GIANTS, CROOKS AND JERKS IN SCIENCE

Giants, Crooks And Jerks In Science

Gordon K. Klintworth

Copyright © 2014 by Gordon K. Klintworth.

Library of Congress Control Number: 2014911757
ISBN: Hardcover 978-1-4990-4198-9
Softcover 978-1-4990-4197-2
eBook 978-1-4990-4199-6

All rights reserved. No part of this book may be reproduced or transmitted in any form or by any means, electronic or mechanical, including photocopying, recording, or by any information storage and retrieval system, without permission in writing from the copyright owner.

Any people depicted in stock imagery provided by Thinkstock are models, and such images are being used for illustrative purposes only.
Certain stock imagery © Thinkstock.

This book was printed in the United States of America.

Rev. date: 08/01/2014

To order additional copies of this book, contact:
Xlibris LLC
1-888-795-4274
www.Xlibris.com
Orders@Xlibris.com
551649

CONTENTS

PREFACE 7

ABBREVIATIONS 11

CHAPTER 1: Introduction 23

CHAPTER 2: Science and Scientists 37

CHAPTER 3: Reactions to Scientific Misconduct 65

CHAPTER 4: Fabrication and Falsification 107

CHAPTER 5: Hoaxes 113

CHAPTER 6: Plagiarism 129

CHAPTER 7: Misconduct in Publications 147

CHAPTER 8: Human Research 167

CHAPTER 9: Misconduct in Research Involving Human Subjects 181

CHAPTER 10: Fraudulent Finances in Research 233

CHAPTER 11: Unethical Research 245

CHAPTER 12: Inappropriate Behavior by Scientists 265

CHAPTER 13: Unintentional Errors in Science 311

CHAPTER 14: Scientific Misconduct Prior to the USA Outbreak of Fraudulent Research in the 1970s and 1980s 365

CHAPTER 15:   Modern American Examples of Scientific Misconduct 399

CHAPTER 16:   Modern Non-American Research Misconduct 457

CHAPTER 17:   Effects of Scientific Misconduct 495

CHAPTER 18:   Correction of Errors in Science 507

CHAPTER 19:   Reasons for Scientific Misconduct 533

CHAPTER 20:   Non-Scientific Frauds and Scandals 559

CHAPTER 21:   Detection of Scientific Misconduct 589

CHAPTER 22:   Prevention of Scientific Misconduct 611

CHAPTER 23:   Investigations of Alleged Scientific Misconduct 627

CHAPTER 24:   COI 637

CHAPTER 25:   Legal Aspects of Research 663

CHAPTER 26:   Epilogue 677

INDEX 685

Tables

Table 1 Chapter 14 Scientific misconduct reported by ORI

PREFACE

I have had a longtime interest in scientific fraud stemming from my days of a medical student when I was appalled to learn what I considered an unbelievable example of stupidity. When cardiac surgery was in its infancy, a major experimental surgical procedure was performed on the heart of a dog in the department of surgery at the University of the Witwatersrand in South Africa. The surgical team achieved a major breakthrough with a new operation. Being enchanted with their achievement, the press was notified, and the information was passed on to the public together with a photograph of the dog. Subsequently the media followed up this newsworthy event, and reporters came to the medical school to obtain another more recent image of the dog that had attained fame. Unfortunately the dog died during the postoperative period, and this created a predicament for the department since an image of the famous dog had previously been distributed. Stupidity won the day, and those involved in the surgery failed to tell the truth and instead decided to perform a paint job on a somewhat similar-appearing dog that would hopefully appear identical to the deceased dog and satisfy the press when they arrived to photograph the dog. Fortunately the truth emerged, and the authentic story became common knowledge; and the Rand Daily Mail in Johannesburg accused Professor William E. Underwood, head of the department of surgery of scientific, fraud. Following this allegation Underwood resigned in 1956, a year before my graduation from the Medical School of the University of the Witwatersrand. He disappeared to Kitwe-Nkana, the second largest city in Zambia (called Northern Rhodesia at the time). When I decided to write a book on scientific fraud, I contacted the University of the Witwatersrand for more information about the incident and was not too surprised to learn that nobody had a recollection of the episode. However, in my readings I discovered that Martin Veller mentioned the scandal in an article on the history of the department of surgery at the University of the Witwatersrand in 2006 [17]. According to Veller, Underwood was involved in pioneering experimental cardiac surgery with Reg Crawshaw, Vernon Wilson, and Paul Marchard. This work was prior to the world's first groundbreaking successful human-to-human heart transplant that

Christiaan Neethling Barnard (1922-2001) performed in 1967 at Groote Schuur Hospital at the University of Cape Town in South Africa.

Being aware of the Underwood scandal, I paid particular attention to episodes of fraudulent research that emerged particularly in the 1970s and 1980s in the USA more than a decade after I embarked on a career as a clinician-scientist. When this apparent epidemic of fraudulent research cropped up, I closely followed the cases of William L. Summerlin (1938-), John Rowland Darsee (c.1948-), Elias A. K. Alsabti, Marc J. Straus, Vijay Soman, John C. Long, Margit Hamosh (1933-2011), Stephen C. Breuning (1953-), Scott S. Reuben, Mark Spector, Robert Slusky, Eric Poehlman (c. 1956), Jan Hendrik Schön, and other dishonest researchers and began collecting everything that I could about fraudulent scientists and other inappropriate behavior by researchers. I desired to learn why people with an apparent honest desire to pursue new knowledge would ruin careers that took to develop. After reading much about scientific misconduct and collecting information on different cases over almost half a century, I thought that the time was ripe to organize my thoughts on the subject into a monograph that would review the subject for all who share my obsession with the topic. Hopefully the text of this book will be valuable for trainees who seek a career in research by drawing their attention to the importance of research integrity and the consequence of not becoming involved in scientific misconduct and other unacceptable behavior.

The definition of the slang term "jerk" was originally used to describe a shamefully naive, fatuous, foolish, or inconsequential person. The errors of researchers with these attributes fall into a category that has been assigned to a jerk. Howard K. Schachman, who fulfilled the role of the ombudsman for the Department of Health and Human Services (DHHS) in the USA, recognized these errors as being different from scientific misconduct. Jerks make unintentional errors because of mistakes created because of their defective background and failure to collaborate with scientists having the appropriate qualifications. Their errors are not because of intentional scientific misconduct caused by fraudulent research but largely because of sloppy research or studies based on inadequate scientific credentials. Schachman directed researchers to distinguish between unintentional errors in research and purposeful violations, namely, the difference between "jerks and crooks." The research of crooks is characterized by the purposeful manipulation or falsification of the research. The distinction between cheating and an ability to synthesize honest established scientific facts has become somewhat blurred. Researchers suspected of research misconduct would undoubtedly prefer to be regarded as a jerk rather than by the more ominous title of crook. The distinction between jerks and crooks stimulated my selection of these terms in the book title. The addition of "giants" was included in the title to stress the research of scientists who reach the highest possible standard. Presidents of the USA who signed into law certain laws that affected research are also listed as another category of giants. For jerks I have broadened the category to include persons whose behavior extended beyond what was appropriate for a professional scientist.

Some famous scientists, even Nobel Prize recipients, deserved this additional label when their behavior was inappropriate, for example, if they failed to acknowledge aptly those who helped them in their research or stole the thunder of subordinates. Examples of such potential jerks are Selman Waksman (1888-1973), John James Rickard McLeod (1876-1935), James Dewey Watson (1928-) and Francis Crick (1916-2004), and Sir Ian Wilmut. Clearly some jerks are extremely intelligent researchers, and a decision as to whether they deserve this title is debatable. I leave the verdicts about specific researchers to the readers.

Numerous books and articles have been written about fraud in science. From an analysis of the stories of those who have tarnished the name of science, I gained the distinct impression that the workings of contemporary science are not adequately understood by the general public and particularly by legislators. For this reason, a brief chapter on science is provided together with one on human research—a topic that has become extremely complicated with numerous rules and regulations. Many of these directives have been established to safeguard human subjects and to maintain high ethical standards in research.

During my more than fifty-two-year-long tenure at Duke University (Duke) as a clinician-scientist, I have had the opportunity to witness research in many departments over the Duke campus. Some of this was as the director of research at the Duke Eye Center (1978-1999) and as the program director of a National Eye Institute (NEI)-funded K12 clinician-scientist training program (2004-2013). I have also followed reports of disgraceful research misconduct in the scientific literature as well as in publications made available to the public. My attention was drawn to the scandal of John Beuttner-Janusch, PhD, after he founded the primate center at Duke (currently known as the lemur center); and as member of the Free Electron Laser Committee, I witnessed the squabbles between John Madey, PhD, and the administration of Duke. The Darsee scandal was also witnessed firsthand as Duke faculty exposed an experimental multicenter trial that he ruined.

The topic of research misconduct has been of considerable interest; and many papers, books and book chapters, and books [1-16, 18] have been written on the subject. Throughout the text I have attempted to document many facts that are within the public domain. Unacceptable behavior by researchers is fortunately uncommon, but it remains much more prevalent than one would desire. It still takes place in a wide variety of disciplines that range from botany to ornithology in different countries. Examples are known to involve graduate students and junior faculty but also distinguished scientists. Other undesirable research is scientifically sound but unethical.

References

1. Broad W and Wade N, Betrayers of the Truth. 1983, New York: Simon and Schuster.
2. Caplan AL, When Medicine Went Mad: Bioethics and the Holocaust (Contemporary Issues in Biomedicine, Ethics, and Society), ed. Caplan AL. 1992) (Editor), Totowa, NJ: Human Press.
3. Dodge M and Geis G, Stealing Dreams: A Fertility Clinic Scandal. 2003, Boston MA: Northeastern University Press.
4. Feder KL, Frauds, Myths, and Mysteries: Science and Pseudoscience in Archaeology, 6th Edition. 1990, New York: McGraw-Hill.
5. Freeman D, Margaret Mead and Somoa: The Making and Unmaking of an Anthropological Myth. 1983, Cambridge, MA: Harvard University Press
6. Freeman D, The Fateful Hoaxing of Margaret Mead. A Historical Analysis of her Samoan Research. 1998, Boulder, CO Westview (Perseus). pp. 291.
7. Garfield B, The Meinertzhagen Mystery: The Life and Legend of a Collosal Fraud. 2007, Washington DC: Potomac Books. 353.
8. Goldberg JS, The "Baltimore" affair. 1986: University of Nebraska Press
9. Gould SJ, The Mismeasure of Man. 1981, New York: W. W. Norton and Company. pp.352.
10. Hixson JR, The Patchwork Mouse. 1976, Garden City, New York: Anchor Press. x, pp. 228.
11. Judson HF, The Great Betrayal: Fraud in Science. 2004, Orlando: Harcourt.
12. LaFolette MC, Stealing into Print. Fraud, Plagiarism, and Misconduct in Scientific Publications. 1992, Berkeley, CA: The University of California Press.
13. Lock S, Wells F, and Farthing ME, Fraud and Misconduct in Biomedical Research 3rd edition. 2001, London: BMJ Books. pp. 268.
14. Miller DJ, Hersen M, and Editors, eds. Research Fraud in the Behavioral and Biomedical Sciences. 1992, John Wiley and Sons: New York.
15. Sabbagh K, A Rum Affair A True Story of Botanical Fraud. 2001: Da Capo Press.
16. Sarasohn J, The David Baltimore Affair: Science on Trial. The Whistleblower, the Accused, and the Nobel Laureate. 1993, New York St.Martin's Press.
17. Veller M, Department of Surgery, University of the Witwatersrand—a brief history. South African Journal of Surgery, 2006. 44(2): 44-51.
18. Wells FO and Farthing MJGE, Fraud and Misconduct in Biomedical Research 4th edition. 2008, London: Royal Society of Medicine Press Limited.

ABBREVIATIONS

5-FU. 5-fluorouracil
AAAS. American Association for the Advancement of Science
AAMC. Association of American Medical Colleges
AAU. Association of American Universities
ABC. ATP-binding cassette, American Broadcasting Company
ACA. Affordable Care Act
ACC. Atlantic Coast Conference
ACE. American Council on Education
ACNP. American College of Neuro-Psycho-Pharmacology
ACSH. American Council on Science and Health
AD. Anno Domini
ADAMHA. Alcohol, Drug Abuse, and Mental Health Administration
AEC. Atomic Energy Commission
AFIP. Armed Forces Institute of Pathology
AGI. Arizona Glaucoma Institute
AhR. Aryl hydrocarbon receptor
AID. U.S. Agency for International Development
AIDS. Acquired immune deficiency syndrome
AIR. American Institutes for Research
ALJ. Administrative Law Judge
ALG. Anti-lymphocyte globulin
AMA. American Medical Association
Am J Med. American Journal of Medicine
ARVO. Association for Research in Vision and Ophthalmology
ATP. Adenosine triphosphate
ATPase. Adenosine triphosphatase
ASM. American Society of Microbiology
ASPD. Antisocial personality disorder
ASU. Arizona State University
AUC. American University of the Caribbean
BALB/c. A stain of mouse

BBC. British Broadcasting Company
BBP. Blood-borne pathogens
BCM. Baylor College of Medicine
BDNF. Brain-derived neurotrophic factor
Bluffton. Bluffton University
BMJ. British Medical Journal
BRCA1. Breast cancer type 1 susceptibility protein
BRCA2. Breast cancer type 2 susceptibility protein
BRU. The initials of a patient with AIDS
BU. Boston University
BUMC. Boston University Medical Center
BUSM. Boston University School of Medicine
BYU. Brigham Young University
c. Circa
CAIF. Coalition against Insurance Fraud
Caltech. California Institute of Technology
Cambridge. Cambridge University
CASPPER. Program for writing papers
CBS. Columbia Broadcasting System
CDC. Centers for Disease Control and Prevention
cDNA. Complementary DNA
CELG. Celgene Corporation
CEO. Chief executive officer
CERN. Conseil Européen pour la Recherche Nucléare
CFC. Chlorofluorocarbon
CFO. Chief financial officer
CFR. Code of federal regulations
CFS. Chronic fatigue syndrome
CGK. The name of a company in South Korea
CGS. Council of Graduate Schools
CIA. Central Intelligence Agency
CME. Continuing Medical Education
CMS. Center for Medicare and Medicaid Services
CMV. Cytomegalovirus
CNRS. Centre National de la Recherche Scientifique
COGR. Council on Government Relations
Columbia. Columbia University
COMS. Collaborative Ocular Melanoma Study
COSEPUP. Committee of Science, Engineering, and Public Policy
COPE. Committee on Publications Ethics
Cornell. Cornell University
CRCT. Criterion references competency test
CRI. Commission on Research Integrity, Committee for Research Integrity

CRISP. Computer Retrieval of Information on Scientific Products (now designated RePORTER)
CRSO. Clinical Research Support Office
CRU. Climate Research Unit
CSIR. Council of Scientific and Industrial Research
CSU. Colorado State University
CT. Computerized tomography
CTMS. Clinical Trial Management System
CTQA. Clinical Trials Quality Assurance
CU. Creighton University, University of Colorado
CU Boulder. University of Colorado-Boulder
CWRU. Case Western Reserve University
DARPA. Defense Advanced Research Projects Agency
DC. District of Columbia
DCAA. Defense Contract Audit Agency
DECREASE. Dutch Echocardiographic Cardiac Risk Evaluation Applying Stress Echocardiography
DEZYMER. A computer program to predict protein sequences
DFG. Deutshe Forschungsgemeinschaft
DFMO. Difluoromethylornithine
DHEW. Department of Health, Education, and Welfare
DHHS. Department of Health and Human Services
DIO. Division of Investigative Oversight
DMSR. Division of Management Survey and Review
DNA. Deoxyribonucleic acid
DRG. Deutsche Forschungsgemeinschaft
DSO. Defence Science Organization
DU. Deakin University
DUHS. Duke University Health System
Duke. Duke University
DUMC. Duke University Medical Center
DUMS. Duke University Medical System
ECG. Electrocardiographs
ED. Department of Education
EdD. Doctor of Education
EEG. Electroencephalography
EEG. Electrocephalogram
EMA. European Medicines Agency
EMBL. European Molecular Biology Laboratory
EMF. Electromagnetic field
Emory. Emory University
EPA. Environmental Protection Agency
ERB. Ethical Review Board

ESF. European Science Foundation
F&A. Facility and Administrative
FAB. Federal Appeals Board of the Department of Health and Human Resources
FACS. Fluorescence-activated cell sorting
FAES. Foundation for the Advancement of Education in the Sciences
FASEB. Federation of the Societies of Experimental Biology
FBAR. Foreign bank account reporting
FBI. Federal Bureau of Investigation
FDA. Federal Drug Agency
FEL. Free electron laser
FFDCA. Federal Food, Drug, and Cosmetics Act
FERPA. Family Education Rights and Privacy Act
FHF. Fred Hollows Foundation
FHPT. Forced hyperventilation provocative test
FOIA. Freedom of Information Act
FQ. A cell line derived from a patent with Hodgkin disease
FRS. Fellow of the Royal Society
FSU. Florida State University
FWA. Federal Wide Assurance
G-6-PD. Glucose 6-phosphate dehydrogenase
GAAP. General accepted accounting principles
GAO. General Accounting Office
GCP. Good clinical practice
GDR. German Democratic Republic
Georgetown. Georgetown University
GLP. Good Laboratory Practices
GMC. General Medical Council
GSK. GlaxoSmithKline
GWI. Gulf war illness
GWS. Gulf war syndrome
GWU. George Washington University
H9. A human cell line
HAART. Highly active antiviral therapy
Harvard. Harvard University
HC. Health Canada
HDL2. High-density lipoprotein 2
HeLa cells. Immortal cells derived from a cervical adenocarcinoma of Henrietta Lacks
HEW. Health, Education, and Welfare
HGH. Human growth hormone
HHS. Health and Human Services
HHMI. Howard Hughes Medical Institute
HIPPA. Health Insurance Portability and Accountability Act

HIV. Human immunodeficiency virus
HL23. A human retrovirus
HMS. USA Department of Health and Human Services
HPLC. High-pressure liquid chromatography
HR. Human resources
HRPP. Human Research Protection Program
HTLV-I. Human T-lymphotropic virus type I
HTLV-II. Human T-lymphotropic virus type II
HTLV-III. Human T-lymphotropic virus type III
HTLV-IIIB. Human T-lymphotropic virus type IIIB
HUT78. A human cell line
HUT102. A human cell line
IBAMA. Instituto Brasileiro do Meio Ambiente e dos Recursos Naturais Renováveis
IBM. International Business Machines
IBT. Industrial Bio-Test Laboratories
ICU. International Cycling Union
ICH. International Conference on Harmonization
ICMJE. International Committee of Medical Journal Editors
IDE. Investigational Device
IDH1. Isocitrate dehydrogenase 1
IDH2. Isocitrate dehydrogenase 2
IEC. Independent Ethic Committee
IEEE. Institute of Electrical and Electronic Engineers
IMT. Industrial Bio-Test Laboratories
IND. Investigational new drug
INPA. Instituto Nacional de Pesquisas da Amazônia
INSERM. Institut National de la Santé la Recherche Médicale
IOC. International Olympic Committee
IOG. Institute of Gerontology
IOM. Institute of Medicine
IOP. Intraocular pressure
IPCC. Intergovernment Panel on Climate Change
IRB. Institutional Review Board
IRS. Internal Revenue Service
IT. Information technology
IQ. Intelligence quotient
IU. Indiana University
IVF. In vitro fertilization
IIIB. Human T-lymphotropic virus type IIIB
JAANP. Journal of the American Academy of Nurse Practitioners
JAMA. Journal of the American Medical Association
JCR. Journal Citation Reports

JEHT. Justice, Equality, Human Dignity, and Tolerance
JGMS. Journal of GLBT Family Studies
JHU. Johns Hopkins University
KAIST. Korea Advanced Institute of Science and Technology
K-INBRE. Kansas IDeA Network of Biomedical Research Excellence
KKG. Kappa Kappa Gamma
KU. Kansas University
LAN. Local area network
LAV. Lymphadenopathy virus
LAV/BRU. A virus isolated from a patient with AIDS
LSD. Lysergic acid diethylamine
LSUHSC-S. Louisiana State University Health Sciences Center Shreveport
LTDD. 2, 3, 7, 8-tetrachlorodibenzo-p-dioxin
MACS. Multicenter AIDS cohort study
MAGIC. Magnetism-based interaction capture
MALDI-MS. Matrix-assisted laser desorption/ionization-mass spectrometry
Mass Gen. Massachusetts General Hospital
MBA. Master of business administration
MBS. National Bureau of Standards
MCG. Medical College of Georgia
MCH. Miami Children's Hospital
MCI. Microwave Communications Inc.
McGill. McGill University
MCSC. Medical College of South Carolina
MD. Doctor of medicine
MEDLINE. Medical literature analysis and retrieval system
MEEIH. Massachusetts Eye and Ear Hospital
Miami. Miami University
MIT. Massachusetts Institute of Technology
MMR. Measles, mumps, and rubella
MOST. Ministry of Science and Technology
mph. Miles per hour
MLA. Modern Language Association
MS. Multiple sclerosis
MSN. Master of science in nursing
MSSM. Mount Sinai School of Medicine
MSU. Michigan State University
MTA. Material transfer agreement
MUSC. Medical University of South Carolina
NACUA. National Association of College and University Attorneys
NAE. National Academy of Engineering
NAS. National Academy of Sciences
NASA. National Aeronautics and Space Administration

NASDAQ. National Association of Securities Dealers Automated Quotations
NASULGC. National Association of State Universities and Land Grant Colleges
NCAA. National Collegiate Athletic Association
NCAB. National Cancer Advisory Board
NCBI. National Center for Biotechnology Information
NCCAM. National Center for Complementary Medicine
NCD. National Coverage Determination
NCHGR. National Center for Human Genome Research
NCI. National Cancer Institute
NCLS. National Conference of Lawyers and Scientists
NCPRE. National Center for Professional and Research Ethics
ND. University of Notre Dame
NEJM. New England Journal of Medicine
NHGRI. National Human Genome Research Institute
NIA. National Institute of Aging
NIAID. National Institute of Allergy and Infectious Disease
NICHHD. National Institute of Child Health and Human Development
NIDA. National Institution of Drug Abuse
NIDCR. National Institute of Dental and Craniofacial Research
NIDDK. National Institute of Diabetes and Digestive and Kidney Disease
NIEHS. National Institute of Environmental Health Sciences
NIGMS. National Institute of General Medical Sciences
NIH. National Institutes of Health
NIMH. National Institute of Mental Health
NHGRI. National Human Genome Research Institute
NHS. National Health Service
NHTSA. National Highway Traffic Safety Administration
NINDS. National Institute of Neurological Disorders and Stroke
NIST. National Institute of Standards and Technology
NNI. National Neuroscience Institute
NOAA. National Oceanic and Atmospheric Administration
NOS. National Organic Symposium
NOU. New Orleans University
NPR. National Public Radio
NSABP. National Surgical Adjuvant Breast and Bowel Project
NSB. National Science Board
NSERC. Natural Sciences and Engineering Research Council
NSF. National Science Foundation
NSTC. National Science and Technology Council
NTTAA. National Technology and Advancement Act
NUS. National University of Singapore
NWU. Northwestern University
NYMC. New York Medical College

NYU. New York University
NYUSOM. New York University School of Medicine
PCBs. Polychlorinated biphenyls
OBE. Order of the British Empire
OCR. Office of Civil Rights
OCRC. Office of Civil Rights Compliance, Office of Corporate Research Collaborations
OESO. Occupational and Environmental Safety Office
OHRP. Office of Human Research Protections
OHSU. Oregon Health Sciences University
OIG. Office of the Inspector General
OMB. Office of Management and Budget
OMK-210. An owl monkey cell line
ONR. Office of Naval Research
OPHS. Office of Public Health and Science
ORA. Office of Research Administration
ORI. Office of Research Integrity
ORIR. Office of Scientific Integrity Review
ORO. Office of Research Oversight
OSHA. Occupational Safety and Health Administration
OSI. Office of Scientific Integrity
OST. Office of Science and Technology
OSTP. White House Office of Science and Technology Policy
OSU. Ohio State University
Oxford. Oxford University
PANAMIN. A Philippine agency meaning both Private Association for National Minorities and Presidential Arm for National Minorities
PBS. Public Broadcasting Service
PCR. Polymerase chain reaction
PCT. Patent cooperation treaty
PECASE. Presidential Early Career Award for Scientists and Engineers
PhD. Doctor of philosophy
PDF. Portable document format
Penn. University of Pennsylvania
Penn State. Pennsylvania State University
PHS. Public Health Service
PI. Principal investigator
Pitt. University of Pittsburgh
PNAS. Proceedings of National Academy of Sciences
PNT. Pneumatic trabeculoplasty
Princeton. Princeton University
Purdue. Purdue University
RA. Responsible author

RAC. Recombinant Advisory Committee
RB. A cell line derived from a patent with Hodgkin disease
RCA. Research Compliance Administration
RCOG. Royal College of Obstetrics and Gynecology
RCR. Responsible conduct of research
RePORTER. Research portfolio online reporting tool
RIO. Research integrity officer
RNA. Ribonucleic acid
RNAi. Short interference RNA
RNase. Ribonuclease
Rochester. Rochester University
RRTA. Rapid response technical assistance
RT-PCR. Real-time polymerase chain reaction
Rutgers. Rutgers University
RY. A cell line derived from a patent with Hodgkin disease
S&P. Standard and Poor
SAGE. Study of Addiction Genetics and Environment
SCNT. Somatic-cell nuclear transfer
SDS-PAGE. Sodium dodecyl sulfate-polyacrylamide gel electrophoresis
SEC. Securities and Exchange Commission
SIDS. Sudden infant death syndrome
SMARTS. Small and moderate aperture telescope system
SMC. Singapore Medical Council
SMU. Silesian Medical University
SNU. Seoul National University
SpR. A cell line derived from a patent with Hodgkin disease
SSAT. Society for Surgery of the Alimentary Tract
SSI. Solid-state imaging system
Stanford. Stanford University
SUNY. State University of New York
SUNY Buffalo. State University of New York Buffalo
SVU. Sri Venkateswara University
TCDD. 2, 3, 7, 8 tetrachlorodibenzo-p-dioxin
Temple. Temple University
THA. Tetrahydroaminocrydine
TNT. Trinitrotoluene
TRDRP. Tobacco-related disease research program
TSU. Tennessee State University
TU. Towson University
Tufts. Tufts University
Tulane. Tulane University
UAB. University of Alabama-Birmingham
UAH. University of Alabama-Huntsville

UC. University of Cincinnati
UC. University of Colorado
UC. University of California
UChicago. University of Chicago
UCI. University of California-Irvine
UC Berkeley. University of California-Berkeley
UConn. University of Connecticut
UCDavis. University of California-Davis
UCL. University College London
UCLA. University of California-Los Angeles
UCSD. University of California-San Diego
UCSF. University of California-San Francisco
UD. University of Delaware
UEA. University of East Anglia in Britain
UH. Houston University, University of Hawaii
UI. University of Iowa
U of I. University of Illinois
UIUC. University of Illinois at Urbana-Champaign
UK. United Kingdom, University of Kansas
U of M. University of Minnesota
UM. University of Michigan
UMD. University of Maryland
UMass. University of Massachusetts
UMMS. University of Minnesota Medical School
UMMS. University of Michigan Medical School
UMN. University of Minnesota
UNC. University of North Carolina, University of Northern Colorado
UPenn. University o f Pennsylvania
UPMC. University of Pittsburgh Medical Center
URL. Uniform resource locator
USA. United States of America
USADA. USA Antidoping Agency
USAAMRIID. USA Army Medical Research Institute of Infectious Diseases
USC. University of Southern California
USF. University of South Florida
USPTO. USA Patent and Trademark Office
UST. University of Science and Technology
UT. University of Texas
Utah. University of Utah
UToronto. University of Toronto
UTSA. University of Texas-San Antonio
UTSWMC. University of Texas Southwestern Medical Center
UVA. University of Virginia

UVM. University of Vermont
UW. University of Wisconsin
VA. Veterans Administration
WFU. Wake Forest University
VUSM. Vanderbilt University School of Medicine
WHICFS. White House Interagency Committee on Fundamental Science
WHO. World Health Organization
WI-38. A human cell line
WITS. Women and infants transmission study
WMU. Western Michigan University
WPI. Whittemore Peterson Institute
WSU. Wichita State University
WUSL. Washington University in St. Louis
XMRV. Xenotropic murine leukemia retrovirus
Yale. Yale University

CHAPTER 1

Introduction

Prior to the late 1970s allegations of scientific misconduct were rarely noted in the public domain and so was dialogue about the subject. Initially when fraudulent research appeared in the 1970s and 1980s the scientific community and the public were taken by surprise and this behavior was considered a rare aberration or at least so it was hoped.

High profile cases at renowned universities in the USA opened a new era in science with considerable adverse publicity in newspapers and on television. The appearance of fraudulent research severely shook the edifice of science and stirred up a hornet's nest. The scientific community received a wakeup call and realized that something needed to be done about this dishonest research regardless of its prevalence. Examples emphasized that major academic centers are not immune to fraud and indeed suggested that some may be more prone to it. Also, researchers that ended up in respectable academic positions like John Darsee and Scott Reuben probably achieved their academic status through continuous deceptive behavior over many years. Abraham Lincoln (1809-1865) the sixteenth president of the USA allegedly stated "You can fool some of the people all of the time, and all of the people some of the time, but you cannot fool all of the people all of the time". While Lincoln did not have dishonest research in mind when making this statement it does indeed apply to those who commit scientific misconduct as over time the truth eventually emerges even if the fraudulent nature of the research does not become evident until many years after the death of the perpetrator.

Most scientists agree that inappropriate research falls into two main categories: scientific fraud and carelessness, which is considerably more widespread.

Time to Expose Cheaters

Most recognized cases of scientific misconduct are presumably discovered during the life of the perpetrator who faced the consequences of the dishonest behavior. This wrongdoing is detected at an early stage especially when whistle-blower draw attention to it. However, the exposure of some questionable research is inordinately slow and does not occur until many years after the death of the scientist. Examples of this occurred with Claudius Ptolemy (c. 90-AD168), Isaac Newton (1642-1727), Gregory Mendel (1822-1884), Louis Pasteur (1822-1895), Cyril Lodowic Burt (1883-1971), and Margaret Mead (1901-1978). The fraudulent work of Ptolemy took two millennia to be exposed.

Definitions

Precise, rigorous, and unambiguous definitions of misconduct in science are essential. After the recognition of frequent widespread research misconduct the scientific community wrestled with definitions related to fraudulent research. Eventually the Office of Research Integrity (ORI) defined research misconduct as "fabrication, falsification, plagiarism or other practices that seriously deviate from those that are commonly accepted within the scientific community for proposing, conducting or reporting research. It does not include honest error or differences in interpretation or judgments of data." Most institutions, professional societies and USA federal agencies eventually accepted the ORI definition with minor modifications [19]. Research misconduct does not usually violate criminal law, but it is a crime against science.

After it became apparent that fraudulent research was not as rare as originally suspected questions were raised about to define it. Congressional subcommittees left these debates primarily to institutions and professional societies. In the considerations of dishonest research fraud was a dominant concern because many cases were characterized by falsified and/or fabricated data or documented experiments that had never been performed. The designation of misconduct in research became preferable to the term fraud in science because of the input of numerous lawyers at discussions on definitions of mistakes in research. The attorneys represented a wide variety of government agencies and in contrast to scientists they desired a restricted definition emphasizing the ambiguities of science and the necessity of researchers to be free to exploit their instinct and considered opinion in deciding on which data to select. Scientists argued that "research misconduct" should be limited to deeds such as concocting data, altering data, as well as the theft of data or ideas without acknowledgment. The original demarcations were more than usually broad and vague in keeping with the purpose of Congress in setting up supervision of federal funds for scientific research. Research misconduct eventually became packaged into "fabrication, falsification and plagiarism". Following years of debate the Public Health Service (PHS) announced "responsibilities of its awardees and applicant institutions for dealing with reporting possible misconduct in science". This document defined research misconduct as "fabrication, falsification, plagiarism, deception or other practices that seriously deviate from those that are commonly accepted within the scientific community for proposing, conducting or reporting research. This delineation was accepted by numerous federal agencies for a decade. It also became clear that mistakes by scientists are not synonymous with "misconduct in research" as researchers are human and many of their errors unintentional and not performed with the intent of deception. They have always been regarded as unintentional "honest" mistakes. Institutions did not consider it appropriate to deal with vandalism under the "serious deviates" clause of the definition of scientific misconduct. In 1992 a panel of the National Academy of Sciences (NAS), National Academy of Engineering (NAE), and the Institute of Medicine (IOM) considered options for the definition of research misconduct. In 1993 the advisory committee on scientific integrity of the PHS advocated an elimination of the phrase "other practices that seriously deviate from those that are commonly accepted within the scientific community" in the definition of misconduct in science being used by the ORI. Their proposal moved closer to the definition recommended by the *NAS* panel.

Various types of dishonest research are carried out deceptively without regard for the truth. Fraudulent research occurs, and questions have been raised about how research misconduct should be defined. The changing of "fraud in science" to "misconduct in science" caused trepidation among scientists. Some actions portrayed at the congressional hearings were aptly labeled as fraud, such as the faking or falsifying of data. However, some confusion existed over the connotation of fraud. A preference of the term scientific misconduct instead of fraud was initiated not by scientists, but by members of the legal profession in PHS and the federal funding agencies preferred that designation because fraud in law imposes a legal liability of having to prove intent and injury to persons. According to Black's Law Dictionary [13] fraud is an intentional perversion of truth for the purpose of inducing another . . . . to part with some valuable thing. Misconduct is a transgression of some established and definite rule of action . . . . Its synonyms are misdemeanor, misdeed, misbehavior, delinquency, impropriety, and mismanagement, but no negligence or carelessness. In the national response to the dishonest research in the USA definitions for this inappropriate behavior needed to be established. For some entities definitions were already in the book. The term misconduct was proposed in preference to fraud to dodge any misunderstanding with the word fraud as used in common law. Howard K. Schachman, a professor in the Department of Molecular and Cell Biology at the University of California, Berkeley (http://www.berkeley.edu/) expressed unease over the imprecision of the phrase "misconduct in science" and how individuals with diverse points of reference construe an assortment of alleged abuses [32]. Researchers underscored the fact that scientific misconduct does not include features intrinsic to the process of science such as error, conflicts in data, or dissimilar explanations or judgments related to data or experimental design. Especially troublesome was the insertion of the vague expression "other practices that seriously deviate from those that are research", because the phrase was open to different interpretations. More importantly it would persuade against unusual exceedingly groundbreaking lines of attack by scientists that have been known to result in major scientific advances. Gifted, imaginative cutting edge research often diverges from the widely used procedures within the scientific population. To include vague terms in the definition of research misconduct possibly commits a breach in the principle of due process. Institutions did not consider it appropriate to consider vandalism under the "serious deviates" clause of the definition of scientific misconduct. The original definitions were more than usually broad and vague consistent with the intent of Congress in establishing oversight of federal funds for scientific research.

During 1994 Samuel C. Silverstein MD, president of the Federation of the American Societies of Experimental Biology (FASEB) objected vehemently to the inclusion of the phrase "or other practices that seriously deviate from those that are commonly accepted within the scientific communities for proposing, conducting or reporting research" on behalf of FASEB [34]. He recommended modifications of the definition to the Commission on Research Integrity (CRI). His objection was based on three reasons: (i) the guideline was imprecise and ambiguous, but had the force of law, (ii) the more original research is the more it approaches a paradigm shift from standard research and the inclusion of the phrase invites misuse, (iii) the phrase implies that the scientific community is a defined entity, but it is not. Over several years expert panels debated the definition of scientific misconduct, but could not reach a consensus. Scientific misconduct like dishonesty covers a wide range of situations that are difficult to define precisely. In this regard they are comparable to pornography, and easy to recognize for what they are.

Mainly for legal reasons definitions are necessary for different kinds of inappropriate behavior by scientists. A definition for what became known as scientific misconduct was needed and the scientific community struggled with the issue for several years. A variety of agencies and organizations debated various proposals. Those that got into the act included the USA Public Health Service (PHS), the DHHS, the *Office of Science and Technology* Policy (*OSTP*), the *National Science Foundation* (*NSF*), the *White House Interagency Committee on Fundamental Science* (*CFS*), *FASEB*, the *NAS*, the *American Association for the Advancement of Science* (AAAS), the *National Science and Technology Council* (NSTC), the CRI and the *National Science Board* (NSB). In April 1996 the NSTC defined research misconduct as the additional wording in the NSCTC definition made certain that wrongdoings do not fall through the cracks. It unambiguously made it evident that destroying the data of a colleague falls under the definition and so does plagiarism during peer review. A proposal by the CRI did no elicit a general consensus and it was criticized for being too-open minded and hypothetically restraining creativeness [18]. By 1997 scientists still failed to reach a consensus on what constitutes research misconduct. After discussing scientific misconduct for almost three and half years a panel of the OSTP finally decided on a definition of the improper conduct under scientific misconduct in October 1999. The misconduct policy appeared in the *Federal Registry* [2].

Loss of Confidence in Science

Scientists have made considerable advances over the past quarter of a century and these have been appreciated by the public, but the deliberate fraudulent behavior and blunders of a relatively small number of researchers have had a detrimental embarrassing effect on honest hardworking scientists. The net effect of intentional and unintentional mistakes by scientists has been a relative loss confidence in the integrity of science in the minds of many members of the public. For example a considerable amount of evidence from different sources, such the retreating of glaciers in different parts of the world, the reduction of snow covered regions in the Northern Hemisphere and the decrease and sea ice worldwide indicates that global warming is a reality [23]. Information supporting global warming is maintained in data sets by the Climate Research Unit (CRU) of the University of East Anglia (UEA) in the United Kingdom (UK) as well as at the National Aeronautics and Space Administration (NASA) and the National Oceanic and Atmospheric Administration (NOAA). Widespread notoriety emerged at the December 2009 convention on climate change in Copenhagen, Denmark following the hacking of electronic records of scientists working at UEA under the direction of Professor Philip Jones (1952—)[22]. Emails disclosed that climate data were not being openly shared and that efforts were being made to not cite particular papers in reports on international climate change. The UEA initiated an investigation into an allegation that their scientists acted inappropriately by manipulating or suppressing climate related data. This inquiry was precipitated by questions raised about the integrity and professional behavior of the UEA scientists. Ralph J. Cicerone, who was president of the *NAS* at the time stressed that this episode damaged the public trust in science [8]. Comparable unpleasant incidents of scientific misconduct in which researchers breached the trust of the scientific community for personal gain seriously also harmed the confidence of the public in the research achievements of scientists and weakened their belief in the generally well-accepted standards of science.

Politicians avoid taking action on sensitive issues and a letter from 256 scientists expressed concern about the political assaults on scientists and especially climate scientists, who could not produce inequitable proof that the earth was heating [14]. The authors of the letter pointed out that there is undeniable "evidence that humans are changing the climate in ways that threaten communities and the ecosystems on which they depend". They pleaded for an end of McCarthy-like intimidation of unlawful action against their collaborators based on insinuation and blame by connection. The integrity of some scientists involved in the consensus reports for the United Nations body on climate change had been questioned [22]. An important point missed by the scientists is the fact that the public has gradually over time been losing confidence in scientists. In early January 2010 the Intergovernmental Panel on Climate Change (IPCC) regretted their inclusion of the unconfirmed forecast that the Himalayan glaciers would no longer exist by 2035 [3, 4].

Incidence and Prevalence

In medical terminology the designations incidence and prevalence are used to indicate how widespread a condition may be as well as the rate of its occurrence. Both terms are significant for scientists and medical doctors who analyze both figures to establish the future course of action and treatment procedures. These interchangeable terms are frequently misunderstood and incorrectly used. A researcher working on a particular disorder often determines both the incidence and prevalence of that disorder in a community. The prevalence of an entity is determined by establishing the ratio of affected cases of it to the population in the region of interest. On the other hand the term incidence refers to the number of new cases of the condition that occurs in a single year. Clearly the ratio of prevalence is invariably larger than that of incidence. It is now well-established that examples of fraudulent research have occurred over a considerable period of time and new instances are constantly being discovered. Many individuals have wondered about how widespread this unacceptable behavior is. In reality we do not know the answer, but the deliberate fraudulent behavior and blunders of a relatively small number of researchers have had a detrimental embarrassing effect on honest hardworking scientists.

Several studies have attempted to gain insight into the prevalence of research misconduct. A study by the small group known as the Acadia Institute found that about 40% of the deans in the major graduate schools in the USA were aware of cases of research misconduct that had been confirmed in their institutions during the past five years [35-37]. Another survey of scientists by the *American Association for the Advancement of Science (AAAS)* disclosed that 27% of them personally knew of cases of research misconduct that had taken place during the past decade [17]. Moreover almost half of the respondents were of the opinion that the fraudulent research was becoming more frequent. It is now well established that more than a casual number of scientists perform dishonest research much more frequently than anyone in the scientific community would like. The precise incidence and prevalence of research misconduct will probably remain unknown.

Several attempts have been made to determine how common research misconduct is, but estimates vary in different studies. In surveys of researchers, around 1-2% of them admit having fabricated, falsified or modified data or results at least once [10]. Fanelli [10] reviewed the findings of twenty-one anonymous academic misconduct surveys of randomly-sampled research scientists and used eighteen surveys in a meta-analysis [5, 9, 12, 15, 16, 20, 21, 24-28, 30, 31, 37-39]. From this analysis, he concluded that 14% of scientists knew a colleague who had falsified data and 2% admitted committing misconduct. The overall anticipated admission rate of fabricating data, the most severe variety of research misconduct, was low (1.9%); but when less harsh versions of misconduct were included, the projected underestimate rate of research misconduct was considerably higher (9.54%). These figures are almost certainly miscalculations as even with promises of anonymity, researchers are unlikely to admit any form of misconduct in their research. Observing others fabricating data elevated the likely rate to 14.12%, and with less crucial witnessed research misconduct, the rate jumped to 28.53%. Judith P. Swazey, president of the Acadia Institute in Bar Harbor Maine, surveyed faculty and graduate students with her colleagues and found that 6% of faculty respondents had knowledge of plagiarism and 9% were aware of data falsification [37]. One study indicated that 27% of scientists encountered 2.5 episodes in ten years [17]. Another investigation gave 0.28% in audits of cancer clinical trials [40]. Research organizations and funding agencies record scientific misconduct in different ways. In the USA, information about cases can be obtained from federal funding agencies under the *Freedom of Information Act*, but this time, consuming chore only reveals an incomplete depiction of the situation. In Europe, Scandinavian countries have national committees that can provide information about the frequency of scientific misconduct, and it is fairly low. Over a period of seven years from 1992, the Danish committee on scientific dishonesty investigated twenty-five allegations but only found four accused researchers guilty.

Several studies suggest a higher incidence than the reported numbers [33, 41]. A 1995 survey of almost three hundred scientists picked at random in Norway found that 22% of respondents were aware of significant violations in ethical research guidelines and 9% admitted that they had individually participated in these serious infringements. Virtually 60% of those participating in the survey knew of less severe misconduct by members of the faculty at their universities. Despite the inherent weaknesses of surveys to assess the incidence and prevalence of scientific misconduct this procedure provides useful information by suggesting that this unacceptable practice of research misconduct may be significantly more common than suspected and this does not make this dishonest behavior inconsequential.

In a survey of science faculty in a major university Price found that one-third of the respondents had suspected a colleague of falsifying data, but that only half of them had ever acted to confirm or correct the impression. More than 50% of the respondents were of the opinion that indisputable fraudulent research would be dealt with in silence by university officials out of the public eye.

The perpetrators of research misconduct have been from a wide variety of universities and research institutions (Table 1). In September 1988, the OIG guesstimated that of 95 research misconduct allegations virtually 50% were validated and had been addressed by NIH grantee institutions. This estimate was consistent with records retained by NIH since 1982 documented that 102 cases of scientific misconduct had been investigated by its grantees and reported. Insufficient data is available to evaluate the gender racial predilections for scientific misconduct, but a substantial number of researchers performing fraudulent research have been males. A small percentage has been women. The gender differences may reflect the ratio of males and females involved in research, but as pointed out by Fang and coauthors, men engage in more risky behavior than women [11]. Research misconduct is not a peculiarity of specific scientific disciplines or of one country or of one generation. It has been documented in a wide variety of subspecialties, but in the clinically relevant studies dishonest research seems to be considerably more prevalent than in basic biomedical sciences. Many wondered whether this new trend was the emergence a modern scourge, but it did not take long for historians of science to point out that fraudulent research and had taken place from the beginning of science and that it was not something unique, not a peculiarity of recent times but it was publicized in the past. In 1983 Broad and Wade [6] considered fraud in science to be widespread as it has been at all times. They also believed that the confidence that scientists had in their own techniques made them especially susceptible to fraud. Government agencies, professional organizations and research institutions consistently affirmed that the primary responsibility for dealing with misconduct allegations should rest with institutions where the research is being performed. Estimates of its prevalence vary in different studies from 27% of scientists encountering two to five episodes over ten years through 0-28%. The prevalence of research misconduct probably lies somewhere between "a few bad apples" and what lies below the "tip of an extremely large iceberg". The latter was predicted by William Broad and Nicholas Wade in *Betrayers of the Truth* [6]. From 1993 new instances of alleged research misconduct in PHS funded research were investigated by the ORI and the number of cases found guilty by that federal office provides an indication, but this number only reflects biomedical research funded by the NIH in the USA. In an extensive review on the incidence of research misconduct Abbott and coauthors draw attention to the observation that the ORI received about one thousand allegations of scientific misconduct between 1993 and 1997. During that period the ORI closed 150 investigations and found that misconduct took place in almost half of these cases [1]. Abbott and colleagues also drew attention to the fact that the NIH funded more than 150, 000 grants during the same five years. During its first year of its existence eighty-six new allegations came to the attention of the ORI. From then onward, the number progressively increased over time and by 2007, the number reached 183. The identified incidence of scientific misconduct among NIH awardees during the five-year period from 1993 to 1997 was 0.67%. During its first fifteen years of its existence the ORI received 1686 allegations of scientific misconduct or almost ten per month. The annual number of allegations received by the ORI between the years 1992—and 2007 was 198, between 2004—and 2006 was 271, in 2007, it was 217 and in 2008, it was 201. The number of retractions of scientific papers by journals also provides an indication of the frequency of research misconduct, but many retracted papers are for unintentional honest" mistakes.

The number of proven cases of scientific misconduct is relatively small compared to the vast number of practicing scientists awarded NIH grants, and the amount of scientific publications released annually. An enormous number of papers generated by research are published weekly and virtually all of them are preserved in the scientific literature forever.

Every year throughout the world almost a million papers are published in 5, 600 professional scientific journals. The massive online database of scientific publications recorded in the Thomson Reuters Web of Science specifies that almost twenty-seven thousand papers were published weekly in 2012. In 1981, the NIH funded nearly eighteen thousand extramural projects. Over time this number increased enormously and approximately 150, 000 grants were awarded by that agency during the five-year period from 1993 to 1997. Since the 1970s fraudulent research has not inevitably become more widespread, but discovered examples have received more publicity. It is not clear whether the cheating has intensified or diminished in recent years or whether it has just received more publicity.

Boundary between Truth and White Lies

The distinction between truth and dishonesty is like day and night, but the boundary between them is blurred like dawn and dusk. It is not always in a person's best interest to always be truthful. A white lie sometimes prevents hurting the feelings of an individual when telling the truth is not socially or politically advisable. In research unintentional mistakes and fraudulent behavior fall into similar zones that are not always black and white. The distinction between scientific misconduct and unintentional sloppy research is not at all times clear-cut and a distinct gray zone exists. As the late American senator Edward Moore "Ted" Kennedy (1932-2009) put it at a Senate hearing regarding the Federal Drug Administration (FDA) and its activities—"inaccurate science, sloppy science, and fraudulent science—these are the great threats to the health and safety of the American people. Whether the science is wrong because of a clerical error, or because of poor technique, or because of incompetence or because of criminal negligence, is less important than the fact that it is wrong" [7].

Fabrication, falsification and plagiarism are indisputably deceitful and unquestionably scientific misconduct. Unintended mistakes are considered "honest" errors of science even though some of them result from sloppy experimentation, poor scholarship and other undesirable performances of research. But what about the gray area where people fudge a little here and there and discard or forget some experimental data that might have significantly influenced the final conclusion? The same applies to the distinction between what is ethical and what is unethical. At both ends of the spectrum, the distinction is obvious, but between these extremes, the boundary is fuzzy and debatable and often a matter of opinion.

Moreover, editors and reviewers of peer-reviewed journals cannot reliably detect fraudulent papers and sometimes allow them to become published. This sinister face of science continued to display its gruesome image as more and more examples of deceitful research were exposed in ever-increasing detail in the 1980s at reputable institutions. In 1980 important cases of dishonest research emerged making reporters in the lay press predict a crime wave and at least two cases were suspected to be on a path toward criminal prosecution.

Failure of Replication in Research

Research findings need to be replicated by others before they become accepted into the general scientific knowledge. One pointer to research misconduct is the failure of others to replicate the results, but failed research replication should not automatically be attributed to research misconduct. Several other potential explanations for a failure to replicate research findings exist and it is difficult to differentiate between honest scientific error and scientific misconduct as an experiment can only be right or wrong. To make sure that the experiments are valid significant studies should ideally be repeated by other members of the research community. As discussed in elsewhere science does not hold the monopoly in untruthful unacceptable behavior, but scientists are expected to not tolerate it because a search for the truth is the pinnacle of its goals. Some members of other professions cheat and fabricate or falsify data or perform other inappropriate actions and understanding of why they do it may help identify the causes of scientific misconduct.

Oversight of Research

The American policy on scientific misconduct was assessed in the 1990s and it became apparent that the scientific community must not only watch over itself, but also be seen as keeping an eye on itself. In the USA for the sake of the public Congress has the responsibility to see that the system is functioning properly and that fraudulent research is being combated. The federal government should clearly intervene when deceitful behavior by academicians is funded by misused federal grants, but the government presumably does not desire to be the ethical overseer of academia. Most scientists do not object when Congress gets involved in allegations of scientific fraud, because it is their right as protectors of taxpayer money to support worthy causes.

Investigation of Research Misconduct

After conducting basic science research at four major institutions, Murray [29] found that any form of scientific misconduct was seriously dealt with by university administrators;, but at one institution selective data reporting, pirating of reagents, and fabrication of results were common. It was Murray's impression that some research communities function like businessmen marketing their results as manuscripts. It is essential that unethical and dishonest researchers be caught and punished.

References

1. Abbott A, Dalton R, and Saegusa A, Briefing. Science comes to terms with the lessons of fraud. Nature, 1999. 398(6722): 13-17.
2. Anonymous, Federal Register, 1999. 64(198): 55615-55808.
3. Bagla P, No sign of Himalayan meltdown. Indian report find. Science, 2009. 326(5955): 924-925.
4. Bagla P, Climate science leader Rajendra Pachauri confronts the critics. Science, 2010. 327: 510-511.
5. Bebeau MJ and Davis EL, Survey of ethical issues in dental research. Journal of Dental Research, 1996. 75(2): 845-855.
6. Broad W and Wade N, Betrayers of the Truth. 1983, New York: Simon and Schuster.
7. Cartwright OJ, Castro C, and Pile E, Senator Ted Kennedy and the SQA: A tribute. Society of Quality Assurance, 2010. 26(1).
8. Cicerone RJ, Ensuring integrity in science. Science, 2010. 327(5966): 624.
9. Eastwood S, Derish P, Leash E, et al., Ethical issues in biomedical research: perceptions and practices of postdoctoral research fellows responding to a survey. Science and Engineering Ethics, 1996. 2(1): 89-114.
10. Fanelli D, How many scientists fabricate and falsify research? A systematic review and meta-analysis of survey data. PLoS ONE, 2009. 4(5): e5738.
11. Fang FC, Bennett JW, and Casadevall A, Males are overrepresented among life science researchers committing scientific misconduct. mBio, 2013. 4: e00640-00612.
12. Gardner W, Lidz CW, and Hartwig KC, Authors' reports about research integrity problems in clinical trials. Contemporary Clinical Trials, 2005. 26(2): 244-251.
13. Garner BA, Black's Law Dctionary (9th edition). 2009, New York City: Thomson Reuters.
14. Gleick PH, Adams RM, Amasino RM, et al., Climate change and the integrity of science.[Erratum appears in Science. 2010 May 14;328(5980):826]. Science, 2010. 328(5979): 689-690.
15. Glick IJ and Shamoo AE, Results of a survey on research practices, completed by attendees at the third conference on research policies and quality assurance. Accountability in Research, 1994. 3: 275-280.
16. Glick JL, Perceptions concerning research integrity and the practice of data audit in the biotechnology industry. Accountability in Research, 1993. 3(2-3): 187-195.
17. Hamilton DP, In the trenches, doubts about scientific integrity. Science, 1992. 255(5052): 1636.
18. Kaiser J, HHS is looking for a definition. Science, 1996. 272(5269): 1735.
19. Kaiser J, A misconduct definition that finally sticks? Science, 1999. 286(5439): 391.

20. Kalichman MW and Friedman PJ, A pilot study of biomedical trainees' perceptions concerning research ethics. Academic Medicine, 1992. 67(11): 769-775.
21. Kattenbraker MS, Health education research and publication: ethical considerations and the response of health educators, Carbondale: Southern Illinois University, Doctoral dissertation 2007: Carbondale: Southern Illinois University.
22. Kintisch E, Stolen e-mails turn up heat on climate change rhetoric. Science, 2009. 326(5958): 1329.
23. Kintisch E, Embattled U. K. scientist defends record of climate center. Science, 2010. 327(5968): 934.
24. List JA, Bailey CD, Euzent P.J., et al. Academic economists behaving badly? A survey on three areas of unethical behavior. Economic Inquiry. 2001 January:162-170.
25. Lock S, Misconduct in medical research: does it exist in Britain? British Medical Journal, 1988. 297(6662): 1531-1535.
26. Martinson BC, Anderson MS, and de Vries R, Scientists behaving badly. Nature, 2005. 435(7043): 737-738.
27. May C, Campbell S, and Doyle H, Letter to editor article. Research misconduct: A pilot study of British addiction researchers. Addiction Research, 1998. 6(4): 371-373.
28. Meyer MJ and McMahon D, An examination of ethical research conduct by experienced and novice accounting academics. Issues in Accounting Education, 2004. 19: 413-442.
29. Murray EJ, Is research fraught with fraud?. The Journal of NIH Research, 1990. 2: 16.
30. Rankin M and Esteves MD, Perceptions of scientific misconduct in nursing. Nursing Research, 1997. 46(5): 270-276.
31. Ranstam J, Buyse M, George SL, et al., Fraud in medical research: an international survey of biostatisticians. ISCB Subcommittee on Fraud. Control Clin Trials, 2000. 21(5): 415-427.
32. Schachman HK, What is misconduct in science? Science, 1993. 261(5118): 148-149.
33. Sekas G and Hutson WR, Misrepresentation of academic accomplishments by applicants for gastroenterology fellowships. Annals of Internal Medicine, 1995. 123(1): 38-41.
34. Silverstein SC, Statement to the commission on research integrity in FASEB Newsletter1994. 6-7.
35. Swazey JP, Ethical issues of artificial and transplanted organs. American Scientist, 1987. 75(2): 192-196.
36. Swazey JP, Louis KS, and Andrerson MS, University policies and ethical issues in research and graduate education:highlights of the CGS deans' survey. CCS Comunicator, 1989. 22: 1-3, 7.

37. Swazey JP, Anderson MS, and Lewis KSKL, Ethical problems in academic research. American Scientist, 1993. 81: 542-553.
38. Tangney JP, Fraud will out—or will it? New Scientist, 1987. 115(1572): 62-63.
39. Titus SL, Wells JA, and Rhoades LJ, Repairing research integrity. Nature, 2008. 453(7198): 980-982.
40. Weiss RB, Vogelzang NJ, Peterson BA, et al., A successful system of scientific data audits for clinical trials. A report from the cancer and leukemia Group B. Journal of the American Medical Association, 1993. 270(4): 459-464.
41. Xi S, Sketches of the American Scientist. 1989: Sigma Xi.

CHAPTER 2

Science and Scientists

Science has existed since at least the days of the ancient Greeks when Euclid (c. 300 BC), Archimedes (c. 287-212 BC), Hipparchus (c, 200-126 BC) and Ptolemy (c. 90-AD168) made names for themselves. Throughout the ages these great scientists and others have been admired for their remarkable discoveries. The most esteemed scientists in the past have included Al-Khwarizmi (5th century AD), Ibn Sina (c. 980) and Al-Biruni of ancient Arabia. Others who have stood the test of time include Leonardi da Vinci (1452-1519), Nicolas Copernicus (1473-1543), Andreas Versalius (1514-1564), Galileo Galilei (1564-1642), Christiaan Huygens (1629-1693), Anton von Leuwenhoek (1632-1723), Robert Hooke (1635-1703), Sir Isaac Newton (1642-1727), Carolus Linnaeus (1707-1778), James Hutton (1726-1797), Antoine Lavoisier (1743-1794), John Dalton (1766-1844), Michael Faraday (1791-1867), Charles Babbage (1791—1871), Charles Darwin (1809-1882), Louis Pasteur (1822-1895), Gregory Mendel (1822-1884), Dmitri Mendeleyev (1834-1907), James Clerk Maxwell (1831-1879), Max Planck (1858-1947), Marie Curie (1867-1934), Ernest Rutherford (1871-1937), Albert Einstein (1879-1955), Alfred Wegener (1880-1930), Niels Bohr (1885-1962), Edwin Hubble (1889-1953), Werner Heisenberg (1901-1976), Linus Pauling (1901-1994), Francis Crick (1916-2004), James Watson (1928—), Rosalind Franklin (1920-1958) and Stephen Hawkins (1942-). The achievements of these geniuses and great scientists have been ideal role models for young individuals embarking on careers in science and their achievements are summarized in a book by Farndon [9].

Wonder has always been at the forefront of scientists and it drives researchers to explore the unknown. The workings of nature was a golden age for British natural philosophers, such as Sir Humphry Davy (1778-1829), Johann Wolfgang von Goethe (1749-1832), Friedrich Wilhelm Joseph Schelling (1775-1854), Sir

Frederick William Herschel (1738-1822) and his sister Caroline Herschel (1750-1848), Mungo Park (1771-1806), and Sir Joseph Banks (1743-1820) the official botanist on Captain James Cook's first voyage on the *Endeavour* who discovered the beauty and terror of science [12].

A wide variety of individuals perform research. At the top end of the spectrum are the geniuses who design innovative imaginative hypotheses which they test experimentally with logical scientific methods that they understand. Below them are well-trained scientists who devote their entire careers to research. Other individuals dabble in research and frequently pursue it without formal training or the necessary knowledge or background to embark on studies that attracts their attention. Other researchers are amateurs who usually tackle a scientific question because of a burning desire to undertake it or because research is expected of them in academia even though their interest in scientific investigations is trivial. The amateur cannot hope to compete with a full-time professional researcher in productivity or in raising funds to support research.

The talent of scientists varies considerably. At the top end of the spectrum are unique imaginative individuals who work at the cutting edge of their discipline and discover truly innovative information. Every now and then major fundamental breakthroughs or unique ideas emerge, but these are rare even with the most ingenious scientists. Scientists of the caliber of Galileo, Darwin, Pasteur and Einstein thought out of the box and made significant advances in our knowledge by not accepting the *status quo*.

Nobel Prizes

A major event in the history of science took place on November 27, 1895, when Alfred Nobel (1833-1896), the wealthy scientist and entrepreneur who made a fortune from 2, 4, 6-trinitrotoluene (TNT), signed his last will and testament establishing the Nobel Prizes. They have been awarded annually in physics, chemistry, physiology or medicine, peace and literature since 1901. Another prize, the Sveriges Riksbank Prize in economic sciences created in the memory of Alfred Nobel, has been awarded at the same time since 1968. The work of the remarkable scientists who earn a Nobel Prize (http://www.nobelprize.org/nobel_prizes/lists/all/). and others who devote careers to understanding various aspects of the natural and physical world in which we live has been admired by the public as well as by other researchers. A noteworthy description of scientists who have won the Nobel Prize is available at the Nobel Prize website (*http://www.nobelprize.org/*) and in publications such as *Over 100 Titles by Nobel*

Laureates and on the Nobel Prizes [3]. Receiving a Nobel Prize has always been regarded as the pinnacle of scientific achievement and the greatest honor that a scientist can receive. The Nobel Prize is only given to living individuals because the statutes of the Nobel Foundation stipulate that the prize cannot to be awarded posthumously. The only exception has been Ralph M. Steinman (1943-2011) who shared the 2011 Nobel Prize in physiology or medicine *(http://www.nobelprize.org/ nobel_prizes/medicine/laureates/2011/steinman.html)* with Bruce A. Beutler *(http:// www.nobelprize.org/nobel_prizes/medicine/laureates/2011/beutler.html)* and Jules A. Hoffmann *(http://www.nobelprize.org/nobel_prizes/medicine/laureates/2011/ hoffmann-lecture.html)*. Steinman died of pancreatic carcinoma three days before the award was announced by the committee, which was unaware of his death. After a consideration of this unfortunate oversight the prize was still awarded as it "was made in good faith."

With research on extremely important subjects scientists move into the arena of potential fame and fortune and psychological factors become tremendously prominent and jealousies bring out undesirable human traits that make life difficult for scientists. This is well-illustrated by the story of how the virus that causes the acquired immunodeficiency syndrome (AIDS) was discovered. In a description of the competition between Robert Charles Gallo (1937—) and other scientists at the *National Cancer Institute* (NCI) over credit for the discovery of the human immunodeficiency virus *(HIV)* John Crewdson offered a revealing look at how research laboratories really work in a highly competitive world [6, 7]. Competition for Nobel Prizes is fierce and most awardees are well deserved, but the awards are sometimes controversial. The awardees of the 2008 Nobel Prize in physiology or medicine included Françoise Barré-Sinoussi (1947—) *(http:// www.nobelprize.org/nobel_prizes/medicine/laureates/2008/barre-sinoussi.html)* and Luc Montagnier (1932—) *(http://www.nobelprize.org/nobel_prizes/medicine/ laureates/2008/montagnier.html)* for discovering the virus that causes AIDS, but Gallo who played an important role in the research was neglected probably because of the scandal over his work that was emphasized by the investigative reporter Crewdson.

When the American virologist John F. Enders (1897-1985) shared the 1954 Nobel Prize in physiology or medicine *(http://www.nobelprize.org/nobel_prizes/ medicine/laureates/1954/enders-lecture.html)* with Thomas H. Weller (http:// www.nobelprize.org/nobel_prizes/medicine/laureates/1954/weller-lecture.html) and Frederick C. Robbins *(http://www.nobelprize.org/nobel_prizes/medicine/ laureates/1954/robbins-lecture.html)* the worthy awardee was debatable. Although Enders received the prize as the father of modern vaccines, many expected the prize to go to Jonas Salk (1914-1995), the creator of the poliomyelitis vaccine and the darling of the public, but Salk was shunned by the scientific community.

A Wakeup Call for American Science

Prior to World War II, scientific research was primarily performed by comparatively small groups of aficionado who by and large had other income to make a living, frequently teaching or if they were physicians, by clinical practice. A watershed in the history of American science took place on October 24 1957. On that day the ego of the USA received a major blow when to the surprise of all Americans the Russians launched *Sputnik 1* into an elliptical low Earth orbit. This unexpected event showed that the Soviet Union, the enemy of the western world at that time, was ahead of the USA at least with regard to conquering space. In 1958 after the wakeup call from the Soviet Sputnik program the USA Congress established its *Select Committee on Astronautics and Space Exploration*. This committee drafted the *National Aeronautics and Space Act* that created the *National Aeronautics and Space Administration* (NASA). It also gave birth to the permanent *House Committee on Science and Astronautics*. In 1974 that committee was renamed the *House Committee on Science and Technology* and later in 1987 the name was changed once again this time to the *House Committee on Science, Space and Technology*. For political reasons the Republican Party renamed the committee *House Committee on Science* in 1994, but this title only lasted until 2007 when the Democratic Party reverted the committee back to the *House Committee on Science and Technology*.

After recovering from the immediate shock of *Sputnik 1* the USA government reacted by starting to pump enormous amounts of money into research and scientific development. Universities started training an increased number of scientists and research laboratories suddenly found that money was readily available in the USA to support research, particularly by federal grants. The scientific enterprise expanded immensely and for a vast number of bright creative young individuals research became a central way in which to make a living. The ability to obtain government support became a function of a researcher's productivity. Because of the difficulty in judging quality particularly in disciplines foreign to persons not familiar with them career advancement became evaluated for the most part by the quantity of publications. From that time onwards until the scandals of scientific misconduct in the 1970s and 1980s decisions related to academic promotions and funding were primarily based on the number of publications. Scientists became aware early in their careers that published papers were essential for their survival and that without them they would perish. More recently the shoe moved to the other foot as publishing, particularly with dishonest coauthors, has the danger of leading to publish and perish. Fallout from the publicity of scientific misconduct was detrimental to the public image of scientists and not appropriate for providing well-informed policies. Different professions value self-regulation and so does the scientific community, but science like other activities supported by the community must interconnect better with the general public and earn and retain their respect and show that research must be performed with absolute honesty. The public pay considerable attention to new discoveries that are cited by the media and the announcements are rightfully expected to be true. There is a clearly a need for the public to become more conversant with science, which has unfortunately lost the understanding and support of the public which it desperately needs for funding. An apparently small number of cheaters in science, such as those described elsewhere in this book, have seriously damaged the public trust and credibility in science. A loss of confidence and integrity in science stemmed from the perceived misbehavior of a relatively small number of the active scientists. Science plays and continues to play a vitally important role in the development of society, and it has made considerable advances over the past quarter of a century. It has also reached a low ebb because fake research and anxieties about research integrity has diminished the public trust in science. When a significant new discovery is documented other scientists generally try to replicate the research, but also for assessing grant applications.

The Review of Journal Submissions and Grant Applications

Although a single person may prepare the initial version of a grant application other co-investigators as a rule participate in the writing of subsequent drafts in which they contribute ideas as well as text that become so comprehensively amalgamated that it unfeasible and impractical to determine who added what when an alliance terminates. Editors often have difficulty recruiting reviewers of papers and numerous reviewers do not impart a full honest disclosure because they do not want to be bothered. With the high degree of contemporary specialization in science it is virtually impossible to find a competent critic with relevant expertise who is not in the same profession.

Importance of Accepting Valid Scientific Data

Because of a lack of trust in the findings of scientists valid conclusions have not always been reached. For example silicone gels have been implanted into human breasts of thousands of women for more than three decades. Dow Corning Corporation was eventually sued by more than 3, 500 women many of whom claimed that the implants caused autoimmune disease. For a long time the implants were thought to be inert, but in 1993 scientists at Dow Corning Corporation, discovered that at least in animals they were a strong irritant of the immune system [11]. However, in 1998 an independent panel of medical experts reported that the incidence of immune-system abnormalities among women with breast implants was not greater than in the general population [8] yet a constant flow of lawsuits over the implants continued despite valid evidence by expert scientist did not substantiate the link between autoimmune disease and the silicone gels. Other later studies came to the same conclusion [22]. A failure to accept the scientific evidence bankrupted the billion-dollar Dow Corning company.

Research Regulations

It is now practically impossible to keep up with the continuously changing detailed regulations that impact on research. A core foundation in science is trust and in collaborative research it is essential that coworkers trust each other in an open, honest, and forthright manner. These regulations are established by numerous federal departments and agencies and are governed by laws. Increasingly more responsibilities are transferred onto the principal investigator (PI).

Building of Scientific Knowledge

All scientific observations establish a foundation on which future generations can build and advance knowledge. Scientists are focusing their research on smaller and smaller entities and often have difficulty communicating about their passionate research with persons not familiar with the topic. The requirements of business or the government often drive research for financial reasons. Scientists commonly focus their research on topics that they enthusiastically pursue, such as a disease that affects them or someone dear to their heart. Ideally science should be pursued for the sake of discovery and knowledge advancement rather than less pungent goals such as power, financial gain, and prestige. Politically or commercially charged research can sometimes stir up emotion that can override scientific meticulousness. The selected topic and the line of attack to study it is determined by the leader of the research group and it is prejudiced by the past experience of that researcher and the person's immediate fascination with the topic. Aside from designing and carrying out experiments laboratory heads are responsible for managing the research and for obtaining funding, which is particularly important when the research is extremely expensive. Research advances for three reasons: technical advances, innovative ideas and new discoveries. Most new knowledge progresses essentially by trial and error through a building process in which the established findings of generations of yesterday and today form the basis of tomorrow's breakthroughs. The sentiment of this well-established truism was aptly provided by the great British scientist Sir Isaac Newton (1643-1727) who stated "If I have seen father than most, it is because I have stood on the shoulders of giants.

Scientific data has no absolute truths and the conclusions are seldom definitive and usually some uncertainty is usually present. The public often decry scientists until they are certain about their conclusions, but public policy decisions are sometimes necessary in the absence of indisputable proof.

Truth and honor are vital components of the scientific enterprise and the accumulation of knowledge depends most importantly on honesty. Trust, integrity, openness, the primacy of knowledge and appropriate professional behavior are the essential foundation on which the scientific enterprise has been built since its origin and it thrives to establish new knowledge that benefits mankind. The entire scientific enterprise depends on a milieu in which research can prosper and for this truthfulness and high ethical standards are essential. Scientific integrity is needed for the advancement of knowledge and for the public trust in science.

Scientists have ethical responsibilities and some research is unacceptable, particularly in the minds of the public. A duty of scientists is to identify the exclude unintentional bias and other sources of error in their experiments, such as control groups and double masked procedures. Bias occurs in research when data is intentionally or inadvertently suppressed. So that knowledge can be passed from one generation to the next new discoveries are traditionally documented in learned peer reviewed publications in the public arena. Papers are an extremely significant component of scientific communication. Unlike some other scholarly endeavors scientists report the vast majority of their work in journals rather than in books.

While addressing the graduates at a high school in California in 1990 the American immunologist and microbiologist John Michael Bishop (1936—), who shared the 1989 Nobel Prize in physiology or medicine (*http://www.nobelprize. org/nobel_prizes/medicine/laureates/1989/bishop-autobio.html#*), with Harold E. Varmus (*http://www.nobelprize.org/nobel_prizes/medicine/laureates/1989/varmus-lecture.html*) aptly stated to this young generation: "Scientists depend upon the truthfulness of their colleagues: Each of us builds our discoveries on the work of others. If that work is false, our constructions fall like a house of cards and we must start all over again. The great success of science in our time is based on honesty."

Value of Doubters in Science

Scientists need to be constantly skeptical about their own research and not become overly enthusiastic about a new observation that makes the researcher rush to print. Unfortunately this mistake is far too common because of the harsh competition in the research arena. In the world of science doubters play a cardinal role and as Bertrand Arthur William Russell (1872-1970), the distinguished mathematician and philosopher of the 20th century stated "Doubt is the essence of science". The organized skepticism of science puts the burden of proof on the original researcher and not on the critic. Conventional wisdom is not always correct and heretics challenge the dogma of the day and sometimes replace old ideas with something new. All top-notch researchers know the importance of being their own devil's supporter when evaluating research data. To criticize their work inspires their coworkers to appreciate the importance of reporting bad as well as good news. In the past, scientific skepticism was not extended to the honesty of an investigator's factual statements, but was directed towards the interpretation of the reported results.

Hypotheses, Theories and Scientific Facts

Specially designed experiments start with a testable hypothesis to explain a specific phenomenon. The hypothesis, which is inherently biased in favor of the researchers own theories, bestows the initiative and motivation for the study and presides over its form. Some methods are selected; others are rejected. It is only this preceding anticipation that provides significance to the activities reported by the scientist in the reports. However, hypotheses are not always necessary to advance knowledge. Indeed Newton one of the most influential scientists of all time frowned upon hypotheses. During his time he stated "hypotheses non fingo" and "hypotheses non sequor" (I do not pursue hypotheses). Scientists have hypotheses, suspicions, and ideas, but they don't know all the answers. Serendipity has contributed to many major scientific discoveries, but usually by persons with a trained mind [24]. They sometimes follow accidental unintentional laboratory errors and when scientists pursue projects by doing the unexpected and thinking out of the box. Notable serendipitous discoveries included the finding of Penicillin, X-rays (Röntgen rays), the Dead sea scrolls, safety glass, nylon, polyethylene, Teflon, Velcro, and the Rosetta stone [24].

Experiments must be repeatable to consistently forecast future findings. Accurate predictions from a hypothesis support the hypothesis, but it still remains questionable and on probation. An objective in testing a hypothesis is not to verify it, but to disconfirm it. If this approach is not adopted the researcher runs the risk of forsaking objectivity and becoming too engrossed in an idea and the researcher may persistently believe in a hypothesis indefinitely when it is unquestionably wrong. A profound commitment to a hypothesis can provide a scientist with intuition as to how to maneuver through much conflicting data, but it can also cause a researcher to follow a path down a blind alley leading nowhere. Researchers often adhere to their beliefs and doggedly try to prove their hypothesis. Sometimes they do not succeed, but they maintain the tradition of always challenging the status quo. Researchers must be prepared to abandon a treasured concept as soon as new data rejects the hypothesis. When the predictions of a hypothesis are not confirmed the hypothesis needs to be abandoned or at least amended. Researchers can support a hypothesis without it being correct, but they must be prepared to exercise self-reflection and critically question their own theories, because if they do not others undoubtedly will. The renowned cell biologist and genius Sydney Brenner (1927—), who shared the 2002 Nobel Prize in physiology or medicine (*http://www.nobelprize.org/nobel_prizes/medicine/laureates/2002/brenner-lecture. html*) with H. Robert Horvitz (1947—) (*http://www.nobelprize.org/nobel_prizes/ medicine/laureates/2002/horvitz-lecture.html*) and John Sulston (1942-)(http:// www.nobelprize.org/nobel_prizes/medicine/laureates/2002/sulston-lecture.html) understood this issue and gave his advice on more than one occasion. According to Brenner "theories should be treated like mistresses. One should never fall in love with them and they should be discarded when the pleasure they provide is over" [10].

When hypotheses are thoroughly and exhaustively tested and questioned by countless scientists, who become convinced that the theories are well-established, they reach the status of established facts. This level is reached when diverse evidence from many scientists supports the theory. Examples of this type include the Darwinian theory of evolution in which all living organisms evolved in the past from lower creatures, the dogma that earth has an age on the order of 4.5 billion years and that its origin dates back to a single event that took place 14 billion years ago according to the "Big bang" theory. Until a genius arrives and can disprove these "facts" they will remain believable.

Theories that include broad domains of investigation may unite different autonomously developed hypotheses collectively into a reasoned, united structure, which may assist in the formation of new hypotheses or put collections of hypotheses into perspective. A logically rigorous method of devising hypotheses does not exist. They stem largely from guesswork or inspiration and not from logic. They are not created by deduction, but specific consequences can be predicted by reasoning from a hypothesis, which can be tested thoroughly by acts designed to test it. William Whewell (1794-1866), a theologian, geologist, philosopher and historian of science, was the first professional scientist to provide a completely logical opinion as to how scientists indeed think when they make their scientific discoveries. A different interpretation of the scientific method sometimes called the hypothetico-deductive interpretation was explained by Karl Popper in the *Logic of Scientific Discovery* [21] and Medawar favors it [20]. It is only in the light of prior anticipations that the scientist documents anything consequential. A significant number of the general public apparently believes that scientists suppress hypotheses that differ from what they believe and also withhold data and manipulate aspects of peer review to prevent disagreement.

Experimental Replication and the Gold Standard of Research

The gold standard in scientific studies is confirmation of the results by impartial investigators. After substantiation of research studies by independent laboratories they become widely accepted. Research findings need to be replicated by others before they become accepted into the general scientific knowledge. Reproducibility is an essential part of scientific investigations and real discoveries get replicated by the original researchers or by others. However replication is not applicable to validating the entire spectrum of scientific research. Complex questions being addressed with new methods, as well as interdisciplinary collaborative research that is often international pose challenges for replication, particularly when the studies are long term and performed at great and are difficult to replicate. Research involving field based studies, primate cognition and behavior, generate colossal databases. An inability of researchers to reproduce results does not necessarily indicate research misconduct. Indeed numerous scientists have reported discoveries that could not be replicated by others. The complexity of modern research makes it extremely difficult for other research teams to confirm the research independently, particularly if subtle technical details are not provided in publications documenting the research. To replicate the findings and meet expected customs it is essential that researchers document the specific experimental conditions. Sometimes the reason has not been determined and the work has just been ignored after it could not be replicated. Unsuccessful attempts at replication are regarded as spurious and are caused by unintentional errors, an unrecognized quirk in the experimental conditions or fraudulent research. This practice of replication promotes truthfulness in research, because it exposes cheaters and questionable conclusions. Regardless of the reason non-replicated research tarnishes the reputation of scientists. Robert K. Merton, who spawned the discipline of the sociology of science, coined the term "organized skepticism" in 1942 for this self-correcting mechanism [17]. He regarded scientific inquiry to be subject to thorough supervision perhaps unlike any other human pursuit. Scientific self-regulation became accepted as a traditional way in which progress is made and cheaters became outcasts.

While replication is an important component of research some research findings are accepted in good faith without replication or accusations of scientific misconduct, or unintentional errors, especially if there is no reason to suspect dishonesty. For example, nobody bothers to replicate extremely time-consuming expensive multi-institutional clinical trials of new drugs and therapies. This point is illustrated by the experiments of D. Carleton Gadjusek that led to to him sharing the 1976 Nobel Prize in physiology or medicine in 1976 with Baruch S. Blumberg (1925-2011) *(http://www.nobelprize.org/nobel_prizes/medicine/laureates/1976/#)* for his discovery of an unusual infectious agent with a very long latent period were not independently replicated, because they were extremely tedious, time consuming, and profoundly expensive. It is essential that the public become convinced that the scientific community is taking all possible steps to maintain the validity of research and in claims that are asserted about new findings. Field-based studies enable researchers to observe living creatures in their natural environment, but the replication of the findings under these conditions is hindered by a difficulty in defining the precise unique conditions under which the new findings are collected [25]. Research on primate cognition and behavior research also face replication barriers, such as small sample sizes, the high cost of research and ethical issues. Moreover scientists working in this arena need to construct assignments that elicit complex cognitive behaviors and captive populations of the species being studied are not always comparable. To replicate studies of this nature a recommendation has been made to create data banks where primary data and videotapes can be deposited, perhaps at the time of publication [28]. Research involving genomics, transciptomics, proteomics, metabolomics, and other types of "omics" has grown exponentially and energized hope for a modern age of personalized medicine. All data created by these analyses have generated gigantic databases within which clinically significant discoveries remain to be unearthed. Hopefully new knowledge derived from such huge collections of information will lead to a better understanding of human disease and spawn novel patient specific therapies. Rigorous replication criteria with high statistical significance is demanded by some of these information arenas [14].

Computer sciences also face difficulties in achieving the gold standard of replication. Sufficient information about computer methods and code are needed so that independent investigators can come to consistent conclusions using the initial raw data. Other investigators are unable to replicate research projects using complicated computer programs without making the code and data about detailed projects available. At least the journal *Biostatistics* encourages authors of accepted papers to submit code and data for posting online to allow the material to be reproduced.

Intuition

Intuition plays an important role in science and it sometimes leads to major advances, but it may escort the scientist down the incorrect path. A superb case in point is in the prejudice that led to triumph in 1920, when the Harvard University (Harvard) (*http://www.harvard.edu/*) astronomer, Harlow Shapley (1885-972), revealed not only that the sun was not at the center of the Milky Way Galaxy, but that the Milky Way was considerably larger than previously appreciated [27] According to the historian, Gingerich some of Shapley's methods may appear particularly arbitrary to current astronomers but some measurements from variable stars, did not seem right to him. If Shapley had performed his analysis using all available data, he might have generated an evenhanded unconvincing report, but with his intuition he made a spectacular leap forward.

The World of Science

Researchers face constant pressure to be outstandingly successful to survive. The world of the scientist is extremely competitive and individuals pursuing a career in science have to compete with classmates for admission to colleges of variable strength and then for graduate school and postdoctoral fellowships as well as faculty positions and subsequently for promotions and tenure. Unless future scientists are fortunate to stem from wealthy families they must also compete for scholarships to help fund the education. The success of researchers entails hard work, discipline and ingenuity and requires complete honesty and the highest standards of objectivity and ethical behavior with regard for all existing creatures. They also have to compete for grants to support their research. A fierce competition also exists between scientists in different laboratories over who gets credit for important discoveries. A major strength of American science is the policy of high quality universities, such as Sanford, Harvard, and MIT, to provide junior members of their faculty with ample freedom to pursue their academic interests. A cultural characteristic of American universities that is of immense importance is that they place self-sufficiency and congeniality on the individual faculty. In law individuals are presumed innocent, unless proven guilty, but in the court of science a researcher is required to prove his/her point and show that the experiment was performed entirely, self-assuredly, and thoroughly with suitable controls.

The Scientific Method

Scientists have shared values and further knowledge using the scientific method, which includes numerous *procedures* for exploring observable facts, attaining new information, or amending and amalgamating past knowledge. The research tools vary from discipline to discipline, but the investigative method is based on collecting discernible, experimental and quantifiable *evidence* conditional on specific principles of logic. Science is much more than an unemotional quest for new knowledge. The basic laws of science stem from observations of nature, experiments in the laboratory, as well as computer and mathematical models. A scientific discovery or the formulation of scientific theory begins with a simple unbiased, naïve observation and generalizations emerge. From a disorderly collection of facts, an orderly theory somehow materializes as pointed out by Mill. He attempted to explain difficulties in the inductive approach to science, such as verification bias and established falsification as a key component in the scientific method ("John Stuart Mill (Stanford Encyclopedia of Philosophy)". plato.stanford.edu. *http:// plato.stanford.edu/entries/mill/#SciMet. Retrieved 2009-07-31).* Prior to Mill deduction in itself was considered powerless as a method of scientific discovery as it only uncovers and explicitly exposes information that is already in the truisms and principles from the process of deduction. Mill considered induction as the essential method whereby science discovers and improves general dispositions, but the theory underlying the inductive method cannot be retained for at least three reasons. A scientific study must be objective so that bias will not influence the analysis of the results, but induction begins with a naïve observation.

Scientists

Those who pursue a career as a scientist do so with enthusiasm and idealism, but scientists are human and express all of the attributes of persons in other professions. Many are highly motivated, self-serving, opportunistic, competitive, far from certain, assertive, and frequently selfish and arrogant. Like all inspired individuals scientists exhibit pride and jealousy and desire credit for their work and do not look lightly on those who snatch glory from them either deliberately or inadvertently. Their most treasured possession is their reputation for truthfulness and objectivity. While the observations of all scientists suffer from bias influenced by past experience they attempt to make objective unbiased observations guided by the evidence regardless of whether it supports or reputes a specific point of view. Like all humans scientists frequently make errors, but are often reluctant to admit them because of a fear that they might blemish their image. Scientists should not accept risks to the integrity of science, whether they arise from within the scientific enterprise or outside of it. Amiability is an essential component of academic research. It is also essential to scrutinize multifaceted problems involving the allotment of data and unusual materials as well as authorship and publication practices. Scientists are extremely judgmental of untested notions, yet the assumption of scientific impartiality is virtually untested. In psychological studies the beliefs and expectations of scientists can be conveyed to human research participants in unintentional ways and this can affect their behavior in the experiment. Many books and articles in the psychology literature document this experimenter effect. The reputations of scientists are built not only in furthering conventional knowledge, but also in proving that accepted agreements are incorrect and that better explanations exist.

Many researchers ignore intentional bias and the degree of it in science remains unknown.

Congeniality is an essential component of academic research.

Research Projects

To solve modern complex scientific issues researchers depend increasingly on multidisciplinary collaborative research teams and the resulting papers have multiple authors often from many countries. From 1990 to 2005, the quantity of international group efforts in research, determined by coauthorship in peer reviewed publications expanded linearly, while the worldwide addresses grew exponentially. Translational research assumes that researchers will extend their studies beyond their own know-how into multidisciplinary collaborative research. After a scientific study is completed the researchers wrap up the study and prepare a manuscript documenting the finding and how the study fits into the big picture of science. The paper is then submitted to a suitable journal hopefully for publication.

An expectation is that scientists will document as precisely, honestly and entirely as possible the observations and findings, as well as the used experimental *methods* so that other scientists can scrutinize and verify the work by replicating the research. To not do so discredits the profession and retards the rate of scientific progress by those who come afterwards.

The resultant paper is submitted to a professional reputable journal. Until the modern electronic age the most efficient method for researchers to disseminate their work was by way of journals produced by professional societies and independent publishers.

For career advancement researchers try to get their papers published in prestigious journals in which the average number of *citations* to articles published in it is higher than other professional journals. These top notch journals are referred to as high impact journals and they receive so many submissions that they are forced to reject the vast majority of papers that they receive. Researchers also strive to maintain a moral and professional investment in furthering knowledge and attempt to ostracize from their ranks those who are guilty of research misconduct. Scientists are expected to pursue their research endeavors faithfully to further knowledge and to be unswervingly objective in making and interpreting observations.

Society in general benefits from those who can think outside of the box and make original thoughts. Organizations hence need to nurture resourcefulness without the undesirable consequences.

The peer review process using well-informed unbiased reviews is important not only for evaluating articles submitted to journals for possible publication. After receiving a manuscript the journal editor sends the paper to two or three experts for review so that its quality can be evaluated and procedural errors and flaws can be detected and flushed out before it appears in print. For a long time the customary practice of professional scientific journals was send submitted papers to respected experts on the same subject so that the validity of the research and the quality and clarity of the manuscript can be judged. This peer review process dates back to the mid-1600s according to historians. Most journals attempt to obtain independent evaluations from at least three reviewers for each article. The reviewers also provide the editor with input related to the suitability of the paper for that specific journal as well as an indication of the relative importance of the paper. Reviewers of manuscripts are not expected to find perfection, but rather to improve the quality of papers and to reject manuscripts that have obvious defects.

To retain a high standard of research journal editors need well-informed reviewers without COI to evaluate articles. Prior to the submission of scientific paper to an appropriate journal it is traditionally reviewed by all authors who are expected to read it carefully for accuracy, clarify and syntax. After being submitted to a journal the manuscript is sent to anonymous reviewers, usually at least three, for evaluation. Having manuscripts reviewed by the peers of a scientist dates back to the *Royal Society of England* which requested guidance from its members regarding which papers to publish in its journal *Philosophical Transactions*. Almost a century later according to Stephen Lock the president of the Royal Society established an external committee to assess manuscripts [19]. The custom of using outside authorities to evaluate scientific material before publishing it did not come in vogue until almost the 19th century. In the 1800s, editors of scientific journals in both Europe and USA relied more or less entirely on their own considered opinion according to science historian John Berman of Ohio State University (OSU) *(http://www.osu.edu/)*. Editors of medical and scientific journals regarded their publications as personal journals and adopted the custom of newspaper and magazine journalism. One of the first journals to use outside referees was the *British Medical Journal* and other journals gradually adopted this policy. As long ago as the 1930s, the *Journal of the American Medical Association* depended upon an undersized journal staff to provide decisions on submitted manuscripts and barely ever used external assistance. Michael M. Sokal PhD a retired professor in the history of science at the Worchester Polytechnic Institute (WPI) *(http://www.wpi.edu/)* pointed out that when James McKeen Cattell was editor of *Science* (1894 to 1945) his son, a Harvard graduate in physiology, passed judgment on many papers that were received by that journal. Following Cattell's death *Science* was taken over by the *American Association for the Advancement of Science* and an external peer review became a standard procedure. The reviewers of manuscripts are anticipated to be recognized experts on the subject and they are expected to identify and exclude factual errors, such as unintentional bias, as with control groups and double blinded procedures that are more appropriately designated "masked" when the eye is involved.

The critiques of the reviewers and editorial board members are communicated to the authors who seriously take them into account so that they can improve their manuscript and remove mistakes. The authors have an opportunity to clarify issues raised by the reviewers. Because of the competition for acceptance is high many papers are rejected and the authors often need to improve their manuscript taking into account the criticisms or try to get the paper accepted by another journal. This often forces researchers to publish in less prestigious journals, which may or may not have a less rigorous review process. Based on the critiques the editor decides whether revisions need to made and whether an additional review is necessary should an acceptable revision be received. Specific professional societies often have their own journals as official publications to provide credibility to the particular society.

If a paper is acceptable, with or without revisions, the editor decides whether or not to publish it. Getting publications in high impact journals can advance the careers of scientists immensely and there is a major incentive to submit papers to such journals.

Papers must not only be found acceptable after peer review, commonly by competitors, and still compete with other articles for space in the top journal of choice before it can be published. Researchers performing mediocre work find it is difficult to get their papers published in respected journals and sometimes can only get them published in deplorable journals that publish garbage. It would be more appropriate for them to improve their study taking into account valid criticisms or alternatively filing the manuscript in the trash can, but authors in this position need to bolster their curriculum vitae for career development.

Funding Agencies

As in other professions the work of scientists needs to be judged and regulated. The work of scientists gets evaluated in papers submitted for publications and in applications for research funding. It also transpires in the awarding of prizes and honors. The Howard Hughes Medical Institute (HHMI) supports the best and brightest scientists on the faculty of many universities and research institutions in the USA. All of these investigators receive their salary and funds for research expenses. Funding agencies, such as the National Institutes of Health (NIH) and the National Science Foundation (NSF) need unbiased opinions without COI that also seem fair to the scientific community at large when grant applications are evaluated for the limited available funds. This has traditionally been performed by the peer review process and COI cannot be completely eliminated. In grant applications to fund their research scientists must also share their best thoughts with competitors. Researchers like businessmen market their results primarily in publications and presentations at scientific meetings as well as in lectures at different institutions. For academic promotions enormous attention is paid to scholarly publications and until recently more attention was paid to quantity rather than quality. Scientists desire to publish their new findings as soon as possible before their competitors and often select journals with a rapid acceptance and publication records. To accelerate the speed of getting publications into the public domain many journals now publish papers online in an "ahead of print" format after satisfactorily passing through peer review and the final approval by the authors of the edited manuscript.

Research Administration

Like research its administration is exceptionally complicated and research institutions must currently possess an infrastructure to take care of the many activities related to the complexity of contemporary science and the imposed regulations. When research grants are awarded universities and other research institutions receive colossal amounts of money in the name of indirect costs (also known as facility and administrative costs) to pay for the costly overhead expenses of the research. These expenses relate to utilities, depreciation of buildings and capital equipment, repair and maintenance of buildings, equipment and libraries as well as part of the expenses of the institutional offices of the president and deans. The indirect costs of grants also pay certain expenses related to the research activities, such as office supplies. The actual amount awarded to universities for this purpose is derived during negotiations between each university and the funding sources and it varies markedly from institution to institution because the cost of doing business differs in different parts of the country. Some institutions receive a much higher percentage of the direct costs of grants than other universities. Since the government started to fund research animosity has existed between researchers and their institution because of the indirect costs of research. Researchers have criticized institutions from receiving this money, amounting to billions of dollars each year, because it was being used for general university support and not directly for research. They felt that the funds would be better allocated to research grants. Members of Congress have also been unhappy about funding this additional cost and have maintained that it was their duty to fund research, but not to sustain universities [26].

Publications

Traditionally credit goes to the discoverer of something new and not to those who independently replicate the findings. It is not a question of how the game is played, but who wins and who loses, because in science winners get virtually all the credit. Honor, position, power and money go to researchers who swiftly publish their novel discoveries and establish priority. Ironically from a practical point of view it is perhaps better to be first author in a suboptimal journal than second in a high impact journal. In the past when committees appointed by the NIH and NSF with special scientific expertise known as Study Sections reviewed grant applications they commonly slashed the budgets with the result that investigators soon learnt to apply for more money than necessary. They hopefully anticipated that the cut in the awarded budget would not be severe enough to hinder the research and there would still be sufficient funds to perform the desired research. Gatekeepers that control the entry of new information into the scientific literature include professional journals, the mentors of students, postdoctoral fellows and other junior researchers as well as the anonymous reviewers of papers submitted to journals. Criteria have been established to judge their quality.

Professional societies started to have their own journals to enhance their authority and to provide a resource for specialized papers of their members. Many such journals now exist. After English became the preferred international language of science, persons brought up in countries where the population is not always eloquent in English, such as China, scientists faced a difficulty in making a name for themselves in their scientific discipline when they were required to publish in high quality journals in a language with which they were not entirely fluent. The quality of publications can also be judged on the caliber of the journals in which the work was published. Another objective measure is the impact factor of the journal which often varies from year to year. In a particular year it is the average number of citations obtained per paper published in that journal during the two previous years. High impact journals are those having their articles cited sooner and more often after publication than journals with a low impact fact. The editors and editorial boards of journals pay considerable attention to the impact factor of journals and battle to increase it for their journals. One trick is to encourage review articles because they are popular and influence the impact factor by being cited more often than the usual original research publication. Some journals with high impact factors do not publish notable original papers and are not necessarily better than other journals with lower impact factors. To achieve a top notch ranking journals need impact factors of greater than 30.

The citation index is a measure of publication quality as it reflects the number of times that a publication has been cited. Several different citation indices are recognized and to obtain them massive world wide databases of peer reviewed papers are searched to identify citations. Of all databases that perform this task the SciVerse Scropus database is the most powerful. Elsevier, a leading publisher of scientific, technical and medical information, developed this database as well as the SciVal Spotlight tool for providing information about the track record of individuals in different disciplines within institutions over time. An annual publication of *Journal Citation Reports* (JCR) by the healthcare and science division of *Thomson Reuters* also keeps track of citations by different journals and is able to identify the top journals in cell biology and biochemistry by their impact factors. However when citation indexes of particular papers are considered other factors need to be taken into account. Some papers elicit many citations because the authors overly cite their own publications and numerous citations may reflect criticisms of a paper rather than the caliber of the reported research. The number of citations is also a sign of the popularity of the subject. Papers on popular subjects are cited more often than those on rare entities.

Another method that has been developed to measure the productivity and impact of the publications of researchers is the h-index also known as the Herfindahl index, Hirsch index and Hirsch number. An index of h indicates that at least that number of a scientist's publications has each been cited and that the other papers have less than h citations (N_p—h) (*http://enwikopedia.org/wiki/H-index*). Unfortunately authors are able to manipulate this index by citing their own articles particularly in dubious journals. An example of someone who has done this is H. B. Kekre a senior professor of computer science at an institution deemed to be a university. He had over 3396 citations in July 2013 and a h-index at the astoundingly high level of 33 *(http://scholar.google.com/citations?user=oiVHiwcAAAAJ&hl=en)*. This index is considerably higher than what would be expected from a combination of all professors at a major university.

Sometimes the direction of a scientist's research suddenly changes based on new discoveries that commonly follow novel concepts and technical advancements.

Until relatively recently scientists performed research on their own or with the assistance of medical students, graduate students, postdoctoral fellows, technicians, and other research support staff. Scientists have now moved away from single researcher investigations into interdisciplinary collaborations have become extremely common and often involve multiple institutions in different countries. Some contemporary research is cumbersome and difficult to perform without refined sophisticated equipment and computer programs. Modern-day research has become extremely complex and is performed more frequently by collaborative multidisciplinary research teams of numerous researchers often with different backgrounds and expertise in which each member of the group contributes to the overall project. It is funded mainly by governmental and other agencies with the support of the public. Some specialized areas are beyond the comprehension of individual investigators who must accept parts of the research on trust.

Research Productivity

Despite the vast amount of total grant money accessible for research competition for research funds is fierce and inadequate to support all worthwhile projects by qualified scientists the large number of researchers. Over the years the total number of professional scientists has increased immensely and today it is estimated that that 90% of all scientists who have ever lived are still alive. Their amazing productivity is reflected in the enormous number of papers published each year. The rate of scholarly publications is at an all time high (*http://www.nlm.nih.gov/medline_cit_counts_yr_pub.html ml*). Throughout the world 40, 000 journals publish about 1 million papers annually, but the number of those that are important is probably relatively small. Most are probably not read after being published and an exceedingly large number are not cited by authors other than those who wrote the paper. Very few scientists have more than one thousand research papers to their credit. In 1992 Kantha [16] surveyed this select few.

Rigid rules about how research must be carried out and how data should be retained do more harm than good.

The NIH funds a vast number of research grants. It allocates $7 billion towards the funding of research for 20, 000 projects a year by 50, 000 biomedical scientists. In the five years after 1992, 49, 815 research grants were active.

Honesty is an essential requirement in research and almost all scientists have the intentions of being honest, but as pointed out by Tina Gunsalus, director of the online National Center for Professional and Research Ethics (NCPRE) at the University of Illinois at Urbana—Champaign (UIUC) "Almost everyone wakes up every day and wants to do the right thing", but because of many potential pitfalls later in the day the goal may be thwarted [1]. A book describing the obligations of being a scientist has been published by the USA National Academy of Sciences (NAS) [5] (*www.nap.edu/catalog/12192.html*).

Outliers are common in scientific experiments and are to be expected from statistical analyses. They should not be ignored as they may need an explanation and an investigation of them may lead to new knowledge. However, some researchers are never fully at ease in accepting them because of a desire for order and some people battle with the issue and may fudge the data because of a psychological need for order. Scientists have a duty to identify and rule out sources of error in their experiments. This includes methods devised to exclude unintentional bias, such as the use of control groups and "double blinded procedures" (called double masked when the research involves the eye and vision). However, intentional bias seems to be ignored by many scientists and the degree of it in science remains unknown.

Research Training

In the past students were required to memorize facts and regurgitate the learnt information in examinations so that the instructor could judge the knowledge of the various students and rank them based on the examination results. Over the years the value of this approach has been questioned. With a new approach Danish students are permitted to use the internet during examinations and this sparked a debate about whether this is cheating and plagiarism or creative innovation *(http://eduratireview.com/2010/10/student-cheating-and-plagarism-or creativity-and innovate . . . 8/4/2011).*

Researchers need to learn how to write and publish scientific papers [13]. Teaching and education involves more than teaching about how to plan and carry out experiments and shed light on the data. Training in research involves more than learning about how to design and perform experiments and interpret data. Genuine face to face instruction by experienced experts and mentors provide oversight and advice that is both necessary and essential. Accepted scientific customs and ethics are now taught in universities and researchers must adhere to accepted standards. Most researchers study the history and philosophy of science while training to become scientists. Philosophers of the stature of Francis Bacon (1561-1626), John Stuart Mill (1806-1873), Bernard Arthur William Russell (1872-1970) and Sir Karl Riamund Popper (1902-1994) have made notable contributions to science and their teachings are still most worthwhile. In the USA individuals involved in research are required to undergo mandatory training at regular intervals on a variety of topics. Some fall under the occupational and environmental safety office (OESO) and occupational safety and health administration (OSHA) and relate to the safety and health of workers in research laboratories, such as the shipment of biological materials, general laboratory safety, blood borne pathogens (BBP), tuberculosis safety training, fire prevention, indoor air control, laboratory waste disposal, occupational exposure to biological pathogens (like human immunodeficiency virus [HIV], hepatitis B, hepatitis C, and tuberculosis), the Health Insurance Portability and Accountability Act of 1996 (HIPPA) privacy and security training, compliance update training, infection control. Online training is available at different institutions, such as *https://ors.duke.edu*. Researchers are required to take courses of the collaborative institutional training initiative (CITI) and pass an electronic examination.

Manuals of the responsible conduct of research such as the committee on science, engineering and public policy (COSEPUP), Ensuring the Integrity, Accessibility and Stewardship of Research Data in the Digital Age, National Academies Press, Washington, DC 2009, On Being a Scientist, National Academies Press, ed.3, Washington, DC 2009 are obtainable (Responsible Science: Enduring the Integrity of the Research Process, Vol. 1, National Academy Press, 1992, 199 pp).

Scientists Generating Financial Gain from their Research

Until the *Bayle-Dole Act* of 1980 academicians considered the worlds of the university and industry incompatible. Serious COI become apparent and they can be diminished if academicians work with, but not for the industry. Since then universities have built significant collaborations particularly with major drug companies. The interaction generates additional funds to support universities and their research activities. It also encourages the transfer of technology from academic laboratories into the private sector as desired by the USA Congress. Aside from being meritorious commercial ties between investigators in universities and those in industry made many scientists wealthy. Because of the vast incomes generated by researchers with federal research grants the congressional watchdog John Dingell asked the Department of Health and Human Services (DHHS) and the Defense Contract Audit Agency (DCAA) in 1993, to embark on an investigation of "profiteering" by academic researchers.

Until recently few scientists were motivated by personal financial rewards. Those with this inclination ran the risk of being ostracized by the scientific community and could lose their credibility as a researcher for life. Particularly in the USA since the *Bayh-Doyle Act* of 1980 many scientists have dreamed of establishing wealth from their research, but very few have achieved this goal. Scientists who have generated considerable wealth from their research include Irving Weissman (1939-), David Sinclair, Stanley N. Cohen (1935-), and Herbert W. Boyer (1936-). Those who have established it have done so by patenting PI and the patent has become profitable after creating a startup company or by developing collaborations with industry.

Industry Academia Collaborations

Collaborations between academia and industry are highly desirable but they are not without their downside. Industrial research is highly competitive and needs to be performed in secrecy if products are to become commercialized. COI are common. Hence mutually beneficial partnerships and collaborations between academia and appropriate industries are highly desirable. In collaborative research within academic and research institutions material is commonly transferred from one institution to another by a Material Transfer Agreement (MTA). This contract stipulates the rights of both the recipient and the provider with regard to the transferred items and their by-products. The subject is extremely complicated and MTAs differ with regard to the terms and conditions. When human material is transferred IRB approval is required. Over the years major scientific advancements have been made for the benefit of mankind.

Patents

Before the 1970s biomedical researchers rarely patented intellectual property (IP). The invention of recombinant DNA (rDNA) took place in the laboratories of Paul Berg (1926-) Stanford University (Stanford) (*http://www.stanford.edu/*), Stanley N. Cohen(1935-) (Stanford) and Herbert W. Boyer (1936-) (University of California San Francisco (UCSF). For his contribution to the discovery of rDNA Berg shared the 1980 Nobel Prize in chemistry (*http://www.nobelprize.org/nobel_prizes/chemistry/laureates/1980/berg-facts.html*) with Walter Gilbert (1932-) (*http://www.nobelprize.org/nobel_prizes/chemistry/laureates/1980/gilberfacts.html*) and Frederick Sanger (1918-) *(http://www.nobelprize.org/nobel_prizes/chemistry/laureates/1980/sanger-facts.html)*. One of the most significant patents was granted to Stanford in 1980 for the process of rDNA. Products of rDNA spawned a new technology which prospered in numerous industries. The life of the patent lasted 25 years until December 1997 and during that time the technology was licensed to 468 companies. Stanford and UCSF accumulated more than $255 million from the licensing revenue and the management of the patents became the gold standard for universities [23]. As a result of their research academicians often generate inventions and other potentially worthwhile intellectual property as a result of their research. Some such achievements are patented, but without their commercialization they will not generate revenue and benefit anyone. For researchers desiring to spawn wealth from their research a crucial step is the patenting of their discoveries known as IP, but for a patent to generate income it must become commercialized and generate royalties. The axiom of "Publish or Perish" has become replaced by "Patent and Prosper" [26].

Society has always encouraged new discoveries and provided inventors with incentives related to patent laws, which undergo changes every now and again. To patent an invention and commercialize it to generates income is time consuming and very expensive. It also has financial risks that outweigh the entrepreneurial ambitions of most academicians. Hence it is not surprising that the vast majority of patents filed by researchers are at non-university research institutions and industry. A major detrimental effect of patents on science is the secrecy needed to protect patent rights. Until the German patent law was changed on February 7, 2002 professors in German universities were exempt from notifying the university that employed them about their inventions. This privilege of the professor (Hochschullehrerprivileg) was an upshot of article 5 of the German constitution relating to freedom of science and research [18]. This benefit was regardless of the fact that the professor was funded by the university and taxpayers. Later the German ministry of science contended that this privilege for the professor not only depleted the country of an economically precious asset, but also inhibited science and technology transfer. After a 2002 patent law change in Germany professors would no longer hold exclusive IP for their inventions.

Reputable institutions share the rights of patents with investigators who participate in their creation. For example, while working in the laboratory of Frank McKeon at Harvard Jiang Yu and Kayoko Kimbara signed an agreement giving Harvard all rights to any discovery or invention made by them during their research fellowships, yet when Harvard patented two genes found by them the university listed both Zhu and Kimbara as co-inventors. Although both Cohen and Boyer received Nobel Prizes they acquired wealth from their patents. Anada Mohan, a scientist working at the General Electric Company, invented a bacterium from the *Pseudomonas sp.* capable of degrading crude oil with the potential of destroying oil spills. General Electric applied to the USA Patent Office (USPTO) for a patent listing Chakrabarty as inventor, but his request was denied because living organisms were deemed non-patentable. However, the Supreme Court ruled that because the specific bacteria did not exist in nature and had indeed been created a patent was possible and indeed one was granted on March 31, 1981.

Other researchers create small companies and hope that they will prosper and enable them to sell shares for millions of dollars, and perhaps even sell the company at a vast profit. Only then does the inventor produce income. Because creativity can generate considerable revenue complex laws and institutional regulations govern patents. The policies governing the distributions of income spawned by patents vary from country to country and within institutions. Employers can claim the rights to an invention by an employee or leave all or part of the rights with the employee.

IP has limited financial value unless it is patented and even then the patent has no significant value until it becomes commercialized and generates vast amounts of money. Because money drives everything universities now encourage their faculty to be entrepreneurial and acquire patents on IP whenever possible. After the *Bayle-Dole Act* universities developed different policies about relationships with industry and they ranged over a spectrum from outright prohibition to qualified approval to an intimate relationship. Even the HHMI differed from universities such as Stanford in policies concerning consultations and IP. Scientists realized that by generating intellectual property they could apply for patents and potentially produce considerable extra income by commercializing the patents. They sometimes form startup companies. After obtaining patents the course for those taking this path was not always smooth and disputes were often generated.

In 1989 the HHMI made Weissman, an immunologist at Stanford one of its investigators, but without knowing that Weissman had already created SyStemix, Inc, in 1988 to develop immunotherapeutics with Stanford's blessing and with the assistance of a Stanford colleague and financial aid from a venture capitalist from New York. By 1991 this company became one of the hottest biotech possessions on Wall Street. This economic accomplishment became known to the HHMI officials after they read about it in *The Wall Street Journal* and they were not happy [2]. Stanford viewed the extension of Weissman's endeavors into the private sector acceptable within the university rules, but in October 1992 the HHMI considered his business a possible conflict of interest. Hence, in October 1992, Weissman resigned his HHMI investigatorship, a position for which many scientists would gladly give their eye teeth. Other HHMI investigators became concerned that they might need to sever their relationship with the HHMI, but they retained their bond with HHMI when they created new companies. David Sinclair PhD is a prominent professor at Harvard University (Harvard) *(http://www.harvard.edu/)* with expertise in genes that slow the aging process as well as the antioxidant resveratrol that is found in red wine, and other grape extracts. In 2006 Sinclair headed an investigation demonstrating that resveratrol could offset the adverse effects of overfeeding laboratory mice. A notable benefit was that it enabled the overfed mice to live more than 12 weeks longer than those not given this grape derived chemical. Despite this beneficial experimental therapeutic effect in mice a therapeutic effect has not been replicated in humans. Because of the great commercial value of his research he made a considerable amount of money from his connections with the pharmaceutical industry. Sinclair created Sirtris Pharmaceuticals in 2004 with the venture capitalist Christoph Westphol MD, PhD. This company was studying resveratrol for potential use as a human drug. When Glaxo acquired Sirtris (currently a division of GlaxoSmithKline PLC [GSK]). Sinclair received more than $8 million. He also received an annual salary of $297, 000 from the company as a consultant. In August 2008 Sinclair joined the scientific advisory board of Shaklee Corp., on a paid position, but he resigned shortly thereafter because this company misused his picture and name in advertisements on Vivix sales web sites giving the impression that he endorsed Vivix, which was alleged to be the world's best anti-aging supplement [30] *(http://online.wsj.com/article* email/SB123025446150734561-1MyQjAxMD14MzIwIy . . . 12/26/2008).

While there may be some truth in health benefits of red wine and resveratrol related to cardiovascular health other researchers jumped on the band wagon. One was Dipak Kumar Das (1947-2013) a professor of surgery and the director of the cardiovascular research center at the University of Connecticut (UConn) *(http://www.uconn.edu/)*. He was charged with fraudulent research related to 26 articles [29]. UConn produced a 60, 000 page document to substantiate the scientific misconduct of this extremely prolific researcher who has 588 articles to his credit in Google scholar. He retired from his academic position in 2012 and while Das was under investigation for research misconduct by the Office of Research Integrity (ORI) he suddenly died at 67 years of age for undisclosed reasons. The cause of his death was not provided in his obituary [4]. His name appears on hundreds of articles that focus on resveratrol. UConn found 145 instances of fabrication by Das over seven years [15].

The traditionally the science of discovery for its own sake has become marginalized by a thrust to perform clinically relevant research that is fundable and publishable in high-impact journals. Human research is extremely important and it needs to be carried out in an efficient ethical way to produce meaningful results. Intellectual curiosity and open debate over research findings and hypotheses is a widely practiced component of science.

Research Resources

By screening vast numbers of professional journals, newspapers, newsletters, books, conferences and other sources a researcher can obtain summaries of important reports relevant to a scientist's research. The National Center for Biotechnology Information (NCBI) *(www.ncbi.nim.nih.gov/)* is an extremely valuable resource on molecular biology information for scientists. One of its most precious assets is a collection of 22 million citations from the biomedical literature known a PubMed with its pushbutton function. Other included databases are Blast, GeneBank, Taxonomy, Nucleotide, and Gene.

Experimentation on Human Subjects and in Animals

Research on human subjects is diverse and influenced by numerous regulations.

Research using animals is extremely important and needs to be performed as humanely as possible with as few animals as necessary. On August 24, 1966 the Animal Welfare Act (Laboratory Animal Welfare Act) was signed into law by Lyndon Baines Johnson (1908-1973) the 36th President of the USA. This law covers the use of live and dead warm-blooded animals and regulates the treatment and care of them in research, as pets and in exhibitions. The Act was originally created for the protection of large animals and pets. All institutions that have facilities for animal experimentation need to be certified by an institutional animal care and use committee (IACUC). At least one member of that committee must be a veterinarian and another one must have no affiliation with the facility.

References

1. Ahearne JF, Honesty: Ultimately, ethics in scientific publishing, as in life, comes down to one word., in For the Record: American Scientist Essays on Scientific Publication. 2011, Sigma Xi, The Scientific Research Society: Research Triangle Park, North Carolina. 1-6.
2. Anderson C, Hughes' tough stand on industry ties. Science, 1993. 259(5097): 884-886.
3. Anonymous, Over 100 Titles by Nobel Laureates and on the Nobel Prizes. 2012.: World Scientific Publishing Co.
4. Anonymous. Orbituary. Dipak Kumar Das. The Hartford Courant. 2013 September 21.
5. COSEPUP (Committee on Science E, and Public Policy), On Being a Scientist. A Guide to Responsible Conduct in Research, Third Edition. 2009, Washington DC: The National Academies Press. 82.
6. Crewdson J. The great AIDS quest. Chicago Tribune. 1989 November 19.
7. Crewdson J, Science Fictions: A Scientific Mystery, A Massive Coverup and the Dark Legacy of Robert Gallo. 2002, Boston, New York, London: Little, Brown and Company.
8. Diamond BA, Hulka BS, Kerkvliet NI, et al., Silicone Breast Implants in Relation to Connective Tissue Diseases and Immunologic Dysfunction. A report by the National Science Panel to the Honorable Sam C. Pointer Jr., Coordinating Judge for the Federal Breast Implant Multi-district Litigation, 1998.
9. Farndon J, The Great Scientists: From Euclid to Stephen Hawkins. 2007, New York: Metro Books.
10. Friedberg EC, Sydney Brenner: A Biography. 2010, New York Cold Spring Harbor Laboratory Press.
11. Hilts PJ. Breast implants found to irritate immune system. The New York Times. 1993 March 20.
12. Holmes R, The Age of Wonder: How the Romantic Generation Discovered the Beauty and Terror of Science. 2008, New York: Panthcon Books a division of Random House, Inc.
13. Huth EJ, How to Write and Publish Papers in the Medical Sciences. Second edition. 1990, Philadelphia: ISI Press.
14. Ioannidis JPA and Khoury MJ, Improving validation practices in "omics" research. Science, 2011. 334(6060): 1230-1232.
15. Jaslow R. Red wine researcher Dr. Dipak K, Das published fake data: UConn. CBS News. 2012 January 12.
16. Kantha SS, Clues to prolific productivity among prominent scientists. Medical Hypotheses, 1992. 39(2): 159-163.
17. Kennedy D, Mixed grill. Science, 2007 317(5842): 1145.

18. Kilger C and Bartenbach K, Patent law. New rules for German professors. Science, 2002. 298(5596): 1173-1175.
19. Lock S, A Delicate Balance: Editorial Peer Review in Medicine. 1985, London: Nuffield Press.
20. Medawar P. Is the scientific paper a fraud ?. The Listener 1963 September 12:377-378.
21. Popper KR, The Logic of Scientific Discovery. 1934: Routledge publishers, Taylor and Francis group.
22. Price JM and Rosenberg ES, The silicone gel breast implant controversy: the rise of expert panels and the fall of junk science. Journal of the Royal Society of Medicine, 2000. 93(1): 31-34.
23. Program. TSUL, Published in Intellectual Property Management in Health and Agricultural Innovation: A Handbook of Best Pracrtices (eds. A. Krattiger, RT Mahoney, L. Nelsen et al.) Oxford, UK and Davis, USA: MIHR and PIP.
24. Roberts RM, Serendipity: Accidental Discoveries in Science. 1989, New York: John Wiley and Sons, Inc.
25. Ryan MJ, Replication in field biology: the case of the frog-eating bat. Science, 2011. 334(6060): 1229-1230.
26. Schachman HK, From 'Publish or Perish' to 'Patent and Prosper' Journal of Biological Chemistry, 2006. 281(11): 6889-6903.
27. Smith RW, The Great Debate, In Cosmology: Historical, Literary, Philosophical, Religious, and Scientific Perspectives, Hetherington, Norris S. (editor) 1993, New York: Garland Publishing Inc.
28. Tomasello M and Call J, Methodological challenges in the study of primate cognition. Science, 2011. 334(6060): 1227-1228.
29. Wade N. University suspects fraud by a reseacher.who studied red wine. The New York Times. 2012 January 11.
30. Winstein KJ. Harvard anti-aging researcher quits Shaklee advisor. The Wall Street Journal. 2008 December 26.

CHAPTER 3

Reactions to Scientific Misconduct

Apprehensions about fraudulent research triggered research institutions, professional societies, government agencies and congressional oversight committees to seek policies that will boost research integrity and improve the excellence of the research milieu. Prior to 1970, known examples of research misconduct in PHS funded projects were extremely uncommon, but over the next two decades incidents of embarrassing fraudulent research popped up at different institutions. This scandalous dishonest research hit the scientific world and the general public like a bolt of lightning, because trust is a cardinal component of science and deceit of this nature had never been suspected. Examples of data fabrication, falsification and plagiarism came to the attention of the media, the public, Congress, academic circles as well as the scientific population [9]. Perhaps even more surprising was the fact that this fraudulent activity took place at prestigious institutions, such as Harvard University (Harvard) *(http://www.harvard.edu/)*, at Yale University (Yale) *(http://www.yale.edu/)*; Massachusetts General Hospital (Mass Gen) (an affiliate of Harvard) and Boston University (BU) *(http://www.bu.edu/)*. The fraudulent research that surfaced was largely funded by the National Institutes of Health (NIH). Some newspaper reporters speculated that an epidemic "crime wave" of scientific fraud had emerged. After realizing that scientific misconduct was not the product of a handful of weirdoes steps were taken to curb this unacceptable behavior. At the time of these scandals institutional and NIH mechanisms for coping with fraudulent science did not exist and the federal government was extremely slow in developing policies and procedures for dealing with it. It was obvious that something needed to be done about the apparent new scourge. Most academic institutions, funding agencies as well the USA federal government did not significantly establish policies and procedures to deal with fraudulent research until 1986. In May 2013 the National Science Foundation (NSF) added intellectual merit and impact to the criteria that reviewers would use to evaluate grant applications.

The apparent upsurge in fraudulent research triggered responses by the USA federal government, professional scientific societies, journal editors, research institutions, universities, federal funding agencies and the scientific community.

In an attempt to bring fraudulent research under control congressional oversight committees sought policies that would fortify the integrity of the research milieu.

Response by the USA Public Health Service

In 1985 after the media continued to report cases of serious research misconduct the USA Congress passed legislation requiring the Public Health Service (PHS) to establish methods of investigating allegations of scientific misconduct entailing research funded by the USA federal government. By July 1986 the PHS embraced required guidelines for research institutions in dealing with research misconduct. Initially the PHS listed individuals accused of scientific misconduct in their alert system, but later abandoned this practice in favor of only providing the names of those found guilty. In a major departure from the authority granted in Section 493 the PHS made an effort to include under its jurisdiction all cases of misconduct stemming from PHS funded research [36]. Years after forms of research misconduct were reported in the media the PHS issued a proposal entitled "Responsibilities of PHS awardee and applicant institutions for dealing with and reporting possible misconduct in science" (http://ori.dhhs.gov/reg-sub-part-a).

The DHHS distributed rules that allowed universities with NIH grant support to design their own procedures for investigating accusations of research misconduct. They included an assurance of due process and were mainly in response to the threat of congressional action if the shortcomings were not rectified. On August 8 1989 the DHHS published new rules in the *Federal Register* for dealing with misconduct in the research laboratory. Institutions getting money from the PHS were required have policies and procedures for dealing with accusations of misconduct and to make declarations equivalent to those currently needed for research affecting humans or animals. Institutions were empowered with the responsibility of preventing, detecting, investigating, reporting, and resolving allegations of scientific misconduct. The rules delineated processes for maintaining order and secrecy in alleged misconduct inquiries including conscientious efforts to re-establish reputations when misconduct allegations are unverified. The rules also protected individuals who brought attention to charges of scientific misconduct in good faith so that their positions and reputations would not be jeopardized. Later rules were about what institutions must do to promote scientific integrity. DHHS retained the "ultimate responsibility and authority for monitoring such investigations and becoming involved in these investigations if appropriate or necessary". Institutions were required to complete a preliminary analysis within 60 days except if additional time was merited. As a rule if a full-blown scrutiny was needed, it should be finalized in 120 days.

Under the *Public Health Service Act* regulations generated by the PHS place requirements on institutions that receive PHS research funding and ORI monitors the institutional compliance with these requirements using both the assurance and the compliance review programs. Every institution that gets research funds from a PHS agency must have on file an assurance form with ORI to guarantee that the institution has policies and procedures for taking care of allegations of scientific misconduct. By 2008 4, 828 institutional assurances were on file with ORI.

The Public Health Service Act

The USA government responded to the fraudulent research by adding a section (Section 493) to the *Public Health Service Act* in 1986 to maintain the expectation that federal research funds would be used in an ethical way to protect the integrity of their use. By concentrating on scientific fraud, it also tried to maintain the equilibrium between suitable government action and the authority of institutions to evaluate and deal with cases of fraudulent research with as little governmental participation as possible. The House committee report that accompanied Section 493 supported an institutional review of both scientific misconduct and fraud. It also stated that institutions need not have a standing committee to investigate cases of misconduct, but should have a mechanism to establish a committee swiftly when the need arose. In a major departure from the authority granted in Section 493 the PHS make an effort to include under its jurisdiction all cases of scientific misconduct stemming from research funded by the PHS [36]. Because of a conflict of interest the subcommittee had difficulty accepting the idea that research institutions could investigate themselves without conceding objectivity and the vivacity of the investigation. Advisors would include senior and junior researchers and representatives of professional societies, clinical research organizations, scientific and medical publications and the NAS.

Congressional Legislation

Legislation was passed setting up a government fraud office. Governmental oversight over the spending of taxpayer money is lawfully authorized and obviously appropriate. In congress, legislators kept an eye on how procedures for investigating cases of scientific misconduct were progressing and they were discontented. Soon after these hearings additional fresh examples of significant scientific misconduct emerged and Mark Straus at BU see chapter 14) and the manner in which the institutions dealt with involving John Darsee at (Harvard) with them cast doubt on the capability of institutions to perform impartial investigations into allegations of research misconduct by members of their own research personnel.

In 1985 after the media continued to report examples of significant research misconduct, the American Congress passed legislation requiring the PHS to establish procedures for investigating allegations of research misconduct funded by federal grants. By 1989 science misconduct was the issue of the day. Some members of congress considered fraudulent research unacceptable and more common than desirable. Aside from proposals for regulations three congressional hearings and a study by the office of the inspector general in the DHHS zoomed in on this subject.

In addition the human resources and intergovernmental relations subcommittee held hearings to assess comparable charges and to delve into public wellbeing and security concerns that might stem from deceptive research. The 1989 rules delineated processes for maintaining order and secrecy in alleged misconduct inquiries including conscientious efforts to re-establish reputations in instances where misconduct allegations are unverified. The rules also protected individuals who bring attention to charges of scientific misconduct in good faith so that their positions and reputations are not jeopardized. To prevent misconduct future rules were also anticipated for authorship practices, the preservation of raw laboratory data, and processes for audits. In 1998 the scientific community was serious about coping with research misconduct, but was received in still recovering from the rough treatment that science received in back to back congressional hearings on research misconduct during the Spring of 1988.

A fallout from the publicity of scientific misconduct was the belief that the work of scientists was often mediocre and inappropriate for establishing well-informed policies. There was clearly a need for the public to become more conversant with science. Scientists were still highly regarded perhaps second to fire fighters in status, but the confidence in them and their prestige depended on recreating confidence in the integrity and credibility of science.

Response by the USA federal government

The Department of Health, Education and Welfare (DHEW) revised and expanded its regulations for the protection of human subjects (45 *CFR* part 46) late in the 1970s. In1980 DHEW split into the Department of Education (ED) and the DHHS (*http://www.hhs.gov/about/hhshist.html*).

Response by USA Presidential Commission

In 1981 a presidential Commission for the Study of Ethical Problems in Medicine and Biomedical and Behavioral Research started holding hearings on the ethical issues in medicine and biomedical and behavioral research. Federal regulations required research institutions to establish written instructions for dealing with accusations of research misconduct.

Congressional Hearings

Administrators of science testified at the congressional sub-committee hearings doubting any important increase in fraudulent behavior existed. A greater awareness and interest in the topic suggested a growing distrust of science and scientists by the public and Congress [7]. Before 1980, known examples of research misconduct in PHS funded projects were extremely uncommon, and performed at prominent universities, but subsequently a handful of extremely well known cases came to the forefront even though they represented only a small fraction of all research and training awards. In response to these reports of biomedical fraudulent research the first congressional hearings on scientific fraud were held between March 31 and April 1 1981 by the Investigations and Oversight subcommittee of the House Committee on Science, Space and Technology chaired by Albert Arnold "Al" Gore, Jr (1948—) (Democrat, Tennessee) [7, 8]. During the prior five years about a dozen cases of scientific misconduct were investigated by the NIH and some had still not yet been resolved [8]. These hearings focused on institutional experiences in dealing with fraudulent biomedical research and the different institutional spokespersons testified. It became apparent during the hearings that the gap between Congress and science was large and with few exceptions neither camp knew or understood much about the workings of the other. During these hearings, delegates of the research community put forward the belief research fraud was extremely rare, and that these uncommon episodes were satisfactorily taken care of by existing self-regulatory procedures. NIH officials, bioethicists, and research directors as well John Long and Philip Felig received invitations to appear before Congress which wanted to know what was happening in biomedical research in the USA. Donald S. Fredrickson (1924-2002), director of the NIH from 1975-1981, emphasized that scientific fraud was uncommon and readily dealt with by the self-correcting nature of scientific inquiry. Fredrickson acknowledged that the findings of numerous experiments had not been confirmed, but he pointed out that noteworthy experiments will be replicated and that each hypothesis will be tested again under many other circumstances. He rebuffed the opinion that individuals accused of scientific misconduct should not receive grant support when accusations had not yet been proven. He also stated that allegations on their own should not hinder the ability of a researcher to obtain ongoing grant support. In the absence of proof the NIH needs to be exceptionally vigilant about black-listing individuals. He pointed out that if the NIH acted carelessly and harshly more damage would be done to science than to an individual.

Philip Handler (1917-1981), president of the National Academy of Sciences (NAS) at the time declared that the fraud issue in science was "grossly exaggerated" and with wishful thinking he felt that the problem should be handled internally by scientists. He also told the congressional committee that he had come across two cases of falsified research at Duke University many years apart. Both researchers left the institution and the scientific community heard no more from them. NIH officials rejected the notion that scientific misconduct was widespread and they announced important alterations in policy. William F. Raub associate director of the NIH, from 1978-1983 pointed out that in the future the results of ongoing investigations of scientific misconduct will be disclosed to individual institutes of the NIH, national advisory councils and boards. Some members of Congress believed that fraudulent research by scientists is more common than desirable. In the middle of an apparent increase in fraudulent research as well as a misuse of federal funds controversial federal regulations related to scientific misconduct were introduced 1980 [41].

By June 1990 Dingell had held three congressional hearings focusing primarily on the *Baltimore Affair* in which the Nobel laureate might have participated in possible fraud. The Dingell congressional sub-committee got involved in the investigations of notable cases, such as the accusations against Teresa Imanishi-Kari and Robert Charles Gallo. The Dingell committee spent a substantial amount of time investigating the manuscript in *Cell* that received substantial attention after O'Toole questioned the validity of some of its contents. This Dingell committee clearly demonstrated that it was not an appropriate place for judging the authenticity of complex experimental findings. During the hearings two days were planned for the examination of the notorious Baltimore case and any person targeted by the Dingell inquiry was strongly advised to obtain the input of a lawyer. Indeed Baltimore employed legal firms from both Washington and Boston. The decade long ordeal of *The Baltimore Affair* reached a high point on May 4 1989 when Baltimore and others testified before the congressional hearings on scientific fraud was approaching an endpoint. A report by the Dingell subcommittee released in May 1989 identified several mistakes in the controversial study. Some entries in the laboratory notebooks contained incorrect dates and were perhaps fabricated. Baltimore was interrogated extensively by the Dingell subcommittee over the paper in *Cell* that he coauthored with Weaver, Imanishi-Kari and Constantini [45]. Baltimore was not happy with the grilling and pointed out that the inquiry into his research activities could intimidate researchers in the USA. In June 1989 Baltimore became aware that the Dingell subcommittee had recruited the Secret Service to evaluate the laboratory notes of Imanishi-Kari. Baltimore stressed "I do not question the right of this subcommittee to conduct this investigation. I do not question the right of this subcommittee to demand accountability for government funding of scientific research, but I do question in the most serious way the manner in which this investigation has been pursued".

Those interviewed involved people from Tufts University, MIT, and the NIH. The content of the paper, which documented complex research involving immunology and serology was beyond the comprehension of many scientists, and the disputed subject matter was clearly beyond the understanding of laypeople and the vast majority on members of Congress.

After the hearings chaired by Gore science in the USA came under the scrutiny of the congressional oversight and investigations subcommittee of the house energy and commerce committee chaired by the powerful terrifying John Dingell (Democrat, Michigan), who was extremely critical of the performance of government agencies, asserting that they cannot satisfactorily deal with issues such as disclosure, notification and protection of whistleblowers. He established a reputation as a class bully for his brutal committee hearings, which had lots of the trappings of a criminal trial with persistent probing of the witnesses. Dingell forced American scientists to take research fraud seriously. He stressed that scientists need to understand that the best way, perhaps the only way, to avoid the threat of 'science police' is for them to show that they have the ability and will to police themselves. Immediately before the congressional hearings by the Dingell subcommittee in May 1989 NIH officials first learnt that the Secret Service had been probing the Imanishi-Kari case for 9-months. Delays in reaching a decision were apparently caused by Healy, and demands by Imanishi-Kari's attorneys to gain access to evidence generated by the Secret Service. The powerful chairman of the subcommittee of the House oversight and investigations got involved in the investigations of notable cases, such the accusations of Teresa Imananishi-Kari and David Baltimore.

Dingell inappropriately pried into investigations by the Office of Research Integrity (ORI).

Imanishi-Kari and Baltimore differed in their interpretations of the experimental findings. O'Toole also argued that the data could be accounted for in another way.

Scientists are assumed innocent until guilt is established, but some law makers felt that this is a legal concept relevant to punishment and inappropriate to the issue of rewarding someone with a research grant. Over and over again members of the USA Congress were astonished at the faith that NIH study sections had in investigators that were subsequently discovered to have falsified their work.

The interest of the USA federal government in scientific misconduct gathered momentum in 1988. New regulations were proposed by three congressional committees in April of that year (the House Committee on Science, Space and Technology, the House Subcommittee on Human Resources and Intergovernmental Relations and a congressional subcommittee of the House committee on Oversights and Investigation) that held independent hearings on fraudulent research. At a congressional subcommittee hearing in April 1988 the accusers, but not the accused, were provided an opportunity to testify. Hence those on whom blame was placed lacked a chance to explain their side of the dispute. Moreover they were not even informed about the hearing. The pursuit of research misconduct by the USA government gathered momentum when two different congressional committees began holding hearings to consider contemporary accusations of research misconduct and investigations of them in biomedical research in April 1988.

Dingell wrote to the NIH questioning whether the Gallo laboratory had acted unprofessionally with regard to the Pasteur institute after reading the long narrative in the *Chicago Tribune* by the reporter John M. Crewdson (1945—) [15], but this was made difficult because of a lack of published procedures in dealing with such allegations [27, 42].

On April 11 1988 representative Ted Weiss held a hearing on scientific fraud and misconduct. On the following day the Dingell subcommittee also held a hearing. A momentous event in *The Baltimore Affair* took place in August 1988 when Dingell subpoenaed loose notebook pages belonging to Imanishi-Kari for a forensic investigation by the Secret Service. Because of this and the unjust pestering of scientists at congressional hearings, more than a few scientists rightfully likened Dingell's actions to the witch hunts of late republican senator Joseph Raymond McCarthy (1908-1957) in the 1950s during the cold war against communism. Dingell resented the comparison, but he established a reputation as the most hostile and controlling member of the USA Congress. His behavior to scientists during the congressional hearings on "Fraud in NIH Grants" in 1988 terrified the scientific community because of his prosecutorial manner. Fraud in NIH grant applications hearings before the subcommittee on oversight and investigations of the committee on energy and commerce [28]. Don S. Doering of the Whitehead Institute at the Massachusetts Institute of Technology (MIT) felt that the Dingell's investigation was a serious encroachment on the free pursuit of science and upon the right for the self-regulation of scientists [22]. Like others Doering believed that scientists, just like other citizens in civil offenses, have the right to be judged by their peers in charges of alleged scientific misconduct. Like many scientists Doering believed that a congressional subcommittee hearing was not the appropriate forum for such inquiries. Next week Dingell took on Baltimore in an inquiry into the process of science frightened many scientists [19].

The House Committee on Science, Space, and Technology, chaired by Representative Robert A. Roe (Democrat, New Jersey) held additional hearings which concluded in June 1989. Roe's committee paid attention to scientific misconduct like the Dingell subcommittee, but in contrast to the Dingell committee, it focused on the big picture of scientific misconduct. This was perhaps an attempt by Roe to reclaim the preeminence his committee had on the subject in the early 1980s.

Prior to the extensively publicized August 1 1991 congressional subcommittee on oversight and investigations hearing a number of leading scientists published powerfully worded views in *Nature* and *The Journal of NIH Research* stating how they thought disputes about research are best decided by those appropriately trained with the relevant expertise to judge disagreements [5].

The congressional hearings considered policy implications of fraudulent research. Representatives of the scientific enterprise testified at these hearings and they maintained without exception that research fraud was rare. They also felt that ample current procedures were available to take care of them and that it was unnecessary to introduce further policies or procedures to guarantee the integrity of biomedical research funded by federal grants. The congressional investigation into the *Cell* paper led Baltimore, numerous scientists, and others into a campaign of lectures, opinion statements and letter writing. Gossip started about the possibility of legislation being brought up in Congress that would remove investigations of fraudulent science out of the hands of both the NIH and universities. It is a matter not only of morality, but also of self interest [21].

After serving as the chairman of the congressional subcommittee investigating fraudulent research Dingell delivered the Shattuck lecture at the annual of the Massachusetts medical society in June 1993. This lecture honored Frederick Cheeves Shattuck (1847-1929), who graduated from Harvard College in 1868 and received his medical degree from Harvard medical school in 1879. Dingell used the lecture to summarize the findings in the congressional subcommittee that he chaired. His views were controversial and several scientists who commented on the lecture were critical of its contents [20] [33] [47]. For example Baltimore pointed out that Dingell's claim that he "looked only at clear-cut cases involving fabrication, falsification and plagiarism" was not true [4].

Baltimore's research was never questioned and he was never accused of scientific fraud, but Dingell asserted that data in the infamous *Cell* paper seemed to be "manipulated" perhaps with an "effort on conceal or to confuse or to deceive". An emotionally charged battle endured when Baltimore, the star witness of the scientific misconduct hearing, took Dingell's symbolic hot seat as an unmistakably hostile witness and faced the powerful terrifying chairman of the subcommittee. In anger the markedly annoyed Baltimore pointed out that he was charged with fraud in the *Boston Globe* [23] story by Peter Stockton Dingell's aide who provided information to the newspaper about the hearing of the Dingell subcommittee in 1988. A surprised Dingle rapidly brought the hearing to a close. The Dingell subcommittee's inquiry into *The Baltimore Affair* took place mainly out of public view, but this case presumably played a cardinal role in hauling this saga on for so long. The Dingell subcommittee employed a staff of investigators, including some who were or had been on the NIH payroll. The fraud busters Stewart and Feder, who were initially the only spokespersons from the NIH on the Dingell subcommittee, received approximately 100 allegations of scientific misconduct each year. This was about 4 times as many as formally received by the NIH. In spite of laboring for 70-80 hours a week only a small number of allegations of research misconduct could be investigated.

They interviewed numerous crucial individuals and evaluated piles of memos, notes of meetings, and other papers as well as telephone conversations between subcommittee staff and federal investigators.

Despite the negative findings of the learned institutions that investigated the allegations against the notorious paper in *Cell* Representative John Dingell (Democrat, Michigan), chairman of the house energy and commerce committee' subcommittee on oversight and investigations stubbornly trailed the case and held three hearings on it between 1988 and 1991. Records of the hearings engrossed virtually all aspects of *The Baltimore Affair* from choice of evidence to its distribution in the public domain. Dingell's participation engendered hostile sentiment between federal administrators and numerous members of the scientific community. A *New York Times* journalist, who happened to be a friend of Baltimore criticized Dingell as a bully and many important researchers in the anti-Dingell camp expressed disapproval of the manner in which the congressional subcommittee mishandled its constitutional responsibility. Congress ignored its need to keep sensitive documents secret and allowed all and sundry the right of entry into ongoing investigations. To Barbara Mishkin, a Washington attorney who specializes in defending clients accused of scientific misconduct the behavior of the subcommittee was extremely inappropriate. Also, the accused scientists did not receive correct due process.

Despite the fact that O'Toole only alleged scientific error in the paper coauthored by Baltimore the congressional hearing discussed fraud and misconduct and the distinction between fraudulent research and honest errors in science became blurred. Many considered the Dingell subcommittee heavy-handed and frightening. It was within their power to subpoena witnesses and records and they made extensive use of this authority.

"Protection of Human Subjects". Title 45, Code of Federal Regulations, Part 46. US Department of Health and Human Services. *http://www.hhs.gov/ohrp/humansubjects/guidance/45cfr46.htm. Retrieved 2008-12-04)/*. During the Dingell hearings the NIH and many universities and medical schools improved their policies for scrutinizing allegations of research misconduct and safeguarding whistleblowers. The issue of how to manage fraud in science was still being debated by June 1989 and appropriate legislation was being contemplated, but a bill had not been formally introduced in the House of Representatives. However, the staff of the Dingell subcommittee distributed draft legislation behind closed doors which they hoped to introduce during the 1989 congressional session. Ironically as pointed out by Barbara Culliton (1989) the very committee that made a case for free and open debate of scientific disputes attempted to conceal the contents of the draft legislation, which contained provisions to transfer some oversight authorization from the NIH to the office of the assistant secretary of health in the DHHS. In 1993 the USA Congress created a twelve member Commission on Research Integrity (CRI) with a mandate to define research misconduct for the DHHS and to recommend ways of improving the practice and oversight of research. The CRI was chaired by Kenneth J. Ryan MD (1926—2002), a Harvard specialist in obstetrics and gynecology and monthly meetings were held from June 1994 until September 1995. The mission of the CRI was primarily to ensure the integrity of science and prevent scientific misconduct rather than to punish those who go astray. The CRI also dealt with whistleblower protection as well as improvements in how institutions deal with scientific misconduct. From the outset the CRI commission considered the entire role of the USA federal government in scientific misconduct. The commission hoped to define science and intended to propose how to promote research integrity and stomp out misconduct. Witnesses that testified before the CRI included respondents in cases of alleged research misconduct, leaders in the scientific community and numerous representatives of professional societies, including Samuel C. Silverstein president of the Federation of American Societies for Experimental Biology (FASEB), a consortium of nine scientific societies embracing a membership of 42, 000 scientists involved in biomedical and related laboratory research.

In 1995 the recommendations of the CRI for dealing with misconduct in science, included a separation of the scrutinizing committee from the one that passes judgment [2].

The CRI contemplated whether or not to include "other practices that seriously deviate" within the definition of misconduct as used by the PHS at that time. The CRI also considered what level of intent should be required for scientific misconduct and who should bear the burden of proof when an assertion of an honest error is made. The CRI also mulled over whether a national regulation should stipulate how long raw data should be retained. The commission also wandered whether statutes of limitation on misconduct claims should be established. After gathering all this information the CRI advised the secretary of the DHHS as well as Congress in ways to potentially improve policies related to these crucial issues. In 1997 the CRI proposed a strict definition of scientific misconduct that applied to all research to assist courts in evaluating complicated cases.

Dingle still intended to release a report on how Tufts University, MIT, and NIH scrutinized the *Baltimore Affair*. Dingell was concerned about whistleblowers and he was under the impression that scientific community was reluctant to protect them. Dingell maintained that he was not trying to regulate how scientists work, but was pursuing the investigation because of concerns that institutions have not taken acceptable action in response to accusations of suspected scientific misconduct. Baltimore and numerous other scientists advised Dingell not to become involved in complex scientific disputes, such as the one precipitated by O'Toole, as he would never understand the details of such highly sophisticated papers. Also Congress was not the place to resolve scientific truth or deceptiveness. Dingell failed to heed this advice, perhaps because of his distrust for scientists and a desire to nail a Nobel laureate to the wall. He was known for its tenaciousness in scrutinizing targets within the range of the subcommittee as it had previously homed in on the Environmental Protection Agency (EPA), the Pentagon, Defense

Department contractors, oil companies and other goals. Dingell commonly targeted well-known individuals so that their grilling would deter wrongdoers who would appreciate the consequences. Dingell learnt this tactic from his father and periodically reminded the public that "My own Daddy used to observe that a few public hangings would help situations to a marked degree". Questions raised by Dingell generated headlines in the newspapers and this publicity boosted Dingell's reputation as someone determined to get scientists. Dingell's attitude to science was surprising as his father John David Dingell Sr (1894-1955)(formerly known as John David Dzieglewicz) supported biomedical research at the NIH during its early days when he served as a member of the USA House of Representatives (1933-1955).

Moreover, his brother James had been employed by the NIH and his wife Deborah Insley Dingell volunteered in the creation of the Children's Inn for young patients and their families at the NIH.

After reading the long narrative by Crewdson in the *Chicago Tribune* [15] questioning whether the Gallo laboratory acted unprofessionally with regard to the Pasteur Institute. Dingell notified the NIH about the concern and the Office of Scientific Integrity (OSI) initiated an inquiry into Gallo, but this was hindered because of a lack of published procedures [16]. Over and over again members of Congress were astonished at the faith that the NIH and its Study Sections had in investigators that were subsequently discovered to have falsified their work.

Response by Universities and Research Institutions

Prior to the 1980s university administrators had little experience in dealing with allegations scientific misconduct. They often looked the other way when misconduct was discovered on their campuses to avoid bringing embarrassment to their institutions. By early 1980s only a handful of research institutions made an asserted attempt to establish written policies and procedures to handle research misconduct. The manner in which academic institutions dealt with John Darsee at Harvard, and Mark Straus at BU brought up uncertainties about the capability of academic establishments to evaluate objectively research misconduct by their own faculty.

Universities differed markedly in the attempts to identify suitable standards for research and in some institutions it remained vague. There was a concern not only about ethical research, but also about the overall quality of research, and about how it may endure in an environment where research integrity is stressed and an excess effort is devoted to scrutinizing mistakes in research. By July 1982 many medical schools in the USA had launched well-defined mechanisms for dealing with someone accused of fraudulent research. Julius R. Krevans MD, the Chancellor of the medical school of the University of California San Francisco (UCSF) stressed the importance of institutions having an established policy to enable them to react promptly and justly when allegations of fraudulent research surface which is almost unfeasible to foresee. Shortly after the Darsee and Straus affairs a number of institutions other than Harvard and Boston University (BU) developed policies and procedures for dealing with scientific misconduct and in establishing guidelines for good research procedures. One recommendation was that scientists should not take on more students in their laboratory than what they can individually supervise. After the Straus affair BU set up a system of random audits to ensure that they would never again be susceptible to unexpected scientific misconduct.

Realizing that the number of publications was a factor leading to research misconduct institutions began to realize that they should place less emphasis on the quantity of publications and more on their quality in the evaluation of faculty for promotion. In September 1985 Donald Kennedy (1931—), president of Stanford University (Stanford) *(http://www.stanford.edu/)* published an account on scholarly authorship which conferred the necessity to elucidate the allotment of accountabilities and recognition for scholarly endeavors and the issues that made determinations of suitable authorship difficult. To help prevent scientific misconduct many institutions introduced the notion in 1988 that faculty promotions and appointments should be based on an evaluation of a limited number of publications. During that year only a handful of academic institutions had precise recommendations for handling allegations of misconduct and gradually guidelines became more widespread. By January 20, 1989, numerous universities still lacked documented procedures for dealing with alleged cases of research misconduct at their institutions.

In December 2009universities expanded the courses available to researchers. Philosophy departments made their courses, which had previously been ignored in the education of researchers, available to graduate students and postdoctoral fellows. Trainees in research needed information on ethics and ethical decisions, yet comparatively modest notice was devoted to incorporating ethical thoughts into the instruction of Responsible Conduct in Research (RCR). The University of Northern Colorado (UNC) *(http://www.unco.edu/)* and the University of Alabama at Birmingham (UAB) *(www.uab.edu/)* used educational videos on RCR to teach students to reach ethical decisions in research. Institutions tried to define suitable responsibilities for government, universities, research institutions, professional societies and scientific journals that could stimulate scientific integrity without generating an unwarrantable authoritarian drain on the activities of scientists. Universities became compelled to scrutinize accusations of scientific misconduct impartially without delay. Universities were required to inform funding agencies and journals about established cases of research misconduct and they needed to be prepared to speak to the media. After wrongdoers were relieved of their positions and obtained employment elsewhere their new employers needed to be notified about their indiscretions.

According to changed policies researchers were permitted to retain copies of the experimental records, but raw research data should not be kept by researchers. It should be retained as long as possible by the institution where the research was performed. Allegations of research wrongdoings need to be investigated regardless of whether the accused remains at the institution or relocates to another university. It is both unethical and unacceptable to make deals with persons guilty of research misconduct.

Universities differ considerably in their endeavors to characterize proper professional criteria for research and sometimes they have stayed indistinct.

Response by Funding Agencies

The director of the National Institutes of Health (NIH) and National Science Foundation (NSF), the two major funding agencies of the USA government, were compelled to investigate accusations of fraudulent research by scientists funded by them and to enforce sanctions when guilt was found. was required to establish a process for the prompt and appropriate response to suspected scientific fraud.

Both the NSF and NIH were powerless to impose sanctions on persons not having grant support from them, even if that person tampered with or sabotaged an experiment of an NSF or NIH funded researcher. The NIH and the NSF established guidelines for mandatory training in Responsible Conduct of Research (RCR). For their long-drawn-out effort in getting a paper published on the Darsee publications in *Nature* [39] Stewart and Feder gained amazing recognition from the NIH and overnight they became the "experts" in scientific misconduct.

National Institutes of Health

In response to the growing number of reports of scientific fraud in the 1970s and 1980s the NIH, which funds most biomedical research in the USA took action to prevent future cases. The NIH devoted much attention to gaining a hold on how to deal with allegations of scientific misconduct, which appeared to be reaching epidemic proportions in these turbulent times. The NIH created a watch-dog organization to monitor and investigate scientific misconduct in all research supported by the PHS.

In November 1980 the NIH established regulations related to scientific misconduct. It placed the responsibility for fraud prevention and detection on institutions and the NIH was empowered to stop all NIH grants at an institution where just a single researcher was discovered to be misusing grant funds or falsifying reports [41]. This threat showed that the NIH was serious about dealing with scientific misconduct and wanted universities to get a clear message that they too needed to deal with the problem even though institutions had no control over those who resorted to this unacceptable behavior. The burden on the institutions was taken extremely seriously, particularly by research universities where a considerable amount of research funding was received. This additional burden for taking care of the research enterprise increased the administrative costs of universities immensely.

The NIH also required all training grants for young scientists to state specifically what training the trainees will receive in research integrity and ethics.

It also became necessary for individuals eligible for federal research grants in the USA to work at institutions that certified that the funded research was performed and monitored in agreement with an extensive list of specifications which included rules of practicing good research. The eagerness for enforcing the good practice varied markedly between institutions.

According to the NIH policy scientists are presumed innocent until unequivocal guilt is established, but some law makers felt that this is a legal concept relevant to punishment and inappropriate when rewarding someone with a research grant. The Straus affair was a notable example of a researcher found guilty of scientific misconduct who retained NIH funding. The rule about rewarding person's only accused of misconduct was also challenged with respect to whether researcher, who admitted falsifying data, would be eligible for future NIH funding should the person apply. However, today the Office of Research Integrity (ORI) now allows persons with a past record of research misconduct to obtain grants provided that specific conditions are met. The NIH took the stand that researchers found guilty of scientific misconduct should not be automatically disqualified from receiving grants, but some congressman felt that researchers who confessed to scientific misconduct were not worthy of a research grant. In 1981 the NIH could bar the awarding of grants to those who were proven guilty of research misconduct, but it still faced the issue of how to deal with research funding in the presence of scientific misconduct allegations.

The performance of research is influenced by the NIH offices that oversee compliance. Before 1988 the NIH lacked procedures for evaluating accusations of scientific misconduct. Robert Charles Gallo and his research team at the National Cancer Institute (NCI), as well as David Baltimore, regrettably became entangled in a quagmire of highly publicized cases when courses of action related to the investigation and prevention of research misconduct had not yet been fully established. James Barnes Wyngaarden (1924—) became director of the NIH in 1982 during the mayhem when fraudulent research funded by the NIH hit the fan. The pressures of his job were unrelenting as well as the aggravation of not having as much influence as credited to the director. Fraudulent research heated up and so did the pressures on the NIH to deal with it. Aside from fraudulent research other issues demanding his attention were anti-intellectualism, animals in research, and fears related of recombinant DNA. Until his resignation from the position of NIH director in April 1989 Wyngaarden was extremely bothered about how scientific misconduct was affecting science. In the Fall of 1988 the

NIH published regulations on research misconduct. For years cases of research misconduct often did not receive attention until prodded by the press or congress. Prior to 1989 the NIH lacked procedures for coping with accusations of scientific misconduct. The NIH took steps to deal with the presumed increase in rogue behavior.

By 1992, the NIH required all grantee and applicant institutions to provide assurances that they had policies and procedures to encourage responsible research practices and research applicants needed to acknowledge their awareness with them and indicate how they intend to store research data accumulated in studies. The training of developing researchers often faced significant gaps in their education and commonly lacked instruction in good research practices and most of them were unaware of the issues that affect professional behavior in the present-day research environment. Universities were compelled to make certain that every student funded by the NIH obtain proper training in RPR, professional and scientific organizations, initiate studies to better understand the dynamics of responsible research compared to fraudulent or sloppy research practices.

Bernadine Healy (1944-2011) replaced Wyngaarden by becoming the new director of the NIH on April 9, 1991 after leaving her position as head of the Cleveland Clinic Foundation's Research Institute. After assuming the directorship Healy faulted many features in the operations of the Office of Scientific Integrity (OSI) including its failure to adhere to established *modus operandi* and breaks in confidentiality. When she testified in a hearing before the Dingell committee Healy steadfastly upheld her position while she attacked the actions of Hadley and the OSI. Her stand provided the lawyers for Imanishi-Kari and Popovic with fodder to defend their clients who could now use of the words of the NIH director to confront the reliability of OSI. The Dingell subcommittee believed that Healy was not only providing critics of either Hadley or the OSI with ammunition to attack this agency, but they also suspected her of inappropriate interference in the OSI.

Many suspected Healy of attempting to undermine ongoing investigations by the OSI. Additional issues fueled the controversy. A major one was the inordinate delay in the issuing of final reports on the two most famous scandals related to accusations of scientific misconduct—the *Baltimore Affair* and the Gallo dispute. These disagreements culminated in the resignation under duress of Suzanne Hadley as the deputy director of the OSI on July 1, 1991. After the adoption of Section 493 of the *Public Health Service Act* in 1993 the NIH published policies and procedures for dealing with possible scientific misconduct without public comment or debate, but Section 3(a) of the *Administrative Procedure Act* explicitly required that OSI procedures be written and published in the *Federal Register* [11]. The OSI had not done this and hence the legal integrity of its findings was unnecessarily put at risk. Indeed a lawsuit was brought against the NIH and a few officials in the PHS by Abbs [26] and these policies of the OSI were declared invalid by a federal court in the case Abbs versus Sullivan (Abbs vs, Sullivan, U.S.

District Court for the Western District of Wisconsin, 90-C-470-C, December 28, 1990) (Hansen and Hansen).

In 1993 Harold Elliot Varmus (1939—), who shared the Nobel Prize in physiology or medicine in 1989 *(http://www.nobelprize.org/nobel_prizes/medicine/laureates/1989/varmus-autobio.html)* with J. Michael Bishop (1936—) *(http://www.nobelprize.org/nobel_prizes/medicine/laureates/1989/bishop-autobio.html)* acquired the directorship of the NIH and held the position until 1999. The NIH jeopardized the funding of any institution that failed to oversee its workers like a watchdog *(http://www.time.com/time/magazine/article/0, 9171, 953258, 00.html#ixzz0f9xRe7s1)*.

Some academic laboratories established rules regarding coauthorship designed to ensure that the name of nobody appears on manuscripts unless it can be satisfactorily justified. Congressional pressure was placed on granting agencies to sanction strayed scientists.

Office of Scientific Integrity

The Office of Scientific Integrity (OSI) lacked a permanent director until Jules V. Hallum was appointed to the directorship in November 1989. He had outstanding credentials and had previously been the chairman of the department of microbiology and immunology at the Oregon Health Sciences University (OHSU) *(http://www.ohsu.edu/xd/)* in Portland Oregon and was also the chairman of the ethical practices committee of the American society for microbiologists. Hallum was disturbed by Dingell's weighty pressure as chair of the House Energy and Commerce committee's subcommittee on Oversight and Investigations overseeing the NIH budget to forcefully make OSI see eye to eye with his committee in investigations of scientific misconduct. When he assumed the leadership of the OSI Hallum was astonished to learn about the Secret Service investigation of Imanishi-Kari. During its first year of existence the OSI resolved almost 60 cases and tried to deal with all allegations of misconduct in a consistent, deliberate, objective, and thorough manner during its first year of existence [24]. The inquiries or investigations followed fundamental, equitable, established procedures.

Initially the OSI lacked sufficient staff and was slow to respond. The OSI was fair and efficient and provided statistics to hold up its assertions. Later it got better and scientists extensively deliberated on the role of outsiders in alleged scientific misconduct investigations, while Congress was particularly interested in how universities dealt with the matter and after the investigations were completed the information transferred to the OSIR for re-examination and proposed penalties. From 1988 the PHS and OSI abandoned the practice of listing individuals under investigation for research misconduct and started to only provide the names of cases found guilty after a complete scrutiny of the evidence. The government had virtuous reasons for withholding information. In about 1991, Charles W. McCutchen, a physicist at the NIH, took legal action against the OSI and a group of officials in DHHS in federal court maintaining that the withholding of the identity of researchers cleared of misconduct prevented the public from determining whether or not the OSI was meticulously investigating cases that came before it. Both McCutchen and his attorney did not expect to win the case, but surprisingly John Helm Pratt (1910—1995), an 82-year old federal district court Judge, who was nominated to the bench by Lyndon Baines Johnson (1908-1973) the 36th president of the USA in 1968, ruled that the PHS must make public the identities of federally examined researchers no matter what the rulings. Other rulings by Pratt over almost three decades helped to shape legal definitions of civil rights and discrimination before he died at the age of 84 years (http://www.nytimes.com/1995/08/14/obituaries/john-h-pratt-84-federal-judge-who-helped-define-civil-rights.html).

The judge considered the public interest of knowing who had been investigated scientific misconduct stronger than the privacy interests of the individuals whose integrity was questioned. In the USA investigations of alleged scientific misconduct do not move rapidly. They are extremely time and effort consuming. They have become increasingly prolonged and may take many years before a final decision is made, particularly to the dismay of scientists in other countries.

From 1989 to June 1992 suspected research misconduct was handled by the understaffed and underfunded OSI in the office of the director of the NIH. The mandate of the OSI was to protect the public interest in science, and not the interests of the NIH.

Hansen and Hansen warned that the developing rules, however well meaning, far exceeded the congressional mandate and may have untoward effects on American science.

The NIH failed to comply with the law and the OSI faced a turbulent course and much criticism. The initial policies and procedures of the OSI were refinements of those published in 1986, but they were revised after 1990. Before an inquiry or investigation was started the procedures were explained to all concerned individuals. The OSI called for each respondent to receive swift and extensive notification about the subject on which the inquiry or investigation was focused. Each accused person was authorized to testify and present all obtainable records to support the integrity of the individual's research. Legal counsel was permitted to escort each respondent. Interviews were taped and transcribed and the interviewees were provided an opportunity to examine and set right the transcription. Every individual found guilty of misconduct was given a copy of the draft report and provided with an opportunity to comment and refute the findings as well as the recommended sanctions. Final OSI reports were transmitted to the assistant secretary of health together with comments. If the misconduct was considered serious enough to justify a recommendation of debarment the accused was permitted to appeal the OSI decision to a debarment officer. Investigations of complex cases were sometimes slow and occasionally extended beyond 120 days, because it was not possible to review even the raw data in a shorter period of time. For example, in one case more than 50 laboratory notebooks needed to be scrutinized and busy scientists with standing commitments at different institutions needed to be interviewed. Many cases had been previously investigated at institutions, but had not been dealt with correctly or had not yielded a decisive conclusion.

The *Baltimore Affair* was but one example among many that triggered congress to probe the ability of universities and the NIH to keep an eye on themselves.

Hallum and Suzanne Hadley, deputy director of the OSI, defended their office just as Pascal, acting director of the ORI, defended the ORI (see The *Journal of NIH Research* September 1990 issue, page 12).

Hadley, encountered difficulties drafting the OSI report on Imanishi-Kari. She got together with the Dingell subcommittee aides on August 1990.

By May 1989 the NIH endorsed the report of its scientific panel maintaining that Imanishi-Kari was not guilty of scientific misconduct, but a separate investigation by the OSI was started during the same year and a harshly worded draft report on the Imanishi-Kari case by a panel of five experts was leaked to the public in 1991 [25]. The confidential draft report on the Popovic case was alo were illegally leaked to the media without authorization. The leaked Imanishi-Kari report concluded that she fabricated data in the controversial 1986 article in *Cell* while at the MIT and the OSI recommended that she be barred from getting research funds for 10 years. The report also maintained that she "repeatedly presented false and misleading information" to investigators and may have falsified other data.

Aides to the Dingell subcommittee did not believe that the OSI was forcefully tracking sufficient cases of alleged scientific misconduct and pressured the agency to instigate more inquiries into dishonest science. The principle deputy general counsel at DHHS, Mr. Robert P. Challow was a relentless critic of the OSI and he contended that the agency did not provide universally acceptable legal procedures for due process. Qui tam lawsuits complicated some OSI investigations.

The leaking of secret documents rekindled uncertainty about the competence of the methods used for dealing with scientific misconduct by the OSI at that time. Because of leaked confidential reports on investigations into accusations of research misconduct the OSI considered methods to strengthen security. Lawyers representing scientists facing allegations of misconduct cast doubt on the fairness and legality of the procedures used by the OSI and attorneys for Imanishi-Kari and Popovic protested to Healy about how the OSI's procedures intruded upon their client's rights under due process. The failure of the OSI to produce a final report on the Imanishi-Kari case frustrated the Dingell subcommittee and each hearing of the Dingell subcommittee initiated a new investigation by the NIH.

The protection of whistleblowers was a concern of Dingell and others. Dingell in particular had been under the impression that scientific community was reluctant to protect whistleblowers.

Office of Scientific Integrity Review

At the same time that the OSI was created the *Office of Scientific Integrity Review* (OSIR) was established in the DHHS under the assistant secretary of health to ensure that all agencies under its control were performing satisfactorily to investigate as well as prevent research misconduct. The OSIR was expected to review all OSI investigations into alleged scientific misconduct and to propose suitable punishments to the assistant secretary. Neither the OSI nor the OSIR were ruled by any course of action and the Gallo investigation stressed the need for better procedures.

In April 1990 the OSIR made public every finalized report of established scientific misconduct on which penalties were imposed. During the first year only three researchers (Phillip Berger, Douglas O. Nelson and Lonnie Mitchell passed through slow federal analyses of the accusations. An investigation of the three cases started in 1987 and did not end until in the summer of 1989. Lyle Bivens, director of the OSIR and a later director of the ORI, stated that misconduct was found in a few more cases.

The USA and some other countries established an organized oversight of research integrity.

On March 18, 1989 the PHS announced in the *Federal Register* its intention to create an OSI under the office of the director of the NIH and an OSIR under the assistant secretary for health of DHHS to deal with scientific misconduct. Guidelines for promoting high ethical standards in research were also provided. By August 11 1989 the two offices dealing with scientific misconduct (the Office of Scientific Integrity (OSI) and the Office of Scientific Integrity Review (OSIR) had been setup. When these rules were published in the *Federal Register* it seemed likely that Dingell, was considering additional legislation on scientific misconduct.

By August 11 1989 the OSI and the OSIR were setup.

In September 1999 a decade later, the PHS proposed regulations in the *Federal Register* defining the responsibilities of recipients of PHS funds.

The PHS authorized the creation of the Office of Scientific Integrity (OSI) within the Office of the Director of the NIH. The OSI was to watch over scientific misconduct investigations conducted by the awardee's institutions and allowed both the OSI and OSIR to perform their own investigations. Some investigated cases lingered on for years and were not resolved within the expected period of time.

By 1991 doubts were raised as to whether OSI provided both the accused and the whistle-blower access to information with sufficient security. Questions were also raised about whether the OSI was sufficiently funded and staffed with people having the appropriate credentials to carry out their required responsibilities. In view of the seriousness of OSI investigations its decisions needed to survive a highly demanding analysis as in the criminal charges against Imanishi-Kari by the USA district attorney's office in Maryland [5]. After receiving comments on its draft reports the OSI issued its final reports with recommendations for sanctions that were independently reviewed. Draft OSI reports were frequently the first time that those accused of misconduct as well as their lawyers had the opportunity to learn of and rebut specific testimony. Barbara Mishkin, a vastly experienced attorney in the defense of scientists accused of misconduct drew attention to the unfairness of the system used by the OSI at that time. She believed that witness testimony should be available to both sides during the entire proceedings. Attorneys representing the accused also desired legal safeguards in the exploratory process as a scientist could be deprived of years of productivity before getting a full hearing.

Other events also darkened the OSI investigations. After confidential OSI reports were finalized the press with time-honored connections to the Dingell subcommittee contacted Hadley for comments. The inspector general traced the leakage of OSI confidential information to a Dingell aide who was empowered to release it by the congressional subcommittee.

Dingell's association with the OSI became more enmeshed after Hadley, the prior deputy director of the OSI, joined the Dingell subcommittee as a full-time employee in 1992 while remaining on the NIH payroll. Hadley was extremely valuable to Dingle because of a combination of access to the NIH and insider knowledge. The OSI investigated many scientists including Robert Gallo.

The investigation of alleged scientific misconduct by the OSI led to power struggles between the NIH, OSI and congress and the resulting disputes were conducive to confrontation rather than solution.

In 1996 William Raub played an important role as scientific advisor to the secretary of DHHS in the scientific misconduct debate and at that time a newly constituted FAB was established and it functions on its own [1]. The original arrangement of having the NIH fund biomedical research and oversee the integrity of the grant recipients as the watchdog under the OSI gave the public the perception that the chickens were being guarded by the fox. Because the USA government was more in favor of purging the NIH of all accountability in managing the oversight of research funded by it the responsibilities of the OSI were transferred from it into the newly created independent ORI in the DHHS under the Office of the Secretary in May 1992 and started to function a month later.

Office of Research Integrity

Eventually the OSI was replaced by the Office of Research Integrity (ORI) an agency independent of the NIH. Challow reiterated some observations that he made in 1990 when he responded to a similar claim by Hallum and Hadley in a September 1990 letter: "Justice . . . involves far more than merely determining the truth. It requires a recognition that at some point conclusions must be reached, even though certain information may not be available. In short, truth without speed is not justice, it is agony." It is noteworthy that it took ORI and its predecessor offices more than 7 years to prepare the very same case for trial [24].

Officials at the ORI and DHHS testified about the intimidating behavior of the Dingell aides despite their lack of authority to browbeat anybody. The Dingell subcommittee's intent in repeated engagements with ORI aides was documented in a deposition by Lyle Bivens.

Lyle Bivens, who had previously directed the OSIR became the director of the ORI in 1993 until he retired in March 1995. David E. Wright the current CRI director became the director of the ORI after serving as Michigan State University (MSU) assistant president for Research Ethics and Standards and its Intellectual Integrity Office (1993-2004). New Director 2012 Because the ORI has more lawyers than scientists on its staff than the prior OSI investigations of allegations of research misconduct by the ORI were performed predominantly by the legal profession. One might have anticipated more scientists being found guilty, but convictions became less common because of the higher required legal standards. The ORI can only find the accused guilty or innocent of scientific misconduct. The Recombinant DNA Advisory Committee (RAC), and the office responsible for protecting the rights of human research participants and animals [7, 8]. The ORI not only handles allegations of misconduct but also promotes education in RCR, scientific misconduct prevention, institutional compliance and information on privacy. For instruction in RCR the ORI has two resources: an online education device for research integrity and image processing. In addition the ORI makes a list of learning objectives, especially for graduate schools and postdoctoral fellows. Laboratory management training is a biennial RCR conference.

Its newsletters function as a remedial tool for retracting defective publications when journals are reluctant to do so because of a fear of litigation. The ORI also provides information on recent respondents that have been found guilty of research misconduct. When misconduct is established the ORI documents the names the guilty parties together with their institutions and the imposed sanctions.

The ORI encourages honesty in the biomedical and behavioral research supported by the PHS at about 4, 000 institutions worldwide.

Like its predecessor the OSI the ORI faced a controversial stormy beginning and its director staunchly defended the ORI just as the OSI director stuck up for the OSI. The ORI was criticized for several shortcomings [44]. Robert P.

Charrow, an American lawyer from Washington DC who served on the task force on scientific misconduct criticized the ORI procedures [10], but Chris B. Pascal acting director of the ORI at the time disagreed with several points made by Charrow.

ORI's budget was a tiny fraction of the multi-billion dollar PHS budget and it was insufficient for its needs. However, the ORI consumed a significant amount of money that could have funded the research of some worthy scientists. By 1996 the ORI appropriation was more than $3.5 million, but that did not include the 4 or 5 attorneys assigned to the ORI each year that were funded by the office of the general council of the DHHS. Charrow maintained that the "the only people with a vested economic interest in seeing that there are large numbers of misconduct cases are those who work at ORI" [10].

Between 1993 and 1997 the ORI received approximately 1, 000 accusations of research misconduct, but had only closed 150 investigations. From 1994 to 2005 the ORI summarized all established cases of scientific misconduct funded by the NIH, but did not provide the identity of the guilty parties or the names of those found not guilty, but from 2005 guilty individuals are specially named. By September 21, 2010 the ORI listed on its website (*http://ori.dhhs.gov/*) all investigated individuals found guilty of research misconduct sponsored by the PHS. The annual average number of sanctioned persons from 1994 to 2010 ranged from 4-14 (mean 7.6). ORI received 17 allegations in 2006. In 2009 fourteen people were found to have participated in research misconduct. A person found guilty of research misconduct may need to re-establish the reputation of colleagues damaged by being innocently associated with the misconduct. The ORI has a research integrity officer boot camp training program and an education and prevention program.

Initially the ORI received 76 former cases and more than 450 unresolved allegations of misconduct [34]. It had closed 172 formal cases since 1992 with 59 cases of misconduct. It reduced the time taken for processing allegations from almost 400 days in 1992 to about 58 days by 1995. Charrow recommended that ORI be required to complete its review of university findings of misconduct in less than 60 days and then immediately advance to trial or let the allegations drop. The ORI considered this time frame impractical as investigations can be extremely complex and prolonged and indeed when the accused hired legal counsel the preparation by the legal defense sometimes took more than 10 months. Unless cases were simple and extremely straight forward there was a tendency to rush to judgment in rapid evaluations of the material. An initial inquiry into the allegation is usually performed by the institution and if it decides to launch a detailed investigation ORI must be notified. The findings of the institution are passed on to the ORI that looks into the matter further. After the scrutiny is completed the matter goes to the division of investigative oversight (DIO) for oversight review. Other duties of the DIO staff include the organization of conferences and workshops on the handling with research misconduct allegations. The ORI tried to resolve cases as soon as possible, but the processing of them seemed extremely slow partly because analyses by the ORI were meticulous.

Gradually the process for investigating accusations of scientific misconduct became smoother and numerous cases were investigated by institutions and the ORI from 1991 onwards each year (less than 100 per year) from 1991 onwards an increased number of institutions investigated suspected scientific misconduct. Evaluations of scientific misconduct accusations by the ORI averaged 185 days, but some cases lasted 2 years. Of 50 cases investigated by universities between 1994 and 1996, ORI closed 19 cases within 6 months. By 1996 the ORI began to triage cases investigated at universities for which reports had been submitted. From 2007, the ORI only investigated cases of research misconduct after institutional investigations had already found evidence of fabrication, falsification, or plagiarism. The ORI has a Rapid Response for Technical Assistance program (RRTA), which interacts with institutional officials and journal editors with concerns about how to manage particular cases of suspected scientific misconduct. Together with the NIH some grants were awarded by the ORI for research into scientific integrity. By 2008 forty-nine groups were supported and some of this research was published in professional journals. Not all accusations of research misconduct referred to the ORI fall under its authority and for a formal investigation to be initiated by the ORI specific criteria must be met. The wrongdoing must be funded by the PHS or involve a grant application to that service and it must fit the ORI definition of scientific misconduct. To reach a conclusion of research misconduct the ORI must establish by a preponderance of the evidence that the accused "knowingly, intentionally, or recklessly" carried out unacceptable research. To hold a scientist guilty of scientific misconduct intentional deception must be shown, but the ORI believes that the researcher should be held responsible if the accused just knew or should have known that a declaration in a scientific paper was false regardless of whether the offense was a few topographical errors or obvious fakery of notebook entries. The ORI does not investigate situations in which virtually indistinguishable expressions are employed to describe techniques or past research. Accusations of plagiarism received by the ORI have often embroiled disagreements between past collaborators. In such cases the ownership of intellectual property is hardly ever obvious and the ORI regards such cases as authorship or credit disputes rather than plagiarism. When sufficient information is available to conduct an investigation into misconduct the ORI may contact the whistleblower for more detailed documentation.

The ORI makes available detailed information about how suspected cases of research misconduct need to be investigated and issues that require attention, such as worries about human subjects, monetary deception, violations of animal rights, and potential unlawful goings-on. The intramural research program of the ORI focuses on research integrity and on how institutions deal with scientific misconduct and/or promote research integrity. One study concentrates on the role of the institutional Research Integrity Officer (RIO) in implementing PHS policies on research misconduct. Others deal with such diverse topics as the effectiveness of institutions in educating their staff on policies related to research misconduct and research integrity, the training and mentoring of PhDs, faculty views on the role of their institutions to promote the development of responsible researchers, evaluating the impact on whistleblowers who report research misconduct. The ORI has permitted persons guilty of scientific misconduct to reenter the research profession despite the loss of their trustworthiness and reputation, provided that they undergo supervised remedial activities for a period of time. Such individuals sometimes needs to be supervised for a few years.

Response by Professional Societies

Professional codes of ethics, established by particular scientific societies, additionally develop some standards for the conduct of research. Societies pointed out that harsh or interfering regulations preside over the details of research conduct may damage the creativity and quality of the research environment, especially if those regulations are unsupported by documented evidence of widespread dishonesty or ethical misconduct by researchers.

The professional societies were appalled to learn about unethical and fraudulent research and steps were taken to eradicate both of these unacceptable types research. Organizations focused sessions on the detection and prevention of unethical and fraudulent research.

The Committee on Science, Engineering, and Public Policy (COSEPUP) representing a joint effect of the National Academy of Sciences (NAS), National Academy of Engineering (NAE), and the Institute of Medicine (IOM) dealt with cross-cutting issues in science and technology. It responded to scientific misconduct by providing reports on research integrity [13, 14].

Many professional societies including the Association of American Medical Colleges (AAMC), the Association of American Universities (AAU), the IOM, and the NAS produced guidelines on how to deal with research misconduct. The recommendations included (i) that a formal investigation of alleged misconduct be undertaken and if identified the funding agency should be notified for potential action, (ii) that allegations reported in good faith that were not substantiated should not be held against the complainant and vengeance against the whistleblower should be prevented, and (iii) that adverse publicity, particularly if the accusations were not justified can be detrimental to the careers of innocent people.

The IOM, a quazi non-governmental organization of the NAS studied the incidence and prevalence of scientific misconduct extensively and considered outright laboratory fraud rare. In September 1987 this institute appointed a 17 member committee chaired by Professor Arthur Rubenstein, of the University of Chicago (UChicago) *(http://www.uchicago.edu/),* to develop proposals for the responsible conduct of research to strengthen professional standards in federal and academic laboratories and to conduct a workshop and develop recommendations that would assist government agencies, professional societies, journals, and universities in formulating policies and procedures to improve the inequity in the quality of biomedical research. The IOM released its final report on *The Responsible Conduct of Research in Health Sciences* on February 15 1989 [12]. An extremely permissive attitude by institutions tends to allow careless and sometimes even fraudulent medical science. The IOM concluded that improved research standards and systematic ways of investigating laboratory responsibility were needed. The report drew attention to the fact that few institutions had precise research guidelines and this permitted shoddy endeavors to be tolerated by a trivial number of researchers who neglect to abide by normally recognized customs.

The combination of peer review and research replication has traditionally protected the scientific literature against research misconduct and unintentional scientific mistakes, but the peer review system depends on trust and data reported by researchers are presumed to be accurate and amenable to replication.

The IOM made numerous recommendations: (i) the NIH should establish an office to promote responsible research and to evaluate investigations of misconduct by institutions, (ii) all institutions conducting medical research should be required to adopt specific policies to promote ethical research practices and investigate misconduct, (iii) the institutes should restrict the number of publications in a grant application and evaluate the researcher's past record based on quality, not quantity, (iv) academic departments should adapt new policies that do not emphasize quantity, (v) scientific journals should develop policies to promote responsible authorship practices, including a system to respond to charges of misconduct, (vi) the medical research community should overhaul the ways it polices itself to control fabricated data, plagiarism and carelessness, (vii) to protect society from research flaws, (viii) a need of the science community to find innovative approaches to encourage ethical research and to deal with cheaters, (ix) a responsibility for insuring that that institutions strengthen oversight mechanisms by removing incentives for research misconduct, (x) each institution should adapt written standards and specify explicit policies on such matters as the recording and retention of research data and an awarding of authorship on research papers, and (ix) the federal government should provide leadership to assist the biomedical community to develop and implement guidelines for responsible research.

The IOM also proposed numerous changes for the NIH, universities and other research centers and for professional and scientific organizations and journals.

Some professional societies, such as the *NAS* and *Sigma Xi* [3] developed materials to educate young scientists on RCR. paper in *American Scientist* [37] [38].

In April 1992, the NAS issued its report *Responsible Science: Ensuring the Integrity of the Research Process*.

Following the AAMC report a survey of almost 500 institutional responses from 1982 to 1984 found that about 25% adopted rules for handling accusations of scientific misconduct and an equivalent number lacked such rules. More than half of respondents were developing courses of action at the time of the survey. The majority of academic institutions believed that their existing policies could cope with fraudulent research on their campuses and that additional guidelines were unnecessary. By the mid 1980s some universities encouraged research integrity.

Numerous professional scientific societies discussed the implications of fraudulent research and recommended procedures and policies for dealing with it. Although different organizations came up with different recommendations most were remarkably similar. The *NAS* felt that the accused needs to be dealt with fairly. A precise definition of scientific misconduct would lead to procedures for dealing with accusations of it that resemble the judicial system and that the targeted researcher be notified if a formal procedure would start.

In January 1982 the AAMC appointed a panel to preserve high ethical standards in research and this committee reiterated the accountability of research teaching staff and their research benchmarks. The AAMC recommended clearly define fraudulent research as a major breach of contract and that those found guilty will be fired and that research institutions establish policies and procedures for promptly coping with accusations of fraudulent research when an incident occurs and that all researchers be familiar with the process. They advocated that confidentiality give way to disclosure if the original review justifies a more comprehensive scrutiny of the case. The AAMC guidelines included procedures for safeguarding whistleblowers who make accusations in good faith from retaliation. The AAMC advised universities to obtain legal advice when establishing policies on dealing with scientific misconduct.

In June 1982 the AAMC became the first professional society to adopt guidelines for dealing with research misconduct. If universities created procedures compatible with the AAMC recommendations, a two-tiered system somewhat comparable to the grand jury system would deal with accusations of research misconduct. Allegations of fraudulent research would initially be considered by the head of the department and the dean would be informed at the outset. To safeguard the accused against unfounded accusations a preliminary inquiry would need to be started in secret. The AAMC was concerned that an investigation into alleged scientific misconduct might damage the reputation of the accused, but the university's accountability to the funding agencies outweighed this apprehension. If sufficient reasons for an investigation exists all coworkers and funding agencies of the accused must be notified.

A public announcement of an allegation during the investigation was viewed as a necessary price to pay. The AAMC report contained a controversial paragraph. "It is appreciated that in these courses of action the reputation of the accused faculty member is placed at risk during the investigation. This is justified since scientists on university faculties need to be held to a more than the usual standard because of their unique favored place in society. The procedures must be fair to the accused and must also be planned to be receptive to the special accountability that the scientific community has to society".

The AAMC advocated that investigations of serious misconduct be conducted by a committee composed of persons not in the department of the accused but by persons from other departments and institutions. The AAMC also recommended that all previous research by persons found guilty of misconduct be reviewed and that institutions and funding agencies with which the person has been previously connected should also be notified of their scientific misconduct. When a person alleged to commit research misconduct is absolved of any wrongdoing official steps should be carried out to completely clear the name of those accused. The AAMC endorsed the belief that that the accused deserved the right to be informed of the identity of the accuser and needed to be able to confront the faultfinder. The accused needs to be fairly dealt with during an investigation of alleged misconduct and should have ample opportunity to defend against the accusations. A legal dispute needs to be anticipated when someone is found culpable of research misconduct as such an adverse decision may be challenged in court and it will be important to determine that fair procedures were used to safeguard the accused against unfounded allegations. Aside from considering how to deal with accusations of scientific misconduct the AAMC recommended that universities be adamant about retaining high moral benchmarks. The AAMC also recommended the establishment of policies on the authorship of papers as well as abstracts and that all designated authors accept the responsibility for the reported research and vouch that they have read the written material. In October 1982, the AAU, which embodies more than 50 of the foremost research universities in the USA, appointed a committee on research integrity chaired by William Danforth, chancellor of Washington University (*http://wustl.edu/*). This committee concluded that informal efforts to cope with scholarly dishonesty were no longer workable as they had been in the past. The AAU recommended that all academic research institutions develop policies and procedures to ensure a high standard of ethical behavior for researchers. The AAU developed the *AAU Framework for Policies and Procedures to Deal with Research Fraud* (ref) on November 4, 1988 with the assistance of numerous professional societies that included an interassociation group representing the Association of Academic Health Centers (AAHC), the Association of American Medical Colleges (AAMC), the American Council on Education (ACE), the American Society of Microbiology (ASM), the Council of Graduate Schools (CGS), the Council on Government Relations (COGR), the Federation of American Societies for Experimental Biology (FASEB), the National Association of College and University Attorneys (NACUA), and the National Association of State Universities and Land Grant Colleges (NASULGC) [17]. In January 1989 when fifty four of the leading universities in the USA belonged to the

AAU this organization published a policy to help universities develop procedures for handling scientific misconduct among their faculty since many lacked any procedure [32]. The AAU reissued its recommendations with a final rule on November 10, 1989 and this was endorsed by numerous professional societies (*www.aau.edu/WorkArea/show content.aspx?id=6360*). The AAU advised that allegations of research misconduct be referred to a senior faculty member and that a preliminary investigation be undertaken in secret perhaps by a committee and that this be completed within 30 days. The AAU also advocated that the formal investigation be completed in less than 120 days and that a written report be submitted to senior university officials and the funding sources. The *AAU* felt that universities had a duty to follow up on all accusations of research misconduct even after the implicated researchers even if the researcher moves elsewhere and it was essential that all charges and investigations be kept confidential.

Response by Professional Journals

After the existence of scientific misconduct became apparent there was a demand for major changes in the way that science gets documented, appraised, and published. The editors of journals found themselves in the line of fire for allowing fraudulent papers to be published after passing through a gateway overseen by journal editors, editorial boards and the peer review process before entering the scientific literature. Certain issues were raised with regard to the peer review process, such as the criteria for authorship [46]. Following the Darsee affair it became apparent that numerous flawed papers slip past the gatekeepers of the professional prose. Obviously the existing peer review process did not block the entry of flawed papers into the literature and journal editors became seriously concerned about how to prevent the products of scientific misconduct from becoming published. Many were concerned with how this could occur without editors and peer reviewers being able to detect the wrongdoings before papers go to press. Editors of journals improved their procedures for publication related to the submission and acceptance of manuscripts. Scientific journals accepted the responsibility of creating procedures for detecting fraudulent papers and developed guidelines on the authorship of scientific papers. Now for an article to be accepted each author is required to identify his/her contribution to the research and COI must be provided together with identity of funding agents and active or pending patents. Journals needed to cease being faint-hearted about withdrawing such papers from the literature and publishing corrections. In the past journals were reluctant to retract flawed papers because of a fear of litigation.

A paper by Edward Huth, editor of *Annals of Internal Medicine,* provided guidelines for the authorship of research papers, case-series analyses, case reports, review articles and editorials based on statements issued by the ICMJE [29].

Journals were sometimes blamed for publishing the flawed information, even though they had no control over the activities in research laboratories where fraudulent behavior takes place.

Symposia on fraudulent and other forms of scientific research were held. Because the survival of a publication through peer review clearly did not guarantee the integrity of research, journal editors questioned what they ought to do to weed out scientific fraud. These included the education of researchers in ethics, responsible science, institutional review boards (IRBs), compliance offices, and the disclosure of COI. Science journals stated to adhere to the Sarbanes-Oxley Act of ethics that was passed in 2002 as a response to scandals in the business profession.

Components of the government became involved largely because of the demand for answerability in all topics supported financially by tax payers. The USA system sometimes allowed allegations of research misconduct to drag on unnecessarily for years even in cases where the accused was not found guilty of the charges that disrupted the person's career.

Following the realization that some images were manipulated in fraudulent publications certain journals began to screen images before accepting papers for publication.

After the discovery of scientific misconduct in publications in the 1970s and 1980s journals tightened their criteria for accepting papers and they took steps to safeguard against the publication of flawed papers.

Committee for Publication Ethics

An important pioneering action related to the combat of fraudulent research publications in journals took place in the UK, when an unofficial small group of intensely perturbed British medical journal editors created the Committee for Publication Ethics (COPE) in 1997. This group campaigned for a national committee to investigate all cases of alleged misconduct in medical research. This proposal was discussed in March 1999 by the General Medical Council (GMC), which establishes professional and research standards for clinicians in the UK.

Over time more and more editorial boards accepted COPE's policies and there are now more than 7, 000 members worldwide covering all academic disciplines. Even major publishing companies, such as Elsvier, Wiley-Blackwell, Springer, Taylor and Francis, Palgrave Macmillan and Walters Kluwer joined on behalf of journals published by them. The COPE code of conduct and best practice guidelines for journal editors provides minimum standards to which its members are expected to adhere. The editors need to be accountable for everything published in their journals. They should "strive to meet the needs of their members and authors, struggle to persistently enhance their journal, have processes in place to ensure the quality of the material they publish, champion freedom of expression, maintain the integrity of the academic record, preclude business needs from compromising intellectual and ethical standards, always be willing to publish corrections, clarifications, retractions and apologies when needed". Many other duties of editors are provided at the COPE website (http:/pubicationethics.org). Among them is the desire to ensure that peer review is fair, unbiased and timely. Also, editors should have systems that ensure that submitted material remains confidential during the review process. COPE recommends that that editors have an obligation to not only reject manuscripts suspected of scientific misconduct, but to also act on the misconduct. COPE solicits problematic cases in which ethical or research misconduct concerns arise and all such cases are listed in a searchable database indexed according to topic. Thus the following topics are listed: author mistakes, authorship, changes in authorship, complaints, consent for publication, copyright breaches, data fabrication, data manipulation, data ownership, editorial decisions, editorial independence, editorial misconduct, falsification, ghost authorship, gift authorship, image manipulation, impact factors, journal mistakes, lack of ethical review, approval, misleading reporting, multiple submissions, overlapping publications, participant confidentiality, participant consent, patient confidentiality, peer-review process, plagiarism, protection of subjects (animals), protection of subjects (human), redundant publication, relation to society/owner, research ethics investigation, retractions, reviewer misconduct, role of publisher, role of sponsor, sanctions for misconduct, self-plagiarism, undeclared conflict of interest (authors), undeclared conflict of interest (editors), undeclared financial support for publication, unethical research, unethical treatments, whistleblowers. By November 2013 more than 400 cases were in the COPE database.

International Committee of Medical Journal Editors

Another important response to the fraudulent publications was the creation of the International Committee of Medical Journal Editors (ICMJE). This group of about 18 general medical journal editors formed international instructions for journals. Unfortunately guidelines need to be backed up by law, but laws are burdensome for the scientist and hinder research. It is also not practical to expect a researcher to be cognizant of all research regulations and keep abreast of them especially since that constantly changes in response to new issues.

Members of ICMJE approved revised uniform requirements for manuscripts in April 2010 (*www.ICMJE.org*), but most journals do not use their procedures. The ICMJE requires that participants in all clinical trials be entered into a public registry (such as clinicaltrials.gov) before they become part of a publication.

By January 1989 the massive computerized database of the Library of Medicine became a link between for scientific publications and retracted papers, letters to the editor and grant support [43].

The Department of Health and Human Services

The Department of Health and Human Services (DHHS) established an appeals board because the e ORI procedures did not originally provide accused scientists with all of their legal rights, including due process with the right to cross examine witnesses and appraise the evidence a federal appeals board (FAB) was set up in the DHHS in 1992 to permit a person found guilty of research misconduct by the ORI with a mechanism to appeal the decision. An additional hearing could take place before a administrative law judge of the DHHS.

Ethics instructors employed ethical theories in the framework of group discussions, projects, and other tasks that compel researchers to reason t in more righteous ways. New federal regulations required research institutions to establish written guidelines and policies for dealing with accusations of research misconduct. They disclosed a need for further procedures to promote high ethical research standards. The secretary of the DHHS published two sets of proposed rules in the 19 September 1988 issue of *Federal Register* and called for public comment. Universities and research institutions were compelled to start an inquiry when an allegation of research misconduct was received and the institution was required to establish whether a more formal investigation was needed. At the start of an investigation, the institution was required to notify the federal funding agency and the agency of the outcome.

In March 1991 the OSI received serious setbacks from its investigations of the cases of Imanishi-Kari and Margit Hamosh. Proposals by the science community and professional societies suggested new administrative entities for dealing with fraudulent research [18].

Several years after a number of PHS cases of fabrication and falsification by researchers were reported in the media the PHS issued a proposal entitled "Responsibilities of PHS Awardee and Applicant Institutions for dealing with and reporting possible misconduct in science".

Annual reports of the ORI deal with a number of issues: (i) the response to research misconduct accusations, (ii) education and misconduct deterrence and (iii) research and research integrity and institutional compliance related to scientific misconduct and privacy. Governmental oversight over the expenditure of taxpayer money is legally mandated and clearly proper. Different institutions and professional societies played a role in establishing definitions for research misconduct. Future rules were anticipated for authorship practices, the preservation of raw laboratory data, and processes for audits.

Response by the Scientific Community

The scientific community began to become organized to ensure that Congress would not get involved in investigating cases of alleged scientific misconduct.

Because of the attitude of the Dingell subcommittee many scientists were concerned that it was likely that Congress would create a watchdog agency to investigate scientific disagreements. Baltimore's enrollment of the scientific community to support him was astonishing and Stephen Jay Gould (1941-2002), a historian of science and an influential writer of popular science, compared the inquisition of Baltimore to the trial Galileo and the journalism department of *Science* aggressively pointed the finger at the accuser while sticking up for the defendants. Expecting Baltimore to face a difficult time, the American geneticist and molecular biologist Phillip Allen Sharp (1944-), who shared the 1993 Nobel Prize in physiology or medicine *(http://www.nobelprize.org/nobel_prizes/medicine/laureates/1993/sharp-autobio.html)* with Richard J. Roberts (1943—). *(http://www.nobelprize.org/nobel_prizes/medicine/laureates/1993/roberts.html)* for co-discovering RNA splicing, instigated a crusade soliciting the support of scientists throughout the nation. Dingell needed to be brought to a halt either by scientists or by his congressional colleagues. The persistent aggravation of Baltimore and other scientists by the Dingell subcommittee had grave repercussions for all researchers. To offset the inappropriate path that the Dingell subcommittee was taking Sharp urged the scientific community to write letters to newspaper editors nationwide and to the entire membership of the Dingell subcommittee.

In September 1988 more than a 100 scientists representing both basic and applied research, government and university officials, professional society offices, journal editors and representatives of the media attended a workshop in Washington DC sponsored by the National Conference of Lawyers and Scientists (NCLS). The meeting evaluated effective misconduct policies and procedures for universities.

After the realization that scientific misconduct was more prevalent than originally anticipated several changes were introduced related to the manner in which research was conducted. The science community was horrified the rough treatment that science received during the two adjacent congressional hearings on research misconduct in 1988 when scientists were serious about coping with scientific misconduct. It was clear that if scientists did not police their profession the government would. Scientists with administrative positions seemed to be attaining an amazing consensus on the limited steps that would be necessary to achieve it.

The scientific community began to become organized to ensure that Congress did not get involved in investigating cases of alleged research misconduct,

Those responsible for investigating alleged scientific misconduct must carry out these duties in secret, but despite the desire to keep this information confidential this cannot always be guaranteed because of anonymous leaks to media that thrive on gossip and items that make headlines. Witch hunts on anonymous and unverified statements need to be avoided. Investigations are often initiated after a whistle-blower reports inappropriate behavior. In the Potti scandal his accusers brought him into open when they drew attention to his falsified credentials in which he claimed that he had been a Rhodes Scholar.

Response to Scientific Misconduct in Non-USA Countries

A scourge of research misconduct hit institutions in Europe and elsewhere after the USA the response in other nations was slower than in the USA. It enabled other countries to benefit from the American debate regarding the definition of scientific misconduct and how to deal with it.

Canada

By 1994 only about half of Canadian universities had policies for investigating scientific misconduct [40], but in that year Canada's three governmental granting agencies the Natural Sciences and Engineering Research Council of Canada (NSERC), the Social Sciences and Humanities Research Council (SSHRC) and the Medical Research Council (MRC) announced that research universities must establish guidelines for dealing with allegations of scientific misconduct by 30 June 1995 or their faculty would be ineligible for government research grants. The proclamation was precipitated after four faculty members were killed by gunfire in August 1992 by Valery I. Fabricant (1940-1992), a *Belarussian* émigré and former professor at Montreal's Concordia University *(http://www.concordia.ca/)*, after asserting that some coworkers had participated in unacceptable consultations and other activities with COI. A committee chaired by Harry William Arthurs (1935—), one of Canada's leading labor law scholars and a former President of New York University (NYU) (*http://www.nyu.edu/*) investigated the deaths of the victims and unearthed some merit in these allegations [35]. The Arthurs committee found that the entire Canadian research community potentially faced similar troubles because of increased competition for smaller amounts of available funds.

Europe

In contrast to the USA the oversight of research in Europe differs extensively from country to country, but the procedures for investigating alleged scientific fraud are basically similar in many countries, even though notable differences exist. In contrast to the USA the European scientific community avoided an obligation to protracted investigations of scientific fraud by attempting to look into the accusations rapidly by limiting the procedural phases. In essence the scrutiny is by a committee of peers and the whistle blower as well as the accused is protected. In Germany and in Scandinavian countries a sanctioned scientist has the ability to appeal the verdict in the courts. Apart from Scandinavia and, to a reduced degree Germany, the United Kingdom (UK) Croatia, and France little or no regulations governed scientific misconduct [6]. In some European countries, such as France and Italy, there was a wide belief that science was honest in their countries and that codes of practice are unnecessary.

Scandinavia. In Scandinavia (Denmark Norway, Sweden, Finland) alleged scientific misconduct is uncommon and it is initially investigated by institutions and not referred by national committees. In Denmark 75% of investigations on alleged research misconduct can be dealt with by the secretariat but some would need "due process, " and because of this the attendance of a judge on the committee, as in the Nordic countries, would be essential. Denmark is apparently the only country that assigns scientific misconduct accusations initially to a national body (the Committee on Scientific Dishonesty), which was created in 1992. It is chaired by a judge, rather than a university appointed person. This procedure was started because of the belief that a university would be disinclined to identify one of its researchers, especially an outstanding faculty member, as a fraud and initially focused on biomedicine but later its scope was expanded in 1999 to include all sciences. In Norway Jarla Ofstad, a former chairman of the Norwegian medical ethics committee, made a strong case against the formation of the Norwegian committee on scientific dishonesty because it would create an excessive bureaucracy focusing on an uncommon problem, but a comparable Norwegian committee was established two years after its Danish counterpart and within the next five years their attention was only drawn to nine examples of which two were significant. It is debatable whether each country in Europe should establish its own national committee to investigate allegations of scientific misconduct rather than have institutions undertake the initial investigations.

France. Claude Griscelli, a former director general of Institut National de la Santé et de la Recherche Médicale (INSERM) in France suggests that when a local institution seems incapable of dealing with an especially potentially serious case of scientific misconduct, a committee with members drawn from other European countries could provide the relevant expertise and not become embroiled in the local academic politics. In France universities are not under the same demands as other countries for codes of scientific practice, but scientific institutions in France also responded to the unacceptable number of cases of scientific misconduct. In February 1999 INSERM circulated guidelines to its institutions and the *Centre National de la Recherche Scientifique* (CNRS) the largest French research organization was still deliberating on the issue at that time.

Germany. In 1997 the Deutshe Forschungsgemeinschaft (DFG) released proposals as to how universities and research institutions should investigate scientific misconduct and nurture ethical research [30] (*www.sciencemag.org*). This main funding agency in Germany for basic research, was profoundly distressed to learn about the Hermann-Brach scandal involving two professors accused of falsifying data in biomedical research and one was formerly a member of the DFG advisory panel [30, 31]. This was a apparently the first known case of scientific misconduct on German land and it was particularly embarrassing to Wolfgang Frühwald, a former DRG president, who had stated in 1995 with some delight that the incentive for a researcher to falsify data to accelerate his career is greater in the USA system than in the German research system, which is tightly controlled by self-regulation. During the same year the Max Planck society also established a modus operandi for dealing with allegations of scientific misconduct in Germany a few months before a series of major fraud scandals shocked the German belief that their country was culturally immune to the "American scourge".

They also suggested that grants not be awarded to universities and research institutions that do not implement ethical research and successful methods to investigate scientific misconduct, but it was questionable whether the DFG could legally deny grants. These proposals were recommended by a 13-member panel after the notorious scandal involving Friedhelm Hermann and Marion Brach. Other recommendations protected whistle-blower, the tightening of coauthorship standards in publications, the elimination of honorary coauthorships, the preservation of research data from publications for 10 years, and the use of qualitative considerations in decisions related to hiring and the awarding of grants. The panel rejected the idea of establishing a government bureaucracy, such as the ORI in the USA to investigate cases of alleged scientific misconduct. At about the same time that the DFG announced its proposal and the extremely prestigious Max Plank society announced its regulations for dealing with allegations of scientific fraud. It will establish a standing committee headed by an outsider and it will recommend sanctions ranging from warning to dismissal. The president of Max Plank society will make the ultimate decision.

In 1998 the University of Freiburg (*http://www.uni-freiburg.de*) in Germany launched rules that necessitate an informal inquiry of suspected research dishonesty, followed if necessary by a formal investigation. A summary of the committee's formal findings are transferred with recommended sanctions to the university President who acts on the information as deemed appropriate. The creation of these rules was aided by Albin Eser, a professor of law at the university, and an appeal phase was intentionally sidestepped because it would lengthen the time required to resolve the accusations as shown in the USA. In another initiative the DFG appointed a commission after allegations emerged indicating that two researchers manipulated data while working at Berlin's Max Delbrück center for molecular medicine in the mid-1990s [31]. Ernst-Ludwig Winnacker (1941—), president of DRG in 1998 felt that fraud can never be excluded in any system, but that it is desirable to make the system as transparent as possible to prevent it. Many institutions savor their independence, but in Germany the DFG has provided a solemn incentive to have a course of action for investigating allegations of scientific misconduct as well as for supporting high-quality scientific practice. Institutions without such procedures by 2002 were ineligible to obtain its funds. This requirement was successful and all universities were soon in compliance. The Herrmann-Brach scandal disclosed the hazard of the German hierarchical research system in which young investigators are afraid to deal with the omnipotent professor. Distrusting institutions with centralized power Germans opposed the formation of a national body to look into allegations of scientific misconduct in 1999.

United Kingdom. The research councils in the UK convinced universities to adopt appropriate policies of practice by making recipients of research grants sign a declaration that their research will be performed in institutions with acceptable rules of practice. Thus the responsibility for potential misconduct was transferred to the grant recipient. In the UK the medical research council (MRC) was adamant that it have its own appeals procedure based on fairness.

Prior to 1997 a private firm called medicolegal investigation was set up in the UK by Frank Wells, a former director of the association of the British pharmaceutical industry to investigate cases of scientific fraud. By 1997 it had referred seventeen cases of serious misconduct to the General Medical Council (GMC), including Dr. John Anderton and was involved in another dozen cases. Towards the end of 1998 the biotechnology and biological sciences research council in the UK unveiled good practice guidelines and other research councils tagged along. In March 2006 the UK established a national research integrity office.

China

A misconduct case in China related to a homegrown DSP chip was dismissed as a hoax [48]. In June 2006 China's Ministry of Science and Technology (MOST) announced reforms to restrain and eradicate scientific misconduct in a 26-point "Recommendations on reforming management of science and technology programs". These included a scheme to measure work performance. Some researchers praised the initiatives, but many were unconvinced that they would lead to substantial differences. MOST prepared to enlarge its database of knowledgeable scientists to review grant applications and appraise projects and to choose reviewers at random to lessen COI. An additional step entailed the establishment of a credit management system to keep performance scores of experts who evaluate grant applications and of institutions and individuals who embark on projects. Scores created by reviewers of grant applications will be considered in reaching forthcoming decisions related to grant applications. According to Shang the principal objective was to increase the "transparency, equity, and fairness" in program management. All non-confidential projects overseen by MOST will be dealt with online with the aid of a searchable public database. Anybody will have the capability of reading the applications, approvals, implementations, and the evaluations of the applications. The opinions of experts who judge the quality of the applications will be kept secret. Applications already must be submitted online, and an online evaluation system will eventually be accepted.

MOST also made use of 100 accounting firms to audit more than 2000 projects in 2004 and 2005, involving an entire funding package of $2 billion. In the past wrongdoers were usually disciplined. MOST has been criticized for its dual role as both "umpire" and "player" in the research administration game. The chair of the physics department at Qinghua University (http://www.china.org.cn) praised the new measures as improvements over past rules, but some observers such as Yu Lu, a theoretical physicist at the Chinese academy of sciences, regarded them too vague.

The Chinese government recognized that science is essential to their nation's transformation and China is expected to expand its budget for science and technology. In China plagiarism is extremely common and it is openly carried out without any embarrassment that this behavior is wrong and unacceptable. Critics hold feeble sanctions for research misconduct and a system that focuses the promotion of faculty on the quantity of publications rather than on their quality.

India

The non-governmental Society for Scientific Values in Delhi is responsible for investigating cases of scientific misconduct in India and because such behavior is thought to be rare in that country the scientific advisor to the Indian government (Rajagopla Chidambaram) believed that a full-time oversight body was not indicated and that alleged cases of scientific misconduct should be investigated by journals and universities.

Japan

The science council of Japan, which advises the Japanese government, published a report in 1998 recommending a new system to enhance good scientific practice at universities. After the first investigation into research misconduct at Tokyo University involved the chemist Kazunari Taira. Following this Tokyo University considered establishing a permanent committee or an office to investigate research misconduct.

In April 2005 RIKEN, a natural sciences institute in Japan, consisting of a collection of about three thousand campuses created a research compliance office, which is authorized to investigate allegations of scientific misconduct. Each participant in RIKEN is required to retain experimental records for five years and the contributions and responsibilities of all authors must be clearly stated.

In 1987 Steven Barlow noted something strange in an article published in Neurology by his former PhD adviser James Abbs. Barlow wrote a letter to UW officials, the NIH and to the editors of two journals on April 9, 1987

References

1. Agnew B, Misconduct in the prosecution of misconduct?. The Journal of NIH Research, 1993. 5: 10.
2. Anderson C, Misconduct panel sets ambitious agenda. Science, 1994. 264(5167): 1841.
3. Anderson MS, Kot FC, Shaw MA, et al., Authorship diplomacy, Chapter 3 in For the Record: American Scientist Essays on Scientific Publication. 2011, Research Triangle Park, North Carolina: Sigma Xi The Scientific Research Society
4. Baltimore D, To the editor. Shattuck lecture—misconduct in medical research. New England Journal of Medicine, 1993. 329(5286): 732-734.
5. Barnes DM, Focus on fraud shifts from the conduct of science to the conduct of investigations. The Journal of NIH Research, 1991. 3: 35-37.
6. Bosch X, A view from Europe on European research oversight. Office of Research Integrity Newsletter, 2009. 18(1): 1.
7. Broad WJ, Congress told fraud issue "exaggerated". Science, 1981. 212(4493): 421.
8. Broad WJ, Fraud and the structure of science. Is fraud a trivial excrescence on the process of science or do the recent cases have deeper roots Science, 1981. 212(4491): 137-141.
9. Broad WJ and Wade N, Science's faulty fraud detectors. Psychology Today, 1982. 16(11): 51-54.
10. Charrow R, Judgments: miscalculated risk: redefining misconduct is unnecessary and dangerous. The Journal of NIH Research, 1996. 8: 61.
11. Charrow RP, Response to Hallum J and Hadley S. The Journal of NIH Research, 1990. 2: 12-14.
12. Committee on the Responsible Conduct of Research NRC, The Responsible Conduct of Research in Health Sciences. 1989, Washington DC: The National Academies Press.
13. COSEPUP, Ensuring the Integrity, Accessibility and Stewardship of Research Data in the Digital Age. 2009, Washington, DC: National Academies Press.
14. COSEPUP, On Being a Scientist: A Guide to Responsible Conduct in Research, Third Edition, . 2009a, Washington, DC: National Academies Press.
15. Crewdson J. The great AIDS quest. Chicago Tribune. 1989 November 19.
16. Culliton B, Inside the Gallo probe. Science, 1990. 248(4962): 1494-1498.
17. Culliton BJ, Bill would set fraud guidelines for scientific publications. Science, 1988. 242(4876): 187.
18. Culliton BJ, Scientists confront misconduct. Science, 1988. 241(4874): 1748-1749.
19. Culliton BJ, Dingell v. Baltimore. Science, 1989. 244(4903): 412-414.

20. Delaney M, To the editor. Shattuck Lecture—misconduct in medical research. New England Journal of Medicine, 1993. 329(10): 734.
21. Dingell JD, Shattuck Lecture—misconduct in medical research. New England Journal of Medicine, 1993. 328(22): 1610-1615.
22. Doering DS, The Dingell Investigation. Letter to Editor. Science, 1989. 244(4910): 1243-1244.
23. Foreman J. A noted researcher disputed work lands in congress. The Boston Globe. 1988 April 10.
24. Hallum J and Hadley S, Scientific misconduct. The Journal of NIH Research, 1990. 2: 12.
25. Hamilton DP, NIH finds fraud in cell paper. Science. 251(5001): 1552-1554.
26. Hamilton DP, NIH sued over misconduct case. Science, 1990. 249(4968): 471.
27. Holden C, New rules on misconduct. Science, 1989. 245(4918): 593.
28. House of Representatives OHC, Second session, April 12, 1988, volume 2-3 US Gro, 1989, Social Science, 1989. pp. 268.
29. Huth EJ, Guidelines on authorship of medical papers. Annals of Internal Medicine, 1986. 104(2): 269-274.
30. Koenig R, Panel proposes ways to combat fraud. Science, 1997. 278(5346): 2049-2050.
31. Koenig R, Panel calls falsification in German case 'unprecedented'. Science, 1997. 277(5328): 894.
32. Norman C, How to handle misconduct allegations. Science, 1989. 243(4489): 305.
33. Onek JN, To the editor. Shattuck Lecture—misconduct in medical research. New England Journal of Medicine, 1993. 329(10): 733-734.
34. Pascal CB, Don't pass judgment on ORI. The Journal of NIH Research, 1996. 8: 16-17.
35. Powell D, Report condemns pressure to publish. Science, 1994. 264(5166): 1662.
36. Register F, Federal Register. 151; 32449.
37. Resnik DB, Gutierrez-Ford C, and Peddada S, Perceptions of ethical problems with scientfic journal peer review: an exploratory study. Science and Engineering Ethics, 2008. 14: 305-310.
38. Resnik DB, A troubled tradition: It's time to rebuild trust among authors, editors and peer reviewers, Chapter 2 in For the Record: American Scientist Essays on Scientific Publication. 2011, Research Triangle Park, North Carolina: Sigma Xi The Scientific Research Society
39. Stewart WW and Feder N, The integrity of the scientific literature. Nature, 1987. 325(6101): 207-214.
40. Stone R, Canadian schools: draft misconduct rules or lose funds. Science, 1994. 266(5191): 1631.

41. Sun M, For future grants ski trip are out. Science, 1980. 210(4471): 746-747.
42. Sun M, Peer review comes under peer review. Science, 1989. 244(4907): 910-912.
43. Unknown. Unknown. The New York Times. 1991 April 1.
44. Unknown, Unknown. The Journal of NIH Research, 1996. 8.
45. Weaver D, Costantini F, Imanishi-Kari T, et al., A transgenic immunoglobulin mu gene prevents rearrangement of endogenous genes. Cell, 1985. 42(1): 117-127.
46. Weissmann G, Science fraud: from patchwork mouse to patchwork data. FASEB Journal, 2006. 20(6): 587-590.
47. Wortis HH, Huber B, and Woodland R, To the editor. Shattuck Lecture—misconduct in medical research. New England Journal of Medicine, 1993. 329(10): 733.
48. Xin H, Scientific misconduct. Invention of China's homegrown DSP chip dismissed as a hoax. Science, 2006. 312(5776): 987.

CHAPTER 4

Fabrication and Falsification

Concocting data is an accepted skill in writers of science fiction and other forms of creative writing, but in science the making up of data or results (fabrication) is taboo and a common form of scientific misconduct in academic and industrial research.

The curriculum vitae of Elias A. K. Alsabti (see Alsabti chapter 5) was filled with falsifications. Falsification is changing data or results and it includes leaving out data points to polish the appearance of a graph.

The term falsification also covers the misrepresentation of a person's one's accomplishments s and qualifications as in a curriculum vitae or grant application. To obtain professional positions and to advance careers, some people pad their résumés with untruthful made up information about research experience and publications. Sometimes they even falsify their academic degrees and honors. Initially this information may be accepted in good faith and the dishonest individual may get away with it, but eventually the truth usually emerges and the credibility of the cheater collapses. After encountering falsified credentials in two applicants for a gastroenterology fellowship, Sekas and Hudson [6] performed a retrospective review of 236 applications submitted for this fellowship within a single year to get an indication as to how widespread this falsification was and found that it was common. One hundred and thirty-eight of the applicants (58.5%) reported research experience during a residency in a USA training program, but research activity could not be confirmed in 47 of these 138 applicants (34.1%). Fifty-three applicants (22.4%) claimed published articles; and 16 of these applicants (30.2%) misrepresented them by including citations of nonexistent articles in journals, articles in nonexistent journals, or articles noted as "in press." Based on this retrospective study, Sekas and Hudson recommended several steps to prevent misrepresentation in fellowship applications.

Paul Arthur Crafton and Other Aliases

Professor Paul Arthur Crafton (John B. Hext, Peter H. Pearse, David Gordon, and others aliases) was an expert impostor who taught courses for which he often had no formal training [1], [2] at seven different colleges and universities (George Washington University (GWU), Towson State University (TU)*(http://www.towson.edu)*, Millersville State University *(http://www.millersville.edu/)*, Shippensburg State College *(http://www.ship.edu/)* the University of Delaware (UD) (http://www.udel.edu/), Rutgers University (Rutgers) *(http://www.rutgers.edu/) and Wagner College (http://wagner.edu/)* over a span of 4 years using different names [7] *(http:www.nytimes.com/1983/05/04/us/around-the nation-impersonation-suspect-had-a top-lecture-rating.html8/8/2010)*. This con-artist with fraudulent credentials from a university in Scotland managed to get faculty teaching positions at different institutions in the departments of mathematics, as well as engineering administration and business administration [2] *(http:www.nytimes.com/1983/03/05/us/around-the nation-impostor-of—professors-said-to-be . . . . 3/5/2010)*. At these institutions the screening procedure was often sophisticated. He was considered "a hardworking, interested and contributing member" of one department of mathematics and computer science. He was fired from different jobs for failing to meet contractual obligations and other reasons. The impostor managed to get his job at Shippenburg State College after answering an advertisement in The Chronicle of Higher Education and going through a rigorous recruitment process which included an interview with a faculty committee, a review of his credentials and references, a telephone interview and the delivery of a lecture at the university. Shippensburg State College eventually exposed the impostor after students pointed out his flaws and constant complaints were received from Fran Lucia, one of the students. Other suspicions about Crafton were raised when a faculty member noticed a journal article by an Australian professor John Bryon Hext after a student had seen the person with this stolen identity teaching in Millersville. After being exposed Shippenburg State College turned the impostor over to the attorney general's office in Pennsylvania. Eventually Crafton was arrested when he was 59 years old [1] and charged with 16 counts of forgery, tampering with records, false swearing, and theft by deception. During an investigation the police found documents in Crafton's apartment connecting him to 34 other aliases.

Misrepresentation of information in professional credentials may point to future dishonest research as happened with Anti Potti who falsely claimed that he had been a Rhodes scholar. Also, after Mark H. Williams was found guilty of scientific misconduct, he was discovered to have previously falsified his curriculum vitae to get scholarly posts in the United Kingdom (UK) and he even forged letters of reference and cited phony credentials and degrees.

Perpetrators of fabrication in science are also frequently guilty of falsification and vice versa.

Fabrication Related to DNA Breakage from Cell Phones

The wide use of mobile cell phones raised concern about the biologic effects of microwave radiation (300 MHz-300 GHz) because of the closeness that the phone is held to the head [4]. Some studies reported in peer-reviewed publications indicated that low intensity electromagnetic fields (EMFs) can cause single-strand DNA breaks [10]. Khurana from the department of neurosurgery at the Australian National University (http://www.anu.edu.au/) pointed out that peer reviewed papers from at least seven countries have shown that low-intensity EMFs can break DNA or modulate it structurally [4]. Vogel responded [8]. Critics of two publications by scientists at the Medical University of Vienna (http://www.meduniwien.ac.at/) questioned the claim that EMFs from cell phones damage DNA because the variation was too low for a biological experiment and the data was too good to be true. Allegations of fabrication were made, and a university investigation concluded that both studies were fabricated and recommended that the papers be retracted. Following this, the technician (Elizabeth Kratochil) who worked on the project resigned despite denying any wrongdoing. Initially, Hugo Rüdiger, the senior author of both papers, agreed to retract the two papers, but then changed his mind and agreed to retract only the 2008 paper [5], which was published after his retirement in October 2007. He did this because he could not guarantee an exclusion of bias by the masking of the experimental and control groups. His change of mind stemmed from his realization that the chair of the ethic committee was a lawyer who had worked for a telecom company. Rüdiger refused to withdraw the other earlier paper [3].

Moreover Rüdiger pointed out that his critics had a conflict of interest and were funded by the cell phone industry. A foremost critic was Alexander Lerchl, a professor of biology at Jacobs University (http://www.jacobs-university.de/) in Bremen, Germany whose research was funded by multiple cell phone operators and manufacturers. He conveyed his concerns to the officials of the Medical University of Vienna as well as to the editor of Mutation Research. Franz Adlkofer a coauthor on both papers would not consent to having the papers retracted and Lerchl continued to attack this research which had a detrimental effect on the mobile phone industry. The Medical University of Vienna was called upon to investigate eight more papers coauthored by the same technician. By August 2008 there was still an ongoing investigation of this controversial work [9]. There are effects of microwaves in microwave ovens and in wireless local area networks (LANs).

Industrial Bio-Test Laboratories

Joseph C. Calandra had been a professor of pathology and biochemistry at Northwestern University (NWU) (http://www.northwestern.edu/) since 1942. In 1953 this entrepreneur founded Industrial Bio-Test Laboratories (IBT) and became its president. Under his leadership IBT developed into a major industrial product safety testing laboratory. In 1966, before the downfall of IBT, Calandra sold the company to Nalco Chemical but remained as its president until 1977. IBT was a highly influential contract laboratory with about 23, 000 studies listed in its portfolio. In 1976 the USA Food and Drug Administration (FDA) discovered that pesticide toxicology reports by IBT did not match the raw data during a standard scrutiny of this company. Because of this discrepancy the Environmental Protection Agency (EPA) demanded an audit of every analysis performed by IBT to justify the registration of different pesticides. Pesticide manufacturers, such as Monsanto, utilized facilities comparable to those of IBT to evaluate the toxicity of their manufactured goods. The inspection of IBT disclosed that some studies on the herbicide Roundup® unacceptable. In October1983, three past officers of IBT were convicted by a USA federal jury of fabricating safety tests in one of the most prolonged criminal trials in USA history.

The analyses needed to be repeated and this cost Monsanto about $6.5 million. Publicity about this scandal in the media harmed the reputation of Monsanto and activists exploited the information to cast immeasurable doubt on the truthfulness of Monsanto's data. All tainted residue analyses were duplicated and unblemished, up-to-date data were acknowledged by the EPA. After the exposure of the falsified tests, the EPA introduced Good laboratory practices (GLP), to safeguard the dependable generation and authentication of all data. The punishment for falsifying EPA data is brutal and includes massive fines and jail terms. The defense of IBT was unfortunately hindered by not helping with the investigation. Both FDA and USA attorney's office revealed that IBT stopped scientists deeply caught up in the scandal that they resisted a federal investigation in 1976 and plotted on their cheating. After the indictment it took several years until the case came to trial on April 13, 1983 in the district court in Chicago. The accused were charged with providing the government with falsified data, mail fraud, and the possession of phony data. The safety reviews of more than 200 pesticides were questioned and many were retested at considerable expense by the pesticide manufacturers. One of IBT's richest customers claimed that it spent $12 million repeating the flawed analyses by IBT.

Craven Laboratories

In 1990, the pesticide business was victimized by other fraudulent analyses of pesticide residues in produces sprayed with insect repellent. These wrongdoings were detected in tests performed at Craven Laboratories in Dallas, Texas. The EPA investigated the matter; and on February 25, 1994, Don Craven, the president of Craven Laboratories, and fourteen former workers at the company were fined or given jail sentences after being found guilty of falsifying pesticide residue tests over a decade.

References

1. Anonymous. Colleges fooled by job applicant. The New York Times. 1983 March 27.
2. Anonymous. Around the nation; Imposter of professors said to be the real one. The New York Times. 2010 March 25.
3. Diem E, Schwarz C, Adlkofer F, et al., Non-thermal DNA breakage by mobile-phone radiation (1800 MHz) in human fibroblasts and in transformed GFSH-R17 rat granulosa cells in vitro. Mutation Research, 2005. 583(2): 178-183.
4. Repacholi MH, Health risks from the use of mobile phones. Toxicology Letters, 2001. 120(1-3): 323-331.
5. Schwarz C, Kratochvil E, Pilger A, et al., Radiofrequency electromagnetic fields (UMTS, 1, 950 MHz) induce genotoxic effects in vitro in human fibroblasts but not in lymphocytes. International Archives of Occupational and Environmental Health, 2008. 81: 755-767.
6. Sekas G and Hutson WR, Misrepresentation of academic accomplishments by applicants for gastroenterology fellowships. Annals of Internal Medicine, 1995. 123(1): 38-41.
7. UPI. Around the Nation; impersonation suspect had top lecture rating. The New York Times. 1983 May 4.
8. Vogel G, Response to Khurana letter.Letter to editor. Science, 2008. 322: 11325.
9. Vogel G, Fraud charges cast doubt on claims of DNA damage from cell phone fields. Science, 2008. 321: 1144-1145.
10. Yao K, Wu W, Wang K, et al., Electromagnetic noise inhibits radiofrequency radiation-induced DNA damage and reactive oxygen species increase in human lens epithelial cells. Molecular Vision, 2008. 14: 964-969.

CHAPTER 5

Hoaxes

A variant of fabrication is the hoax in which something is done without malice for deception or mockery as a practical joke. Through the ages, many individuals with a sense of humor have generated hoaxes sometimes affecting one or more of the sciences. Hoaxes may also be designed to make an academic point [40-42]. In this regard, Sokal stressed the sloppiness of post-Modernist thought [45]. The perpetrators have often not concealed their identity, and the hoaxes have not involved taxpayer's money or impacted adversely on their fellow scientists.

The Ape Man of Waterton

"Squire" Charles Waterton (1782-1865), an eccentric English naturalist and explorer, was alleged to have descended from eight saints (Vladimir the Great, Saint Anna of Russia, the Holy Martyrs Boris and Gleb, Saint Stephen of Hungary, Saint Margaret of Scotland, Saint Mathilde, Saint Thomas) as well as Humbert III of Savoy and a number of European royal families [50]. During the reign of King Henry VIII of England, the Waterton aristocratic family refused to convert from Catholicism to the new Protestant religion. This obstinate adherence to their religious beliefs resulted in the confiscation of a considerable part of the Waterton estates. In 1802 the Squire went to British Guiana to assume responsibility for his uncle's estates, and a decade later he began to delve into the hinterland of South America and even reached Brazil allegedly with bare feet in the rainy season. In 1825 Waterton documented his travels in a tome [52] in which he maintained that he came across an ape man in the rainforest and explained lightheartedly that he had killed the beast and carried the head and shoulders out of the jungle back to England, where he preserved the creature according to Waterton's personal technique of taxidermy. A sketch of the ape man is illustrated in the frontispiece of his book. In reality Waterton took the head and shoulders of a red howler monkey and fashioned its facial appearance to produce a humanoid appearance. In addition to this frolic, Waterton combined parts of two completely entirely dissimilar animals into a single taxidermic animal. Waterton's hoax damaged his reputation as a genuine naturalist, but he did not care.

The Great Moon Hoax of 1835

Sir John Frederick William Herschel (1792-1871) was the son of Sir Friedrich Wilhelm Herschel (1738-1823), a German musician who immigrated to England and made numerous contributions to astronomy. He was a renowned scientist in the midnineteenth century and the subject of a stunning hoax forever dubbed the Great Moon Hoax of 1835 [17].

Like his father, Sir John was also fascinated with astronomy and made several original discoveries related to the heavenly bodies. He created the Julian day system and named four moons of Uranus and seven moons of Saturn. In November 1833 he embarked on a private expedition to South Africa to catalogue the stars and other objects in the skies overlooking the Southern Hemisphere. While there Richard Adams Locke (1800-1871), a British-born journalist, published a series of six articles in the Sun, a New York newspaper starting on August 25, 1835, contending that they were reprints from the Edinburgh Journal of Science (the Edinburgh Courant) documenting the actual discoveries made by Sir John with his astonishing telescope while in South Africa [25]. The author of the narrative of the six articles was a factious Dr. Andrew Grant who was stated to be a traveling companion of Sir John. The first article started with a discussion on what changes were being made in the construction of the telescope that would be appropriate for the mission. Even His Majesty King William IV of England agreed to pick up the tab for its creation. Part of Sir John's mission was to observe the crossing of the sun by Mercury, and this could not be achieved in the Northern Hemisphere because the transit would occur during the night, whereas in South Africa it would be daytime! After the long sea voyage, the site for the telescope was selected "aided by several companies of Dutch boors." Late in 1835 the Halley comet was due back, so Locke made use of the fact by reporting that Sir John has spotted it that year before anyone else. In reality Sir John had difficulty pinpointing the comet in Cape Town. Sir John's telescope allegedly disclosed a forest on the moon and in the shade of the woods herds of brown quadrupeds resembling the bison. Aside from these living animals, four flocks of large-winged creatures were detected, and they bore no resemblance to any known birds. Other livestock seen on the moon were goats, unicorns, and bipedal tail-less beavers. There were even bat-winged humanoids who built temples. A survey of the shores of Mare Nubrium revealed a sea with brilliant white beaches and rocks that seemed to be composed of green marble. During the publication of this stunning hoax about Sir John and his discoveries, the circulation of the Sun increased immensely from 8, 000 to 19, 360, making it briefly one of the newspapers with the largest circulations. Rival newspapers were obviously not happy with the success of the Sun based on the story of these remarkable findings by Sir John and looked into the narrative, and so did two professors at Yale University (Yale) *(http://www.yale.edu/)* who wanted to inspect the "original" material. Suddenly Locke confessed, and the game was over. After the hoax was exposed, it was appreciated more for its entertainment than as indignation of fabricated journalism. In June 1835 two months prior to the Great Moon Hoax, Edgar Allan Poe (1809-1849) published his own moon hoax entitled "The Unparalleled Adventure of One Hans Phaall" in the Southern Literary Messenger [33]; but because of its satirical nature, this early science fiction story was not as successful as the Great Moon Hoax. Locke upstaged Poe, who regarded Locke as a genius for contriving his hoax that fooled more than one in ten gullible readers. Reprints of the hoax spread, and eventually a copy reached Sir John who took the matter in stride. Nevertheless, Poe's story established him as one of the first science fiction writers [48].

Minie Ball Pregnancy

Dr. LeGrand G. Capers (1834-1877) was the chief surgeon connected to the Cuttashaws battery during the Vicksburg campaign in the war between the American states. He submitted a concocted story to the American Medical Weekly as a joke and was surprised to get it published [7, 8]. His report focused on a revolutionary bullet known as the minie ball that was invented in the 1840s by the French army captain Claude-Étienne Minié (1804-1879) and modified by another French captain Henri-Gustave Delvigne (1800-1876) for a muzzle-loaded rifle. According to his November 21, 1874, report, during the battle of Raymond, Mississippi, a Confederate soldier was shot in the scrotum by a Yankee soldier with a minie ball. After penetrating a testicle of the Confederate soldier, the bullet went through the private parts of a young virgin who was standing very nearby and she delivered a baby 278 days later. Prior to the birth, the women's hymen was found to be intact, implying that she had not undergone sexual intercourse. Further evidence that this pregnancy resulted from the implantation of sperm was the removal a deformed minie ball from the newborn baby's skin. The honorable Southern gentleman married the Southern maiden for making her pregnant in this unorthodox manner. The article was quoted in 1959 as reality in the New York Journal of Medicine [15] *(http;/www.suite101.com/article/famous-minie-ball-pregnancy-a15052).*

Piltdown Man

The most famous so-called hoax is the Piltdown Man, but this conspiracy may have had an ulterior motive and not been intended as an innocent practical joke. Indeed the fabrication set back research on paleontology for many years. Charles Dawson (1864-1916) was a practicing solicitor in Sussex, England, and spent his free time working on his hobbies of geology and anthropology. His past achievements included the discovery of the fossils of the first Mesozoic mammals in the United Kingdom (UK), and he provided the British museum (natural history) with a notable number of fossils. On December 18, 1912, Dawson and Arthur Smith Woodward (1864-1944), the British museum's leading paleontologist, made an important announcement to the geological society of London that would shake the world and impact science for more than half a century. They declared that the remains of an early human fossil had been found in a gravel pit in the village of Piltdown in East Sussex, England. This fossil consisted of fragments of a skull and jaw bone and became known technically by the Latin name *Eoanthropus dawsoni*. Because of where the remains were found, the bones were attributed to Piltdown Man, but it was also referred to as Sussex Man and Dawn Man [46]. This remarkable find occurred while the British empire was still expanding and little was known about the antiquity of ancient man. The discovery was a distinct feather in the cap of Dawson and Woodward but also for the nation establishing its primacy in this arena, because it appeared to be older than what the French or Germans had found. From far and wide, Piltdown Man became accepted as the earliest human fossil.

Human evolution was still a novel idea, and the views of Charles Darwin (1809-1882) had not yet become entrenched. The finding of Piltdown Man occurred at a time when discoveries of preglacial human ancestral remains were being made throughout Europe but sadly for the British not on their land. The Heidelberg jaw had been discovered in 1907 on the European continent. Simple stone artifacts (eoliths) were a subject of debate as to whether they were merely a product of natural abrasion or a precursor to the more refined stone instruments of ancient man (paleoliths). Pleistocene sites dating back to between 10, 000 and 2 million years ago were being discovered in Britain, but earlier Pliocene sites (2 to 5 million years ago) were extremely disappointing to British fossil hunters. Paleontologists were convinced that early human remains would be found in the gravel beds of parts of Europe as well as southern England.

Many experts of the day including Frank O. Barlow (a staff member of the British Museum of Natural History), Sir Grafton Elliot Smith (1871-1937), Woodward, Arthur S. Underwood (a dentist), Sir Arthur Keith (1866-1955), William Plane Pycraft (1868-1942), and Sir Ray Lankester (1847-1929) (a zoologist) examined these bones; and all were convinced that they were the fossilized remains of an early human ancestor. The combination of the bone fragments and the teeth supported extremely well the prevailing belief that brain development was an early attribute of human evolution. Scientists debated about the controversial findings in Piltdown Man for more than four decades. In 1923 Franz Weidenreich (1873-1948), a German anthropologist, correctly concluded that the cranium was from a modern human with filed-down teeth, but the "experts" had deaf ears to the truth [27].

On November 21, 1953, Joseph S. Weiner, Kenneth Page Oakley (1911-1981), and Sir Wilfrid Edward Le Gros Clark (1895-1971) startled the world with the declaration that Piltdown Man, the delight of British science, was a forgery fabricated by one or more unknown persons and that the experts had been hoodwinked. Their paper in the *Bulletin of the British Museum of Natural History (Geology)* [53] provided stunning evidence that the most important human fossil remains found in Britain that fooled the experts was a forgery made to deceive the scientific community. Later Weimer documented this forgery in a book titled The Piltdown Forgery in 1955[54] and in another later book published in 2003 [55]. Piltdown Man suddenly became the most famous paleontological "hoax" in history. It is estimated that this forgery was the focal [4, 16, 28, 35, 43] point of more than 250 papers, and numerous books have been published on Piltdown Man [51]. The fraud of Piltdown Man consumed an unnecessary vast amount of time and effort by talented scientists. The announcement also appeared in the *Times* on November 23, 1953 [49], and questions were raised in the British houses of parliament. Piltdown Man was in reality a human skull of medieval age. The mandible was not human but belonged to a juvenile female orangutan and appeared to be about 500 to 600 years old. Its owner must have died somewhere in Borneo or Sumatra where the orangutan only lives naturally. The molars and canine tooth were those of a chimpanzee, and to give the impression that their shape was compatible with a diet suited to humans, they had been filed. To provide the bones with an appearance of age, they were stained with a solution of iron and chromic acid.

The hinge of the jaw bone (condyle) was seemingly deliberately broken so that it did not articulate accurately with the skull. Fragments of the skull were indeed human but of extraordinary thickness and construction. Other bones discovered in the Piltdown gravel dated back about a half a million years to the early or late Pleistocene. Flaked flint stones were also present at Piltdown, but they were also being found in vast numbers in different sites of southern England. Some were characteristic of British fossils, but other appeared to come from the Mediterranean region. Although the Piltdown Man is widely regarded as a hoax even today, this is too mild a term for this piece of fraud. It set back research on human evolution significantly. Moreover, Piltdown Man threw scientists off track because of the belief that in human evolution the brain enlarged before the jaw changed in response to new types of food. Because of Piltdown Man, important significant fossils found in South Africa during the 1920s were ignored or received too little attention. The interpretation of objective evidence for the reconstruction of human evolution ended in a blind alley. After Piltdown Man was discovered, the scientific community regarded it as the missing link between apes and humans, but it gradually lost its validity after the discovery in 1924 by Professor Raymond A. Dart (1893-1988) in South Africa of the Taung child designated *Australopithecus africanus*, which was published in the following year in Nature [12]. This early potential precursor of *Homo sapiens* lived 1-4 million years ago. Perhaps because Piltdown Man, became widely accepted it took many years before the Taung skull became appreciated as a member of the ancestral hominoids from which *Homo sapiens* evolved. Later incomplete bony remains of Lucy were discovered and fossilized footprints belonging to *Australopithecus afarensis* dating back to approximately 3.7 million years. A most important breakthrough in human evolution was the discovery of pieces of jaw and teeth from *Ardipithecus ramidus*, the oldest-found fossil species of primitive hominids. More than a hundred specimens of this species have been thoroughly studied, and the findings suggest that body and brain sizes of both males and females do not differ much from each other. These primitive hominids were more primitive than Australopithecus and characterized by tiny faces and abridged canine and a premolar complexes. They were existing chimpanzees. *Ardipithecus ramidus* is the last-known common precursor of humans and African apes. Over the decades, a considerable amount of research has been performed by numerous scientists of multiple nationalities on the bones of this presumed human ancestor [19]. An analysis of the 4.4-million-year-old fossil called Ardi shows that early humans left the trees before departing the forest and getting much smarter January 25, 2010.

Once exposed the most devastating effect of the Piltdown fraud was its effect on the general public that lost confidence in science. In particular it produced much skepticism in the eyes of creationists who doubted Darwin's theory of evolution.

Who was the culprit to shame British science by creating a fraud that fooled their experts for many years? Was the Piltdown Man story merely a joke that got out of hand? Nobody confessed to the crime, and even today the perpetrator remains uncertain despite investigations by many scholars. If Piltdown Man was a hoax, why did the culprit not come clean and state that the incident was only done as a joke? After the realization that the Piltdown Man was a fraud, an attempt was made to identity of the fraudster, and a list of suspects was established with reasons for and against each potential offender continues to be debated.

Dawson, the discoverer of the skeletal parts and one of the main excavators of the Piltdown site, was regarded as the prime suspect, and for a long time he was suspected of acting alone. Miles Russell made an extremely strong case blaming Dawson as the sole perpetrator. Dawson participated in at least 38 obvious other archeological hoaxes during a decade or two prior to Piltdown [37]. However, authorities who have studied the fossils in considerable detail believe that he was aided by a professional paleoanthropologist. Dawson, despite his enthusiasm about geology and paleontology, remained an amateur in these branches of science. Dawson died 37 years before it was established that the Piltdown Man was a fraud.

Aside from making commendable contributions to science as an amateur, he also had an unscrupulous side. He plagiarized an unpublished historic document of Hastings Castle, Sussex. He also seemingly purchased his stylish house on the grounds of Lewes Castle by professing to carry this out on behalf of the Sussex Archaeological Society. During his interactions with the British museum, Dawson became acquainted with Woodward. While searching for fossils near Hastings, Dawson met Father Féllix Pelletier, a Jesuit priest, and Pierre Teilhard de Chardin, a seminary student who were also hunting for old bones. Other suspected fraudsters include the neuroanatomist Sir Grafton Elliot Smith and the British geologist and anthropologist William Johnson Sollas (1849-1936). Martin Alister Hinton (1863-1961), a British zoologist, was implicated in 1970 after a trunk left in storage at the natural history museum in London was found to contain carved teeth stained like those in Piltdown Man.

Another person implicated in the fraud is Sir Arthur Ignatious Conan Doyle (1859-1930), the talented writer and creator of Sherlock Holmes. Doyle loved practical jokes, adventure, and danger; and outstanding evidence implicating him as the perpetrator of Piltdown Man has been summarized by Winslow and Meyer [56]. They embarked on a path leading to Doyle because of several similarities between Piltdown Man and an earlier hoax brought about by the peculiar English naturalist "Squire" Charles Waterton (1782-1865), described in his book Wanderings in South America [52]. Doyle was familiar with Piltdown Man; and in a note to Woodward, Dawson stated, "Conan Doyle has written and seems excited about the skull. He has kindly offered to drive me in his motor anywhere." Doyle had an interest in fossils and was extremely excited when dinosaur remains were found in a project mining for coal in Kent. Winslow and Meyer pointed out that a former neighbor of Doyle collected specimens from the Malay Peninsula while working there as a magistrate. Several fossil remains found in the Piltdown gravel pit were identified as coming from the Mediterranean, and two years before the discovery of Piltdown Man, Doyle visited one of the few archeologists (Joseph Whitaker), who had often visited the Ichkeul region of the Mediterranean. Winslow drew attention to several similarities between Doyle and Waterton, the joker who created an ape man. Doyle knew both Dawson and Woodward before the fossils were found, and he lived 7-8 miles from the gravel quarry at Piltdown. Moreover, he apparently visited it openly in 1912. Dawson was delighted to discover Doyle's interest in the fossils and pointed this out to Woodward in a note. Another potential clue pointing to Doyle relates his abiding interest in human jaws that apparently developed after Doyle moved into to a dentist's house containing many casts of human jaws. The perpetrator of Piltdown Man undoubtedly had access to a large collection of skulls. Winslow and Meyer pointed out that Doyle had the opportunity to obtain all the fragments found at Piltdown. In his rendering of the fraudulent nature of Piltdown Man they speculated that similarities between Doyle's book The Lost World [13] might have been inspired by the Piltdown episode, and characters in that popular book make statements such as "If you are clever and you know your business, you can fake a bone as easily as you can a photograph." It also states that a practical joke "would be one of the most elementary developments of man." According to Winlow and Meyer, when Dawson learned about the imminent publication of The Lost World, he wrote to Woodward, pointing out that "Conan Doyle is writing a sort of Jules Verne book. I hope someone has sorted out his fossils for him!" In The Lost World, ape men are discovered in the upper Amazon, near where Waterton's red-haired nondescript ape man allegedly lived. Doyle documented a plateau in The Lost World that was "as large perhaps as Sussex, which could have been lifted en bloc with all of its contents." Doyle completed the novel in December 1911 and then sent it to the Strand, and it started to come out as a serial in April 1912—eight months before the discovery of Piltdown Man. The ideas that spawned The Lost World were unmistakably in Doyle's mind long before Piltdown became recognized as a site to be remembered for years to come (http://www.talkorigins.org/faqs/homs/piltdown2003.html). Experts who studied the evidence against the potential perpetrators of Piltdown Man in considerable detail considered Sir Arthur Keith the most likely coconspirator with Dawson [47]. In the August 1980 issue of Natural History, Dr. Stephen Jay Gould (1941-2002) provided a solid case for incriminating the French Jesuit scholar Pierre de Chardin Teilhard (1881-1955), who assisted Dawson at Piltdown digs, as an accomplice to Dawson based on circumstantial evidence. Letters implicating de Theilard were written between him and Oakley, one of the three British scientists who exposed the fraud in 1953. Also, elephant and hippo bones found in the Piltdown quarry were thought to come from Malta and Tunisia that de Teilhard visited between 1905 and 1908. Gould also suggested that Teilhard was motivated by his belief "that evolution moved in an intrinsic direction representing the increasing domination of spirit over matter" [32].

Dawson's coexcavator Woodward visited Piltdown but was considered above reproach since he was the leading paleontologist at the British museum, and although he visited Piltdown, he did not find the original bones and came into the study mainly to authenticate the findings because of his expertise. Several others have been suspected of being the hoaxer, perhaps in cahoots with Dawson.

Piltdown Chicken

Because some dinosaurs possess birdlike features, such as 3-toed feet, hollow bones, and forked bones in front of the breastbone (wish bones), many scientists believe that dinosaurs spawned birds about 150 million years ago. Others maintain that birds arose separately. The difficulty in distinguishing birds from dinosaurs gave rise to the controversial belief that some dinosaurs were flightless birds. The *Haplocheirus* genus of alvarezsauroid theropod dinosaurs seemed to settle this dispute as it lacks some birdlike features. It predates all other members of the genus by about 63 million years and is about 15 million years older than the oldest-known bird Archaeopteryx. The theory that birds evolved from dinosaurs gained considerable support when a fossilized potential missing link showed a dinosaur with birdlike plumage. Such a fossil made its way to Utah after being smuggled out of China where it was found. Stephen Czerkas, a dinosaur enthusiast with no scientific qualifications, purchased it at a gem and mineral sale. On October 15, 1999, the fossil named *Archaeoraptor liaoningensis* was displayed jubilantly by the National Geographic Society at a press conference, and it was written up in National Geographic the following month. Xu Xing, a prominent paleontologist and major researcher in dinosaur fossils at China's Institute of Vertebrate Paleontology and Paleoanthropology in Beijing, disputed the interpretation of the fossil and maintained that it was a combination of two fossils: the body and head of a birdlike animal and the tail of a dissimilar dinosaur. Xing had also found a mirror image of part of the specimen with a tail in a private collection in China. Paleontologists were only temporarily ecstatic, because, like the bones discovered in the Piltdown quarries as described above, the fossil was apparently the product of a hoax allegedly created by a Chinese farmer who joined bird parts with the tail of a meat eater. A computerized tomography (CT) scan of the fossil by the National Geographic Museum confirmed Xing's impression by revealing abnormalities in the reconstruction. The *U.S. News & Report* coined the term "Piltdown chicken" for the hoax, likening it to Piltdown Man. This fraudulent fossil illustrates the danger of rushing into print before a peer review by competent scientists in the relevant discipline. In March 2000 the *National Geographic* published an admission of its error. Despite this hoax, the concept of a dinosaur origin for birds lingers on and an astonishing new dinosaur fossil with feathers [44].

Together with his colleagues, Xing unearthed at a 160-million-year-old site in northwestern China an almost complete fossilized skeleton of a new genus of dinosaurs (*Haplocheirus sollers*). This fossil was estimated to be roughly 15 million years older than the most primitive identified bird (Archaeopteryx). Alvarezsauroids, a group of dinosaurs once regarded as flightless birds, were initially collected in the late 1920s but not fully characterized until the 1990s. Their astonishing discovery of feathered dinosaurs raised the question of why feathers would evolve in flightless creatures [39]. Perhaps they provided insulation, camouflage, or helped attract a mate. A paper in *Nature* in January 2010 reported the presence of eumelanin and pheomelanin types of melanin containing organelles in dinosaurs [2], suggesting that the structures are feathers and not collagen fibers.

The Holy Oak Pendant

An item that attracted considerable attention in American archeology was the Holy Oak pendant, which consisted of a whelk shell with a rudimentary drawing of a mammoth or mastodon. It was allegedly found in Delaware by Hilborne T. Cresson AM, MD (-1894) of the Peabody Museum of Harvard University *(http://www.harvard.edu/)* in 1864 the same year that a fragment of a mammoth ivory was found in France. For reasons that remain obscure, Cresson only brought it to the attention of a cynical group of archeologists in 1889; and soon thereafter, he exposed it to the general public. Its discovery was used as evidence of human antiquity in North America. Cresson was not highly regarded by the archeological community during his time, and for stealing artifacts, he lost his job at the Peabody Museum. He subsequently committed suicide in New York in September 1894 while in a mentally deranged state with delusions and hallucinations [1]. A relatively recent analysis of the pendant with accelerator mass spectrometry disclosed that it was only slightly more than 1, 000 years old [20] and was obviously made long after the woolly ancient animal became extinct in North America. In a major article in 1976, John C. Kraft and Jay F. Custer raised the plausibility of the pendant being a fraud [11].

Tasaday Hoax

In 1968 the Philippine government created a Private Association for National Minorities or Presidential Arm for National Minorities (PANAMIN) to protect cultural minorities. The head of the agency Manuel Elizalde Jr., the eldest son of a Filipino millionaire, learned about a small primitive prehistoric "Stone Age" society of forest inhabitants known as the Tasaday from a local hunter. Soon after becoming aware of them, Elizalde notified the media, and this led to much excitement; and a path toward them through the forest was started. Shortly thereafter, the passage to the Tasaday was blocked by guards, and only a select handful of extremely important visitors were permitted to reach them. In 1971 Western scientists confirmed the existence of this primitive society living on the Philippine island of Mindanao. The individuals had somehow maintained their primitive lifestyle while being isolated from the rest of the Philippine population. With 20-20 hindsight, the prehistoric community was apparently a fraud. Despite leaving the bodies of their deceased beneath leaves in the forest, nobody had found places with compost or bones. Also, even though the Tasaday had allegedly existed at their jungle cave for many years, evidence of garbage or human waste was not detected. The caves were extremely clean despite an expectation that a Stone Age tribe would have had garbage. Also, how would such a small tribe have avoided the consequences of consanguinity? In addition, it seemed strange that the Tasaday would not have stumbled upon the modern village located just a three-hour walk away while rummage around for food. Several months later, the *American Broadcasting Company* (ABC) aired on its television program 20/20 "The Tribe that Never Was." Through an interpreter, two young Tasaday men informed the interviewer that they were not Tasaday. Two years later, during another documentary made by the British Broadcasting Company (BBC) the same two persons from the Tasaday documented in the 20/20 program together other Tasaday and acknowledged they had not told the truth to the interviewers but just what they had been told to say. For this obedience, they were promised cigarettes, clothing—anything they wanted (http://www.tasaday.com/). Gradually disbelief about the reality of this "Stone Age" primitive society began to accumulate, and eventually in the 1980s, it became disclosed that this story was a sophisticated hoax; but the evidence for the existence of this "Stone Age" population remained controversial [14, 21, 58]. The Tasaday communicated with each other in a language that differed from the bordering tribes, and linguists considered it most probably derived from the Manobo languages of neighboring tribes about two centuries ago [29] (Reid, Lawrence A., "Another Look at the Language of the Tasaday," *(http://www.aa.tufs.ac.jp/~reid/Tasaday/Papers/pdffiles/tas1.pdf)* [34]. In March 1972 reporters from Associated Press, the *National Geographic Society*, and other media met with Elizalde and the Tasaday at a secluded cave home site. A report on this meeting was published in the *National Geographic* by Kenneth MacLeish and John Launois [26].

The occasion was highlighted with a photograph of a Tasaday boy climbing vines. On December 1, 1972, the National Geographic released a documentary titled "The Last Tribes of Mindanao." By then, eleven anthropologists had performed field studies on the Tasaday, but each study lasted less than six weeks. As stated by Elizalde, 24 remaining Tasaday did not share wives and did not commit adultery or practice divorce. The diet of the Tasaday was also suspicious. Allegedly they existed entirely on forgeable items, such as wild fruit, palm pith, forest yams, tadpoles, grubs, and roots; but such a diet was estimated to contain insufficient calories for survival, or they should have been emaciated. Dietitians and health care personnel advocated additional research, but they were prohibited from approaching the Tasaday community. An anthropologist witnessed soldiers sneaking cooked rice to the Tasaday, but after reporting this, he too was forbidden from performing research on the Tasaday. Before sealing the Tasaday reservation to sightseers in 1976, PANAMIN financially supported endeavors to find, visit, and study, the Tasaday; and most financial aid to protect them came from the Elizalde and his family, but a smaller amount was received from the Philippine government. The Tasaday community became effectively isolated after visitors were prohibited from contacting them during 1976. Between 1987 and 1990, Elizalde maintained that he spent more than $1 million standing up for the Tasaday against accusations of being a hoax. In 1986, after Ferdinand Marcos (1917-1989) was deposed as president of the Philippines, Oswald Iten, a Swiss anthropologist and journalist, escorted by Joey Lozano (a journalist from South Cotabato) and Datu Galang Tikaw (a member of the T'boli tribe to serve as the main translator even though he spoke no Tasaday) made an unlawful exploration to the Tasaday caves, spending approximately two hours with six Tasaday [23]. After coming back from the forest, Iten and Lozano testified that the caves were abandoned and maintained that the Tasaday were only members of local tribes and that they were maintained to mimic a Stone Age way of life because of Elizalde. The Tasaday continued to elicit much attention in television and radio programs. The uproar about the Tasaday provoked studies among scholars, politicians, and businessmen. After spending almost a year with the Tasaday and adjoining linguistic populations, Lawrence A. Reid, an emeritus professor in the department of linguistics at the University of Hawaii (UH) *(http://www.hawaii.edu/)*, presumed that they "probably were as isolated as they claim, that they were indeed unfamiliar with agriculture, that their language was a different dialect from that spoken by the closest neighboring group, and that there was no hoax perpetrated by the original group that reported their existence." Despite much evidence suggesting the existence of the Tasaday as a prehistoric primitive society, it was a fabricated hoax, and not everyone was convinced that they were not indeed a primitive tribe. Parts of Reid's research findings on the Tasaday are available on the Tasaday website together with other material [24] *(http://www.aa.tufs.ac.jp/~reid/Tasaday/index.html)*.

The Nipple and Cello Hoaxes

Musicians that play the guitar or the cello have been targeted by hoaxes. On April 27, 1974, the *British Medical Journal* contained a letter to the editor from a physician under the heading of "Guitar Nipple" [10]. It mentioned that three girls aged between 8 and 10 had recently been seen with mastitis in one breast characterized by a slightly inflamed cystic swelling about the base of the nipple. Questioning revealed that all three were learning to play the classical guitar, which requires close attention to the position of the instrument in relation to the body. In each case a full-sized guitar was used and the edge of the sound box pressed against the nipple. Two of the patients were right-handed and consequently had a right-sided mastitis while the third was left-handed with a left-sided mastitis. When the guitar playing was stopped, the mastitis subsided spontaneously. The author asked whether any other doctors have come across this condition. This letter came to the attention of Dr. Elaine Murphy who thought the report was a fake, and it incited the following response by her: "SIR, though I have not come across 'guitar nipple' as reported 'by Dr. P. Curtis [27 April, p. 226], I did once come across a case of 'cello scrotum' caused by irritation from the body of the cello. The patient in question was a professional musician and played in rehearsal, practice, or concert for several hours each day. I am, etc. J. M. Murphy" [31]. To Murphy's surprise, the journal published the note, and the hoax remained unexposed for 35 years later until Murphy, now a baroness and a member of the British House of Lords, pointed out in a letter with her partner in crime that their report was meant as a joke [30]. Despite the disapproval of dishonesty in science, this harmless prank has had no significant repercussions [22, 38, 57].

Sokal Hoax

In 1996 the physicist Alan Sokal participated in an elaborate hoax to fool the editors of the leftist journal Social Text. He submitted an essay containing at least six types of nonsense [42]. To the embarrassment of the editors of the journal, the paper was published without spotting the gibberish. Three weeks later, in the journal Lingua Franca, Sokal reported that the earlier paper was a hoax [41]. Stanley Fish, executive director of the Duke University Press, which publishes social text, was not amused by Skolal's sense of humor [18]. Humanists and social scientists were made most uncomfortable by the hoax. They lashed out in ways that sometimes made them appear even more ridiculous than the editors, who allowed the contribution to be published [3]. Later Sokal elaborated further on the subject in a textbook [40].

Boy Clone

David M. Rotvik (1944-) wrote a nonfiction book In His Image: The Cloning of a Man [36]. It allegedly documented that an eccentric aged millionaire Max had solicited Rotvik's aid to have a genetic clone of himself created. In the book Rotvik claimed that he located a gynecologist ("Darwin") who took the millionaire to a distant tropical site where Max possessed rubber plantations, nutmeg trees, and rice paddies and had built a hospital. Numerous Oriental women living in the region contributed their eggs and uteruses for the cloning experiment. In December 1976 a cloned child was born to the surrogate mother. The book was published in 1978, and it infuriated J. Derek Bromhall, an Oxford University geneticist, who sued the author and J. B. Lippincott, the publisher, for $7 million for defamation. Bromhall, who had developed the cloning technique in rabbits, claimed that the book was a hoax and Rotvik had used his name in the book without his permission. Scientists disputed that a human clone had been created [9]. After the lawsuit was instigated, the author acknowledged that three insignificant individuals in this nonfiction book were fictitious. Judge John P. Fullam, the USA District Court judge who heard the case, ruled the story a fraud and a hoax [5]. The publisher agreed that the book was a fraud and hoax and agreed to an out-of-court settlement in which provided Bromhall with $100, 000 [6].

References

1. Anonymous. Think Dr. Cresson was insane; The suicide an accomplished scholar, Well known in Philadelphia. The New York Times. 1894 September 8.
2. Benton MJ, Evolutionary biology: new takes on the red queen. Nature, 2010. 463(7279): 306-307.
3. Bérubé M, Post hoax, ergo propter hoax American Scientist, 2009. 97: 60-61.
4. Blinderman C, The Piltdown Inquest. 1986, Buffalo: Prometheus Books.
5. Broad WJ, Saga of boy clone ruled a hoax. Science, 1981. 211(4485): 902.
6. Broad WJ, Court affirms: boy clone saga is a hoax. Science, 1981. 213(4503): 118-119.
7. Capers LG, Unknown. American Medical Weekly, 1874. 1(19): 233-234.
8. Capers LG, Unknown. American Medical Weekly, 1874. 1(21): 212.
9. Culliton BJ, Scientists dispute book's claim that human clone has been born. Science, 1978. 199(4335): 1314-1316.
10. Curtis P, Letter: Guitar nipple. British Medical Journal, 1974. 2(5912): 226.
11. Custer JF, Kraft JC, and Wehmiller JF, The holly oak shell. Science, 1989. 243(4888): 151-152.
12. Dart RA, Australopithecus africanus. The man-ape of South Africa. Nature, 1925. 115: 195-199.
13. Doyle AC, The Lost World. 1912, London, England: Hodder and Stoughton.
14. Dumont JP, The Tasaday, which and whose, toward the political economy of an ethnographic sign. Cultural Anthropology, 1988. 3: 261-275.
15. Eger C, Famous minnie ball pregnancy: facts on the war urban legend. Woman impregnated when hit by bullet that shot civil war soldier's testicles. Military History, 2007.
16. Feder KL, Frauds, Myths, and Mysteries: Science and Pseudoscience in Archaeology, 6th Edition. 1990, New York: McGraw-Hill.
17. Fernie JD, The great moon hoax. American Scientist, 1993. 81: 120-122.
18. Fish S. Professor Sokal's bad joke. The New York Times. 1996 May 21.
19. Gibbons A, Breakthrough of the year. Ardipithecus ramidus. Science, 2009. 326(5960): 1598-1599.
20. Griffin JBea, A mammoth fraud in science. American Antiquity, 1988. 53: 578.
21. Hemley R, Invented Eden: The Elusive, Disputed History of the Tasaday. 2007, New York: Farrar, Straus and Giroux.
22. Herr HW, 'Cello scrotum' hoax. BJU International, 2009. 103(11): 1585.
23. Iten O, Die Tasaday: Ein Philippinischer Steinzeitschwindel. Neue Zurcher Zeitung. Zurich, 1986: 77-89.
24. Laurie L, About the Tasaday. 1999: Tokyo University of Foreign Studies.
25. Locke RA, The great moon hoax of 1835 1835.

26. MacLeish K and Launois J, Stone age cavemen of Mindanao. National Geographic, 1972. 142(2): 219-249.
27. McCort JJ, Franz Weidenreich; 1873-1948. New England Journal of Medicine, 1957. 257(14): 670-671.
28. Millar R, The Piltdown Men. 1972, New York: Ballantine Books.
29. Molony CH, The Tasaday language: Evidence for authenticity? In Thomas N. Headland (ed.), The Tasaday controversy: Assessing the evidence, 107-16), American Anthropological Association Scholarly Series, 28 ed. Headland TN. 1992, Phillippines: American Anthropological Association.
30. Murphy E and Murphy JM, Cello scrotum confession. Murphy's lore. British Medical Journal, 2009. 338: b288.
31. Murphy JM, Letter: Cello scrotum. British Medical Journal, 1974. 2(5914): 335.
32. O'Toole T. Piltdown Hoax said to involve Jesuit scholar. Washington Post. 1980 July 16.
33. Poe EAP. The Unparalleled Adventure of one Hans Phaal. Southern Literary Messenger. 1935.
34. Reid LA, The Tasaday language: a key to Tasaday prehistory. In Thomas N. Headland (ed.), The Tasaday controversy: Assessing the evidence, 180-193. American Anthropological Association Scholarly Series, 28 1992, Washington DC: American merican Anthropological Association
35. Roberts NK, From Piltdown Man to Point Omega: The Evolutionary Theory of Teilhard de Chardin. Studies in European Thought. 2000, New York: Peter Lang.
36. Rotvik DM, In his Image: The cloning of a Man. 1978, Philadephia and New York: J.B. Lipinncott.
37. Russell M, Piltdown Man: The Secret Life of Charles Dawson and the World's Greatest Archeological Hoax. 2003, Stroud: Tempus Publishing.
38. Shapiro PE, "Cello scrotum" questioned. Journal of the American Academy of Dermatology, 1991. 24(4): 665.
39. Sloan C, Feathers for T.Rex? National Geographic, 1999. 196(5): 98-107.
40. Sokal A, Beyond the Hoax: Science, Philosophy and Culture. 2008, Oxford: Oxford Press.
41. Sokal AD, A physicist experiments with cultural studies. Lingua Franca, 1996. 4: 62-64.
42. Sokal AD, Transgressing the boundaries: towards a transformative hermeneutics of quantum gravity. Social Text, 1996. 46/47: 217-252.
43. Spencer F, Piltdown: A Scientific Forgery. 1990, Oxford: Oxford University Press.
44. Stone R, Paleontology. Bird-dinosaur link firmed up, and in brilliant Technicolor. Science, 2010. 327(5965): 508.
45. Thomas E, Professor critiques post-midernist thinkers. The Chronicle, 1996.

46. Thomson KS, Piltdown man; the great English mystery story. American Scientist, 1991. 79: 194-201.
47. Tobias PV, Piltdown: An appraisal of the case against Sir Arthur Keith. Current Anthropology, 1992. 33: 243-293.
48. Tresch J, Extra! Extra! Poe invents science fiction! The Cambridge Companion to Edgar Allan Poe, ed. Hayes KJ. 2002: Cambridge University Press. pp.1155.
49. Unknown. Unknown. The Times. 1953 November 21.
50. Walker JW, The Burghs of Cambridgeshire and Yorkshire and the Watertons of Lincolnshire and Yorkshire. The Yorkshire Archeological Journal, 1931. 30: 314-419.
51. Walsh JE, Unraveling Piltdown: The Science Fraud of the Century and its Solution. 1996, New York: Random House.
52. Waterton C, Wanderings in South America, the North-West of the United States, and the Antilles, in the Years 1812, 1816, 1820 and 1824: with Original Instructions for the Perfect Preservation of Bird &c. for Cabinets of Natural History. 1825, London: J. Mawman.
53. Weiner JS, Oakley KP, and Le Gros Clark WE, The solution to the Piltdown problem. Bulletin of the British Museum of Natural History (Geology), 1953. 2: 141-146.
54. Weiner JS, The Piltdown Forgery (First edition). 1955, London: Oxford Univesity Press.
55. Weiner JS, The Piltdown Forgery: the Classic Account of the Most Famous and Successful Hoax in Science. 2003, Oxford: Oxford University Press.
56. Winslow JH and Meyer A, The perpetuator at Piltdown. Science. Vol. 83. 1983: American Association for the Advancement of Science. 32-43.
57. Wyner LM, Cello scrotum, 1974-2009: history of a medical hoax. Medical Problems of Performing Artists, 2010. 25(3): 130-132.
58. Yengoyan AA, Shaping and reshaping the Tasaday: A question of cultural Identity—a review article The Journal of Asian Studies, 1991. 50(3): 565-573.

CHAPTER 6

Plagiarism

Plagiarism and the special variety known as self-plagiarism are variants of scientific misconduct involving publications. The term plagiarism is derived from the Latin word plagiaries meaning kidnapper, seducer or plunderer and the first known recorded use of the word plagiarism was by the Roman Poet Marcus Valerius Martialis (36-41 AD—102-104 AD), known in English as Martial, who accused the poet Fidentius of the crime. The designation is defined in different ways by different people and different dictionaries, but in the 1620s the term was used for "literary thief". Yet a workable definition of plagiarism in scientific misconduct was still needed in 1993. Plagiarism is a subset of the misappropriation of IP [52] and the lifting of extensive pieces of unattributed pieces of text from the writings of other individuals. Accusations of plagiarism are sometimes made between former collaborators. The unlawful use of ideas or exclusive techniques achieved by accessed confidential communiqués, such as in the review of a manuscript text or a grant application is theft of IP. For eons society has condemned plagiarism and considered it graver than the stealing of a material item and certainly more than borrowed text. Plagiarism includes considerable unattributed copying of text about it word for word or almost verbatim without acknowledgment giving a causal reader the impression that the work was original. The plagiarism of ideas is often not easy to prove, but when the evidence leads to an unequivocal conclusion that a person deliberately appropriated the work of others most scientists support forceful prosecution and strong penalties.

There is a fine distinction between self-plagiarism and a comprehensive reprocessing of information from earlier publications by an author into a later manuscript. It is understandable why a scientist from a non-English speaking country with imperfect skills in English would become involved in self-plagiarism.

Considerable reuse of written passages by veteran writers with a complete appreciation of the English language is unforgivable and ought to be forbidden.

It includes considerable verbatim or almost word for word copying of text without acknowledgment giving a causal reader the impression that the work was original.

Some professional societies consider plagiarism as the copying of wording, a table, a figure, or data from published or unpublished work of another without acknowledgment. Plagiarism not only comprises the lifting of sentences or phrases from a text without credit, but also embraces the illegitimate use of ideas, data, and explanations and unique techniques acquired during the evaluation of privileged communications, such as grant applications or in the review of manuscripts for journals. Plagiarism has no place in society and plagiarists are constantly being exposed on the Internet *(http://www.famousplagiatists.com)* exposes major plagiarists, including literary giants.

In academic and legal circles the definition of plagiarism are not identical. The law protects society against plagiarism. As pointed out by the legal expert Robert P. Charrow [40] the USA constitution provides Congress with the power to "secur (e) for limited Times to Authors . . . . the exclusive Right to their respective Writings" (Article 1, §8, cl.8, U.S. Constitution). Under this authorization Congress has passed different copyright laws, one of which was the amended Copyright Act of 1976. The latter distinguishes between "expression", which it protects against, and "ideas". Unwaveringly the Supreme Court has taken the position "that ideas and facts, regardless of their origin, may not be copyrighted and are part of the public domain available to every person" [37]. The Supreme Court clarified the copyright protection of the telephone directory white pages [62]. To paraphrase the writings of another person is completely legal under the copyright laws even without acknowledgment. Moreover, Congress has ratified under the "fair use" doctrine that limited quantities of another individual's published text may be copied word for word without giving credit to the source. In essence many pieces of text that the academic community would designate as obvious plagiarism do not legally represent plagiarism under the copyright laws. In academia plagiarism it is an unmitigated mortal sin and the commonest form of scientific misconduct. It is also the most frequent variety of misappropriation of IP. All scientists as well as the general public have an interest in making sure that thefts of IP are exposed and appropriately penalized. Here scientists are not alone because this dishonest behavior is extremely widespread and has been detected in journalism, literature, politics, theology and religion, pop fiction/nonfiction in addition to science and medicine. It is found in the writings of authors, poets, biographers, novelists, and occasional reporters. Historical novels have always been popular even when written by swindlers or plagiarists who frequently cut and paste the text from the works of others. LaFollette's book Stealing into Print [54] focuses on plagiarism and covers misconduct in academic fields other than science.

IP refers to various achievements resulting from intellectual creations. It includes discoveries that may be inventions of commercial value that could lead to the application of patents. Many legal principles governing IP have evolved over centuries, but it was not until the nineteenth century that this term was introduced and in the late twentieth century that the designation became commonly used at least in the USA.

As with other types of research misconduct authors of plagiarized text do not believe that they will get caught. Some are extraordinarily careless and on other occasions it is the advertent borrowing of text without giving credit to the writer of the original collection of words. NIH held a conference on plagiarism on 21 and 22 June 1993.

Plagiarism is extremely common and the literature contains numerous examples particularly before copyright laws protected IP. In his book on plagiarism Thomas Mallon [55] points out that plagiarists often attribute their misappropriation of contributions by others to unintentional confusion between text lifted from particular writers and their own handwritten notes. Now the standard excuse is mistaking electronic files with personal writings. As pointed out by Mallon the sanctions for plagiarism is markedly indefinable, but in recent times plagiarists have faced litigation. The situation is complicated especially since good writers appropriate and emulate the works of others. Charges of plagiarism or inadequate documentation have faced biographers, who have published texts that extend earlier biographies or histories. Cases of this nature bring up the issue of where the line between plagiarism and errors in citation is drawn. John Milton (1608-1674), William Shakespeare (1564-1616), Alexander Pope (1688-1744), John Keats (1795-1821), Charles Dickens (1812-1870), Thomas Stearns Eliot (1886-1965), the Pulitzer Prize winning author Doris Cones Goodwin (1943-), Rudyard Kipling (1865-1936), François Rabelais (c. 1494-553), Alexandre Dumas (1802-1870), Victor Hugo (1802-1885), Herman Melville (1819-1891), Herbert George (H. G.) Wells (1866-1946), Samuel Langhorne Clemens (Mark Twain) (1835-1911), as well as others have been accused of appropriating pieces of text from the works of others without acknowledgment. Other famous plagiarists include Dr. Martin Luther King, Jr. (1929-1968), Benjamin Franklin (1705-1790), the child psychologist Bruno Bettelheim (1903-1990), Dan Brown (1964—), Stephen E. Ambrose (1936-2002), the renowned American writer John Gardner (1933-1982) who generously borrowed portions of the text of others in a 1977 biography of Geoffrey Chaucer [49]. The Modern Language Association (MLA) stresses the importance of using quotation marks to signify quotations that are given verbatim.

Jacob Epstein. A year after graduating from Yale University (Yale) *(http://www.yale.edu/)* Jacob Epstein, a son of Jason Epstein the editorial director of Random House, and Barbara Epstein who coedits The New York Review of Books, spent a year in London trying to master the craft of a novelist and jotted down relevant information in notebooks including passages of his literary idols, such as Martin Amis, who wrote some wonderful phrases. While still a neophyte he borrowed much from the literature just like a potential artist copies the works of masters in the fine arts. He then published his first book, Wild Oats [47] and this comic novel was widely acclaimed. So much so that in October the following year a paperback edition of 110, 000 copies was released. This 31 year old British author was accused of plagiarism by Amis, the son of Kingsley Amis a renowned novelist, because more than 53 large pieces of text within the book were lifted verbatim from his own first novel The Rachel Papers [31]. Epstein admitted guilt for unintentionally using some of Amis's phrases and the discovery of his plagiarism worried him causing him to develop insomnia. The episode generated a transatlantic plagiarism debate between the two famous literary families, but Epstein clearly "committed literary suicide" [55].

Christopher Sawyer-Lauçanno. The American poet Edward Eastlin Cummings (1894-1962) popularly known as E.E Cummings or e.e. Cummings was the subject of at least two biographies. One by Richard S. Kennedy was published in 1980 [53]. Another authored by Christopher Sawyer-Lauçanno [65] was published a quarter of a century later and it received glowing reviews until it was discovered to contain many passages that repeated or directly duplicated parts of the Kennedy book [56, 74]. After being confronted about the plagiarism Sawyer-Lauçanno denied the charges stating that it was a mistake in which he may have missed some places that should have been documented. He corrected those oversights in a paperback edition of the book.

Stephen Edward Ambrose. During his life Stephen Edward Ambrose (1936-2002) was an American historian and a professor of history at the University of New Orleans (UNO) *(http://www.uno.edu/)*. Throughout his life he authored numerous best sellers on American history and he was the biographer of two American presidents (Dwight David Eisenhower [29] and Richard Milhous Nixon [28, 30]. As pointed out by David Plotz [64] *(http://www.slate.com/id/2060618/)* he passed off the elegant prose of some victims as his own. Analysts of his publications found him to be a master in plagiarism, falsification, and an inaccurate portrayal of historical facts. As an overly productive author he wrote many books and developed a reputation of running a history-book mill. At one time he delivered eight books in five years with the aid of his five children.

Karl-Theodor zu Guttenberg. Baron Karl-Theodor Maria Nikolaus Johann Jacob Philipp Franz Joseph Sylvester von und zu Guttenberg (1971—) lived in a castle in Bavaria which his family owned from 1482. This German aristocrat was an extremely popular charismatic politician. He had many titles and did not really need any more, but because a doctoral degree is a political asset in Germany he decided to get a PhD at the University of Bayreuth *(http://www.uni-bayreuth. de/).* This also provided him with two additional titles: Doctor Cut-and-Paste or Doctor Googleberg. In 2011 while serving as the German defense minister his political career collapsed *(http://news.bbc.co.uk/2/hi/programmes/from_our_own_ correspondent/9410282.stm)*, *(http://online.wsj.com/article/SB20001424052748704 50600457617397076502052 8.html).* Guttenberg was forced to resign from the government because of plagiarism in his doctoral thesis. He took this strategic move, to enable the German chancellor Angela Merkel to resist calls from the opposition to relieve her defense minister of his cabinet position. His PhD thesis included not only notes prepared by his research assistants but also extensive portions of text lifted word for word from newspaper articles and journals without citing their source. Because of this unacceptable behavior the baron requested the university to retract his doctoral degree *(www.spiegel.de/international/germany/0, 1518, 747402, 00.html).* Guttenberg attributed his errors to excessive commitments, namely his position as a parliamentarian and his duties as a husband and father, while writing the thesis. Neither Guttenberg's plagiarism nor his resignation apparently affected his popularity. Some have predicted that he might one day become a future chancellor of the German nation. Time will tell whether or not this will occur.

Annette Schavan. Aside from Guttenberg Annette Shavan the German education and research minister faced allegations of plagiarism in parts of her doctoral thesis in educational science that was published in 1980 when she was 24 years old. A doctoral degree was at one time regarded as an asset for German politicians, but with modern methods of picking up plagiarized text it seems to becoming a liability as six other German politicians have had PhDs revoked for using the material of others in their theses [35] *(http://scim/ag/Schavan).*

Alex Haley. Alex Haley the author of Roots: The Saga of an American Family [51] received many awards for his famous book, including a Pulitzer Prize. In 1978 he came under attack because of plagiarism. Another writer Harold Courlander (1908-1996) filed a lawsuit claiming that Haley lifted substantial pieces of text from his book The African [44] that was published in 1967. Following a five week trial in the USA district court of the southern district of New York Haley admitted inserting minor pieces of lifted text in his award winning book. The slipup cost Haley more than $500, 000.

Penelope Ann Douglass Conner Gilliatt. Michael Mewshaw (1943—) wrote an article about Henry Graham Greene (1904-1991) the English author, playwright and literary critic, but tried unsuccessfully to get it published until The Nation and London Magazine accepted it in 1977 [68]. To his disgust Mewshaw discovered that the New Yorker in 1979 contained an article by Penelope Ann Douglass Conner Gilliatt (1932-1993) the English novelist and film critic that was lifted almost verbatim from well-expressed words that he had written in his paper that was rejected by the New Yorker and other magazines. Eventually Mewshaw penned an entire book on Greene that included numerous grips such as the plagiarism of his work by Gilliat [57].

Leonard Mosley. Leonard Mosley (1913-1992) filed a lawsuit against Ken Follett (1949—) the Welsh thriller-novelist in the federal district court in Manhattan for filching from his book titled The Cat and the Mouse [58], but he later voluntarily dropped his suit and received no payment or other concessions [32].

Elias A. K. Alsabti. A plagiarist with considerable audacity was Elias A.K. Alsabti, an alleged Iraqi student claiming Jordanian citizenship. He claimed that he was a blood relative of the Jordanian Royal family when he came to the USA in 1977 for postgraduate medical education. His studies in the USA were allegedly paid for by the brother of King Hussein of Jordan (1935-1999), his Royal Highness Crown Prince Hassan bin Talal (1947—) the forty-second generation direct descendant of the Prophet Mohammad. However, the Jordanian embassy in the USA pointed out that he had no connection to the royal family. Alsabti performed research on cancer and was an extremely skilled liar, who acquired a remarkable portfolio that made it possible for him to trick his way into several outstanding research laboratories. Because of his personality, apparent royal connection and astounding curriculum vitae he managed to con positions in several American hospitals and universities, including the MD Anderson Hospital and Tumor Institute. Even though he lacked a medical degree Alsabti maintained that he graduated from medical school in Jordan, but so far the written proof of this remains to be seen. One day while in the USA, a PhD degree appeared by his name, and he claimed that he received this doctorate in cancer immunology, but he was only known to have worked in the USA in nondegree fellowships. While in the USA, Alsabti communicated with Paul S. Tien, who opened a branch of the American University of the Caribbean (AUC) *(http://www.aucmed.edu/)*. Alsabti received credit for clinical rotations in a Houston hospital and eventually received the MD from AUC.

He lifted entire articles from published journals that were commonly esoteric and replaced the authors with other names including his own and submitted the papers to other little-known journals. During the decade between 1970 and 1980, he managed to get 60 phony articles on a variety of subjects published in journals in different parts of the world. Most of these papers were published in 1979 and reprint requests were often assigned to an address at the Royal Scientific Society in Jordan. Papers by Alsabti appeared in a wide variety of Journals (Acta Haematologica [16], Journal of Cancer Research and Clinical Oncology [8, 9] Neoplasma [10, 11, 17, 24], European Surgical Research [12] Oncology [1, 3, 13, 14, 22, 23], Urologia Internationalis [19, 20], Journal of Surgical Oncology [5, 21], Gynecologic Oncology [27], British Journal of Urology [2]. Alsabti even published papers in the Japanese Journal of Experimental Medicine [4, 6, 7, 18, 25, 26], and the Japanese Journal of Medical Science and Biology [15]. The Journal of Clinical Hematology and Oncology received nine articles from Alsabti and six were published. When the editor of that journal (Ekkerd Grundmann) was informed about the plagiarized paper on "tumor dormancy" that was published in the Journal of Clinical Hematology and Oncology Broad received the curt relay of "we never print a retraction" [39].

Despite the established plagiarism of Alsabti, only one of his publications was retracted by the journal that published it. To some of his colleagues, Alsabti hinted that he would eventually return to Jordan as the director of a prominent cancer institute. Three years after coming to the USA, Alsabti's career collapsed when his plagiarism was discovered and most, if not all, of his scientific contributions were found to be fraudulent and in reality the work of others. Three reviews by Alsabti (two on tumor dormancy) in different journals were lifted in their entirety from a research grant of E. Frederick Wheelock, professor of microbiology at Jefferson Medical College in Philadelphia. Wheelock initially hesitated to incriminate Alsabti in dishonest acts, but he evicted him from his laboratory after two young researchers proved that Alsabti was fabricating data. After Wheelock discovered that Alsabti was publishing his work he brought the matter to the attention of the National Cancer Institute (NCI), which proposed that Wheelock alert a broad scientific community. Initially Wheelock wrote to Alsabti commanding him to retract the papers, but this did not occur. Letters were hence written to several different journals (Nature, Science, Journal of the American Association and Lancet) [71] with suggestions on how to avoid similar episodes. Alsabti also worked for Herman Friedman at Temple University (Temple) *(http://www.temple.edu/)* in 1977 to whom he introduced himself as an MD from Baghdad. The published manuscripts of Alsabti had numerous coauthors and many were strange and appeared fictitious [39]. For example, they included KA Saleh, AM Taleb, and AS Talat, who have never published papers without him. In more than one paper the reprint address was the Albaath Specific Protein Reference Unit in Baghdad, Iraq. In one paper Alsabti claimed that he was the director of that unit. William Broad was unable to find anyone that had heard of that unit [39]. Three different researchers drew attention to the unequivocal plagiarism of Alsabti, who reproduced virtually every word in least five papers that had previously been published by others [39]. At the time of Alsabti journal editors often sent manuscripts by surface mail to reviewers without first determining whether they were prepared to review a particular paper. One such paper was inadvertently sent to Jeffery Gottlieb from the European Journal of Cancer despite the fact that he had died in July 1975. The package remained in the mailbox until Alsabti opened it, made some minor alterations, changed the title and added fabricated coauthors (Omar Naser Ghalib and Mohammed Hamid Salem) and then mailed it with the original illustrations to the Japanese Journal of Medical Science and Biology which surprisingly published it [15]. When the original author, Daniel Wierda, who had been a graduate student at the University of Kansas (KU) *(http://www.ku.edu/)* discovered the paper, shortly after it was published in the same journal coauthored with his PhD advisor [72] he hit the roof. When the dishonest exploits of Alsabti became well established His Royal Highness Crown Prince Hassan of Jordan put an end to the funds that supported him.

Multiauthored Textbooks

Plagiarism is extremely common in successful multiauthored textbooks that get published in multiple editions. For new editions of such textbooks, the editors need authors to revise and update chapters. Frequently the author of a chapter declines to accept this role, and the editor has difficulty recruiting a new author who lacks the time, or the incentive, to prepare a completely new chapter even though much of the chapter content has stood the test of time and does not need to be rewritten. To rewrite such a chapter would also impose a heavy burden on the editor and publisher, because the wording of the existing chapter has often been extensively edited. The new author commonly lifts major portions of the chapter verbatim.

Kenneth L. Melmon. The sixth edition of William's Textbook of Endocrinology was released in 1981; but plagiarism charges against Kenneth L. Melmon, chairman of the department of medicine at Stanford University (Stanford)*http://www.stanford.edu/)*, did not surface until William W. Douglas, a pharmacologist at the School of Medicine at Yale, detected parts of his personal writings in Goodman and Gilman's The Pharmacological Basis of Therapeutics in the chapter prepared by Melmon in the William book. Douglas uncovered this academic misconduct in February 1984 while going through the literature seeking new information to bring his chapter in the Goodman and Gilman book up to date for the next edition. Subsequently, other authors involved in the episode were infuriated. After discovering that neither the editors nor publisher of the Goodman and Gilman book was aware of the incident Douglas informed Dominick Purpura, dean of the medical school and he passed the issue on to the Stanford Medical School's committee on ethical scientific performance and the accusations of plagiarism against Melmon were investigated [60]. Authors of replicated text in Melmon chapter of William's book confirmed that their consent was neither requested nor provided. Moreover the Macmillan publishing company, which published the Goodman and Gilman text, discovered no documents to signify that permission was requested from the publisher or that the publisher had granted its blessing. Some onlookers considered it implausible that a publisher would provide consent for such a vast amount of material to be reproduced from their extremely renowned textbook. Melmon reproduced fifteen pages of copyrighted material from eight chapters by four discrete authors in the sixth edition of Goodman and Gilman's The Pharmacological Basis of Therapeutics [50], a highly popular textbook (now in its twelfth edition), for which he was an associate editor and dropped them into a chapter he had written for the sixth edition of Williams Textbook of Endocrinology [73]. Melmon used the replicated material without acknowledgment and evidently without authorization, but Mellon claimed he grudgingly integrated the text at the persistence of the editor after being guaranteed that consent had been given. Moreover, Mellon asserted that his submitted 73-page chapter included major credits where they were indicated. He also pointed out that he was "stunned" to find that these attributions were missing from the published version.

Melmon's lawyer claimed that a draft of the manuscript with Melmon's notations assigned attribution for the incorporated material. Melmon sent a letter of apology to everybody affected in the incident and offered to forego his royalties for the Williams book. In retrospect Melmon believed that this regrettable situation was the result of a mix of different events associated with a collapse in communication as well as his failure to personally examine the galley proofs of his chapter himself. There was also editorial pandemonium after the unexpected sudden death on November 1979 of Robert Williams, editor of the Textbook of Endocrinology from a heart attack before the new edition was completed. Furthermore while his book chapter was in preparation, Melmon was overcommitted and moving to Stanford from the UCSF. The scandal resulted in the resignation of Melmon from his chairmanship at Stanford [61]. In retrospect, Melmon realizes that he should not have depended so much on other individuals.

Sherbert H. Frazier. After serving as the director of the National Institute of Mental Health (NIMH) Dr. Sherbert H. Frazier, a well-respected academic psychiatrist, became psychiatrist-in-chief at the McLean hospital of Harvard University (Harvard *http://www.harvard.edu/*) for more than a decade. He was also a professor at Harvard Medical School. When he gave speeches, Frazier commonly cited considerable work of others without giving acceptable credit. This was too much for Paul Scatena of the University of Rochester (Rochester) *(http://www.rochester.edu/)* to swallow after reading reviews by Frazier and running into problems going back to the original sources of the cited papers. Scatena noted that sizeable sections of Frazier's writings had been lifted from papers written by other authors. He also found examples in which cited references varied from paper to paper and that the primary source was not always condensed as correctly as it should have been. Scatena notified the dean of Harvard Medical School (Daniel C. Tosteson) about the flaws in Frazier's writings. This led to an investigation of plagiarism by Harvard and after being found guilty of plagiarism he was forced to resign from the university in December 1988 despite Frazier's appointment with tenure [46].

The exposure of this plagiarism gave Harvard another black eye shortly after the Darsee scandal and this university's unbending severe response in the Frazier case was to demonstrate to the world that Harvard was going to be tough on persons guilty of scientific misconduct. Tosteson notified the Harvard faculty in a letter that plagiarism had been identified in four review articles written by Frazier between 1966 and 1975. There was also careless scholarship in three of his publications, but evidence of fabrication was not detected. The biomedical literature database known as PubMed lists 37 papers by Frazier but none have been retracted.

Raghunath Mashelkar. Raghunath Mashelkar (also known as Ramesh Mashelkar) a chemical engineer and director general of the Council of Scientific and Industrial Research (CSIR) in India until his retirement in 2007 found himself identified as a plagiarist because as had relied on others to write Intellectual Property and Competitive Strategies in the 21st Century appearing under his name [34]. As an extremely busy individual he relied on other researchers to provide him with the content of a book. Unfortunately his helpers did not rewrite the material and nor did he in a chapter coauthored ironically on intellectual property. The book contains a long section copied verbatim from an article written by Darrell A. Posey (1947-2001) and in Graham Dutfield [33, 69]. After the plagiarism was discovered by Poser, the 2006 Indian edition of the book inserted the copied text in quotation marks and a footnote indicated the source.

Chaturvedi Sunil. Plagiarism is commonly detected by the author of the original work as in the case of Chaturvedi Sunil and colleagues. While reading an article in the Indian Journal of Ophthalmology [42]. Gail M. Seigel, a cell biologist, discovered a major segment of a review article that she had written in the 2001 copy of Digital Journal of Ophthalmology [66] extensively copied including images of an ocular melanoma [38, 41]. After notifying the editors of the two journals, Barun K. Nayak, the editor of the Indian Journal of Ophthalmology, retracted the paper [42] and pointed out that the four authors (Chaturvedi Sunil, Mittal Sanjeev, Bahadur Harsh, and Methrotra Amar Nath) from the Himalayan institute of Medical Sciences in India were banned from publishing in Indian Journal of Ophthalmology until further notice [59].

Douglas Mann. Five pairs of papers authored or coauthored by Douglas Mann, a cardiologist at Washington University *(http://wustl.edu/)*, were exposed for plagiarism by Harold "Skip" Garner [45].

Lee S. Simon. Lee S. Simon, a rheumatologist at the Harvard-affiliated Beth Israel Deaconess medical center and an associate clinical professor of medicine at Harvard, wrote a thirty-two-page review article on new treatments for rheumatoid arthritis [67]. Many pages in this review were almost indistinguishable from passages in an article previously published by Roy Fleischmann of the University of Texas Southwestern *(http://www.utsouthwestern.edu/)* in Expert Opinion on Drug Safety [48]. After *déjà vu* detected this situation, Harvard investigated the matter, and this led to Simon's resignation in early March 2009 [43]. After realizing that parts of the paper were plagiarized, the publisher retracted Simon's paper [45].

Rudolph A. Weiner. Another example of plagiarism detected by *déjà vu* was by Rudolph A. Weiner, a bariatric surgeon at Krankenhaus Sachsenhausen Hospital in Frankfurt Germany who published a paper on obesity research [70]. This paper duplicated essentially one-third to a half of what Daniel Herron another surgeon in New York City had documented [45]. When confronted with this unethical behavior Weiner brushed aside the incident stating that the paper was a review article and a coauthor had written the introduction. That coauthor died suddenly, but the paper in Surgical Technology International was not retracted.

Other

Aside from the aforementioned individuals some other prominent scientists have been accused of plagiarism or of not reporting their findings accurately and as honestly as they probably should have. Some of them have even admitted the plagiarism. A website lists famous plagiarists in science and medicine *(http://www.famousplagiarists.com/scienceandmedicine.htm)*. Jorge Bucay (an Argentinean psychiatrist) plagiarized almost verbatim sixty pages of philosopher Mónica Cavallé. Vishwa Jit Gupta, a professor at Punjab University *(http://www.pu.edu.pk/)*, not only plagiarized the work of others, but also fabricated geological findings in the Himalayas. Mostafa M. Iman lifted the work of others over a twenty-year period. The astronomer Jose-Luis Ortiz and his student Pablo Santos-Sanz hacked into a publically accessible log of the SMARTS (small—and moderate-aperture research telescope system) telescope and claimed the discovery of ten Neptunian objects before the genuine discoverers announced their findings. The British psychiatrist Raj Persuad pilfered the work of the American psychologist Thomas Blass. Aside from plagiarism, some of these cheaters also falsified data (Gupta and Iman). The misuse of ideas can end up in litigation.

Plagiarism in Research Grant Applications
Lonnie Mitchell

Lonnie Mitchell, a psychologist at Coppin State College, was found guilty of using plagiarized material in a grant application to the National Institute of Mental Health (NIMH) for a minority access to research careers research training grant [63]. Because of this indiscretion he was required to certify about integrity, honesty, and reliability of any Public Health Service (PHS) grant application he submitted in the subsequent 5 years. Moreover both Mitchell and his department chairman (Jesusa Wilson) were barred from participating on PHS advisory committees for 3 years.

Joseph Abbott

Those involved in the writing of grant applications need to examine more closely the use of boilerplate texts in preparing grant applications and particularly interpersonal relationships with past and present collaborators. The use of identical material in different grant applications is a sensitive topic where charges of plagiarism can arise. The wording used in an earlier grant application with collaborators can be considered plagiarism according to a clash between cardiologists at the University of California, San Francisco (UCSF) *(http://www.ucsf.edu/)*[36]. A grant application submitted to investigate the effect of antismoking procedures on the heart precipitated a dispute between Joseph Abbott and David Siegel, another UCSF cardiologist, in which Siegel, a prior collaborator, made an accusation of plagiarism. The dispute revolved around Abbott's proposal, to study whether nicotine patches used to help stop smoking might be injurious to individuals with heart disease. In the summer of 1997 a faculty panel at UCSF as well as an external mediator found Abbott guilty of plagiarism in the grant application. Abbott asserted that components in the second grant were his personal studies and the material was collected before the submission of either grant application. Siegel, his faultfinder, disagreed. The case brought in renowned experts as witnesses such as Donald Kennedy, former Stanford president, and Drummond Rennie, editor of the Journal of the American Medical Association (JAMA). Rennie testified as an expert witness for UCSF and argued that federal regulations permitted universities to establish their own standards; Kennedy testified as an expert witness for Abbott. Kennedy found it disturbing that a distinguished researcher and university would attempt to condemn someone on this type of charge. It was Rennie's opinion that the reuse of hefty pieces of a grant application without the permission of Siegel, the original principal investigator (PI), was undoubtedly outside the realm of acceptable grant writing and unquestionably plagiarism. Abbott appealed the plagiarism ruling and in 1996 he was given a hearing at the American arbitration association. In July 1997, the arbiter (Golden Gate university law professor J. Lani Bader) upheld UCSF's action believing that Abbott was guilty of plagiarism for copying major portions of Siegel's NIH grant application without acknowledgment and for submitting it to the Tobacco-Related Disease Research Program (TRDRP) without Siegel's awareness or approval. Surprisingly Scheinman the PI of the TRDRP grant application was not charged with any wrongdoing by USCF apparently because Abbott was allegedly the only author of the application; however the PI of a grant application is accountable for the entire contents of the proposal, no matter what this person's role was in its composition. The dispute resulted an arbitration hearing in which Siegel denied being informed by Abbott of the submitted TRDRP grant application. Moreover Siegel maintained that he would not have agreed to such an application shortly after his NIH proposal. Nevertheless the grant application submitted to TRDRP contained large sections identical to those in the NIH application, but without Siegel listed as a participant. As required the TRDRP application stated that a similar submission was under consideration at NIH. Adding more fuel to the fire the TRDRP application was funded, but not Siegel's application to the NIH. During the hearing Paul Friedburg, former dean of academic affairs at the University of California San Diego (UCSD) *(http://www.ucsd.edu/)* school of medicine and former chair of the committee on research integrity of the American Sssociation of Medical Colleges testified as an expert witness. He pointed out that persons who have played a part in a grant application are normally deemed to have a right to reprocess material in the proposal, even in the absence of the PI's consent.

During the hearing Melvin Cheitlin, chief of cardiology at San Francisco general hospital pointed out that when Abbott and Scheinman submitted a request to begin enrolling subjects for the project resemblances were noted between the TRDRP and NIH applications. Cheitlin notified Siegel and attempted unsuccessfully to negotiate a resolution to the conflict. The USCF fired Abbott in April 1994 apparently because of financial issues related to the grant that paid his salary. Abbott claimed that Scheinman and he had good reason for leaving Siegel's name off the TRDRP proposal because the basic ideas in both proposals were derived from research performed by them before they collaborated with Siegel. Abbott had been in charge of a smoking cessation clinic at UCSF. Before collaborating with Siegel, Abbott submitted an unsuccessful grant application to investigate whether nicotine patches were harmless in patients with cardiac disease. Abbott maintained that much of the wording in the initial Scheinman grant and from Abbott's earlier grant ended up in the application submitted to the NIH by Siegel. The testimony by Siegel at the hearing clashed with Abbott's statements. He maintained that more than 90% of his proposal was derived from new information, including his formerly reported classification of cardiac arrhythmias, which played an important role in both grant applications. Evaluators of the Abbott-Siegel dispute could not reach a consensus as to whether plagiarism took place. Clearly permission is needed if large pieces of another person's words are used and acknowledgment should be given. In retrospect if Abbott did get Siegel's permission he should have documented it in writing. This case was an institutional blunder by a prominent university.

The Office of Research Integrity (ORI) took the stand that an assumption of implied consent exists between past collaborators in the use of products of the collaboration. ORI does not regard cases such as this as plagiarism, but authorship disputes and lets universities handle them. According to Alex Parrish, Abbott's attorney, the UCSF should not have accused Abbott of plagiarism as that university professes to use the PHS definition of plagiarism. The ORI's refusal to deal administratively with authorship disputes does not signify that one coworker will never plagiarize the work of a collaborator. If the decision of this case is upheld by the general scientific community, significant problems will arise with grant writing that involving collaborators. As pointed out in chapter 2, the preparation of grant applications commonly involves the input of several individuals. Surprisingly the arbitrator found Abbott guilty despite not being the PI on the relevant grant.

References

1. Alsabti EA, Serum alphafetoprotein in bladder carcinoma. Oncology, 1977. 34(2): 78-79.
2. Alsabti EA, Serum alphafetoprotein (AFP) in bilharziasis. British Journal of Urology, 1978. 50(2): 134-135.
3. Alsabti EA, Paraproteinemia in normal family members of eight cases with primary intestinal lymphomas in Iraq. Oncology, 1978. 35(2): 68-72.
4. Alsabti EA and Safo M, Fetal proteins in various tumors. Japanese Journal of Experimental Medicine, 1978. 48(4): 283-285.
5. Alsabti EA, Serum immunoglobulins in breast cancer. Journal of Surgical Oncology, 1979. 11(2): 129-133.
6. Alsabti EA, Lymphocyte transformation in patients with breast cancer and the effect of surgery. Japanese Journal of Experimental Medicine, 1979. 49(2): 101-105.
7. Alsabti EA, Histopathological subtypes of Hodgkin's disease in childhood in Iraq. Japanese Journal of Experimental Medicine, 1979. 49(5): 319-324.
8. Alsabti EA, Tumor dormancy: a review. Journal of Cancer Research and Clinical Oncology, 1979. 95(3): 209-220.
9. Alsabti EA, Hodgkin's disease in childhood. Journal of Cancer Research and Clinical Oncology, 1979. 95(1): 75-81.
10. Alsabti EA, Tumor dormancy (a review). Neoplasma, 1979. 26(3): 351-361.
11. Alsabti EA, Serum immunoglobulins in acute myelogenous leukemia. Neoplasma, 1979. 26(5): 611-615.
12. Alsabti EA, Prognostic value of urinary fibrinogen degradation products in bladder carcinoma. European Surgical Research, 1979. 11(3): 185-190.
13. Alsabti EA, In vivo and in vitro assays of immunocompetence in bronchogenic carcinoma. Oncology, 1979. 36(4): 171-175.
14. Alsabti EA, Serum lipids in hepatoma. Oncology, 1979. 36(1): 11-14.
15. Alsabti EA, Ghalib ON, and Salem MH, Effect of platinum compounds on murine lymphocyte mitogenesis.[Retraction in Shishido A. Japanese Journal of Medical Science and Biology. 1980 August;33(4):235-7; PMID: 6765630]. Japanese Journal of Medical Science and Biology, 1979. 32(2): 53-65.
16. Alsabti EA and Hammadi M, Inherited bleeding syndromes in Jordan. Acta Haematologica, 1979. 61(1): 47-51.
17. Alsabti EA and Kamel A, Carcinoembryonic antigen (CEA) in patients with malignant and non-malignant diseases. Neoplasma, 1979. 26(5): 603-609.
18. Alsabti EA and Muneir K, Serum proteins in breast cancer. Japanese Journal of Experimental Medicine, 1979. 49(4): 235-240.
19. Alsabti EA and Saffo MH, Plasma levels of CEA as a prognostic marker in carcinoma of urinary bladder. Urologia Internationalis, 1979. 34(5): 387-392.

20. Alsabti EA and Safo M, Prognostic value of bone marrow acid phosphatase in prostatic carcinoma. Urologia Internationalis, 1979. 34(5): 350-355.
21. Alsabti EA, Safo MH, and Shaheen A, Lymphocytes subpopulation in normal family members of patients with alpha-chain disease. Journal of Surgical Oncology, 1979. 11(4): 365-374.
22. Alsabti EA, Safo MH, and Shaheen A, Lymphocyte subpopulation in primary lung cancer. Oncology, 1979. 36(4): 176-179.
23. Alsabti EA and Saleh K, Colony-stimulating and colony-forming cells in peripheral blood as prognostic markers in acute myelogenous leukemia. Oncology, 1979. 36(4): 180-183.
24. Alsabti EA and Shaheen A, The prognostic value of serum innumoglobulin levels in Hodgkin's disease. Neoplasma, 1979. 26(3): 329-333.
25. Alsabti EA and Talat AS, Periodic acid-thionin Schiff/potassium hydroxide/periodic acid-Schiff (PAT/KOH/PAS) approach in the detection of adenocarcinoma. Japanese Journal of Experimental Medicine, 1979. 49(3): 209-212.
26. Alsabti EA, Taleb AM, Raji HM, et al., Clinical trials in vaccination with leukemia associated antigens in acute myelogenous leukemia. Japanese Journal of Experimental Medicine, 1979. 49(3): 157-168.
27. Alsabti EA, The immunostatus of untreated cervical carcinoma. Gynecologic Oncology, 1980. 9(1): 6-11.
28. Ambrose SE, Nixon: The Education of a Politician, 1913-1962. 1987, New York: Simon and Schuster.
29. Ambrose SE, Eisenhower: Soldier and President. 1990, New York: Simon and Schuster.
30. Ambrose SE, Nixon: Ruin and Recovery, 1973-1990. 1991, New York: Simon and Schuster.
31. Amis M, The Rachel Papers. 1973, New York: Knopf.
32. Anonymous. Mosley drops Follett suit over 'Key to Rebecca'. The New York Times. 1981 January 15
33. Anonymous, Misconduct. Science, 2007. 315(5818): 1475.
34. Anonymous, Speaking of intellectual property. Science, 2007. 315(5816): 1205.
35. Anonymous, German research minister faces plagiarism allegations Science, 2012. 336(6082): 656.
36. Baringa M, UCSF case raises questions about grant idea ownership. Science, 1997. 277(5331): 1430-1431.
37. Bell R, Impure Science: Fraud, Compromise and Political Influence in Scientific Research, . 1992, New York: John Wiley and Sons.
38. Bhattacharjee Y, Misconduct. Science, 2006. 315: 1233.
39. Broad WJ, Would be acdemician pirates papers. Five of his papers are demonstrable plagiarisms, and more than 55 others are suspect. Science, 1980. 208(4451): 1438-1440.

40. Challow RP, A rose is a rose is . . . . plagiarism. The Journal of NIH Research, 1993. 5: 58.
41. Charatan FB, Psychologist wins damages over theft of research. British Medical Journal, 1997. 315(7107): 501-504.
42. Chaturvedi S, Mehrotra AN, Mittal S, et al., The conundrum of lenticular oncology. Indian Journal of Ophthalmoloy, 2003. 51(4): 297-301.
43. Cooney E. Harvard Medical School resignation. White Coat Notes. 2009 March 11.
44. Corlander H, The African. 1967, New York: Crown Publishers.
45. Couzin-Frankel J and Grom J, Scientific publishing. Plagiarism sleuths. Science, 2009. 324(5930): 1004-1007.
46. Culliton BJ, Harvard psychiatrist resigns. Science, 1988. 242(4883): 1239-1240.
47. Epstein J, Wild Oats. 1980, New York City: Little Brown and Company.
48. Fleischmann R, Safety and efficacy of disease-modifying antirheumatic agents in rheumatoid arthritis and juvenile rheumatoid arthritis. Expert Opinion on Drug Safety, 2003. 2(4): 347-365.
49. Gardner J, The Life and Times of Chaucer. 1977, New York: Alfred A. Knopf.
50. Gilman AG, Goodman, Louis S, and Gilman A, Goodman and Gilman's The Pharmacolological Basis of Therapetics. 1980, New York: Macmillan Publishing Company. pp. 1843.
51. Haley A, Roots: The Saga of an American Family. 1976: Doubleday.
52. Judson HF, The Great Betrayal: Fraud in Science. 2004, Orlando: Harcourt.
53. Kennedy RS, Dreams in the Mirror: A Biography of E.E. Cummings. 1980, New York: Liveright Publishing.
54. LaFolette MC, Stealing into Print. Fraud, Plagiarism, and Misconduct in Scientific Publications. 1992, Berkeley, CA: The University of California Press.
55. Mallon T, Stolen Words—The Classic Book on Plagiarism. 2002, New York: Harcourt, Inc.
56. Mason W. Make it newish: E.E. Cummings, plagiarism and the perils of originality. Harper's Magazinne. 2005:92-102.
57. Mewshaw M, Do I Owe You Something?: A Memoir of the Literary Life. 2003 Baton Rouge, Louisiana: Louisiana State University Press. pp. 226.
58. Mosley L, The Cat and the Mouse. 1958, London: A. Barker.
59. Nayak B, Notice of retraction. Indian Journal of Ophthalmology, 2006. 54(2): 75-76.
60. Norman C, Stanford investigates plagiarism charge. Science, 1984. 224(4644): 35-36.
61. Norman C, Melmon resigns Stanford chairmanship. Science, 1984. 224(4655): 1324.

62. O'Connor J, Rehnquist CJ, White, et al., Feist Publications, Inc. v, Rural Telephoe Service Co., 499 U.S. 340 1991.
63. Palca J, Scientific misconduct cases revealed. Science, 1990. 248(4953): 297.
64. Plotz D. The plagiarist. why Stephen Ambrose is a vampire. Slate Magazine. 2002 January 11.
65. Sawyer-Lauçanno C, E.E. Cummings: A Biography, 2005, Sourcebooks. Inc.: Naperville, IL.
66. Seigel GM, The enigma of lenticular oncology. Digital Journal of Ophthalmology, 2001. 7(4).
67. Simon LS, The treatment of rheumatoid arthritis. Best Practices and Research: Clinical Rheumatology, 2004. 18: 507-538.
68. Slavitt DR. Settling the score. A review of do I owe you something? A memoir of the literary life by Michael Mewshaw. The New Criterion. 2003 October.
69. Unknown, Unknown. Science, 2003: 23.
70. Weiner RA, Pomhoff I, Schramm M, et al., New advances in laparoscopic treatment of morbid obesity. Surgical Technology International, 2004. 13: 79-90.
71. Wheelock EF, Plagiarism and freedom of information laws. Lancet, 1980. 1(8172): 826.
72. Wierda D and Pazdernik TL, Suppression of spleen lymphocyte mitogenesis in mice injected with platinum compounds. European Journal of Cancer. 15(8): 1013-1023.
73. Williams RH, Textbook of Endocrinology, Sixth Edition. 1981: W.B. Saunders company.
74. Wyatt E. E. E. Cummings scholar is accused of plagiarism The New York Times. 2005 April 16.

CHAPTER 7

Misconduct in Publications

Unfortunately many authors are ignorant of the numerous forms of misconduct involving publications. The sins of publishing include plagiarism, self-plagiarism, ghostwriting, patchwriting, duplicate publications, the cosubmission of papers to more than one journal, the misquotation of citations, authorship abuses, image manipulation, the submission of papers without the approval of all authors, abuse of the confidential peer review of manuscripts for journals, and the repetitive publication of short-term research results (salami publications).

Because the number of publications, rather than their quality played an important role in academic advancement prior to the scandals of scientific misconduct in the 1970s and 1980s, researchers commonly used a variety of strategies to enhance their curriculum vitae, such as the awarding of honorary authorships, salami publications, and duplicate publications.

Citations

It behooves authors to personally verify the accuracy of citations in their manuscripts and to only quote those that they have personally read. When this practice is violated, citation errors are extremely common and frequently become propagated. Some incorrect quotations cite the wrong reference; others provide misinformation on author names, article title, journal name, and the precise reference. Mistakes in this category commonly reflect the fact that the original manuscripts have not been read and are mistaken citations from earlier papers that have been copied. Errors of this type are made mainly by junior authors supervised by mentors who fail to check the references of the mentee who writes the bulk of manuscripts that they coauthor.

Inaccurate Statements

Errors in publications are common and are most often unintentional careless mistakes produced by researchers under pressure to publish their findings before competitors or to enhance grant applications to fund further research. Inaccurate statements in papers may be construed as scientific misconduct as in the paper authored by Mikulas Popovic and his collaborators. The Office of Research Integrity (ORI) found Popovic guilty of research misconduct because statements about certain experiments were not true. Some experiments were stated to have not been done when they had been carried out. Also, a table in one of the papers mentioned 10%, but this was not substantiated by the data. Popovic had also maintained that some fluids initially manifest signs of virus replication before being mixed together, despite this not being true. Likewise, the text of the paper in Cell authored by Weaver and colleagues [32] did not reflect the truth. With regard to this paper, the ORI found Theresa Imanishi-Kari guilty of research misconduct. However, in both of these instances, the federal appeals board (FAB) of the Department of Health and Human Resources (DHHS) overruled the ORI and cleared Popovic and Imanishi-Kari of misconduct.

Abuse of the Peer Review Process

An abuse of the peer review process during which manuscripts are evaluated for potential publication in journals is another form of research misconduct involving publications. Well documented examples of this are illustrated in chapter 14 by Vijay Soman of Yale University (Yale) *(http://www.yale.edu/)* and C. David Bridges of Baylor College of Medicine (BCM) *(http://www.bcm.edu/)*. In November 1978, Helena Wachslicht-Rodbard, a Brazilian researcher at the National Institutes of Health (NIH), submitted a manuscript with Jesse Roth MD, head of the diabetes laboratory at the NIH with whom she worked and who was the senior author. The paper titled "Insulin Receptors in Anorexia Nervosa" documented research performed by her and colleagues and it was submitted to the New England Journal of Medicine (NEJM). As usual the editor of that journal submitted the paper to experts for a peer-review of its scientific reliability. After an inordinately long period of two and one-half months, the paper was returned to Wachslicht-Rodbard, informing her that if she revised the paper taking into account the comments of three anonymous reviewers, it would be reconsidered by the NEJM. One reviewer recommended rejection instead of revision. Without delay Wachslicht-Rodbard began revising the manuscript; and a few days later, Roth asked her to review a paper by Soman and Felig, which the editor of the American Journal of Medicine had requested an appraisal. While reading it, she was astonished to discover numerous pieces of text identical to what she had written with Roth and submitted to the NEJM. It also included a formula Wachslicht-Rodbard had created for quantifying receptor sites per cell. Only her colleagues at the NIH, the editors of the NEJM and the three reviewers of her paper had seen this confidential information. After scrutinizing the critiques of her paper, she deduced correctly that Felig was the reviewer who recommended that her paper be rejected. In reality, after Felig received her paper from the NEJM, he asked Soman to provide the initial remarks on the manuscript; and after reading it, he provided an unfavorable critique. Felig failed to regard it inappropriate to pass the paper on to Soman for an initial review despite it being sent to him for a confidential review even though the two of them had a virtually indistinguishable project approaching conclusion. Felig even failed to notify Arnold S. Relman MD (1923-), the NEJM editor at the time, of his extremely strong academic conflict of interest. Also he did not receive authorization to delegate the review process to a subordinate, perhaps because it was a common practice to pass manuscripts to trainees and other laboratory personnel for review.

Toward the end of February1978, Relman phoned Felig and informed him that he had discovered that Felig and Soman had submitted a paper on the identical subject as Wachslicht-Rodbard and colleagues after he had reviewed her manuscript. Relman was concerned not only that Felig's appraisal may have been affected by a COI, but also about the resemblance of the papers. Felig reassured Relman that his evaluation of the paper was solely on its merit.

After revising the paper Wachslicht-Rodbard resubmitted it to the NEJM together with a letter to Relman, accusing Soman and Felig of plagiarism showing photocopies of the initial paper with an indication of the portions lifted. Felig had difficulty accounting for the copied text, but informed Roth, a friend of his from childhood, that their study had been performed on their own initiative and he maintained that his research had been finished prior to the arrival of the Wachslicht-Rodbard manuscript. Roth did not doubt this explanation and recommended that he and Felig compare the two manuscripts during a forthcoming conference at the NIH. While not extensive the plagiarism, was clearly evident, but by now Wachslicht-Rodbard was convinced that the Felig-Soman manuscript might have been completely fabricated and based on their paper. Roth and Felig devised a plan to correct the situation. Because the priority for scientific discovery is extremely important Felig would indicate that the Wachslicht-Rodbard study came first and the Felig-Soman paper would not be published until after the Wachslicht-Rodbard study appeared in print. Soman and Felig would withhold their paper as long as uncertainty persisted about the autonomy of the research. Felig contacted Wachslicht-Rodbard and informed her that he was embarrassed to learn about what has transpired and delineated the plan to correct the situation. Shortly thereafter Roth wrote to Felig summarizing a proposal to settle the dispute. Roth asked Wachslicht-Rodbard to cosign the letter, but she refused. Even after her improved manuscript was published [28] Wachslicht-Rodbard remained adamant that Felig and Soman, had committed a serious wrong that warranted punishment and that Roth, her supervisor, was attempting to muzzle her. Further ramifications of the Felig-Soman scandal are covered in elsewhere.

C. David Bridges PhD, a professor of ophthalmology and biochemistry at BCM plagiarized the work of members of the Robert R. Rando laboratory at Harvard University (Harvard) *(http://www.harvard.edu/)* from a paper that he was reviewing during August 1986 on the regeneration of visual pigments [5]. Bridges not only stole ideas and data from the paper, but also the protocols and experimental techniques and falsely dated important laboratory records back to May 1986 claiming that his original data was stolen. He also created some dubious data and was suspected of not performing some of the research. The accusations of improprieties came to the attention of the BCM in the fall of 1987 and Bridges resigned with an effective date of October 15, 1987. An ad hoc committee at BCM concluded on May 24, 1988 that Bridges had acted improperly mainly in attempting to establish priority for his own research. An NIH panel found Bridges guilty of serious scientific misconduct and suggested harsh penalties [7], but after learning about the panel's finding congressman John Dingell wrote to James B. Wyngaarden (1924—) the then director of NIH indicating that an inspector general of the DHHS had found Bridges guilty of the charges raised against him on March 7 1991 and the debarring official recommended that he be debarred from contracting or subcontracting with the federal government and from eligibility or involvement in grants from the federal government for three years. *(http://www.hhs.gov/dab/decisions/dab1232.html)*. The FAB of the DHHS found "that Bridges improperly used information from a draft scientific manuscript sent to him for peer review." The plagiarized manuscript from the Rando laboratory was eventually published in the Proceedings of the National Academy of Sciences USA [3].

Multiple authorship

Today numerous publications have multiple authors and the various authors receive credit for the publications. When fraud is detected in publications the different coauthors should also accept the blame for allowing their names to be linked to such papers. Authorship in papers carries a responsibility. Because it is not always feasible to exclude liability from authors more recently journals have required each author to stipulate the specific role of each author.

Authorship Abuses

As pointed out in chapter 2 publications in journals are how researchers document most of their discoveries. Because of this journal editors, the gatekeepers to the scientific literature realized that they could potentially detect and block dishonest research before it is published, but papers with fraudulent data, unintentional errors, or the reporting of sloppy research cannot always be readily detected through the peer review process during which submitted papers are evaluated. Articles containing such mistakes are sometimes published in reputable journals after slipping through the safety net of reviewers, editorial boards, and the editor-in-chiefs.

Because the number of publications is an important criterion in the evaluation of academicians for promotion many individuals add the names of colleagues to manuscripts as coauthors. Some individuals receive credit for research papers in which they have barely participated, or in which their role was often trivial and hardly justified. Moreover the heads of laboratories and even of departments have included their names in papers in which they did not closely oversee the work sufficiently well to exclude fabrication of falsification. Many scientific papers have multiple coauthors and all of them take credit for publications listing them among the authors. They should also rightfully accept blame for such papers when flaws are detected. Multiauthored articles should not be submitted for publication without the consent of all contributors. Many contemporary research publications have multiple authors from different disciplines and often from numerous countries. When such papers contain mistakes all authors are presumably not guilty of research misconduct or sloppy research, but because the authors allowed their names to be associated with the manuscripts and consented to the submission of the paper their careers become tainted by association.

In academic institutions it is not uncommon for graduate students, research fellows and other junior researchers to provide honorary authorships to colleagues who return the favor on their papers. Particularly in the era when publish or perish was rampant this practice was not only widely practiced but also approved by faculty supervisors, because it increased the productivity of each laboratory participant at least in their list of publications. Particularly in the past it was a common practice in multiauthored papers for the senior or first author to provide authorship to persons who did not contribute significantly to the writing of the manuscript or in performing the research that is being documented. For example, some might just provide a reagent for the study, such as an antibody, a tissue specimen or the use of a piece of equipment. The awarding of a coauthorship was clearly more appealing to a researcher than just an acknowledgment for a trivial contribution. Many journals fueled this practice by not allowing acknowledgments at the end of articles, mainly because of space restrictions.

Different types of honorary authorship are added to publications for unjustified reasons: gift, surprise, ghost, courtesy, prestige, legitimizing, and coercive. A gift authorship is usually given by a more senior investigator to a junior person to enhance the person's career, but the senior researcher may ask to have his/her name left off. A surprise authorship is received by someone who did expect an authorship and just discovered that he/she was an author in a publication without participating in the study. The legitimizing author is added to increase the probability of getting the manuscript published. The names of well-known scientists are commonly appended to the authentic writers in an attempt to secure publication. Honorary authorship is also provided to individuals in multicenter publications for inappropriate reasons. Several studies have shown that the name of institution(s) as well as those of the authors influences the outcome of the review process of scientific papers. Manuscripts from prestigious institutions with a renowned author often obtain lenient reviews. Hence, inexperienced authors sometimes suffer from the misconception that the inclusion of the name of a prominent expert among the authors will have a major impact on getting a manuscript excepted by a journal. Some authors provide honorary authorships to prominent individuals of their manuscripts to enhance the probability of getting a paper published. In an attempt to achieve this goal at least one prominent coauthor has been given an honorary coauthorship. An even more despicable misconduct is for the author to be included without the person's knowledge or consent as documented in chapter 14 with Elias A.K. Alsabti. However, in the days when publish or perish was rampant individuals with unsuspected honorary authorships seldom objected to publication because it increased their bibliography. This type of honorary authorship commonly went to a more senior person in a hierarchical system such as the departmental head. In Germany, Japan, and some other countries, coauthorship was provided to the departmental chairs as a requirement for submission to a journal by a faculty member. This policy was started because the reputation of a department was dependent on the quality and quantity of publications coming from it and the departmental chair had the responsibility of overseeing all manuscripts published from the department. Endorsement by such a prominent individual also presumably influenced journal reviewers to provide favorable critiques that would result in an acceptance of the article. In such cases the chairperson read the paper and made editorial changes, but was not necessarily involved in the research being reported. The chair served as a gatekeeper controlling the content being released into the public domain, but this practice also increased the number of publications of this person immensely. Honorary authorship has sometimes gone to colleagues who reciprocated with their papers so that the publication list in the curriculum vitae of the entire group would be enhanced. Students or fellows would commonly be included in authorships from laboratories as a guarantee for working in the laboratory even if they were not successful in producing their own publications during that time commitment. These individuals needed publications for career advancement.

Because of a real concern by editors of peer reviewed journals regarding honorary authorships journals took steps to discourage inappropriate authorships. Numerous journals required each author to sign a declaration acknowledging accountability for the entire content of the article written under his/ her name. To decrease this practice most journals currently have specific prerequisites for authorship and mandate that each author must certify with a written or electronic signature from a list of eligibly criteria what their contribution was to the research as outlined on web sites *(www.sciencemag.org/about/authors)*. In many cases this probably has no significance other than the ethical issue, but when one of the authors was guilty of scientific misconduct the reputations of all authors were affected. Guidelines for being included or excluded as an author have been recommended. Genuine authors have made significant scholarly contributions to the project and are responsible for at least one component of the study. Authorship is not considered defensible for just providing space to a researcher to perform a project or for making a trivial contribution, such as submitting an occasional sample for analysis or providing a chemical reagent.

Contemporary collaborative research commonly involves laboratories from different institutions and sometimes different countries. In recent years academic institutions, funding agencies and publishers have been looking at novel ways of dealing with attribution *(http://projects.iq.harvard.edu/attribution_workshop)* and numerous publishers currently expect the contributions of all authors to be disclosed. In an analysis of six major medical journals in 2008 a report in the British Medical Journal (BMJ) [35] found honorary authors in 25% or research reports, 15% of review articles, and 11% of editorials.

Aside from the inappropriate inclusion of individuals as coauthors, persons worthy of authorship are sometimes consciously or subconsciously excluded from the authorship or given an inappropriate position among the listed authors, despite having performed a significant part of the research or written a major part of the paper. Examples of this are papers incorporating research data from studies performed by students or research personnel [34].

Patchwriting

Many scientists, particularly those with difficulty writing papers in English, resort to a variant of plagiarism known as patchwriting in which the essential blueprint of a paper is lifted, but not the actual data. Patchwriting and the recycling of parts of former publications are undeniably instances of inappropriate scholarship, but some consider that an exemption should be allowed for descriptions of the methods since they often include extremely complex techniques that are frequently difficult to rephrase. Numerous examples of lifted text exist, particularly from the introduction and scientific design of published papers, in which the authors insert details from their own experiment. With patchwriting authors repeat a study, but in a different population and report the findings by inserting them into a previously published text. The International Journal of Radiation Oncology, Biology, Physics published a paper by Odilia Popanda and coworkers of the German Cancer Research Center in Heidelberg on breast cancer in 2003 [23]. In 2005 Wei-Dong Wang and his colleagues in China lifted more than 95% of this paper in a manuscript published in the more prestigious Clinical Cancer Research using patchwriting. Wang and colleagues organized their results using the template from the paper by Popanda and coauthors but gave the results for nasopharyngeal cancer rather than breast cancer [30, 31]. The paper by Wang and coworkers was retracted four years later [31] and Wang pointed out that patchwriting was used because his English and that and his coauthors were not good enough to meet the language requirements of Clinical Cancer Research [6].

Authorship Disputes

Traditionally the order of authorship names has significance. The first position is usually reserved for the name of the person who has performed most of the work, but this position is commonly given to a student, postdoctoral fellow or junior faculty member regardless of effort to boost the person's career. The last position is usually reserved to the head of the laboratory, or the funded investigator. With projects involving significant contributions from multiple individuals many journals now allow more than one person to be designated as the first author or the senior author. After papers are written a dispute commonly occurs about the order of the author names. This drawback can be readily prevented as authorship is negotiable and the order of the names can be established in an open discussion between the research team that contributes to the study before the manuscript is written. Standards vary in international collaborations.

Some departments with a powerful leader had a relatively large staff doing the research and the writing of papers derived from it. In Belgium for example Jules François MD, PhD (1907-1984) was an extremely successful renowned ophthalmologist with power over that profession in his country, while he was director of the Ophthalmology Clinic of the University of Ghent *(http://www.ugent.be/en)*. During his widely acclaimed academic career he authored more than 1, 500 scientific papers and wrote or edited countless books allegedly numbering more than 400. François generally claimed position number one in the authorships and often failed to give appropriate billing to the people who did most of the work. The faculty in his department resented this, but knew that they had to live with the reality that they could not change the system.

Another type of honorary authorship is a "coercive authorship" that is inappropriately forced on the legitimate authors by a senior member of the faculty who notifies a junior colleague that a more senior individual must be included as an author despite contributing little or nothing of substance to the manuscript. A variant of the coercive authorship is the research team member who insists on being the first or senior author of a publication in which the person has neither actually carried out the research nor supervised it or contributed significantly to the writing of the paper. Honorary coauthorships of different types are unethical unpardonable, unacceptable and in reality fraudulent. Some universities, such as Washington University in St. Louis (WUSL) *(http://wustl. edu/)* regard honorary authorships as scientific misconduct *(http://wusrl.edu/ policies/authorship.html)*. Moreover researchers are ill-advised to accept the authorship of a study with which they are not intimately involved as it can result in considerable humiliation if flawed data in it gets questioned leading to the retraction of the paper. Despite concerted efforts to bring honorary authorships to an end they remain common today [12]. At present many journals now require all authors to certify their contribution to submitted papers and to vouch for the validity of the paper and to take responsibility for the content of the paper should it be challenged for intentional or unintentional mistakes and need to be retracted. The latter will hopefully discourage individuals from accepting honorary authorships.

To author a paper carries responsibilities. All coauthors should provide appropriate input into the manuscript based on their expertise and should approve the final product. Customarily, in small studies from a single laboratory this duty has fallen on the shoulders of only one person typically the senior author who should personally evaluate the raw data, especially when created by graduate students and postdoctoral fellows under their supervision. This person should also verify that the data chosen for publication has been properly portrayed in figures and tables. Over time it has become progressively impracticable for a single person to oversee all contributions in multi-authored, multidisciplinary collaborative publications, involving input from individuals with diverse know-how. Today more than half of the publications in important scientific journals have authors from more than a single country. When research is a collaborative effort involving multiple subspecialties it beholds the different senior authors to be accountable together for the entire content of each publication and particularly with regard to their areas of expertise.

Ghostwriting

The early drafts of speeches for presidents and busy high officials as well as the final polished versions of these scholarly communications have typically been written by official speech writers. Likewise some scientists are too busy to report all of their discoveries and many have a poor command of English and are unable to express themselves clearly. For such researchers others often write their papers for a coauthorship or a payment for compensation. Such ghostwriting appears to be widespread but its scope remains uncertain. Many consider ghostwriting reasonable and morally ethical under certain conditions, but a line in the sand needs to be drawn between what is acceptable and what is not. Some ghostwriting is clearly fraudulent and unethical, such as that sponsored by drug companies in which the names of reputable individuals are added as authors to scientific papers that document information that helps market the manufacturer's products. A feature common to some clinical news magazines is ghostwritten articles written for industry supporting their products adjacent to advertisements promoting them. Dozens of companies draft scientific papers on behalf of scientists who merely take credit for the article as an author without contributing anything of significance to its content.

Editors of newsmagazines find that advertisers are reluctant to place their advertisements adjacent to articles unless they can be promised favorable publicity or the absence of information promoting a product of a competitor. Ethical editors of such magazines do not allow industry to dictate the content. Unethical ghostwriters breach the public trust. Drug companies often employ outside firms to draw up documents aimed at marketing their drugs and getting physicians to take the role as the official authors. Members of the university faculty are responsible for the integrity of their work including papers derived from their research. Some institutions prohibit ghostwriting and advise faculty to maintain records of their participation in the preparation of manuscripts. The USA federal government is getting involved in ghostwriting. Charles E. Grassley (Republican senator of Iowa) [25] has started putting pressure on the NIH, the federal agency that supports much of the biomedical research in the USA, to block medical ghostwriting.

Industry-sponsored ghostwriting is known to date back until at least the end of the twentieth century and information about this despicable unethical behavior became known in various ways and particularly from leaks from pharmaceutical company employees and lawsuits. For example the Wyeth Pharmaceutical company hired the medical writing company DesignWrite to prepare an estimated 60 articles that were favorable to hormonal drugs produced by the company [25]. These papers appeared in medical journals between 1998 and 2005 promoting the beneficial effect of drugs, particularly in relation to competing medications, and they deemphasized the risk of breast cancer. However, later studies showed that certain hormonal therapies put postmenopausal women at an increased risk of breast cancer as well as heart disease. While some authors of papers related to the Wyeth ligation contributed significantly to the writing of the papers the role of others was minor. The main unethical issue was the fact that the authors failed to disclose that Wyeth paid ghostwriters to prepare articles favorable to the company for medical journals in which doctors took credit that they endorsed the particular manufactured drugs [25]. This ghostwriting benefits the involved doctors by both saving them time as well as enhancing their curriculum vitae, but the practice also aided pharmaceutical companies because the articles provided positive information related to their drugs. At one time

GlaxoSmithKline (GSK), one of the largest drug companies in the world, used a program called CASPPER, which helped physicians in many aspects of paper writing from topic development to manuscript submission [11]. The program was designed to fortify product positioning and hopefully it enabled it to triumph over the competition's antidepressant runaway successes with which Paxil was competing, namely Eli Lilli's Prozac, and Pfizer's Zoloft. CASPPER was drawn to the public's attention, during a wrongful death lawsuit brought by the Los Angeles Law Firm Baum Hedlund PC on behalf of hundreds of former users of the GlaxoSmithKline drug Paxil *(http://www.nytimes.com/2009/08/19/health/reserch/19ethics.html?r=1 (8/20/2009).*

Image Manipulation

Images constitute an important component of many modern scientific publications. Prior to the modern electronic era of image processing drawings of photographs were used to make a point. Photographs needed to be accurate and flawless, but in early days of photography photographs were sometimes manipulated to obtain a more esthetically pleasing image or obtain an image that was difficult to obtain with the existing cameras. The famous Charles Darwin made use of this, but became severely criticized in recent times despite the fact he merely wanted to illustrate a point. In modern times with the advent of digital image processing it is perhaps not surprising that dishonest researchers manipulate images in their manuscripts to distort the truth. Scientific misconduct involving fraudulent images led to the discovery of the fraudulent research by Amitav Hajra and H-H Hsu, who was the first author of a paper by an all-Taiwan group of researchers on microbiology in the prestigious journal Cell [15]. A research team challenged the prevailing view on how the DNA in bacteria is transcribed, but documented their data with fraudulent images. The flaw was not detected until unnamed detectives drew attention to image manipulation in the paper involving several dozen western lanes and posted the exposure on an electronic bulletin board [37]. By December 2006 Cell had not yet published a retraction, but the National Chung Hsing University in Taiwan *(http://www.nchu.edu.tw/)* set up a committee of distinguished academicians to investigate the alleged image manipulation and they recommended that the paper be retracted. This may not be the only example of image manipulation by Chang and his colleagues as bulletin boards have challenged the images in another online publication by them in the Journal of Biological Chemistry [14].

After learning about the fraudulent publications by Jon Sudbø the NEJM reevaluated two of his papers and noted that two photomicrographs in a 2001 paper [27] which were denoted to represent two separate patients and stages of disease were indeed the same photomicrograph but at different magnifications.

Salami Publications

The word salami, which dates back to ancient times, is perhaps best known for a sausage made of air-dried meat made from one of a variety of animals. The English term is the plural form of the Italian salame that is derived from Sale (salt) and—ame as a collective noun for all kinds of salted (meats). Meats treated in this way are forced into animal gut with an elongated and thin shape, then left to undergo fermentation. The sausage is cut into small pieces perpendicular to the long axis of the sausage and this divides it into pieces comparable to what some scientists do in documenting their research. Rather than reporting a solid complete single study some researchers pad their résumés by dispersing the findings of a single experiment into multiple short term salami publications. This was a response by creative researchers to the need of multiple publications for career development and for progress reporting required by funding agencies. Unfortunately the custom of documenting research findings in little fragments presents untruthful researchers with an opportunity to publish dishonest research that others are unlikely to try and replicate. It has been recommended that researchers who break up their research findings into numerous manuscripts of this type should be censured rather than rewarded.

Duplicate Publications

Another violation of journal policies is the simultaneous submission of articles by the same authors to different journals. To repeat the publication of a scientific article more than once is a form of scientific misconduct. It is also wrong, unethical, deceptive and an infringement of copyright law. When caught the culprits of duplicate publications typically express regret and willingly consent to withdraw the article (s). Liou and coworkers submitted a paper to Developmental Biology in March 1993 [17] and were advised on October 6, 1993 that it had been accepted and forwarded the required copyright form to the publisher. During the inordinate delay in hearing from Developmental Biology they sent a virtually identical copy of the paper containing one table and several figures describing the same experiment to Investigative Ophthalmology and Visual Science and that journal published essentially the same paper a second time. When this was discovered the editor of Investigative Ophthalmology and Visual Science withdrew the paper [18] and pointed out that Liou and colleagues dishonestly signed the standard copyright transfer notice indicating that the paper was original and not under consideration by another journal and had not been previously published other than as an abstract. Pleading ignorant of the ethics of the scientific publication and copyright law Liou apologized and voluntarily agreed to withdraw the article published in Investigative Ophthalmology and Visual Science [2]. Miguel Roig from the department of psychology at St. John's University *(http://*

www.stjohns.edu/) echoed the opinion that any form of covert duplication of data is a threat to scientific integrity [24]. While journal editors are concerned with duplications in publications and redundancy as in "salami publication" [29], these practices fortunately do not damage the integrity of the content of science.

It seems reasonable that a scientist should be able to quote his/her own work or the work of others provided that due credit is given and that it is made clear what is being repeated. Translations are an understandable form of approved duplication [20]. The Vienna born Canadian endocrinologist Hans Hugo Bruno Seyle (1907-1982) was a famous scientist who performed pioneering research on the biological response of stress. At the age of 22 he became a doctor of medicine and chemistry in Prague and two years later he was at John Hopkins University (JHU) *(http://www.jhu.edu/)* followed by a stint at McGill University (McGill) *(http://www.mcgill.ca/)*. He then joined the French university in Montreal (Université de Montréal) *(http://www.umontreal.ca/)* where he had an astounding 40 assistants to help him with his research on 15, 000 laboratory animals. Seyle was extremely productive and published 1, 700 articles, 15 monographs and 7 books. To help publicize his research Seyle had many of his manuscripts translated into a wide variety of languages.

Duplicate papers with different authors are highly suggestive of plagiarism. Publications that appear in the literature more than once by the same authors are usually autoplagiarism and a variety of scientific misconduct, but some such cases are probably acceptable if the duplicate publication is in response to a legitimate request. For example journal editors occasionally plan a composite of previously published review articles with the approval of the publisher. When this is done with the acknowledgment of the original source it can probably be considered acceptable, but to publish the data more than once deceptively under the pretense that it has not been published before is clearly a no-no.

In the past some papers written by a group of individuals were sometimes published more than once with the same data, but with a different order of author names because separate authors desired a certain number of first author papers in their biosketches for promotion. This is a way that squabbles were settled among individuals working in the same laboratory. In more recent times this issue has become resolved by allowing more than one person to be the first or senior author with a statement that the individuals contributed equally to the work.

To speed up the publication date of submitted articles some authors have submitted manuscripts to multiple journals under the misconception that one journal may review the paper quicker than the others and it will appear in print much more rapidly. Moreover, if one journal rejects it another may accept it and in many instances this has indeed happened. A particularly unacceptable practice is to report the same information more than once in different journals. Such publications have been duplicated on numerous occasions to bolster the publication output of the author(s) or to publicize the research findings of the group. A repeat publication of material in publically available media is frowned on particularly when it is not cited in the later article or if the material has previous appeared in book chapters and symposia. Theodore Friedmann, a pediatric researcher and prominent geneticist at the University of California at San Diego (UCSD) *(http://www.ucsd.edu/)*, published a paper in the Proceedings of the National Academy of Sciences USA (PNAS) [36] in 1987 shortly after identical data was reported in Somatic Cell and Molecular Genetics. The two papers documented a technique for inserting genes into rat liver cells and the method became widely used. Particularly disturbing in this instance neither article referred to the other. Moreover, the six coauthors were unaware that the paper had been submitted to more than one journal and they did not realize it until the papers appeared in print. Because of this episode of research misconduct in which Friedmann violated the publication rules of the PNAS this society prohibited him from publishing in its journal for three years [13]. According to the San Diego Union Friedmann admitted his judgmental blunders in not citing the first paper in the second one and his failure to consult with his coauthors. He also informed the newspaper that he had nothing to gain by the double publication and had no ulterior motive. When the dean of UCSD, Paul J. Friedman, learned about the incident he initiated an internal inquiry and following it decided that no further action was necessary. Indeed, he felt that the journal's punishment was particularly harsh.

Even allegations of inappropriate duplicate publications can affect the accused. Charges of scientific misconduct prevented Stephanie Dimmeler (1967—) a cardiologist at the Wolfgang Goethe University *(http://www.uni-frankfurt.de/)* in Frankfurt am Main from receiving a $2 million Leibniz prize from the Deutsche Forschungsgemeinschaft (DFG) when the award was first announced because she was accused of misconduct for making mistakes in three publications and for publishing the same information in two different papers with dissimilar captions. The prize was only awarded after a university committee found that she had merely made an innocent mistake and found her not guilty of scientific misconduct [4]. The mistakes concerned false figures that had been prepared by a postdoctoral fellow. They were acknowledged and corrected as an erratum [26].

Review Articles

Editors of journals often solicit reviews from authors who have previously written papers on specific topics making it virtually inevitable that some authors will duplicate text in their publications even if the reviews paraphrase the same information. Many scientists include duplications of their own material within their papers and see nothing wrong with this autoplagiarism, but when the content of papers use the work of others without proper attribution it is clearly plagiarism and dishonest. Moreover the infringement of copyright may also be a setback if it has been transferred to the original publisher and it usually has been. A second submission of a published paper with or without a signed declaration that the information has been published elsewhere is out-and-out fraud. Authors who are transparent and seek editorial permission when reusing material should not fear inclusion in déjà vu if they can readily document integrity.

Text-matching software has been used to detect similarities between manuscripts. Numerous papers showing up in Déjà vu reflect papers that either repeat, or seem to repeat their own past published work. While some scientists frown upon this behavior others consider review articles and reports that document incremental advances part of the acceptable norm, especially when the work is related to funded research that demands periodic published reports if funding is to be continued. High similarities between papers may be justifiable, but they are more often illegitimate. Those that are ethical include conference proceedings, clinical trial updates and errata [9]. Unethical practices include substantial reproductions of other studies without proper acknowledgment.

Acknowledgments

Acknowledgments at the end of manuscripts usually give credit to grant support and to individuals who have contributed in some way to the reporting of the project. Nearly all journals do not permit acknowledgments of secretarial help and most other persons because of space restraints. Statements attributed to particular persons in the acknowledgments are not always accurate and sometimes authors inaccurately credit others without their permission. This is usually done to enhance hopefully the reliability of content in the paper.

Autoplagiarism

Couzin-Frankel and Grom [6] suggest that "repetitious reviews and incremental reports are part of an accepted tradition", but some scientists, such as John Loadsman of the Royal Prince Alfred Hospital in Camperdown Australia [19] feel that it misleading and indeed misconduct if an author includes material in a later publication without making it clear that the information has been previously published [19]. The International Committee of Medical Journal Editors (ICMJE) finds this questionable and points out that it has a detailed section on overlapping publications in its, "Uniform requirements for manuscripts submitted to biomedical journals: Writing and editing for biomedical publication" *(www.icmj.org)*.

Publication of Controversial Papers

Some controversial papers become published even though the editor of the relevant journal is aware of significant flaws in the research. Jacque Benveniste was able to get a manuscript published with his colleagues in the high impact journal Nature in June 1988 [8] documenting that water could retain information about the shape of highly diluted molecules. The publication was permitted largely because Benveniste was a prominent scientist and certain conditions were required after its acceptance. The reason why the conditions were not required prior to publication is obscure, but the defective paper stimulated debate.

In 1986 after becoming editor of Geophysical Research Letters Alex Dessler proclaimed that he would print audacious contentious articles because of the freedom of expression and inquiry and he published a short original report by Lewis Frank and JB Sigwarth [10]. Space scientists were extremely critical of this publication and of the editor's decision to publish it. Opposition to this editorial policy expanded over time and many readers considered Dessler accountable for turning up a dispute. Several associates of Frank were disenchanted by the society's contentious reaction rather than by the thesis of Frank, who maintained that all wording to reject his thesis is for some solid physical confirmation One of Frank's most severe opponents, a member of the National Academy of Sciences, maintained that Frank had made many notable accomplishments in the past, but that he had got off track.

Scientific Paper

Sir Peter Brian Medawar (1915-1987), a one-time director of the British National Institute of Medical Research shared the 1960 Nobel Prize in physiology or medicine *(http://www.nobelprize.org/nobel_prizes/medicine/laureates/1960/medawar-bio.html)* with *Sir Frank Macfarlane Burnet (1899-1985) (http://www. nobelprize.org/nobel_prizes/medicine/laureates/1960/burnet-bio.html#)*. Three years later he proposed that the research paper was a fraud and stressed that it provided an entirely misleading account of the thought processes involved in scientific discoveries [21] and made this point again in Saturday Review almost a year later [22]. He accepted the notion that the usual scientific paper accurately reports the facts, but he contended that the thought processes involved in the research were misrepresented and indeed fraudulent because the expected unorthodox manner of presenting the material exemplified a false conception, even a mockery, of the character of scientific thought. The usual convention is to start research papers with an introduction followed by a description of previous work, methods, results, discussion, and conclusions. The outcome of these findings is that generalizations grow and take shape as if by spontaneous generation. Medawar blamed John

Stuart Mill (1806-1873) on the existing notion that scientific discovery results from the inductive method of thinking logically. Medawar brought up three philosophic arguments as to why the inductive method cannot be maintained and should be discarded from the standpoint of the scientific paper. Firstly, an unbiased observation does not exist, because there is always a subjective factor of interpretation or experience. Secondly, the putting together of a scientific idea and its presentation or confirmation are two completely different ideas. Mill on the other hand considered that induction was the "operation of discovering and proving general propositions". Medawar pointed out that to conform to the accepted tradition of writing a paper the author pretends that the mind is an empty receptacle into which information flows from the external world for no apparent reason. The entire assessment only comes during the discussion when the absurd pretense is taken up questioning whether the gathered information has any meaning. Medawar's exaggerated assertion has an element of truth which basically regarded the existing style of reporting scientific discovery as an inductive process. He believed that the inductive format of the scientific paper should be abandoned and recommended that hypothesis belonged at the beginning of the discussion and not toward the end where scientific journals customarily insert it. Some journal editors have modified the format of papers in their journal by inserting the discussion at the beginning of the article as recommended by Medawar. Medawar also stated that "The scientific facts and scientific acts should follow the discussion, and scientists should not be ashamed to admit, as many apparently are that the hypotheses appear in their minds along uncharted byways of thought; that they are imaginative, inspirational and are indeed adventures of the mind. What, after all, is the good of scientists reproaching others for their neglect of, or indifference to, the scientific style of thinking they set such great store by, if their own writing show that they themselves have no clear understanding of it?" [1].

References

1. Anonymous, The "fraudulent" character of scientific papers. Canadian Medical Association Journal, 1964. 91: 1324
2. Anonymous, Notice of duplicate publication. Precocious IRBP gene expression during mouse development.[Retraction of Liou GI, Wang M, Matragoon S. Investigative Ophthalmology and Visual Science. 1994 March 35(3):1083-8; PMID: 7510271]. Investigative Ophthalmology and Visual Science, 1994. 35(8): 3127.
3. Bernstein PS, Law WC, and Rando RR, Isomerization of all-trans-retinoids to 11-cis-retinoids in vitro. Proceedings of the National Academy of Sciences of the United States of America, 1987. 84(7): 1849-1853.
4. Bhattacharjee Y, Follow-up: delayed payment. Science, 2005. 309(5733): 379.
5. Bridges CD and Alvarez RA, The visual cycle operates via an isomerase acting on all-trans retinol in the pigment epithelium. Science, 1987. 236(4809): 1678-1680.
6. Couzin-Frankel J and Grom J, Scientific publishing. Plagiarism sleuths. Science, 2009. 324(5930): 1004-1007.
7. Culliton BJ, NIH sees plagiarism in vision paper. Science, 1989. 245(4914): 120-122.
8. Davenas E, Beauvais F, Amara J, et al., Human basophil degranulation triggered by very dilute antiserum against IgE. Nature, 1988. 333(6176): 816-818.
9. Errami M and Garner H, A tale of two citations. Nature, 2008. 451(7177): 397-399.
10. Frank L and Sigwarth JB, Detections of small comets with a ground-based telescope. Journal of Geophysical Research—Space Physics, 2001. 106(A3): 3665-3683.
11. Goldstein J. CASPPER, GlaxoSmith Kline's friendly ghostwriting program. Wall Street Journal. 2009 August 20.
12. Greenland P and Fontanarosa PB, Editorial. Ending honorary authorship. Science, 2012. 337(6098): 1019.
13. Hamilton DP, PNAS bars papers from UC geneticist. Science, 1990. 249(4969): 622.
14. Hsieh C-F, Chang B-J, Pai C-H, et al., Stepped changes of monovalent ligand-binding force during ligand-induced clustering of integrin alphaIIB beta3. Journal of Biological Chemistry, 2006. 281(35): 25466-25474.
15. Hsu H-H, Chung K-M, Chen T-C, et al., Role of the sigma factor in transcription initiation in the absence of core RNA polymerase.[Retraction in Hsu HH, Chung KM, Chen TC, Chang BY. Cell. 2007 January 12;128(1):211; PMID: 17218266]. Cell, 2006. 127(2): 317-327.
16. Knight B, Latters, Multiple authorship. Science, 1997. 275(5299): 461-465.

17. Liou GI, Wang M, and Matragoon S, Timing of interphotoreceptor retinoid-binding protein (IRBP) gene expression and hypomethylation in developing mouse retina. Developmental Biology, 1994. 161(2): 345-356.
18. Liou GI, Wang M, and Matragoon S, Precocious IRBP gene expression during mouse development.[Retraction in Investigative Ophthalmology and Visual Science. 1994 July;35(8):3127; PMID: 8045705]. Investigative Ophthalmology and Visual Science, 1994. 35(3): 1083-1088.
19. Loadsman J, Plagiarism: transparency required. Science, 2009. 325(5942): 813.
20. Long TC, Errami M, George AC, et al., Responding to possible plagiarism. Science, 2009. 323(8919): 1293-1294.
21. Medawar P. Is the scientific paper a fraud ?. The Listener 1963 September 12:377-378.
22. Medawar PB. Is the scientific paper fraudulent? Saturday Review 1964 August 1:42-43.
23. Popanda O, Ebbeler R, Twardella D, et al., Radiation-induced DNA damage and repair in lymphocytes from breast cancer patients and their correlation with acute skin reactions to radiotherapy. International Journal of Radiation Oncology, Biology, Physics, 2003. 55(5): 1216-1225.
24. Roig M, Plagiarism: consider the context. Science, 2009. 325(5942): 813-814.
25. Singer N. Senator moves to block medical ghostwriting. The New York Times. 2009 August 18.
26. Stubbard J. Stephanie Dimmeler cleared of scientific misconduct. Medical News Today. 2005 July 12.
27. Sudbo J, Kildal W, Risberg B, et al., DNA content as a prognostic marker in patients with oral leukoplakia.[Retraction in Curfman GD, Morrissey S, Drazen JM. New England Journal of Medicine. 2006 November 2;355(18):1927; PMID: 17079770]. New England Journal of Medicine, 2001. 344(17): 1270-1278.
28. Wachslicht-Rodbard H, Gross HA, Rodbard D, et al., Increased insulin binding to erythrocytes in anorexia nervosa: restoration to normal with refeeding. New England Journal of Medicine, 1979. 300(16): 882-887.
29. Wager E, Fiack S, Graf C, et al., Science journal editors' views on publication ethics: results of an international survey. Journal of Medical Ethics, 2009. 35(6): 348-353.
30. Wang W-d, Chen Z-t, Li D-z, et al., Correlation between DNA repair capacity in lymphocytes and acute side effects to skin during radiotherapy in nasopharyngeal cancer patients.[Retraction in Wang WD, Chen ZT, Li DZ, Cao ZH, Pu P, Sun SL, Shen XP. Clinical Cancer Research 2009 May 15;15(10):3642; PMID: 19390114]. Clinical Cancer Research, 2005. 11(14): 5140-5145.

31. Wang W-D, Chen Z-T, Li D-Z, et al., Retraction: Correlation between DNA repair capacity in lymphocytes and acute side effects to skin during radiotherapy in nasopharyngeal cancer patients.[Retraction of Wang WD, Chen ZT, Li DZ, Cao ZH, Sun SL, Pu P, Chen XP. Clin Cancer Res. 2005 Jul 15;11(14):5140-5; PMID: 16033828]. Clinical Cancer Research, 2009. 15(10): 3642.
32. Weaver D, Albanese C, Costantini F, et al., Retraction: altered repertoire of endogenous immunoglobulin gene expression in transgenic mice containing a rearranged mu heavy chain gene.[Retraction of Weaver D, Reis MH, Albanese C, Costantini F, Baltimore D, Imanishi-Kari T. Cell. 1986 April 25;45(2):247-59; PMID: 3084104]. Cell, 1991. 65(4): 536.
33. White B, Letters, Multiple authorship Science, 1997. 275(5299): 461-465.
34. White B and Knight J, Letters. Multiple authorship: letters from. Science, 1997. 275(5299): 461-465.
35. Wislar JS, Flanagin A, Fontanarosa PB, et al., Honorary and ghost authorship in high impact biomedical journals: a cross sectional survey. British Medical Journal, 2011. 343: d6128.
36. Wolff JA, Yee JK, Skelly HF, et al., Expression of retrovirally transduced genes in primary cultures of adult rat hepatocytes. Proceedings of the National Academy of Sciences of the United States of America, 1987. 84(10): 3344-3348.
37. Xin H, Scientific misconduct. Online sleuths challenge Cell paper. Science, 2006. 314(5806): 1669.

CHAPTER 8

Human Research

A mission of numerous medical schools and hospitals includes the performance of clinically relevant research, and such research sometimes leads to novel new methods and devices for the therapy of patients. Such research has also generated extra revenue to medical doctors and hospitals when it involves clinical trials with pharmaceutical companies. Human research is extremely important, and it needs to be carried out in an efficient ethical way to produce meaningful results. In these research projects, the subjects commonly face risks, and these are listed in the approved written informed consent form. Institutions also face potential billing risks. Many research projects on human subjects take place in hospitals and other health care facilities where patients are billed for therapeutic activities, whereas study expenses are paid for by research grants or by sponsors of the research, such as pharmaceutical companies. It is extremely important that insurance companies are not billed for expenses that are more appropriately paid for from other sources. Some institutions place human studies into two major categories (billing risk studies and no-billing risk studies). Billing risk studies are those in which some research activities are carried out during conventional health care visits. In such instances the study team needs to review and approve all charges before bills are sent to insurance companies and other third parties for payment. As an additional precaution, all billing risk studies are required to have an institutional approved charge assignment grid before the inquiry can begin. For a clinical trial to meet the national coverage determination (NCD) requirements to be paid for by Medicare, the trial needs to have a therapeutic purpose and enroll diagnosed beneficiaries related to Medicare. It should not solely evaluate toxicity or the pathophysiology of disease. It must also be supported by the National Institutes of Health (NIH), Centers for Disease Control and Prevention (CDC), Agency for Healthcare Research and Quality (AHRQ), Center for Medicare and Medicaid Services

(CMS), Department of Defense (DOD), Veterans Administration (VA), or have an investigational new drug (IND) number or an IND exemption.

Human Research Projects

Aside from learning how the human body works, research on human beings hopes to understand the causes and basic mechanisms whereby humans develop disease. All types of human research require appropriate training, and a particularly important type of research is the development of new and better methods of treating human disease. According to the NIH, clinical research falls into three categories: epidemiological and behavioral studies, patient-orientated research, and outcomes research and health services research. Approaches to human research include observational studies, record reviews, database or repository studies, and interactive studies with human subjects, such as the gathering of specimens, such as deoxyribonucleic acid (DNA), for genetic studies.

Interventional research entails the testing of investigational drugs, devices, and biologics in different types of clinical trials, which may be randomized (such as trials evaluating investigational drugs and comparative evaluations of different treatments). Prospective and retrospective clinical reviews and specific case reports commonly take place. Some research by psychologists and social psychologists involves human subjects. In some research projects, information and material on human subjects is stored in databases or repositories.

Institutional Review Boards

Adverse publicity about unethical abused research involving human subjects during the early twentieth century, such as the Nazi crimes that were exposed during the post-World War II doctors' trial, the Tuskegee syphilis study, and the biological warfare studies of Unit 307 of the Japanese Imperial Army, shocked the world and led to oversight of human research. By the 1960s it became apparent that research needed to be overseen by independent institutional review boards (IRBs); and the PHS established them in 1966 to officially approve, monitor, and review biomedical and behavioral research entailing human subjects. IRBs are also known as independent ethics committees (IECs) and ethical review boards (ERBs). IRBs are defined and controlled by title 45 of the code of federal regulations (CFR) part 46 (45CFR46) *(http://ohsr.od.nih.gov/guidelines/45cfr46.html)*. All research on human subjects must be approved by an IRB or equivalent committee. In the USA IRBs are empowered with regulations from both the Office for Human Research Protections (OHRP) in the Department of Health and Human Services (DHHS), and the Food and Drug Administration (FDA) approves and requires input into modifications in planned research prior to the approval or disapproval of research involving human subjects. These complex regulations relate to situations outlined in the National Research Act. IRBs oversee the scientific, ethical, and regulatory aspects of human research and protect human subjects from unethical and cruel research.

The IRB establishes what human research is and for reviewing and approving all human research projects. The entire protocol and all pages of the written informed consent form and past versions of it must be reviewed by IRBs. Before the onset of any human research, IRB approval must be obtained; and after a study has received IRB approval, the protocol can only be changed if an amendment receives IRB approval. Approval for a project is provided for no longer than a single calendar year, and if the research lasts longer, a renewal application must be submitted to the IRB no more than 60 days and no less than 45 days before the study expiration date. The IRB may declare a study exempt from IRB oversight. Depending on the level of risk studies that require IRB, oversight may be reviewed by a fully convened IRB or undergo an expedited review. When the risks are minimal, an expedited review can be conducted by an IRB chair, vice chair, or a knowledgeable IRB member designee. The IRB protects not only private information about persons but also confidential data. The IRB needs to authorize the entry of information into databases as well as the removal of the data. Typically a study is allocated to an IRB member who serves as the primary reviewer. This person appraises all the pertinent material related to that research project involving human subjects including the entire research protocol, all pages of the written informed consent, and past versions of the protocols. The IRB member attempts to resolve all concerns about the protocol with the principal investigator (PI) prior to the next scheduled meeting of the IRB. The primary reviewer summarizes the project to the rest of the IRB at a regular meeting with recommendations. When a study is completed, files related to it need to be brought to closure. If a project receives IRB approval, the approval is valid until the date referred to in the letter of endorsement. Investigators and research support staff are obligated to inform the IRB of any suspected or obvious noncompliance, and this may be done anonymously. Certain research consent forms are not permitted to be included in the patient's medical records. Depending on the risks to human subjects, the IRB performs different types of reviews.

A vast number of human studies are performed, and these are controlled by specific FDA regulations. When a scientist plans to perform research on non-English-speaking individuals, it is essential that the subject understands the project and the contents of the written informed consent. To ensure this, an interpreter needs to communicate with the subject and the consent form needs to be translated into that person's home language. The deciphered document also needs to be interpreted into English to make sure that nothing has been lost in translation. Moreover, the translators must be approved by the IRB. The complexity of doing research on non-English-speaking subjects makes many researchers reluctant to undertake projects that involve persons who do not speak their language *(http://www.hhs.gov/ohrp/)*.

The IRB is officially responsible for approving, monitoring, and reviewing all biomedical and behavioral research that entails human subjects. The IRBs themselves are regulated by the Office for Human Research Protections (OHRP). Additional requirements are designated by title 21 of the CFR part 56 for IRBs that oversee clinical trials involving new drug evaluations. IRB approval is needed for all research supported directly, or indirectly, by the DHHS (formerly the Department of Health, Education, and Welfare [DHEW]). It must oversee the research that researchers are performing on them. In the USA the IRBs are empowered with regulations from both the OHRP in the DHHS and the FDA to approve, require modifications in planned research prior to approval, or disapprove research involving human participants.

IRBs are defined and controlled by title 45 of CFR46 (http://ohsr.od.nih.gov/guidelines/45cfr46.html).

Belmont Report

The Belmont report was an outgrowth of the National Commission for the Protection of Human Subjects of Biomedical and Behavioral Research. It was drafted at the Belmont Conference Center in Elkridge, Maryland, following an intensive four-day period of discussions held in February 1976. The final report released on 30 September 1978 was called the Belmont Report: Ethical Principles and Guidelines for the Protection of Human Subjects of Research. Since 1978 the principles of clinical research have been based on the Belmont report, which was later published in the Federal Register on 18 April 1979 [1] *(http://ohsr.od.nih.gov/guidelines/belmont.html)*. A two-volume appendix to the Belmont report contains statements by prominent citizens who aided the commission in its assignment. It is available for purchase [2]. The Belmont report stresses three main ethical components respect for persons, beneficence, and justice. All human research subjects should be protected and treated as autonomous agents with courtesy and respect. They should voluntarily consent to participate, or not take part, in studies after becoming truthfully fully informed about the details of the research study. The researchers must not be deceptive.

By beneficence, patients must be protected from injury from different potential sources, and harm must be minimized while the benefits are maximized. Reasonable, nonexploitative, and well-considered procedures should be ensured and administered fairly and equally. People who are not likely to benefit from research participation are not automatically excluded from studies *(http://ohsr.od.nih.gov/guidelines/belmont.html)*.

The Belmont report is the basis for regulations that protect humans in research, and it continues as an obligatory tribute for IRBs that oversee research performed on human subjects to make certain that projects meet acceptable ethical standards *(http://ohsr.od.nih.gov/guidelines/belmont.html)* by maintaining privacy and confidentiality to protect patients from potential harm, which may be psychological, social, and criminal or civil liability. Research should be scientifically aligned with the academic interests and missions of the faculty and the institution. It should be financially transparent and accountable for any costs of the research and should be conducted by researchers with appropriate qualifications, skills, and training and compliant with applicable regulatory requirements and institutional standards.

Clinical Trials

A vast number of human studies are clinical trials, and these are controlled by specific regulations of the FDA. Different regulations apply to IND and investigational devices (IDE). Detailed records are expected to ensure that the past history and clinical observations are accurately documented. The records need to permit a restoration of the events related to a study, and they must be stored in agreement with institutional policies and the regulations of the external agency involved in the research.

Clinical trials are a crucial way in which new therapies are evaluated by pharmaceutical companies, and they require significant effort by the PIs and other research support staff. In clinical trials, the studies need to meet the expectation of the sponsoring industry. They are an essential part of the FDA's procedure for evaluating scientifically the safety and efficiency of new drugs. A vast number of clinical trials have been performed and continue to be carried out. Study teams involved in clinical trials are obligated to publish their findings within a year of the completion date. An institutional office of clinical research (OCR) assists investigators in entering relevant information into clinicaltrials.gov.

By April 30, 2014, a registry of 32, 963 clinical trials in 160 countries was available on the Internet (clinicaltrials.gov). The work volume and the complexity of the regulations governing this research have created numerous highly responsible jobs. The Food and Drug Amendment Act of 2007 (FDAA) requires clinical trials to be registered at the website clinicaltrials.gov within twenty-one years of the first enrolled subject. The International Committee of Medical Journal Editors (ICMJE) also requires that prospective clinical trials be registered to the same site or another public registry prior to the enrollment of the first participant. Moreover, within a year of the clinical trial's primary completion date, a summary of the study findings must be entered into clinicaltrials.gov.

A key member of the research team is the clinical coordinator, who works closely with the rest of the clinical trial team to safeguard patient's safety and ensure that all components of the designed research are performed correctly according to the required industry standards and guidelines. The clinical coordinator needs outstanding clerical skills and must be detail oriented.

Regulations Related to Human Research

Regulations related to human research are complicated and human research needs to be overseen by independent institutional committees. Research projects must comply with the regulations of ORHP, FDA, and the IRB. Subjects should enter research studies voluntarily after being provided with adequate information.

Human research is exceptionally complex and abundant regulations created by the DHHS, the IRB, the ORHP, the FDA, the Health Insurance Portability and Accountability Act (HIPAA), the Office of Civil Rights (OCR), the Human Research Protection Program (HRPP), and state laws preside over it. State laws are not identical. For example, the age at which a minor becomes a legal adult and who can serve as a legally authorized representative vary in different states.

Additional regulations and policies reign over the Veterans Administration (VA) and various institutions. Other policies and procedures listed in the code of federal regulations (CFR), such as CFR38, require mandatory training. The CFR is a major FDA regulation; and it consists of numerous parts, including CFR50, CFR54, CFR66, CFR312, CFR803, and CFR812. Title 21 in the CFR, which deals with foods and drugs, has a part 56 that relates to additional requirements for IRBs that oversee clinical trials of drugs involved in new drug applications. Additional regulations and policies reign over research performed in VA facilities and various institutions. The CFR title 21 deals with foods and drugs. It has a part 56 that relates to additional requirements for IRBs that oversee clinical trials involved in new drug applications. The CFR is a major FDA regulation; and it consists of numerous parts, including CFR parts 50, 54, 66, 312, 803, and 812. CFR protects human subjects, and it is qualified by several federal agencies. The CFR regulations include definitions of certain terms of human participants and human research and safeguards for pregnant women, fetuses, neonates, children, students, disabled individuals, prisoners, and other groups of people. Some regulations in human research are governed by CFR46, a federal policy for the protection of human subjects that is qualified by several federal agencies. It is also known under the aegis of the "common rule, " and it defines research as "a systematic investigation designed to develop or contribute to generalizable knowledge." The regulations safeguard various groups of people, such as pregnant women, human fetuses and neonates, children, students, disabled individuals, and prisoners. These regulations include descriptions of both human research and human subjects.

Drugs developed for potential therapeutic use by pharmaceutical companies in the USA are required by the FDA to be tested in distinct phases of clinical trials. Only those drugs that successfully complete all phases are approved by the FDA for precise applications. Once a drug is approved, medical doctors are permitted to prescribe them for the specific approved application but also for unapproved uses designated off-label use at their discretion despite the lack of valid scientific evidence to support its administration in these other circumstances. Doctors usually do this out of desperation to help patients because they hope that the drug may work for some theoretical reason. Some doctors justify this practice by stating that people are not machines and that they need to practice the art of medicine as well as the science.

Manufacturers, however, are forbidden by the FDA from marketing drugs for medical conditions other than the specific approved uses. Pharmaceutical companies can only promote drugs for the treatment of diseases after the FDA is convinced that the drug is not only beneficial in the treatment of the other disease but is also safe. Members of the medical profession often report success in a small number of patients treated with the off-label use of drugs. To justify the positive results of some off-label drugs, it has been hypothesized that a drug may work in one person but not in another, but this may be a placebo effect. Billions of dollars are spent annually by drug companies to convince doctors to prescribe their drugs even for off-label applications, and strict rules govern the manner by which this is achieved. The regulations are intended to make certain that doctors receive dependable information. Pharmaceutical companies as well as industries that specialize in devices or other useful products need professional advice from the medical profession and other appropriate specialists.

Research on human subjects is highly regulated by several administrative offices that facilitate the research and ensure compliance with institutional and government regulations. The required infrastructure for research includes offices of oversight research administration, clinical research support, patient revenue management, clinical trials billing, compliance and investigational drug service, effort reporting, posting and publication, patient billing, research contracts, clinical trials quality assurance (CTQA) audits, clinical trials management system (CTMS), site-based research unit (SBR), Office of Research Integrity (ORI), IRB, Clinical Research Support Office (CRSO), Office of Corporate Research Collaborations (OCRC), misconduct, study approval process, Human Resources (HR), payroll, procurement, Office of Research Administration (ORA), and the Office of Science and Technology (OST).

In research projects, human participants commonly face risk, and these are listed in the IRB informed consent. Institutions also confront risk that needs to be protected against.

These offices assist study teams with contract negotiations, grant awards, billing, study approvals, compliance, conflict of interest reporting, investigational product handing, and other matters. Other regulations for human research are controlled by the International Conference on Harmonization (ICH), Good Clinical Practice (GCP), which was adopted as guidelines by the FDA; the DHHS; HIPAA; OCR; National Coverage Determination (NCD); the Office of Inspector General (OIG). The DHHS protects human research by regulations in the CFR title 21 and CFR title 45 (the common rule) by way of several federal agencies. One program that protects human subjects in research is the Human

Research Protection Program (HRPP), which is a collective endeavor to review, approve, and facilitate human research to ensure the safety, rights, and well-being of human subjects.

HIPAA protects humans against the disclosure of their protected health information (PHI). The CTR56 relates to additional requirements for IRBs that oversee clinical trials involved in new drug applications. The National Coverage Determination (NCD) is a countrywide entity that originated in the OIG to ascertain whether Medicare is responsible for payments for particular items or services. Data collected during human research teams needs to managed and stored electronically appropriately in an approved secure format so that it can be handled and transferred when indicated. To be reproducible, all human studies need to be accurately documented regardless of their nature. Audits sometimes take place to make sure that the studies are carried out in compliance with the relevant regulations and the study protocols. The PI, or director of a research project, has the responsibly of making sure that a study is carried out efficiently as planned. A major concern in human research is the protection of the welfare and safety of the subjects. The research is carried out in an extremely regulated environment. Some clinical research crosses international boundaries, and the ICH has provided the research population with guidelines for GCP with regard to how clinical trials should be performed so that accurate credible data can be collected while protecting the rights, confidentiality, and integrity of the human participants. These guidelines cover such diverse topics as to how the trials should be designed, conducted, monitored, analyzed, and reported.

To perform research on human subjects the investigator needs to pass through an approval process at certain institutions. This process starts with an evaluation of the research question (raised by the investigator or sponsor) by the sponsoring clinical research unit (CRU) to determine study feasibility and funding. Depending on the funding sources, the investigator may work with the office of research administration or the office of corporate research collaboration or grant and industry funding. After CRU feasibility has been established, the protocol is reviewed by an IRB for oversight or exemption from oversight. For institutional approval, consent is needed from specialty committees, the institutional OCR, and other sources. A fund code needs to be identified and billing grid may be necessary. A clinical research study passes through three distinct phases (start, active, and close out). During the startup of the study, authorizations are obtained. Participants are registered during the active phase during which data is gathered and evaluated. Papers are also written and submitted for publication during the active phase. During this stage, study subjects are followed by means of enrollment logs, and an eye is kept on grant billing. If changes in the research protocol are deemed necessary, they need to be submitted for the approval of the IRB. Prior to the closure of a human research project, obligations to sponsors and the institution need to be met. These include the reporting of results and the requirements of clinical trials that need to be registered with ClinicalTrials.gov. In some medical centers, research on human subjects is organized into therapeutic segments-designated clinical research units (CRUs), which oversee the research. The CRUs are responsible under the leadership of a medical director (who is responsible for the unit and for making sure that studies fall within the goals of the department), research practice manager, and finance practice manager. The financial analysts in many CRUs report to the finance practice manager. Clinical research coordinators and the research staff report to the research practice manager while obtaining daily direction from the PI. Potential participants in research projects should be introduced to the study by a known member of their health care, and their recruitment requires it to be approved by the IRB.

Study Teams. Most human research involves a team made up of many people with different backgrounds and areas of expertise. The members of the team have different duties and include such diverse persons as the PI, subinvestigators, a clinical research coordinator (CRC), clinical trials assistant, regulator coordinator, research analyst, residents, fellows, biostatistician, and a data technician/analyst. Statisticians play an important role in human research and particularly in the study design to that precise analyzable data can be gathered. To ensure the data is consistent with integrity, it is essential that technicians/analysts collect and evaluate data is an identical manner.

Data collected during human research needs to appropriately managed and stored in an approved secure format so that it can be handled and transferred when indicated. To be reproducible, all human studies need to be accurately documented regardless of their nature. During as well as after human studies audits sometimes take place to make sure the studies are carried out in compliance with the relevant regulations and the study protocol. A wide variety of human research projects take place. They include retrospective chart reviews and industry-sponsored clinical trials. To document the research, GCP standards and federal regulatory requirements need to be followed.

With respect for persons, individuals are to be treated as autonomous agents that exercise their right of privacy and to have private information remain confidential. Beneficence is to do no harm and to maximize the benefits and minimize harm. An understanding of the rules and regulations that are involved in human research helps researchers protect human subjects.

The mission of numerous medical schools, hospitals, and medical centers includes clinically relevant research; and these institutions maintain their research reputation by not using humans as experimental guinea pigs. Human research includes genetic studies involving the analysis of DNA, and epidemiologic studies.

Clinical trials are a crucial way in which new therapies are evaluated. In these research projects, human participants commonly face risks.

To be a PI most universities and research institutions require the researcher to hold a faculty appointment or to be authorized by the department chair. The PI has the responsibility for conducting the entire research study in accordance with the institutional regulations. Some responsibilities and duties may be delegated to members of a research team with the necessary skills provided that they have had appropriate training. The delegation of authority needs to be documented in a log. Members of the study team need to be familiar with all regulations that apply to the research study and must ensure the safety and well-being of all human participants during the study.

For human subject protection (HSP), clinical research personnel must undergo repetitive training usually every two years. This is usually done with online courses such as those given by the modules at the collaborative IRB training initiative (CITI) website. For certification, a set of stipulated modules must be taken. The modules cover such diverse topics as the vulnerable population for minors. All individuals listed as key must compete a required number of modules. Prior to performing research on human subjects, all employees are required to pass a health review with employee occupational health and wellness.

In conducting studies on human subjects, research teams are responsible for understanding the involved laws, regulations guidelines, and institutional policies and procedures. The penalties for not adhering to the requirements range from litigation to removal of institutional research privileges. This is in contrast to the HIPPA definition, which includes both living and deceased individuals. HIPPA protects human subjects for privacy and protects PHI related to physical and mental health. All research on living or deceased persons may lead to the collection of information that contains PHI. When PHI becomes stored or transmitted across electronic devices, it becomes sensitive electronic information (SEI). This material should not be stored on unencrypted devices. For some research, the collected data is de-identified by removing the 18 HIPPA identifiers (name, address, telephone numbers, all elements of dates related to an individual, fax numbers, e-mail address, Social Security number, medical record number, health plan beneficiary number, account number, certificate/license number, any vehicle or other device serial number, device identifiers or serial numbers, web URL, internet protocol address numbers, finger or voice prints, photographic images, and any other characteristics that could uniquely identify the individual). Social Security numbers may only be stored by the research team with regard to payments if they are registered with the information security office, but they must be redacted as soon as possible. OESO training is needed every year. HIPAA regulations are important in retrospective chart reviews. 45CFR outlines the basic provisions for IRBs, informed consent, and assurances of compliance.

Data Management

To perform research on human subjects, it is important to first establish a plan that will answer the intended question. The data elements need to be described, and the collected research data must be valid and truthful with integrity. Specific measurement techniques need to be defined so that they can be consistently recorded. Data collectors need to be trained and the research data ought to be recorded at the time that it is gathered in agreement with the research protocol and with the required HIPAA regulations. If the study team is using de-identified data, HIPAA identifiers need to be removed, and a key needs to be established. If anonymized data are used, HIPPA identifiers are removed so that no key can re-identify the data.

Research Documentation

Documentation should permit the reconstruction of the events related to the performance of specific studies. A complete accurate documentation of raw data in research studies is essential and the details about the research methods and their findings should be reported in professional journals so that the work can be replicated by other researchers. Documentation also provides evidence that the rights and welfare of human research subjects have been protected. Moreover, in the USA, federal regulations insist that researchers write down their research studies using GCP standards. A top-notch record of investigations is also extremely valuable when auditors scrutinize the research in the presence of allegations of research misconduct. Raw data should be stored at the institution where the research was performed for as long as required by federal and institutional regulations. Copies of raw data may be retained by the researcher even when the person relocates at another university or research institution.

Record Reviews

Before starting a review of medical records in the health system approval by the IRB is essential. Initially the researcher might perform a feasibility study by carrying out a review preparatory to the research (RPR) in which PHI is reviewed but not recorded. Such a RPR is helpful if patients need to be screened to determine eligibility for a specific study. Should the investigator desire to review and document PHI, appropriate authorization that is compliant with HIPAA regulations is needed. The query tool DEDUCE is available to the research community to build cohorts and extract data from computerized health system records. DEDUCE can detect a possible population. The use of it requires training, access to eBrowser, an approved IRB protocol, and a RPR number from the IRB.

Informed Consent

The wording of the informed consent form needs IRB approval before a study starts. Human subjects participating in research need to be completely cognizant about the study so that they can make an informed decision as whether or not to participate. Information about the study needs to include not only its duration but also the risks and benefits of participating. This material should be provided not only orally but also in writing. The person obtaining the consent must be a trained member of the study team who has been delegated to this task. The consent form should be printed on watermarked pages. The original consent form remains in the study file, and copies go to the subject and to the medical center records if the subject is patient at the medical center where the research is being performed. Throughout the study, participants should be notified about new information about the project.

Human research is tremendously valuable, and it must be performed in an efficient ethical manner so that significant new information is derived. All projects involving human subjects must be approved by an IRB. Human research is exceptionally complex and must comply with abundant regulations created by the DHHS and the FDA that preside over it. For the informed consent to be valid, the human subject must sign a copy form and initial every page as well as in indicated places.

Human Protection

The HRPP is a cooperative endeavor of all who partake in the conduct, review, approval, and facilitation of human research and work together to make certain that the rights and welfare of human subjects are protected by improving systems/processes/communications/quality of research. It makes certain the safety of the rights and well-being of human subjects by enhancing systems/ processes/ interactions/quality of research. The HRPP at different institutions is accredited by the nationally acknowledged Association for the Accreditation of Human Research Protection Programs (AAHRP). IRBs adhere to the demanding standards of the AAHRP for protecting human research subjects.

References

1. Anonymous, Part IV. Protection of Human Subjects: Notice of Public Comment, Department of Health E, and Welfare.Office of the Secretary, Editor 1979. 23192-23197.
2. Publication D, DHEW Publication No. (OS) 78-0013 and No. (OS) 78-0014: USA Government Printing Office, Washington, D.C. 20402

CHAPTER 9

Misconduct in Research Involving Human Subjects

Publicity about unethical abused research involving human subjects during the early twentieth century shocked the world. First there was the Tuskegee syphilis study and some years later the war crimes related to twelve World War II trials were revealed by the USA war crimes tribunal. A prime participant in the horrific activities was Karl Brandt (1904-1948), the personal physician of Adolf Hitler who was also the Reich commissar for heath and sanitation. The tribunal was officially referred as the USA versus Karl Brandt et al., but it was also known as the Doctors' Trial [114]. Twenty of the defendants were medical doctors and they were accused of inhumane unethical human experimentation and mass murder under the pretext of euthanasia. The defendants were Nazi officials. A prime participant in the horrific activities was Brandt who was also the Reich commissar for heath and sanitation. He was sentenced to death by hanging *(en.wikipedia.org/wiki/Doctor's_Trial)*.

Less well known than the Nazi crimes against humanity were the experiments performed by the secret biological and chemical warfare research and development Unit 731 of the Imperial Japanese Army during the second Sino-Japanese war (1937-1945) and World War II in the Pingfang district of Harbin in Manchuria (now Northeast China). Unit 731 was under the command of General Shiro Ishii until the end of World War II, and during his leadership, thousands women and children died while experiments were carried out on them [4, 126, 169, 204]. The vast majority of the victims were Chinese (70%), and a high percentage of them were Russian (about 30%). The test subjects used by Unit 731 and its affiliated units (Unit 1644 and Unit 100) included a diverse population composed of political prisoners, captured bandits, common criminals, anti-Japanese partisans, individuals arrested for mistrustful actions as well as infants, pregnant women, and the elderly. After infecting subjects with pathogenic microorganisms responsible for the plague, smallpox, botulism, typhus, anthrax, cholera, tularemia, syphilis, gonorrhea, and other infectious diseases, vivisection was carried out without anesthesia. In other experiments, bombs were loaded with the dangerous infectious agents, and infected clothing or foods were dropped by low-flying airplanes. Poisoned food and candies were also distributed to unsuspected victims from the air. Even plague-infested fleas were sprayed on thousands of unsuspected victims, leading to the death of thousands of people. Aside from being responsible for at least 580, 000 Chinese deaths, 1, 700 Japanese are estimated to have accidentally killed by their own biological weapons [59]. Unit 731 consisted of eight divisions, and some participating scientists were arrested by the Soviet forces and were tried during the Khabarovsk War Crime Trials. Other researchers in Unit 731 surrendered to the USA Forces and were excluded from the potential of being tried for war crimes, because the USA considered that the knowledge gained by them in biological warfare was of great value and should be retained in intelligence channels [59]. The USA concealed the unethical activities of Unit 731 by denying the existence of its biological warfare as claimed by the Soviet Union and branding the Soviet allegations as Communist propaganda. The USA did not want other nations and particularly the Soviet Union to acquire information on biological warfare *(http://news.bbc.co.uk/2/hi/programmes/correspondent/1796044.stm—Unit 731: Japan's biological force).*

Research Misconduct in Clinical Trials

Patients are recruited into clinical trials because they are healthy or suffer from particular diseases and meet very specific criteria, but sometimes they are dishonestly recruited by someone who falsifies their eligibility for a particular trial.

Ornithine Transcarboamylase Clinical Trial. Jesse Gelinger (1981-1999) was the first person known to have participated in a clinical trial using gene therapy. This young man suffered from an X-linked genetic disease known as ornithine transcarboamylase deficiency. Affected individuals have severe liver disease and are unable to catabolize ammonia the byproduct of protein metabolism. Usually, the disease is fatal at birth, and because Gelinger disease was less severe than usual, it was thought to result from a less than usually severe spontaneous mutation. On September 13, 1999, Gelinger participated in a clinical trial at the University of Pennsylvania for infants with severe ornithine transcarboamylase deficiency. On the first day, he was injected with an adenovirus vector containing the corrected gene. Four days later, he died of a massive immune response. After his death, the Federal Drug Administration (FDA) investigated the situation and found several violations of conduct. Gelinger replaced another child in the trial, but he should have been ineligible because of high blood levels of ammonia. Also, the university had failed to report serious side effects in two patients who had undergone gene therapy. The consent form for participating was defective because it did not mention that monkeys had died following a similar therapy. Another major drawback in this clinical trial was the fact that both the university and James M. Wilson, MD, the director of the institute for human gene therapy at Penn, both had a financial interest in the research. Despite participating in the clinical trial, he did not receive complete knowledge of what was involved. There were major deficiencies in how the trial was conducted, and the documentation of the data was defective and so were the reporting requirements. The Gelinger tragedy set back research on human gene therapy for years.

Barry Garfinkel. In 1986, the Swiss-based pharmaceutical company Ciba-Geigy selected the University of Minnesota (U of M) *(http://www1.umn.edu/)* as one of five USA sites to evaluate the company's antidepressant drug Anafranil (clomipramine hydrochloride). The investigation was under the direction of Barry Garfinkel, MD, an outstanding child psychiatrist at the U of M. According to the protocols for the clinical trial, patients were required to have weekly psychiatric evaluations and to be monitored for adverse side effects as well as benefits. Instead of adhering to the required strict protocol, Garfinkel created his own protocol and executed it as he saw fit without notifying the drug company, the FDA, or the university. In March 1989, the clinical coordinator of this trial, Michelle Rennie, blew the whistle on Garfinkle and reported him to Ciba-Geigy, the FDA, and the university. The university investigated the matter and found irregularities but no intentional offenses. A few months later, the drug company pointed out that it would not use the U of M data because of significant unearthed problems. The FDA performed its own investigation, but this was delayed because of the university's refusal to share its observations with the FDA. Eventually through a subpoena, the FDA substantiated widespread misconduct. The USA attorney's office also investigated the allegations, and this led to the conviction of Garfinkel on mail fraud and making false statements in documents about the clinical research. After being convicted on five counts of felony, Garfinkel was sentenced on November 19, 1993, in the USA district court for Minnesota to six months in a federal correctional facility and six months at home detention with work release [61]. In addition, he was fined $214, 000 and required to perform 400 hours of community service [53]. The FDA considered the verdict too lenient and permanently debarred him from providing services in any capacity to a person that has an approved or pending drug product application *(http://findarticles.com/p/articles/mi_m1370/is_v28/ai_15330335/)*[54]

Werner Bezwoda. When Werner Bezwoda treated breast cancer patients with drugs followed by a bone marrow transplant in South Africa, he allegedly found them to live longer than similar patients on standard chemotherapy. Seventy-five patients were treated twice with a high-dose drug cocktail, designed to kill the cancer cells. The drug also annihilated the bone marrow, and to counteract this adverse effect, the patient's own bone marrow cells were transplanted after each episode of chemotherapy. A set of other patients were only given a low dose of the chemotherapy. Surprisingly, Bezwoda maintained that survival in the high-dose group was on average twice as long without a relapse. A similar therapeutic approach was successful against testicular cancer and some types of leukemia. The findings were published [19] and widely cited in professional journals as well as in the lay press. In May 1999, Bezwoda presented his findings on a clinical trial of 134 patients with advanced breast cancer at a meeting of the American Society for Clinical Oncology in Atlanta Georgia. The patients remained at high risk for metastases following tumor removal. Bezwoda raised high hopes in the cancer community, and his new approach triggered considerable enthusiasm, and a major clinical trial based on his protocol was proposed. However, these hopeful findings turned out to be too good to be true. Others could not confirm Bezwoda's findings [156], and the significant inconsistency was bewildering. Three other similar studies reported at the same 1999 meeting of the American Society for Clinical Oncology documented that a high-dose regimen was not beneficial to a standard therapy. In December 1999, at a meeting at the NCI, several American oncologists decided to take a closer look at the Bezwoda data, and a seven-member panel was sent by the NCI to South Africa on January 25, 2000. The NCI panel was flabbergasted by their findings, which were published on March 10, 2000 on the web page of the Lancet *(www.thelancet.com)* and in the printed version of the journal [200]. Bezwoda was only able to produce fifty-eight records of patients treated with high-dose chemotherapy, seventeen fewer than mentioned at the professional meeting in Atlanta. Most of the patients should not have been recruited into the study based on Bezwoda's protocol. Even more bothersome records on the seventy-nine alleged controls were nonexistent. Also, there was no documentation of informed consent by any patient; and the ethics board of the University of the Witwatersrand (Wits) *(http://www.wits.ac.za/)* in Johannesburg, South Africa, had no record of reviewing the clinical trial before it began. After the attention of Peter Cleaton-Jones, chair of the university's committee for research on human subjects, was informed of these findings, he launched a probe on January 31, 2000. The health professional council of South Africa, the authority to rescind medical licenses, launched its own investigation. On March 10, 2000, officials at Wits ended an inquiry set off by USA experts. It concluded that Bezwoda did not accurately report his observations and neglected to get approval for the clinical trial before undertaking it [56]. As a consequence of his research misconduct, Bezwoda resigned with an effective date of March 31, 2000; but the university beat him to the gun and fired him on March 10 [57]. Ten years after Bezwoda published his fraudulent research, the American Society of Clinical Oncology officially retracted Bezwoda's article on April 26, 2011 [170]. The work of Bezwoda diminished dreams of a novel approach to breast cancer therapy, and the episode tarnished the public impression of clinical trials, reinforcing a common misconception that medical scientists cannot be trusted.

John Anderson. John Anderson, a senior nephrologist in Edinburgh, Scotland, participated in a clinical trial to evaluate the efficacy of a drug manufactured by the pharmaceutical company Pfizer. The project lasted fifteen months and cost £42, 000 (about $70, 000). A reviewer of this trial detected concerns about Anderson's report on the study. Hence, a private company called Medicolegal Investigation looked into the matter and discovered that Anderson falsified data related to electrocardiographs (ECGs) and magnetic resonance images (MRI) for participating patients. During an investigation into the faked data, Anderson claimed that Dr. Shaffick was responsible for falsifying the data, but this allegation was problematic because this person did not exist. For this research misconduct, the GMC struck Anderson from the UK medical register [45].

Clinical Trials in Investigational Devices. Studies on the evaluation of the efficiency of investigational devices (IDEs) are sometimes seriously flawed because of inadequate or inappropriate controls. New devices for treating diseases are usually patented and sometimes used in legitimate clinical trials, but sometimes they are not appropriately evaluated before they are widely used in therapy that is charged to the patient. In an example of this type, the Arizona Glaucoma Institute (AGI) started to use a new therapy for open-angle and pigmentary glaucoma in 1997 called pneumatic trabeculoplasty (PNT), which is not an established safe and useful method of treating glaucoma. This surgical procedure was stated to decrease intraocular pressure (IOP) and lessen or eradicate the requirement for medication. A study by Stephen Barrett claimed that PNT not only did not achieve its alleged goal, but that it placed patients at serious risk for significant complications [17].

The patent described this IDE as being comprised of a vacuum source and a vacuum applicator (an eye ring or eye cup) united by a tube. By employing a vacuum source to the eye after placing the device on the front of the eye, the applicator becomes securely attached to the eye. The patented IDE had been approved by the FDA for making the eye stable during refractive surgery, but it was not approved for treating glaucoma; and in 1997, the AGI offered a new treatment for glaucoma using the device in PNT.

In 1999, the Arizona medical board closed the AGI clinic and acted unfavorably to Dr. Leo Bores. The sale of the IDE in the USA was obstructed because it did not receive FDA approval, but in 2004, the device became certified by CERMET, an agency with lower standards than the FDA; and this allowed it to be marketed in approximately twenty-five countries. Advocates of PNT have claimed that three publications document its effectiveness, including one study of 177 patients, but the reported clinical trials related to the PNT device were flawed [14]. In the trials, patient selection was not randomized, the background IOP measurements were not established appropriately, and no steps were taken to impede the investigators from being aware of whether the eyes had or had not been treated. Control groups not submitting themselves to PNT were not used. A third purportedly affirmative study entailed thirty-seven patients from two clinics in Italy [28]. The choice of the eye was based on which one had the uppermost IOP originally. The findings were not judged against those of equivalent patients getting standard therapy. To establish the effect on the other eye, the authors compared the IOP after PNT with the IOP when the project started. The FDA warned Ophthalmic International President Richard Smith that during their assessment of their company, they found that two vacuum fixation devices with suction rings were delivered to the AGI for taking care of individuals with glaucoma using a PNT device.

ADVANTAGE Study. A special type of pseudoscientific clinical trial known as a seeding trial is conducted by pharmaceutical companies not to answer a scientific question, but rather to fulfill marketing objectives, which are concealed from the medical profession, the public and institutional review boards (IRBs). For a long time pharmaceutical companies had been suspected of using seeding trials, but conclusive evidence of its existence was only established in documents obtained in lawsuits related to the ADVANTAGE (Assessment Differences between Vioxx and Naproxen To Ascertain Gastrointestinal Tolerability and Effectiveness) study by Merck and company, Inc [62]. Seeding trials by drug companies face several unacceptable components. Firstly, it is scientifically invalid to have a division of a drug company plan and carry out a study in which they have a conflict of interest. Moreover it is unethical and not feasible to provide informed consent without divulging the real purpose of the study. The physicians and patients who participated in the ADVANTAGE study were provided with the objectives of the study, but they were not told the genuine reason for the trial or of the vital function of Merck's marketing division. The ADVANTAGE study was a deceptive pseudoscientific clinical trial designed primarily for marketing reasons rather than to test a clinical hypothesis. The original findings in this trial, which was conceived by the marketing division of the Merck company, were published in the Annals of Internal Medicine in 2003 [73]. In 2003 at about the time of the ADVANTAGE study Merck started its VIGOR (Vioxx Gastrointestinal Outcomes Research) trial to settle the issue of whether Vioxx had less gastrointestinal toxicity. The reason for ADVANTAGE was not to establish a new indication for its use or to achieve postmarketing supervision. It is noteworthy that while many of the authors were employed by Merck and Company, Inc, the first author and many others had academic or other clinical positions. A stated aim of the study was to compare the gastrointestinal safety of Vioxx and naproxen (an older and less expensive analgesic) in patient) with osteoarthritis. ADVANTAGE did not pose any new clinical questions, but was designed to connect future potential prescribers precisely when Vioxx was coming to market. After being linked to cardiovascular side effects Vioxx was withdrawn from the market in 2004 and Merck faced two lawsuits (Cona v. Merck and Co., Inc, and McDarby v Merck and Co, Inc. [62]

ENHANCE was another probable seeding trial. It began at about the same that the FDA approved ezetimibe, but was not to obtain a new drug indication.

In 2003 the large pharmaceutical company Merck performed a biased unscientific clinical trial to evaluate the value of the painkiller Vioxx (rofecoxib) and published the findings. The alleged objective of the trial was to appraise the gastrointestinal safety of Vioxx in comparison to the older and less expensive analgesic naproxen in individuals with osteoarthritis. The research team headed by Harlan Krumholz of Yale University (Yale) *(http://www.yale.edu/)* pointed out that the Merck Advantage study failed to pose novel study questions, and was in effect planned to involve future prescribers when Vioxx entered the market. The Yale team reached their conclusions after assessing correspondence by Merck that they acquired by court order, as well as in records filed against Merck in two lawsuits against the drug company following the removal of Vioxx from the market in 2004. In response to criticisms of the ADVANTAGE study Merck claimed that the clinical trial was the brain child of their marketing division. The executive director for Global Center for Scientific Affairs at Merck Research Laboratories, Jonathan Edelman, stated in an open letter that the trial answered an important clinical question by concluding that concomitant use of aspirin did not impact on the gastrointestinal toxicity profile of Vioxx and naproxen. The study was carried out to mimic legitimate research, but reality it was primarily designed to market Vioxx rather than to test a clinical hypothesis. An objective of the project was to market Vioxx to doctors using a marketing ploy in a pseudoscientific form of research designed to target selected doctors. Attention to the flaws in this Merck's ADVANTAGE study of Vioxx were exposed in 2008 by Hill and his coauthors in AIM [62].

But the team maintained that it was during this legal process that they uncovered evidence that the trial was a seeding study whilst reviewing court documents. An editorial accompanying the study, coauthored by the editor of AIM Harold Sox, suggested that seeding studies are common in the pharmaceutical industry. The editorial stated :" The company flatters a physician by selecting him because he is 'an opinion leader' and incorporates him in the research team with the title of 'investigator.' Then, it pays him good money: a consulting fee to advise the company on the drug's use and another fee for each patient he enrols."

The New England Journal of Medicine became the first most important medical journal to make it obligatory for authors of unique papers to provide their financial connections with industry related to the content of their articles [127] and six years later that journal provided its editorial policy with regard to editorials [128]. A paper on the benefit of the antidepressant drug nefazodone by Keller and twenty eight other coauthors illustrates extensive COI by the authors in connection with Bristol-Myers Squibb the drug company that manufactured the reported drug [70]. From 2000 numerous researchers in medical schools have had COI because of relationships with industry. Some have been because of financial connections that influence biomedical investigations. Because of the frequency of COI some journal editors, such as those of the New England Journal of Medicine [3] and The Journal of the American Medical Association [77] have written editorials on the subject. Those who have discussed the matter have commonly stressed the need for disclosure about both financial interests and sponsorships, but dishonor has humiliated researchers who have ignored this advice. Such scandals have not been particularly extraordinary because for many years journals passed over the importance of making disclosures obligatory. Accusations of abused clinical trials by academic researchers because of COI commonly appear in the popular newspapers. Particularly in the past clinical researchers one clinician scientists became enmeshed with the drug industry through grant support and other financial arrangements. The researchers frequently served as paid consultants, advisory board members and on speaker bureaus promoting specific products. Some even agreed to be listed as authors on ghost-written publications reporting research on which they had participated in little or no way.

William K. Summers. Dr. William K. Summers, a psychiatrist in private practice in California hypothesized that Alzheimer disease was due to a defect in the cholinergic neuronal pathways of the brain. He discovered and patented Tacrine, now known as Cognex®, which became the first Federal Drug Administration (FDA)-approved drug for the treatment of Alzheimer disease. In 1986, he published the first report detailing the treatment of Alzheimer disease for an average of one year with Tacrine [162] claiming that tetrahydroaminocrydine (THA), an experimental drug was beneficial against Alzheimer disease by vastly enhancing the mental capabilities of people with this incapacitating disease. The article precipitated a gigantic response. Firstly it provoked an enormous demand for the drug by persons caring for patients with this common debilitating degenerative disease of the brain and it triggered a major federal study on the effectiveness of THA. By February 12, 1988 the study was still incomplete, but numerous reputable scientists pointed out that Summers's research practice was unacceptable and that his conclusions were unwarranted. The data from the disputed research was investigated by the FDA and Dr. Frances Kelsey, who monitors compliance with FDA regulations in clinical trials, found Summers guilty of repeatedly or deliberately violating regulations pertaining to the proper conduct in the clinical testing of investigational drugs. She listed fourteen findings that cast serious doubt on the conclusions of the Summers study [72]. One of them was the fact that he did not take into account that eight patients in the clinical trial were already being treated with medications that effect mental functioning. Kelsey sent a letter to Summers informing him that he could either concur with the FDA's findings or could reply in writing or attend an informal conference at the FDA. Summers was also notified that the FDA was considering barring him from testing new drugs because of skepticism about a paper written by him. If Summers could not reach an agreement with the FDA he would be offered a regulatory hearing to determine his future as an investigator for the agency.

On May 4, 1989 the FDA issued a restriction on Summers for direct personal involvement in specific aspects of study conduct and reporting. The restriction was removed on July 24, 2007 [72].

Study of Natural Course of Syphilis

The despicable unethical Tuskegee syphilis study caused the creation of the Office for Human Research Protections (OHRP) within the Department of Health and Human Services (DHHS) to watch over ethical aspects of human research chiefly in National Institutes of Health (NIH) funded clinical studies. The prime responsibility of the OHRP is to implement the) for institutional review boards (IRBs) that precisely reiterate FDA guidelines related to regulations given in Code of Federal Regulations (CFR) Title 45 Part 46 (45CFR46) for IRBs that precisely reiterate FDA guidelines related to clinical research carried out by drug companies aside from other relations. Institutions that partake in DHHS funded research must have a contract with OHRP institutional review boards (IRBs) , provides guidance on research ethics, and advises the related to ethical oversight called "Federal-Wide Assurance" (FWA). Moreover, the OHRP instructs secretary of health on issues relevant to medical ethics.

National Surgical Adjuvant Breast and Bowel Project

An enormous National Surgical Adjuvant Breast and Bowel Project (NSABP) involving tens of thousands of patients and more than 400 collaborating clinical centers across North America brought together different trials assessing treatments of bowel and breast cancer using surgery and anti-cancer drugs. The project involved tens of thousands of patients. The NSABP was headed by Dr. Bernard Fisher, a prominent surgeon and cancer researcher at the University of Pittsburgh (Pitt) *(http://www.pitt.edu/)*, from 1969 until March 1994. A study of many women with early breast cancer by Fisher found similar survival rates with 5 and 10 year follow-ups after partial mastectomy or lumpectomy when compared to total mastectomy [48, 49].

An extremely important major clinical trial costing $68 million was started in May 1992 to evaluate whether the hormone-like drug tamoxifen could prevent breast cancer in healthy women with a high risk of this cancer. In this masked study half of the cases received tamoxifen for 5 years and the other half were given a placebo. The study design was to monitor participants for 7 years to ascertain whether the treated group had less breast cancer than the controls. This was one of the most ambitious and controversial clinical trials supported by the National Cancer institute (NCI). The trial was contentious as critics alleged that tamoxifen had not been adequately studied since it was first synthesized in 1966 and that it might lead to lethal tumors, but this did not made a significant dent on the principal scientific advisors of the NCI. The possibility of this drug causing cancer was supported by several reported deaths late in 1993 from endometrial cancer among breast cancer patients taking tamoxifen.

The vast NSABP multicenter trial ran into major difficulties when Dr. Roger Poison of St. Luc's Hospital in Montreal falsified some information related to 14 patients that he recruited. The fraud included adding follow-up data on a patient 2 years after her death. The study's organizer and NCI officials knew about the suspected fraud for at least two years, but did not publish a re-evaluation of the findings. On learning about the alleged fraud the NCI initiated a colossal audit on all studies in the NSABP. Fisher crawled at a snail's pace publishing corrections and the trial was temporarily stopped while the NCI tried to fix the harm caused by the scientific misconduct. The Office of Research Integrity (ORI)investigated the allegations and ORI appeared in the federal register in 1993. In January 1994 Dorothy MacFarlane, an ORI investigator received a deposition from congressman John Dingell, chairman of the congressional subcommittee on oversight and investigations insisting on learning why the ORI was not investigating Fisher for reporting, but failing to publicize, fraud by a doctor in this extremely large clinical trial for which Fisher was responsible. Two months later word about the corrupted data in his clinical trial became widely known and the ORI started investigating Fisher.

The public's confidence in breast cancer therapy was panic-stricken when news about falsification in this contentious study was first reported by John Crewdson in the Chicago Tribune on March 13, 1994 [35]. Federal officials audited all documents at Pitt, where the study was coordinated. They also checked original records at many hospitals in the USA and Canada. 1, 843 women participated in a clinical trial showing that partial mastectomy and radiation are as effective as total mastectomy [1].

On May 5, 1994, an NCI board of scientific counselors unanimously agreed to restart the trial, but they requested the NSABP to provide participants with additional detailed information about risks and as an extra safeguard they required each participant to be screened annually for endometrial carcinoma [79]. The fraud allegations against a member of the colossal multicenter research project directed by Fisher damaged Fisher's reputation immensely and caused him to lose his job. Fisher and his deputy Carol Redmond was forced to resign their leadership roles in NSABP [2]. From the standpoint of the patients this was serious. About 11, 000 healthy women aged from 35 to 78 years been recruited into the tamoxifen trial after it was halted. NSABP had anticipated enrolling an additional 5, 000 women before the end of 1994. By putting the project on hold investigators were concerned that an enthusiasm for it would wane and that the dispute about the risks of administering this powerful drug to healthy women would be rekindled. Fisher was the focus of a federal investigation in which he was acquitted of all allegations in the spring of 1997 [7]. Because he tried unsuccessfully to restore his previously held excellent reputation, Fisher sued both the NCI and Pitt charging them of denying him due process and infringing on his right to free speech and free association. Six days before the case was due to go to trial a settlement was reached in which Fisher and his attorneys were to be paid $2.75 million by the university and Fisher was to receive a public apology from the university and the NCI was made to make an announcement commending his contributions to cancer research [69]. According to Robert Charrow, an attorney in Washington DC, the $2.75 million settlement was the largest for a first amendment defamation with which he was familiar [69].

Misconduct in Uveal Melanoma Clinical Trial. A national multicenter Collaborative Ocular Melanoma Study (COMS) under the leadership of Stewart L. Fine MD, chairman of the department of ophthalmology at the Scheie Eye Institute of the University of Pennsylvania (UPenn) *(http://www.upenn.edu/)* assessed whether iodine 125 bachytherapy was more effective than enuclucleation in prolonging the lives of persons with rare malignant melanomas of the choroid. The study ran into difficulties in 1993 when key data was inappropriately altered in the records of two of the 43 participating sites. The dates that photographs of the tumors were taken were altered by non-physicians in the study, but fortunately this did not affect the integrity of the data [47]. An employee at one participating site was fired and another one at a different institution agreed to resign. The ORI probed possible falsification or fabrication in this clinical trial. The ORI recommended that the National Eye Institute (NEI) not publicize the problematic cases in this clinical trial to avoid adverse publicity like that which emerged after the breast cancer clinical trials. After the completion of the ORI investigation a letter to a medical journal was anticipated [157].

Marc J. Straus. After working as a clinical associate at the NIH Marc J. Straus, a senior cancer researcher accepted a faculty position at Boston University Medical Campus (BUMC) *(http://www.bumc.bu.edu/)*, where he received a $1 million grant over three years to support his research on the response of patients to complex drug treatments intended to control a virulent small cell lung cancer. The disease was usually fatal within three months of diagnosis, but research conceived by Straus supposedly caused remissions in 93% of cases. While at BUMC Straus evaluated new cancer therapies from a large computerized database of approximately 200 patients collected from the eastern cooperative oncology group by the NCI. In 1978 prior to the congressional hearings on fraudulent research five members of Straus's research team informed BUMC officials that they had detected a wide variety of irregularities in the database, ranging from changes in a patient's birth date, details about therapy and the results of laboratory investigations that had never been performed. In addition one patient was stated to have a tumor when one was not present. When BUMC officials learned about the accusations a new standard was set on how to deal with scientific misconduct after the Darsee scandal. It took them only ten days to investigate the allegations and force Straus's resignation. A detailed study of the medical records by BUMC disclosed nothing to suggest patient mistreatment or unacceptable care, but a FDA official believed that some of these defects had grave clinical implications. Patients were put in danger and the data would generate poorly founded conclusions. Despite being legally responsible for the false data, Straus was adamant that he was the target of a conspiracy devised by a few peeved associates among his staff of about 40 individuals. Straus acknowledged submitting fabricated and falsified data to federally funded programs repeatedly in his progress reports.

After Straus was compelled to resign from BUMC he relocated at New York Medical Center in Valhalla where he submitted another grant application to the NCI. Despite BUMC's rapidity in getting rid of Straus the NIH awarded him a $1.32 million research grant 22 months later. The 26 member National Cancer Advisory Board (NCAB) approved this application, but NCI officials failed to notify the advisory board about Straus' falsified and fabricated data. Later the board became aware of them and censured NCI officials for not notifying them. When word about this faux pas reached the Dingell subcommittee it was extremely critical of NIH officials for their prolonged investigations into scientific misconduct and for continuing to fund scientists accused of data falsification.

William F. Raub PhD, deputy director of the NIH, maintained that Straus was presumed innocent until proven guilty. FDA and NIH officials investigated the charges against Straus between June 1979 and March 1982, but did not bring criminal charges against him. Following the congressional inquiry a visit to Valhalla by the NIH found that Straus had disobeyed several conditions of the award and in April 1982 his research grant was terminated. Shortly thereafter on May 17th 1982 Straus became the first scientist to be debarred for receiving federal research funds because of scientific misconduct [27]. Straus reached a settlement with the government when he realized that federal regulations penalize PIs unfairly, even if they are unaware of wrong doings. An agreement was signed by Straus, the NIH, and the FDA after Straus admitted that he was responsible for the falsified research data performed at BUMC even though he did not personally commit the misconduct in the oncology studies. Aside from not being able to receive financial aid from the DHHS for four years Straus was prevented from studying novel investigational drugs. After being accused of falsifying data Straus spent four years engaging in an endless battle with inadequate resources against the USA government for a just and complete peer review of his research. Three years after leaving BUMC, Straus filed a $33 million conspiracy suit in a federal court in Boston against five members of his research team claiming that they had falsified data, abused patients and conspired to blame these acts on him [26]. The defendants remained steadfast in their claim that Straus ordered most of the fabrications. The Straus affair set a clear precedent for the responsibility of a senior scientist regarding the behavior of subordinates, even when unethical acts are carried out by them without others being aware of them. In giving testimony to the president's commission Straus pointed out that a senior investigator must rely on the integrity of people under his control. The Straus scandal raised the question of whether a senior investigator should continue to receive federal funds during an unresolved allegation of scientific misconduct.

Michael Briggs. A collaborative arrangement between Schering AG (West Berlin in West Germany) and Wyeth (Philadelphia in Pennsylvania in the USA) manufactured oral contraceptives containing the synthetic "triphasic" hormone evonorgestrel. Following a marketing campaign by these drug companies this second generation synthetic progestogen was before long being taken by more than 15 million women throughout the world (more than any other contraceptive). The popularity of this contraceptive was the direct result of the research of Professor Michael Briggs a British scientist and a former senior staff member of the Schering pharmaceutical company. He was an adviser to the World Health Organization (WHO) and was an internationally recognized expert on contraceptives. He fabricated data on the long-term side effects of oral contraceptives in 1981 mainly while a professor of human biology in the endocrinology department at Deakin University (DU) *(http://www.deakin.edu.au/)* in Australia. Fraudulent research by Briggs concluded that the protracted use of oral contraceptives placed women at risk of both heart and arterial disease. His research culminated in the largest published survey of these risks and he extensively relied on the drug company advertising to promote the pill's safety claims. The drug companies which manufactured the contraceptives and funded most of the research were Wyeth and Schering AG. Briggs presented his research findings at special international symposia with expenses paid by the two pharmaceutical companies. Schering also paid for the cost of the publications which were handed out to members of the medical profession globally. The drug companies denied familiarity with the fraud, but started encouraging physicians to contemplate Femodene (ethinylestradiol and gestodene) as a new oral contraceptive. A number of products, including Logynon (ethinylestradiol and levonorgestrel) and Trinordial (ethinylestradiol and levonorgestrel) in the UK and Tri-Levlen (ethinylestradiol and levonorgestrel) and Triphasil (levonorgestrel, ethinylestradiol) in the USA received government licenses based on research by Briggs over more than a decade. After an investigation in Australia, Europe and the USA, Briggs was traced by the Sunday Times to Marbella Spain where he was living with his wife, who coauthored some of his scientific publications [23-25]. Briggs admitted serious deceptions in his research. He pretended organizing studies on the effects of oral contraceptives but did not do them. He maintained that the investigations were not arranged by him, but by another person whom he refused to identify. Briggs also claimed that he had performed studies on female beagles, but such studies were never carried out. His scientific papers documented the results of complex biochemical tests that allegedly used sophisticated equipment, which did not exist at DU, where he performed the research. In one paper Briggs maintained that he was unaware of the country where the data were analyzed. The phony research findings of Briggs entered the medical literature though books and journal articles and gradually nearly all major researchers in contraceptives cited Briggs. His published research was distributed to doctors by the Wyeth and Schering pharmaceutical companies that disavowed any awareness of the fraudulent research and the two drug companies pointed out that they were not responsible for overseeing the research even though they provided financial support for it. Even after Briggs's research became disputed Wyeth continued to publicize his data. Doubt about the validity of Briggs's research was raised by his ability to recruit vast numbers of women into the study with considerable rapidity and little effort that other researchers found impossible to emulate. After the validity of his work was challenged attempts to initiate an inquiry into the matter were blocked by a legal injunction. Since 1985 the drug companies were aware that Briggs's research was in dispute and in that year an international meeting of contraceptive research specialists in Berlin held an emergency discussion about Briggs after growing incredulity at the speed with which he produced highly complex data on the participants of the clinical trials. Many experts became convinced that Briggs's research was questionable, but focused on papers he wrote while at DU, which contained astonishing flaws. At a meeting in Chicago in September 1986 senior staff of the two drug companies involved in the project urged the Sunday Times to stress that the public was protected through research by other scientists. The FDA stated that it would launch an investigation into the continued use of Briggs's work in promotions to doctors. A second investigation in 1986, ended with the resignation of Briggs [38, 154]. If the research had been done in the USA, the study would have been stringently checked, but these controls did not extend into foreign countries [39]. Primary care physicians in the UK were advised to change more than 2 million women from their existing oral contraceptives restricting the man-made after research indicated that they contained most health risks among contraceptives [39]. The hormone levonorgestrel (Microgynon, Eugynon, Ovran, Ovranette, Trinordial, and Logynon) safety of levonorgestrel containing contraceptives circulated among both researchers. Concerns about doctors and an investigation by the Sunday Times divulged that Briggs, fabricated papers disproving a link between levonergestrel and heart disease. Vital publications contending that the contraceptives were relatively safe were fabricated. The gap created in the literature by the exposure of the fraudulent research needed to be filled by research which disclosed that users of the pills were more prone to heart and other cardiovascular disease than other available brands. Because of the fraudulent research by Briggs the NIH funded a £500, 000 project under the leadership of Professor Victor Wynn, an expert on contraceptive safety. Random samples of more than 30 components of the blood were evaluated by a research team at the Cavendish clinic in north London on over 1, 400 healthy women taking nine dissimilar formulations of the oral contraceptives. The Wynn research group determined complex alterations in the blood, which have become recognized as important markers in the predisposition to cardiovascular disease. Over a wide range of parameters Wynn's assessment disclosed that oral contraceptives containing levonorgestrel put women at a greater risk of cardiovascular disease in contrast to the claims of Briggs. The most predictive marker found by him was a drop in the blood high density lipoprotein 2 (HDL2), which is regarded as protective against cardiac disease in women. The contraceptive which performed best in the study was the desogestrel, containing Marvelon (ethinylestradiol and desogestrel), rather than the suspected levonorgestrel. Extremely close behind were contraceptives with small quantities of norithisterone, such as Brevinor (ethinylestradiol and norethisterone), Minovlar (ethinylestradiol and norethindrone acetate) and Loestrin (ethinylestradiol and norethindrone acetate). However, contraceptives containing levonorgestrel are well-liked as they effectively thwarted "breakthrough bleeding" and apparently often brought about a sense of happiness and an increased libido.

Scott S. Reuben. Scott S. Reuben, MD, was a prolific pioneer in postoperative pain management working at the Baystate Medical Center, in Springfield Massachusetts on one the campuses of Tufts University (Tufts) *(http://www.tufts.edu/)*. He was an internationally renowned anesthesiologist and former chief of acute pain management at this respected academic medical center. After a year-long investigation by Maystate Medical Center Reuben, was found guilty of one of the biggest known examples of academic fraud. Many of his papers were retracted (15). Reuben was charged with fabricating no less than twenty-one health-related projects over thirteen years. Conceivably many more of the 72 papers created by Reuben were also fraudulent. The fraud, which included clinical trials without patients, was detected following questions about two studies that had not been approved for human research. How Reuben was able to deceive his colleagues for almost two decades remains a mystery.

A disturbing aspect of Reuben's research was his indisputable COI involving drugs that he tested for the pharmaceutical company Pfizer. Also, of note his studies disclosed greater benefits of the drugs than what other investigators found and indeed some other investigators found no significant benefit with some of his positive drugs. One of Reuben's reported studies in Anesthesia and Analgesia is the only one supporting the use of a particular expensive drug combination for spinal surgery patients [136]. The Reuben scandal provoked several responses. The editor-in-chief of Anesthesia and Analgesia published a letter alerting its readers that he received a summary of the Reuben investigation and that fabricated data in the referenced article was created under the sole control of Reuben. He also stated that most, if not all, of the listed articles are being retracted by the relevant journals. In May 2009 Anesthesia and Analgesia formally retracted 19 articles by Reuben and two editorials [153, 202] discussed the medical and treatment implicationsn.

Reuben exposed a major flaw in the peer-reviewed process that evaluates the scientific contributions of researchers. Reuben may have been responsible for one of the furthermost instances of fraudulent research over 5 decades and his fraudulent behavior was likened to Bernard Lawrence Madoff (1938-) [22] *(http://junkfoodscience.blogspot.com/2009/03/one-of-biggest-cases-of-academic-fraud.html)*. He maintained that Vioxx and Celebrex had analgesic effects. These two therapeutic agents were subsequently removed from the global market because of worries about an augmented risk of cardiovascular complications. The majority of Reuben's papers were retracted [122-125, 126, 127, 128, 129-131, 130-135, 137-144], Reuben urged the FDA in writing not to impede the consumption of numerous analgesics investigated by him. In support of their safety and efficacy he even had the audacity to quote his own fabricated research. Aside from supposedly falsifying data, Reuben seemingly also carried out forgery. Evan Ekman, MD, an orthopedic surgeon in Columbia, South Carolina, was partly accountable for drawing attention to flaws in Reuben's research. He pointed out that he was listed as an honorary coauthor on at least two manuscripts that were eventually retracted [135, 139, 141, 142] in spite of playing no part in the studies or in the writing of the papers. Reuben published fabricated studies and included his alleged finding of the enormous efficiency of a combination of drugs. Ruben was regularly paid by pharmaceutical companies. Pfizer sponsored Reuben's research from 2002 to 2007, and as a member of Pfizer's speaker's bureau, he spoke on behalf of the drug company. Aside from the apparent greed that tainted many medical research activities, something even more sinister convinced Reuben to jeopardize human lives by falsifying experimental data in research involving human subjects. The tale of the Reuben fiasco established itself in history for several reasons. Reuben is a prime example of the inability of the peer reviewed process to detect fraudulent publications. This medical researcher published at least 21 papers containing fabricated data between 1996 and 2009. None of the journal editors or reviewers questioned the integrity of his manuscripts. Amazingly throughout his career Reuben by no means reported a negative study and his results were invariably extremely robust when other researchers did not succeed in detecting notable differences.

Paul H. Kornak. Qualified clinical coordinators are in high demand and often get hired without undergoing a detailed scrutiny of their past. Paul H. Kornak [8] had been conflicted of mail fraud in 1992 but still obtained employment as a research coordinator by falsifying a declaration of employment to the Stratton veterans administration (VA) medical center in Albany New York with the responsibility of organizing, coordinating, implementing, and directing all research in their oncology research program. At that hospital he was the site coordinator for four clinical trials. Between 1999 and 2002, Kornak defrauded the sponsors of the clinical trials by repeatedly submitting false documents about the participants in the studies and enrolling those who did not meet the criteria for enrollment. Kornak's inappropriate behavior caused in the death of a study subject when he failed to perceive a substantial and unjustifiable risk that death would occur when he knowingly and willfully made and used documents falsely stating and representing the results of blood chemistry analysis.

Kornak was indicted on 48 criminal charges. He pled guilty to three of them and admitted his dishonest handling of the research records and pleaded guilty to criminally negligent homicide. Kornack was sentenced to 71—months in prison. This case strongly emphasizes that a clinical trial involving powerful therapeutic agents can have serious consequences to participants that are inappropriately enrolled by ignoring the inclusion and exclusion criteria outlined in the study protocol. Aside from naming individuals found guilty and the institutions that led the investigations the effective dates and durations (2 months to debarment for life) of the sanctions were provided as well as information as to whether the individuals were serving on committees are listed. The same penalty was imposed upon Kornak based on criminal convictions of making and using a materially false statement, mail fraud, and criminally negligent homicide.

Because of his criminal convictions, admitted serious offenses, and criminally negligent homicide Kornak was barred for life from participating in federal agency transactions by the PHS. The penalty for his inappropriate behavior was considerably more severe than usually inflicted on individuals found guilty by the ORI in cases of research misconduct. In addition to the PHS sanctions, Kornak was excluded from all federal non-procurement transactions for an indefinite period by the office of personnel management.

Other Defective Research Involving Clinical Trials. Defective research involving clinical trials has the potential of harming patients and the FDA has identified numerous cases of this type. By September 20, 2010 this agency listed 87 individuals from 1993-2010 who had been debarred for 5-10 years and often permanently in drug product applications. The following is a public list of persons debarred pursuant to sections 306(a), (b)(1), and (b)(2) of the Federal Food, Drug, and Cosmetic Act (21 U.S.C. 335(a), (b)(1), and (b)(2) as published in the Federal Register: Eduardo Caro Acevedo, Anthony W. Albanese, James Michael Anthony, Mohammed Azeem, Kun Chae Bae, Norma D. Banks, Padam C. Bansal, Baldev Raj Bhutani, Richard L. Borison, David J. Brancato, John W. Bushlow, Anne L.Butkovitz, Maria Anne Kirkman Campbell, Charles Y.Chang, Laverne M. Charpentier, Dulal C. Chatterji, Steven F. Colton. John D. Copanos, Robert Ray Courtney, Charles G. Dicola, Kanubhai C. Desai, Mary L. Donnelly of Green Bay City, Michigan (not to be confused with Dr. Mary Sandra Donnelly), Robert Elbert, Rami Elsharaiha, Scott Feuer, Robert A. Fiddes, Gena R. Finelli, Robert A. Fogari, Barry D. Garfinkel (discussed elsewhere in this chapter), Premchand Girdhari, Wallace Gonsalves, Jr., Kim C.Hendrick, Hedviga Herman, James A. Holland, Liaquat Hossain, Amirul Islam, Sanyasi Raju Kalidindi, James T. Kimball, Niaja Kane, George J. Kindness, Walter S. Kletch, Edwin Kokes, Paul H.Kornak (discussed above), Constantine I. Kostas, Elaine Yee-Ling Lai, Patrick J. Lais, Susan M. Long, Muhammad Z. Mannan, Jay Marcus, Rajaram K. Matkari, Gary D. Mays, Arnold S.Mendell, Allyn M. Norman, Roy C. Page, Andrew Morris Daphne Pai, (also known as Daphne Lau), Ashok Patel (of Upper Saddle River, New Jersey, USA not to be confused with Dr. Ashokkumar Bhailaibhai

Patel (also known as Ashok B. Patel), Craig H. Petrik Renee Peugeot, Kumar Prasad, Nandlal Rana, Thomas M. Rodgers, Jr., Patrick T. Ryan, Suhas V. Sardesai, Frederick Jay Shainfeld. Gloria H. Schetlick, Robert Shulman, Harry W. Snyder, Jr., Edmund J. Striefsky, Mark B. Perkal, Abu Quamruzzaman, Jacob H. Rivers, Juan Manuel Rodriguez, Robert E.Sacher, Mary E. Sawaya (also known as Marty Sawaya), Atul Shah, Dilip Shah, Satish R. Shah. Michelle Lynn Torgerson, Jan T. Sturm, Thomas Ronald Theodore, Mohammad Uddin, Brian Ullom, Jason Vale, Raju Vegesna, Kevin Xu, and Seth M.Yoser.

Misconduct in Other Human Research

Jon Sudbø. Jon Sudbø, a pioneering Norwegian oncologist at the University of Oslo's Norwegian radium hospital, reported that long-term treatment with nonsteroidal anti-inflammatory drugs reduces the risk of oral cancer, but renders individuals to greater risks of death from cardiac disease. The oral cancer community was captivated by Sudbø's research [159, 160]. Sudbø claimed that lesions in the mouth with an abnormal number of chromosomes (aneuploid) were at high risk for oral cancer and an aggressive form of the malignant tumor. Unluckily for Sudbø his fraudulent study was exposed by Camilla Stoltenberg of the Norwegian Institute of Public Health who discovered that 250 of the 908 individuals in the primary data set study had the same date of birth. After this remarkable finding was revealed Sudbø acknowledged that he invented some data and a five member investigative panel headed by Anders Ekbom of the Karolinska institute in Stockholm found that most of his research was invalid [80, 116]. On learning about this fraudulent paper the New England Journal of Medicine (NEJM)reevaluated two earlier papers by Sudbø and noted that two photomicrographs in a 2001 paper [158] which were denoted to stand for two separate patients and stages of disease were indeed the same photomicrograph, but at different magnifications. To make the similarity of the two photomicrographs less evident Sudbø overlay them with patched histograms of ploidy. Until the papers were formally retracted the NEJM issued an expression of concern [36]. Surprisingly these studies had numerous coauthors who should have detected the fraud. An explanation of their role in the manuscripts has not been provided, but for the paper in the Lancet [161] the authors verified that they approved the final report. Despite the misconduct of Sudbø other oral pathologists, such as Edward Odell at King's College London *(http://www.kcl.ac.uk/)* in the UK and Richard Jordan at the University of California, San Franscico (UCSF) *(http://www.ucsf.edu/)*, believe that Sudbø might have been on the right track with his aneuploidy theory [34] and some of them have confirmed his findings.

Anil Potti. In 2003, Duke University (Duke) *(http://duke.edu)* created an Institute of Genome Sciences and Policy (IGSP). During that year Anil Potti came to Duke University Health System (DUHS) for training in hematology and oncology. Potti was an impressive highly regarded bright respected researcher and he had no difficulty getting a junior faculty position at Duke under the supervision of Joseph R. Nevins PhD, a renowned scientist, because of his impressive credentials. Nevins is currently the Barbara Levine professor of breast cancer genomics, molecular genetics and microbiology at Duke was previously James B. Duke professor of genetics and chairman of the department of molecular genetics and microbiology at Duke. This brilliant cancer researcher was a Howard Hughes Medical Institute (HHMI) investigator from 1986-2004 and he was involved in cutting edge research based on the sound scientific concept that genetic profiles of individuals with certain types of cancer may respond to particular forms of treatment. The research related to the potential development of tailored personalized chemotherapy for cancers characterized by specific biological markers and the expression of a vast number of genes. It dealt with ways of employing genomic information to prognosticate the response to cancer therapeutic agents. It had extensive major clinical ramifications for the therapy of cancer.

Unfortunately lax hiring practices permitted Potti to be hired without his curriculum vitae being verified. Potti's padded résumé claimed a past receipt of numerous awards including the prestigious Rhodes scholarship. Potti maintained that for his alleged Rhodes scholarship he worked under the mentorship of a specific scientist, in Australia, but this person did not know him, moreover all recipients of this scholarship are required to study at Oxford University *(http://www.ox.ac.uk/)*. After attention was drawn to this questionable scholarship the foundation that awards Rhodes scholarships could not substantiate that he had ever received this award. The falsified information in Potti's résumé was listed on federal grant applications. Without an anonymous leak the major lies in his curriculum vitae might never have been exposed. Hence it was not surprising that Keith Baggerly PhD and Kevin Coombes PhD, professors of bioinformatics and computational biology at the MD Anderson Cancer Center found flaws in his research [15]. The basic defect in the research by Potti and his coworkers is described in detail in the videolecture by Keith Baggerly, who first noticed flaws in the research by this Duke team. Researchers who falsify their resumes are apt to cut corners in their research and even commit research misconduct. If this red flag in his resume had been checked it would have alerted DUHS to potential dishonest behavior by Potti *(http://videolectures.net/cancerbioinformatics2010_baggerly_irrh/)*. Baggerly warned Duke Officials that data made public at a website by Potti failed to match raw data in the public domain. This information was not provided to the investigating panel because of a fear that it might bias the review, but in retrospect the Duke administration realized that this was a mistake. The Potti case made Duke change its policies with regard to moving from research studies to clinical trials, particularly in topics that involve complex sophisticated cutting edge research. Information about the Potti case obtained through the USA Freedom of Information Act by *Nature* [129] disclosed that a review conducted by Duke into the research of Potti, Nevins and Dr. William T. Barry (a coauthor of several papers) was completed in December 2009. The review panel substantiated the research using data made available by Potti, but it did not correspond to the initial raw data. Moreover their efforts to verify his results were impeded by the absence of data or software scripts [166]. Duke conducted an inquiry into Potti's research and then to obtain a totally neutral objective evaluation of his alleged flawed research the Institute of Medicine (IOM) was asked to lead an investigation [13]. The identities of the distinguished members of the committee were originally kept secret. The IOM report issued on March 23, 2012 criticized how Duke handled the flawed research of Potti. It also felt that the university failed to oversee his research that involved more than 100 patients. From July 2010 until his resignation from Duke on November 19, 2010 Potti was on paid administrative leave during an investigation of research misconduct lasting more than three months. The resignation of Potti was not an admission of scientific misconduct, but he accepted "full responsibility for a series of anomalies in data handling, analysis and management". Potti hoped to avoid potential sanctions if he had been found guilty of research misconduct [20, 21, 42]. For his involvement in flawed research Potti in essence became a fugitive. After resigning from Duke he moved to South Carolina and by June 9, 2011 Potti was practicing oncology at the Coastal Cancer Center in Myrtle Beach, South Carolina [12]. Because that state would not accept his medical license he moved to Missouri where he obtained a medical license on February 1 2011, but on March 6 2011 the Missouri State Board of Registration for the Healing Arts revoked it. Potti moved again and when last heard from he was in North Dakota where he had apparently completed an internship and residency in internal medicine before originally coming to Duke. After leaving his position in North Carolina Potti attempted unsuccessfully to block adverse publicity that was high on the Google list of entries that appeared when a search was conducted on his name. Duke did not explain whether or not Potti made deliberate or irresponsible errors in his research, but Nevins retracted two papers because of a lack of confidence in some of the findings based on a flawed method [11, 164]. The first retraction was for a paper published in the *Journal of Clinical Oncology* in November 2010 [124]. The second publication in *Nature Medicine* [121, 122] that was cited 467 times by January 2013 was retracted when Nevins found that he could not reproduce the method. In January 2011, the Duke University Chronicle reported [9] that a third paper on the prediction of the response to breast cancer authored by Potti and numerous coauthors was retracted by the *Lancet Oncology* [21]. A third paper published in the *Journal of Clinical Oncology)* [67] Other payers were retracted because the authors were unable to replicate their experimental findings. Eventually numerous papers by Potti were either retracted, but some were corrected with errata [21, 118-124]. Many other papers in different scientific journals authored by Potti were retracted or received an erratum [21, 122, 123, 125]. By August 2013, several papers by Potti's papers had been retracted, and the retractions acknowledged that key findings in the studies could not be reproduced and that multiple samples appeared to be duplicated in the training and validation datasets. Three clinical trials (two on lung cancer and one on breast cancer) were started based on the work of Potti and his collaborators. These trials were first suspended in the autumn of 2009 when Baggerly and Coombes expressed concern about Potti' research but the Duke IRB permitted them to resume until early 2010. They were suspended again in July 2010 after the allegation that Potti misrepresented his academic record. They were stopped in November 2010 and a Duke official allegedly stated that the trials should never have been conducted. All patients in these trials were fortunately receiving customary or extensively accepted cancer therapies and the IRB found no concerns to warrant the termination, so the trials were restarted. Later it became suspected that the patients in the clinical trials may have been at some risk and could sue because the trials were conducted under deceptive pretenses. Not long thereafter it a legal firm in Raleigh, North Carolina that specializes in personal injury (Henson-Fuerst) started producing advertisements on the local television stations inviting participants in breast and lung cancer clinical trials at Duke to contact them and a YouTube video was posted as well as a notice on its website requesting information [168]. On September 7, 2011 the Raleigh law firm filed a lawsuit in the Durham county superior court on behalf of 8 patients (only 2 were still living) who had participated in clinical trials interrelated with the research of Potti and his coworkers [155]. The lawsuit was based on an allegation that the patients participated in fraudulent and dangerous clinical trials and unnecessary chemotherapy. On legal advice the lawsuit targeted not only Potti and Nevins, but also Duke, DUHS, the Duke private diagnostic clinic, Oncogenomics, Inc a company formed by Nevins, Michael Cuffe (vice-president for medical affairs at Duke), Sally Kornbluth Duke University Prowest and John M. Harrelson (chairman of one of the IRBs at DUHS). Duke's priorities in the Potti affair were the lives of the patients in the clinical trials and the institution's academic integrity. As part of the fallout of the Potti affair Duke was requested to repay the American Cancer Society that portion of a grant that had been awarded to Potti in 2007 [60]. This was because he had falsified an application by stating that he had been a Rhodes scholar. After Potti's credentials and research techniques were questioned the pharmaceutical company Eli Lily and Cancer-Guide Diagnostics put an end to their affiliations with him in July 2010 [10]. Under the USA Freedom of Information Act, the journal *Nature* obtained information on the reported investigation of Potti, William Barry, and Nevins that was released in December 2009. Although the flaws in the study to which Baggerly had drawn attention where known to Duke administrators, they were not passed on to the investigating panel because of fear that they might bias the review [167].

By August 2010 Duke neared the 'final stages' of naming external review body in Potti case[151]

Andrew J. Wakefield. Dr. Andrew J. Wakefield (1956—) was the son of an upper middle class neurologist and family practitioner and the father of four children. In November 1988 he became a member of the Royal Free Hospital after training in Toronto, Canada. In February 1998 as a 41 year old physician he authored a five page article with a dozen other doctors in the extremely reputable British journal called the Lancet [184]. This infamous unethical human study impacted millions of children. It documented 12 children with developmental disorders and bowel problems who were admitted to the Royal Free Hospital in north London between July 1996 and February 1997. The children all had a prior history of exposure to the measles, mumps and rubella (MMR) three in one vaccine which had been effectively employed from the early 1970s in the USA and late 1980s in the UK. Almost all children received the vaccine at one year of age and this virtually eliminated these infections.

The Wakefield paper attracted considerable public attention at a press conference as well as in a video news-release. It provoked a prolonged assault on the MMR vaccine supported partly by additional papers on the same subject by Wakefield [171-195, 197-199] which led to a breakdown in the public trust of the vaccine beyond comparison. The study stirred up a hornet's nest, because it implicated adverse effects by the widely used MMR vaccine that was used to prevent extremely common serious childhood infections. Wakefield concluded that the MMR vaccine caused autism and that he could not in good conscience recommend the vaccination of children. During the study, Wakefield obtained blood samples from children and compensated each of them with a small amount of money equivalent to $8 for their contributions.

The scandal caused thousands of unnecessary deaths in children because their parents were too afraid to vaccinate them with a vaccine that had already been established as the only safeguard against serious childhood illnesses. With Wakefield's anti-MMR crusade vaccination rates using MMR plummeted below the level required to maintain measles under control. The alleged complications of the MMR vaccine reported by Wakefield created a health crisis that lasted more than a decade after beginning in the UK before spreading to the USA and other countries. After a crumpling of the anti-MMR movement that Wakefield promoted vaccination levels in children started to bounce back. In the USA, the vicious anti-MMR vaccine pressure group emerged following a November 2000 appearance of Wakefield on the Columbia broadcasting system (CBS) network's 60 minutes in which he spoke of an epidemic of autism. The fear of MMR vaccination because of Wakefield's research caused a reappearance of measles outbreaks and at times this led to lethal outcomes or disabling brain damage. Numerous parents were thrust into anguish thinking they were personally responsible for one or more of their children developing autism because of consenting to the implicated vaccination. The tragic fraud of Wakefield initiated epidemics of preventable infectious diseases as well fear and guilt particularly in the minds to parents who became involved in decisions as to whether or not to have their children vaccinated with the MMR vaccine. Aside from its effect on MMR the issue generated grave concerns over the vaccination of children for other diseases. The paramount cause of the anxiety was the remark in the paper that the parents of two-thirds of the dozen children pointed the finger at MMR for the abrupt beginning of a mixture of inflammatory bowel disease and "regressive autism". Particularly alarming was the onset of behavioral signs within only 14 days of the vaccination. In the USA the Wakefield-Barr agreement strengthened the allegation by American attorneys that thimerosal, a mercury-based preservative, in vaccines, was the reason for the exponential growth in the number of cases of autism.

The report by Wakefield and his coauthors was subsequently discredited because Wakefield was found to have acted unethically in conducting the research. On March 6, 2004, ten of the original authors of the paper by Wakefield and colleagues made a "retraction of an interpretation, " not a retraction of the factual content of the paper [81]. The coauthors who removed their names from the manuscript argued that there were not enough facts to make a link to the MMR vaccine, but the likelihood was still present. These authors pointed out that the data were insufficient to establish a causal link between the MMR vaccine and autism. In a letter to the editor of the Lancet published on April 17, 2004, Wakefield and two colleagues pointed out that almost 6 years had passed since Wakefield divulged in the Lancet that he had agreed to evaluate some children for the legal aid board before the 1998 publication [196].

Despite the retraction later papers continued to cite the notorious papers of Wakefield. He also emphasized that the initial paper was not a scientific study but a clinical report and that these laboratory studies received financial support from the legal aid board (later called legal services commission). Wakefield denied allegations made against him in the *Sunday Times* report of February 22, 2004 [40] which included "claims of misappropriate patient referral, inappropriate use of legal aid funding, lack of ethics approval, unmerited clinical investigation, and keeping secret for 6 years the involvement of the legal aid board in a separate study. All of these claims have been investigated and we know that they were unfounded and vigorously deny them."

The report by Wakefield and colleagues related to only 12 children and when the observations were first reported they could have been the first sign of a coming epidemic of shocking dimensions. Observations on small numbers of patients occasionally draw attention to important new medical discoveries, as exemplified by the beginning of the retinopathy of prematurity (formerly called retrolental fibroplasia) and the adverse side effects of administrating thalidomide during pregnancy. At a press conference in 1998, Wakefield recommended that the MMR vaccine be embargoed, and replaced by vaccines that immunize against single viruses at annual intervals rather than by all three at the same time. Wakefield avowed that he could not endorse a continuation of the administration of the three vaccines in combination, until the matter was settled. Wakefield's findings were never reproduced by anyone else. According to the American Academy of Pediatrics countless studies refuted Wakefield's link of the MMR vaccine to bowel disorders and autism. Notwithstanding an assessment by the WHO and the Institute of Medicine (IOM) that Wakefield's allegations against MMR were not valid.

The London *Sunday Times* designated Brian Deer the task of exploring the MMR crisis, and he exposed a disgraceful fraud of astonishing proportions. Deer performed a major investigation of considerable public health importance in the UK for the newspaper and also for the Channel 4 television network in the same country. He doggedly looked into allegations that the MMR vaccine caused a frightening novel syndrome in children with anomalies affecting both the bowel and brain. He found that far from stemming from the truth the public distress signal completely lacked a scientific foundation. Deer discovered that Wakefield had been paid to generate proof against the MMR vaccination and at the same time was preparing astounding commercial plans to generate money on the public panic. Wakefield also altered and misreported information on the unnamed children described in his paper in the Lancet, which gave the impression that the kids (11 boys and one girl) were not precisely as described. The impression was given that they were a customary series of children with developmental disorders and gastrointestinal symptoms, requiring treatment at a London hospital. Many of their parents attributed setbacks on a vaccine that they had all received. Deer ascertained that the children (aged between 2½ and 9½) had been enlisted because of MMR crusades. Prior to the referral of any of the dozen children to the Royal Free Hospital Wakefield was hired by Richard Barr, a lawyer who dreamed of starting an exploratory class action suit against the manufacturers of MMR. In contrast to the usual medicolegal case in which doctors provide guidance and judgments as expert witnesses, Wakefield reached a profitable deal with Barr to perform both clinical and scientific investigations. The purpose was to unearth support for a recently recognized "new syndrome, " meant to be the focus of future legal action on behalf of 1, 600 families, enlisted by way of tales in the media. For Wakefield this task formed a dreadful COI. When Deer drew the public's attention to it in February 2004 [40], pandemonium broke loose in the UK. Money for Wakefield came through Barr from a UK legal aid fund administered by the government to provide the poor with an entrée to justice. For his time Wakefield billed $150 an hour through his wife's company. After being pushed under the Freedom of Information Act, the UK legal services commission disclosed that the expenses were £435, 643 (about $750, 000 USA).

When hospitalized, most of the parents were clients of Barr or had connections with him. None of the 12 children resided in London. Two went to a doctor's office, 280 miles from the Royal Free Hospital. Three were patients at a different hospital clinic. One flew in from the USA! The assertion in the Lancet paper that autism suddenly followed MMR vaccination was a deception. The parents were expected to hold MMR responsible for the disease of their children, since that was why they came to Royal Free Hospital. The actual number of families blaming MMR was not eight, as reported in the Lancet. It turned out be 11 of the 12 and subsequently all parents. In June 1996 Wakefield received £55, 000 but failed to notify the Lancet as he should have. An astounding £18 million of taxpayers' money in the UK would ultimately be divided amongst a group of doctors and lawyers, toiling under the leadership of Barr and Wakefield in an attempt to establish that MMR brought about an alleged new syndrome, which would later be dubbed "autistic enterocolitis". In a confidential letter, six months before the Lancet report, Barr reminded Wakefield "I have mentioned to you before that the prime objective is to produce unassailable evidence in court so as to convince a court that these vaccines are dangerous."

Investigations by the Sunday Times and the television station Channel 4 exposed an additional outrageous conflict of interest on the part of Wakefield. In June 1997, almost nine months before Wakefield announced at a press conference that single vaccines were probably safer than the MMR vaccine he had filed a patent on products, including a single measles vaccine, which simply had little hope of triumph unless support for MMR was destroyed. Confidential documents indicated that a group of companies planned to raise venture capital from professed inventions—including a vaccine, testing methods, and strange potential miracle cures for autism. One business was subsequently awarded £800, 000 from the legal aid fund, for now-discredited data which he supplied. Three months after his original paper in the *Lancet* Wakefield wrote that only he had agreed to assist in the evaluation of a small number of children on behalf of the legal aid board. Wakefield also maintained that no attempt was made to hide the fact that the viral study was taking place prior to the publication of the notorious 1998 paper in the *Lancet*. Responding to a letter in the Lancet on May 2, 1998 by Wakefield and his colleagues [183] Richard Horton (1961-) the editor of the *Lancet* at the time drew attention to reasons why the interpretation of Wakefield and his coauthors were not accepted [6, 64].

Horton pointed out that the journal did not accept their interpretation [65], which was only 3 months after the 1998 paper in the *Lancet*. This letter was written in response to one from Dr. Rouse [149]. The investigative reporting of Deer achieved something that other journalists had not previously been able to do and perhaps will never be permitted to do again. He managed to identify the specific children in Wakefield's original provocative 1998 article, and to gain access to their unique medical records. Piercing barriers of confidential medical documents, Deer learned that nothing was found by the hospital staff to incriminate MMR, yet Wakefield had again and again altered and misstated diagnoses, clinical histories as well as the signs and symptoms of the children creating the impression that he had discovered an important new syndrome. On the last day of 2006 Deer pointed out in the Sunday Times [41] that Wakefield received more than eight times his reported annual salary and had a monetary reason to set off the MMR alarm and make it last as long as possible.

Time and time again the pathology department at the Royal Free Hospital reported no abnormalities in bowel biopsies from the children. Some of the kids manifest features of autism before being vaccinated and some did not even suffer from autism. Aside from these factual errors in the Lancet paper, other unethical issues were present. Research on human subjects is overseen by national and international benchmarks, principally the Helsinki declaration. No IRB approved the project that Wakefield undertook for the lawyer, Barr. Without such approval reputable medical journals would not normally have accepted a paper dealing with a clinical study. Wakefield claimed that he had obtained approval to perform numerous unpleasant procedures, including anesthesia, ileocolonoscopies, lumbar punctures, brain scans, electroencephalograms (EEGs), and X-ray examinations, over a five-day period on these children. They also swallowed drinks with radioactive isotopes. However the ethics committee of the Royal Free Hospital was not truthfully notified about the details of the proposed project and did not approve the invasive procedures.

Wakefield's business enterprise was based on the supposition that both the inflammatory bowel disease and autism were caused by an unrelenting live measles virus infection caused by MMR. Yet complex analyses performed in Wakefield's own laboratory failed to detect measles virus in the intestine or blood of the children.

In July 2007 a panel of the GMC of the UK consisting of three doctors and two lay persons began a hearing on Wakefield. After a 197-day inquiry, the longest professional misconduct hearing in the UK to date, the GMC announced on January 28, 2010 that it had found major defects in Wakefield's research. In addition invasive unnecessary diagnostic procedures that were not in the paramount interest of the youngsters were carried out on the youngsters and family financial resources were misused. Wakefield failed to disclose COI, but he was also entailed in lawful accusations hostile to the vaccine makers. Wakefield was labeled as a dishonest, unethical, irresponsible, and callous individual. The research that he reported in the *Lancet* in 1998 [184] was dishonestly portrayed and carried out on developmentally disordered children without ethical approval. Five days later, the *Lancet* took the bold step of officially retracting the controversial paper because of numerous errors in it after Wakefield refused to do so insisting that that the study was authentic. The UK's GMC's fitness to practice panel concluded that Wakefield had provided false information in the report and acted with a callous disregard for the children in the study the council deliberated on whether Wakefield was guilty of serious professional misconduct. Less than a week later the council's conclusive finding of misconduct caused him to lose his medical license. Aside from the amoral issues the authors had carried out an invalid study because it was not a randomized double-masked study as required to obtain acceptable data for a justifiable conclusion. After being discharged from the Royal Free Hospital in October 2001 Wakefield obtained a job with an annual salary of $280, 000.

Despite being aware that his charges against MMR were groundless, Wakefield still promoted his assertions from a contentious business in Austin, Texas designated Thoughtful House. In February 2010 Wakefield was still practicing in Austin. He also started, and later ditched with some £1.3 million costs, a two-year libel lawsuit. He received financial support from the medical protection society, which defends doctors against grievances from patients. Wakefield insisted that accusations against him were unfounded and unjust *(hhttp://briander.com/nmr/lancet-summary.htm)*, but by January 2012 Time Magazine had him listed as one of the "Great Science Frauds".

Mark H. Williams. Dr. Cameron Bowie, emeritus director of public health in Somerset England, was forced to retract a highly praised paper on a fundamental community based scientific contribution that advanced health care policy significantly in the UK with cost-effectiveness for the disabled. Bowie coauthored the paper with Dr. Mark H. Williams in 1993. Despite the fact that Williams over and over again denied fabricating data, Bowie could not confirm, certain aspects of the study, and could not presume that the published data was valid [203]. The concepts within it became encompassed in national examinations of doctors specializing in public health. Williams was found guilty of scientific misconduct casting a dark cloud over the paper documenting the unmet requirements for ways to permit 181 harshly disabled adults to live by themselves. According to the manuscript affected individuals were interrogated and studied in 1989 and 1990. The survey revealed that health professionals improperly supervised such needs but with the coordinated aid of a social worker the care could effectively be enhanced at a nominal expense. Three years after the paper appeared in print Bowie was astonished to hear that Williams, who had relocated to a neighboring university was accused of falsifying statistics in a different research project and was confronting disciplinary action. After the paper was questioned Bowie looked for the questionnaires and computerized particulars related to the study but with no avail. Bowie visited all of the patients listed by Williams as being in the study and found that none of them had been interviewed. After his dishonesty was established Williams's medical license was revoked by the GMC.

Mark Williams. Another Mark Williams also held an appointment in public-health medicine and participated in fraudulent behavior. This doctor was a senior lecturer at the University of Bristol *(http://www.bris.ac.uk/)* in the UK. In planning to attend a 1994 conference in Sweden he submitted an abstract of his intended presentation with suspicious statistics. The chairman of Williams's department, professor Stephen Frankel, did not challenge his explanation, but in 1995 when Williams prepared a manuscript for the *British Medical Journal* Frankel became doubtful about the integrity of the paper and reanalyzed the data. Williams was confronted and admitted his wrongdoing [44]. All that he had done was then carefully scrutinized. He was found to have falsified his qualifications for his initial university appointment in 1991 and again a couple of years later for an upgrade in his position. According to his credentials he received an MD and PhD from the University of Cambridge *(http://www.cam.ac.uk/)*. He attributed the false credentials to a typist's mistake and acknowledged that the statement related to the statistics was ambiguous. As soon as his fraudulent behavior was unearthed he resigned from the University of Bristol and accepted a temporary job as a locum tenens in a different city. This position came to an abrupt end when Williams was struck off the register of practicing physicians of the GMC.

Malcolm Pearce. Dr. Malcolm Pearce a British gynecologist, who was a senior lecturer in the obstetrics and gynecology department at St George's hospital medical school in south London published fabricated work in two papers in the August 1994 issue of the British Journal of Obstetrics and Gynaecology for which he was an assistant editor. The head of the department where Pearce worked was Professor Geoffrey Chamberlain, who was not only the editor of the *British Journal of Obstetrics and Gynaecology*, but also the president of the Royal College of Obstetricians and Gynecologists. One paper documented the successful reimplantation of the embryo from an ectopic pregnancy leading to the birth of a live baby [113]. The other paper was a preposterous clinical trial of treating recurrent miscarriages in 191 women with polycystic ovary syndrome [112]. Ideally a reasonable editor might have doubted the validity of Pearce's report and wondered how he could have recruited 191 women with the extremely uncommon association of recurrent spontaneous abortions with polycystic ovary syndrome over three years when major tertiary medical centers barely encounter one or two such cases each month. After an elapse of 9 months a whistle-blower pointed out that the patient with the reimplanted pregnancy did not exist. Attention was also drawn to the unlikelihood that Pearce could have recruited 191 patients with polycystic ovary syndrome and recurrent abortions. The fabricated publications of Pearce slipped past the peer review process because of unethical practices by the *British Journal of Obstetrics and Gynaecology*. The report of birth after the ectopic pregnancy escaped peer review because case reports submitted to *the British Journal of Obstetrics and Gynaecology* did not undergo peer review at that time. Soon after a scrutiny by the medical school confirmed the fraudulent nature of the two publications by Pearce he was fired and the GMC struck him from the register of medical doctors in the UK for his misconduct [33, 45]. The Pearce scandal was Britain's highest profile case of research misconduct and it attracted considerable international attention, because obstetricians and gynecologists had attempted for a long time to reimplant embryos from ectopic pregnancies without significant success.

Pearce created the most prominent high profile scandal on scientific misconduct in the UK and it was dealt with exceptionally well by the head of medical school at St. George's hospital, where Pearce was employed while protecting the rights of the accused as well as the whistle-blower [45, 74, 75].

Admirably as a reaction to the Pearce disgrace the Royal College of Obstetricians and Gynecologists authorized a report into the part of their journal in this scandal, and it put into practice the recommendations. Another clue that something was amiss was the wide variety of complex evaluations that the patients underwent, including karyotyping of both partners. Another inappropriate aspect of the Pearce publications was the bestowing of honorary authorships on publications reporting research that had not been performed. To add insult to injury Pearce added Chamberlain's name to the paper on the ectopic pregnancy survival as the most senior author despite being told twice to remove it. It was within Chamberlain's rights to coauthor the paper if he so desired, but having his name associated with this research misconduct was a major embarrassment. Following this scandal Chamberlain resigned his position as president of the Royal College of Obstetricians and Gynecologists and of editor of that society's *British Journal of Obstetrics and Gynaecology.*

Misconduct in Clinical Trials

Charles J. Glueck. Even distinguished citizens and scientists can run into trouble with research involving human subjects as Charles J. Glueck MD, a member of one of the most well-known families in Cincinnati discovered. This NIH funded researcher was a nationally and internationally renowned cholesterol researcher. He had also been dean at the University of Cincinnati (UC) *(http://www.uc.edu/)* for 18 years and headed both the general clinical research center and the lipid research center of UC. At least one renowned colleague of Glueck portrayed him as extremely energetic, assertive, hardworking investigator who appeared to be driven to publish his research findings. In June 1986 when forty-eight—year-old Glueck submitted the findings of a study to Pediatrics comparing the results of two treatments for extremely high serum cholesterol levels in 73 children with hereditary hypercholesterolemia. One treatment was a low-fat, low-cholesterol diet; the other used cholesterol lowering drugs combined with a diet. While the results of the study were in press and due to appear in Pediatrics two anonymous whistle-blower at UC drew attention to flaws in the soon to be reported study [55]. After an investigation by UC into accusations of serious research misconduct by Glueck, the paper was retracted in 1987 [55]. Glueck composed a letter referring to the study. Like many other researchers who are found to report scientifically unsound research Glueck stubbornly refused to accept the seriousness of his research mistakes. In November 1986 Glueck provided a 303 page rebuttal in is defense. He acknowledged some mistakes and inconsistencies in the paper, but he considered them random and inadvertent and maintained that the charges against him stemmed from mistakes and distortions of the reported research. During the investigation Glueck claimed that notebooks holding study data were in storage and he was obligated to depend on incomplete information in patient records. As with other overcommitted administrators who take on the burden of research Glueck attributed flaws in his research to a harsh overcommitted weekly work load of 70-80 hours. Aside from clinical and managerial responsibilities Glueck was the principal investigator (PI) on five major studies.

Glueck followed variable aged children for more or less five years. The report concluded that both of the evaluated therapeutic options lowered plasma cholesterol efficiently without undesirable side effects. A UC committee investigated the allegations and concluded that no meaningful interpretation could be made from this nonprospective study. The committee also detected evidence of serious research misconduct by Glueck in the published paper. Glueck selected his data subjectively and the paper contained both inaccuracies and contradictions. The study was sloppily carried out and some data came from earlier studies by Glueck between the years 1970 and 1977. None of Glueck's protocols treated children with diet alone and hence could not have been part of any prospective study. Glueck incorrectly claimed that he had not submitted any prospective protocols for follow up studies after 1978. The UC inquiry discovered that the 39 normal children in the control group were selected from a private pediatric practice when the manuscript was being written and data on them were not collected as defined in the study protocol. An inferred degree of accuracy in the study proposal was not backed up by infuriated evidence. UC notified the NIH about questionable aspects of the Glueck research. On February 9 1987 Glueck resigned from UC and became employed by the Jewish Hospital in Cincinnati. The reported findings of Glueck and colleagues were of potential clinical importance and relevant to the safety and efficacy of lowering cholesterol levels in childhood.

Just after the paper appeared in print officials at UC notified the editor of Pediatrics that UC wished to disassociate itself from the paper. The editor of Pediatrics, Jerold Lucey, was displeased that he was not informed about the questionable article prior to its publication when he could have delayed publishing it. At least some of the flaws should have been detected by peer review. Important discrepancies were found between initial patient charts and tables in the manuscript, and uncertainties existed about dates that the therapeutic methods began and ended. Because of these faults the committee asserted that the assumptions bridging the effectiveness of these treatments were not justified. After the UC evaluation the raw data was appraised by three researchers from other institutions, namely Robert Bare Linna of Yale University (Yale) *(http://www.yale.edu/)*, Justin Shawnfield of Washington University (UWSL)*(http://wustl.edu/)* and Robert V. Furetz of the University Chicago (UChicago)*(http://www.uchicago.edu/)*. This external committee concurred that not being prospective, the findings in the study could not be meaningfully interpreted. The NIH assessed these concerns and agreed that the paper was unacceptable and failed to achieve expected standards. The NIH urged Glueck to retract promptly the Pediatrics manuscript or publish an elucidation and be barred from receiving federal funding for 2 years and from serving on peer review committees for five years [63]. In a letter to the editor coauthored with Peter Laskarzewski, Margot J. Mellies and Tammy Perry Glueck withdrew his controversial paper in Pediatrics [55] claiming that it was seriously flawed with unintentional errors. In their statement of retraction the authors pointed out that the committee that reviewed the patients charts found errors and inconsistencies in some of the lipid data, but not in the height and weight data. They also stated that when they corrected the errors they found them to be nonsystematic and random and that the clarified data led to the same conclusion as the original paper.

Philip A. Berger. In May 1987 Stanford University (Stanford) (http://www.stanford.edu/) announced that Dr. Philip A. Berger, a distinguished researcher on schizophrenia in charge of two clinical research centers at its institution had resigned effective December 1, 1987 and that he had been given an immediate leave of absence. Prior to his departure questions had been raised about Berger's professional conduct concerning private patients. Berger was one the nation's most productive scientists investigating the biochemical basis of mental diseases and he had been the recipient of a large 10 year grant from the National Institute of Mental Health (NIMH). Money was being returned to the NIMH because an audit confirmed that some costs to the grant could not be justifiably charged to the grant. However, there was no misappropriation of money for Berger's private use. Largely because of concerns about the salary that Berger's wife was receiving while spending much time working at home an audit of Berger's NIMH grant was ordered by the dean of Stanford school of medicine. Berger spent a considerable amount of time on consultations and this interfered with his ability to supervise the people under his control [16]. Berger was the first individual to be barred for three years by the Office of Scientific Integrity Review (OSIR) from receiving federal funds from grants or contracts. He was found to be responsible of deviations from accepted practices in the conduct and reporting of science and for not using valid controls. Some so-called normal controls were categorized in a different study as having memory loss. In particular, he erroneously documented that some people in certain studies were drug free while in reality they were receiving medication.

Promotion of Off-Label Drugs by Pharmaceutical Companies. Billions of dollars are spent annually by drug companies to convince doctors to prescribe their drugs, and strict rules govern the manner by which this is achieved. The regulations are intended to make certain that doctors receive dependable information, but unfortunately some drug manufacturers and their staff find loopholes to circumvent rules. Individuals in the pharmaceutical industry sometimes misuse scientific evidence to promote their products and occasionally medications are made available to the gullible public using distortions of the truth. Sales representatives commonly lie to doctors about the effectiveness of drugs and in the past doctors were often paid vast amounts of money by pharmaceutical companies to promote their drugs and even to get drugs approved by the FDA. Large drug companies that commit federal crimes can afford to pay colossal fines and still flourish despite the adverse publicity. It is a small price to pay for doing business when powerful drugs have a very large market. Insurance companies and Medicaid programs are key forces encouraging a fraudulent marketing of drugs. The financial reward to a whistle-blower who draws attention to the fraud is a major preventive force of drug company fraud. If evidence disclosed by this person leads to a successful prosecution the whistle-blower receives 50% of the money received from the fine. Usually the whistle-blower works, or has worked, for the pharmaceutical company, but unfortunately loyal happy employees are not going to bite the hand that feeds them. The traitor is more likely to be a disgruntled employee with an axe to grind, and this may lead to an expensive time-consuming investigation in which the allegations of fraud may be unfounded. It is illegal for drug companies to sell FDA approved drugs for reasons other than the FDA approved reason, but doctors may prescribe them for other reasons at the discretion of the medical doctor. Prescribing drugs for this other reason is referred to as off-label and anecdotal case studies are often reported by members of the medical profession on a small number of patients treated with the off-label use of drugs. To justify the negative results of some off-label use of drugs it has been pointed out that a drug may work in one person but not in another, but this may be a placebo effect. The illegal practice of off-label was promoted by marketing by paying doctors to attend so-called educational meetings at luxuriant restaurants and resorts.

It is illegal for drug makers to pay doctors directly to prescribe specific products. Federal rules also bar manufacturers from promoting unapproved or off-label uses for drugs, but medical doctors may prescribe these drugs at their discretion. Pharmaceutical companies can get around marketing exclusions by paying doctors to lecture on their drugs, and during the question-and-answer session, they can bring up potential off-label uses of the drugs [115]. Many medical doctors as well as the drug industry regard promotional lectures on particular drugs as a useful teaching exercise for participants of health care. Opponents of the system regard the payments to the lecturers as well as the educational sessions to the medical profession that frequently take place at luxurious restaurants as camouflaged kickbacks that persuade doctors to prescribe potentially hazardous drugs. The issue is especially relevant to psychiatry, because mental disease is poorly understood, and its therapy is frequently based on trial-and-error, and off-label prescriptions are common. An analysis of Minnesota records disclosed that from 1997 to 2005, more than a third of the licensed psychiatrists in Minnesota accepted payments from pharmaceutical companies, including the eight presidents of the Minnesota Psychiatric Society. The psychiatrist receiving the most from drug companies was Dr. Annette M. Smick, who resided outside Rochester, Minnesota, and she received more than $689, 000 from drug makers between 1998 and 2004 [58]. At one time, Smick was delivering so many lectures sponsored by drug companies as an educational service to her colleagues that she found it difficult to make time to treat patients in her clinical practice. Steven S. Sharfstein, a past president of the American Psychiatric Association, was of the opinion that psychiatrists had become too friendly with the manufacturers of drugs. As an example of the influence of drug companies on therapy, he pointed out that Lexapro, a product of Forest Laboratories, became the most commonly prescribed antidepressant in the USA; even though less expensive alternative medications, such as generic adaptations of Prozac, were available. Due to marketing, expensive drugs were being prescribed rather than generic versions.

Neurontin. The drug Neurontin (gabapentin) was approved by the FDA only as an antiseizure medication for epileptics in 1994 and later in 2003 also for pain from Herpes zoster. However, some literature distributed to doctors hinted at a beneficial effect in other disorders. Initially Parke-Davis owned Neurontin, but it was later acquired by the pharmaceutical giant Pfizer. In the USA alone almost 12 million prescriptions were written for Neurontin, but most of the revenue generated by this drug is not from its FDA approved application, but for the treatment of other disorders. By 2002 94% of Neurontin's sales were for off-label uses. A lawsuit by a whistle-blower and former salesman, David Franklin, accused Parke-Davis of making most of its profit from Neurontin by marketing it to doctors for unapproved uses. Franklin alleged that internal documents showed that the drug company promoted it for untested uses. In 2004 the Warner-Lambert division of the Pfizer drug company paid a $430 million fine after admitting guilt for illegally marketing Neurontin to treat ailments for which it was not approved and to settle charges of Medicaid fraud and other crimes [152] *(http://www.npr.org/display_page/features_920362.html)*. However, Warner-Lambert claimed that it was effective in a long list of maladies, including pain, bipolar disorder, attention deficit disorder, and alcohol detoxification. Some employees of the pharmaceutical company allegedly dubbed the drug as being effective against a "snake oil" list. The confession did not hurt Pfizer as sales of the blockbuster drug Neurontin rose significantly and major insurers and most state Medicaid programs continued to pay for Neuronin as before. Snigdha Prakash, a reporter for National Public Radio (NPR) *(http://www.npr.org/display_page/features_920362.html)* detailed how a single company endeavored to get around the off-label rules with Neurontin. From 1994 until 2000 the pharmaceutical company paid doctors up to $230, 000 to permit sales representatives to attend sessions with patients.

Other Blockbuster Drugs. In September 2009 Pharmacia and Upjohn another subsidiary of the Pfizer was fined $2.3 billion for illegally promoting off-label uses for four of its drugs Bextra (a painkiller), Geodon (an antipsychotic), Zyvox (an antibiotic) and Lyrica (an antiepileptic) in the largest health care settlement in history [150]. Bextra had only been approved for the treatment of menstrual pain, rheumatoid arthritis and osteoarthritis, but Pfizer pleaded guilty of promoting off-label uses of it, as in pain relief after knee-replacement surgery. An investigation into this inappropriate behavior by the drug company was initiated by six whistle-blower lawsuits. The first involved John Kopchinski, a former Pfizer sales representative, who was fired after questioning Pfizer's marketing of Bextra. From the fine $102 million will be divided among the whistle-blowers and Kopchinski was expected to receive $51.5 million [152].

Charlatans

John Taylor. Perhaps the most famous known charlatan was the British eye surgeon John Taylor (1703-1772) [163]. He claims to have been the personal eye surgeon of such famous people as King George II of Great Britain and Ireland (1683-1760), the pope of the Catholic Church, and various members of the European royal families. Taylor removed the cataracts of the renowned composers Johann Sebastian Bach (1685-1760) and George Frederic Handel (1685-1759) by breaking them into small pieces. The surgical procedures on both the latter distinguished musicians were not successful, and as with hundreds of other patients of Taylor, they ended up becoming blind; and Bach actually died during the postoperative period of a suspected infection. While his surgery left much to be desired, particularly since local anesthesia was not available and postoperative infections were not understood, Taylor had considerable talent in self-promotion and in womanizing persons of the female gender. When he traveled in Europe, he used a coach painted with images of eyes together with the motto "Qui dat videre dat viver"(He who gives sightgives life). Taylor sought publicity, and before he moved to a new location, information about his itinerary was advertized beforehand so that large crowds would show up. Taylor took special steps to prevent being caught with adverse postoperative results. He insisted that operated eyes be covered for a week before the bandage be removed, and by this time, he had already moved to another town.

John Romulus Brinkley. John Romulus Brinkley (1885-1942), who later changed his name to John Richard Brinkley, was an extremely controversial medical doctor *(en.wikopedia.org/John_R._Brinkley)*. He was born into a humble family and when he was 21 years old he married his first wife and the two of them roamed the country masquerading as Quaker doctors and promoting patent medicines. Later he registered at the unaccredited Bennett Medical College in Chicago shortly before the famous 1910 report by Abraham Flexner (1866-1959) on the appalling state of medical education in the USA and Canada [50]. By 1912 Brinkley attempted to get credits from his Bennett Medical College education transferred to another medical school, but Bennett College refused because of his unpaid debts. Not to have his career held back Brinkley purchased a diploma from the Eclectric Medical College in Kansas City. Eventually Brinkley became exceptionally wealthy and even campaigned unsuccessfully for governor of Kansas on two occasions. He proposed a simple hypothesis to treat the common human male ailment of impotence. Today a successful therapy is available with Viagra and its derivatives, but at the time of Brinkley nothing was effective. Because the goat was regarded as the horniest animal Brinkley thought that the transplantation of goat testes into a male subject would cure the ailment. To carry out this therapy he opened clinics in many parts of the USA. Being an expert in both advertising and in radio transmission the public readily learned about his treatment. The medical profession was not impressed with this extremely popular natural showman and in many states his medical license was revoked. Like many of the rich and famous he lost his fortune. He was almost penniless when he died at 56 years of age in San Antonio Texas. His downfall was caused by numerous lawsuits brought against him for fraud, malpractice and wrongful death.

Andrew McNaughton Jr and Ernesto Contreras. Scientific information is sometimes ignored because of a public disbelief in the advice of those who are knowledgeable about the criteria for determining acceptable therapies. In the 1920s Dr. Ernst Krebs, Sr, an unknown general practitioner in San Francisco put forward the notion that an extract of apricot seeds reduced rodent tumors [37]. In 1936, Krebs warned that the predominant chemical in the apricot seeds seemed to be amygdalin and that it was too capricious and too risky for human consumption. More than two decades later his son, Ernst Krebs, Jr., maintained that he had created a safer amygdalin derivative which he named "laetrile", even though the active component of most laetrile sold has been identified as amygdalin.

This controversial anticancer drug received some publicity because of the Canadian adventurer, Andrew McNaughton, Jr., and the establishment of a clinic at the USA-Mexican border by a Mexican pathologist Dr. Ernesto Contreras. When Dr. John Richardson, an activist of the John Birch Society, was arrested in 1972 for selling laetrile in violation of California law an uproar was started with arguments coming from the sophisticated political organization of the far right aided by the manpower of numerous Americans overinvolved in extreme food faddism that quickly spread across the country.

To restrict the use of laetrile was considered a violation of an individual's freedom of choice in deciding a personal form of cancer therapy. This roused much support among the general public in whom post-Watergate sentiments of anti-establishment, anti-government, and anti-regulation were reaching their zenith.

Contreras and Richardson made huge amounts of money, as evidenced by bank deposits of $2.5 million for one individual in as short a period of time as 27 months (1973 to 1975) [5]. It is extremely economical to manufacture laetrile and modest consideration is needed for purity and sterility. It was sold at more than a dollar per tablet or 10 dollars per ampoule. Laetrile has been exhaustively scrutinized but it failed to obtain FDA approval for human therapy under the provisions of the Federal Food, Drug, and Cosmetic Act. In numerous animal tumor models the drug has over and over again failed to reveal a reproducible benefit. Human data has been exceptionally mediocre and anecdotal with inadequate recordings of the cancer and hardly ever dependable quantified therapeutic results. While under the weight of considerable political pressure the NCI performed a retrospective analysis of human experiences with laetrile to establish if an ethical or scientific good reason could be found to initiate a prospective human clinical trial of laetrile [46]. Despite finding no definitive evidence to support an anticancer effect of laetrile, the NCI proposed a prospective clinical trial of laetrile. Such a trial was unreasonable in terms of the accepted FDA standards, because new drug studies in humans normally require safety and effectiveness to be first determined in experiments on animals. Perhaps because laetrile met FDA standards of purity and sterility the FDA made an exception to its rule on animal experiments.

Peter Nixon. Dr. Peter Nixon of the Charing Cross hospital in London made a name for himself for his weird hyperventilation hypothesis that caused AIDS, post-traumatic stress disorder, Gulf War syndrome (GWS), premenstrual tension, and other entities. Nixon published information about his hypothesis in medical journals [51, 52, 71, 108, 109, 111, 117, 145-148]. In 1988 Nixon's opinions were endorsed by Neville Hodgkinson, the medical correspondent of the Sunday Times in a front page section of this reputable newspaper. The story maintained that Nixon had discovered a flawless successful therapy for chronic fatigue syndrome (CFS). Patients identified with the hyperventilation syndrome were treated with diazepam to induce sleep, followed by physiotherapy and breathing training. In 1994 Channel 4, a television station in the UK, took a bold step in exposing Nixon's fraudulent medical practice to the public in a program designated Preying on Hope in which Nixon was branded a charlatan, incompetent to practice medicine. Channel 4 clandestinely videotaped a discussion between Nixon and a terminally ill patient known to have AIDS, but attributed to overbreathing with a recommendation of a cure with valium or diazepam and "two weeks of sleeping." Channel 4 ascertained that he manipulated the findings on a breathing test using the capnograph by instructing the patient to become "hugely angry, frustrated, and trapped ". After exposing Nixon Channel 4 faced what many individuals fear in exposing fraudulent research. Nixon sued the investigative journalist Duncan Campbell and his television production company for £2 million in a libel suit. He confronted the defendants for five weeks at the Royal Courts of Justice in London. Unfortunately for Nixon his career was flawed and the exposure of it was supported by solid evidence. In testimony he divulged mistakes in professional publications coauthored by him that seemed to be "more than an honest slip of the pen." After being cross-examined about papers that Nixon had written in the Journal of the Royal Society of Medicine [51, 108, 109, 111, 117, 146-148]. During cross-examination Nixon acknowledged the presence of mistakes in some of them and even admitted that he did not write some of the articles listing him as a coauthor.

Evidence against him was overwhelming and the truth prevailed causing him to withdraw his charges of libel. In doing so he consented to pay £765, 520 in expenses to Channel 4 toward their legal costs. Nixon was not allowed to use health explanations for terminating his litigation. He also came to an understanding not to start further litigation against Channel 4, the producer and Campbell, or his production company if they repeated their accusations. In addition Nixon consented to have the entire documentation in the libel case forwarded to the GMC for further action if he failed to voluntarily retire from practicing medicine. However, soon after the court issued its verdict Nixon was anticipated to retire from the practice of medicine. The application of different standards in comparing patients and controls made the differences between the patients and the controls falsely more noticeable. Nixon performed diagnostic tests without obtaining written informed consent or providing information about the risks. Moreover Nixon failed to obtain the approval of the Charing Cross ethics committee. The medical defense union that financed the legal case of Nixon, was left with a bill estimated at almost £2 million [43]. Throughout the court case, the charges made by Channel 4 were validated and Nixon even admitted some of them during cross examination.

He admitted having "no honest grounds" for assertions that a "hypnotic challenge" test for diagnosing hyperventilation syndrome was trustworthy. He also conceded that he came to the wrong assumption in asserting that his test, in which affected individuals were instructed to consider anger or other pessimistic emotions, was more successful than the customary forced hyperventilation provocation test (FHPT). Nixon seemed to use dissimilar percentage criteria in the two groups. Nixon agreed that it "looked rather suspicious" that three of 27 patients documented in the Journal of the Royal Society of Medicine were excluded from the study when it was subsequently reported again in the American Journal of Clinical Hypnosis [32]. He admitted that this nullified the findings documented in the latter journal, which were not mentioned in the prior manuscript. Nixon submitted several communications to the British Medical Journal [78, 82-87, 94, 99, 100, 102, 104, 105] and The Lancet [18, 29-31, 43, 68, 76, 88-93, 95-98, 101, 103, 106, 107, 110, 201] *(http://duncan.gn.apc.org/nixon1.htm)*.

Other Dishonest Human Research

Aside from researchers discussed in this chapter fraudulent research involving human participants has been performed by numerous other scientists, including Cyril Lodowic Burt, Vijay Soman, Stephen E. Breuning, Roxana Gonzalez, William Griffith McBride, Scott Monte, and Kugler and colleagues.

References

1. Altman LK. Federal officials to review documents in breast cancer study with falsified data. The New York Times. 1994 March 27.
2. Anderson C, Breast cancer. How not to publicize a misconduct finding. Science, 1994. 263(5154): 1679.
3. Angell M, Is academic medicine for sale? New England Journal of Medicine, 2000. 342(20): 1516-1518.
4. Anonymous, Khabarovsk War Crime Trials. Materials on the Trial of Former Servicemen of the Japanese Army Charged with Manufacturing and Employing Biological Weapons. 1950, Moscow: Foreign Languages Publishing House.
5. Anonymous, Laetrile: The political success of a scientific failure. Consumer Reports, 1977. 42: 444-117.
6. Anonymous, Ethical watchdog's just desserts. Nature, 1991. 349(6308): 355.
7. Anonymous, Fisher absolved of scientific misconduct. Science, 1997. 275(5305): 1407.
8. Anonymous, Case Summary Paul H. Kornak. Federal Register, 2006. 71(37): 9555-9556.
9. Anonymous. Potti Investigation. Journal retracts paper based on withdrawn work. The Chronicle. The Independent Daily of Duke University. 2011 January 31.
10. Anonymous, Potti Investigation. Report shows review of Potti research was based on flawed data. The Chronicle. The Independent Daily of Duke University, 2011.
11. Anonymous. Potti Investigation. Report shows review of Potti research was based on flawed data. The Chronicle. The Independent Daily of Duke University. 2011 January 12.
12. Anonymous. Fifth Potti paper sees retraction. The Chronicle. The Independent Daily of Duke University. 2011 August 22.
13. Anonymous, IOM. Omics committee offers blueprint to prevent other Duke-style disasters. The Cancer Letter, 2012.
14. Avalos Urzúa G, Bores LD, and Livecchi JT, Pneumatic trabeculoplasty. A new method to treat primary open-angle glaucoma and reduce the number of concomitant medications. Annals of Ophthalmology., 2005. 37(1): 37-46.
15. Baggerly KA and Coombes KR, Deriving chemosensitivity from cell lines: forensic bioinformatics and reproducible research in high-throughput biology. Annals of Applied Statistics, 2009. 3(4): 1309-1334.
16. Barinaga M, NIMH assigns blame for tainted studies. Science, 1989. 245(4920): 812.

17. Barrett S. Pneumatic trabeculoplasty (PNT) has not been proven safe and effective for treating glaucoma. Quackwatch 2006; Available from: www.quackwatch.com.
18. Bethell HJ and Nixon PG, Rehabilitation after myocardial infarction. The Lancet, 1972. 1(7740): 49.
19. Bezwoda WR, Seymour L, and Dansey RD, High-dose chemotherapy with hematopoietic rescue as primary treatment for metastatic breast cancer: a randomized trial.[Retraction in Journal of Clinical Oncology. 2001 June 1;19(11):2973; PMID: 11387377]. Journal of Clinical Oncology, 1995. 13(10): 2483-2489.
20. Bonnefoi H, Potti A, Delorenzi M, et al., Validation of gene signatures that predict the response of breast cancer to neoadjuvant chemotherapy: a substudy of the EORTC 10994/BIG 00-01 clinical trial.[Retraction in Bonnefoi H, Potti A, Delorenzi M, Mauriac L, Campone M, Tubiana-Hulin M, Petit T, Rouanet P, Jassem J, Blot E, Becette V, Farmer P, Andre S, Acharya CR, Mukherjee S, Cameron D, Bergh J, Nevins JR, Iggo RD. Lancet Oncology. 2011 February;12(2):116; PMID: 21277543]. Lancet Oncology, 2007. 8(12): 1071-1078.
21. Bonnefoi H, Potti A, Delorenzi M, et al., Retraction—Validation of gene signatures that predict the response of breast cancer to neoadjuvant chemotherapy: a substudy of the EORTC 10994/BIG 00-01 clinical trial. [Retraction of Bonnefoi H, Potti A, Delorenzi M, Mauriac L, Campone M, Tubiana-Hulin M, Petit T, Rouanet P, Jassem J, Blot E, Becette V, Farmer P, Andre S, Acharya CR, Mukherjee S, Cameron D, Bergh J, Nevins JR, Iggo RD. The Lancet Oncology. 2007 December;8(12):1071-1078; PMID: 18024211]. The Lancet Oncology, 2011. 12(2): 116.
22. Borrell B, A medical Madoff: anesthesiologist faked data in 21 studies. A pioneering anesthesiologist has been implicated in a massive research fraud that has altered the way millions of patients are treated for pain during and after orthopedic surgeries. Scientific American, 2009.
23. Briggs MH and Briggs M, Preliminary studies on metabolic effects of a continuous-dose, progestogen-only, oral contraceptive. African Journal of Medical Sciences, 1972. 3(2): 105-115.
24. Briggs MH and Briggs M, Molecular biology and oral contraception. New Zealand Medical Journal, 1976. 83(562): 257-261.
25. Briggs MH and Briggs M, Plasma lipoprotein changes during oral contraception. Current Medical Research and Opinion, 1979. 6(4): 249-254.
26. Broad WJ, Straus defends himself in Boston. Science, 1981. 212(4501): 1367-1369.
27. Broad WJ, Researcher denied future U.S. funds. Science, 1982. 216(4550): 1081.

28. Bucci MG, Centofanti M, Oddone F, et al., Pilot study to evaluate the efficacy and safety of pneum(atic trabeculopasty in glaucoma and ocular hypertension. European Journal of Ophthalmology, 2005. 15(3): 347-352.
29. Carruthers M, Nixon P, and Murray A, Letter: Safe sport. The Lancet, 1975. 1(7904): 447.
30. Catchpole LA and Nixon PG, A hand clamp for plastic tubing. The Lancet, 1959. 2(7098): 329.
31. Catchpole LA and Nixon PG, A labour-saving plate for the Melrose heart-and-lung machine oxygenator. The Lancet, 1959. 2(7098): 329.
32. Conway AV, Freeman LJ, and Nixon PG, Hypnotic examination of trigger factors in the hyperventilation syndrome. American Journal of Clinical Hypnosis, 1988. 30(4): 296-304.
33. Court C and Dillner L, Obstetrician suspended after research inquiry. British Medical Journal, 1994. 309(6967): 1459.
34. Couzin J, Cancer research. Fake data, but could the idea still be right? Science, 2006. 313(5784): 154.
35. Crewdson J. Fraud in breast cancer study. Chicago Tribune. 1994 March 13.
36. Curfman GD, Morrissey S, and Drazen JM, Expression of concern: Sudbo J et al. DNA content as a prognostic marker in patients with oral leukoplakia. New England Journal of Medicine 2001;344:1270-1278 and Sudbo J et al. the influence of resection and aneuploidy on mortality in oral leukoplakia. New England Journal of Medicine 2004;350:1405-1413. New England Journal of Medicine, 2006. 354(6): 638.
37. Cuthbert M, Freedom From Cancer: The Amazing Story of Vitamin B17, or Laetrile. 1976, California: Seal Beach Press.
38. Deer B. The pill: professor's safety tests were faked. The Sunday Times. 1986 September 28.
39. Deer B. Research reveals birth bill risk for 2M British women. The Sunday Times. 1988 September 18.
40. Deer B. Focus: MMR the truth behind the crisis. The Sunday Times. 2004 February 22.
41. Deer B. MMR doctor given legal aid thousands. The Sunday Times. 2006 December 31.
42. Doherty T and Tracer Z. Cancer researcher resigns, accepts blame for mistakes. The Chronicle. The Independent Daily of Duke University. 2010 November 22.
43. Dyer C, Cardiologist admits research misconduct. British Medical Journal, 1997. 314(7093): 1501.
44. Dyer C, Doctor admits research fraud. British Medical Journal, 1998. 316(7132): 645.
45. Dyer O, Consultant struck off for fraudulent claims. British Medical Journal, 1995. 310(6994): 1554-1555.

46. Ellison NM, Byar DP, and Newell GR, Special report on Laetrile: the NCI Laetrile Review. Results of the National Cancer Institute's retrospective Laetrile analysis. New England Journal of Medicine, 1978. 299(10): 549-552.
47. Fine SL and Hawkins BS, The investigators' perspective on the collaborative ocular melanoma study. JAMA Ophthalmology, 2007. 125(7): 968-971.
48. Fisher B, Bauer M, Margolese R, et al., Five-year results of a randomized clinical trial comparing total mastectomy and segmental mastectomy with or without radiation in the treatment of breast cancer. New England Journal of Medicine, 1985. 312(11): 665-673.
49. Fisher B, Redmond C, Fisher ER, et al., Ten-year results of a randomized clinical trial comparing radical mastectomy and total mastectomy with or without radiation. New England Journal of Medicine, 1985. 312(11): 674-681.
50. Flexner A, Medical Education in the United States and Canada. A Report to the Carnegie Foundation for the Advancement of Teaching. Bulletin No.4. 1910, New York City: The Carnegie Foundation for the Advancement of Teaching. pp.346.
51. Freeman LJ, Conway A, and Nixon PG, Physiological responses to psychological challenge under hypnosis in patients considered to have the hyperventilation syndrome: implications for diagnosis and therapy. Journal of the Royal Society of Medicine, 1986. 79(2): 76-83.
52. Freeman LJ and Nixon PG, Pain and the heart: discussion paper. Journal of the Royal Society of Medicine, 1988. 81(2): 100-102.
53. Garfield BD, Denial of hearing; final debarment order, in Federal Register1997. 15713.
54. Garfinkel B, CF No.400 (http://www.hhs.gov/dab/decisions/cr-400, htm). 1995.
55. Glueck CJ, Laskarzewski P, Mellies MJ, et al., Letter to the editor. Dr. Glueck and coauthors' letter of withdrawal. [Retraction of Glueck CJ, Mellies MJ, Dine M, Perry T, Laskarzewski P. Pediatrics 1986 August 78(2):338-348; PMID: 3226270]. Pediatrics. 80 (5):766 1987.
56. Grady D. Breast cancer researcher admits falsifying data. The New York Times. 2000 February 5.
57. Hagmann M, Scientific misconduct. Cancer researcher sacked for alleged fraud. Science, 2000. 287(5460): 1901-1902.
58. Harris G, Carey B, and Roberts J. Psychiatrists, children an drug industry role. The New York Times. 2007 May 10.
59. Harris SH, Factories of Death: Japanese Biological Warfare 1932-1945 and the American Cover-Up. 1994, New York: Routledge a member of the Taylor and Francis Group. pp. 297.
60. Havele S. Duke repays $437k of ACS research grant. The Chronicle. The Independent Daily of Duke University. 2010 December 7.

61. Henkel J, Psychiatrist sentenced for research fraud. University of Minnesota child psychiatrist Barry Garfinkel. FDA Consumer, 1994. 28(3).
62. Hill KP, Ross JS, Egilman DS, et al., The ADVANTAGE seeding trial: a review of internal document. Annals of Internal Medicine, 2008. 149(4): 251-258.
63. Holden C, NIH moves to debar cholesterol researcher. Science, 1987. 237(4816): 718-719.
64. Horton R, MMR-responding to retraction. The Lancet, 2004. 363(9417): 1328.
65. Horton R, MMR-responding to retraction. Editor's reply. The Lancet, 2004. 363(9417): 1328.
66. Horton R, The lessons of MMR. The Lancet, 2004. 363(9411): 747-749.
67. Hsu DS, Balakumaran BS, Acharya CR, et al., Pharmacogenomic strategies provide a rational approach to the treatment of cisplatin-resistant patients with advanced cancer.[Erratum appears in Journal of Clinical Oncology 2010 June 1;28(16):2805]. Journal of Clinical Oncology, 2007. 25(28): 4350-4357.
68. Ikram H and Nixon PG, Cardiac puncture. The Lancet, 1964. 2(7370): 1178.
69. Kaiser J, Fisher wins $2.75 million settlement. Science, 1997. 277(5331): 1425.
70. Keller MB, McCulloch JP, Klein DN, et al., A comparison of Nefazodone, the cognitive behavioral-analysis system of psychotherapy, and their combination for the treatment of chronic depression. New England Journal of Medicine, 2000. 342(20): 1462-1470.
71. King JC, Rosen SD, and Nixon PG, Failure of perception of hypocapnia: physiological and clinical implications. Journal of the Royal Society of Medicine, 1990. 83(12): 765-767.
72. Kolata G. U.S. is considering disciplinary action in Alzheimer's study. The New York Times. 1988 February 12.
73. Lisse Jeffrey R, Perlman M, Johansson G, et al., Gastrointestinal tolerability and effectiveness of rofecoxib versus naproxen in the treatment of osteoarthritis: a randomized, controlled trial. Annals of Internal Medicine, 2003. 139(7): 539-546.
74. Lock S, Misconduct in medical research: does it exist in Britain? British Medical Journal, 1988. 297(6662): 1531-1535.
75. Lock S, Lessons from the Pearce affair:handling scientific fraud. British Medical Journal, 1995. 310(6974): 1547-1548.
76. Lum LC and Nixon PG, EndorphIns, I presume—or hyperventilation? The Lancet, 1981. 1(8212): 160.
77. Lundberg GD and Flanagin A, New requirements for authors: signed statements of authorship responsibility and financial disclosure. Journal of the American Medical Association, 1989. 262(14): 2003-2004.

78. Macaulay AJ and Nixon PG, Parascending: a safer alternative to hang gliding. British Medical Journal, 1977. 1(6072): 1352.
79. Marshall E, Tamoxifen: hanging in the balance. Science, 1994. 264(5165): 1524-1567.
80. Marshall E, Panel:extensive Sudbø fraud. Science, 2006. 313(5783): 29.
81. Murch SH, Anthony A, Casson DH, et al., Retraction of an interpretation. The Lancet, 2004. 363(9411): 750.
82. Nixon P and Bethell HJ, Letter: Presentation of myocardial infarction. British Medical Journal, 1974. 1(5903): 326.
83. Nixon P, Letter: doubts and lignocaine. British Medical Journal, 1975. 2(5962): 87-88.
84. Nixon P, Predicting illness by numbers. British Medical Journal, 1976. 2(6035): 586-587.
85. Nixon P and Weinstock N, Return to work after coronary artery surgery for angina. British Medical Journal, 1979. 1(6158): 265.
86. Nixon P, Haemodynamic monitoring in the intensive care unit. British Medical Journal, 1980. 280(6224): 1187.
87. Nixon PG, Recurrent myxoedema and goitre attributed to potassium iodide. British Medical Journal, 1957. 1(5021): 748-749.
88. Nixon PG, A simple flow-meter and safety device for the extracorporeal circulation. The Lancet, 1959. 2(7107): 830.
89. Nixon PG, A pump for use in open heart surgery. The Lancet, 1959. 1(7082): 1074-1075.
90. Nixon PG, Grimshaw VA, and Wooler GH, Clinical observations on vasomotor reflexes in relation to blood-volume in open-heart surgery. The Lancet, 1960. 2(7166): 1429-1431.
91. Nixon PG and Snow HM, A multipurpose cardiac catheter. The Lancet, 1960. 2(7150): 582.
92. Nixon PG, The arterial pulse in successful closed-chest cardiac massage. The Lancet, 1961. 2(7207): 844-846.
93. Nixon PG, A training device for extracorporeal circulation. The Lancet, 1961. 2(7209): 966.
94. Nixon PG, Hepburn F, and Ikram H, Simultaneous recording of heart pulses and sounds. British Medical Journal, 1964. 1(5391): 1169.
95. Nixon PG, Ikram H, and Morton S, Infusion of dextrose solution in cardiogenic shock. The Lancet, 1966. 1(7446): 1077-1079.
96. Nixon PG, Pulmonary oedema with low left ventricular diastolic pressure in acute myocardial infarction. The Lancet, 1968. 2(7560): 146-147.
97. Nixon PG, Taylor DJ, and Morton SD, Left ventricular diastolic pressure in cardiogenic shock treated by dextrose infusion and adrenaline. The Lancet, 1968. 1(7554): 1230-1232.
98. Nixon PG, Taylor DJ, Morton SD, et al., A sleep regimen for acute myocardial infarction. The Lancet, 1968. 1(7545): 726-728.

99. Nixon PG and Bethell HJ, Coronary deaths—how unexpected? British Medical Journal, 1971. 4(5785): 486-487.
100. Nixon PG, Letter: The heart during sleep. British Medical Journal, 1974. 3(5924): 174.
101. Nixon PG, Letter: Impending heart-attacks. The Lancet, 1976. 1(7949): 34.
102. Nixon PG and Dighton DH, Letter: Meditation or methyldopa? British Medical Journal, 1976. 2(6034): 525.
103. Nixon PG, Diagnosis of chest pain. The Lancet, 1977. 1(8012): 646.
104. Nixon PG, Prognosis after myocardial infarction. British Medical Journal, 1979. 2(6204): 1583.
105. Nixon PG, Hypnosis. British Medical Journal, 1979. 1(6172): 1212-1213.
106. Nixon PG, "Total allergy syndrome" or fluctuating hypocarbia? The Lancet, 1982. 1(8268): 404.
107. Nixon PG and Pugh S, Coronary artery bypass surgery. The Lancet, 1983. 1(8318): 242.
108. Nixon PG, Freeman LJ, Nixon S, et al., Problems of patients' dependency on doctors. Journal of the Royal Society of Medicine, 1987. 80(6): 394-395.
109. Nixon PG and Freeman LJ, The 'think test': a further technique to elicit hyperventilation. Journal of the Royal Society of Medicine, 1988. 81(5): 277-279.
110. Nixon PG and King JC, Overpromotion of the dietary theory of coronary disorders. The Lancet, 1989. 2(8675): 1334-1335.
111. Nixon PG, The broken heart-counteraction by SABRES. Journal of the Royal Society of Medicine, 1993. 86(8): 468-471.
112. Pearce JM and Hamid RI, Randomised controlled trial of the use of human chorionic gonadotrophin in recurrent miscarriage associated with polycystic ovaries.[Retraction in British Journal of Obstetics and Gynaecology. 1995 November;102(11):853; PMID: 9063864]. British Journal of Obstetrics and Gynaecology, 1994. 101(8): 685-688.
113. Pearce JM, Manyonda IT, and Chamberlain GV, Term delivery after intrauterine relocation of an ectopic pregnancy.[Retraction in British Journal of Obstetrics and Gynaecology. 1995 November;102(11).853; PMID: 9063864]. British Journal of Obstetrics and Gynaecology, 1994. 101(8): 716-717.
114. Pellegrino E, The Nazi doctors and Nuremberg: Some moral lessons revisited. American College of Physicians, 1997. 127(4): 307-308.
115. Perry S. Thousands of doctors get payments from drug firms, investigation reveals. Is your doctor one among them? MINNPOST. 2010 October 19.
116. Pincock S. Lancet study faked. Investigation to probe all research by scientist accused of fabricating results from 900 reserach participants. The-Scientist.com (http://www.thescientist.com/news/display/22952). 2006 January 16.

117. Pinney S, Freeman LJ, and Nixon PG, Role of the nurse counsellor in managing patients with the hyperventilation syndrome. Journal of the Royal Society of Medicine, 1987. 80(4): 216-218.
118. Potti A, Bild A, Dressman HK, et al., Gene-expression patterns predict phenotypes of immune-mediated thrombosis.[Retraction in Blood. 2011 October 20;118(16):4497; PMID: 21856862]. Blood. 107(4): 1391-1396.
119. Potti A and Carson PJ, Comparison of rectal and infrared ear temperatures in older hospital inpatients. Journal of the American Geriatrics Society. 48(8): 1023-1044.
120. Potti A, Mukherjee S, Petersen R, et al., Retraction: A genomic strategy to refine prognosis in early-stage non-small-cell lung cancer. New England Journal Medicime 2006;355:570-80.[Retraction of Potti A, Mukherjee S, Petersen R, Dressman HK, Bild A, Koontz J, Kratzke R, Watson MA, Kelley M, Ginsburg GS, West M, Harpole DH Jr, Nevins JR. New England Journal Medicine. 2006 August 10;355(6):570-80; PMID: 16899777]. New England Journal of Medicine. 364(12): 1176.
121. Potti A, Dressman HK, Bild A, et al., Genomic signatures to guide the use of chemotherapeutics.[Erratum appears in Nature Medicine. 2008 August;14(8):889].[Erratum appears in Nature Medicine. 2007 November 13(11):1388]. Nature Medicine, 2006. 12(11): 1294-1300.
122. Potti A, Dressman HK, Bild A, et al., Genomic signatures to guide the use of chemotherapeutics.[Erratum appears in Nature Medicine. 2007 November;13(11):1388].[Erratum appears in Nature Medicine. 2008 August;14(8):889].[Retraction in Potti A, Dressman HK, Bild A, Riedel RF, Chan G, Sayer R, Cragun J, Cottrill H, Kelley MJ, Petersen R, Harpole D, Marks J, Berchuck A, Ginsburg GS, Febbo P, Lancaster J, Nevins JR. Nature Medicine. 2011 January;17(1):135; PMID: 21217686]. Nature Medicine, 2006. 12(11): 1294-1300.
123. Potti A, Mukherjee S, Petersen R, et al., A genomic strategy to refine prognosis in early-stage non-small-cell lung cancer.[Erratum appears in New England Journal of Medicine. 2007 January 11;356(2):201-2].[Retraction in Potti A, Mukherjee S, Petersen R, Dressman HK, Bild A, Koontz J, Kratzke R, Watson MA, Kelley M, Ginsburg GS, West M, Harpole DH Jr, Nevins JR. New England Journal of Medicine. 2011 March 24;364(12):1176; PMID: 21366430]. New England Journal of Medicine, 2006. 355(6): 570-580.
124. Potti A, Pharmacogenomic strategies provide a rational approach to the treatment of cisplatin-resistant patients with advanced cancer. Journal of Clinical Oncology, 2010.
125. Ramar K, Hauer D, Potti A, et al., Ecthyma gangrenosum and chronic lymphocytic leukaemia.[Erratum appears in Lancet Infect Dis. 2003 Mar;3(3):177]. The Lancet Infectious Diseases, 2003. 3(2): 113.

126. Rapoport DC, Terrorism and Weapons of the Apocalypse. In Ludes, James M., Sokoloski, Henry (Eds). Twenty-First Century Weapons Proliferation: Are We Ready ? 2001, Portland Oregon: Frank Cass Publishers.
127. Redman AS, Dealing with COI New England Journal of Medicine, 1984. 310(18): 1182-1183.
128. Redman AS, New "information for authors"—and readers. New England Journal of Medicine, 1990. 323(1): 56.
129. Reich E, Samuel Cancer trial errors revealed. University officials admit data withheld from review panel before misconduct charges arose. Nature, 2011. 469(7329).
130. Reuben SS, Connelly NR, and Maciolek H, Postoperative analgesia with controlled-release oxycodone for outpatient anterior cruciate ligament surgery.[Retraction in Shafer SL. Anesthesia and Analgesia. 2009 April;108(4):1350; PMID: 19299812]. Anesthesia and Analgesia, 1999. 88(6): 1286-1291.
131. Reuben SS and Reuben JP, Brachial plexus anesthesia with verapamil and/ or morphine.[Retraction in Shafer SL. Anesthesia and Analgesia 2009 April;108(4):1350; PMID: 19299812]. Anesthesia and Analgesia, 2000. 91(2): 379-383.
132. Reuben SS, Fingeroth R, Krushell R, et al., Evaluation of the safety and efficacy of the perioperative administration of rofecoxib for total knee arthroplasty.[Retraction in Journal of Arthroplasty. 2010 January ;25(1):172; PMID: 20117681]. Journal of Arthroplasty, 2002. 17(1): 26-31.
133. Reuben SS, Makari-Judson G, and Lurie SD, Evaluation of efficacy of the perioperative administration of venlafaxine XR in the prevention of postmastectomy pain syndrome.[Retraction in J Pain Symptom Manage. 2009 May;37(5):941; PMID: 19422087]. Journal of Pain and Symptom Management, 2004. 27(2): 133-139.
134. Reuben SS, Rosenthal EA, Steinberg RB, et al., Surgery on the affected upper extremity of patients with a history of complex regional pain syndrome: the use of intravenous regional anesthesia with clonidine.[Retraction in Journal of Clinical Anesthesia. 2009 May; 21(3):237; PMID: 19475772]. Journal of Clinical Anesthesia, 2004. 16(7): 517-522.
135. Reuben SS and Ekman EF, The effect of cyclooxygenase-2 inhibition on analgesia and spinal fusion.[Retraction in Heckman JD. Journal of Bone and Joint Surguery—American Volume. 2009 April;91(4):965; PMID: 19339584]. Journal of Bone and Joint Surgery—American Volume, 2005. 87(3): 536-542.
136. Reuben SS, Interscalene block superior to general anesthesia. Anesthesiology, 2006. 104(1): 207; author reply 208-209.
137. Reuben SS, Buvanendran A, Kroin JS, et al., The analgesic efficacy of celecoxib, pregabalin, and their combination for spinal fusion surgery.

[Retraction in Shafer SL. Anesthesia and Analgesia. 2009 April;108(4):1350; PMID: 19299812]. Anesthesia and Analgesia, 2006. 103(5): 1271-1277.
138. Reuben SS, Buvanendran A, Kroin JS, et al., Postoperative modulation of central nervous system prostaglandin E2 by cyclooxygenase inhibitors after vascular surgery.[Retraction in Anesthesiology. 2009 March;110(3):689; PMID: 19218855]. Anesthesiology, 2006. 104(3): 411-416.
139. Reuben SS, Ekman EF, Raghunathan K, et al., The effect of cyclooxygenase-2 inhibition on acute and chronic donor-site pain after spinal-fusion surgery. [Retraction in Neal JM. Regional Anesthesia and Pain Medicine. 2009 March-April;34(2):184; PMID: 19288610]. Regional Anesthesia and Pain Medicine, 2006. 31(1): 6-13.
140. Reuben SS, Pristas R, Dixon D, et al., The incidence of complex regional pain syndrome after fasciectomy for Dupuytren's contracture: a prospective observational study of four anesthetic techniques.[Retraction in Shafer SL.Anesthesia and Analgesia. 2009 April;108(4):1350; PMID: 19299812]. Anesthesia and Analgesia, 2006. 102(2): 499-503.
141. Reuben SS and Ekman EF, The effect of initiating a preventive multimodal analgesic regimen on long-term patient outcomes for outpatient anterior cruciate ligament reconstruction surgery.[Retraction in Shafer SL. Anesthesia and Analgesia 2009 April;108(4):1350; PMID: 19299812]. Anesthesia and Analgesia, 2007. 105(1): 228-232.
142. Reuben SS, Ekman EF, and Charron D, Evaluating the analgesic efficacy of administering celecoxib as a component of multimodal analgesia for outpatient anterior cruciate ligament reconstruction surgery.[Retraction in Shafer SL. Anesthesia and Analgesia 2009 April;108(4):1350; PMID: 19299812]. Anesthesia and Analgesia, 2007. 105(1): 222-227.
143. Reuben SS, Buvenandran A, Katz B, et al., A prospective randomized trial on the role of perioperative celecoxib administration for total knee arthroplasty: improving clinical outcomes.[Retraction in Shafer SL. Anesthesia and Analgesia. 2009 April;108(4):1350; PMID: 19299812]. Anesthesia and Analgesia, 2008. 106(4): 1258-1264.
144. Reuben SS, Steinberg RB, Maciolek H, et al., An evaluation of the analgesic efficacy of intravenous regional anesthesia with lidocaine and ketorolac using a forearm versus upper arm tourniquet.[Retraction in Shafer SL. Anesthesia and Analgesia. 2009 April;108(4):1350; PMID: 19299812]. Anesthesia and Analgesia, 2009. 95(2): 457-460.
145. Rosen SD, King JC, and Nixon PG, Magnetic resonance muscle studies. Journal of the Royal Society of Medicine, 1988. 81(11): 676-677.
146. Rosen SD, King JC, Wilkinson JB, et al., Is chronic fatigue syndrome synonymous with effort syndrome? Journal of the Royal Society of Medicine, 1990. 83(12): 761-764.

147. Rosen SD, King JC, and Nixon PG, Hyperventilation in patients who have sustained myocardial infarction after a work injury.[Retraction in Rosen SD, King JC, Nixon PG. Journal of the Royal Society of Medicine. 1997 July;90(7):416; PMID: 9303950]. Journal of the Royal Society of Medicine, 1994. 87(5): 268-271.
148. Rosen SD, King JC, and Nixon PG, Hyperventilation in patients who have sustained myocardial infarction after a work injury.[Retraction of Rosen SD, King JC, Nixon PG. Journal of the Royal Society of Medicine. 1994 May;87(5):268-71; PMID: 8207722]. Journal of the Royal Society of Medicine, 1997. 90(7): 416.
149. Rouse A, Autism, inflammatory bowel disease, and MMR vaccine. The Lancet, 1998. 351(9112): 1356.
150. Rubin R. Pfizer fined $2.3B for illegal marketing. USA Today. 2009 September 3.
151. Rushton T. Duke nears 'final stages' of naming external review body in Potti case. The Chronicle. The Independent Daily of Duke University. 2010 August 23.
152. Schmit J. Drugmaker admitted fraud, but sales flourish. USA Today. 2004 August 17.
153. Shafer SL, Tattered threads. Anesthesia and Analgesia, 2009. 108(5): 1361-1364.
154. Smith. Renewed attack on pill researcher. The National Times 1986 July 25-31.
155. Spector J. Patients sue Duke over Potti trials. The Chronicle. The Independent Daily of Duke University. 2011 September 9.
156. Stadtmauer EA, O'Neill A, Goldstein LJ, et al., Conventional-dose chemotherapy compared with high-dose chemotherapy plus autologous hematopoietic stem-cell transplantation for metastatic breast cancer. Philadelphia Bone Marrow Transplant Group. New England Journal of Medicine, 2000. 342(15): 1069-1076.
157. Stone R, NIH confronts new fraud allegations. Science, 1994. 264(5159): 647.
158. Sudbo J, [DNA ploidy analysis—a possibility for early identification of patient with risk of oral cancer]. Lakartidningen, 2001. 98(45): 4980-4984.
159. Sudbo J, Kildal W, Risberg B, et al., DNA content as a prognostic marker in patients with oral leukoplakia.[Retraction in Curfman GD, Morrissey S, Drazen JM. New England Journal of Medicine. 2006 November 2;355(18):1927; PMID: 17079770]. New England Journal of Medicine, 2001. 344(17): 1270-1278.
160. Sudbo J, Lippman SM, Lee JJ, et al., The influence of resection and aneuploidy on mortality in oral leukoplakia.[Retraction in Curfman GD, Morrissey S, Drazen JM. New England Journal of Medicine. 2006

November 2;355(18):1927; PMID: 17079770]. New England Journal of Medicine, 2004. 350(14): 1405-1413.
161. Sudbo J, Lee JJ, Lippman SM, et al., Non-steroidal anti-inflammatory drugs and the risk of oral cancer: a nested case-control study.[Retraction in Horton R. The Lancet. 2006 February 4;367(9508):382; PMID: 16458751]. The Lancet, 2005. 366(9494): 1359-1366.
162. Summers WK, Majovski LV, Marsh GM, et al., Oral tetrahydroaminoacridine in long-term treatment of senile dementia, Alzheimer type. New England Journal of Medicine, 1986. 315(20): 1241-1245.
163. Taylor J, The Life and Extraordinary History of Chevalier John Taylor. 1761, London: M, Cooper.
164. Tracer Z. Nevins requests retraction of key genomics research paper. The Chronicle. The Independent Daily of Duke University. 2010 November 22.
165. Tracer Z. Scientists call for experimental reproducibility. The Chronicle. The Independent Daily of Duke University. 2010 October 6.
166. Tracer Z. Scientists call for experimental reproducibility The Chronicle. The Independent Daily of Duke University. 2010 October 6.
167. Tracer Z. Health care companies cut ties with Potti. The Chronicle. The Independent Daily of Duke University. 2010 September 14.
168. Tracer Z. NC law firm investigates Potti trials. The Chronicle. The Independent Daily of Duke University. 2011 February 11.
169. Tsuchiya T, The Imperial Japanese Experiments in China, In The Oxford Textbook of Clinical Research Ethics, edited by Ezekiel J. Emanuel, Christine C. Grady, Robert A. Crouch, Reidar K. Lie, Franklin G. Miller, and David D. Wendler. 2011, Oxford, England: Oxford University Press.
170. Vickers A and Christos P, Re: "Bezwoda: evidence of fabrication in original article". Journal of Clinical Oncology, 2000. 18(22): 3875.
171. Wakefield A, A statement by Dr Andrew Wakefield. The Lancet, 2004. 363(9411): 823-824.
172. Wakefield AJ and Gordon EM, A huge renal cyst presenting in childhood. A case report and review of the literature. Journal of the Royal Society of Medicine, 1989. 82(7): 443-445.
173. Wakefield AJ, Sawyerr AM, Dhillon AP, et al., Pathogenesis of Crohn's disease: multifocal gastrointestinal infarction. The Lancet, 1989. 2(8671): 1057-1062.
174. Wakefield AJ, Sankey EA, Dhillon AP, et al., Granulomatous vasculitis in Crohn's disease.[Erratum appears in Gastroenterology 1991 August;101(2):595]. Gastroenterology, 1991. 100(5 Pt 1): 1279-1287.
175. Wakefield AJ, Sawyerr AM, Hudson M, et al., Smoking, the oral contraceptive pill, and Crohn's disease. Digestive Diseases and Sciences, 1991. 36(8): 1147-1150.
176. Wakefield AJ, Fox JD, Sawyerr AM, et al., Detection of herpesvirus DNA in the large intestine of patients with ulcerative colitis and Crohn's disease

using the nested polymerase chain reaction. Journal of Medical Virology, 1992. 38(3): 183-190.
177. Wakefield AJ, Pittilo RM, Sim R, et al., Evidence of persistent measles virus infection in Crohn's disease. Journal of Medical Virology, 1993. 39(4): 345-353.
178. Wakefield AJ, More LJ, Difford J, et al., Immunohistochemical study of vascular injury in acute multiple sclerosis. Journal of Clinical Pathology, 1994. 47(2): 129-133.
179. Wakefield AJ, Ekbom A, Dhillon AP, et al., Crohn's disease: pathogenesis and persistent measles virus infection. Gastroenterology, 1995. 108(3): 911-916.
180. Wakefield AJ, Hudson M, and Pounder RE, Leukocyte-endothelial cell interactions: potential targets for mesalazine in Crohn's disease. Advances in Experimental Medicine & Biology, 1995. 371B: 1307-1311.
181. Wakefield AJ and Pounder RE, Measles virus in Crohn's disease. The Lancet, 1995. 345(8950): 660.
182. Wakefield AJ, Sim R, Akbar AN, et al., In situ immune responses in Crohn's disease: a comparison with acute and persistent measles virus infection. Journal of Medical Virology, 1997. 51(2): 90-100.
183. Wakefield AJ, Autism, inflammatory bowel disease, and MMR vaccine. The Lancet, 1998. 351(9112): 1356.
184. Wakefield AJ, Murch SH, Anthony A, et al., Ileal-lymphoid-nodular hyperplasia, non-specific colitis, and pervasive developmental disorder in children.[Retraction in The Lancet. 2010 February 6;375(9713):445; PMID: 20137807]. The Lancet, 1998. 351(9103): 637-641.
185. Wakefield AJ, MMR vaccination and autism. The Lancet, 1999. 354(9182): 949-950.
186. Wakefield AJ and Montgomery SM, Autism, viral infection and measles-mumps-rubella vaccination. Israel Medical Association Journal, 1999. 1(3): 183-187.
187. Wakefield AJ, Montgomery SM, and Pounder RE, Crohn's disease: the case for measles virus. Italian Journal of Gastroenterology & Hepatology, 1999. 31(3): 247-254.
188. Wakefield AJ, Anthony A, Murch SH, et al., Enterocolitis in children with developmental disorders.[Retraction in Am J Gastroenterol. 2010 May;105(5):1214; PMID: 20445528]. American Journal of Gastroenterology, 2000. 95(9): 2285-2295.
189. Wakefield AJ and Montgomery SM, Measles virus as a risk for inflammatory bowel disease: an unusually tolerant approach. American Journal of Gastroenterology, 2000. 95(6): 1389-1392.
190. Wakefield AJ and Montgomery SM, Measles, mumps, rubella vaccine: through a glass, darkly. Adverse Drug Reactions & Toxicological Reviews, 2000. 19(4): 265-283; discussion 284-292.

191. Wakefield AJ and Montgomery SM, Immunohistochemical analysis of measles related antigen in IBD. Gut, 2001. 48(1): 136-137.
192. Wakefield AJ, Enterocolitis, autism and measles virus. Molecular Psychiatry, 2002. 7 Suppl 2: S44-46.
193. Wakefield AJ, The gut-brain axis in childhood developmental disorders. Journal of Pediatric Gastroenterology & Nutrition, 2002. 34 Suppl 1: S14-17.
194. Wakefield AJ, Puleston JM, Montgomery SM, et al., Review article: the concept of entero-colonic encephalopathy, autism and opioid receptor ligands. Alimentary Pharmacology & Therapeutics, 2002. 16(4): 663-674.
195. Wakefield AJ, Measles, mumps, and rubella vaccination and autism. New England Journal of Medicine, 2003. 348(10): 951-954; author reply 951-954.
196. Wakefield AJ, Harvey P, and Linnell J, MMR-responding to the retraction. The Lancet, 2004. 363(9417): 1327-1328.
197. Wakefield AJ, Harvey P, and Linnell J, MMR—responding to retraction. The Lancet, 2004. 363(9417): 1327-1328; discussion 1328.
198. Wakefield AJ, Ashwood P, Limb K, et al., The significance of ileo-colonic lymphoid nodular hyperplasia in children with autistic spectrum disorder. European Journal of Gastroenterology & Hepatology, 2005. 17(8): 827-836.
199. Wakefield AJ, Stott C, and Krigsman A, Getting it wrong. Archives of Disease in Childhood, 2008. 93(10): 905-906; author reply 906-907.
200. Weiss RB, Rifkin RM, Stewart FM, et al., High-dose chemotherapy for high-risk primary breast cancer: an on-site review of the Bezwoda study. The Lancet, 2000. 355(9208): 999-1003.
201. Wetherill JH and Nixon PG, Spontaneous cessation of ventricular fibrillation during external cardiac massage. The Lancet, 1962. 1(7237): 993-994.
202. White PF, Kehler H, and Liu S, Perioperative analgesia: what do we know?. Anesthesia and Analgesia, 2009. 108(5): 1364-1367.
203. Williams MH and Bowie C, Evidence of unmet need in the care of severely physically disabled adults. British Medical Journal, 1993. 306(6870): 95-98.
204. Williams P, Unit 731: Japan's Secret Biological Warfare in World War II. 1989, New York: Free Press.

CHAPTER 10

Fraudulent Finances in Research

In the past research grants were awarded to institutions, just as they are today, and the principal investigator (PI) performed the research, which did not necessarily relate to what was described in the grant application. A major difference between grants and contracts was conceived. For a long time researchers and some government officials believed that a research grant was a gift to the recipient like a prize for past work and that minimal strings were attached as to how the funds were to be used. With grants the investigator was free to carry out whatever research the grantee wished to do after the award was made. Indeed the proposed research in the application had often already been performed so that there would be a guaranteed positive result for the progress report and ultimately for a renewal application to continue the research.

In sharp contrast to grants, contracts provided money to perform a specific piece of research and there was often no vigorous peer-review equivalent to what grant applications went through, but there was no flexibility about how the funds could be used. Business oriented officers at the National Institutes of Health (NIH) did not share this viewpoint.

Clinician-scientists in clinical departments would be paid from research grants, but would spend much effort on clinical activities as well and would often receive bonuses from income generated from the clinical activities. When federal research funds were abundant in research institutions in the 1950s and 1960s departments often had more money than needed. It became a common practice for departments to sometimes pool research funds derived from different grants to support a wide variety of funded and unfunded research projects as well as other activities. The research by those without grants was funded by the excess money coming from colleagues skilled in grantsmanship. At that time investigators typically applied for one or more research grant and each application came with detailed declarations about the time that each member of the research team would devote to the project. The budget also included the cost and justification for equipment, consumable supplies, and miscellaneous other items needed to perform the research. Each grant application was reviewed by a committee of the researcher's peers who assumed that the project was a distinct unit that would be investigated as documented in the proposal.

A regulation of the Office of Management and Budget (OMB) (FMC 73-8J-7(d) required that the investigator certify each month how the money was in fact spent. The institution employing the researcher was also required to provide an annual report of disbursement with comparable information. In the vast preponderance of cases both the investigator and the institution officially stated that the funds were expended as outlined in the initial research plan. But in reality the investigator was frequently juggling time between more than one research grant and contracts, as well as additional other commitments, such as teaching, administration, travel to meetings and in the case of clinician-scientists patient care. The same was true for collaborators, technicians, and students. With numerous activities going on simultaneously an accurate recording of effort was difficult if not impossible. The financing of research activities and its bookkeeping was likewise complicated and beyond the understanding of most researchers.

Starting in the 1960s Lawrence H. Fountain (1913-2002) (Democrat, North Carolina) embarked on a comprehensive exploration of how research grants, especially from the NIH were spent. The committee chaired by him found substantial waste and negligence. During the 1970s more than a few examples of dishonesty with research funds were discovered. Some examples are provided in this chapter. By 1977 certain government auditors and investigators were of the opinion that some financial activities related to grant management violated federal guidelines and spent taxpayer's money in a way that was not originally intended.

Leonard Hayflick. Leonard Hayflick PhD (1928—), a distinguished biologist was appointed to Stanford University (Stanford) *(http://www.stanford. edu/)* school of medicine in 1968 as professor of medical microbiology and as professor of anatomy at the University of California, San Francisco (UCSF) *(http://www.ucsf.edu/).* He studied the aging process and acquired fame for his discovery that human cells cannot divide an unlimited number of times in vitro [3]. He formed a company called Cell Associates and entered into a sizable contract with the pharmaceutical company Merck and company which could have potentially generated $1 million for Cell Associates. He allegedly sold cells designated WI-38 (named for the Wistar Institute) for a profit of $67, 000 after developing them with federal financial support. The cell line was originally established in July 1962 by Hayflick from lung tissue acquired from a human female fetus obtained at a Swedish hospital when he worked at the Wistar Institute in Philadelphia. He supposedly kept the income despite charging the cost of human cell production to the NIH, except for postage and freight and resigned under threat of dismissal [10]. Unfortunately by the time of the financial dispute the cells were contaminated with bacteria and virtually useless for applications, such as rubella virus vaccine production.

Reports of the Division of Management Survey and Review (DMSR) indicate that allegations of wrongdoing by some researchers or institutions are vindicated, but others have turned up significant misuse of federal funds. For example a single investigator studying abortion and unwanted children at the American Institutes for Research (AIR), a nonprofit organization performing behavioral and social science research in Washington, was in charge of a five year grant of $200, 000 from the National Institutes of Mental Health (NIMH). Financial support was also obtained from the National Institute of Child Health and Human Development (NICHHD) and from the Ford foundation. The DMRS maintained that the scientist falsified charges to the government for the compensation of consultants on the project who were in the former country of Czechoslovakia and purportedly would have been arraigned by their government if they were paid openly. To overcome this difficulty the investigator deposited the payments in several different bank accounts until they amounted to more than $100, 000. The DMSR referred the matter to the Internal Revenue Service (IRS) and law enforcement agencies. The investigator eventually persuaded the authorities that the consultants were not only real, but had worked on the project. A trust was established by the USA district court of Columbia to hold the money for any Czech providing proof of deserving it. AIR was the whistle-blower in this case and it fired the investigator.

Management of Federal Grants and Contracts. Individuals who misappropriate federal grant funds should be prosecuted like other thieves as pointed out in detail in Science by Shapley [7]. The management of federal grants and contracts was becoming intensely scrutinized by different agencies of the government, including the NIH and the Government Accountability Office (GAO). At issue was the precise manner in which the funds were spent. This ranged from unequivocal fraud to creative accounting that fudged the accounts. "Time and effort reports" and "monthly certifications" and other aspects of grant management were investigated. In 1977 the intergovernment relations and human resources committee chaired by Fountain was concerned that the reforms accomplished during the 1960s may not have been continued as required. The issue surfaced in 1976 because of two incidents. One involved Phin Cohen, a former assistant professor of nutrition at Harvard University (Harvard) *(http://www.harvard.edu/)* claimed that he had been forced to sign blank forms, assuring how the money in his NIH grant had been consumed. The department of nutrition in the school of public health at Harvard filed them in with unconnected items and dispatched them to the government. The NIH investigators discovered that the allegations were true, and they unearthed significant accounting drawbacks on two additional Harvard grants that they evaluated. To offset the misspent items on these grants Harvard was required to return $132, 349. In 1977 auditors from the Department of Health, Education and Welfare (DHEW) began to audit all federal funds, totaling about $400 million, obtained by Harvard. Harvard failed to renew the appointment of Cohen after he made the aforementioned allegations so he sued the president and fellows of Harvard college claiming that his first and fifth amendment rights were violated by not renewing his appointment in retaliation for his complaints about the expenditure of federal grant monies by Harvard school of public health. The case was initially dismissed, but it went to the USA court of appeals, almost seven years later and the decision of the lower court was upheld (729 F2d 59:Phin Cohen, M., d., Plaintiff, Appellant, v. President and fellows of Harvard college, et al., Defendants, Appelles. Argued Feb. 10, 1984. Decided March 13, 1984.

Phillipe Shubick.Another episode concerned the Eppley Institute for Research in Cancer and Allied Diseases at the University of Nebraska *(http://www. unomaha.edu/)* medical Center in Omaha, Nebraska, under the directorship of Phillipe Shubick (1921-2004) which was awarded more than $18 million from the National Cancer Institute (NCI) to evaluate chemicals for potential carcinogenic properties between 1973 and 1976. According to the GAO report (iD:11531) that was published on June 2, 1981 these contracts with NCI were extended not using standard procedures *(www.legisorm.com?score_gao/show/id/1003.hml)*. Eleven projects for instance were authorized only by a verbal consent from NCI and some had already started. A large number of laboratory animals (about 50, 000) propagated at $1.75 per animal were evidently killed and not used by the Eppley Institute. Moreover, some equipment, materials, and animals purchased from the government funds were used for industrial research contracts performed at the Eppley Institute. Particularly disturbing was the fact that Shubick, director of the Eppley Institute was a member of the National Cancer Advisory Board (NCAB) which had oversight responsibilities for the NCI. Individuals responsible for overseeing the spending of federal research funds were concerned about whether the money was being used in the expected manner. In 1977 Joseph Califano, HEW secretary, instructed the HEW to review all NCI contract awards. He issued a NCI directive on May 18, 1977 and delivered an order setting sights on eradicating waste, abuse and mismanagement.

Others

Two researchers at Brandish University *(http://www.brandeis.edu)* in Waltham Massachusetts were awarded a grant from the National Institute of General Medical Sciences (NIGMS), but departed for Israel apparently with equipment worth about $6, 000 that was purchased from the grant. The DMSR required the cost of the equipment and part of the salaries of the two investigators to be refunded to the government. The DMSR also found circumstances in which the same applications were submitted by institutions to separate parts of the NIH or in which the salaries of investigators were charged to incorrect accounts. The

DMSR also found cases in which institutions deliberately intended to misuse the approved use of research funds.

Money used to enter the departments and laboratories of research institutions at irregular and sometimes unpredictable times. Some funds arrived before projects were able to start; others arrived after they began. Some accounts had abundant funds, whereas others were too lean for what they were expected to support. In practice charges were initially allocated temporally for items like salaries, supplies, equipment to a fund code that was handy at the time. Monthly financial reports intended to be signed by a person familiar with the progress of the projects often signed the statement, despite knowing that this was not true.

This practice, known as pooling, was widespread and provided some choice to institutions regarding governmental bookkeeping rules. A danger of this practice was that a project could be deprived of money that was awarded to support it. The OMB considers this dishonest certification of expenditures to the federal government inexcusable and laws and such false reports to the government are punishable. The agency sponsoring the research was responsible for enforcing the law. The use of funds for purposes other than those in the awarded grant damaged the peer review process for selecting worthy research projects to fund. Another objection that researchers had to the financial recordkeeping was the fact that they were teachers and thinkers, not certified public accountants. The financial recordkeeping gave at least one scientist the impression that the government was attempting to purchase research in a comparable manner to which it buys missiles or the way that safety pins or diapers are bought. Federal auditors were aware of this point of view and maintained that they had been hearing it for more than decade. One federal auditor pointed out that the academic community regarded attempts to identify amounts of efforts spent on funded research as an unjustifiable interference in research. By charging non-research related expenses to grants some investigators initially benefitted by not having to pay these expenses out of personal funds, but when such financial fraud became detected the consequences were serious and embarrassing. In accordance with the federal rules of evidence, a lawyer is permitted to cite a conviction involving dishonesty or false statement by a witness provided that less than ten years has lapsed since the time of the conviction.

James Allen. An example of these fraudulent finances involved Dr. James Allen, who had an international reputation as a pathologist for his research on the health effects of 2, 3, 7, 8-tetrachlorodibenzo-p-dioxin (TCDD), an extremely toxic contaminant in Agent Orange the defoliant used in the Vietnam War. It is also a constituent of herbicides, such as 2, 4, 5-trichorophenoxyaceticacid (2, 4, 5-T) and silvex. In a lapse of judgment Allen dipped into a government grant for $900 to pay for two skiing trips [1]. In March 1978 Allen went to a toxicology conference in San Francisco and stopped over in Utah for relaxation and skiing. While in Utah he anticipated interviewing individuals at the University for postdoctoral positions, but failed to do so. Nonetheless, for the reimbursement of his travel expenses he falsely included the costs for a detour into Park City,

Utah, claiming that that portion of the trip was to interview postdoctoral fellows. During the same month Allen flew to Colorado before attending a conference in New Orleans. He intentionally established the notion that he planned to meet a researcher in Colorado and to interview candidates when seeking reimbursement for his travel expenses. In a later trip with his son Allen visited Telluride, Colorado where his wife owned property. Allen gave the impression that he visited Colorado State University (CSU) *(http://www.colostate.edu/)* in Fort Collins, by presenting the receipts for airline tickets for him and his son. For the price tag of his son Allen used a postdoctoral researcher's name on an expense account. In the fall of 1978 a woman working as Allen's assistant wrote to the personnel office of the University of Wisconsin (UW) *(http://www.wisc.edu/)* alleging that he had repeatedly violated federal grant regulations. The charges were investigated by a university committee and found to be valid and the matter was passed on to the dean of the UW medical school. Allen maintained that there had been a misunderstanding of the federal guidelines. The dean proposed a meeting with both Allen and the whistle-blower to resolve the issue, but the former worker, evidently sensed that such a dialogue would be strenuous and informed senator Edward William Proxmire (1915-2005) (Democrat Wisconsin) who forwarded the grievance to the inspector general of the DHEW. Following an initial investigation the matter was referred to the department of justice and criminal charges materialized. In October 1979 Allen pleaded guilty to unlawfully taking $892 from the NIGMS postdoctoral training grant for which he was the director. As part of a plea arrangement Allen gave a three page affidavit documenting his dishonest actions.

After the dust settled Allen received a criminal conviction with 6 months' probation, and a fine of $4, 000 on 27 November 1979. He was also forced to resign from the University of Wisconsin-Madison *(http://www.wisc.edu/)* in 1980 when 52 years old with a seriously tarnished reputation and much of his life's research became discredited. Following his guilty verdict he admitted that he had shamed his family, his university and his state. Allen's conviction occurred at an unfortunate time because he was to be an important witness for the Environmental Protection Agency (EPA) that was going to hold hearings concerned with banning of 2, 4, 5-T and silvex. Because Dow Chemical Co, a most important producer of 2, 4, 5-T was against the ban its staff devoted considerable effort to damage the reputation of Allen's credibility and integrity based on his criminal conviction. The conviction of Allen for his fraudulent use of funds in a federal training grant was used to discredit him as a witness [1]. Some passionate environmentalists, loathing herbicides, believed that Allen's whistle-blower was in cahoots with the pesticide industry and were afraid that Allen's conviction would influence the conclusion of the EPA hearing. In March 1979, the EPA provisionally stopped most uses of 2, 4, 5-T and silvex after human exposure to TCDD was shown to increase the risk of spontaneous abortions in an epidemiologic study [8]. The hearings banning TCDD were scheduled to begin on February 13 1980 and could make an everlasting ban, Allen was scheduled to testify on the amplified threat of miscarriages in monkeys exposed to TCDD. The EPA already had concerns about Allen's laboratory practices and credible arguments for questioning the validity of his research existed. Toxic polychlorinated biphenyls (PCBs) had been detected in tissue samples from his laboratory animals in significant quantities raising grave concerns about the dependability of Allen's research. Allen's indiscretions with his travel expenses became fodder for the lawyers of Dow Chemical Co, who needed to find defects in his evidence. Allen's confession of guilt in the theft of funds from the USA federal government added a significant burden to his attempt to be a reliable witness in the hearing about whether TCDD should be banned. The conviction of Allen was a serious blow to his ability to get other grants, but Allen maintained that the National Institute of Environmental Health Sciences (NIEHS) comforted him that his indiscretions would not affect his ability to obtain future research grants. Allen's research was widely cited in claims of impaired health in Vietnam veterans stemming from exposure to Agent Orange as well as by the EPA and others.

In 1980 new vague and powerful regulations were introduced to deal with the misuse of federal funds awarded by the Department of Health and Human Resources (DHHR) The new rule, called the debarment regulation, was an excessive reaction to an apparently uncommon problem and it could punish entire institutions and honest researchers. The senior NIH attorney, William Metterer, was the main proponent of the regulation and considered it a stick to make institutions prevent their researchers from misappropriating research funds. The debarment rule could be imposed on persons convicted of a criminal offense related to a federal grant, unsatisfactory performance, and any other activity regarded as serious. Of greater concern to institutions was the fact that they could be debarred if they knew about the offense of an individual or should have known about it. Debarment was less likely to be imposed on institutions taking corrective action. Metterer felt that the debarment would rarely be imposed, but that the secretary should have the power to debar under extraordinary occurrences that are unpredictable [9].

Wasim Siddiqui and James M. Erickson, Wasim Siddiqui, an internationally renowned malaria researcher, was accused of using various methods to divert university research funds under his control as chairman of the department of tropical medicine for his personal use or benefit [5]. At least $130, 000 had supposedly been embezzled from such funds from 1984 to 1987 by unlawful bookkeeping trickery. When the investigation into Siddiqui began in 1987 the Agency for International Development (AID) looked into allegations of sexual harassment by James Erickson the chief technical officer for the malaria vaccine program [6]. Erickson was removed from this position and placed on administrative leave with pay despite claims that he was a victim of a feud. This provoked a legal clash with Erickson pointing the finger at AID for incompetence. Erickson, who directed the program from 1983-1987 pleaded guilty to filing false statements [2].

On September 14, 1989, Siddiqui was indicted by a grand jury for embezzling these funds and Susan Lofton his assistant was arraigned as a coconspirator. According to the indictment, $8, 639 was charged in unrelated expenses to a fund code intended to pay for an Asia and Pacific conference on malaria in Honolulu in 1985. In addition the accusation also charged that another $8, 500 of unspent conference funds eventually ended up in Siddiqui's private bank account and so did some checks from the travel agency taking care of the project. Siddiqui also apparently used $30, 000 intended for researchers in his department for his own personal expenses. In an attempt to cover up the incriminating inappropriate paper trail Siddiqui allegedly backdated a letter and convinced the comptroller at an outside organization that managed charges for the conference to sign a letter in 1988 allowing transactions that occurred three years earlier. Unfortunately for Siddiqui his indictment was announced on the exact day that the USA AID notified the University of Hawaii (UH) *(http://www.hawaii.edu/)* that it was renewing Siddiqui's malaria research grant of $1.65 million for 3 more years. Because of Siddiqui's indictment AID was adamant that his name be taken away from the project until the allegations were settled.

Misuse of Funds in Federal Training and Research Funds.

Whistle-blowers have drawn attention to the inappropriate use of federal grant funds.

These cases have resulted in major settlements between the USA Department of Justice and numerous universities and medical centers. Examples of the misuse of training grants have involved Wilfred van Gorp and Dr. New

Wilfred van Gorp. Cornell *(http://www.cornell.edu/)* Weill Medical College was involved in at least three whistle-blower cases involving fraudulent finances. Dr. Wilfred van Gorp, a clinical neuropsychologist and former faculty member of Cornell was awarded a research training grant from the NIH to train postdoctoral fellows committed to a research career in the neuropsychology of human immunodeficiency virus/acquired immunodeficiency syndrome (HIV/AIDS). The grant was funded with money specifically allocated by Congress to support research on HIV/AIDS. One trainee Daniel Feldman brought suit against van Gorp and Cornell under the federal False Claims Act designed for whistle-blowers. Feldman alleged that the awarded funds were used for improper reasons. According to the accuser the trainees were required to participate in the clinical care of an excessive number of private patients with medical conditions unrelated to AIDS. About 160 clinical patients were examined by the trainees over five years of support by the NIH grant, but only three patients were HIV-positive. Rather than dealing with HIV-patients, the trainees frequently assessed "medicolegal" cases in New York, where van Gorp often testified as an expert witness for the defense of criminal defendants. In this costly trial of several hundred thousand dollars in legal fees Feldman initially raised concerns to van Gorp and Cornell during his fellowship from 1998-1999. Following his fellowship Feldman forsake his original career goals and in 2003 he joined the pharmaceutical industry.

Feldman raised other allegations in both the grant application and its progress reports, such as listed courses that were never taught, and that the trainees neither met key faculty listed on the grant, nor received the breadth of HIV-research that they should have based on the description of the training program. Feldman also maintained that the activities of the trainees were not as described in the grant application, but a skillfully written to influence continued funding.

By July 22 2010 when van Gorp was at Columbia University (Columbia) *(http://www.columbia.edu/)*, the jury of a federal court found Cornell and van Gorp guilty of submitting false claims to the NIH *(http://www.prnewswire.com/ news-releases/federal-appeals-court-upholds-fraud-verdict-against-cornell-university- medical-school for-misuse-of-nih-aids-funding-168818716.html)* from 1991 and 2001 for payment from a grant to train postdoctoral fellows in a research grant in the neuropsychology of HIV/AIDS. Over the five-year tenure of the grant, van Gorp and Cornell were found to have intentionally submitted progress reports with false or fraudulent statements to the NIH. Judge William H. Pauley III, who presided over the trial, awarded damages at $855, 714 (three times what the government had paid in research grants). In addition $32, 000 was awarded in statutory penalties and more than $635, 000 was granted in attorney' fees to Feldman. By bringing the fraud to the government's attention Feldman was awarded a portion of the settlement known as the "relator's share." The case was appealed, but on September 5, 2012 the second court of appeals upheld the judgment of the lower court.

Kyriakie Sarafoglow

In some institutional grant applications universities have applied for financial support for certain programs, but have unethically spent the money on something else. Dr. Kyriakie Sarafoglou, a pediatric endocrinologist, blew the whistle on what she considered an inappropriate use of a five-year $23 million NIH grant by Cornell University (Cornell a *(http://www.cornell.edu/) (http://www.ahrp.org/ infomail/05/08/17.php)*. The grant with Dr. New as PI, was awarded to support a center for research on children's diseases yet several of the projects just existed on paper and many of the research participants did not have the diseases that the grant was supposed to support. Much of the money was used to treat adults. Sarafoglou interpreted this grant as a fraudulent use of taxpayer money because the money was being used to take care of patients at the New York-Presbyterian Hospital. She complained to university officials about her concerns, but because she failed to elicit a satisfactory response she filed a qui tam civil lawsuit against Cornell and several of its faculty. Prosecutors charged Cornell with fraudulent claims and eventually without admitting any wrongdoing Cornell agreed to pay $4.4 million in a settlement [11, 12]. Cornell claimed that universities are provided considerable discretion as to how they spend the funds in federal research grants. While there might be some truth in this policy a number of universities have not only faced civil lawsuits, but agreed to pay millions of dollars in settlements during the early 2000s. This abuse of the system has led to prosecutions by the justice department and it has been costly for the Mayo clinic ($6.5 million)(2005), Harvard University ($2.4 million)(2004), Johns Hopkins University ($2.6 million)(JHU), the University of Alabama at Birmingham (UAB) ($3.4 million)(2005), Northwestern University (NWU) ($5.5 million and some other institutions. Stanford University (Stanford) *(http://www.stanford.edu/)* was humiliated in 1994 because of its misuse of the indirect costs awarded for the overhead charges of research.

Misuse of Federal Funds Intended to Support the Indirect Costs of Research, An investigation disclosed that Stanford charged inappropriate items to the indirect costs provided to the university to support the overhead of federal research funded grants. Dingell challenged Donald Kennedy (1931-), President of Stanford at the time to explain "How fruitwood commodes, chauffeurs for the university president's wife, housing for dead university officials, retreats in Lake Tahoe and flowers for the president's house are supportive of science". An investigation disclosed that Stanford had charged certain inappropriate items to the indirect costs provided to the university to support the overhead of federal research funded grants. Kennedy tried to explain the university's indirect cost recovery in an interview by reporter Stone Philips on the American Broadcasting Company (ABC) news show 20/20, but was scorned for charging flower arrangements and a yacht as indirect costs of research. Philips had been fortified with information provided by the Dingell subcommittee. An inadequately prepared Kennedy struggled to provide satisfactory responses to the questions, such as why the government was billed by Stanford for his wedding reception in 1987.

The Stanford scandal on indirect costs led to the resignation of Kennedy as president of Stanford and brought into the open the issue of whether other universities were also misusing similar funds intended to support the overhead expenses of research [4].

Inappropriate Use Departmental Funds. In August/September 1993 Elwin Fraley, a former head of the urology department at the University of Minnesota (UMN) *(http://www1.umn.edu/)* resigned after a university investigation had concerns about his use of department funds. He used the funds from patent fees and borrowed money without written agreements to employ his wife and children and to purchase luxury cars *(http://www.mndaily.com/1996/05/22/urologic-surgery-department-gets-permanent-leader 6/22/2011)*.

References

1. Broad WJ, Ski trips cost researcher his job. Science, 1980. 207(4432): 743-744.
2. Brown G. Aid malaria unitact store gain credibility as probe continues. The Scientist. 1990 March.
3. Hayflick L and Moorhead PS, The serial cultivation of human diploid cell strains. Experimental Cell Research, 1961. 25(3): 585-621.
4. Jaroff L, Nash M, and Thompson D. Crisis in the labs. Time Magazine. 1991 August 26:45-51.
5. Marshall E, Malaria researcher indicted. Science. 245(4924): 1326.
6. Marshall E, Crisis in AID malaria network. Science, 1988. 241(4865): 521-523.
7. Shapley D, Research management scandals provoke queries in Washington. Science, 1977. 198(4319): 804-806.
8. Smith J, EPA halts most use of herbicide 2, 4, 5-T. Study finds miscarriages occurred among Oregon women just after the spraying period there. Science, 1979. 203(4385): 1090-1091.
9. Sun M, For future grants ski trip are out. Science, 1980. 210(4471): 746-747.
10. Wade N, Hayflick's tragedy: the rise and fall of a human cell line. Science, 1976. 192(4235): 125-127.
11. Wysocki BJ. Cash injection. As universities get billions in grants, some see abuses; Cornel doctor blows whistle over use of federal funds, alleging phantom studies; defending a star professor. The Wall Street Journal 16 August 2005.
12. Wysocki BJ. Universities misappropriate federal grants. Cornell doctors blows whistle alledging phantom studies. The Wall Street Journal. 17 August 2005.

CHAPTER 11

Unethical Research

For science to progress and retain the confidence of the public it must go hand in glove with ethics. The only protection that researchers have against plagiarism in grant applications and articles submitted for publication is the high expected ethical standards of science, but the trust in this honor system is very rarely violated because of numerous safe guards and the severe consequences for breaches (Vijay Soman and David Bridges).

Notorious unethical abuses in research have been carried out by a small percentage of scientists, such as those performed between 1932 and 1972 by the USA Public Health Service (PHS) in rural Alabama and Guatemala as discussed below and by Nazi physicians during the third Reich. Dishonest unethical behavior is not peculiar to research. It also occurs in other human endeavors as discussed in chapter 19.

Dealing with Serious Infectious Agents

Hopefully everyone knows that it is desirable to strive for honest ethical behavior, but the standards of morality change over time and depend on the circumstances facing research endeavors. Most will agree that the boundary between unethical and moral behavior is blurred.

Throughout history major epidemics of serious lethal infections have occurred. A prime example is bubonic plague (caused by Yersinia pestis) that killed millions of people in Europe during the fourteenth century. Other devastating infections have been smallpox, rabies, tuberculosis, syphilis, and the acquired immunodeficiency syndrome (AIDS) of recent years. These infections have lacked effective treatments and have sometimes provoked individuals to resort to measures that some would regard as unethical. But desperate times need frantic means.

Smallpox. In May 1721 a ship arrived in Boston harbor containing passengers with smallpox. Despite the ship being quarantined an epidemic of smallpox struck the small population of Boston. At that time little was known about how to control this extremely infectious and often lethal disease. The Reverend Cotton Mather (1663-1728) an influential New England Puritan minister, who was particularly well-remembered for his role in the Salem witch trials was a citizen of Boston at the time and he had been told by a slave named Onesimus about an ancient way to prevent or subdue the disease in Africa. Before he was forcibly removed from Africa into slavery as a child Onesimus had been inoculated in Africa from pustules of a patient with small pox. Mather encouraged physicians to try this method but initially his suggestion fell upon deaf ears. One medical doctor, Zabdiel Boylston (1676/1679-1766), a great-uncle of President John Adams (1735-1826), eventually became courageous to try the method on his only son Thomas and two slaves after much urging by Mather. The three of them developed a mild illness after pus from a smallpox sore was applied to small wounds on the skin of the subjects. The select men of Boston forbade Boylston to repeat the experiment partly because of a fear that inoculation would spread smallpox rather than prevent it. Citizens, particularly from some religious groups as well as most other physicians, were outraged and some even urged the authorities to prosecute Boylton for murder. Eventually despite fierce opposition some found inoculation acceptable and Boylton inoculated about 180 more people and it became apparent that the mortality was much lower in the inoculated individuals (2.5%) than in the persons who contacted smallpox naturally (14%). In 1726 Boylston published the results of his experiments that probably would never have been permitted by an institutional review board (IRB) today [2]. Although some may regard his experiments as unethical he faced the need to save lives at a time when a vast number of individuals in his community were suffering from an epidemic of a dreadful disease with a high mortality and no acceptable treatment. Vaccination was a standard method of treating smallpox in the early eighteenth century, but it had serious risks. A breakthrough in the therapy of smallpox did not occur until the 1770s when it became common knowledge that young women (milkmaids), who milked cows, were immune to smallpox. Their protection was thought to be due to their exposure to the pus in blisters of cowpox. Although others had developed vaccines against smallpox with cow pox [33]. It was only after Edward Anthony Jenner (1749-1823), the father of immunology, that the first effective preventive therapy for smallpox became established.

Syphilis. One of the most appalling and embarrassing unethical clinical studies that left permanent stains on the American nation was carried out under the leadership of John Charles Cutler, MD (1915-2003) a senior surgeon and acting chief of the venereal disease program in the PHS in Tuskegee, Alabama *(http://www.cdc.gov/tuskegee/timeline.htm)*. In 1932 when the study started the natural history of syphilis was unknown and there was no effective treatment for this venereal disease. To determine the innate progression of untreated syphilis the disease was followed on six hundred poverty-stricken, African-American male sharecroppers, who lived in the countryside of Macon County in Alabama. The study violated numerous ethical standards. The participants were infected with *Treponema pallidum* the causal agent of syphilis under the untruthful assurance that they were getting free health care from the USA government [19]. Informed consent was never acquired as data was collected. The Tuskegee study furthered knowledge of syphilis and it is virtually impossible to discard this important information. The enrolled participants were 399 men who acquired syphilis before the project began and 201 were disease free at the start of the study. The incentive for participation was free medical care, meals, and at no cost funeral insurance. At no time during the study were participants informed that they had syphilis. The men were informed that they were being treated for "bad blood" an expression used in their community to portray a number of illnesses, including syphilis, anemia and fatigue. A disclosure of details about the flaws in the research project by a whistle-blower resulted in extremely important alterations in USA law and in the introduction of regulations safeguarding participants in clinical research.

As a result of the Tuskegee syphilis experiment research on humans now require informed consent and an IRB pays particular attention to this and makes the principal investigator (PI) modify it as the situation changes over time. Specific exemptions are permitted for some USA federal agencies which can remain secret by executive order *(http://www.hhs.gov/ohrp/policy/ohrpregulations.pdf)*. Researchers involved in the Tuskegee syphilis study thwarted their participants from a right of entry into syphilis therapy programs to which others in the community had access [19]. After 1940 during the study an effective treatment for syphilis with penicillin emerged that often resulted in a cure, but the researchers withheld this antibiotic and continued the project without using it or informing the patients about its availability and effectiveness. Another unethical aspect of the study was its prolongation over four decades until 1972. The project ultimately came to an end when its existence was leaked to the press. Aside from the overwhelming ethical issues in the Tuskegee project extremely serious casualties occurred. Many men died from syphilis, and those wives who contracted the venereal disease gave birth to offspring, with congenital syphilis [13] *(http://select.nytimes.com/gst/abstract. html?res=F40616F6345A137B93C4AB178CD85F468785F9)*. The infamous Tuskegee study was indisputably the most notorious biomedical research study in USA history [21] *(http://muse.jhu.edu/cgi-bin/resolve_openurl.cgi?issn=1049-2089&volume=17&issue=4&spage=698&aulast=Katz)*. The study led to the creation of the Office for Human Research Protections (OHRP) and the Belmont report. On May 16, 1997, President William Jefferson "Bill" Clinton (1946—) formally apologized to the Tuskegee study participants on behalf of the USA at a ceremony in their honor. If the syphilis experiment was not enough to embarrass the USA and particularly the PHS another scandal involving that agency surfaced in

October 2010. While inspecting the documents of Cutler, a government researcher involved in the Tuskegee study, Wellesley college's historian Susan Reverby discovered records of a somewhat comparable experiment performed on syphilis in Guatemala [37, 41] *http://www.boston.com/news/science/articles/2010/10/02/wellesley_professor_unearths_a_horror_syphilis_experiments_in_guatemala/).* After Cutler's death on February 8, 2003, his involvement in the controversial and unethical medical experiments regarding syphilis was found not only in Tuskegee, but also in Guatemala. From 1946 to 1948, American doctors intentionally infected prisoners, soldiers, and patients in a mental hospital with syphilis and, in some cases, gonorrhea, with the support of Guatemalan health ministries and officials. Without signing an informed consent 696 men and women were exposed to syphilis. When these individuals contracted the disease they were treated with antibiotics, but it remains uncertain whether infected persons were cured [28] *(http://www.guardian.co.uk/world/2010/oct/01/us-apology-guatemala-syphilis-tests.).* In October 2010, the USA formally apologized to Guatemala for conducting these experiments [11] *(http://www.reuters.com/article/idUSTRE6903RZ20101001).*

Unethical Research Involving In Vitro Fertilization and Human Cloning

The creation of offspring with cloning has been an extremely controversial scientific challenge, particularly because it places scientists in the role of trying to play God. Research on in vitro fertilization (IVF) was extensively talked about and in 1980 it was disallowed by NIH supported researchers. Making it feasible to fertilize human eggs with IVF was a major scientific breakthrough that enabled infertile couples to give birth to living babies. This translational research that took laboratory work into the clinic was of such profound importance that one of the developers of the technique the British scientist Sir Robert Geoffrey Edwards (1925-2013) was awarded the Nobel Prize for physiology or medicine in 2010 *(http://www.nobelprize.org/nobel_prizes/medicine/laureates/2010/edwards.html).* Unfortunately an earlier death of Patrick Steptoe (1913-1988) his collaborator and a pioneer in laparoscopy and IVF prevented him from becoming a Nobel laureate because this treasured prize only goes to living scientists. The Edwards-Steptoe team was responsible for the birth on July 25, 1978 of Louise Brown, the first test-tube baby. The subject of IVF has been divisive and many religious people disapprove of it and believe that scientists have gone beyond the line by playing God. Scientists who have devoted much effort to the subject have faced difficulties because of the banning of embryonic research and because of confusing regulations that need lawyers to interpret.

Severino Antinori. Severino Antinori, an Italian gynecologist and embryologist, graduated from the University of Rome La Spienza in 1972. He then embarked on a contentious career in reproductive biology and infertility and established his own clinic in Rome near the Vatican. At first he offered IVF to menopausal women and one of his patients Rosa Della Cortes gave birth at 63 years of age in 1994. After it became known that another of his patients was pregnant when 62 years old Josephine Quitavalle from the Christian prolife organization in the UK known as Comment on Reproductive Ethics (CORE) (corethics.org) criticized Antinori severely and pointed out how difficult it was for someone as old as a grandmother to mother a child.

Antinori then developed an interest in human cloning in 1998, while working with Panayiotis Zavos, who ran a fertility clinic in Lexington Kentucky. He maintained that reputable scientists need to take the initiative to clone humans before other undesirable individuals develop the technology to replicate themselves. In 2001 Antinori and Zavos announced that they would try to clone a human within a year as a therapeutic procedure for fertility in an unspecified Mediterranean country where he had permission. Antinori took the provocative stand of arguing that human cloning should be offered to infertile couples when other methods failed. The hope was to inject genetic material from the father into a human egg stripped of its own genetic content, which would then be inserted into the uterus of the woman if an embryo began to develop. In theory the offspring would essentially be a genetic replica of the father. According to an Italian newspaper Antinori claimed that 1, 500 couples were eager to participate in this research and in January 2002 he announced that he had induced pregnancy in three women with this technique [25] and another woman was pregnant following IVF [10]. In a non-fiction book In His Image: The Cloning of a Man David M. Rotvik claimed that human beings were successfully cloned [39], but this turned out to be a hoax [6].

In 1993 congress opened the door for financing research on embryos, but before any projects were approved Harold Elliot Varmus (1939—), the director of the NIH at the time and a the recipient of the Nobel Prize in physiology or medicine in 1989 *(http://www.nobelprize.org/nobel_prizes/medicine/laureates/1989/varmus-autobio.html#)* with J. Michael Bishop *(http://www.nobelprize.org/nobel_prizes/medicine/laureates/1989/bishop-autobio.html#)* for the discovery of the cellular origin of retroviral oncogenes, requested counsel from an advisory panel of experts about what should be permitted. A member of the panel was Mark R. Hughes, who was a National Institutes of Health (NIH) funded molecular geneticist searching for disease producing mutations in DNA from embryos generated by IVF, but the panel recommended limited research on embryos toward the end of 1994 [26]. William Jefferson "Bill" Clinton (1946—) the forty-second president of the USA in 1994 prohibited the use of federal funds in embryonic research. Later the Republican Congress made it illegal to use federal funds for research in which human embryos are destroyed, discarded, or knowingly subjected to risk of injury or death greater than that allowed under other laws. Hughes was recruited to the NIH in 1994 when policies related to embryonic research were in a state of uncertainty and on June 12 1995 Varmus and other top NIH officials at the NIH told Hughes explicitly that he could not use NIH resources for the DNA analysis to screen embryos generated by IVF to determine if they contained mutations in disease producing genes Hughes wanted to make sure that they were fit to be implanted in the recipient mother's uterus. In June 1997 Varmus in his capacity as director of the NIH was required to testify before the House investigative panel for three hours answering questions related to Hughes. In 1996 Hughes was accused of breaching a federal ban on embryo research [26] and Varmus and top NIH officials were criticized by the lawmakers for negligent supervision. The congressional subcommittee presumed that Hughes defied the ban on embryo research during 1995 and 1996. Varmus verified Hughes's infringement of the embryo research ban as well as other rules created to protect human subjects and was adamant that he and other NIH officials had been oblivious to the alleged misconduct of Hughes because of his thorough steps to conceal it. When Hughes's clandestine prohibited research was unearthed the NIH promptly severed ties with Hughes on October 21, 1996 to make certain that comparable deviances would not recur. Varmus stressed that he did not become aware of Hughes's research until the scandal hit the newspapers in January 1997; three months after officials of the National Human Genome Research Institute (NHGRI) had broken all ties with Hughes because of his obvious transgressions [27]. According his lawyer Hughes never intended to violate the ban on embryo research, and only did so because the NIH had not made it clear that the federal rules forbade him from doing as needed for research on preimplantation genetic diagnoses. Regrettably no written memos were sent to Hughes or NIH staff about the rules and Hughes continued to obtain deoxyribonucleic acid (DNA) from single cells removed from embryos at IVF clinics and to analyze them in the laboratory at a suburban hospital near NIH. This technique was later shown to be valuable in families prone to retinoblastoma, a malignant cancer of the retina caused by a mutation in the Rb gene. Persons carrying this defective gene could pass it on to half of their eggs or spermatozoa. By permitting the fertilization of the egg with the husband's sperm an embryo could be allowed to develop to the 8 cell stage. One of the cells can then be removed and analyzed for the mutation known to cause retinoblastoma in that family. If the defective gene is detected the remaining 7 cells are discarded, but if it is not detected the embryo can be implanted in the future mother and allowed to develop further as the risk of retinoblastoma has been eliminated. This important practical application of DNA analysis in early human embryos was carried out effectively by faculty at New York-Presbyterian Hospital/Weill Cornell Medical Center who used a negative Preimplantation Genetic Diagnosis (PGD) for retinoblastoma to give birth to the world's first baby without a predisposition to this potentially lethal genetically determined eye cancer. A description of the first case that was evaluated in this way was published in the January 2004 issue of the American Journal of Ophthalmology [55].

University of California Irwin Fertility Clinics. IVF spawned fertility clinics in many parts of the Western world. In 1990 the University of California Irwin (UCI) *(http://uci.edu/)* Medical Center in Orange county, California lacked a fertility clinic, but operated such clinics at Garden Grove and Saddleback Memorial Medical Center in Laguna Hills. More than a decade ago fertility doctors at these clinics stole eggs and embryos from patients. The university vowed that it would notify the women who may have been victims, because its administrators believed that this was the ethical thing to do. But others were not convinced and believed that the criminals should just be punished without putting the families though more unnecessary stress. Regardless the UCI conceded in January 2006 that it had been unable to get in touch with at least 20 couples related to this scandal in a timely fashion. Lawsuits were filed mostly on the grounds of fraud, concealment, and emotional stress. Attorneys representing the victims of these fraudulent cases sued the University of California system, mainly in 2003. The UCI Medical Center admitted that it failed to contact patients and in an attempt to reduce the total amount of money that it could be compelled to compensate victims UCI contended that the statues of limitations had expired. When ethics and money come into conflict unfortunately money usually wins as it did in this scandal. In lawsuits filed in 2002 and 2003 Byron Beam, UCI's attorney admitted that UCI failed to contact two-thirds of the patients. Letters were mailed to the relevant people and the university even tried to find some of them using private investigators. It was only after more than 15 years that some victims of the scandal were shocked to discover that stolen human eggs and fertilized embryos had been successfully implanted unethically and illegally into other women who gave birth to living babies. The scandal produced so much adverse publicity that many fertility clinic patients became aware of it and hearings were held by the state of California. UCI paid as many as 137 legal settlements amounting to more than $24 million. The disgrace spoiled the reputation of UCI which overlooked earlier forewarnings and tried to cover up this extremely unethical behavior. Two doctors at the Center for Reproductive Health at UCI, Ricardo Asch and Jose Balmaceda, were indicted by a federal grand jury on mail fraud and tax evasion but fled to Mexico to avoid prosecution [56].

Some of the patients were told that their eggs were not viable and fertility drugs were administered to them. Clinical records of one patient disclosed that seven of her eggs were removed and implanted into another woman who delivered twins. Knowledge about this disgrace had an incredible effect on some subjects. One patient likened the incident to biological rape, and dreamt most nights about the children that might be hers. She had a nervous breakdown and started taking medication for severe depression and anxiety. She also gained weight and distrusted doctors. One couple spent $30, 000 on fertility treatments and the woman underwent three attempts at IVF. A book entitled Stealing Dreams, a

Fertility Clinic Scandal documents in detail this fertility scandal [7], which caused a major setback in medical schools and hospitals from 1995.

Cecil Jacobson. In the days when regulations regarding the qualifications of persons adequately trained and competent to maintain a fertility clinic did not exist, men and women frantic to create a child flocked to the fertility clinic of Dr. Cecil Jacobson in the town of Vienna in Virginia. He pioneered the technique of amniocentesis and established an excellent reputation as a geneticist. The doctor regarded himself as the baby maker and his patients trusted him immensely, but in reality he informed many women that they were pregnant when they were not. False pregnancies were created by treating the patients with hormones that simulate the first part of pregnancy. Others anxious to conceive consented to being inseminated by a genuine lawful sperm donor, but instead of this they received clandestinely introduced semen of the doctor. Sometimes this led to pregnancies and at least seven children are known to have been fathered by Jacobson. This was substantiated after several of his patients pleaded for genetic paternity tests. In 1989 at hearings before a committee of the Virginia Board of Medicine patients emotionally testified how they were shown sonograms of their apparent babies and given photographs of their fetuses and then to be informed a few weeks later that the fetus had died. The Virginia Board of Medicine withdraw Jacobson's medical license after finding sufficient evidence to implicate his disgusting behavior. Following this he gave up his practice and moved to Provo, Utah. In November 1991 he was indicted and faced 53 felony charges. In defense of his concealed sperm donations Jacobson's attorney argued that this was performed to provide the patient with a clean good sample in this era of human immunodeficiency virus/acquired immunodeficiency disease syndrome (HIV/AIDS) [8].

Unethical Cancer Research

Cornelius R. Rhoads. Cornelius R. Rhoads (1898-1959) was a prominent American pathologist and a pioneering cancer researcher with a superb reputation as a scientist. He became director of the Memorial Sloan-Kettering Cancer Institute and he even made the cover of TIME Magazine on June 27, 1949. In 1979 because of an anonymous donor the American Association of Cancer Research established the Cornelius P. Rhoads Memorial Award to be given in his honor to a promising researcher under 40 years of age. But the reputation of Rhoads became mortally damaged after the turn of the millennium because of deadly human experiments in 1931 when he worked at the Rockefeller Institute for Medical Research (now Rockefeller University) *(http://www.rockefeller.edu/)* [51]. By 2002 it became known than secret unethical medical experiments by Rhoads deliberately infected several Puerto Rican patients with cancer cells. The subjects were not notified about the nature of the experiments and did not provide informed consent. These studies outraged Puerto Ricans *(http://www.*

puertorico-herald.org/issues/2002/vol6n44/PROutragMedExp-en.html). The secret experiments might have remained concealed if Rhoads had not put pen and paper in a letter that became notorious. This letter was written when Rhoads was at Presbyterian Hospital in San Juan, but he never mailed it. It was inadvertently discovered by Edwin Vázquez, a biology professor, while preparing a lecture. In the communication Rhoads belittled Puerto Ricans as "the dirtiest, laziest, most degenerate and thievish race of men to inhabit this sphere". In addition Rhoades wrote "What the island needs is not public health work but a tidal wave or something to totally exterminate the population". Rhoads then made an appalling declaration: "I have done my best to further the process of extermination by killing off eight and transplanting cancer into several more"[51]. Investigations into the documented statements in the letter took place; and one by Jay Katz, an emeritus professor of law, medicine, and psychiatry at Yale University (Yale) *(http://www.yale.edu/)*, concluded that there was no evidence that Rhoads actually killed patients or transplanted cancer cells. According to one colleague, Rhoads got drunk at a party on the night before the letter was written and found his car stripped with flat tires" [51]. Susan Lederer, an associate professor of medicine at Yale has written about this case [24]. In 2003 the American Cancer Research Association removed Rhoads's name from the award.

Chester Milton Southam, In the early 1950s many thought that cancer was caused by a virus or an immune system deficiency. A major scientific breakthrough occurred in 1951 when George Otto Gey (1899-1970) established immortalized cells from an adenocarcinoma of the cervix of Henrietta Lacks (1920-1951) an African American woman that became known as HeLa cells. The story of their discovery and the many ethical violations related to the original and subsequent research are well documented in an excellent book authored by Rebecca Skoot [40]. When the American virologist and oncologist Chester Milton Southam (1919-2002), the chief of virology at Sloan-Kettering Institute for Cancer Research, injected HeLa cells into rats tumors grew in some of the rodents. These cells were widely used at the time in many experiments using cell culture systems. During his career Southam performed numerous experiments with tumors and HeLa cells [22, 29, 43-50]. While transferring HeLa cells form from vial to vial Southam had the terrifying thought that cancer may develop in persons inoculated with the cells or a product of them during vaccine production or scientists using HeLa cells might be at risk of cancer if accidentally injected during laboratory work. To test this potential danger Southam injected a suspension of about five million HeLa cells in saline into the forearm of a woman hospitalized with leukemia in February 1954. He tattooed the site of injection to follow the reaction. Within hours an acute inflammatory reaction developed at the injection sites. Subsequently HeLa cells were injected into about a dozen other patients with cancer. The subjects of this experiment were told that their immune system was being tested, but they were not informed that the injection contained cells derived from the HeLa cancer cells. Local tumors formed in some cases and these were excised so that tumor formation could be verified. Some nodules were followed to determine if the immune system would reject them and in some cases it apparently did, but others evidently recurred after excision and in one patient HeLa cells metastasized to the lymph nodes. All of these patients already had cancer before the experiment began so Southam decided to find out how healthy individuals would react. To obtain volunteers he placed an advertisement in the Ohio State Penitentiary newsletter in May 1956. It stated: Physician seeks 25 volunteers for cancer research. There was a superb response that reached 150 within a short time, but the vulnerable prisoners were unable to provide informed consent. The inmates of this penitentiary had established their reputation as cooperative participants in research projects needing their services, such as a study on the lethal infection of *Franciscella tularensis* the pathogenic bacterium that causes tularemia. In June 1956 injections of HeLa cells into this defenseless population were started. Multiple injections were administered to each prison volunteer and all of them fought off the cancer suggesting that a vaccine for cancer may one day become a reality. Following the penitentiary project Southam injected HeLa cells into hundreds of other healthy patients as well as those with cancer. Furthermore some individuals were inoculated with living cancer cells. Individuals were not informed about the true nature of the injections as he did not want to unnecessarily frighten them about the dreaded disease of cancer. His cancer research eventually moved to the Jewish Chronic Disease Hospital in Brooklyn, New York where an arrangement was made with the director of medicine, Emanuel Mandel, to use the hospital patients in cancer studies. Doctors on Mandel's staff were expected to inject a certain number of patients with cancer cells, but three young physician refused because of the similarity of this work to what the Nazis did in World War II.

The genomic and transcriptomic structure of a HeLa cell line was determined by Lars Steinmetz and colleagues at the European Molecular Biology Laboratory (EMBL) [23]. This private genetic data could disclose private information about the family of Henrietta Lacks. Because of the ethical implications related to the study of HeLa cells an agreement was made with the Lacks family by Hudson and Collins [18].

Psychological Behavioral Studies

Social psychologists have been struck by differences in the behavior of Caucasians who live in the southern USA compared to those in the northern states. Those living in the South were considered to have a distinct culture known as the "culture of honor". It was characterized by an avoidance of unintentional offense to others and a reputation for not tolerating inappropriate behavior by others and a propensity for violence. Different hypotheses were proposed to explain this observation. Some social scientists believe that it stems from the qualities of the early settlers and is inheritable and rooted in the herding societies in Northern England, the Scottish borders, and Irish border region [9]. Others consider the culture of honor oversimplified and that the violence committed by Southerners to result from social tensions [38]. One theory maintains that toughness is an attribute developed to protect one's possessions against predators. In the early 1990s Richard Nisbett and Dov Cohen, two social psychologists attempted to perform experiments to test the validity of the culture of honor [30]. They were familiar with the culture of Harlan Kentucky in the nineteenth century and in the English borderland centuries before, but wanted to determine if this culture still existed today. For their studies they recruited a group of students and insulted them to witness their responses. Presumably this was done without an informed consent or the approval of an IRB as the study needed to catch the participants by surprise and no reasonable IRB would have permitted the study on students. In the social sciences building at the University of Michigan (UM) *(http://www.umich.edu/)* young men were invited to enter a classroom one at a time to fill in a questionnaire and then walk down a long narrow passage lined with filing cabinets and to then return to the classroom. For half the men this was all as they were controls. When the others walked down the hallway someone pulled out a drawer of a filing cabinet and almost blocked the passageway and when the young men tried to squeeze by the blocker slammed the drawer closed, appeared annoyed jostled the passing student with his shoulder and in a low audible voice insulted the student by calling him an "asshole". Nisbett and Cohen recorded all kinds of data on the insulted students including levels of testosterone and cortisol before and after the insult. Most of the students from the northern USA treated the episodes with amusement. The Southerners were however angry. The study has many inherent weaknesses including the fact the participants from the UM were not necessarily representative of the intended populations for an evaluation of "culture of honor". Nevertheless the findings were published in Culture of Honor: The Psychology of Violence in the South [30] and the subject was propagated in a later book by Malcolm Gladwell [12]. Understanding of violence is important if it is to be controlled and if more guns are not the cause of violence in the South gun control is unlikely to be the solution.

Studies Performed During the Nazi Regime

Nazi Anatomical Studies

In 1938 an unusual mutually beneficial agreement was made between senior anatomists at the University of Vienna and the Nazi regime. Corpses were needed for teaching and research in the anatomy department and the Gestapo needed to dispose as quietly as possible of large numbers of bodies from the prisoners that they shot or guillotined. So many corpses were transported to the university and special arrangements were needed to move the bodies in an exclusive streetcar nicknamed "Death Transport". When the medical school morgue reached its capacity executions were officially postponed. In addition at least one anatomist, Hermann Stieve (1886-1952) apparently manipulated the timing of the executions by wanting to study terrorist female prisoners waiting unduly to die [54]. According to a report by the university that was released in 1998 at least 1337 corpses of Nazi victims were accepted by the Vienna Medical School.

Germany's Anatomic Society held its first symposium on "Anatomy of Third Reich" on September 29, 2010 at the University of Würzburg. A series of three studies published in the Fall of the prior year by the anatomist Sabrine Hildenbrandt in the journal Clinical Anatomy traces the customs of German anatomists prior to and during the Third Reich when unclaimed bodies from hospitals, psychiatric facilities and prisons were used [14-16]. Between 1933 and 1945 at least 16, 000 civilians were executed in German prisons and by 1942 the bodies of every prisoner executed for high treason was turned over to anatomists. Those corpses that were beheaded probably affected the students emotionally and at least one young researcher gave up anatomy after dissecting the body of an executed political prisoner whom he had personally known [34]. As a response to the unethical activities in anatomy during the Third Reich the German Medical Council recommended in 2003 that German universities eliminate from their anatomical collections all histological slides, bones and other specimens derived from Nazi victims. Numerous German universities did just that and buried the remains in places of honor. Starting in 1933 Eduard Pernkopf an anatomist at the University of Vienna toiled for many hours over two decades dissecting structures for a four volume anatomical atlas which was published with the title Topographical Anatomy of Man [32]. He worked closely with artists who painted many of the details. The book was widely acclaimed and according to the New England Journal of Medicine it was "an outstanding book of great value to anatomists and surgeons" [42]. Until now the illustrations are unrivaled by any other text. Regrettably Pernkopf and several of his artists were fervent Nazis as disclosed by David Williams of Purdue University [53]. Moreover, an investigation by Vienna University *(http://www.univie.ac.at/en/)* established in 1998 that Perkopf's department of anatomy received the bodies of executed prisoners from the Gestapo and the assize court of Vienna. These revelations have complicated the ethical issue of what to do with this morally defiled scientifically precious Magnum Opus.

Other Nazi Projects

Various Nazi research projects have been exposed over the years, including the use of psychiatrists in the selection of mentally ill patients for euthanasia [35]. Nonethical biomedical research was performed in the 1940s by German scientists on inmates in Nazi concentration camps and many subjects were killed in the name of research. Over two dozen distinct projects were carried out in concentration camps at Auschwiz, Dachau and Ravensbruck and the experiments yielded some valuable scientific information. The issues investigated included the efficacy of electric shock treatment for depression, the effect of high—altitude decompression on the human body, the efficacy of sulfanilamide in the therapy of gunshot wounds, the influence of stress on ovulation and menstruation and the physiological response of the human body to extreme cold [3]. During the Nazi years much research at universities and other institutions was performed mainly by incompetent individuals on unfriendly malnourished usually sick subjects. Some made its way into professional publications even if it was blatantly immoral. Nazi research on hypothermia is cited in medical journals and textbooks from the 1950s to at least the 1980s [36]. In the 1980s American researchers cited Nazi research on phospholine gas exposure as well as the effect of stress on menstruation. Eventually the Environmental Protection Agency (EPA) banned the use of Nazi data [52].

Victims of Atomic Bombs In Japan

Aside from the ethically atrocious Nazi and Tuskegee studies numerous other morally questionable biomedical investigations have been performed. Current knowledge about the effect of high-dose radiation on humans stems partly from studies on the victims of the two atomic bombs that were dropped on the Japanese cities of Hiroshima and Nagasaki to bring World War II to a rapid conclusion. These studies included the determination of the cancer incidence of various atomic bomb survivors, but surely nobody will find fault in such follow-up studies, even though many challenge the ethical decision by Harry S. Truman (1884-1972), the thirty-third president of the USA, to drop the bombs.

Research on Workers in Uranium Mines

What we know about the ill effects of radon gas stems partly from research on Native-Americans who worked on uranium mines in the 1950s and 1960s, while not being informed of the risks. Over a period of four decades of Communist rule the government of the German Democratic Republic (GDR) secretly supplied the Soviet Union with uranium for atomic bombs [20]. An estimated 450, 000 workers involved in this clandestine activity participated in the mining and processing of uranium. Before 1993 with the assistance of the German government researchers began scrutinizing the vast collection of detailed medical records on the huge number of individuals exposed to different levels of radiation in the former GDR before 1993. How this epidemiological information on the effects of radiation and environmental toxins on the innocent populations in Eastern Europe relates to the Native-American studies remains debatable.

Studies on Chemical Weapon Development

Suspect data collected by military scientists in arsenal research is another issue of concern. In weapons research as recently as 1975 scientists working for the USA military exposed so-called human volunteers to agents in chemical weapons [31]. A panel from the National Academy of Sciences (NAS) investigated research on military chemical warfare in 1993 and concluded that the rights of tens of thousands of American military personnel were violated in experiments related to chemical weapons between 1942 and 1975. Moreover the subjects were dishonestly recruited using lies and half-truths. Many volunteers developed severe disabilities or died prematurely. Nevertheless most knowledge about mustard gas and lewisite stems from this research. The results of these despicable studies are regularly cited without comments on the questionable conditions under which the data was assembled.

Unethical Misuse of Information in Electronic Databases

In 2006 the NIH established an electronic database of genotypes and phenotypes called dbGaP for confidential information garnered in NIH funded genetic studies on human diseases. It was established to enable all federal grantees to share the enormous quantity of genetic data that is accumulating exponentially as a sequel to federal grants. NIH funded researchers are required to deposit information about phenotypes and genotypes into this database. Authorized individuals are permitted to gain access to this material and to use it in genetic analyses. Investigators that receive permission to access the raw data on dbGaP must sign a memorandum of understanding promising not to submit a paper for publication using an analysis of the dbGaP material before a restricted embargo of 12 months. The embargo provides the first opportunity of analysis to the submitter of the data. Some scientists see merit in dbGaP, but others consider that it is too generous and a temptation to predators without an investment in the data. On September 18, 2009, Constance Holden (1941-2010) [17] drew attention to a paper that had been retracted following a genome data breach. Laura Bierut, a psychiatrist, performing genetic research on alcoholism and other addictions submitted the genetic data that she had acquired in a Study of Addiction: Genetics and Environment (SAGE) to dbGaP as required by NIH regulations, but before the restriction period for its use was over she discovered that a research team led by Heping Zhang had unethically violated the intent of dbGaP and published a paper using the data in a report documenting addiction in women of European origin in association with changes in the PKNOX2 gene. Despite the embargo having an end date of September 23, 2010 a paper by Zhang et al. appeared online in the prestigious Proceedings of the National Academy of Sciences (PNAS) on August 31, 2010 [5]. Several points about this case are particularly disturbing. Firstly, the unethical paper was sponsored by a member of the NAS (Burton Singer) the most esteemed society in the USA for scientists. Secondly, the culprits were a team of investigators from Yale one of the top universities in the USA. Bierut was extremely gracious on learning about the genome data breach and pointed out that the incident was probably an unintentional act, despite its chilling effect on investigators with data in dbGaP. Without delay she emailed numerous colleagues as well as NIH officials and the PNAS. In contrast to other unacceptable acts by scientists reported elsewhere in this monograph the response in this case was swift. NIH froze access to dbGaP by Zhang and his colleagues and the PNAS retracted the paper. A review of how the breach occurred was started and Zhang, who signed the required agreement to dbGaP data, warrants further sanctions.

How to Deal with Unethical Research Studies

A dark cloud hovers over the biomedical community regarding the issues of whether and how to use ethically tainted data. Useful scientific information can and sometimes has been collected under the most heinous situations. Such unethically obtained data has raised the question of whether papers reporting such morally unacceptable data should be cited in publications. Caplan [4] feels that ethically defensible rules can be established for handling such tainted data and biomedical scientists should follow the rules. Those responsible for unethical research clearly should not receive accolades, awards and prizes. We have learned not to condone unethical research now and in the future and everyone agrees that unethical research should be condemned and not tolerated. Many studies performed in the past under unethical conditions yielded valuable new information. Some ethicists question how contemporary biomedical scientists should handle such knowledge. Should present-day publications cite unethical research? Those deeds, however, have taken place and one cannot brush the horrors of past historical facts under the carpet and pretend that they never occurred solely because they were unethical. Caplan [4] raises the question of whether data derived from unethical situations should be published and whether it should be referenced. Tainted unethical data falls into three categories according to Caplan: (i) already published and cited research data, (ii) data obtained from the immoral government policies such as the uranium records of the GDR, and (iii) data yet to be produced in immoral experiments. He recommends that the third category be treated as taboo and be kept beyond the mainstream of science. Angell [1] stressed that editors and reviewers of journals and periodicals must agree not to permit the publication of papers based on immoral means, regardless of the importance of their findings. Scientists have an obligation to determine whether tainted data are the only achievable basis for obtaining the required information.

References

1. Angell M, Editorial responsibility: protecting human rights by restricting publication of unethical research, in The Nazi Doctors and the Nuremberg Code, Annas GJ and Grodin MA (Eds), ed. Code TNDatN. 1992 New York: Oxford University Press.
2. Boylston Z, An Historical Account of the Small-Pox Inoculated in New England. 1726, London: S. Chandler.
3. Caplan AL, When Medicine Went Mad: Bioethics and the Holocaust (Contemporary Issues in Biomedicine, Ethics, and Society), ed. Caplan AL. 1992) (Editor), Totowa, NJ: Human Press.
4. Caplan AL, How should science handle data from unethical research? The Journal of NIH Research, 1993. 5: 22, 24, 26.
5. Chen X, Cho K, Singer BH, et al., Retraction. PKNOX2 gene is significantly associated with substance dependence in European-origin women.[Retraction of Proceedings of the National Academy of Sciences of the United States of America. doi: 10.1073/pnas.0908521106]. Proceedings of the National Academy of Sciences of the United States of America, 2009. 106(40): 17241.
6. Culliton BJ, Scientists dispute book's claim that human clone has been born. Science, 1978. 199(4335): 1314-1316.
7. Dodge M and Geis G, Stealing Dreams: A Fertility Clinic Scandal. 2003, Boston MA: Northeastern University Press.
8. Elmer-Dewitt P. Scandals: The cruelest kind of fraud. Time. 1991 December 2:27.
9. Fischer DH, Albion's Seed: Four British Folkways in America. 1989, Oxford, England: Oxford University Press.
10. Follain J and Foggo D. Professor Severino Antinori: Mother-to-be too old at 66. The Sunday Times. 2009 17 May.
11. Fox M. US apologizes for '40s syphilis study in Guatemala. Reuters. 2010 1 October.
12. Gladwell M, Ourliers: The Story of Success. 2008: Little, Brown and Company.
13. Heller J. Syphilis victims in U.S. study went untreated for 40 Years; Syphilis victims got no therapy. New York Times. 1972 26 July
14. Hildebrandt S, Anatomy in the Third Reich: An Outline, Part 3. The science and ethics of anatomy in national socialist Germany and postwar consequences. Clinical Anatomy, 2009. 22: 906-915.
15. Hildebrandt S, Anatomy in the Third Reich: An outline, Part 2. Bodies for anatomy and related medical disciplines. Clinical Anatomy, 2009. 22: 894-905.

16. Hildebrandt S, Anatomy in the Third Reich: An outline, Part 1. National socialist politics, anatomical institutions, and anatomists. Clinical Anatomy, 2009. 22: 883-893.
17. Holden C, Scientific publishing. Paper retracted following genome data breach.[Erratum appears in Science. 2009 Oct 16;326(5951):366]. Science, 2009. 325(5947): 1486-1487.
18. Hudson K and Collins F, Biospecimen policy: family matters. Nature, 2013. 500(7461): 141-142.
19. Jones J, Bad Blood: The Tuskegee Syphilis Experiment. 1981, New York: Free Press.
20. Kahn P, A grisly archive of key cancer data. Science, 1993. 259(5094): 448-451.
21. Katz RV, Kegeles SS, Kressin NR, et al., The Tuskegee legacy project: willingness of minorities to participate in biomedical research. Journal of the Health Care for the Poor and Underserved, 2006. 17(4): 698-715.
22. Kikuchi K, Reiner J, and Southam CM, Diffusible cytotoxic substances and cell-mediated resistance to syngeneic tumors: in vitro demonstration. Science, 1968. 165(3888): 77-79.
23. Landry JJM, Pyl PT, Rausch T, et al., The genomic and transcriptomic lanscape of a HeLa cell line. G3 Genes, Genomes, Genetics, 2013.
24. Lederer SE, "Porto Ricochet": Joking about germs, cancer, and race extermination in the 1930s. American Literary History, 2006. 14(4): 720-746.
25. Lyman EJ. Italy's Antinori says he's cloned three people. United Press International. 2002 May 9.
26. Marshall E, Human embryo research. Clinton rules out some studies. Science, 1994. 266(5191): 1634-1635.
27. Marshall E, Varmus grilled over breach of embryo research ban. Science, 1997. 276(5321): 1963.
28. McGreal C. US says sorry for 'outrageous and abhorrent' Guatemalan syphilis tests. Experiments in 1940s saw hundreds of Guatemalan prisoners and soldiers deliberately infected to test effects of penicillin. The Guardian. 2010 October 1.
29. Moore AE, Rhoads CP, and Southam CM, Homotransplantation of human cell lines. Science, 1957. 125(3239): 158-160.
30. Nisbett RE and Cohen D, Culture of Honor: The Psychology of Violence in the South. 1996, Boulder, Colorado: Westview Press.
31. Pechura CM and Rall DPE, Veterans At Risk: The Health Effects of Mustard Gas and Lewisite. 1993, Washington, DC: National Academy Press.
32. Pernkopf E, Pernkopf Anatomy: Atlas of topographic and applied human anatomy. (Volume 2: Thorax, abdomen and extremities.) Third edition. Edited by Warner Platzer, translated by Harry Monsen. 1989, Baltimore: Urban and Schwarzenberg.

33. Plett PC, [Peter Plett and other discoverers of cowpox vaccination before Edward Jenner] (in German). Sudhoffs Archiv, 2006. 90(2): 219-232.
34. Pringle H, The Master Plan: Himmler's Scholars and the Holocaust. 2006: HarperCollins.
35. Pringle H, Confronting anatomy's Nazi past Science, 2010. 329(5989): 274-275.
36. Pross C, Nazi Doctors, German Medicine and Historical Truth, in The Nazi Doctors and the Nuremberg Code, Annas GJ and Grodin M (Eds.). 1992, New York: Oxford University Press.
37. Reverby SM, "Normal exposure: and inoculation syphilis; a PHS "Tuskegee" doctor in Guarmala, 1946-1948. Journal of Policy History, 2011. 23: 6-26.
38. Roth R, American Homicide. 2009, Cambridge MA: The Belknap Press of Harvard University Press.
39. Rotvik DM, In his Image: The cloning of a Man. 1978, Philadephia and New York: J.B. Lipinncott.
40. Skoot R, The Immortal Life of Henrietta Lacks. 2010, New York: Broadway Paperbacks.
41. Smith S. Wellesley professor unearths a horror: Syphilis experiments in Guatemala. US apologizes for performing unethical study in 1940s. Boston Globe. 2010 October 2.
42. Snell RS, Pernkoft Anatomy: Atlas of topographic and applied human anatomy. New England Journal of Medicine, 1990. 323(3): 205.
43. Southam CM, Goldsmith Y, and Burchenal JH, Heterophile antibodies and antigens in neoplastic diseases. Cancer, 1951. 4(5): 1036-1042.
44. Southam CM and Goettler PJ, Growth of human epidermoid carcinoma cells in tissue culture. Cancer, 1953. 6(4): 809-827.
45. Southam CM, Management of disseminated neoplastic disease: the direct effects of cancer. Bulletin of the New York Academy of Medicine, 1957. 33(10): 720-734.
46. Southam CM, Areas of relationship between immunology and clinical oncology. American Journal of Clinical Pathology, 1974. 62(2): 224-242.
47. Southam CM, Prospects for tumor immunotherapy in man, with emphasis on its use as an adjunct to surgery for primary operable cancer. Journal Medical Libanais—Lebanese Medical Journal, 1975. 28(1): 43-66.
48. Southam CM, Laboratory models of minimal residual cancer; development and preliminary immunotherapy studies. Recent Results in Cancer Research, 1976(54): 206-217.
49. Southam CM, History and prospects of immunotherapy of cancer: an introduction. Annals of the New York Academy of Sciences, 1976. 277: 1-6.
50. Southam CM, Bower EB, Currie VE, et al., Specific adoptive and passive immunotherapy be parabiosis for syngeneic mouse and rat tumors. Annals of the New York Academy of Sciences, 1976. 277: 505-515.

51. Starr D, Ethics. Revisiting a 1930s scandal, AACR to rename a prize. Science, 2003. 300(5619): 573-574.
52. Sun M, EPA bans use of Nazi data. Science, 1988. 240(4848): 21.
53. Williams DJ, The history of Eduard Pernkopf's Topographisiche Anatomie des Menschen. The Journal of Biocommunications, 1988. 15(2): 2-12.
54. Winkelmann A and Udo S, Hermann Stieve's clinical-anatomical research on executed women during the "Third Reich". Clinical Anatomy, 2009. 22(2): 163-171.
55. Xu K, Rosenwaks Z, Beaverson K, et al., Preimplantation genetic diagnosis for retinoblastoma: the first reported liveborn. American Journal of Ophthalmology, 2004. 137(1): 18-23.
56. Yoshimo K. UC Irvine fertility scandal isn't over. Los Angeles Times. 2006 January 20.

CHAPTER 12

Inappropriate Behavior by Scientists

Scientists who have isolated human genes or identified mutations in genetically determined diseases have often patented their discoveries. Crucial to this research are the affected individuals who provide written informed consent for their DNA and that of their dependent children to be analyzed. With knowledge of the genetic defect tests can be developed to identity the precise diagnosis of numerous genetically determined diseases. Also, carriers of the diseases can be detected even when the affected person is still *in utero*. For almost three decades the USA patent and trademark office (USPTO) awarded patents to the discoverers of specific human genes. This enabled researchers to profit from patents on the genes by forming biotechnology companies or by licensing the patents to companies that commercialized them.

Patents on BRCA1 and BRCA2 Genes. Patents on human genes were controversial and many contended that DNA is part of nature and not an invention subject to the regulations of patent laws and for more than a decade a debate over the issue lingered on. Certain mutations in the BRCA1 and BRCA2 genes predispose to an increased risk of both breast and ovarian cancer. By holding the patents for these two genes *Myriad Genetics Inc* had a monopoly on tests for mutations in these genes and made a fortune by charging more than $3, 000 a test. Moreover numerous women could not afford the genetic test because of the high cost. Other researchers and companies were unable to develop additional tests because the other genes were patented. Eventually a legal case brought by Molecular Pathology and others against *Myriad Genetics. Inc* reached the USA Supreme Court which unanimously reached a decision on June 13, 2013 declaring that patents held by *Myriad Genetics Inc* for the *BRCA1* and *BRCA2* genes were invalid [86]. The Supreme Court ruled that human genes were part of nature and that genomic DNA, natural phenomena and abstract ideas could not be patented, but that synthetically created DNA (cDNA) was subject to the laws of patents *(https://www.aclu.org/files/assets/12-398_8njq.pdf)*. Previously patents on other genetic information also had adverse effects on many innocent uninvolved bystanders. For example IP squabbles over genetic patents threatened the health of millions. The Centers for Disease Control and Prevention (CDC) in the USA and the NIH claimed ownership of the Indonesian influenza genes and by doing so prevented others from using them as a potential source for vaccines against a potential bird influenza pandemic. A fight over ownership of influenza genes blocked efforts to track this deadly infection.

Petr Táborský. In 1988 when Petr Táborský was a chemistry student at the University of South Florida (USF) *(http://www.usf.edu/)* he carried out research as a laboratory assistant on a project supported by *Florida Progress Corporation* [49]. The research was unsuccessful. According to Táborský a dean in the college of engineering gave him permission to carry out an extension of the unproductive *Florida Progress Corporation* sponsored research experiments on his own using a different approach with the hope of using the data in his thesis for a master's degree [88]. The student's independent research at the USF culminated in a noteworthy finding pertaining to the removal of ammonia from contaminated water [88] *(http://www.theatlantic.com/issues/2000/03/press3.htm)*.

In addition USF selfishly asserted that Táborský did not deserve any compensation for his discovery. Táborský disagreed and based on his laboratory records he successfully applied for a patent [44] *(http://www.cptech.org/ip/csm.html)*. Despite being a naïve student unaware of laws related to IP and interactions between universities and industry and the implications of creating patents, USF sued Táborský for the theft of IP and in 1990 he was found guilty of stealing trade secrets and $20, 000 worth of research material from USF. This was the earliest case of IP theft tried in a USA court. The USF as well as the *Florida Progress Corporation* maintained that despite Táborský's discovery the invention listed in his patent belonged to them as it was created in their laboratory. For the theft of what he regarded as his own research Táborský was initially sentenced to a maximum security state prison, but after two months he was transferred to a minimum-security work-release center [11]. He rejected an order by the judge to shift ownership of his patent to the USF. An offer of clemency by Lawton Mainor Chiles (1930-1998), the 41st Governor of Florida (1991-1998) was rejected by Táborský because he considered such an action equivalent to an admission of guilt [11]. Táborský established fame by being the first person in the USA to be convicted in a criminal court for theft of IP [44] *(http://www.cptech.org/ip/csm.html)*.

J. Milton Harris. In November 1990 Drs. J. Milton Harris and Sedaghat Herati of the University of Alabama in Huntsville (UAH) *(http://www.uah.edu/)* filed a patent application for "The preparation and use of polyethylene glycol propionaldehyde." In 1992 Harris created a company called *Shearwater Polymers, Inc.* and a decade later it was sold to *Inhale Therapeutics, Inc.* and renamed *Nektar Therapeutics, Inc.* and it also became known *Nektar Therapeutics AL, Inc.* Harris was accused by UAH of secretly patenting an important drug formulation. In June

1993 Shearwater entered into a license agreement with UAH for the technology embodied in the patent. A few months later a patent was issued and the rights of it were assigned to UAH according to the university's patent policy. A year after the formation of *Shearwater Polymers, Inc.* UAH established a license agreement with this company for the technology represented in the polyethylene glycol propionaldehyde patent. The profitable cancer therapeutic drug Neulasta manufactured by the large drug company *Amgen*, which uses the Harris technology also generated revenue for Harris and UAH. Legal action was brought on when *Nektar Therapeutics* contacted UAH in May 2005 and proclaimed a one-sided termination of all disbursements under the license arrangement of 1993. To settle a legal dispute Harris and *Nektar Therapeutics* agreed to pay the university $25 million. The defendants paid UAH an initial $15 million and for the next decade an extra payment of $1 million per year was paid. After the initial complaint was filed UAH found almost 30 other patents that were unknown to the institution that Harris had filed while connected with the university. UAH maintained according to the terms of its institutional patent policy Harris was required to divulge and allocate these patents to UAH, but had not done so. Hence UAH modified its grievance with additional claims against Harris on August 3, 2005. The Harris case illustrates the determination of universities to enforce their patent policies even though the creativity of their faculty is responsible for generating the IP that leads to patents

(*http://www.uah.edu/legal/pdf_files/UAH%20Settles%20Intellectual%20 Property%20Litigation.pdf*) [105].

Blood Test for Acquired Immunodeficiency Syndrome. Potential financial rewards from patents profoundly affect the behavior of scientists. Members of university faculty often obtain patents on their IP, but they frequently do not always acknowledge that the discovery of their patented IP occurred at an institution that provided resources and the intellectual environment that made the financially profitable research possible. They often are not keen on sharing the reaped rewards with institutions. When researchers establish patents in universities and research institutions they usually generate them in a partnership with the institution where the scientist performed the work.

Patented commercial tests have been developed to identify individuals with numerous genetically determined diseases, acquired immunodeficiency syndrome (AIDS) and other disorders. The Pasteur Institute applied for a USA patent on a blood test for AIDS in December 1983, but for unclear reasons a patent was not issued. Seventeen months later on May 29, 1985 the USPTO issued DHHS a patent for the human immunodeficiency virus (HIV) blood test based on the research performed in the Robert Charles Gallo (1937—) laboratory. Because of this apparent discrimination the Pasteur Institute went on the offensive. Its officials attempted to convince their counterparts at the National Institutes of Health (NIH) that a theft had taken place, but after unsuccessful discussions they took the matter to court. In 1987 the dispute was resolved by mutual consent when an agreement was reached to split equally the credit and the royalties for the finding of the virus that causes AIDS. A USA patent was provided to the French and the agreement was signed by both Gallo and Luc Montagnier at the highest possible level for the USA and France by Ronald Reagan (1911-2004), the 40th President of the USA and Jacques Chirac (1932—), who was prime minister of France at the time. This seemingly brought the clash between the Gallo and Montagnier laboratories to an end, but truce was brief until the John Crewdson article in the *Chicago Tribune* in 1989 [40] in which research misconduct in the Gallo laboratory was suspected. An analysis of the disputed situation in 1992 by an independent law firm concluded that the USA HIV blood test patent rested on a secure foundation, while the initial French patent application was less certain. An evaluation by a law firm also concluded that even if Gallo made untruthful statements during the quarrel over the patent, the patent would only be worthless if it could be verified that the pronouncements were meant to mislead, but Gallo's disputed statements were not *prima facie* false and could be open to different interpretations. Because Gallo and Mikulas Popovic were suspected of lying about their findings on the HIV virus to government officials in their application for an HIV blood test patent the Office of the Inspector General (OIG) investigated these researchers in November 1991 and issued a report indicating that they had stolen the virus sample that Montagnier sent to Gallo in 1983 from the Pasteur Institute and made use of it to develop a blood test. This report became widely known after the publication of the June 19, 1989 article by Crewdson, which reignited the decade long troubles of Gallo and Popovic.

The patent dispute about the acquired immunodeficiency syndrome (AIDS) virus blood test began in 1983, when USA and European patent applications were put on record for the research performed in the laboratory of Montagneier. The USA federal government filed a patent application for the Gallo blood test a year later in 1984, but the USPTO only issued the patent in 1985 without reaching a decision on the Pasteur Institute application. Several months later a meeting between representatives of DHHS and the Pasteur Institute discussed the contention that the Pasteur Institute warranted priority with regard to the patent. Not only did the Pasteur Institute submit a patent application first, but one of its researchers, Montagnier was the first person to isolate the virus responsible for AIDS. To add insult to injury Gallo's blood test for AIDS was developed using Gallo's viral isolate named HTLV-IIIB (and later referred to as IIIB), which was strikingly similar to the virus that Montagnier isolated from patient BRU and labeled LAV before being sent to Gallo. On behalf of the DHHS the National Cancer Institute (NCI) reviewed the research of Gallo to determine whether the patent royalties should be split between the Americans and the French. This issue provoked claims and lawsuits in different courts in the USA by the Pasteur Institute. The clash between the American and French researchers created headlines in newspapers and the battle became expensive for both sides. The patent was first challenged in 1985 by the Pasteur Institute attorneys who claimed that Gallo had created the diagnostic procedure with LAV/BRU that he had siphoned off from the Pasteur institute from which his laboratory generated the blood test. After a long unpleasant heated discussion an agreement was reached between the USA government and the Pasteur Institute in March 1987. Both Montagnier and Gallo joined forces on each other's patents and DHHS and the Pasteur Institute agreed to split the royalties equally. Moreover an understanding was reached to transfer 80% of the royalties into a newly created French and American AIDS Foundation devoted to supporting AIDS research in developing countries. Soon after its establishment the French and American AIDS Foundation pledged twenty five percent of this revenue to another organization, the World AIDS Foundation, which already supported AIDS research in developing countries. Gallo's admission that his isolate IIIB was indeed identical to LAV/BRU was well established before the 1987 agreement. The dispute was re-evaluated by a committee of specialists at the National Institutes of Health (NIH).

Now and then members of the faculty conceal the existence of their patent(s) and their involvement with biotechnology companies that are commercializing the patent. Some do this because of a lack of loyalty to their institution or the belief that their employee played little or no role in their creation of the IP. Others do it for greed. When researchers attempt to deprive institutions of a portion of the generated revenue that they regard as their due nasty disputes sometimes arise rise between institutions and researchers over patents and between different institutions and many such examples have been documented.

The dispute over the AIDS/HIV patent royalties was expected to be resolved at the annual meeting of the board of the French and American AIDS Foundation on July 11. 1994, which was set up by the Reagan-Chirac accord to watch over royalties from the blood tests [35, 61], but an unpleasant prolonged squabble between France and the USA erupted again with regard to who should get credit and financial reward for unearthing the cause of AIDS. The Pasteur Institute demanded a larger portion of the royalties derived from the HIV blood test. In 1994 Michael Astrue, the top ranking DHHS lawyer contended that the Pasteur Institute did not have compelling legal reasons to receive a larger percentage the royalties from the HIV blood test, but he recommended that the USA willingly give up part its royalties expected to amount to $2 million to the Pasteur Institute, because the damaging USA relations with France was dampening international scientific collaboration. Nevertheless, Harold Varmus, the NIH director at the time insisted that the royalty agreement remain unchanged. Prior to the report of the DHHS's OIG Varmus, believed that there was not a valid reason for rearranging the HIV blood test royalties, but following the OIG report Epstein the attorney for the Pasteur Institute indicated that he might sue on behalf of his client. When the virus was sent to Gallo from the Pasteur Institute a contract pledged not to use it for money-making endeavors. Because of its use in the development of the HIV blood test charges were brought against Gallo and conflicting understandings emerged.

Greenberg versus Miami Children's Hospital. Twenty five years ago the father of two children with the rare genetically determined Canavan disease, caused by a genetic deficiency of aspartoacylase, convinced Reuben Matalon a pediatrician with expertise in genetics, to develop a molecular probe for the diagnosis of this disease that affects 1 in 6, 400 Ashkenazi Jewish children. The gene was found by Matalon and his colleagues in 1993 [71] and he developed a genetic test, which was patented in 1997 by Miami Children's Hospital (MCH) where Matalon was employed at the time. During the following year the test was licensed to identify the lethal mutations responsible for Canavan disease. The commercial test stemmed from data obtained without informed consent from the blood and tissue of affected patients. Because of the patent the Canavan Foundation was compelled to cease offering genetic screening because it would need to pay royalties and meet licensing terms. While comparable situations have arisen with other disease producing genes the control of the gene had not been challenged, but with regard to the Canavan disease gene four parents and three non-profit groups took the matter to court and sued Matalon and the MCH [82]. The case was settled out-of-court with a sealed resolution. Other research groups anticipated disputes over patents and arranged to share the control of the patents.

Andrew J. Wakefield. In June 1997, almost nine months before Andrew J. Wakefield announced at a press conference that single vaccines were probably safer than the measles, munps, rubella (MMR) vaccine he filed a patent on products, including a single measles vaccine, which had little hope of triumph unless support for MMR was destroyed. Confidential documents indicated that a group of companies planned to raise venture capital from professed inventions—including a vaccine, testing methods, and strange potential miracle cures for autism.

Peter Seeburg. Because the University of California San Francisco (UCSF) (*http://www.ucsf.edu/*) held a patent on genetically engineered human growth factor it became involved in a bitter legal dispute with Genentech, a pioneering biotechnical company that was breaking new ground in San Francisco. During the trial Peter Seeburg, a neurochemist, testified on behalf of UCSF stating that he moved to Genentech in 1978 and used growth hormone DNA that he had assisted in cloning while a postdoctoral fellow at UCSF and that he made the vectors necessary for Genentech to market the extremely profitable drug, Protropin. This drug allegedly generated $2 billion by 1999 [22] and UCSF acknowledged that during Seeburg's move to Genentech unequivocal university regulations barred him from taking the samples with him.

This money generator naturally perked the ears of UCSF which wanted a piece of the pie. Early in 1999 the disagreement came to trial and after 8-weeks the jury reached an impasse with 8 to 1 supporting the university. In April 1999 Seeburg testified that he together with a former colleague at Genentech [59] concealed the true origin of the samples. This acknowledgment triggered an investigation into scientific misconduct by the Max Planck Society Germany's foremost research establishment. This censure came after a prolonged confrontation over the rights to a patent involving genetically engineered human growth hormone in court between UCSF and Genentech. In 1999 Seeburg informed *Science* that he may donate part of the $17 million that he and four former UCSF scientists will each receive as part of the settlement to a research foundation or a charity. By November 1999 Seeburg was director of the Max Planck Institute for medical research in Heidelberg, Germany. In December 1999 twenty years after Hubert Markl the President of the Max Planck Society officially reprimanded Seeburg for publishing factious data in a 1979 paper in *Nature* [22, 59]. Seeburg was insignificantly affected and none of his papers were retracted or corrected. His position at the Max Planck institute was not changed, but a report of the scientific misconduct investigation ended up in his personnel file [60]. A retrial in the UCSF-Genentech dispute was scheduled, but prior to that date Genentech consented to pay UCSF $200 million [21]. Part of the settlement was a $50 million donation towards a research building at UCSF and $85 million was split between each of 5 ex-UCSF scientists who first cloned the human growth hormone. Howard Goodman and two of his former postdoctoral fellows (Peter Seeburg and John Shine) are named on the patent as co-inventors.

John Madey. John Madey, a brilliant physicist, invented the tunable free electron laser (FEL) when he was a graduate student at the California Institute of Technology (Caltech) *(http://www.caltech.edu/)* and soon thereafter he refined it at Stanford University (Stanford) *(http://www.stanford.edu/)*. He was then recruited to Duke University (Duke)*(http://duke.edu/)* at considerable expense in 1989 largely because he was a potential Nobel laureate and Duke desperately needed a faculty member who could win this coveted prize. The Nobel laureate count at research institutions is an objective measure of the best universities in the world. The discovery of the FEL had considerable potential to lead to a Nobel Prize, because intense light can be tuned into a wide range of wave lengths of potential therapeutic and military value, including the uncovering of concealed nuclear and chemical weapons.

Rumor has it that the price tag for Duke's recruitment package for Madey was higher than what had ever been offered for any faculty recruit in the history of university. The FEL facility was extremely well-funded beginning in 2000 with $5.4 million from the nuclear treaty program and another $49 million arrived in 2002 for basic research budgeted at more than $3 million a year. Madey was unquestionably a big catch and a feather in the cap for Duke. The original FEL, designated Mark III, was built at Stanford by Madey and it was brought by him to Duke in 1992. Madey subsequently hired Vladimir Litvinenko to manage a Russian FEL in the Duke resource that functions in the ultraviolet wavelength. The Duke FEL facility was promoted as a medical program to build advanced lasers, delve into materials, and especially to create novel surgical and other therapeutic products. Regrettably the medical potential of tunable lasers did not materialize and this goal was probably overestimated. Moreover a serious conflict emerged over the balance between priorities of research related to physics and to medical applications of the FEL. A dispute over how the laboratory should be managed culminated in the removal of Madey in 1997 as director of the Duke FEL Laboratory and as principal investigator (PI) of the grant that funded the research. Madey was adamant that Duke wrongly removed him as chief of the research center. The university offered Madey the title of chief scientist at the defense department-funded center, but he rejected the offer being unhappy with the administrative change. A difference of opinion over the future course of the laboratory was partially responsible for bringing both machines in the FEL laboratory to a standstill over 4 months. Madey considered locating the FEL elsewhere, but Duke resisted and became involved in a long-drawn-out bitter dispute with him [79] over the use of patented tools in academic research. After the departure of Madey Duke continued to use his lasers and other equipment, but without his permission. This infuriated Madey who asserted that this was an infringement of his patents. He sued Duke and a lower court ruled in favor of the university asserting that the use of these devices at the university was not a patent infringement and that scientists at Duke were not engaged in commercial projects. Duke's position was aided by the professed research exemption clause stemming from an 1813 case. The latter put forward the notion that investigators could freely make use of patented pieces of equipment "for amusement, to satisfy idle curiosity, or for strictly philosophical inquiry". Madey was not satisfied with the lower court decision and took the case to a federal court of appeals [1] (Madey v. Duke University, No.01-1567, Federal Circuit Court of Appeals, 3 October 2002). This higher court passed judgment in favor of Madey, who joined the faculty of the University of Hawaii (UH) *(http://www.hawaii.edu/)* in 1998 and demanded that Duke forward his FEL to him at his new Hawaiian laboratory [78]. In 2003 Duke and other institutions were not happy with the 190-year-old policy in which academic scientists could get free use of patented tools in academic research. In 2007 after a nine-year ownership battle with Duke Madey finally received guardianship of his unique FEL at the UH where he had obtained a faculty appointment in 1998 [4] *(http://archives. starbulletin.com/207/01/21/news/story06.html)*. The transfer of this FEL was no small task as it consisted of more than one thousand pieces *(http://archives. starbulletin.com/207/01/21/news/story06.html)*. After the departure of Madey from Duke the university recruited a Nobel Prize winner on its faculty by hiring Peter Agre *(http://www.nobelprize.org/nobel_prizes/chemistry/laureates/2003/agre-bio.html)* who shared the 2003 Nobel laureate in chemistry with Roderick McKinnon *(http://www.nobelprize.org/nobel_prizes/chemistry/laureates/2003/mackinnon-lecture.html)*. Agre did not obtain his Nobel Prize for work performed at Duke, but in 2012 Robert J. Lefkovitz became Duke's first home grown Nobel Prize recipient when he shared the 2012 chemistry Nobel Prize *(http://www.nobelprize.org/nobel_prizes/chemistry/laureates/2012/lefkowitz-facts.html)* with Brian K. Kobilka *(http://www.nobelprize.org/nobel_prizes/chemistry/laureates/2012/kobilka-lecture.html)*.

John Bennett Fenn. John Bennett Fenn (1917-2010) *(http://en.Wikopedia. org/wiki/John_Fenn)* obtained a PhD from Yale University (Yale) *(http://www. yale.edu/)* and subsequently joined its faculty and rose to the rank of full professor while becoming a distinguished chemist. When Fenn reached 70 years of age he was required to retire according to the policies of Yale. He was given the title of emeritus professor and an office, but could no longer perform his own research at the university. A couple of years after Fenn retired from Yale the university inquired about the possible money-making worth of his research. Fenn feigned its trivial value. Nevertheless he continued to complete his electrospray research some distance away from that institution. His cutting edge research on liquid chromatography mass spectrometry earned him a share of the 2002 Nobel Prize in chemistry *(http://www.nobelprize.org/nobel_prizes/chemistry/laureates/2002/ fenn-lecture.html)*, and he split half for the Prize with Koichi Tanaka (1959—) another pioneer in mass spectrometry *(http://www.nobelprize.org/nobel_prizes/ chemistry/laureates/2002/tanaka-lecture.html)*; the rest of the award went to Kurt Wüthrich (1938—) *(http://www.nobelprize.org/nobel_prizes/chemistry/ laureates/2002/wuthrich-lecture.html)* for "his development of nuclear magnetic resonance spectroscopy for determining the three-dimensional structure of biological macromolecules in solution". Under the *Bayh-Dole Act* of 1980 Fenn believed that he had the intellectual rights on his invention of electrospray technology so without the knowledge of officials at Yale he personally patented the technology and sold the licensing rights to *Analytica*, a company to some extent owned by him. Another company expressed an interest in the electrospray technology developed at Yale and contracted the university which discovered that Fenn held the patent. Yale obviously wanted a percentage of the royalties generated and this precipitated a dispute between Fenn and Yale, but attempts to settle it were unsuccessful so Fenn filed a lawsuit against Yale and the university countersued. An out of court settlement could not be reached and in 2005 the USA district Judge Chrisopher Droney ruled in favor of Yale and awarded the university $545, 000 in royalties and $500, 000 in legal fees [84]. Despite winning the case Yale received bad publicity because of the manner in which it dealt with a Nobel laureate who had been associated with Yale for 68 years [73].

Gerald P. Schatten. Gerald P. Schatten a professor at the University of Pittsburgh (Pitt) *(http://pitt.edu/)* was a coauthor of papers on the high profile fraudulent stem cell research of the disgraced Hwang Woo-suk. Despite an admission that he did not really contribute anything to those publications of Hwang on which he was a coauthor Schatten attempted in April 2004 to patent Hwang's technology to create embryonic stem cells with two Pitt researchers (Calvin Simerly and Christopher Navara). Schatten's patent application to the USPTO was drawn to public attention by the *Pittsburgh Tribune Review* in its online Saturday edition. It was noteworthy that his patent application Schatten failed to include Hwang with whom he had previously worked. Korean scientists considered this move outrageous particularly as Hwang has submitted an application for a Patent Cooperation Treaty (PCT) application in December 2003 and had reported the discovery of the method in *Science* in 2004 [69]. However the immediately humiliated breakthrough was pronounced in a manuscript reported in the journal *Science* in June 2005 [70].

Craig B. Thompson. In 2002 the philanthropist Leonard Abramson and his family donated $100 million to the University of Pennsylvania (UPenn) *(http://www.upenn.edu/)* for cancer research. In exchange the university named a cancer research institute "The Leonard and Madlyn Abramson Family Cancer Research Institute" after the generous contributors. The Abramson family was one of the largest donors to UPenn and they stipulated that the institute would own part or all of the discoveries made at the institute. In 2007 Dr. Craig B. Thompson (the current President of the renowned Memorial Sloan-Kettering Cancer Center in New York City), Dr. Cantley and Tak W. Mak of the University of Toronto (UToronto) *(http://www.utoronto.ca/)* created the biotech company *Agios Pharmaceticals*, which focused on developing new treatments for cancer. Thompson's research suggested that metformin, a drug useful in the treatment of diabetes mellitus, might be useful in reducing the risk of cancer [30]. Thompson coauthored some papers involving the isocitrate dehydrogenase 1 (*IDH1*) and isocitrate dehydrogenase 2 (*IDH2*) genes [55, 102] that were the focus of the privately owned *Agios Pharmaceticals*. With a series C round of funding by a syndicate that included a venture capitalist, public investment funds *Agios Pharmaceuticals* raised $78 million. Thompson later formed the drug company *Celgene Corporation (CELG)* that invested in *Agios Pharmaceuticals*. These companies were created when Thompson was employed at the Leonard and Madlyn Abramson family cancer research institute for 12 years. Towards the end of 2012 *CELG* increased its investment in *Agios Pharmaceuticals* to $150 million. Thompson' former employee the Abramson institute filed a $1 billion lawsuit against him and his two companies on December 13, 2011 in the USA district court in Manhattan. After he became President of the prestigious Memorial Sloan-Kettering cancer center, Thompson was accused of using valuable IP that he developed while working at the Abramson research institute to start a biotechnology company that turned his discoveries into revenue without sharing any of it with his prior employee. According to the lawsuit the Abramson Institute maintained that it was not aware of Thompson's involvement with *Agios Pharmaceuticals* until 2011 and that Thompson concealed his involvement with the biotechnology company and allegedly deprived the Abramson cancer institute of the IP that it owned. Thompson considers the allegations in the lawsuit unfounded [18].

Failure to Give Credit when Credit is Due

Technicians and other research support staff are paid to perform various duties and are not usually expected to direct or interpret the research or to write or coauthor papers reporting significant observations. As a rule they do not anticipate receiving more than their salary for what they do, but they appreciate recognition for their small role in furthering scientific information. Indeed, although times are changing, they did not usually receive acknowledgement at the end of papers stemming from their contributions because most professional journals do not want to acknowledge research support staff. Coauthorships were often provided for those with notable technical contributions.

It is not uncommon for junior researchers, who play an important role in new discoveries, to complain about their mentors, supervisors or others for stealing credit for their work. Receiving a Nobel Prize has always been regarded as the pinnacle of scientific achievement and the greatest honor that a scientist can receive. While most awardees are well deserved some are especially controversial. Major discoveries by some scientists bring out the worst in them when they fail to give appropriate credit to others that unearth the same finding independently or contribute immensely to the discovery that made them famous.

John James Rickard Macleod. Persons in this category include John James Rickard Macleod (1876-1935), who was awarded the 1925 Nobel Prize mired in controversy for physiology or medicine (*http://www.nobelprize.org/nobel_prizes/medicine/laureates/1923/macleod-lecture.html*) with Frederick Grant Banting (1891-1941) (*http://www.nobelprize.org/nobel_prizes/medicine/laureates/1923/banting.html#*) despite the fact that the discovery of insulin was a product of teamwork involving Banting, Macleod, Charles Best (1899-1978), James Bertram Collip (1892-1965), and probably others. Before World War I ended the structure of the pancreas and the link between the pancreatic islets of Langerhans and diabetes mellitus had been established. Insulin (pancrein) had been isolated by a method patented by Nicolae Paulescu, a professor of physiology at the University of Medicine and Pharmacy in Bucharest Romania *(http://www.umf.ro/)* [85], but no clinical value came from his work. The separation of insulin from the pancreas was difficult largely because of its degradation by pancreatic exocrine enzymes. At 32 years of age Banting, a surgeon living in the town of London in Western Ontario, Canada had a brilliant innovative thought about how the pancreatic internal secretions could be isolated. He hypothesized that the pancreatic enzymes needed for digestion broke down the islet secretions. To test his hypothesis he approached Macleod, a distinguished physiologist at the UToronto, but Macleod was skeptical about whether it would work. Nevertheless, Macleod reluctantly provided Banting with Charles Best, who was an undergraduate student assistant at the time, eight dogs and laboratory space. Macleod apparently also offered worthwhile ideas to facilitate the research. Macleod then went to Scotland on vacation in 1921 and during that time Banting solved the insulin mystery with Best. After occluding the pancreatic ducts of dogs with a suture pancreatic secretion accumulated above the suture and insulin could be isolated. After two months of relentless effort an extract of the pancreas was injected into a dog whose pancreas had been excised. During the following hour the dog's blood-sugar plummeted by 40%. The dog recovered from coma and its health improved and eventually walked within the laboratory. A few months later Macleod received a request for abstracts for the annual meeting of the *American Physiological Society* so he suggested that Banting and Best report the insulin discovery. Banting requested that Macleod's name be associated with the research paper to draw attention to it and Macleod went along with the suggestion. Banting and Best started drafting a manuscript in late November 1921 documenting their research findings until that time. Because Macleod assisted in finalizing the manuscript Banting asked him if he would like to be a coauthor, but Macleod expressed his appreciation for the offer but declined the invitation even though it was in his rights to have his name included. Macleod stated that it was their project and "he did not wish to fly under borrowed colors". After receiving daily injections of an extract of impure insulin in January 1922 the health of a 14-year-old 65 pound diabetic boy rapidly improved, but the preparation provoked a marked allergic reaction and this treatment was discontinued. The findings generated headlines throughout the world. The historically important landmark manuscript that affected the lives of millions was published in the February 1922 issue of the esteemed *Journal of Laboratory and Clinical Medicine* [20]. It contained minor factual errors and other flaws that would be seriously criticized for not meeting contemporary standards. The assertion that pancreatic extracts invariably reduced the blood and urine sugar was not true. Some figures in their graphs disputed others and/or contradicted laboratory records. The portrayal of the final experiment was especially substandard. Because Banting and Best had difficulty purifying the pancreatic extract, Macleod allowed Collip to join the research team and without delay he prepared extracts from beef pancreas and in less than a month obtained a sufficiently pure pancreatic extract to use in clinical trials.

When the isolation of insulin became possible considerable interest was aroused by both the medical profession and the general public because insulin dramatically prolonged the lives of patients with diabetes mellitus. Prior to this a diagnosis of juvenile diabetes was a kiss of death with a lethal outcome usually within a year. Insulin injections soon started saving lives of millions of human beings worldwide.

Soon thereafter a dispute arose about who should receive credit for the discovery of how to isolate insulin. Macleod clandestinely organized an assessment of the research when it became apparent that a Nobel Prize was in the offering for the insulin research. It was suggested that Macleod bask in the success of his underlings. In the late fall of 1922 rumors about a potential Nobel Prize in medicine or physiology going to scientists at UToronto began when prominent scientists, such as Schack August Steenberg Krogh (1874-1949) the Danish physiologist and 1920 Nobel Laureate in physiology or medicine (*http://www.nobelprize.org/nobel_prizes/medicine/laureates/1920/krogh-bio.html#*) visited Toronto. He learnt all that he could about the insulin discovery and his visit was the talk of the town. In September 1922 an intense quarrel between Banting and Macleod broke out because the British physiologist Sir William Bayliss (1860-1924) was quoted by

The Toronto Star as expressing the opinion that credit for the isolation of insulin should go exclusively to Macleod. In a letter to the *Times* Macleod corrected this misconception and pointed out that Bayliss was in error in stating that "the idea of preparing insulin from pancreas sometime after ligation of the pancreatic duct originating with me. As a matter of fact, this is particularly the part of the work that originated with Banting" [7]. It became evident that the contribution of Best was being lost in the Banting and Macleod conflict. Even though Banting and Best deserved complete recognition for the original discovery, Banting was only prepared to acknowledge Macleod for functional studies on insulin. In September 1922 in an unpublished written text about the discovery of insulin Macleod pointed out that both Banting and Best warranted complete credit for the initial research. This astonishing declaration remained hidden in the archives of UToronto for longer than half a century. The committee that selects Nobel Prize winners thought that Banting and Best should share the award, but Macleod diverted the triumph of his underlings [99]. Later Krogh nominated Banting and Macleod for the 1923 Nobel Prize in physiology or medicine for their discovery of insulin and on August 25 1923 a secret ballot by nineteen distinguished members of the Nobel council voted to accept this recommendation. Banting was infuriated when he heard this news on 28 October and was ready to reject the Prize because of what he thought was injustice, but colleagues controlled his outrage stressing his obligations to Science and his country. After becoming appeased Banting proclaimed that would divide the money and the glory with Best. About a week later Banting appealed to Collip to share the riches of the Prize.

Because of the controversy regarding the 1925 Nobel Prize the President of UToronto avoided reigniting the dispute in the mid 1950's by inappropriately using the power of his office to block knowledge about Macleod's unpublished document. After painstakingly researching the history of the discovery of insulin the Canadian historian Michael Bliss (1941—) of the UToronto was fortunate to gain access to important information in Nobel archives. He released a book on the subject in 1982 [28].

According to Rolf Luft (1914—2007) a past chairman of the Nobel committee for physiology or medicine the awarding of the Prize to Macleod was one of the sorriest blunders of commission in the record of this prestigious award. Further newspaper comments precipitated additional conflict and greater hatred between Banting and Macleod. This led to a bitter battle of words.

Selman Waksman. Selman Waksman (1888-1973) was an undergraduate student at Rutgers College, now known as Rutgers University (Rutgers) *(http://www.rutgers.edu/)*, and then attended the University of California, Berkeley (UC Berkeley) *(http://www.berkeley.edu/)* where he obtained a PhD in biochemistry in 1918. Later Waksman returned to Rutgers where he joined the faculty in the departments of biochemistry and microbiology. Whilst at Rutgers Waksman was an outstanding biochemist and microbiologist whose research focused mainly on soil living organisms. During his career his laboratory unearthed more than a few antibiotic generating microorganisms that became valuable in the therapy of infections. Waksman coined the term antibiotics and became wealthy from income derived from the patents of the more than fifteen antibiotics that he discovered. Waksman studied the *Streptomyces* family of microorganism including *Streptomyces griseus* from which streptomycin was isolated. His wealth enabled him to fund a foundation for microbiological research and he created the Waksman Institute of Microbiology at Rutgers in Piscataway, New Jersey (USA).

In 1952 Waksman was awarded the Nobel Prize for "his discovery of streptomycin, the first antibiotic effective against tuberculosis" *(http://www. nobelprize.org/nobel_prizes/medicine/laureates/1952/waksman-lecture.html#)*. However, it was appreciated that his selection for this coveted award was for his numerous scientific contributions and not only for that specific discovery, but after becoming a Nobel laureate Waksman became recognized as the only discoverer of streptomycin. His graduate student Albert Schatz (1922-2005) tenaciously challenged this assumption [92] and in 1950 initiated a lawsuit in opposition to Waksman for credit as the co-discoverer of streptomycin, Later Schatz became lawfully recognized as the co-discoverer of this enormously valuable antibiotic and collected a portion of the royalties derived from streptomycin. Waksman and Rutgers reached an out of court settlement and for a part of the income derived from streptomycin royalties. A noteworthy essay on the role of Schaltz in the streptomycin discovery has been written by Milton Wainwright [101]. In early 1942 Schatz worked as a laboratory aide at Miami hospital while drafted by the army in World War II. After witnessing youthful soldiers perish from penicillin resistant infections he became inspired to seek soil organisms that might hamper the growth of penicillin-resistant bacteria. He detected some hopeful microbes and sent them to Waksman for additional evaluation. After his release from the army in 1943 for health reasons Schatz retuned to graduate school, where he persisted in his research on soil bacteria in the Waksman laboratory at Rutgers. A patent for streptomycin lists both Waksman and Schatz. Although Waksman is listed first on the Patent, Schatz was the first author of a scientific paper on the subject [92]. Shortly after the discovery of streptomycin Schatz's PhD thesis on this antibiotic was submitted. Eventually Schatz received monetary reward and "legal and scientific' recognition as co-discoverer of streptomycin [6]. In 1994 Rutgers went even further in awarding Schatz the Rutgers medal for his research in the development of streptomycin.

Ian Wilmut. Another scientist, who originally claimed undue personal credit for the work performed by colleagues, was the English embryologist professor Ian Wilmut, director of the medical research council center for regeneration at the University of Edinburgh *(http://www.ed.ac.uk/)*. He was most well-known for his role as the leader of the team at the Roslin Institute outside of Edinburgh that cloned the celebrated Finnish Dorset lamb in 1996 named Dolly in honor of Dolly

Parton [32]. This cloned product was the first mammal to be generated from an adult somatic cell [58, 93], in 1999 for his contributions to embryonic development [50] Wilmut was awarded Order of the British Commonwealth (OBE) (*http://www.timeshighereducation.co.uk/story.asp?storyCode=146682&sectioncode=26.*) and in December 2007 it became known that he was to be knighted by Queen Elizabeth II in the traditional 2008 new year honors [15]. Past staff members of the Roslin Institute blamed Wilmut for acknowledging excessive credit for his role in the creation of Dolly and they considered him to be a "self-confessed charlatan" with an apparent lack of adequate scientific understanding" [75]. These colleagues of Wilmut unsuccessfully petitioned the Queen to deprive him of his knighthood [37, 39] (*http://www.telegraph.co.uk/news/uknews/1512377/I-didnt-clone-Dolly-the-sheep-says-prof.html*) (*http://www.telegraph.co.uk/news/uknews/1512377/I-didnt-clone-Dolly-the-sheep-says-prof.html. Retrieved 2009-04-16*). After the creation of Dolly Wilmut received credit in headline news around the world as the mastermind behind the creation of this lamb and he basked in the glory of this scientific breakthrough without giving appropriate credit to other members of the research team that cloned Dolly. It was only after a decade that he admitted that his contribution in the giant scientific step was "not trivial", but mainly supervisory [39]. Wilmut was made to admit the truth when Dr. Prim Singh, a 45 year old molecular biologist, accused him of racial discrimination, brow beating and attempted theft of IP. These allegations brought Wilmut before an employment tribunal hearing and during questioning under oath Wilmut pointed out that he neither developed the technology, nor performed the crucial experiments. In describing his role Wilmut acknowledged that Dr. Keith Campbell played the major role (66%) in the cloning of Dolly. Wilmut maintained that he only assumed the lead role among the research team because of an earlier arrangement with Campbell.

James Dewey Watson and Francis Crick. The co-discovery of the structure of desoxyribonucleic acid (DNA) by James Dewey Watson (1928-) (*http://www.nobelprize.org/nobel_prizes/medicine/laureates/1962/watson-bio.html*) and Francis Crick (1916-2004) (*http://www.nobelprize.org/nobel_prizes/medicine/laureates/1962/crick-bio.html*) in 1953 rocked them into sharing the Nobel Prize in physiology or medicine in 1962 with Maurice Hugh Frederick Wilkins (1916-2004) (*http://www.nobelprize.org/nobel_prizes/medicine/laureates/1962/wilkins-bio.html*). Tragically the British biophysicist and X-ray crystallographer Rosalind Elsie Franklin (1920-1958), who made critical contributions to the understanding of the fine molecular structures of DNA, ribonucleic acid (RNA), viruses, coal and graphite [56] (*http://profiles.nlm.nih.gov/ps/retrieve/Narrative/KR/p-nid/186*) failed to be included among the 1962 Nobel laureates because she died at the age of 37 years on April 16, 1958 from complications of ovarian carcinoma and according to tradition the Nobel Prize is only awarded to living individuals. The DNA work achieved the most fame because DNA plays essential roles in cell metabolism and genetics, and the discovery of its structure helped scientists understand how genetic information is passed from parents to children. Franklin was best known for her crucial observations on X-ray diffraction images of DNA that led to the Watson-Crick discovery of its double helix structure. According to a letter dated 31 December 1961 from Crick to Jacques Lucien Monod (1910-1976) the French biologist who received the Nobel Prize in physiology or medicine in 1965 (http://www.nobelprize.org/nobel_prizes/medicine/laureates/1965/monod-bio.html) Watson and Crick used the data of Franklin to interpret their findings. This letter was discovered in the archives of the Pasteur Institute by Doris Zeller and it was reprinted in *Nature* [106]. At the opening of the King's college Franklin-Wilkins building in 2000 Watson confirmed this historic fact in formulating the Watson and Crick's 1953 hypothesis regarding the structure of DNA [103]. Franklin's X-ray diffraction image of DNA confirmed that this macromolecule had a helical structure and the data related to her observations were shown to Watson without her authorization or awareness. Her precise interpretation of the data provided precious insight into the DNA structure. Despite this Franklin's scientific contributions to the discovery of the double helix are often overlooked. Unpublished drafts of manuscripts by Franklin reveal that she had determined the DNA helix by herself as well as the sites of the phosphate groups in the molecule. She did not receive the acknowledgement that was her due because her crucial research was published third among three papers on DNA in *Nature*. The papers were led by the Nobel Prize winning one by Watson and Crick and it failed to give adequate credit for her role in their hypothesis [12, 103] (http://www.nature.com/nature/nature/dna50).

Theft of Intellectual Property and Trade Secrets

Bin Han. In recent times Bin Han a forty year old Chinese scientist, who graduated from Xi'an Jiaotung University *(http://www.xjtu.edu.cn/en/)* in China in veterinary studies, immigrated to the USA and became a naturalized citizen. He worked in the department of ophthalmology at the University of California Davis (UCDavis) *(http://www.ucdavis.edu/)* for thirteen years until he was fired on May 13, 2002. A week later Han was arrested, jailed and charged with stealing trade secrets, possession of stolen property and embezzlement. In particular he was charged with stealing vials containing protein used in ophthalmological studies with the intent of taking the materials to China and to profit from them [74]. Han was indicted for clandestinely keeping half of the material from forty vials of protein gels used in corneal transplant research belonging to the university in his home freezer. He was also accused of the more serious crime of economic espionage, but these allegations were dropped by the prosecutors. During an investigation the police found a one-way plane ticket to China for Han backing up the belief that he planned to skip the country and not return, but Han maintained that he planned to visit his mother who was in poor health in China. He also claimed that he merely picked up the gels from a neighboring company and had not yet found sufficient time to deliver them to the laboratory. In August 2002, a jury found Han not guilty of stealing the vials. On learning about this case Asian-Americans in particular were appalled and some were outraged claiming that they were being unjustly targeted at USA laboratories. The charges were ultimately deceased to a solitary misdemeanor accusation of petty theft and embezzlement. Because the gels were readily available in China they had a trivial value of less than $400. After a jury passed a no-guilty verdict Han stated that he might fight to get his job back. This case had similarities to the one charging Dr. Wen Ho Lee (1939—), a physicist who worked at the Los Almos National Laboratory for the University of California (UC) with theft of classified material. In 1999 Lee was indicted by a grand jury for stealing secrets about the USA arsenal for the People's Republic of China, but the federal government could not prove these accusations.

The production of monoclonal antibodies involving human-mouse hybrid cells was first documented in 1973 [94], but this work was controversial and questionable [31]. Eventually this technology became widely accepted after a seminal paper by Kohler and Milstein [72] and in 1984 Georges J. F. Köhler *(http:// www.nobelprize.org/nobel_prizes/medicine/laureates/1984/kohler-lecture.html)*, *César Milstein (http://www.nobelprize.org/nobel_prizes/medicine/laureates/1984/ milstein-lecture.html)* and Niels Kaj Jerne *(http://www.nobelprize.org/nobel_prizes/ medicine/laureates/1984/jerne-lecture.html)* shared the Nobel Prize in physiology or medicine for the discovery of how to generate monoclonal antibodies.

Hideaki Hagiwara. Ivor Royston, an oncologist and an associate professor at the University of California in San Diego (UC San Diego)*(http://www.ucsd. edu/)* created hybrid cell lines by fusing a specific *antibody*-producing B cell with a *myeloma* (B cell cancer) cell. These so-called hybridomas were produced with Hideaki Hagiwara, a young postdoctoral visiting Japanese researcher in his laboratory. This was done by fusing lymphocytes from the mother of Hagiwara, who had cervical carcinoma, with a human cell line. In July 1982 Royston was surprised to learn that Hagiwara had taken this promising new cell line, which was of potential scientific and commercial value, back to Japan without his authorization. The Japanese scientist did this to treat his moribund mother and indeed he injected the monoclonal antibody into his mother before the antibodies had been thoroughly tested. Initially he even injected the antibody into himself, his father and three volunteers to rule out adverse effects. Soon after the removal of the cell line the professor and his past fellow became entangled in a dispute over ownership of the hybridoma and who should receive recognition for the research. The dispute was accentuated by the fact that the hybridoma was derived from lymphocytes obtained from Hagiwara's mother and Hagiwara insisted that he rightfully owned the cells for this reason. At the time of this dispute the creation of hybridomas and monoclonal antibodies was something new, but the scientists responsible for creating the hybridoma presumed that the synthesized monoclonal antibodies might be able to treat cancer. The Royston-Hagiwata monoclonal antibody reacted with cancer cells from the cervix, lung, colon and prostate, but not with normal cells such as fibroblasts and those in blood. Royston was concerned about the theft and became increasingly apprehensive that Hagiwara had applied for a hybridoma patent in Japan and had taken the cells to use them commercially. Royston was concerned that UC San Diego might have lost its opportunity to patent the hybridoma. The law regarding the ownership of the cell line remained unexplored in 1983, but this difference of opinion was eventually settled [97]. While it remains debatable whether scientists that merely fuse cells should take ownership of the product or whether the individual who contributed the cells should also be considered to have ownership rights, especially if the cells had commercial value legally the issue depends on who owns the patent.

Jiang Yu Zhu and Kayoko Kimbara. After receiving an undergraduate degree from the prestigious Beijing University *(http://www.china.org.cn/)* Jiang Yu Zhu, a Chinese citizen, obtained a PhD in biochemistry from Temple University (Temple) *(http://www.temple.edu/)* before accepting a postdoctoral position in 1997 with Frank McKeon, a cell biologist at Harvard University (Harvard) *(http://www.harvard.edu/)* Medical School. In 1998 his Japanese born wife, Kayoko Kimbara accepted her doctoral degree from Tokyo University *(http://www.u-tokyo.ac.jp/en/)* before also joining the McKeon laboratory late that year. While at Harvard the husband-wife team sought novel agents to aid in the avoidance of organ transplant rejections by screening genes and proteins. In 1999, Kimbara identified two useful genes of potential commercial value. While in the McKeon laboratory the post doctoral duo signed an agreement giving Harvard all rights to any discovery or invention made by them during their research fellowships, yet when Harvard patented the two genes the university listed both Zhu and Kimbara as co-inventors. The duo started to toil long hours and supposedly rejected an opportunity to have significant dialogue with McKeon, who became aware that they were no longer sharing information about other potentially valuable genes discovered by them.

McKeon and his colleagues at Harvard evidently had a sufficiently favorable impression of Zhu to commend him for a job at the University of Texas at San Antonio (UTSA) *(http://utsa.edu/)* where he worked after leaving Harvard. Zhu hoped to commercialize his research and supposedly forwarded without McKeon's authorization gene products to a company in Japan in December 1999, with the intent of generating antibodies to them. Zhu also allegedly made plans to transport more than 30 boxes containing a wide variety of items to Texas from Harvard clandestinely in the final days of 1999. In June 2000, Harvard recuperated items valued at about $300, 000. After leaving the McKeon laboratory many items not taken by Zhu and Kimbara were discovered to be spoiled or labeled incorrectly.

Because the theft of IP from university laboratories in the USA had become prevalent and a matter of concern law enforcement agencies attempted to curtail it [83] and in June 2002 Zhu and Kimbara were arrested by the Federal Bureau of

Investigation (FBI) and jailed in California pending extradition to Massachusetts because of a fear that the researchers might flee the country. They faced up to 25 years in prison and a $750, 000 fine for supposedly plotting to steal Harvard-owned trade secrets and for transporting university property across state lines. The two accused carried out experiments late into the night unbeknownst to their supervisor McKeon.

Some scientists were sympathetic to the accused and were concerned about a governmental overreaction and criticized the FBI's approach to the investigation because of their lack of appreciation that postdoctoral fellows often spend long hours on their research and often work at unusual times. Moreover, it is not uncommon for them to take materials related to their research with them when they change jobs and try to find assistance with industry. Harvard endorsed Zhu's application for his next position. When their annual appointments in Texas were not renewed Zhu and Kimbara moved to California, where Zhu worked for 18 months as a postdoctoral research fellow at UC San Diego in the laboratory of Jean Wang, who believed that young researchers deserve a second chance. She offered him a position after he willingly drew attention to his difficulties at Harvard. After coming to California Kimbara received a postdoctoral fellowship at the Scripps Research Institute in La Jolla. Wang and other researchers conjectured that Zhu and Kimbara might have been afraid to ask McKeon for authorization to transfer materials to others. The Zhu-Kimbara case might encourage postdoctoral fellows and their mentors to discuss the ownership of their work and material transfer agreements (MTA).

Hiroaki Serizawa. Hiroaki Serizawa, an assistant professor of biochemistry and molecular biology, working on Alzheimer disease was fired by the University of Kansas (UK) *(http://www.ku.edu/)* after being under "increased supervision" since 2001 for allegedly conspiring to steal the products of research from the Cleveland Clinic. In 2003 he admitted lying to FBI agents in September 1999 about recently gotin touch with with Takashi Okamoto when he was employed at the Cleveland Clinic and about how many containers with pilfered genetic material were being stored for Okamoto. For this indiscretion Serizawa was fined $500 and sentenced to three years probation and 150 hours of community service. He entered a guilty plea to circumvent accusations of financial spying. The USA government claimed that Okamoto took several hundred research specimens from the clinic to the Serizawa laboratory for storage, and replaced the contents of the vials with tap water. For these accusations Okamoto was charged with conspiracy, economic spying and the transport of stolen material across state lines. The justice department intended to request his extradition from Japan. David Zapol and Diana Laird, two professors at Stanford University (Stanford) *(http://www.stanford.edu/)* hoped to obtain a pardon for Serizawa in addition to raising money to assist the researcher reimburse his legal expenses *(http://www.usatoday.com/tech/news/2003-06-03-gene-theft_x.htm)*.

Terrorism

Bruce Ivins. Senator Patrick Leahy of Vermont, Senator Tom Daschle, and the *New York Post* received letters contaminated with anthrax spores at about the same time in October 2001. Daschle was Senate minority leader at the time. These scary terrorist attacks precipitated a major investigation by the USA justice department. Within weeks the spores were analyzed by X-ray microanalysis, transmission electron microscopy and other techniques at the former Armed Forces Institute of Pathology (AFIP) in Washington, DC and at the Sandia National Laboratories in Albuquerque, New Mexico. A high percentage of the anthrax spores on the letters contained silicone and oxygen and differed from those of the usual *Bacillus anthracis*. These attributes implied that the spores contained silica so that they would readily float and have an enhanced killing power. An analysis of anthrax specimens from almost 200 laboratories, did not have this conspicuous silicone content.

After a thorough investigation the FBI found the terror attack to be the sole work of Bruce Ivins (1946-2008) a mentally deranged scientist in the USA army. The case against him was based on DNA analyses indicating that the anthrax came from a flask managed at the USA Army Medical Research Institute of Infectious Diseases (USAAMRIID) in Fredrick, Maryland, where Ivins worked for thirty five years. Evidence linking the anthrax in the letters to spores under the control of Ivins in his laboratory was overwhelming. Ivins worked more than his usual number of hours during the nights in his laboratory for months before the attacks. As in most government investigations skeptics were not convinced that the evidence against Ivins was irrefutable [26]. Just before federal investigators were about to make the charges against him public, Ivins committed suicide on July 29, 2008. After his suicide thousands of pages of material supporting the belief that Ivins was the terrorist and that he acted alone were released. After his death an analysis of his personal records by different panels disclosed inadequacies that should have prevented the USAMRIID from hiring him [25]. A high level of security clearance is demanded from employees working at USAMRIID, but for incompletely understood reasons Ivins slipped through the net. His psychiatric records should have instinctively prohibited him from being employed in a biocontainment facility. Over many years Ivins divulged his obsessions and criminal activities to mental health professionals, but this privileged client information was kept private and Ivins did not disclose it to his employer. A therapist concerned about Ivins in the 1990s notified the police. USAMRIID failed to follow up red flags that would have pointed to his psychiatric disorder. Because of the lessons learnt in the anthrax fiasco background checks for personnel working with hazardous pathogens and toxins were improved in 2008 and consist of recurrent drug tests and psychological as well as other evaluations. Some procedures may have prevented the anthrax attacks, such the reporting of unusual behavior by fellow workers, a requirement of at least two workers to be present during unusual work hours and video monitoring of biocontainment regions. After his death the department of justice requested an analysis of Ivins' psychiatric records and this was performed by nine experts from medicine, psychiatry and other disciplines. This panel, chaired by the psychiatrist Gregory Saathoff of the University of Virginia (UVA) *(http://www.virginia.edu/)*, evaluated three decades of mental health records by Ivins and found him deeply disturbed with obsessions and criminal thoughts. The committee ratified the government's implication of Ivins and concluded that "Dr. Irvins was psychologically disposed to undertake the mailings; his behavioral history demonstrated his potential for carrying them out; and he had the motivation and the means". His abnormal mental behavior apparently started as an undergraduate student after being rejected by a member of the Kappa Kappa Gamma (KKG) sorority. Ivins developed a long time obsession with the KKG following this rejection and he broke into buildings of the sorority on several occasions. Moreover the mailbox from which he allegedly mailed the letters contaminated with *Bacillus anthracis* was located less than 200 feet from the KKG office at Princeton University (Princeton) *(http://www.princeton.edu/)*. A report after Ivins' death attributed his mental behavior to a disturbed childhood in which he was physically abused and ridiculed in public and in which his mother hit and stabbed his father.

Sexual Perversions

Daniel Carleton Gajdusek. Daniel Carleton Gajdusek (1923-2008) was one of the most eminent scientific intellects of the 20th century and the brilliant son of a poor east European immigrant. He was born to become a famous scientist. His scientific interests began under the inspiration of his maternal aunt, an entomologist, when he was only 5 years old *(http://nobelprize.org/nobel_prizes/ medicine/laureates/1976/gajdusek-autobio.html)*. From his childhood he knew that he wanted to be a scientist and was strongly influenced by the biographies of Louis Pasteur (1822-1895) [23] and Marie Curie (1867-1934), who shared the 1903 Nobel Prize *(http://www.nobelprize.org/nobel_prizes/physics/laureates/1903/marie-curie.html)* with her husband Pierre Curie (1859-1906) *(http://www.nobelprize.org/ nobel_prizes/physics/laureates/1903/pierre-curie-lecture.html)* and Henri Becquerel (1862-1908) (http://www.nobelprize.org/nobel_prizes/physics/laureates/1903/ becquerel-lecture.html) [42] as well as Paul de Kruiff's *Microbe Hunters* [43]. His career could not have been better launched than in *the way that it was. He graduated from the University of Rochester (Rochester) (http://www.rochester.edu/)* in 1943 majoring in biophysics and three years later he obtained a MD from the school of medicine at Harvard University (Harvard) *(http://www.harvard.edu/)*. He then performed research at the California Institute of Technology (Caltech) *(http://www.caltech.edu/)* under the supervision of Max Ludwig Henning Delbrück

(1906-1981), who was the 1969 Nobel laureate in physiology or medicine *(http://www.nobelprize.org/nobel_prizes/medicine/laureates/1969/delbruck-bio.html#)*, and then carried out research at Harvard under John Franklin Enders (1897-1965) the father of modern vaccines, who received the Nobel Prize for physiology or medicine in 1954 *(http://www.nobelprize.org/nobel_prizes/medicine/laureates/1954/#)*. As a medical student children and their medical disorders fascinated Gajdusek and offered him a greater challenge than adult medicine so he embarked on a career in pediatrics and became board certified in that specialty.

In 1951, he was drafted from Ender's laboratory to the Walter Reed Army Medical Service Graduate School as a young research virologist. He spent time at the Pasteur Institute in Iran working on rabies, plague, arbovirus infections, and scurvy. He also studied a variety of infectious diseases in the USA and Australia. In 1954 he was a visiting investigator at the Walter and Eliza Research Institute in Australia with Sir Frank Mcfarlane Burnet (1899-2011), who shared the 1960 Nobel Laureate in physiology or medicine *(http://www.nobelprize.org/nobel_prizes/medicine/laureates/1960/burnet-lecture.html)* with Peter Brian Medawar (1915-1987) *(http://www.nobelprize.org/nobel_prizes/medicine/laureates/1960/medawar-lecture.html)*. From there he embarked on studies in child development and disease patterns in Australian aborigines and New Guinean populations. This exposure to the *crème de la crème* in science throughout his formative research years effected Gajdusek immensely as evidenced by his numerous original discoveries during his career.

The ground-breaking research of Gajdusek advanced the understanding of certain brain diseases immensely. He documented the first medical description of a peculiar incurable non-inflammatory disorder of the central nervous system called Kuru, which was spread by a cannibalistic custom of eating the organs of deceased persons. Together with his collaborator Clarence J. Gibbs Jr (1924—) he discovered that Kuru, Creutzfelt-Jacob disease, scrapie and some other degenerative disorders of the nervous system were caused by an exceptionally extraordinary filterable infectious agent with a long incubation period after the subject became infected. Heat, radiation, or formaldehyde did not destroy the pathogen unlike other infectious agents. Subsequently Dr. Stanley Ben Prusiner (1942—), the 1997 Nobel laureate in physiology or medicine *(http://www.nobelprize.org/nobel_prizes/medicine/laureates/1997/prusiner-lecture.html)* found that the unusual infectious agent was a protein rich infectious particle for which he coined the term prion from a combination of the words *protein* and *infection*. After carrying out research in New Guinea Gajdusek returned to the NIH in 1958, when 35 years old and he remained working there until his retirement in 1997.

In 1958 during the cold war the FBI and the Central Intelligence Agency (CIA) began investigating Gajdusek because of his possible ties to the communist party. He had circulated pamphlets containing Russian political propaganda from the Soviet Union between 1953 and 1958. He was also in contact with a medical doctor from communist China. On October 25, 2009, the Fredrick County newspaper *Fredrick Newpost* reported FBI findings from Gajdusek's journal entries that were released under the *Freedom of Information Act*.

In 1976 Gajdusek shared the Noble prize in physiology or medicine *(http:// www.nobelprize.org/nobel_prizes/medicine/laureates/1976/gajdusek-autobio.html#)* with Baruch Samuel Blumberg (1925-2011) *(http://www.nobelprize.org/nobel_ prizes/medicine/laureates/1976/blumberg-lecture.html)* for his remarkable discovery that degenerative diseases of the nervous system can be caused by an unusual infectious agent in the absence of an inflammatory reaction. Throughout his life, Gajdusek compared himself to the Pied Piper, who led children, but not rats, out of the small town of Hamelin in Germany. This strong attraction explains in part why he imported 56 children, mostly boys, during the 1960s into the USA from Micronesia and New Guinea which he adopted creating a large family. He took care of them and their schooling expenses. For this he was applauded until it was realized that he had a clandestine reason. In his personal diaries Gajdusek again and again stressed his necessity or longing to be encircled first and foremost by preadolescent and adolescent young boys. These desires were the eventual cause of his downfall to become the only Nobel laureate convicted as a child molester. Gajdusek frequently documented erotic, seductive, sensual and equivalent adjectives in descriptions of children and their behavior. He had a long documented record of suspected sexual abuse and confessed to having sexual intercourse with under aged young men cared for by him. Over the years the Maryland State Police was notified about allegations of child abuse but victims were reluctant to testify. Eventually Gajdusek was charged in April 1996, when he was 72 years old, of molesting a teenage boy who he had imported from Micronesia and unofficially adopted. A federal judge forbade Gajdusek from having unverified contact with minors. He pleaded guilty in 1997 and was sentenced to 5 years of probation and a year in the Fredrick County Allen Detention Center after molesting the teenage boy. When he received his freedom Gajdusek moved to Europe spending time in different cities until he ultimately died of in Tromsø, Norway at 85 years of age *(http://www.guardian.co.uk/science/2009/feb/25/carleton-gajdusek-obituary)*[91].

W. French Anderson. A sexual perversion also caused the downfall of W. French Anderson MD (1936—) a 1963 graduate of Harvard and a prior employee of the NIH. This prominent American scientist was a pioneer in gene therapy research. On July 30, 2004 he was arrested on allegations of three counts of lewd acts upon the underage daughter of an employee. Despite numerous pleas for clemency he was convicted to 14 years in prison on July 19, 2006 for molesting this young girl [14]. This indiscretion caused his resignation from the Institute for Genetic Medicine at the University of Southern California (USC) *(http://www. usc.edu/)* in September 2006 [27].

Politically Insensitive Remarks

Ward LeRoy Churchill. Freedom of speech is a cherished right protected by the constitution of many countries. It is also treasured by academic communities. Members of university faculty are free to express their opinions on any subject no matter how abhorrent the statement, even if everyone disagrees with what is stated. The faculty member cannot be fired for making politically insensitive statements especially if tenured. However, a detailed evaluation of the person's academic record may disclose information that had previously been ignored, and which does not withstand critical review. Such information may contain the seeds for research misconduct allegations and the person can be discharged for research misconduct as Ward LeRoy Churchill (1947—) discovered when he evoked the wrath of the university's administration and precipitated an investigation into his academic record. Churchill is an American writer and political activist. In 1990, he was hired by the University of Colorado at Boulder (CU Boulder) *(http://www.colorado.edu/)* as an associate professor despite the fact that he lacked the doctoral degree typically needed for that academic rank. The next year he surprisingly received tenure in the communications department and in 1992 Churchill received an honorary doctorate of humane letters from Alfred University (AU) *(http://www.alfred.edu/)*. In 1996 when a new department of ethnic studies was created at the CU, Churchill was moved to that department and promoted to full professor. In 2002, he became chairman. His work concentrated on the historical treatment of political rebels and Native Americans by the USA government. His writings were controversial and characterized by provocative allegations frequently composed in an offensive style. Many claims made by Churchill related to his past experience and ancestry could not be substantiated. He stirred up a tempest on September 12 2001 with an offensive essay titled "On the Justice of Roosting Chickens". In the document, which subsequently expanded the text into a book [34] he compared some victims in the terrorist's attacks on the World Trade Center in the USA on September 11th 2001 to American abuses of other countries by international law. His comparison extended to Adolf Eichmann, who annihilated European Jews during World War II as part of Adolf Hitler's extermination plan, by referring them to "little Eichmanns", which was derived from a speech delivered by the human rights activist Malcolm X (1925-1965) in which he associated the killing of John Fitzgerald "Jack" Kennedy (1917-1963), the 35th President of the USA, to the hostility for which Kennedy was responsible as "merely a case of chickens coming home to roost." This controversial analogy precipitated an increased interest in the research of Churchill, which had up to this time been frowned upon by intellectuals. During the tempest that Churchill whipped up with his 9/11 remarks, Churchill resigned as chairman of the ethnic studies department in January 2005 shortly before his term as chair was to run out. Some of his critics asserted that his compositions misrepresented their work. The university determined that it could not fire Churchill for his unacceptable remarks, but in March 2005 administrators at the CU launched an investigation into seven accusations of misconduct in his research. On May 16, 2006 a committee on research misconduct at CU concluded that Churchill was guilty of numerous counts of plagiarism, fabrication, and falsification. He also failed to follow the accepted practice for reporting. The CU investigation found that "his academic publications were nearly all works of synthesis and reinterpretation, drawing upon studies by other scholars, not monographs describing new research based on primary resources".

The committee was unanimous about Churchill's serious and repeated academic misconduct, but it was split on the proper degree of penalty to impose. One member of the committee recommended that he be fired; the other four recommended he be suspended. Churchill appealed against the committee's report. A five member panel of the CU's privilege and tenure committee felt that only five of the seven findings of misconduct comprised offences for which the accused could be fired. Three members of the panel recommended demotion and one year's suspension without pay; two favored dismissal. Bill Owens (1950—) the Governor of Colorado from 1999-2007 stated that Churchill had blemished the university's reputation and should resign. Over and over again Churchill refused to concede any misconduct and asserted that he would delay any further steps until the CU administrators acted, but if fired he threatened to sue. The regents of CU fired Churchill on July 24, 2007 for research misconduct and for failure to comply with standards for listing the names of authors with a vote of eight to one.

The firing of Churchill provoked an accusation by certain academicians that he was fired because of the ideas that he articulated. Churchill filed a lawsuit against CU for illegal termination of employment in the Colorado state court the following day contending that his dismissal was revenge for communicating disliked political opinions. On April 1, 2009, the Colorado jury initially found that Churchill had been inappropriately sacked and awarded him a mere $1 in damages. Following an appeal on July 7, 2009, Judge Naves found that CU was entitled to quasi-judicial immunity under the law and negated the jury verdict and decided that CU did not owe Churchill any monetary payment. Moreover the judge disallowed Churchill's appeal for reinstatement at CU. The *Denver Post* reported that Churchill became politically radicalized as a consequence of his experiences in Vietnam (*http://en.wikipedia.org/wiki/Ward_Churchill*).

Failure to Maintain Database Security

As mentioned in an earlier chapter much complicated contemporary human research data are stored not on laptop or personal computers, but on institutional servers because of the vast about information and the need to have it regularly backed up. The complexity of maintaining such vast databases is beyond the capability of an average scientist. Individuals who specialize in Information Technology (IT) are needed to take care of the databases and their security. For 16 years Bonnie C. Yankaskas PhD, an associate professor of radiology and adjunct professor of epidemiology and prominent cancer researcher at the University of North Carolina (UNC) ran the Carolina mammography registry, which received and analyzed data from radiologists across the state of North Carolina with the goal of improving the detection of breast cancer. Since its beginning the registry had received more than $8 million in federal grants and it brought considerable money to the UNC in the name of indirect costs. One would have expected her institution to have assumed the responsibility for overseeing and maintaining the security of such databases, but in 2009 Yankaskas found that neither her university nor her department took care of this when officials of the school of medicine discovered that the server storing the data on human participants had been infiltrated. Although personal information is not known to have been removed all participants were notified at a cost of $250, 000. A security breach caused personal information to become disseminated on 180, 000 women who had had mammograms. In a kneejerk reaction the university cracked down on Yankaskas and tried to fire her [53]. It later demoted her with a severe decease in salary. After an uproar by the faculty and an appeal by Yankaskas much consideration to the matter was received in publications in higher education [80] *(http://chronicl.com/article/Chapel-Hill-Researcher-Fights/124821/)*. Subsequently the university scaled back on its sanctions and after a protracted expensive course it agreed to repay Yankaskas $175, 000, which was less than half of her legal costs. Under the settlement her academic rank of full professor was reinstated at her prior annual salary of $178, 000, but soon thereafter she planned to retire *(http://en.wikipeda.org/wiki/The_414s)*.

Failure to Meet Research Obligations

In the early 1990s an unusual disease emerged in veterans of the first Gulf war (1990-1991) between the forces of the USA and its allies against Iraq. Affected individuals developed an array of symptoms (pains, dizziness, headaches, loss of concentration and depression). These manifestations also affected troops fighting in Afghanistan. The term Gulf war illness (GWI) or Gulf war syndrome (GWS) was coined for this disorder and its cause remains uncertain. The ailment was suspected of being due to an exposure to one or more neurotoxic chemicals as in fires or pesticides [10, 13]. The neurotoxin sarin was also suspected. A clue to the possible cause was the finding of lower than normal aryl esterase activity in affected veterans, but this observation had not been replicated and was hence questionable. Research on GWI is clearly needed and the University of Texas Southwestern Medical Center (UTSWMC) *(http://www.utsouthwestern.edu/)* was awarded a multimillion dollar contract with the epidemiologist Robert Haley as

PI. Inordinate amounts of money were awarded in this contract obtained outside of the peer reviewed mechanism and as PI Haley was the recipient of much of this money. The USA department of veterans affairs (VA) stopped funding a $75 million contract for this research study after two years alleging severe performance deficiencies [81]. Little progress emerged from the research effort and the researchers missed deadlines and refused to share important data. The VA had already obligated $32 million to the research group and had spent $8 million for laboratory equipment to perform magnetic resonance imaging (MRI) scans, but they had been done on relatively few Gulf war veterans and there were concerns about consent forms and data access particularly since a collaborative research agreement required medical data to be shared with the VA. Without notifying and obtaining the permission of the VA the UTSWMC changed the consent form for participants baring the VA from getting identifiable data [38].

Failure to Obey the Law or Research Rules/Regulations

Marc van Roosmalen. The eccentric Brazilian Dutch-born biologist Marc van Roosmalen (1947—), a specialist in primatology was hired by the Instituto Nacional de Pesquisas da Amazônia (INPA) in 1986 after having performed studies in Suriname and French Guiana. While working in South America he identified some new species of monkey as well as other animals and plants. Aside from his research van Roosmalen was an activist in a quest to safeguard the Brazilian rainforest. For his research he received honors. He received the title of Officer in the Order of the Golden Arch from Prince Bernard of the Netherlands and *Time magazine* made Van Roosmalen a "Hero for the Planet" in 2000 [98] (TIME.com: Heroes for the Planet, 03/06/00—Marc van Roosmalen)(*http://www.time/time/ reports/environment/heroes/heroesgallery/0.2967.rormalen*). Like many scientists van Roosmalen loathed onerous rules and regulations and this displeased colleagues in the INPA when he failed to complete the necessary paperwork for his studies. Some associates even questioned his methods for identifying novel monkey species based on investigations of solitary orphans. They also disapproved of his solicitation of donations with offers to name new species after donors. By 2002 he was in trouble with the Instituto Brasileiro do Meio Ambiente e dos Recursos Naturais Renováveis (IBAMA). It fined him for unlawfully transporting orchids and monkeys out of an unexplored region of the Amazon basin. Shortly thereafter he was dismissed from INPA for illegally exporting DNA from local specimens. To add insult to injury his younger son discovered that he was having an affair with a young Brazilian woman and reported this to his mother bringing the van Roosmalen marriage to an abrupt end. But this was not the end of his troubles. In 2007 the Brazilian government took him into custody for unlawfully keeping orphan monkeys in a place of safety at his home in the Amazon and for the embezzlement of Brazilian funds in the public domain. The latter allegation was based on the disappearance of a scaffolding tower. For this misappropriation and biopiracy he was sentenced to 14 years in prison [17, 104]. He was incarcerated in the infamous Raimundo Vidal Pessoa Penitentiary and during his imprisonment he asserts that he saw two murders. Fortunately for van Roosmalen an appeals court in Brasilia the capital of Brazil overruled all of the allegations against him, except one, and only imposed a year jail sentence before releasing him. Two alleged attempts to murder him sent him into hiding after his release from prison in August 2007 [51].

Charles Gald Sibley. Jonathan M. Marks (1955—), an American biological anthropologist at UNC at Charlotte has drawn attention to the indiscretions of Charles Gald Sibley, PhD (1917-1998) (*http://personal.uncc.edu/jmarks/dnahyb/Sibley%20revisited.pdf*), who was the director of the Peabody Museum at Yale. He was a controversial ornithologist who influenced the classification of birds immensely and altered the understanding of avian evolution by using modern techniques of molecular biology. Partly because of personality conflicts Sibley had few collaborations with researchers other than Jon E. Ahlquist [2]. In 1973 Sibley's career received a fatal blow when he was arrested on six counts of smuggling the eggs of birds in danger of extinction out of the UK. Rather than fight the allegations the emotionally immobilized Sibley decided to plead no contest and pay a fine. Two of the charges were valid but others in this egg-white incident were questionable. For example one specimen came from a legally obtained parrot born in captivity. Another specimen was derived from a bird with a bogus name. For his violation of the *Lacey Act* Sibley was required to resign from the British union of ornithologists. On July 13, 1974 the *New York Times* presented a front page story about Sibley having to pay a fine of $3, 000 for violating the *Lacey Act* of 1904 [54] which requires Americans to respect the wildlife laws of all countries and not to import illegally collected animals. Sibley anticipated being fired from his job as the director of Peabody Museum at Yale and indeed following this scandal Kingman Brewster Jr (1919-1988) president of Yale did not renew his appointment as the director of the Peabody Museum. By the 1980s Sibley's research employed DNA hybridization and he solved previously unresolved questions related to phylogeny of birds and primates [96]. For his scientific advancements Sibley was elected to the membership of the National Academy of Sciences (NAS) in 1986. Aside from not obeying the law four years later Sibley confessed that the phylogenetic observations had been accomplished by data manipulation not discussed in his papers. These exploitations comprised replacing controls across the studies and shifting spots on a scatter plot into the regression line depicting them [95]. His unacceptable data manipulation would probably have failed a junior researcher like an undergraduate student. The undisclosed nature of the data analysis and the strength of the conclusions in his publications deceived both reviewers and readers. His publications have not yet been retracted. Scientists dealing with the concerns of Sibley have regarded his data adjustments as bad judgment errors, but insufficiently severe to challenge his membership in the *NAS*, but he became a *persona non grata* with the *NSF*. Senior members of the molecular evolutionist community did not coauthor Sibley's publications.

Illegal Drug Manufacture and Transportation

The illegal manufacture of drugs within a research laboratory can get a scientist into unexpected trouble with the authorities.

John Narfarian. A clandestine section of the transplant program manufactured and sold the anti-rejection drug for more than two decades without obtaining the required federal approval. The UMMS was not appropriately overseeing its researchers who became involved in commercial endeavors.

The university feared federal sanctions and reacted with an iron rod against not only members of the transplant program, but also against the UMMS with management changes, financial audits and stringent oversight. Nils Hasselmo (1931-), the president of the university from 1988-1997 got involved in the disgrace and the heads of several individuals started rolling. In September 1992 steps were taken to fire Richard Condie, the ALG program's director, who was also accused of diverting $174, 000 from income derived from the drug to his personal bank account. Condie maintained that he received this money for legitimate consulting fees.

In 1971 John Narfarian, a transplant surgeon at the University of Minnesota (U of M) *(http://www1.umn.edu/twincities/)* medical school and was chair of the surgery department until 1995. He obtained permission from the FDA to manufacture antilymphocyte globulin (ALG) and to use it experimentally. This investigational new drug (IND) blocks the immune system from rejecting the transplanted tissue and by 1993 more than 50, 000 transplant patients had been treated with it, which was only obtainable from the transplant program at the U of M medical school. Sales from the drug produced more than $60 million by 1993, but because the drug was not authorized for commercial use almost all sales were illegal. In addition interstate sales of INDs were forbidden until 1980 and thereafter it was permitted under specified conditions, but only with FDA approval. A scandal emerged at the Minnesota medical school in 1993 related to its transplant program, ranked as one of two in the nation [5]. With regard to the manufacture and sale of ALG a clandestine section of the transplant program was discovered to have manufactured and sold the anti-rejection drug for more than two decades without obtaining the required federal approval. The university was not appropriately overseeing its researchers who became involved in commercial endeavors. The *Minneapolis-St. Paul Star Tribune* initially questioned the drug sales and this led to investigations by the university and the federal government. Not only was there a FDA probe, but also investigations by the FBI, the internal revenue service (IRS) and the local USA attorney's office. Officials of the transplant program had been informed on five occasions over 19 years that they could not sell their ALG. Surprisingly it took a long time for the FDA to crack down on the program. In 1992 the clinical trials in the ALG program were abruptly stopped for various reasons including the fact that reports on many of the 100, 000 transplant patients were never filed and informed consent was not always obtained. Moreover the FDA was not informed about three deaths and some other side effects related to the clinical trials. Investigations looking into possible prosecution for criminal charges were performed by the FDA and federal law enforcement agencies. In anticipation of this happening Minnesota medical school hired lawyers and blamed the ALG program for misleading university officials regarding the legal status of the drug and for not informing them about the FDA concerns. The university feared federal sanctions and reacted with an iron rod against not only members of the transplant program, but also against the medical school with management changes, financial audits and stringent oversight. The President of the university, Nils Hasselmo, got involved in the disgrace and the heads of several individuals started rolling. In February 1993 Najarian was asked to resign as head of the surgery department and Hasselmo removed that department's oversight of the ALG program. By June 1993 the vice president for health science at M of U as well as the dean of the medical school had resigned. In October 1993 steps were started to deprive Najarian of his tenure so that he could be fired, but by December the proceedings were on hold to provide him with an opportunity to defend himself against criminal charges. In December 1993 Condie resigned from the university. Not only did the university target specific members of the faculty, but it allowed the university's general counsel to hastily withdraw an application for a competitive renewal of a grant from the NIH to support 20 faculty researchers involved in the ALG program. This action by the university infuriated the medical school faculty, who considered it a violation of both due process and academic responsibilities in research. A month later the application was resubmitted with minor changes. The ALG scandal raised several issues. For example why did the university not take the necessary steps to make the ALG sales legal? ALG was the first most important drug developed in a university and many of the shortcomings of the involved researcher can probably be attributed to inexperience in running companies. Also, according to some observers the FDA was not adequately prepared to deal with a new situation of this type. Some suspect that the FDA was prodded into action by pharmaceutical companies that sell immunosuppressant drugs with which ALG competes. In the late 1990s the university paid the USA government $32 million to settle allegations that it profited illegally from the sale of ALG [16]. Najarian was acquitted on criminal charges related to the sale of ALG [36] and after his retirement the university honored him by establishing an endowed chair in his name [16].

John Buettner-Janusch. The distinguished American physical anthropologist John Buettner-Janusch PhD (1924-1992), commonly known as "B-J", was an authority on the evolution of simians and humans (*http://en.wikipedia.org/wiki/ John_Bueettner-Janusch*). He initiated the use of molecular methods in primate evolution and his text on the origins of man was well received [29]. After obtaining a PhD at the University of Chicago (UChicago) (*http://www.uchicago.edu/*) he joined Yale and remained there until he was recruited to Duke University (Duke) (http://duke.edu/ where he founded the renowned Duke Primate Center (currently known as the Duke Lemur Center). He left Duke to become the chairman of the anthropology department at New York University (NYU) *(http://www.nyu.edu/)*, where he earned one of the highest salaries at the university. He had an eccentric personality and this became particularly evident after the death of his wife and long-time coworker Vina in 1977. The career of Buettner-Janusch came to an abrupt end when the finger was pointed at him for harboring an illegal drug business in his research laboratory. In this facility he allowed his assistants to manufacture the psychedelic drug lysergic acid diethylamine (LSD) (known colloquially as acid) and methaqualone a sedative-hypnotic drug widely referred to as Quaalude. Clifford Jolly, a professor in the department of anthropology, who had collaborated with Buettner-Janusch for several years investigating genetic markers in the blood of primates, blew the whistle on his chairman. Over a period of time Jolly accumulated specimens, notebook pages and photographs implicating Buettner-Janusch. Late in the evening of May 17, 1979, the laboratory of Buettner-Janusch at NYU was invaded by six narcotics agents from the federal government. Within the laboratory they discovered methaqualone as well as residues of a precursor of LSD. Pleas of innocence were rendered, but eventually Buettner-Janusch was indicted and convicted largely on the testimony of several graduate students that implicated him. He was sentenced to five years in prison by the federal Judge Charles L. Brieant in 1980, but was released on parole in 1983. Prior to this during World War II he spent a short stint in jail as a conscientious objector. The drug production did not bring Buettner-Janusch back to reality as he mailed boxes of Godiva chocolates just prior to Valentine's Day in 1987 to Judge Brieant, as well as to Dr. J. Bolling Sullivan, an associate professor of biochemistry at the Duke marine biology in Beaufort, North Carolina. A Valentine's card signed with a question mark accompanied each gift making the receiver ponder over the identity of the secret admirer. The gifts were not sent out of love as they had been injected with atropine and pilocarpine. Three individuals became seriously ill after eating a chocolate or two: the Judge's wife, Virginia, Sullivan's wife, Ashley, and Sullivan's daughter Ann. The individuals who became ill after eating the tainted candy fortunately recovered. Because Judge Brieant's wife collapsed after eating the chocolates, the candy was analyzed and discovered to contain the poisons that Buettner-Janusch had introduced into the boxes. Sullivan did not understand why poisoned chocolates were sent to him. Two decades earlier when Buettner-Janusch was director of the Duke primate center from 1965-1973 Sullivan was a graduate student at Duke and their association was pleasant. The last contact that the two had with each other was in 1970 or 1971 when Sullivan was the recipient of animal tissues from the primate center. The prosecutor in the case (James R. DeVita) maintained that on February 12, 1987 Buettner-Janusch mailed four boxes of poisoned bonbons from a post office and that authorities fortunately intercepted the other two boxes of venomous candy meant for other former academic colleagues. Once accused Buettner-Janusch lacked a feasible defense as candy containing boxes also had his fingerprints. Eventually in June 1987 he admitted inserting atropine and pilocarpine into the chocolates and mailing them with the purpose of killing or injuring the federal Judge and Sullivan his previous learned associate [76] (*http://www.nytimes.com/1987/06/10/ nyregion/professor-pleads-guilty-in-poisoned-candy*—... *1/28/21010*). In the absence of extenuating circumstances Buettner-Janusch was sentenced at the age of 63 years to two consecutive prison stretches of 20 years [8], but he died after serving 6 years. In his final days Buettner-Janusch refused to eat and was force-fed.

Paul Frampton. Paul Frampton PhD a distinguished professor of physics and astronomy at UNC Chapel Hill was arrested and convicted in Argentina in November 2012 after being found with 4 pounds of cocaine in a suitcase. He was hoodwinked into unsuspectingly transporting the drug after first being enticed to Bolivia for a guaranteed encounter with Denise Milani the 2007 Miss Bikini World. Despite being allured by the attractive model the 68 year old Frampton did not actually get to meet her. The shocked Frampton received a sentence of 4 years and eight months and expected to serve it under house arrest [89] (*http:www.charlotteobserver.com/2012/11/22/3682130/argentine-court-convicts-unc-ch.html*).

Premature Research Publicity

It is extremely common for researchers to anticipate not yet known research results when submitting abstracts to a scientific meeting. Such publicity seeking scientists make their new discoveries known to the public even when the data is preliminary and has not survived peer review with acceptance in an satisfactory journal. This is often done at an open forum, such as a professional meeting, to claim priority in new breakthroughs. Unfortunately members of the media jump on new discoveries and commonly report such findings to the general public before they have been validated. Such studies are commonly over interpreted and the scientific facts are misstated. Because the truth becomes distorted under these conditions the reputation of science suffers and trust in it deteriorates. Clearly it is desirable not to release new research findings to the media until after they have passed peer review and are pending publication. Unfortunately some professional societies have press offices and allow interviews with researchers to learn about potential newsworthy discoveries. The timing of such announcements often coincide with professional scientific meetings at which societies allow press releases that make new discoveries available to the general public. Unfortunately the release of information that has not undergone peer review often turns out to be inaccurate and misinformation becomes disseminated. Examples are half truths on Himalayan glaciers and the effects of climate change.

These actions include a failure to cite competitors in scientific publications, and the exclusion of rival programs and review groups, as well as a refusal to provide information or reagents to researchers not in favor [57, 100].

Felonies

Like other citizens scientists can perform serious crimes.

Maintaining Unsafe Work Environment. On December 29, 2008 Shebarbano "Sheri" Sanji a 23-year of research assistant was fatally burnt in a fire in the laboratory of Patrick Harran a professor of chemistry at the University of California, Los Angeles (UCLA) *(http://www.ucla.edu/)*, while performing research using highly flammable chemicals. Sanji died 18 days after the fire. On December 27, 2011 three counts of felony were filed by the Los Angeles district attorney's office against Harran and UCLA for not maintaining a safe work environment (*http://scim.ag/safelabs*). The district attorney's office accused the charged parties of violating the occupational safety regulations of California, which require employees to not only undergo proper safety training, but also to wear suitable protective equipment. Hearings were still taking place in December 2012 to determine whether Harran should stand trial on felony charges for occupational safety hazards in this case (Benderly, Beryl Expert testimony continues on third day of Harran hearing. Science Carrers Blog. January 23, 2013 *(blogs.sciencemag.org/sciencecareers/beryl-lieff-beri/).*

Syred Zaki Salahuddin. Aside from the difficulties that Gallo faced in the investigations of this own research other members of his staff got on the wrong side of the law. Syred Zaki Salahuddin, a biologist in the Gallo laboratory of the NCI was suspended without pay while a grand jury determined whether he hid a financial interest in a company [9]. In September 1990, Salahuddin admitted guilt to 2 felony charges, one for directing laboratory contracts to a biotechnology company that employed Saki's wife and the other for receiving an illegal gratuity from the identical company.

Prem S. Sarin. Prem S. Sarin, PhD a chemist and former deputy director of the Gallo laboratory, was later relieved of all duties by the NCI without pay until an investigation of accusations of criminal COI were concluded. Charges critical of Sarin surfaced in 1991 at a hearing before the Dingle committee [63]. Sarin had relationships with two pharmaceutical companies and in July 1992 a federal jury convicted him on three felony counts entailing the creation of fictitious statements to the government and the misappropriation of money from the NIH. The federal jury determined that Sarin unlawfully obtained $25, 000 from the German pharmaceutical company Degussa/ASTA Pharma for evaluating the company's vaccines in the Gallo laboratory and later he perjured himself about his earnings on the NIH annual financial disclosure form. Sarin was also found guilty of providing false statements on a NIH financial disclosure form when he failed to list a $4, 000 consulting agreement with another pharmaceutical company. The money was deposited in Sarin's personal bank account and it was used for his private expenses. Sarin maintained that he deposited money into the account held by the NIH entitled Foundation for the Advancement of Education in the Sciences (FAES) that is used for miscellaneous expenses. Sarin claimed that he had not intended to harm or defraud anyone, but admitted serious errors in judgment. The three felonies could carry fines of up to $750, 000 and a prison sentence of 20 years [62]. Aside from the above the Dingell committee learnt that two subjects died in a phase 1 clinical trial performed with an AIDS vaccine by Picard and colleagues including Gallo [41, 87]. The adverse reaction should have been reported according to NIH regulations, but was not. Dingell criticized Gallo for this oversight but apparently did not appreciate that Gallo as only one of many participants in this multi-institution study, primarily as a source of supply. Moreover Daniel Zagury the senior author was acquitted of all allegations.

Processing Material Illegally

Vincent Dammai MPhil PhD, a 40 year old assistant professor in the department of pathology and laboratory medicine at the Medical University of South Carolina (MUSC) *(http://academicdepartments.musc.edu/)* was arrested by the FBI on December 22, 2011. Shortly thereafter Dammai was placed on administrative leave and locked out of his office and laboratory. Also named in the 15-count indictment were Francisco "Dr. Frank" Morales, Alberto Ramon, and Lawrence Stowe from Texas [19] (*http://www.postandcoourier.com/news/2011/dec/29/musc-doctor-indicted/?print*). They were accused of harvesting by unapproved procedures stem cells without oversight from umbilical cords and illegally selling them for use in patients with various incurable diseases, including cancer, autoimmune diseases, Parkinson disease, multiple sclerosis, and amyotrophic sclerosis. The stem cell therapy was not administered under a FDA approved protocol and it offered false hope [45]. Between January 2007 and April 2010 more than $1.5 million was apparently generated from patients that Morales met in the USA and treated in Mexico. Earlier in 2011 Fredda Bryan of Scotsdale Arizona, who created a company called Global Laboratories, pleaded guilty of selling umbilical cord derived stem cells for non-research purposes without FDA approval. During 2009 and 2010 more than one hundred and eighty vials containing umbilical stem-cells were sold by Global Laboratories [45] *(http://www.postandcoourier.com/news/2011/aug/27/inquiry-includes-prof-at—musc/)*.

A Failure of Principal Investigators to Take Responsibility for Errors in Their Research

The answer to this major unresolved question in biology would permit researchers to design custom proteins. With his postdoctoral mentor Frederic Richards of Yale University (Yale)*(http://www.yale.edu/)* Hellinga developed a computer program called DEZYMER to predict protein sequences [67]. Later Hellinga discovered how to create an enzyme from a simple protein and his novel innovative accomplishments excited the biochemistry community. His pioneering breakthrough was the type expected of future Nobel laureates. For his remarkable achievements Hellinga received one of five introductory $2.5 million five year Director's Pioneer awards from the NIH in September 2004. A month later he received the Feyman Prize from the Foresight Nanotech Institute and the following year Duke University award of Hellinga a distinguished professorship (James B. Duke professor of biochemistry).

By 2007 the enthusiasm about Hellinga darkened when John Richard, a chemical biologist and mechanistic enzmologist at the State University of New York at Buffalo (SUNY Buffalo) *(http://www.buffalo.edu/)*, who was a past collaborator and coauthor on two papers with Hellinga found that his research laboratory was unable to replicate the behavior of the engineered enzyme as claimed by Hellinga despite adhering to the methods recorded in papers by Hellinga in *Science* [24] and the *Journal of Molecular Biology* [66]. Richard was recognized as one the best physical organic chemists in the world. He had superb credentials and had studied with giants in enzymology, such as Perry Frey, Bill Jencks and Irwin Rose. An inability to reproduce the documented work raised suspicions about the integrity of the research by Hellinga.

In December 2010 more than three years after Hellinga's research findings could not be replicated an investigation into potential scientific misconduct by Hellinga was concluded by Duke after a confidential unbiased review of all relevant incidents. From a public relations perspective the university missed an excellent opportunity to enhance its image with the public and to make it clear that it supports and protects its students. The media do not look lightly on institutions that appear to protect their faculty by not performing transparent investigations of their alleged indiscretions and then not issuing a formal summary of the findings. It should have been particularly important for the university to issue a statement on Hellinga as his research affected many individuals including an innocent graduate student. If Hellinga did nothing intentionally wrong Hellinga will lose nothing if the report clears his name. If the published scientific studies that he was forced to retract were caused by honest mistakes why was this not brought out into the open to restore trust in Hellinga and the university.

Homma W. Hellinga

Another scientist whose research could not be replicated and who has been accused of research misconduct is Homme W. Hellinga PhD. He is a brilliant biochemist and was enthralled with how a series of amino acids encode the function of a protein from the time of his first publication in the high impact journal *Nature* [65]. The answer to this major unresolved question in biology would permit researchers to design custom proteins. Wth his postdoctoral mentor Frederic Richards of Yale University (Yale) *(http://www.yale.edu/)* Hellinga developed a computer program called DEZYMER to predict protein sequences [67]. Later Hellinga discovered how to create an enzyme from a simple protein and his novel innovative accomplishments excited the biochemistry community. His pioneering breakthrough was the type expected of future Nobel laureates. For his remarkable achievements Hellinga received one of five introductory $2.5 million five year Director's Pioneer awards from the NIH in September 2004. A month later he received the Feyman Prize from the Foresight Nanotech Institute and the following year Duke University honored Hellinga with a distinguished professorship (James B. Duke professor of biochemistry) [64]. In 2007 enthusiasm about Hellinga darkened when John Richard, a chemical biologist and mechanistic enzmologist and professor of chemistry at the State University of New York at Buffalo (SUNY Buffalo)(*http://www.buffalo.edu/*). Richard was unable to replicate the behavior of the engineered enzyme as claimed by Hellinga despite adhering to the methods recorded by Hellinga in *Science* [24] and the *Journal of Molecular Biology* [66]. Richard had collaborated with Hellinga in the past and had coauthored two papers with him. Richard was recognized as one the best physical organic chemists in the world. He had superb credentials and had studied with giants in enzymology, such as Perry Frey, Bill Jencks and Irwin Rose. Rose shared the 2004 Nobel Prize in chemistry (*http://www.nobelprize.org/nobel_prizes/chemistry/laureates/2004/rose-lecture.html*) with Aaron Ciechanover *(http://www.nobelprize.org/nobel_prizes/chemistry/laureates/2004/ciechanover-cv.html)* and Avram Hershko *(http://www.nobelprize.org/nobel_prizes/chemistry/laureates/2004/hershko-lecture.html)*. Hellinga asked the university to initiate a formal investigation of her. An inability to reproduce the documented work raised suspicions about the integrity of the research by the adorned scientist [64], but no evidence of research misconduct was discovered. Initially Hellinga blamed Mary Dwyer, a former graduate student of his for the errors in a paper and asked the university to initiate a formal investigation of her. This took place and Dwyer was acquitted of all transgressions in February 2008, but the two papers under dispute were retracted [46, 47]. In October 2009 a former postdoctoral student of Helliga published results that contradicted another two of his papers. Following Dwyer's absolution graduate students in the department of biochemistry at Duke University (Duke) *(http://duke.edu/)* delivered notarized copies of a petition to eight administrators of the university asking for a formal investigation of Hellinga for publishing Dwyer's data even with her objections and that he engaged in "baseless and malicious, or reckless misconduct charges against her". Months later, on July 24 2008, Hellinga wrote to *Nature* apologizing for the errors for which he was personally responsible and pointed out that the university had granted his request for an additional investigation into the matter [68]. By requesting a probe himself, Hellinga turned out to be not only the accused, but also the plaintiff. The Hellinga affair brought forth an uproar from scientists annoyed by a disappointment in him for not giving an explanation for his experimental faults, and graduate students were infuriated by his readiness to make a scapegoat of Dwyer.

To the scientific community at large the university's response was regarded as mediocre and lackadaisical because it allowed Hellinga to be apparently investigated on his own terms. The Hellinga scandal was particularly disturbing to graduate students, because of his responsibility to guide and mentor Dwyer and oversee her research including the content of manuscripts written by her particularly when he was a coauthor. To turn on his student and accuse her of academic misconduct because of the failure of others to replicate the results for which he had in the past received prestigious awards seemed unreasonable. The incident could frighten any student if this behavior is tolerated by any university.

This took place and Dwyer was acquitted of all transgressions in February 2008, but the two papers under dispute were retracted. In October 2009 Hellinga's former postdoctoral student published results that contradicted another two of his papers. Following Dwyer's absolution, before Hellinga's personal call for an investigation, graduate students in biochemistry delivered notarized copies of a petition to eight Duke administrators asking for a formal investigation of Hellinga for publishing Dwyer's data even with her objections and that he engaged in "baseless and malicious, or reckless misconduct charges against her". Months later, on July 24 2010, Hellinga wrote to *Nature* apologizing for the errors for which he was personally responsible and pointed out that the university had granted his request for an additional investigation into the matter. By requesting a probe himself, Hellinga turned out to be not only the accused, but also the plaintiff. The case was complicated by the fact that Hellinga was accused of misconduct in mentoring Dwyer and neglecting to supervise her appropriately where he was responsible. The Hellinga affair brought forth an uproar from scientists annoyed by a disappointment in him not giving an explanation for his experimental faults, and graduate students were infuriated by his readiness to make a scape goat of Dwyer. The Hellinga affair was particularly disturbing to graduate students, because of his responsibility to guide and mentor Dwyer and oversee her research including the content of manuscripts written by her and particularly when he was a coauthor. To turn on his student and accuse her of academic misconduct because of the failure of others to replicate the results for which he had in the past received prestigious awards seemed unreasonable. The incident could frighten any student if this behavior became tolerated by any university. In December investigation 2010 more than three years after Hellinga's research findings could not be replicated an into potential scientific misconduct by Hellinga was concluded by Duke after a confidential wide-ranging thorough allegedly unbiased inquiry into all relevant incidents under its research misconduct policies and procedures which abide with federal law. The findings were discussed with Hellinga and appropriate steps were taken. Unfortunately the results were kept secret and a dark cloud remained over the reputation of the university because of its lack of an open policy in such investigations. In February 2008 Hellinga retracted papers in

Science [48] and *Nature*. Despite the retraction of two important papers a failure to issue an official statement left the matter up in the air because the results of misconduct investigations are only made public if a federal agency finds research misconduct. In May 2008 another paper in the *Journal of Molecular Biology* was retracted [3]. The scientific community deserves feedback in cases publicized as much as the Hellinga case. In December 2010 the university concluded its investigation after two years, but as pointed out in *The Chronicle* the university's secret investigation did not bring the matter to closure [33]. The public rightfully expected the outcome of a high profile case of alleged scientific misconduct, such as this one, which reached the public domain following its publicity in *Nature*. If misconduct by Hellinga breached university bench marks but not federal standards the university does not have a legal obligation to make the findings of its investigation public, but from a public relations perspective the university missed an excellent opportunity to enhance its image with the public and to make it clear that it supports and protects its students. The media do not look lightly on institutions that appear to protect their faculty by not performing transparent investigations of their alleged indiscretions and then not issuing a formal summary of the findings. It should have been particularly important for the university to issue a statement on Hellinga as his research affected many individuals including an innocent graduate student. The faculty and student body at Duke needs to informed whether Hellinga did anything wrong and if he did what steps are being taken to rectify the errors. If Hellinga did nothing intentionally wrong he will lose nothing by having his name cleared. If the published scientific studies that he was forced to retract were caused by honest mistakes why was this not brought out into the open to restore trust in Hellinga and the university [90].

Sabotage of Research by Laboratory Personnel

A rare variety of dishonest research is the sabotaging of a colleague's laboratory research endeavors. It is debatable whether this type of behavior should or should not be included under the federal definition of scientific misconduct. One incident was investigated and ruled upon by the Office of Research Integrity (ORI) in 2011. In this case Vipul Bhrigu, a postdoctoral fellow at the University of Michigan (UM) *(https://www.kmich.edu)*, acknowledged exterminating over and over again the cultured cells of Heather Ames, a graduate student working in his laboratory, with ethanol after foul play was proven with concealed video cameras [77]. The State of Michigan prosecuted Bhrigu for the meticulous destruction of property and Judge Elizabeth Pollard Hines sentenced him to more than $30, 000 in fines and restitution. The motivation for this strange crime was a desire to slow down the research of the student. A somewhat similar case of sabotage involved the research of Magdalena Koziol at Yale University (Yale)(www.yale.edu/). Koziol assumed that someone was sabotaging her research when the eggs of her fertilized transgenic zebrafish kept dying [52]. The perpetrator of the crime was Polloneal Jymmiel Ocbina and his actions were documented on secret cameras. When sabotage is carried out it not only frustrates the honest researcher, but also causes much time to be lost and the targeted scientist's research productivity suffers and this can affect the individual's career.

Other Inappropriate Behavior by Scientists and Institutions

Aside from plagiarism, fabrication and falsification that fall under the umbrella of scientific misconduct other activities by scientists and other research personnel are unacceptable and inappropriate whether they occur in the laboratory or elsewhere. They include gossip, a failure to report observed research misconduct, the selective interpretation of data, leaving out points that do not fit a hypothesis, laying false trails to mislead competitors, a failure to share reagents, and other material derived from research funded by federal agencies. Other unacceptable behavior by researchers include the keeping of sloppy notebooks, and tossing away data, and inappropriate studies on animals especially if not certified by an institutional animal care and use committee (IACUC). Various forms of harassment, violation of government research regulations, exploitation of students and technicians, the misuse of human subjects, the mishandling of hazardous research materials, the misuse of recombinant DNA materials, the suppression of manuscripts or adversely effecting grant applications by competitors. Some of these unacceptable actions are punishable wrongdoings covered by existing rules and laws that govern civilized society regardless of whether the guilty are penalized.

Senior individuals are generally more likely to establish sexual liaisons and bend the rules to get jobs done for their lovers. They are also more likely to author articles promoting the discoveries of biotechnical firms in which they have an interest, and to torpedo grant proposals of their competitors hiding under the shroud of animosity and to perform morally offensive behavior that does not fall under the accepted definition of scientific misconduct.

References

1. Madey v. Duke University, No.01-1567, Federal Circuit Court of Appeals, 3 October 2002
2. Ahlquist JE, Charles G. Sibley. A commentary on thirty years of collaboration Auk, 1999. 116(3): 856-860.
3. Allert M, Dwyer MA, and Hellinga HW, RETRACTED: Local encoding of computationally desighned enzyme activity. Journal of Molecular Biology, 2008. 378(3): 759.
4. Altonn H. Laser beaming anew at UH. Star-Bulletin. 2007 Sunday 21 January.
5. Anderson C, Scandal scars Minnesota Medical School. Science, 1993. 262(5141): 1812-1813.
6. Anonymous. Dr. Schatz wins 3% of royalty, named co-finder of streptomycin; key figures in streptomycin discovery suit. The New York Times. 30 December 1950.
7. Anonymous. Gives Dr. Banting credit for insulin. The Toronto Daily Star. 1922.
8. Anonymous. Former Duke Prof sentenced to 40 years in candy case. The Durham Morning Herald 1987 July 15.
9. Anonymous. AIDS researcher suspended over finances. The New York Times. 1990 May 2.
10. Anonymous, Ethical watchdog's just desserts. Nature, 1991. 349(6308): 355.
11. Anonymous. Disputes rise over intellectual property rights. National Public Radio. 1996 30 September
12. Anonymous, Double helix: 50 Years of DNA. Nature Archives, 2003.
13. Anonymous, Research Advisory Committee on Gulf War Veterans' Illnesses. United States Department of Veterans Affairs. Retrieved 2012-05-09, 2004.
14. Anonymous, In the courts. Researacher convicted. Science, 2006. 313(5786): 137.
15. Anonymous. Dolly creator heads Scots honours. BBC News. 2007 Saturday December 29.
16. Anonymous. U honors surgeon Najarian with endowed chair. StarTribune. 2007 November 27
17. Anonymous, Three Q's. Science, 2007. 317(5843): 1303.
18. Anonymous, Cancer researcher sued. Science, 2012. 335(6064): 19.
19. Anonymous, FBI investigates stem cell case Science, 2012. 335(6064): 19.
20. Banting FG and Best CA, The internal secretion of the pancreas. Journal of Laboratory and Clinical Medicine, 1922. 7: 464-472.

21. Barinaga M, Biotech patents. Genentech. UC settle suit for $200 million. Science, 1999. 286(5445): 1655.
22. Barinaga M, Biomedical Patents. No winners in patent shoot out. Science, 1999. 284(5421): 1752-1753.
23. Bellis M, Louis Pasteur: Biography of Louis Pasteur 1822-1895.
24. Benson DE, Conrad DW, de Lorimier RM, et al., Design of bioelectronic interfaces by exploiting hinge-bending motions in proteins. Science, 2001. 293(5535): 1641-1644.
25. Bhattacharjee Y, Anthrax investigation. Army missed warning signs about alleged anthrax mailer. Science, 2011. 332(6025): 27.
26. Bhattacharrjee Y, Science in Irvins case not iron clad NRC says. Science, 2011. 331(6019): 835.
27. Bhatttacharjee Y, In the courts. Science, 2007. 315(5814): 919.
28. Bliss M, The Discovery of Insulin. Twenty-fifth Anniversary Edition, Second edition. Second ed. 2007, Chicago, IL: University of Chicago Press.
29. Buettner-Janusch J, Origins of Man: Physical Anthropology. 1966: John Wiley and Sons.
30. Buzzai M, Jones RG, Amaravadi RK, et al., Systemic treatment with the antidiabetic drug metmorfin selectively impairs p53-deficient tumor cell growth. Cancer Research, 2007. 67(14): 6745-6752.
31. Cambrosio A and Keating P, Between fact and technique: the beginnings of hybridoma technology. Journal of the History of Biology, 1992. 25(2): 175-230.
32. Campbell KH, McWhir J., Rirchie W.A., Wilmut, I., Sheep cloned by nuclear transfer from a cultured cell line. Nature, 1996. 380(6569): 64-66.
33. Cho J. After Hellinga review closes, admins quiet. The Chronicle. 2010 13 December.
34. Churchill W, On the Justice of Roosting Chickens. Reflections on the Consequences of U.S. Imperial Arrogance and Criminality. 2003, Oakland California: AK Press.
35. Cohen J, AIDS research. U.S.-French patent dispute heads for a showdown. Science, 1994. 265(5168): 23-25.
36. Compton J. Najarian verdict passes on-year mark. Minnesota Daily 1997 February 21.
37. Connor S. 'Dolly' scientist should be stripped of his knighthood, colleagues tell Queen. The Independent. 2008 February 1.
38. Couzin J, Texas earmark allot millions to disputed theory of gulf war illness. Science, 2006. 312(5774): 668.
39. Cramb A. I didn't clone Dolly the sheep says prof. The Telegraph. 2006 March 8.
40. Crewdson J. The great AIDS quest. Chicago Tribune. 1989 November 19.
41. Crewson J. U.S. agency faulted in AIDS vaccine study. Chicago Tribune. 1961 July 16.

42. Curie M, Madam Curie Translated by Vincent Sheean. 1937, New York: Da Capo.
43. de Kruiff P, Microbe Hunters. 1926, New York: Houghton Mifflin Harcourt.
44. DeQuine J. Volatile mix of corporate cash and academic ideals. The Christian Science Monitor. 1996 July 11.
45. Dudley R. Stem-cell inquiry includes professor at MUSC. The Post and Courier 2001 Saturday August 27.
46. Dwyer MA, Looger LL, and Hellinga HW, Computational design of a biologically active enzyme.[Retraction in Dwyer MA, Looger LL, Hellinga HW. Science. 2008 February 1;319(5863):569; PMID: 18239106]. Science, 2004. 304(5679): 1967-1971.
47. Dwyer MA, Looger LL, and Hellinga HW, Retraction.[Retraction of Dwyer MA, Looger LL, Hellinga HW. Science. 2004 June 25;304(5679):1967-71; PMID: 15218149]. Science, 2008. 319(5863): 569.
48. Dwyer MA, Looger LL, and Hellinga HW, Retraction.[Retraction of Dwyer MA, Looger LL, Hellinga HW. Report. Computational design of a biologically active enzyme. Science. 2004 June 25;304(5679):1967-1971; PMID: 15218149]. Science, 2008. 319(5863): 569.
49. Edition NPRM, Disputes Rise Over Intellectual Property Rights, 1996.
50. Editorial. Queen's Birthday Honours. Times Higher Education 1999 June 18.
51. Enserink M, Biologist's legal battle ends. Science, 2008. 322(5907): 1451.
52. Enserink M, Scientific misconduct. Sabotaged scientist sues Yale and her lab chief. Science, 2014. 343(6175): 1065-1066.
53. Ferreri E. Breach costly for researcher, UNC-CH. News and Observer. 2011 May 9.
54. Ferretti F. Fining of bird scholar stirs colleagues. The New York Times. 1974 July 13.
55. Figueroa ME, Abdel-Wahab O, Lu C, et al., Leukemic IDH1 and IDH2 mutations result in a hypermethylation phenotype, disrupt TET2 function, and impair hematopoietic differentiation. Cancer Cell. 18(6): 553-567.
56. Franklin R, The Rosalind Franklin Papers, The Holes in Coal: Research at BCURA and in Paris. 1942-1951.
57. Friedmann T, Ethical duties of scientists, their institutions, and the guild of science. The Journal of NIH Research, 1992. 4: 19-22.
58. Giles J and Knight J, Dolly's death leaves researchers woolly on clone ageing issue. Nature, 2003. 421(6925): 776.
59. Goeddel DV, Heyneker HL, Hozumi T, et al., Direct expression in Escherichia coli of a DNA sequence coding for human growth hormone. Nature, 1979. 281(5732): 544-548.
60. Hagmann M, Scientific misconduct. Researcher rebuked for 20-year-old misdeed. Science, 1999. 286(5448): 2249-2250.
61. Hamilton DP, What next in the Gallo case? Science, 1991. 254(5034): 944-945.

62. Hamilton DP, More woes for Gallo. Science, 1991. 251(4990): 152.
63. Hamilton DP, NIH takes heat for lax investigation. Science, 1991. 251(4999): 1305.
64. Hayden EC, Chemistry: Designer debacle. A high-profile scientist, a graduate student and to major retractions. Erika Check Hayden reports on a case that has rocked the chemistry community. Nature, 2008. 453(7193): 275-278.
65. Hellinga HW and Evans PR, Letters to Nature. Mutations in the active site of Escherichia coli phosphofructokinase. Nature, 1987. 327(6121): 437-439.
66. Hellinga HW, Caradonna JP, and Richards FM, Construction of new ligand binding sites in proteins of known structure. II. Grafting of a buried transition metal binding site into Escherichia coli thioredoxin. Journal of Molecular Biology, 1991. 222(3): 787-803.
67. Hellinga HW and Richards FM, Construction of new ligand binding sites in proteins of known structure. I. Computer-aided modeling of sites with pre-defined geometry. Journal of Molecular Biology, 1991. 222(3): 763-785.
68. Hellinga HW, In the wake of two retractions, a request for investigation. Nature, 2008. 454(7203): 397.
69. Hwang WS, Ryu YJ, Park JH, et al., Evidence of a pluripotent human embryonic stem cell line derived from a cloned blastocyst.[Erratum appears in Science. 2005 December 16;310(5755):1769].[Retraction in Kennedy D. Science. 2006 January 20;311(5759):335; PMID: 16410485]. Science, 2004. 303(5664): 1669-1674.
70. Hwang WS, Roh SI, Lee BC, et al., Patient-specific embryonic stem cells derived from human SCNT blastocysts.[Erratum appears in Science. 2005 December 16;310(5755):1769].[Retraction in Kennedy D. Science. 2006 January 20;311(5759):335; PMID: 16410485]. Science, 2005. 308(5729): 1777-1783.
71. Kaul R, Gao GP, Balamurugan K, et al., Cloning of the human aspartoacylase cDNA and a common missense mutation in Canavan disease. Nature Genetics. 5(2): 118-123.
72. Köhler G and Milstein C, Continuous cultures of fused cells secreting antibody of predefined specificity. Nature, 1975. 256(5517): 495-497.
73. Labowsky M, Mann M, Meng C-K, et al. Yale's patent lawsuit was terrible mistake. Yale Daily News. 2005 February 25.
74. Lawler A, Researcher acquitted of lab theft charge. Science, 2002. 297(5586): 1483.
75. Lister D. Honour for creator of Dolly the sheep 'is insult to science'. The Times. 2008 February 1.
76. Lubasch AH. Professor pleas guilty in poisoned candy case. The New York Times. 1987 June 10.

77. Maher B, Research integrity: sabaotage!. Nature, 2010 457: 516-518.
78. Malakoff D, Intellectual property. Academia gets no help from U.S. in patent case. Science, 2003. 300(5626): 1635-1637.
79. Malakoff D, Intellectual property. Universities ask supreme court to reverse patent ruling. Science, 2003. 299(5603): 26-27.
80. Mangan K, Chapel Hill researcher fights demotion after security breach. The Chronicle of Higher Education, 5 October 2010.
81. Marshall E, Biomedical research. VA pulls the plug on disputed study of Gulf War illness. Science. 325(5946): 1324-1325.
82. Marshall E, Families sue hospital, scientist for control of Canavan gene. Science, 2000. 290(5494): 1062.
83. Marshall E and Normile D, Intellectual property. Alzheimer's researcher in Japan accused of economic espionage. Science, 2001. 292(5520): 1274-1275.
84. Moran K. Nobelist loses to Yale in lawsuit. Yale Alumni Magazine. 2005.
85. Murray I, Paulesco and the isolation of insulin. Journal of the History of Medicine and Allied Sciences, 1971. 26(2): 150-157.
86. Perkel JM. Gene patents decision: everyone wins. Last week's supreme court decision to invalidate patents on human genes was a win for patients, independent researchers, and even the wider biotech industry. The Scientist. 2013 June 18.
87. Picard C, Giral P, Defer MC, et al., AIDS vaccine therapy: phase 1 trial. The Lancet, 1990. 336(8708): 179.
88. Press E and Washburn J. The University as Business (part 3). The Atlantic Monthly—digital edition. 2000 March.
89. Price J. Argentine court convicts UNC-CH physics professor of drug smuggling. UNC-CH physicist jailed for drug smuggling. Charlotte Observer. 2012 November 22.
90. Reddy R. Questions linger about Hellinga case. Investigation of James B. Duke prof still pending. The Chronicle. 2010 Tuesday 20 April.
91. Richmond C. Obituary. Carleton Gajdusek: Nobel prizewinner who first described the prion disease kuru. The Guardian. 2009 Wednesday February 25.
92. Schatz A, Bugie E, and Waksman SA, Streptomycin, a substance exhibiting antibiotic activity against gram-positive and gram-negative bacteria. 1944.[Erratum appears in Clinical Orthopaedics and Related Research 2005 December ;441:379]. Clinical Orthopaedics and Related Research, 1944(437): 3-6.
93. Schnieke AE, Kind AJ, Ritchie WA, et al., Human factor IX transgenic sheep produced by transfer of nuclei from transfected fetal fibroblasts. Science, 1997. 278(5346): 2130-2133.

94. Schwaber J and Cohen EP, Human x mouse somatic cell hybrid clone secreting immunoglobulins of both parental types. Nature, 1973. 244(5416): 444-447.
95. Sibley CG and Ahlquist JE, The phylogeny of the hominoid primates, as indicated by DNA-DNA hybridization. Journal of Molecular Evolution, 1984. 20: 2-15.
96. Sibley CG and Ahlquist JE, Phylogeny and Classification of Birds. A Study of Molecular Evolution. 1990, New Hanover: Yale University Press.
97. Sun M, Scientists settle cell line dispute. Science, 1983. 220(4595): 393-394.
98. Unknown. Hero for the Planet. Time. 2000.
99. Wade N, Nobel follies. Science, 1981. 211(4489): 1404.
100. Wager E, Field EA, and Grossman L, Good publication practices for pharmaceutical companies: why we need another set of guidelines. Current Medical Research and Opinion, 2003. 19(3): 147-148.
101. Wainwright M, Streptomycin: discovery and resultant controversy. History and Philosophy of the Life Sciences, 1991. 13(1): 97-124.
102. Ward PS, Patel J, Wise DR, et al., The common feature of leukemia-associated IDH1 and IDH2 mutations is a neomorphic enzyme activity converting alpha-ketoglutarate to 2-hydroxyglutarate. Cancer Cell. 17(3): 225-234.
103. Watson JD and Crick FHC, A structure for deoxyribose nucleic acid. Nature, 1953. 171: 737-738.
104. Weldeab S, Lea DW, Schneider RR, et al., 155, 000 years of West African monsoon and ocean thermal evolution. Science, 2007. 315(5829): 1303-1307.
105. Williams MH and Bowie C, Evidence of unmet need in the care of severely physically disabled adults. British Medical Journal, 1993. 306(6870): 95-98.
106. Zallen DT, Correspondence. Despite Franklin's work, Wilkins earned his Nobel. Nature, 2003. 425(6953): 15.

CHAPTER 13

Unintentional Errors in Science

Accusations of scientific misconduct are serious and can damage the reputations of innocent scientists who are acquitted of the charges. When research that is assumed to be dishonest is observed it is best to report the questionable research to the head of the suspect's department, who should be able to provide a preliminary judgment as to the possibility that the accusation is valid. Then the dean of the medical school needs to be immediately notified. If the research involves human participants the Institutional Review Board (IRB) must be immediately informed as the suspected scientific misconduct may put partakers in the research at risk without the prospect of any gain to themselves or society. The IRB needs to take steps to safeguard against potential concerns related to participants.

While a case can be made for allowing universities and research institutions to investigate alleged scientific misconduct at their own institutions the committees appointed to investigate the accused have sometimes been questionable because of a COI in favor of the institution. In the John Rowland Darsee scandal Harvard University (Harvard) Medical School appointed a committee of eight to look into the matter and five members were of the Harvard faculty. Not only was the individual being scrutinized, but the institutional environment was not completely without blame as pointed out by an independent NIH committee that also looked into the matter. It is not surprising that Harvard's so-called "blue ribbon committee" absolved Harvard in the Darsee case of fakery [19].

While most unintentional errors are inconsequential, some are not entirely without ill effect if they are not discovered prior to publication. If detected after publication, the papers rightfully need to be retracted to the embarrassment of all authors. During the determination of the cause of the error, blame often gets placed on someone involved in the research. The scapegoat is often a junior member of the research team such as a graduate student or a technician. Some errors lead to serendipitous new discoveries.

In common with funding agencies in other countries, the NIH and NSF depend on academic institutions to initially investigate accusations of dishonesty. When the university confirms the allegations in federally funded research its findings are transferred to the funding agency and in the USA to the Office of Research Integrity (ORI), which looks into the matter further when indicated. If the ORI finds the accused guilty of misconduct sanctions are imposed.

Allegations of scientific misconduct take a major toll not only on the accused, even if the person is innocent, but also on the individuals involved in investigating the accusation. The scientists selected to evaluate the accusations are required to make a major commitment because of the serious nature of such allegations that can affect the careers of those involved. The time commitment is often massive and not trivial. A single case can embroil participants for years. Some individuals with experience in such cases vow that they will never do it again because the time commitment ruins their own careers. Also, tactics attacking the credibility of individuals in an investigation of scientific misconduct has convinced some individuals to never partake in another similar activity.

Some individuals charged with research misconduct, particularly in the USA, hire defense lawyers, who may be particularly aggressive and challenge the integrity of the person making the accusation. In other defense tactics the lawyer sometimes challenges the legality of members of the investigating committee or university officials. In some prolonged cases pre-agreed time limits are often wavered. Lawyers may even raise questions about the whistle-blower having inappropriate personal relationships. A threat of litigation has become another defense against accusations of scientific fraud. In the defense of a client in California some time ago a private investigator was hired to construct a psychological profile of the dean overseeing the investigation. The dean was labeled as paranoid based on a letter he had written to the accused.

William R. Brinkley, a past president of the Federation of American Societies of Experimental Biology (FASEB) observed a long running misconduct case at his own institution, Baylor College of Medicine (BCM) in Texas, when he served as dean. The tough stand taken in the USA against researchers who commit scientific misconduct has given birth to a new type of fugitive. Some individuals whose dishonest careers in research have been exposed in the USA have sought positions in other countries. A typical example is Kimon J. Angelides, a 44—year neurobiologist and tenured professor of cell biology and biochemistry, who lost his position at BCM in March 1996 for allegedly fabricating or falsifying research data in five scientific papers and in NIH grant applications. Angelides blamed the accusations against him on people with axes to grind. In August 1996 he filed lawsuits against BCM and its officials for defamation, blacklisting, wrongful termination, and breach of contact. The trial date was set for February 1997. The ORI was notified, but had not reached a decision by September 1996. Prior of his stint at BCM McGill University (McGill) *(http://www.mcgill.ca/)* identified false academic credentials in his curriculum vitae. These included four or five unpublished papers and a false statement that he obtained an undergraduate degree from Harvard, when in reality he received his undergraduate degree from Lawrence University in Wisconsin (UW) *(http://www4.uwm.edu)* [48]. In the early 1970s and 1980s when the first apparent epidemic of scientific misconduct emerged in the USA the rights of the accused where not appropriately provided, but by the late 1990s researchers, defense attorneys and scientific societies stressed the importance of making sure that the accused receive due process. Now the burden of investigating cases requires that the accused obtain all necessary legal protection of the accused and individuals volunteering to serve on panels looking into allegations of unacceptable research face hazards. The pendulum may have swung far too much in the wrong direction facilitating an environment in which a fraudulent scientist might go scot-free or even unchallenged. Now the accused receive perhaps more protection than they deserve. Even Brinkley who devoted much effort to ensuring that researchers accused of scientific misconduct receive utmost due process thinks that perhaps they are now providing too much protection. Reports on how to deal with allegations of scientific misconduct have been released by numerous professional societies [84].

Because institutions are notoriously weak at policing themselves, several bodies have been purposely established in different countries to carry out this mission. In the USA the ORI serves this necessary function after an initial institutional investigation. In the Nordic countries and Austria central committees on scientific dishonesty become informed of all alleged scientific misconduct. They also teach good research practice and guide whistle-blowers. The committees investigate the allegations of scientific misconduct, monitor all accusations and publish reports. In some countries, such as the United Kingdom (UK), a central committee seems more desirable as a single body would gain relevant experience and proficiency than smaller committees.

Representative Examples of Unintentional Errors in Research

Unintentional Communication of Information to Human Subjects

Not all errors in research fall under the designation of "scientific misconduct" or "inappropriate behavior". Some are "honest mistakes" that are unintentionally made, but the distinction between accidental errors in science and scientific misconduct is not always clear-cut and some whistle-blowers have accused researchers of misconduct when unintentional "honest "errors have been made. Unintentional errors include mistakes in recording findings in laboratory notebooks, mislabeling figures, undetected mistakes in proofreading of manuscripts, and literature citations.

In psychological studies on human subjects the beliefs and expectations of scientists can be conveyed to participants in unintentional ways and this can affect their behavior in the experiment. Many books and articles in the psychology literature document this experimenter effect. Those that commit scientific misconduct sometimes also make unintentional mistakes that accentuate the errors. A prime example is the complex contemporary genomic research that involves special computer programs and complicated bioinformatics. This added to the downfall of Anil Potti in 2010 (page 14).

Unexpected Contaminated Vaccine

A vaccine against *Corynebacterium diphtheriae*, the causal agent of diphtheria, was produced in horses, but in 1901 thirteen children died after being inoculated with horse-serum-derived diphtheria antitoxin. An investigation into the tragic incident from which the diphtheria antitoxin had been derived developed tetanus from which it contaminated the diphtheria antitoxin. This calamity caused the USA Congress to pass the Biologic Control Act of 1902, providing the government with regulatory power over the production of antitoxins and vaccines. This episode is documented in a book written by Margaret B. Liu and Kate Davis titled Lessons from a horse named Jim [106].

Even after an experiment apparently yields valid results, the conclusions can only be regarded as provisional until the experiment has been replicated, because something unnoticed may have influenced the, making the study nonreplicated.

Carelessness

Some investigators are careless and fail to acknowledge unintentional errors. This can become particularly disturbing when the investigators are arrogant and ignore questions about their research and Anti Potti. Inattention makes the errors of some lackadaisical individuals obvious.

Lack of Appropriate Training

Unintentional mistakes are commonly made by researchers that lack the appropriate training. Such errors are caused by ignorance, time pressure to document new observations before competitors, and sloppy work. Mistakes in scientific judgment, inappropriate techniques, statistical errors, and ignorance about existing information can lead to incorrect conclusions. Scientists are human and sometimes honest individuals make unintentional errors in their research, but these mistakes become corrected by the person who makes them or by persons who follow up the defective studies often with different more refined methods.

The analysis of research data often spawns different opinions and the principal investigator (PI) sometimes comes to the incorrect conclusion, especially if wedded to a mistaken hypothesis. Research performed by persons with an inadequate educational background or who are ignorant about the relevant technology commonly result in unintentional errors. Defective controls, incorrect preconceived notions, inadequate sampling and statistical errors, inappropriate techniques, poor observations are also sources of unintentional mistakes in research. Some research carried out by persons with a good intention, may be incompetent because the researcher used methods that the investigator was not qualified to use or interpret. Some such studies are now and then carried out sloppily or inadequate attention is paid to the writing of the resultant manuscript.

Researchers that discover honest mistakes in their studies after a publication of the results should retract the paper with an indication to the journal editor as to why the paper is being retracted. The researcher should realize that this is the thing to do and that the retraction will not tarnish their reputation. An admission of "honest mistakes" does not carry the stigma of retraction because of scientific misconduct.

Unintentional Errors Due to Time Pressure

The speed of light in a vacuum (299, 792, 458 meters per second) is an important constant in physics and it is believed that nothing can travel faster. But this long accepted belief was challenged by the finding that electrically neutral subatomic particles called neutrinos can travel faster. A team of 170 members working with the Oscillation Project with Emulsion-tRacking Apparatus (OPERA) particle detector fired neutrinos from the Conseil Européen pour la Recherche Nucléare (CERN) laboratory in Geneva, Switzerland to Gran Sasso, Italy and they arrived faster than if they had travelled the distance at the speed of light. They were faster by 60 billioniths a second! [31]. To be accepted the work needed to be replicated and the Fermilab in the USA decided to prove or disprove the conclusion of the experiment. If the CERN experiment could withstand the test of time it may have an effect on Einstein's theory of relativity, but the study could not be replicated by other physicists [32].

Controversial Cutting-Edge Research. Scientists doing cutting-edge research on controversial topics may find their work questionable if it is not adequately documented, especially if persons working in the laboratory make allegations of improprieties.

Unintentional Errors due to Observer Bias and Self-Deception

Bias and self-deception can influence the interpretation of scientific studies regardless of whether it is intentional or unintentional, and this can lead to false conclusions in research. This was particularly common in the past, and statistical analyses and randomized clinical trials were designed to avoid this.

Sloppy Science

By 1977, the number of shoddy and unrepeatable articles in scientific journals was increasing at an alarming rate [17]. Sloppy research is commonly performed by students, residents, and fellows often because they are instructed or expected to participate in research as part of their training. They are commonly just enhancing their credentials to achieve a career goal, and their supervisor is often a clinician who needs to coauthor publications for advancement, but who is not really motivated toward research. Sloppy research is also performed by established scientists as exemplified by the kidney cancer vaccine research of Alexander Kugler (see below), the immunological research of Theresa Imanishi-Kari and the AIDS research of Popovic. Research also often becomes sloppy when the investigators are under pressure to obtain grant support or to submit an abstract for a professional meeting.

Defective Software

Geoffrey A. Chang, a promising young crystallographer who received the Presidential Early Career Award for Scientists and Engineers (PECASE) at the White House in 2000, made errors in his research because of flaws in software that he used to generate the molecular structure of crystallized proteins embedded in cell membranes. Researchers [51] at the Swiss Federal Institute of Technology in Zurich raised serious doubt about the 2001 Chang paper while determining the structure of an ATP-binding cassette (ABC) transporter in *Staphylococcus aureus* named Sav1866. Surprisingly, it was significantly different from the structural information that Chang and colleagues had previously published on another ABC transporter named MsbA isolated from the bacterium *Escherichia coli* [27]. According to Google Scholar, that paper had been cited 715 times in other publications by April 2013. After learning that his research had been questioned, Chang looked into the matter and was dismayed to learn that the software program that he used and obtained from another laboratory was defective and flipped two columns of the data. After realizing that they published erroneous structures, Chang and his coworkers retracted five papers [28]: three in Science [27] (Chang, G. Pornillos, J etc. Science 310 (5756): 1950-1953, December 23, 2005) [28], Reyes, CL, Chang G. Science 308, 1028 2006; one in Molecular Biology on Vibrio cholera in 2003 [29]; and one in the Proceedings of the National Academy of Sciences USA in 2004 [110]. Chang was extremely apologetic about the unintentional mistake, which caused some researchers to waste time and effort [116]. Because of Chang's high esteem, other researchers who contradicted Chang's 2001 paper had difficulty getting their finding published and grant applications that did not agree with the retracted papers received a tough time under peer review [129].

Failure to Abandon the Wrong Hypothesis

Some researchers believe deeply in their hypotheses and commonly defend them vigorously even if the theories are off beat, often long before their peers consider them plausible. To diligently follow the wrong theory when objective evidence refutes it is a major weakness of some researchers. This has caused the downfall of numerous scientists or research teams. An avid belief in the wrong hypothesis can turn out to be a pipe dream. With blinders on their eyes, some researchers do not evaluate relevant data objectively. When creative scientists think outside of the box and become enmeshed in their experiments, they sometimes arrive at incorrect conclusions. Rather than paying heed to the flaws to which other draw attention, they become so convinced in the validity of their hypotheses that they fail to consider alternative explanations for the experimental findings. Among the pressures of the research environment is the challenge that faces researchers when their work is questioned. When an error in the research of a scientist is pointed out, there is a human tendency to go on the defensive and defend the work even when evidence refutes the existing hypothesis.

In addition to deliberate bias, some scientists make a serious conscious or unconscious mistake of pursing their research, blindly ignoring all evidence that does not support their hypotheses because they are determined to prove that their theories are correct. Along the path of testing hypotheses, researchers are kept on going by what is virtually devotion. It is not easy for them to tread beyond their personal persuasions and envision the situation by the use of the eyes of an unbiased bystander. Dispassionate individuals pursuing a hypothesis are more likely to interpret data objectively than passionate researchers. Unfortunately, some scientists become so engrossed in their hypothesis that they lead themselves up the wrong futile path. While retaining a passion over their hypotheses, scientists must remain sufficiently detached and objective to prevent becoming trapped in an obsession about their hypotheses. Researchers that become enmeshed in their hypothesis and lose objectivity sometimes believe firmly in a hypothesis that is obviously wrong in the eyes of others.

Insect Pheneromes

L. B. Hendry, an assistant professor at Pennsylvania State University (Penn State) *(http://www.psu.edu/)*, typified someone with a strong belief in a theory that was wrong (cite the New York Times January 23, 1977). In 1975, he contended that the chemical sexscent known as pheromones that insects generate was determined by what they eat [86]. Extraordinary effort was taken to make the theory known in scientific communities and in the print media. His new theory created much attention, but others could not confirm this hypothesis, and its originator would not admit that he was wrong. A study by a team of researchers from Cornell University (Cornell) *(http://www.cornell.edu/)* contradicted the work about a year later, and this led to a two-year-long controversy. Miller and colleagues from Cornell identified the sex pheromones of the oak leaf roller *(Archips semiferanus)* as a specific blend (67:33) of trans-11—and cis-11-tetradecenyl acetates. In the female insects, the blend did not vary when they ate a semisynthetic diet. Also, males responded only to the 67:33 blend. In additional reported studies by David M. Hindenlang and Joseph K. Wickmann, two authors of the initial paper agreed with the Cornell group [89]; and in essence, they retracted the original paper, putting the dispute to rest [5].

The Pursuit of Absurd Hypotheses

Scientific articles are not only published in peer-reviewed journals but also in numerous other publications and magazines intended for the lay public. The latter sometimes provoke scientific studies. A prime example of a passionate conviction to a false hypothesis occurred when Stanley Pons, chairman of the Chemistry Department at the University of Utah (Utah) *(http://www.utah.edu/)*, teamed up with the electrochemist Martin Fleischmann from the University of Southampton *(http://www.southampton.ac.uk/) in England.*

Paul Brodeur and Cancer from Power-Field Lines

The provocative writings by Paul Brodeur (1931-) from 1989 to 1992 in the New Yorker [20-25] raising the possibility of a link between electromagnetic fields from power-field lines and cancer stimulated scientific inquiry. The issue raised was of considerable public concern; and several colossal epidemiological studies were performed in Canada, the United Kingdom (UK), and the USA to obtain statistically valid results. The research was not put to rest until eight years later at a cost of $25 billion according to the administration of William "Bill" Clinton, the forty-third president of the United States [119] *(http://discovermagazine.com/2000/oct/featblunders)*. These expensive studies could have been avoided if those tackling the project had been aware that cancer-inducing agents break chemical bonds in DNA, and this energy has the wavelength of ultraviolet light or smaller. Longer wavelengths and particularly those of power lines that are miles long cannot break chemical bonds by any stretch of the imagination.

Scientific Failures Because of Omission of Crucial Tests

The Hubble space telescope made to obtain the sharpest images in outer space has undoubtedly been one of the greatest scientific achievements. It was made at a cost of $1.5 billion over nearly twenty years by about five thousand skilled craftsmen. After it reached space, the mirror was found to be defective, making it impossible to focus light correctly [172]. A human blunder by Perkin-Elmer Corporation ground the most important mirror to the incorrect curvature. A procedural mistake failed to detect this error in construction in a prelaunch test before the telescope left earth. Fortunately, astronauts were able to install corrective optics that made it possible for the telescope to disclose amazing images of celestial bodies that nobody could have predicted. Aside from the unintentional error in the construction of the mirror, dishonesty was also a component in the construction of this fantastic human achievement. On more than one occasion, some National Aeronautics and Space Administration (NASA) managers kept a truthful assessment of the cost from the USA Congress [66] (www.americanscientist.org). This deception was carried out to keep this expensive project active. As stated by Zimmerman [172], the mirror debacle was the direct outcome of dishonesty in budgeting and scheduling.

Scientific Fiascos Because of Technical Failures

Challenger

The second Space Shuttle orbiter of the NASA, called the Challenger (NASA Orbiter Vehicle Designation: OV-099), was built by Rockwell International's Space Transportation Systems Division in Downey, California. Following its maiden flight on April 4, 1983, it successfully accomplished nine missions. In January 1986, Roger Boisjoy, an engineer with Morton Thiokol Inc., brought up technical worries that he thought should delay a further blast-off of the Challenger; but no attention was paid to the advice. Shortly after its launch for the tenth time on January 28, 1986, the Challenger exploded, tragically killing the entire crew of seven. An analysis of this ill-fated event was found to be primarily the result of a blemish in a low-tech O-ring intended to prevent the escape of hot gases from the booster engines of the spaceship. NASA lift-off officials had been warned by their engineers that the rings could crack in cold weather, but the risk was ignored. Aside from the technical difficulties that plagued Challenger's explosion, disappointments also stemmed mainly from managerial blunders and budgetary constraints rather than scientific errors.

Galileo

The $1.3 billion space probe Galileo departed Earth in 1989 with a planned meeting with Jupiter in 1995. A microprobe was expected to enter the merciless Jovian atmosphere with the remainder of the spacecraft passing between the planet moons. Unfortunately, the main data transmitting antenna of Galileo would not open completely. The Galileo had low-gain and high-gain antennae, and for unknown causes, the high-gain antenna did not completely become set up. In an investigation of the possible explanations, it was realized that Galileo remained in a storeroom for a period of time after the Challenger disaster, and some hypothesized that the system became damaged or that the lubricants evaporated. Luckily, Galileo had a low-gain antenna that could convey data back to Earth, but unfortunately not at the desired resolution. Galileo carried a 1, 500-mm (59-inch) focal length narrow-angle telescope like the one used in Voyager. Together with accessories that included an image sensor, the camera formed a solid-state imaging subsystem (SSI).

Lack of Appropriate Controls

Controls are an important component of scientific investigations and some unintentional errors in science result from a lack of appropriate controls or an inadequate sample size. If controls are not used or are inappropriately chosen, incorrect interpretations can result. In 1984, an analysis of a 6.3-ounce meteorite found in Antarctica disclosed flecks of polycyclic aromatic hydrocarbons, magnetite, and iron sulfide. Because the rock was thought to have reached Earth from Mars, NASA scientists declared in 1996 that the finding of amino acids indicative of life suggested the existence of bacteria on the Red Planet 3.6 billion years ago [115]. But this conclusion was reached on insufficient evidence without appropriate controls [97]. Subsequent studies showed that traces of the amino acids within the Martian rock were also present in the surrounding Antarctic ice. A further blow to the theory of life on Mars was the detection in other rocks, such as those on the moon, of similar "confirmation" of life.

Failure to Consider Other Factors in the Interpretation of the Data

A failure to consider all relevant factors in the interpretation of research data can result in wrong conclusions. The American astronomer and science communicator Carl Edward Sagan (1934-1996) published a manuscript in 1983 with coauthors, forewarning that a nuclear war could contaminate the atmosphere with so much dust that all sunlight would be blocked, creating a climatic change like the one that terminated the survival of dinosaurs [70]. Skeptics disagreed with the conclusion and pointed out that this article had significant unintentional flaws because of a variety to factors that had not been considered. For example, the dust would need to reach the highest levels of the atmosphere so that rain would dissipate it. A few years later, Sagan together with previous authors admitted that their original estimates of temperature were incorrect [146, 147].

Errors from Lack of Adequate Communications

Frequent communication between members of collaborative research teams is essential, especially when complex scientific endeavors are pursued. If this is not carried out adequately, scientific mistakes can occur and they may. According to NASA officials, the failure of this space craft was a direct result of using the metric Newton in its guidance. Regrettably, Lockheed-Martin built the Orbiter to be guided in the English units of poundals.

Mars Climate Orbiter and Polar Lander

A prime example of a communication failure was the tumbling of the $125-million Mars Climate Orbiter into oblivion near Mars in September 1999. An even more expensive fiasco was the loss on December 3, 1999, of the $185-million Mars Polar Lander (Mars Surveyor '98 Lander) after multiple attempts to get in touch with it after the failure of space radio antennas. NASA officials speculated that the trip was destined to failure because one line of missing computer code created a signaling crisis in the landing legs.

An allegation of scientific misconduct was raised against two professors (Dan Feldheim and Bruce Eaton) and a graduate student (Lna Gugliotti) when they performed cutting-edge research at the University of North Carolina at North Carolina State University (NCSU) in Raleigh, North Carolina. Several investigations into the incident took place, including one by the National Science Foundation (NSF). A federal investigation found them guilty of falsifying groundbreaking research that they published in Science [83] In the Science paper, an RNA sequence known as RNA palladium (Pd 17) was reported to create metallic hexagons with an edge length in the micrometer range in an aqueous solution. According to the inspector general of the NSF, the researchers "recklessly falsified their work" [118]; and this investigative arm of the NSF has apparently recommended that the paper be retracted and that the authors be barred from acting as reviewers, advisors, or consultants for the NSF for three years. A supervisor is also expected to be required to certify that the researchers follow federal rules in writing precise documents. An internal investigation of the charges of misconduct was conducted by NSCU, and it also concluded that the research reported in Science contained deceptive data and was not performed according to acceptable scientific standards. NCSU did not find intentional errors of research misconduct but careless unintentional mistakes. Gugliotti submitted the work to the university in her PhD thesis in 2006. The research was supported by the NSF, the USA Department of Energy, and the W. M. Keck Foundation.

Because of significant flaws in the paper by Gugliotti and colleagues, Stephan Franzen, another professor of chemistry at NCSU, battled for years to get the published data rectified; and almost a decade after the paper was published in Science, the manuscript had still not been corrected [118]. No retraction or errata had taken place, and the editor of Science had not expressed an expression of concern. During structural studies on crystalline Pd hexagonal particles, Franzen and coworkers [73] found that RNA Pd 17 was not required for the formation of the hexagonal particles. In the face of experimental evidence that their studies were flawed, Gugliotti and her coauthors persisted in denying their faulty data in a way typical of jerks.

Representative Scientists Performing Unintentional Research

Robert Charles Gallo. A prime example of unintentional errors in science is illustrated in the saga about Robert Charles Gallo (1937-) and the discovery of human immunodeficiency virus (HIV), the causal agent of the acquired immunodeficiency syndrome (AIDS) in the Gallo laboratory. That laboratory demonstrated the drama of scientific discovery in the face of competition. It also exemplified the inevitable disputes that result when money is involved because of patents. In the early 1980s Gallo was attempting to discover and isolate the causal agent of AIDS and by the end of the year the Gallo team detected reverse transcriptase, an enzymatic marker for a retrovirus, in two patients from hundreds of blood samples. Because of his expertise in virology and particularly in retroviruses it was not surprising that Gallo moved his research efforts into finding the etiologic agent of this scourge that was infecting and killing millions of people worldwide.

Professor Luc Antoine Montagnier of the Pasteur Institute in Paris detected reverse transcriptase in a viral isolate known as LAV/BRU in February 1983 and requested reagents from Gallo so that his Pasteur Institute team could characterize the virus further. With Gallo's reagents, Montagnier demonstrated that his LAV/BRU isolate from a patient with AIDS having the initials BRU differed from both human T-lymphotropic virus I (HTLV-I) and human T-lymphotropic virus I (HTLV-II). In February 1983 the Gallo team found reverse transcriptase in two additional samples, but no other features of HTLV-1 indicating a new incompletely understood virus had been discovered. The Gallo team failed to publish early experimental findings suggesting the existence of this novel retrovirus. Instead their initial publications focused on the likelihood that the AIDS virus belonged to the HTLV1 and HTLVII family of retroviruses.

On May 20, 1983 Science published a series of papers on AIDS and it included two from the Gallo laboratory [77, 80] and one from the Pasteur Institute [11]. In July 1983, Montagnier sent Gallo a sample of a virus named lymphadenopathy virus (LAV) that had been isolated at the Pasteur Institute in 1982 from a patient with AIDS. Its genetic code had not been determined, and researchers at the Pasteur Institute did not know whether it caused AIDS. In September, Gallo received a second sample of the LAV/BRU virus and confirmed that it was a retrovirus. The recognition of this fact spawned a dispute among the scientific AIDS establishment that became more provocative over time. The Gallo laboratory was carrying out research on AIDS and other retroviruses before they received LAV.

In April 1984, Gallo and his colleagues at the National Cancer Institute (NCI) announced that they had identified the virus responsible for AIDS and had developed a blood test to detect it. Shortly thereafter on May 4, 1984, they published four papers in Science [78, 128] and provided persuasive evidence that a virus was the causal agent of AIDS. They also outlined the procedure for detecting the AIDS virus in blood.

They named the virus human T-lymphotropic virus type III (HTLV-III), but it later became known as the human immunodeficiency virus (HIV) [76]. The papers documented how the Gallo team raised copious amounts of retrovirus. From 1984 Gallo became widely acclaimed as the discoverer of the cause of AIDS, but he was attacked by opponents who maintained that he stole this prestige from Montagnier of France. Within months, it became obvious that their HTLV-III which they designated HTLV-IIIB (IIIB), bore a striking resemblance to the 1982 French viral LAV/BRU isolate at the Pasteur Institute.

In 1985, while evaluating the Pasteur Institute's claims, the Department of Health and Human Services (DHHS) extensively reviewed the accusations and insinuations and discovered them to lack merit. The inquiry no longer appeared to be one for a legal point of view but rather an investigation by an oversight committee from the National Academy of Sciences (NAS). It appeared that the entire ad hoc process was devised to protect the NIH from congressional criticism. Without published procedures the integrity and fairness of the decision making process was ensured and the NIH was shielded from unjustifiable congressional pressures.

The research career of Gallo was full of controversy and he had a reputation for his superego. In 1988, the NIH began investigating Gallo after he was accused of identifying the causal agent of AIDS in a tissue sample from the Pasteur Institute. This allegation had a devastating effect on Gallo a famous scientist who had made major contributions to our understanding of a ravaging plague that was infecting and killing millions of people worldwide.

The initial investigations into Gallo focused on allegations that his laboratory had misappropriated the AIDS virus (designated LAV) from the French research groups well as disagreements about a patent and disputes with other scientists. The allegation about misappropriation was discarded early in the scrutiny. In 1990 Science devoted an enormous description of the Gallo scandal [7, 43-46, 124]. Later analyses proved that although separate strains of HIV differ in genetic structure LAV/BRU and IIIB, which Gallo used to develop HIV, were essentially the same. The French believed that the two strains stemmed from the fact that they had forwarded samples from the patient coded BRU to Gallo in both July and September 1983. The French concluded that LAV/BRU must also have been the IIIB isolate detected in the Gallo laboratory.

On November 19th 1989 John Crewdson, an investigative reporter for the Chicago Tribune published a 16 page article [37] resurrecting the uncertainty about whether the Gallo AIDS virus had its origin in the French isolate. For his research on Gallo Crewdson filed more than 100 requests from the NIH for information about Gallo's laboratory under the USA Freedom of Information Act (FOIA). This high number of requests virtually brought the Gallo laboratory to a standstill. Crewdson delved into the likelihood that Gallo or his coworkers had stolen their HIV isolate from samples received by them from Montagnier. Crewdson did not discover a "smoking gun", but resorted to an imaginative cautiously written indirect line of reasoning using the style of journalism which he used at the beginning of his career while reporting on Watergate for the New York Times. This journalistic bulldog seldom made factual mistakes, and when he did they were as a rule unimportant. Crewdson was bothered by dishonesty and conclusive rights and wrongs rather than with the most up-to-date research discoveries or petty credit disputes. After interviewing more than 150 people Crewdson concluded that the AIDS-causing virus had first been discovered by the Montagnier research team.

In spite of this, the NIH still looked into inconsistencies in the seminal 1984 publication by Gallo and his coworkers in October 1990. The NIH also acquired original samples of other isolates from patients with AIDS that were established in the Gallo laboratory. The NIH desired to pin down IIIB and requested samples of cells from the Gallo laboratory from 1983 to 1984 when the research was being performed. Gallo cooperated fully with the NIH investigation. The Crewdson article prompted a 10—month long investigation, which culminated in the verdict that Gallo was not guilty, but after this article Crewdson fanatically pursued Gallo for more than three years. Crewdson behaved as if he was a renowned virologist rather than Gallo. Had the timing of the Gallo story not coincided with the wide publicity of numerous cases of scientific misconduct at prominent institutions and the congressional hearings precipitated by them, Crewdson would probably not have been allowed to launch such an in-depth scrutiny of a prominent American scientist, who was devoting a major effort in the public interest to find the cause of a deadly disease that was killing millions of people worldwide.

Not unexpectedly representative John Dingle, insisted that the NIH look into the allegations raised against Gallo in the Crewdson newspaper report. The NIH concentrated its attention on accusations raised by Crewdon and to ensure that its inquiry was independent it was performed by a committee of federal scientists, and the NIH invited the NAS and the Institute of Medicine (IOM) to recommend an outside panel of unconcerned authorities to watch over the investigation by an internal panel.

Crewdson's article precipitated an Office of Scientific Integrity (OSI) investigation and this was continued by the Office of Research Integrity (ORI) which replaced it. Almost continuously over four years the OSI and ORI investigated the Gallo laboratory. Initially the OSI concluded that there was insufficient evidence to implicate misappropriation rather than unintentional contamination. The Crewdson article precipitated another OSI investigation of Gallo and Howard Streicher, a researcher in the Gallo laboratory spent a considerable amount of time responding to questions brought up by the OSI.

Shortly thereafter Montagnier reiterated the incrimination and NIH officials were forced to appoint a panel of independent scientists to reexamine the situation.

Crewdson maintained that Gallo should expect journalists to scrutinize his research thoroughly, because as a NIH scientist he was an employee of the taxpayers. In sharp contrast to most journalists Crewdson showed no respect for Gallo and considered the NIH to be the least analyzed component of the government. Crewdson's long article did not blatantly focus on allegations of corruption, but on issues like potential theft, fraud, and an apparent cover up that became apparent during his scrutiny of Gallo. Crewdson suggested that the only AIDS causing virus yielding productive experimental data was LAV, which Montagnier sent to Gallo in 1983. The article by Crewdson provoked the scrutiny of the NIH, DHHS, and a congressional subcommittee chaired by John Dingell. These different groups unearthed evidence that infuriated administrators at the Pasteur Institute. The latter became convinced that the French scientists had received the short end of the stick in the HIV blood test patent agreement. Some support for this view stemmed directly from the Gallo laboratory. The Gallo blood test for AIDS was patented by the USA government. On February 1, 1978 the research laboratories of John Minner and Gallo joined forces in a productive collaboration for a couple of years Adi Gazdar, a 41 year old pathologist, became a member of the NIH after obtaining a medical degree from London University. Gazdar provided two cell lines (HUT78 and HUT102) that became renowned to Gallo's colleague, Frank Ruscetti, who had isolated the first human retrovirus with him in 1979. To considerable applaud Gallo discovered the first human retrovirus (HL23 known), but the announcement was premature and some months soon after he was compelled to acknowledge that the virus was just a contaminant of his cultures.

Some activities relevant to the AIDS/HIV research of the Gallo laboratory were ignored in the narrative about the big picture. The cardinal papers on the cause of AIDS from the Gallo laboratory described the use of a H9 cell line, but did not mention the origin of this cell line and nor did the papers acknowledge the Minner research group. The failure to acknowledge other researchers who contribute to a scientist's research is not only unethical, but perhaps a form of scientific misconduct. Years later it became known that H9 was cloned from a cell line named HUT78 that was created by Gazdar and Bunn in the Minner laboratory and given freely to the Gallo laboratory, which renamed it H4 to conceal l its origin and deprive Minner's laboratory of any potential royalties derived from the AIDS blood test.

Initially Gazdar was not acknowledged for any part in Gallo's exciting research on AIDS, but later he pleaded for credit since he had his finger in the Gallo newsworthy pie that was attracting international acclaim. For years Gazdar speculated about whether the H9 cell line mentioned in the Gallo laboratory papers may have come from a cell line that he had created and given to the Gallo laboratory. Gallo considered Gazdar's request for recognition "a pathetic joke" as cell lines do not get patented and only scientific achievements warrant credit.

Eventually with DNA sequencing, IIIB and LAV/BRU were found to be remarkably alike. This energized the disagreement between the Gallo and Montagnier laboratories, reinforcing the belief that Gallo may have unethically pilfered the Montagnier virus.

For allegedly stealing the AIDS virus from the Pasteur Institute Gallo was accused of scientific misconduct. After being investigated by the OS1 (the precursor of ORI), but was cleared of accusations in October 1989, but nevertheless pushed on with a scrutiny of data supporting the seminal paper published by Gallo and his former colleague and cell-biologist, Popovic in 1984. By December 1990, scientific integrity had become a major public issue in the USA and the OSI pronounced that it had created a scientific panel to assist its investigation of the controversial aspects of Gallo's 1984 paper in Science.

Subsequently in 1990 the NIH investigated allegations that Gallo had helped himself to the virus containing specimen, but in the fall of 1990, the NIH exonerated Gallo of stealing anything from Montagnier's laboratory. The NIH found that Gallo already had several viral isolates from patients with AIDS in his laboratory when the LAV/ BRU specimen arrived from Paris, and there was no reason to steal the French virus. Culture contamination was the most probable explanation for the striking similarity between the viruses. Investigators apparently tried to determine whether there was a contamination or whether IIIB and LAV/BRU were different were independently isolated viruses. After Gallo was absolved of accusations of stealing the AIDS virus from the Pasteur Institute [47] the NIH attempted to determine from which particular patient the virus was isolated. Almost from the beginning when the origin of the Gallo IIIB virus was questioned, especially since the nucleotide sequence of IIIB was astonishingly similar to that of the virus isolated at the Pasteur Institute named LAV/BRU.

Members of the Gallo laboratory, including Jean-Claude Chermann a former collaborator of Montagnier revisited the specimens and raw data and analyzed three of the five of LAV/BRU specimens that Gallo obtained from the Pasteur Institute in 1983, which had remained frozen and saved at the NIH. The nucleotide sequence in IIIB and LAV/BRU specimens differed by 1%. It seemed more likely that the inconsistency between LAV/BRU and Gallo's IIIB reflected the rapid mutation rate of the virus that both Montagnier and Chermann had observed when the AIDS virus was cultivated in different media.

Gallo published a 352 page autobiography in 1991 [79] recounting his recollection of the discovery of the AIDS virus and the clash with the French that set off the NIH inquiry. While primarily interested in seeking the truth Crewdson seeped beneath Gallo's skin making him feel persecuted and in his book [79] Gallo expressed the opinion that he had become the object of Crewdson "who somehow turned from analysis to assignation".

In February 1991, Gallo published fresh findings with colleagues in Nature [75] contending to settle the issue of whether the AIDS virus identified in his laboratory was really imported from France. But his intended message was not convincingly transmitted and the paper only intensified the uproar and deepened the concerns. The Gallo-Montagnier dispute that was expressed in the public domain tarnished the intentions and trustworthiness of scientists. The unease persisted after Gallo admitted that the virus he identified was isolated through inadvertent contamination from a sample sent to him.

Michael Epstein considered the existing agreement of unfair. This was not a trivial academic quibble because it involved millions of dollars generated annually in royalties. Epstein wanted a greater percentage of the income for the Pasteur Institute because Gallo had admitted in 1991 that the AIDS virus he used to create his blood test seemingly came from blood samples contaminated with the LAV/BRU virus provided to the Gallo laboratory by the Pasteur Institute.

In October 1991, an NIH investigation of the Gallo laboratory documented the history of the Popovic's concoction of blood samples allegedly obtained from ten patients with AIDS. In his response to allegations of falsification, he pointed out that they represent either divergences in scientific explanations or outcomes of his inaccurate use of English [85].

The final OSI report did not consider Gallo guilty of scientific misconduct but was judgmental of his failings as head of his laboratory and as senior author of the publications stemming from it. The OSI passed the Gallo case on to the ORI, which replaced it in May 1992. Initially, the ORI found Gallo guilty of a single episode of scientific misconduct in a statement in the 1984 papers in Science. The authors wrote that the French virus designated LAV "has not yet been transmitted to a permanently growing cell line"; hence, its characterization was challenging. The accusation that Gallo stole the virus from Montagnier was explored but dropped by both the ORI and the former OSI. After Popovic won his appeal, the ORI without delay dropped the charge of scientific misconduct against Gallo.

Mikulas Popovic. The Czech cell biologist Mikulas Popovic came to the USA in 1980 and worked in the Gallo laboratory. He grew vast quantities of HIV in a most unconventional manner by pooling specimens from ten different patients with AIDS into a single soup and by doing so he played an important role in establishing the cause of AIDS. Thinking that viruses might not be making sufficient reverse transcriptase, he used an unorthodox method of pooling viruses from different sources into a solitary culture. Popovic hoped to culture HIV in this manner and indeed his unrushed method ended up being successful. Popovic continued trying to obtain continuous replication of the viruses in a human cell line. From such a soup a virus designated IIIB proliferated and was exploited to develop an AIDS blood test.

While the Gallo laboratory was under investigation, Popovic lost his job at the NIH and he joined the faculty of the Primate Research Institute of the New Mexico State University *(http://www.nmsu.edu/)* as head of the department of virology and immunology. He later obtained temporary work at the Karolinska Institute in Sweden and was still working there in 1994. After being found not guilty by the FBA Barbara Mishkin, one of Popovic's lawyers appealed to the board on behalf of her client asking the DHHS to reimburse his legal fees, which amounted to more than $250, 000. In October 1990 the Gallo laboratory had other accessible viral isolates that could have been used, but he was growing IIIB in vast amounts. The Gallo research team had a computerized database on every AIDS blood sample that had been received for testing. Their records indicated that he received blood samples from patients with AIDS not only from the USA, but also from France and Switzerland and if IIIB and LAV/BRU were identical both laboratories could have received specimens from the same infected patient. In August 1991 Science [148] mentioned that the draft NIH report about the dispute censured Gallo, but did not officially reprimand him or find him guilty of scientific misconduct. It found Popovic guilty of scientific misconduct because of inaccurate statements in the paper published in the May 1984 issue of Science [128], documenting the isolation of the AIDS virus by the Gallo laboratory and providing proof that the virus caused AIDS. Several times Popovic incorrectly entered not done into the notation of a table when it was clearly incorrect when experiments had been performed. The Science paper documented that reverse transcriptase was produced continuously for five months, but the raw data substantiating this was not available. M. G. Sarngadharn, a coauthor of the Science paper, informed OSI that he did not save the data.

The OSI found several flaws in Popovic's research and in his explanation of them. He was found to be untruthful about his effort in growing a permanent culture of IIIB. His account of a viral pool that generated "continuous" virus was deceptive because Popovic needed to re-feed the pool with virus samples twice. The OSI attempted to deal with the source of the IIIB virus and sequences of the nucleotides in the viruses viral samples were performed from 1983 to 1984 in a masked fashion to Roche Molecular Diagnostics [149] so that their DNA sequences could be established without knowing their source prior to analysis. The company verified the presence of dissimilar viruses in the cultures that Popovic claimed that he included in his pool in November 1983 and January

1984. An isolate ascertained to be the basis for both LAV/BRU and IIIB was not found. During 1991 OSI concentrated on suspect data in one of the four ground breaking papers published by the Gallo research team in Science on May 4, 1984 [78, 128, 132, 133]. In August 1991 the OSI was apparently ready to withdraw misconduct charges against Popovic [85], but it became apparent that Dingle was seriously considering charging Gallo for perjury because of an internal NIH memo prepared in June 1991 by Suzanne Hadley objecting to what she regarded as "false" and "incomplete and misleading" statements in three documents signed under oath by Gallo, namely the HIV blood test patent, the patent on Popovic's technique for endlessly generating AIDS virus and a statement signed on November 8, 1986 as a component of the USA government's justification opposed to the patent lawsuit. Gallo had asserted that Popovic employed the cloned H9 cell line to produce virus, but his laboratory records revealed that H9 did not exist until cloned by Popovic on January 19, 1984.

By November 1991 Dingle was planning to initiate an investigation of NIH and DHHS officials, because he suspected them of plotting to cover up falsehoods in the AIDS research of Gallo during the long-drawn-out clash over patent rights of the HIV blood test fight between the USA government and the Pasteur Institute in 1989.

Gallo and Popovic were both found guilty of misconduct by the ORI for supposedly making false statements in their papers. Later the ORI characterized Popovic's errors as "relatively minor". The ORI asserted that Gallo did not establish in appropriate time the origin of the cell line in which his research team cultivated the virus causing AIDS and that this error deprived suitable people of due credit and held back attempts by other investigators wanting to grow the virus. According to the ORI Gallo tried to place excessive control on the use of his cell line.

The ORI found Popovic guilty of scientific misconduct on November 4 1993, but two weeks later the federal appeals board (FAB) of the DHHS, consisting of three lawyers overruled the ORI verdict and cleared Popovic of charges of scientific misconduct. The FAB considered each of Popovic's explanations plausible and it criticized the ORI for misquoting and misparaphrasing key sentences in questionable papers, such as in the one Science paper that caused Popovic much grief. In their assessment of the ORI's case against Popovic the FAB issued a powerfully worded criticism of the government for its enormous endeavor over four years to uncover scientific misconduct by Popovic and by implication his past chief Gallo. Despite the substantial effort noteworthy evidence of wrongdoing was not found.

The FAB made it clear that the ORI must meet high standards of proof and that it had failed to even prove that there were errors in the paper in Science [3].

According to the ORI the Gallo laboratory had grown LAV in a permanent cell line. The AIDS virus proved tremendously grueling to culture. The FAB dismissed some ORI charges, because of insufficient evidence.

The FAB held hearings in November, 1993 on the charges against Gallo. In its consideration of the Gallo and Popovic cases the FAB pointed out that to find someone guilty of research misconduct the ORI must not only substantiate that statements by the accused were untruthful, but also that the accused meant to deceive. The FAB also ruled that when alleged misconduct does not involve fabrication, falsification or plagiarism the ORI must not only prove that the behavior gravely deviates from the standards in the community, but also that any levelheaded researcher would regard the actions as misconduct when it was performed. The latter requisite would be difficult to meet in the Gallo case. According to the FAB, none of the assumed misrepresentations would have influenced the paper's conclusions, and the FAB felt that the ORI had not proven that the paper in Science contained untruthful statements about the data and certainly no deliberate falsification [2]. For various reasons, the FAB also rejected some of the testimony from ORIs expert witnesses.

Embarrassed by having the decisions of the ORI in the Popovic case overturned the ORI protested that its case was hampered by the FAB's prerequisite of having to prove intent and "materiality" of questionable statements in the entire paper [4]. On June 10th 1994 Office of the Inspector General (OIG) expressed uncertainty about whether Gallo and Popovic found the AIDS virus on your own initiative and its findings were transferred to the USA attorney in Maryland for potential criminal action, but after mulling over possible charges, such as perjury, obstruction of justice, mail fraud, and conspiracy to defraud the government the attorney's office decided that it would not act against Gallo and Popovic because it would be hard to persuade a jury beyond a reasonable doubt that these two scientists acted with criminal intent, moreover many of the suspected illegal activities took place before the 5 year statute of limitations [34].

Although the Nobel Foundation sometimes awards controversial Nobel Prizes it rightfully awarded one for the discovery of the AIDS virus, but it snubbed Gallo in favor of Montagnier, who shared the 2008 Nobel Prize in physiology or medicine *(http://www.nobelprize.org/nobel_prizes/medicine/laureates/2008/ montagnier-lecture.html#)* with his AIDS collaborator Françoise Barré-Sinoussi *(http://www.nobelprize.org/nobel_prizes/medicine/laureates/2008/barre-sinoussi-lecture.html)* and Harald zur Hausen (the discoverer that human papilloma virus causes cervical cancer) *(http://www.nobelprize.org/nobel_prizes/medicine/ laureates/2008/hausen-lecture.html)*.

Judy Mikovits. Whilst performing research on a chronic debilitating health disorder with persistent fatigue known most commonly as chronic fatigue syndrome (CFS) Judy Mikovits established international fame in 2009 for her research on this syndrome. During her research, she came to an unintentional incorrect conclusion while working as a biochemist at the virtually unknown nonacademic Whittemore Peterson Institute (WPI) for Neuro-Immune Disease in Reno, Nevada. Collaborative research by her team with distinguished scientists at both the NCI and the famous Cleveland Clinic in Ohio linked CFS to a little known retrovirus of mice called xenotropic murine leukemia retrovirus (XMRV) [107, 108] in attributing CFS to an infection of the patients' blood cells by a XMRV [108, 109]. The research reached the headlines of newspapers in October 2009 and initially Mikovits was highly praised, because the novel discovery offered hope to the many sufferers of CFS and the finding of the cause offered promise in the search for a cure. But later renowned colleagues ridiculed the work. Unfortunately multiple laboratories, including the laboratory of the original authors failed to confirm the XMRV association [138]. A careful review of the original paper disclosed numerous weaknesses in the study and indeed parts of the report were partially retracted by the authors [137]. In their paper the authors omitted important information that was relevant to their findings (J. Cohen Scienceinsider (4 October 2011) *http://scim.ag/_Mikovits)*. When it became apparent to the majority of the authors of the paper by Lombardi and coworkers needed to be retracted are the authors could not agree on the wording of the statement to accompany the retraction. Part of the original paper by Lombardi and colleagues was later retracted [137]. Other laboratories tried to replicate this potentially important study, but were unsuccessful. Bruce Alberts editor-in-chief of Science took the unusual initiative and retracted the manuscript because of a loss of confidence in the validity of the study [1]. The retraction of the papers infuriated many patients with CFS who had hoped that the cause of their mysterious disease had finally been detected and that potential therapy and perhaps even cures were around the corner. Their trust for the scientific community went through the back door. As negative data accumulated it became apparent that some of the positive results represented a laboratory contamination and the cause of CFS had not been found [36]. The emotional pressure that faces a scientist in this position can probably only be appreciated by those who have been in a comparable position, but apparently unrelated to the anguish of the XMRV fiasco Mikovits was fired for insolence and insubordination by Annette Whittemore, president of WPI, who founded the institute with her husband in 2007. A little over a month later WPI filed a civil lawsuit against Mikovits with allegations of breach of contract, misappropriation of trade secrets and other issues [35]. WPI contended that after being fired Mikovits took her notebooks together with data containing flash drives, and a laptop that did not belong to her. An attorney representing Mikovits immediately denied the allegations and pointed out that several employees had keys to her office and laboratory. Aside from the civil suit Mikovits became enmeshed in a simultaneous criminal case. She is alleged to have directed Max Pfost, one of her research assistants, to enter her previous office unlawfully to repossess research notebooks, a laptop, flash drives and correspondence. The missing material apparently included confidential information, such as trade secrets, patented inventions, and pending patent applications. After an investigation by the police an arrest warrant was issued. On November 18, 2011 the researcher was taken into custody by the Ventuta County police in Nevada and she was locked up at the Todd Road Jail. A paper headed by Shch-Ching Lo of the Federal Drug Administration (FDA) initially seemed to support this view with a different group of viruses, but the authors were not able to replicate the work in further studies. All of the authors signed a retraction of the paper *(http:scim.ag/XMRVpaper2)*.

Ronald Dorn. In the mid-1980s Ronald Dorn a geoscientist at Arizona State University (ASU) *(http://www.asu.edu/)* created a controversial method of dating ancient stone carvings, but this was an unintentional error and not scientific misconduct. He extracted minuscule amounts of organic material from a thin layer beneath a natural varnish on the surfaces of rocks and submitted the specimens to an accelerator mass spectrometry laboratory to determine the quantity of radioactive carbon-14. This isotope decays at a known rate and the amount provides an indication of rock age. In the mid 1990s a controversy arose about the reliability of the technique because some stone artifacts in the Southwestern USA were found to be several thousands of years older that suspected based on other analytical methods used to date prehistoric stone carvings. Dorn stopped using this dating technique in 1996 when he appreciated flaws in his original method. Beck and colleagues [12] found coal and charcoal grains of markedly different ages in a rock sample obtained from Dorn. In rocks prepared in laboratories other than the Dorn laboratory the grains were not identified. Because of these discrepancies the National Science Foundation (NSF) that funded some of Dorn's research initiated investigations into scientific misconduct and so did the ASU on the assumption that Dorn had doctored the specimens by adding carbon grains. In October 1999 the investigations failed to disclose misconduct [113]. After being cleared of the charges against his honesty Dorn sued the 8 authors of their article in Science in 1998 that was critical of him for deformation [12, 49]. By rushing to judgment prematurely unjust accusations of scientific misconduct backfired on the whistle-blowers. The case rested on whether the critical authors stepped beyond a line acceptable in academic dialogue.

Adriaan van Maanen. The astronomer Adriaan van Maanen (1884-1946), a contemporary of Harlow Shapley (1855-1972), working in California, declared that he had discovered a substantial relatively nearby internal spin in the Andromeda and other nebulae with many others. Edwin Hubble (1889-1953) rightfully disagreed and maintained that the nebula could not be spinning at the calculated rate as that would mean that portions of the galaxy were traveling faster than light. Others accepted Hubble's view, even though he avoided publishing them [87]. Since making his observations, other experts in astronomy scrutinized the raw data of van Maanen in detail, trying to determine how he arrived at the wrong conclusion. Eventually, it was thought that observer bias affected van Maanen's interpretation of his research, but nobody alleged that he intentionally falsified his recordings. He deciphered the data in a manner that verified what he desired to discover. The intellectual bias of van Maanen contradicted his investigative talents and led him to the wrong conclusion [112].

Theresa Imanishi-Kari. To establish the inference for immune regulation, David Baltimore began to collaborate with David Weaver, his postdoctoral fellow, and Theresa Imanishi-Kari because of her considerable expertise and experience in the relevant laboratory techniques. Questions under consideration included the issue of whether the transgene from the BALB/c mouse influenced the expression of immune cell synthesis in the black mouse. The answer appeared to be in the affirmative. Different possible explanations for the reported observations in the Cell paper existed, and the data did not decisively prop up one hypothesis. Imanishi-Kari thought that she had detected initial confirmation for a regulatory effect.

Niels Kaj Jerne (1911-1994), the Danish immunologist who shared the 1984 Nobel Prize in physiology or medicine *(http://www.nobelprize.org/nobel_prizes/medicine/laureates/1984/jerne.html)* with Georges J. F. Köhler *(http://www.nobelprize.org/nobel_prizes/medicine/laureates/1984/kohler.html)* and César Milstein *(http://www.nobelprize.org/nobel_prizes/medicine/laureates/1984/milstein.html)*, had proposed a contentious theory that the immune system is presided over by a self-regulating network depending on the production of antibodies to its own antibodies.

The decade-long investigation into allegations made by Margot O'Toole against Imanishi-Kari in the notorious paper in Cell [151] will undoubtedly go down in history as the most botched case of unintentional scientific mistakes caused by sloppy research. Perhaps no other publication has ever undergone a more detailed analysis. Without doubt, this article is one of the most comprehensively scrutinized research papers in history. In this case, chaos resulted when many individuals became involved in the investigation of an academic dispute that could have been easily resolved with rapidity if reasonable steps had been taken by competent scientists. This paper, with unintentional errors caused largely by sloppy work, caused considerable misery especially for Imanishi-Kari and Baltimore. The incident extensively disrupted their lives, but the authors were eventually absolved of any allegations of scientific misconduct. While scrutinizing the case, members of Congress did not appreciate the difference between unintentional scientific errors and fraudulent research. Moreover, Dingell saw an opportunity to attack a Nobel laureate in a campaign against unacceptable federally funded scientific research and hoped to nail a big fish to the wall. For the Dingell committee to put an outstanding scientist like Baltimore through such an ordeal was cruel and totally unnecessary, but perhaps Baltimore brought much of his trouble on himself because of his uncompromising attitude. After a long drawn-out evaluation of the paper, it was ultimately settled when the federal appeals board (FAB) of the DHHS found Imanshi Karo not guilty of scientific misconduct. Prior to the paper in Cell that stirred up a hornet's nest, Imanishi-Kari had only authored one peer-reviewed paper [18]. During the investigation of her research integrity, Imanishi-Kari still managed to effectively perform research and publish scientific discoveries [90, 93, 94, 105, 151, 153].

At almost each phase of the investigation of the notorious paper by Weaver and colleagues that was published in the high-impact journal Cell on April 25, 1986 [151], the whistle-blower, scientists, federal officials, politicians, and the media mishandled the scrutiny. The bungled inquiry shed light on how institutional and federal procedures for investigating alleged scientific misconduct should and should not be conducted by Weaver and colleagues[151]. The paper documented some unanticipated observations on how the immune system rearranges its genes to generate antibodies against antigens for the first time in genetically altered mice. It is noteworthy that the first author who should have been responsible for the paper did not get blamed for the errors that appeared under his name. A hornet's nest was stirred up by O'Toole, a young strong-minded postdoctoral fellow of Imanishi-Kari at MIT who could not replicate certain experiments described in the paper. She naively accused her mentor of falsifying the data and discovered what she regarded as "serious misstatements "in the laboratory records. O'Toole found misstatements in the published manuscript that were not in agreement with the laboratory records and brought this to the attention of the authors, who conceded some of the errors but declined to put forward an amendment. Prior to its publication in Cell, O'Toole went over the research paper by Weaver and colleagues, which cited some of her relate unpublished experiments. O'Toole, who naively turned Imanishi-Kari in for alleged faked data in 1986, was severely criticized by other scientists. O'Toole was adamant that she went over all data with Imanishi-Kari before reporting it to Herman Eisen, who was exclusively accountable for the investigation into scientific misconduct at the MIT. In assessing the quarrel over the published paper, Eisen discussed the matter extensively with different colleagues. O'Toole claimed that Eisen "did not even look at that data, " perhaps wishing to convey the impression that a thorough examination of raw notebook data is essential to settle disputes among investigators as in this case. As Eisen pointed out, the task of reviewing a vast number of unedited laboratory notebooks containing raw data is tremendously time-consuming and disruptive on the activities in the laboratory. Individuals may differ at what stage such a harsh process needs to be initiated, but Eisen felt that it was logical to hold back on a detailed analysis of laboratory records until allegations of scientific fraud were made. Initially, the heart of O'Toole's dispute with the authors was their explanation of what she considered weaknesses in their data. Fraud was not put forward in O'Toole's original memorandum, and she specifically rejected such a charge when questioned by Eisen. Later she assumed that some hybridomas had not even been subcloned and that data to back up some published findings did not exist. According to Eisen, she thought three considerations were relevant to the explanation of the experimental data: (i) an overlooked low-level expression of the transgene in many hybridomas from these mice, (ii) a high frequency of idiotype-positive hybridomas from normal mice of the same stain, and (iii) heterodimer formation involving disparate classes of immunoglobulin heavy chains, one from the transgene and the other from an endogenous gene [42, 71].

Later O'Toole was ready to drop the matter; but Charles Mapiethorpe PhD, a former graduate student of Imanishi-Kari, who obtained his doctoral degree from MIT in 1985, opted to push the issue [40]. O'Toole photocopied seventeen pages from Imanishi-Kari laboratory notebook and brought them to the attention of David Baltimore, a famous, brilliant, talented, and scrupulously hardworking scientist with an incredible scientific creativity. In 1970, when thirty-seven years old, Baltimore codiscovered reverse transcriptase [10], an enzyme important in the replication of a class of viruses that enables genetic information to be converted from RNA to DNA. For this work, Baltimore shared the 1975 Nobel Prize in physiology or medicine *(http://www.nobelprize.org/nobel_prizes/medicine/laureates/1975/baltimore.html)* with Howard Martin Telmin (1934-1994) *(http://www.nobelprize.org/nobel_prizes/medicine/laureates/1975/temin.html)* and Renato Dulbecco (1914—) *(http://www.nobelprize.org/nobel_prizes/medicine/laureates/1975/dulbecco.html)*. Baltimore was an eloquent spokesperson for the scientific enterprise.

Walter Stewart and Ned Feder of the NIH became aware of the dispute brought up by O'Toole, and these two self-appointed scientific fraud busters increased pressure on Baltimore by inspiring questions from reporters, becoming involved in what became known as the Baltimore Affair. They were convinced that fraud must be involved and began working on the Baltimore Affair and other cases of research misconduct with the staff of the Dingell subcommittee, even though both were completely incompetent in immunology and to neither was an immunologist qualified to appraise the science. Based on the seventeen photocopied pages acquired by O'Toole, the fraud buster duo prepared a lengthy report assaulting the Cell manuscript and trying to find data to prop up their flawed assumptions. Not surprising, Baltimore declined to cooperate with the Stewart and Feder appraisal of the science in the Cell article. They had already made up their minds about the conclusion with an approach of verdict first, confirmation later.

In 1987, Baltimore appealed to the NIH, which funded the research, to examine the paper officially. Stewart and Feder testified at the congressional hearings and asserted that MIT and Tufts had not satisfactorily dealt with O'Toole's charges. The duo was adamant that they should be included in the NIH review panel, notwithstanding their declared prejudgment, and seemingly persuaded the staff of the Dingell subcommittee that they had come to the right verdict.

Perhaps because of the outstanding reputation of Cell and the fact that Baltimore, a 1975 Nobel laureate in physiology or medicine, coauthored the paper, the manuscript elicited much attention in the media. The Baltimore Affair hit the front pages of newspapers mainly because scientific misconduct was still a topic of wide interest. The high profile of this incident together with other examples of scientific misconduct during the previous decade led not only to congressional hearings but also to an aggressive scrutiny of the Cell study. The questioned research was not performed in Baltimore's laboratory, and he was not personally accused of scientist misconduct. In retrospect, the paper was the product of sloppy research and a failure to detect errors in the manuscript prior to its publication.

The paper described how transplanted genes could stimulate a recipient's immune system to make particular antibodies. If validated, the breakthrough would have made noteworthy progress in our comprehension of immunology.

Investigations into allegations of scientific misconduct were carried out by the Massachusetts Institute of Technology (MIT) *(http://www.mit.edu/)* and Tufts University (Tufts) *(http://www.tufts.edu/)* in Boston where different parts of the experimental studies were carried out. The Boston institutions concluded that the dispute was related to data interpretation and not to data misrepresentation. At the outset, the accusations should have been investigated more earnestly at the university level, but MIT and Tufts responded too casually and informally; and later, both the MIT and Tufts were censured by Congress for the manner in which those institutions investigated the alleged errors. In retrospect, their investigations of potential scientific misconduct appeared more like conversations than serious investigations. Their investigative procedures should have provided the accused with a much earlier opportunity to challenge the allegations.

The case should never have been permitted to last as long as it did. In reviewing the disagreement between O'Toole and Imanishi-Kari, some mistakes were found in the paper, but evidence of scientific misconduct was not detected. The NIH, which funded the research, also appointed a committee to evaluate the paper. Because of concerns raised during the congressional subcommittee hearing chaired by Dingell in April 1988, another panel was appointed by the NIH to review all of the evidence. In 1998, it announced finding significant mistakes in the manuscript but nothing to suggest research misconduct to justify reopening of the investigation.

A couple of years later, a draft report on Imanishi-Kari by the NIH found her guilty of significant research misconduct for repetitively showing a table of untrue and deceptive information. On the Dingell committee's instructions, the Secret Service performed a nine-month forensic analysis of Imanishi-Kari's laboratory records apparently with the intent of terrifying the accused. The Secret Service discovered that one figure in the paper was a composite of different exposures of a single auto radiogram. Since the publication of the paper in Cell, none of its conclusions have been found to be incorrect.

It was inappropriate for the accused to wait nine years for a chance to cross-examine the accuser. The process needed to move much faster. DHHS should provide institutions greater responsibility in the investigations. A necessity of the government is not only to accelerate the investigation of alleged scientific misconduct but also the need to establish faith in its capability to investigate and act against research misconduct.

Even though O'Toole dropped out of science following her important role as a whistle-blower in the Cell paper, she received a subpoena to testify before the Dingle subcommittee, but she was a reluctant witness.

Baltimore pointed out that errors in science are not examples of research misconduct and should not be treated as such. Congress was undoubtedly an inappropriate place to resolve scientific disagreements.

The Dingell committee traced the drama of the Imanishi-Kari case back to 1984 when Frank Constantini PhD of Columbia University (Columbia) *(http://www.columbia.edu/)* collaborated with Baltimore in generating the genetically engineered mouse that was at the heart of the disputed paper in Cell. The transgenic mouse was created by transferring the gene encoding mu immunoglobulin that Imanishi-Kari characterized from a BALB/c mouse to a C59BL black mouse [82]. For reasons that are difficult to understand, the Dingell subcommittee demanded copies of Constantini's laboratory records, even though he was not previously involved in the dispute. Dingell even wanted copies of communications between Constantini and Baltimore, Imanishi-Kari, and all other members of the research group involved in the study documented in the Cell paper.

In May 1988, Baltimore communicated his version of the saga in a letter to a vast number of scientists. In it, he stated that the experimental data were interpreted differently by Imanishi-Kari and him. The adverse publicity received by Imanishi-Kari affected her research funding. The ongoing investigation with the associated congressional pressure undoubtedly contributed to the cancellation of her grant. She submitted an unsuccessful grant application to the American Cancer Society grant to support her research on the gene regulation of the immune system. However, she received financial support from the NIH until William Raub, acting director of the NIH, suggested to Anthony Fauci, the director of the National Institute of Allergy and Infectious Diseases (NIAID) in March 1989, that he terminate this grant; and funding of this grant ended during the following month. When the competitive renewal of another grant of Imanishi-Kari's from the NIAID was submitted in July 1989, the NIH chose not to renew it; instead, it extended the existing grant for three-month periods.

In March 1989, an NIH grant for one of Imanishi-Kari's funded projects came up for a review, and the NIH denied the grant extension, especially because of concerns about allegations raised by the Secret Service. This was an unusual drastic step by the NIH, which had never previously ended grant funding by a researcher under investigation for wrongdoing.

The Imanishi-Kari case was followed especially by representative Dingell largely because Baltimore labeled O'Toole a disgruntled postdoctoral fellow of Imanishi-Kari.

After a major investigation, Baltimore was finally forced in 1991 to retract the paper when the NIH inappropriately concluded that data in the Cell paper had been falsified by Imanishi-Kari. It was only in that year that Baltimore conceded that fraud might be involved in the Cell paper after the Office of Scientific Integrity (OSI) published its report. Eventually, Baltimore apologized for his role in the affair, but the harm to science had already been done. In 1991, Baltimore conceded that it was appropriate for the Dingell committee to watch over the disbursement of federal funds allocated for science. Scientists cannot expect to request more money for research and then to be left alone.

The toll on Imanishi-Kari and Baltimore created by the allegations against them extended far beyond any case of suspected scientific misconduct in memory.

Baltimore's response to the hearing partly accounted for the manner in which it was pursued. He gave the impression of being more intent on defending the research in the Cell paper rather than in trying to reach the truth about the behavior of his coauthor. As acknowledged some time later by Baltimore in a written communication to the NIH, "the better course would have been to suspend further comment on the matter until I had a full opportunity to review and digest all of the new information. In good conscience, I feared a rush to judgment, and I accorded my colleague the benefit of the doubt. I now recognize that I was too willing to accept Imanishi-Kari's explanations and to excuse discrepancies as mere sloppiness."

When the Whitehead Institute was set up as an affiliate of MIT in 1982, Baltimore was chosen as its director because of his outstanding academic reputation, and he established an extraordinary record while serving in this position. Because Baltimore was regarded as one the most outstanding scientists and science administrators in the world, Rockefeller University *(http://www.rockefeller.edu/)* recruited him as its president with the hope of restoring that university to its prior status as the nation's leading biomedical research institute, where major scientific discoveries are made. Senior faculty were against the choice of Baltimore because of the fog over his collaborative research, becoming more disturbed by his behavior and what they thought this told them about his ability to lead the university. He failed to take the warnings of O'Toole seriously, and his defense of Imanishi-Kari was stubborn even when concerns about the validity of her work became apparent. Some faculty was bothered about recruiting Baltimore before the allegations were resolved. They believed that he had not dealt with the humiliation appropriately, and they advised the trustees not to hire him, but the advice was ignored, especially since Baltimore transformed his participation in the scandal into an affair célèbre by convincing other scientists to unite with him in condemning the participation of Congress in scientific matters.

The discovery convert of defective critical research data in the Cell paper injured Baltimore's reputation and provoked annoyance and apprehension at Rockefeller University, when he was tapped to become its president on July 1, 1990, despite opposition by some faculty because the Imanishi-Kari scandal was still unresolved at the time and there was fear that it would damage the university. The hiring of Baltimore still affected the university, which customarily depended less on federal research grants than most leading universities. Indeed, Baltimore's handling of the situation led to considerable discontent by Rockefeller faculty when he was considered for that distinguished position. Being plagued by the adverse publicity of the controversial paper in Cell and the hostile environment at Rockefeller University, Baltimore did not survive long in the presidency of that university and resigned after only holding the position for eighteen months [88]. The adverse publicity of the Baltimore Affair played an important role in Baltimore's abandonment of the presidency of Rockefeller University. He then returned to the MIT before becoming the president of the California Institute of Technology (Caltech) *(http://www.caltech.edu/)* in 1997 [95].

A later investigation by a NIH committee exonerated all authors of the Cell paper while focusing on some technical details in the notorious paper. The NIH committee found distinct errors in the manuscript, which were rectified in a communication to Cell and totally made clear in a second letter that appeared in the May 19, 1986, issue of Cell. The NIH desired to bring closure to the investigation, but it did not and additional critiques were provoked. O'Toole claimed that the facts were inadequately analyzed by the NIH panel. The three NIH panelists were called to defend their report before Congress.

On May 14, Susan Hadley testified to the Dingell subcommittee about Imanishi-Kari's fitness to hold a public health service research grant. The Imanishi-Kari notebooks contained changed dates, altered pages, and out-of-the-ordinary tapes created by gamma ray counters; and she defended these errors, which affected about a third of the relevant raw data, by admitting that she was disorganized and kept sloppy notebooks.

NIH has distinct policies with regard to terminating the grants of investigators that have been approved by peer-reviewed study sections and NIH councils. Researchers being investigated are permitted to retain legal counsel, correct interview transcripts, refute initial drafts of investigative reports, and make remarks on proposed sanctions. In May 1986, the dean of MIT asked Eisen to conduct an official review of O'Toole's grievances. One complaint concerned a particular agent called Bet-1 that did not work the way that Imanishi-Kari claimed in the Cell paper.

On September 9, 1986, Baltimore confirmed that the Bet-1 antibody did not work as portrayed in the manuscript. He also gave two reasons why the paper in Cell should not be retracted: (i) Bet-1, good or bad, was not a critical assumption of the article; and (ii) and it would be not be easy to retract the paper because Weaver would need to be recognized as the senior author despite the fact that he was an innocent bystander without any role in the allegedly tainted data. A forensic analysis of the gamma counter output tapes by the Secret Service was provided during the congressional hearing.

In a letter to the editor of Cell on November 18, 1988, Imanishi-Kari, Baltimore and their coauthors admitted that there were three incorrect declarations in their 1986 paper [91], which became a major target of congressional hearing [39, 40]. The authors pointed out that the antibody called Bet-1, which they originally claimed to be highly specific in its reactivity for IgM (A), was an overstatement as this reaction was not exclusive since it cross-reacted with other molecules. They also acknowledged a couple of mistakes in one of the tables. After establishing that the paper contained no serious mistakes, both the NIH and the review panel requested additional alterations be submitted to Cell, including the substitution of data in Table 2 and a dialogue of the importance of the error about Bet-1 specificity. They also conceded blunders in the second table and reiterated that their errors did not impinge on the paper's conclusions.

Congress cultivated a remarkable interest in scrutinizing federally financed research in 1989 when a House subcommittee held hearings on a disputed research paper during the course of overseeing the NIH. Dingell even requested copies of all correspondence between both Baltimore and Imanishi-Kari and the American Cancer Society from 1982. Dingell was especially concerned about the letter dated January 25, 1985, from Imanishi-Kari that allegedly contained data that conflicted with the notorious Cell paper. As part of an inappropriate overkill, the Dingell subcommittee recruited the Secret Service to scrutinize the laboratory notebooks of the immunology researcher Imanishi-Kari related to the infamous paper in Cell. During an evaluation of the Imanishi-Kari data performed by the Secret Service at the request of Dingell, large parts of one of her notebooks were found to not be genuine.

The Secret Service discovered changed dates in the notebooks that had been entered at times later than indicated. Secret Service imprinted Imanishi-Kari's notes and tapes from detectors that record radioactivity. To establish when the data was produced, these records were compared with other notebooks from MIT researchers. Despite claims about being one of the best document inspectors in the world, the Secret Service acknowledged that it had never encountered such a complex and huge assignment in analyzing sixty notebooks of data. The congressional subcommittee and Secret Service staff met behind closed doors with NIH officials in April 1989 to reveal their discovered forensic evidence. The Secret Service found that at least some tapes and data in Imanishi-Kari's raw data did not seem to have been produced when the experiments were stated to have been performed and were entered after O'Toole requested to examine them. Data sheets may have been torn out and taped back into Imanishi-Kari's notebooks at a later date. Forensic experts also testified that some alterations in laboratory records were probably performed in an attempted cover-up. The analysis of Imanishi-Kari's notebooks by the Secret Service was not completely independent as Stewart would usually determine what was important to evaluate. At that time, he was outspoken and convinced that fraud had taken place because of the many errors. A confidential meeting between Secret Service officials, William Raub the NIH's deputy director at the time, and other NIH officials took place on 14 July 1989. At this meeting, one NIH official was concerned about the inference of Dingell's aides. Dingell's interest in the Baltimore Affair stemmed from the finding by the FAB that the Cell paper was "rife with errors of all sorts." Despite outcries of injustice, the subcommittee obtained sworn open testimony from all pertinent parties, including Baltimore and Imanishi-Kari. Dingell officially did not take part in misconduct investigations, because he was replaced as chairman of the subcommittee when the Republicans regained control of Congress in 1994.

In 1996, the hullabaloo over the Baltimore Affair was expected to affect other attempts to upgrade procedures for dealing with accusations of scientific misconduct [74]. Dingell was dissatisfied with the letter of correction and wrote to Otis R. Bowen, secretary of the DHHS, contending that the rectification was in an effort to forestall the finding of an NIH panel scrutinizing the notorious paper in Cell. Dingell also accused that NIH of conspiring with Baltimore by recommending that all authors of the publication write to the editor of Cell.

Dingell also requested an inquiry into how information in the 15 July 1988 copy of Science [41] was leaked, claiming that the NIH review panel had not detected evidence of fraudulent research.

The behavior of the Dingell subcommittee was hard to imagine, particularly with regard to a manuscript that was mainly beyond reproach except for a handful of highly technical details that did not influence the conclusions. When the congressional hearings focused on the validity of experimental data and errors within them, the Baltimore Affair became a target of inquiry, but this seemed inappropriate because the subject matter was complex and too difficult for most scientists, let alone members of congress to understand. It was undoubtedly beyond the comprehension of politicians with little, if any, knowledge about contemporary immunology.

In 1991, an additional federal investigation took place to ascertain who was aware of suspected fraudulent research. At this time, Baltimore expressed regret to O'Toole for referring to her as a disgruntled postdoctoral fellow. His denial to speak out promptly and under protest reflected what some regarded as conceit and poor reasoning. The episode undoubtedly had a detrimental effect on promising young scientists who admired this Nobel-winning scientist [38, 104, 152].

The Baltimore Affair provided the scientific community with numerous important lessons on how allegations of scientific misconduct should not be handled. After O'Toole drew attention to possible flaws in the renowned paper, Baltimore made significant errors in judgment. He criticized the whistle-blower and tenaciously defended Imanishi-Kari without keeping an open mild and leaving the inquiry into the allegations to independent investigators. Baltimore stubbornly refused to admit that data reported by a colleague in a paper that he coauthored might have been fraudulent, and his inconsiderate handling of the whistle-blower provoked public suspicion about scientific integrity. Baltimore made the tactical mistake of becoming involved in a struggle between science and politics, even though most politicians have little or no knowledge of science. He had an inappropriate rather confrontational attitude toward the congressional subcommittee on oversight and investigations and especially toward its chairman. The grilling that Baltimore experienced could happen to any scientist, and if others received the same type of treatment, science would surely come to a standstill. Fans of Baltimore were rightfully disgusted by the unfounded endeavor to link this well-respected distinguished scientist with possible fraud. Worse, from the standpoint of scientists, the Baltimore Affair became a major part of the congressional investigation; and this led to apprehension about the greater federal regulation of science. Entire books have been written about the Baltimore Affair [81, 131], most notably by Daniel J. Kevles, the Stanley Woodward professor of history at Yale University [98].

An investigation by the NIH lasting almost a year did not detect scientific misconduct or grave conceptional mistakes. Copies of a draft NIH report were sent to all coauthors of the Cell paper on November 11, 1988. In an extraordinary step, James B. Wyngaarden (1924-), the director of the NIH from 1982 to 1989, instructed the authors to clarify the controversial technical mistakes. In announcing his decision, he added, "It is unfortunate that despite the growing challenge to the validity of the research the coauthors apparently did not undertake a comprehensive review of their data until they met the NIH scientific panel." Wyngaarden signed his final report on January 30, 1989. In 1989, Imanishi-Kari and colleagues published a report in Cell, correcting some of the original data as required by the NIH [92, 121].

O'Toole pointed out that her charges against Weaver and coauthors' paper had not changed since the onset of the disagreement [71, 121]. In the July 18, 1991 issue of Nature, Paul Doty (1920-), emeritus professor of biochemistry at Harvard, reprimanded both Baltimore and Imanishi-Kari for many "lapses in scientific standards" [68].

Like many other whistle-blowers before and after her, O'Toole paid a price for accusing her mentor of publishing defective research. For several years after her accusation, she was unable to find work in science but eventually became a staff scientist at Genetics Institute in Cambridge, Massachusetts *(http://www.manta.com/c/mt1r0g0/genetics-institute)*. She will probably go down in history as precipitating the most infamous case of unproven scientific misconduct.

At that time of the disputed paper in Cell, the Office of Scientific Integrity (OSI) existed for dealing with suspected scientific misconduct, and it started investigating the case.

In 1993, Bruce Singal, Imanishi-Kari's former attorney, argued that the ORI investigation of the case should be disposed of completely, because it was tainted by unwarranted congressional meddling. Singal referred to a historic 1966 appeals court verdict, which read, "To subject an administrator to searching examination as to how and why he reached his decision in a case still pending before him sacrifices the appearance of impartiality-the sine qua non of American judicial justice" [74].

In a letter to Cell going first to Wyngaarden for his endorsement, Baltimore stated that if further clarification of the paper seemed needed, the authors would react fittingly. However, it took the authors of the Cell paper a long time to concede that some errors in it should be rectified, and Baltimore disputed the necessity for more clarification and stated that it is was untrue to imply that they failed to go over the data together.

The paper in Cell by Weaver and coauthors was the topic of hearings by two NIH committees and the USA Congress.

After a gap of two years, O'Toole testified at a congressional hearing.

In June 1988, almost a year after Baltimore asked NIH to step into the dispute, the NIH, which funded the research, appointed a panel of three distinguished scientists (Joseph M. Davy, Hugh McDermott of Stanford University (Stanford) *(http://www.stanford.edu/)*, and Ursula Storb of the University of Chicago *(http://www.uchicago.edu/)* to investigate the questionable data.

At least ten NIH officials evaluated the research by its own panel of experts. They endorsed many of O'Toole's allegations about the Cell paper and agreed that the mistakes were serious, but reported that the principal assertion of the study was upheld by unpublished subcloning experiments performed in June 1985. However, the panel recommended that these experiments be documented as a substitute to the published material. McDermott and Storb regarded alternative scenarios plausible for some allegations, but they concurred with the other panelists regarding the significance of the Secret Service's analysis that radiation counter tapes with raw data could not have been created when claimed by Imanishi-Kari. The dissenters wanted to examine the gas chromatograms used in the analysis, but the Secret Service refused to provide them with access to them. The authors disagreed and in response only submitted a small sample of the subcloning data.

When O'Toole received a copy of the report by the NIH panel in May 1986, she informed the NIH that she had been told that these subcloning experiments had never been performed.

The NIH began another investigation under the poorly organized OSI, and its preliminary report of March 1991 considered Imanishi-Kari guilty of scientific misconduct, because she fabricated research data in the manuscript in which she was a key author. The OSI criticized the managing of the case by Baltimore and blamed him for sweeping aside accusations of fraud and found it difficult to understand why he continued to stand up for Imanishi-Kari. Baltimore stressed that he needed to scrutinize the facts in more detail before reaching a conclusion as to whether fraud had or had not been committed; but in 1991, in response to the NIH report, Baltimore was forced to retract the controversial paper in Cell that he coauthored after the NIH concluded that data in the paper had been falsified. The NIH report was rapidly leaked to the press and incorrectly became widely regarded as the definitive decision. The contention was that Imanishi-Kari faked her results and that Baltimore paid insufficient attention to the allegations. The NIH probe took much longer than anticipated, and its findings were announced after an evaluation by two NIH committees and the congressional hearings over three years [72]. The USA attorney sat on the NIH report until the middle of 1992 before deciding not to prosecute.

The ORI completed its final report in November 1994 and found Imanishi-Kari guilty of nineteen counts of scientific misconduct, and it recommended that she be banned from getting federal grant support for ten years. Not surprisingly, Imanishi-Kari demanded a hearing before the federal appeals board (FAB) of the Department of Health and Human Services (DHHS); and starting in 1995, the case against her was reviewed by two lawyers (Judith A. Bellard and Cecilia

Sparks Ford) and Julius S. Younger, an immunologist and distinguished service emeritus professor at the University of Pittsburgh School (Pitt) *(http://www.pitt.edu/)* of Medicine. This was the first opportunity that Imanishi-Kari had to meet her accusers head-on and cross-examine them. Before reaching its verdict, the FAB reviewed seventy original notebooks, a 6, 500-page hearing transcript, thousands of pages of statements, including those from prior investigations, and other materials. The FAB reviewed all charges of scientific misconduct related to the 1986 paper in Cell against Imanishi-Kari; and on June 21, 1996, it unequivocally cleared her of all charges of scientific misconduct [96]. The 175-page report of the FAB on this investigation dated June 21, 1996, is available on the Internet *(http://www.hhs.gov/dab/decisions/dab1582.html)*. This was the second occasion that an ORI verdict had been overturned by the FAB. The first one involved Mikulas Popovic (discussed elsewhere in this chapter). Imanishi-Kari was unfairly prosecuted but paid a high price for sloppy recordkeeping. She moved from MIT to Tufts in 1986 and was delighted to finally see the end of this ordeal. "This was a victory for me and to my fellow scientists it demonstrates definitely that there is something wrong with the misconduct process." As anticipated, the group that accumulated evidence against Imanishi-Kari including Peter Stockton, Susan Hadley, and another ex-Dingell investigator, were disappointed with the verdict. O'Toole, the whistle-blower who started the inquiry, was also embittered.

As pointed out by Baltimore, it was extremely distressing that so much time, money, and energy was spent to resolve a scientific dispute that he considered evident from the beginning. Thousands of dollars of taxpayer money were consumed evaluating the data and circumstances related to this publication. The complexity of the reported research made the article because of concerns incomprehensible to scientists unfamiliar with the specific area of expertise reported.

Alexander Kugler. Attempts have been made to treat metastatic renal cell carcinoma with a vaccine, and one study by the German urologist Alexander Kugler and his coworkers with numerous flaws reported glowing results in a clinical trial using a kidney cancer vaccine [101]. In November 2002, a committee of the University of Göttingen *(http://www.uni-goettingen.de/)* in Germany found evidence of sloppiness and other irregularities in a controversial research paper by Kugler, reporting success in experimental studies with a cancer vaccine [18]. The University of Göttingen did not find the coauthors responsible for this unacceptable behavior; but later, Rolf-Hermann Ringert, a urologist at the university and the senior author of the disputed paper, was barred from applying for grants from the Deutsche Forschungsgemeinschaft (DFG) for eight years after the agency determined that Ringert shared responsibility for "misrepresentations" in the paper in which he was the lead author. Moreover, he was the head of the department in which the research was performed. The DFG decision was not open to appeal. The authors made several incorrect statements and erroneous presentations of primary data, results, and conclusions; but despite finding flaws, a German inquiry detected no evidence of fraud [18]. In September 2003, the contentious paper by Kugler and colleagues was unanimously retracted [102].

Leo V. DiCara. Bodily functions such as heart rate, blood pressure, skin temperature, sweat gland activity and muscle tension are under the control of the endocrine and autonomic nervous systems. When changes take place in these physiological activities normal people commonly become consciously aware of them. The possibility of a conscious control of them became a testable hypothesis in the 1960s by biofeedback and Neal Miller of the Rockefeller University *(http://www.rockefeller.edu/)* became an early investigator into the topic and he recruited Leo V. DiCara to also study the validity of biofeedback. DiCara became a productive researcher in the field and published numerous papers in support of his hypothesis [26, 52-64, 139, 140, 142, 143, 144, 155, 156, 157-160]. He created considerable excitement with his unique discoveries in biofeedback. DiCara was highly regarded at Rockefeller University and rose though the academic ranks to full professor. When at that level he was admitted to medical school. While working with Miller DiCara changed an entire field of investigation. He demonstrated robust visceral (autonomic) learning in rats while they underwent an acute pharmacological paralysis for 2-4 hours. The experiments involved more than 2, 000 rats that were paralyzed with curare. The documentation of his research is eloquently described in scientific publications. DiCara firmly believed in the wrong hypothesis. Over many years he clearly influenced associates about the validity of both his hypothesis and his research. Doubts about his research were only raised after Barry Dworkin took over the Dicara project in the Miller laboratory and could not replicate the experiments [69]. The findings could also not be replicated by others. Despite the fact that his work could not be replicated by others none of his papers were retracted. While gossip led some to suspect research misconduct formal allegations were not made against Dicara and his publications merely became ignored by those in the know. A failure of others to replicate his work is highly suggestive of research misconduct, but the quality of DiCara's publications suggests that he may have become wed to the wrong hypothesis and lost his objectivity in his research. Without retractions some later generations still believe them especially since the renaissance of alternative medicine in the therapy of certain human diseases. Because of concerns about an inability of others to replicate his results Dicara is alleged to have committed suicide, but an obituary has not been identified.

Stanley Pons and Martin Fleischmann. They thought that they had discovered cold fusion, namely the reaction in which the nuclei of light elements combine, forming larger elements as they release energy. Prior to their report, the fusion of atomic nuclei required considerable power and temperature, such as what the sun is capable of generating. On March 23, 1989, five years after starting to work on cold fusion, these two well-respected scientists announced that they had set off nuclear fusion on a tabletop in simple electrochemical cells containing palladium and platinum electrodes immersed in deuterium oxide (heavy water) on a tabletop. Deuterium is a stable isotope of hydrogen containing a neutron and a single proton in its nucleus. Pons and Fleischmann maintained that the nuclei of deuterium were packed so intimately together in the palladium cathode that they fused, releasing energy. When a current passed through the cells, deuterium ions became absorbed into the palladium. Two deuterium atoms would fuse as one, generating heat as well as neutrons, tritium, and helium, the essential products of the deuterium-deuterium fusion. In one of these experiments, Pons and Fleischmann discovered a hole in the beaker and in the floor of the chemistry building, which they attributed to heat generated in the process of cold fusion; but surprisingly, the radiation expected of a nuclear reaction was not detected. The findings were at odds with contemporary knowledge about deuterium-deuterium fusion.

The amazing alleged discovery by this distinguished couple created considerable interest as it put forward a potential source of generating an inexpensive almost boundless source of energy. An extravaganza followed the declaration that cold fusion had been achieved, but despite attempts by many scientists in several hundred laboratories, others could not reproduce their results. This led to the denunciation of the claim and stirred up public annoyance as well as their ridicule in the press. Regardless of whether a new concept stems from original thoughts or unintentional or intentional errors, others almost invariably jump on the band wagon and provide support. The cold fusion hypothesis gained support from John Bockris and Rusi Taleyarkhan.

John Bockris. In April 1989, Bockris of the Texas Agricultural and Mechanical University (Texas A & M) *(http://www.tamu.edu/)* found that the concentration of a tritium within electrochemical cells increased 10, 000-fold overnight in experiments performed in his laboratory. This remarkable event could be repeated and took place in six separate cells within a week. In June 1989, Bockris and Robert Huggins of Stanford persuaded the State Legislature of Utah that cold fusion had been verified and warranted a $5 million investment. Nine months later at the First Annual Cold Fusion Conference, supporters of cold fusion drew attention to the fact that emissions of tritium are indisputable evidence of a nuclear reaction. However, almost from the start of Bockris's scientific confirmation, investigators accustomed to his experiments intimated that the data were possibly overly too good to be true and too straightforward. Also, scientists were puzzled as to how Bockris managed to achieve success within a month of the initial cold fusion proclamation, when no other researcher was able to replicate the experiment using the technique of Bockris. Did Bockris really elicit cold fusion or had he discovered a new phenomenon that indicated that our knowledge of contemporary nuclear physics required revision? Were the electrochemical cells contaminated or did the research findings reflect something more sinister? More or less from the beginning, tritium was suspected of being inserted into the experimental system by human hands. Some scientists sounded an alert about the possibility of fraud and pointed out that it would need to be excluded before the findings of Bockris could be accepted. The researchers in the Bockris laboratory as well as the Texas A & M administration paid insignificant attention to these concerns. Rather than undertaking reasonable constructive actions to investigate and protect against fraud, Bockris and his coworkers presented explanations why they considered fraud improbable, now and then overstating their case.

The success of Bockris did not simply come out of the blue by an independent scientist. He was a longtime colleague of Fleischmann and attempted to duplicate the claims of cold fusion in early April 1989. The Bockris laboratory created numerous electrochemical cells and embarked on seeking evidence of the extra heat that Pons and Fleischmann felt was needed to explain a nuclear reaction. In July 1989, Charles Martin, another Texas A & M electrochemist, attempted to persuade Bockris that if someone believed that the cells had been spiked, it was Bockris's job to run cells in a way that would exclude this possibility. Martin suggested that Bockris run his cells in Martin's locked laboratory with very restricted right of entry. Bockris failed to accept this suggestion, nor did he confine the cells running in his laboratory. Bockris replaced the member of his research team, who sampled the cells for tritium, with two other researchers Ramesh Kainthla and Omo Velev, originally assigned to heat measurements. His initial article on the tritium work was published with colleagues in the Journal of Electroanalytical Chemistry [123]. Bockris brushed aside the possibility that anyone could have hampered with the experiments "because of positive results from the Cylotron Institute to which entrance is prohibited except by the usual personnel at the Institute." Those familiar with that institute were not persuaded by this argument as any graduate student could have entered it, and studies in the Bockris laboratory were performed by four dedicated graduate students. It seemed extremely likely that the amazing conclusion from Bockris's research on cold fusion was not valid and most likely due to scientific dishonesty. A detailed investigation was incompletely performed largely because no whistle-blower existed, and no credible suspect was identified. The experience of Texas A & M with this case illustrates how difficult it can be to investigate potential scientific misconduct. Universities need to define the fine line between dishonesty and unintentional honest scientific errors. To demand that somebody be recognized as accuser or whistle-blower is too restrictive. That is what the Texas A & M administrators let happen. It is important that universities take heed of the fact that a fine line exists between honest unintentional errors and misinterpretations that are not scientific misconduct. At a conference at Brigham Young University (BYU) *(http://home.byu.edu/)* in October 1989, a gathering of scientists considered the pros and cons of cold fusion, and many were not prepared to abandon the idea but wanted to distance themselves from Pons and Fleischman [126]. An internal review of cold fusion at Texas A & M found no direct evidence of scientific fraud [127]. Texas A & M administration had barely done anything about the possibility of fraud and had only asked a few preliminary questions. The Bockris cold fusion case illustrates how justifiably raised questions of potential scientific misconduct can blemish the name and career of scientists and institutions if the allegations are not fully investigated. Without an institutional analysis, the work remained unresolved and open to continuous rumor and gossip. Moreover, it harms the reputation of science when possible cases of fraud are not investigated. In this particular high-profile case, tens of millions of dollars and thousands of scientific man-hours were devoted in the search for the implausible cheap form of energy. Also, it gives the impression of a conspiracy or coverup. Gary Taubes has reviewed the cold fusion saga during the 1990s [144, 145].

Rusi Taleyarkhan. Rusi Taleyarkhan, another scientist who became entrapped in the cold fusion saga as a copycat, was found guilty of scientific misconduct. He initially triggered a dispute after his research team documented the generation of nuclear fusion with an uncomplicated tabletop system at the renowned Oak Ridge National Laboratory in Tennessee; and together with his colleagues, he conveyed the impression that hydrogen atoms become replaced with deuterium atoms, which formed as bubbles and then swelled and collapsed, after firing a pulse of ultrasound and neutrons at a cylinder of acetone [134, 142]. The heat and pressure at the center of the collapsing bubbles allegedly fused the deuterium atoms together, releasing nuclear byproducts and a surplus of energy. In 2004, Taleyarkhan moved to Purdue University (Purdue) *(http://www.purdue. edu/)*, where he attempted to repeat the initial bubble fusion results. He was suspected of scientific misconduct and was investigated three times. The third enquiry found him guilty of two counts of misconduct in research by his own laboratory. According to the panel's testimony, Taleyarkhan's postdoctoral assistant Yiban Yu performed the bubble fusion experiments; and Taleyarkhan described the findings in a paper that was submitted to Science, which rejected it and so did Physical Review Letters. In contrast to most complicated experiments, this paper was written by a solitary author. Several years later, Taleyarkhan recruited Adam Butt, a master's degree candidate, to proofread the paper and to confirm certain calculations. He was then named as a coauthor, and the paper was published in Nuclear Engineering and Design [162]. The panel investigating Taleyarkhan considered this virtual honorary joint authorship unfavorably and as evidence of scientific misconduct as it gave readers the false impression that both authors took part in the study. They labeled Taleyarkhan for another count of misconduct for a paper published in Physical Review Letters in which Taleyarkhan and colleagues cited the Nuclear Engineering and Design paper as independently verifying bubble fusion [143]. The panel did not consider some allegations against Taleyarkhan as scientific misconduct, but it severely criticized his behavior and some of his scientific procedures. Another six-member Purdue panel chaired by Mark Hermodson, a biochemist, was set up in March 2007 after complaints were received by the Inspector General of the Office of Naval Research (ONR), which funded some of Taleyarkhan's research. The panel found him guilty of research misconduct [114, 136] and tendered its report to ONR in April 2008, which was officially accepted; and it entered the public domain on July 18, 2008. ONR indicated that it would keep the case open until Purdue takes remedial action to thwart similar future incidents [135].

Jacques Benveniste. Famous scientists can become enmeshed into research where they lose their objectivity and become influenced by bias. The French immunologist, Jacques Benveniste (1935-2004), was working as the head of the renowned Unit 200 at the Institut National de la Sante la Recherche Medicale (INSERM) in France where he directed research in immunology, allergy, and inflammation. Benveniste was most well-known for papers published in the late 1970s on platelet-activating factor and its relationship to histamine [14, 15, 30]. He was a notable example of how a prominent scientist with an excellent track record can go astray because of self-deception and bias. Benveniste astounded scientists with a report in the high-impact journal Nature in June 1988, documenting that water could retain information about the shape of highly diluted molecules [50]. His research team claimed that when water contained a specific antibody, a memory of it persisted as if the antibody was present after being diluted. They documented the degranulation of human basophils by anti-IgE antibody at extremely low dilutions, supporting the concept of homeopathy and how it works. This manuscript drawing attention to this remarkable discovery of "the memory of water" provoked an international uproar, especially because different researchers failed to duplicate the results of Benveniste [122, 154].

Because of the excellent scientific reputation of Benveniste, the editor of Nature, Sir John Maddox (1925-2009), a physicist, came to a curious agreement to publish Benvenste's highly questionable paper on two conditions, which Benveniste accepted. First, the findings must be confirmed by independent laboratories; and second, after publishing the paper, the authors must demonstrate the experiments in his laboratory in front of a team of witnesses selected by Nature. The required replication was allegedly dealt with by adding an additional group of investigators from different laboratories in Canada, France, Israel, and Italy. The article was accompanied by an editorial labeled "When to Believe the Unbelievable" [6] and an editorial disclaimer. It was also accompanied by an editorial reservation that readers of this article may share the incredulity of the many referees [9]. Shortly after this controversial article was published, Nature arranged for independent investigators to observe repetitions of the experiments. Maddox visited the Beneviste laboratory at INSERM to evaluate the laboratory records and attempt to replicate his research under appropriate conditions with James Randi, a former magician, and the American fraud buster Walter Stewart. The group was unable to replicate Beneviste's results despite the mutual aid of Beneviste's own research group. The Nature committee investigating the Benveniste paper judged that the data were not critically evaluated and that weaknesses in the research were ineffectually documented [111]. The report of this committee tarnished not only the reputation of the authors but also the standing of INSERM and more importantly French scientists in general. The scientific council of INSERM wanted this project brought to an end; but not desiring to suppress innovative investigations, Philippe Lazar, the director of the institution, gave Beneviste six months to determine if bias or other errors accounted for his conclusions. According to Lazar, Nature dealt badly with the controversial paper and made Benveniste and his coauthors undergo an exceedingly public humiliation.

Because of Benveniste's refusal to retract the assertions in his paper, the Nature site visitors published a critique of his initial paper in July 1988 of Nature, pointing out that his experiments were "statistically ill-controlled" and that the laboratory personnel were not conversant with the notion of sampling blunders. Benveniste's technique for obtaining control values was fallacious, and "no substantial effort" was made to exclude systematic error, including observer bias. His interpretation was clouded by the exclusion of measurements in conflict with the claim. Blood that failed to degranulate was "recorded but not included in analyses prepared for publication." From time to time, the experiment failed totally for intervals of months. Steps to avoid contamination were inadequate, and the source of blood used in the studies was not controlled. The research also disclosed a significant COI. Two coauthors received their salaries under a contract between INSERM and the French company Boiron et Cie. The experiments seemed particularly triumphant when performed by investigators paid by a French company that marketed homeopathic medicines. Indeed, one individual in the research group was an active homeopath. All experiments carried out under adequately controlled conditions did not produce the alleged results. The Nature committee concluded that the research team was self-delusional and saw only what they thought they wanted to observe.

Serious doubt was expressed about these findings because molecules of the original antibody could not be present at the highly reported dilutions. Benveniste thought that the configuration of the molecules in water assumed a biologically active configuration, which a journalist designated water memory. Benveniste remained firmly committed to the hypothesis supported by his research; and many years later, he asserted that this active biological property could be transmitted to other samples of water. In July 1989, Benveniste reported in the French newspaper Le Monde that five research teams had confirmed his work on "the memory of water." Two were in France, two in the USA, and one in the Soviet Union. At the same time, Benveniste was placed on probation by INSERM after his laboratory was assessed during a routine evaluation. Benveniste had an inconsiderate attitude toward the explanations of his scientific findings and especially in communications to the public. Benveniste ridiculed the "mockery of scientific inquiry" by the Nature team and alerted other scientists about the dangers of "Salem witch hunts or McCarthy-like prosecutions" that will kill science. In a letter to Nature in 1994,

Benveniste disputed the study by Nature-appointed committee claiming that did not truly pursue his procedures. Yet Benveniste refused to retract his contentious paper by defending it in letters to Nature [16], explaining that the protocol used by the Nature committee was not the same as the one that he used. Nevertheless, Benveniste's reputation was harmed, and he was unable to obtain external funding after his fiasco. This forced him to fund his own research as he remained convinced about his findings. In 1991, the French Academy of Sciences was prepared to publish his latest research findings obtained under the supervision of a statistician but did this under the heading of "right of reply" since its proceedings had previously published a study crucial of the memory of water. In 1997, Benveniste founded the company DigiBio to "develop and commercialize applications of digital biology." Seven years later, he died after open heart surgery at the age of sixty-nine years. When most of the scientific community dismissed the claims of Benveniste as incredible, he received public backing from the Welsh physicist Brian David Josephson, who shared the 1973 Nobel Prize in physics *(http://www.nobelprize.org/nobel_prizes/physics/laureates/1973/josephson.html)* with Leo Esaki *(http://www.nobelprize.org/nobel_prizes/physics/laureates/1973/esaki.html)* and Ivar Giaever *(http://www.nobelprize.org/nobel_prizes/physics/laureates/1973/giaever.html)*. This stimulated many strange experiments including one leading to a 1997 paper alleging that water memory could be transferred over phone lines and two more papers in 1999 and an additional paper on remote-transmission in 2000. The USA Defense Advanced Research Projects Agency (DARPA) was fascinated by the idea that biological interactions could be digitized. Dr. Wayne Jonas, a homeopath working as the director of the USA National Center for Complementary and Alternative Medicine (NCCAM), was hence asked to try to replicate the reported findings. This self-sufficient experiment in remote-transmission by the USA Department of Defense failed to detect any effect. Some positive experiments were documented, but only when a Benveniste researcher ran the equipment. Benveniste acknowledged an experimenter effect and offered various explanations. A group of researchers found no reproducible digital signals (FASEB Journal in 2006) [65] *(http://en.wikipedia.org/wiki/Jacques_Benveniste)*.

References

1. Albert B, Retraction. Science, 2011. 334(6063): 1636.
2. Anderson C, Popovic is cleared on all charges; Gallo case in doubt. Science, 1993. 262(5136): 981-983.
3. Anderson C, ORI faces high hurdle in Gallo case. Science, 1993. 262(5136): 982.
4. Anderson C, Scientific misconduct. Hearing process proves a challenge for ORI.[Erratum appears in Science 1994 January 14;263(5144):159]. Science, 1993. 260(5115): 1714.
5. Anonymous, Diet-pheromone link gets a second look. Chemical and Engineering News, 1977. 55(4): 24.
6. Anonymous, When to believe the unbelievable. Nature, 1988. 333(6176): 787.
7. Anonymous, Viral etiology of AIDS and the Gallo probe. Science, 1990. 249(4968): 465-466.
8. Anonymous, Experimental multiple sclerosis vascular shunting procedure halted at Stanford. Annals of Neurology, 2010. 67(1): A13-15.
9. Anonymous., Editorial reservation: readers of this article may share the incredulity of the many referees Nature, 1988. 333(6176): 818.
10. Baltimore D, Discovery of the reverse transcriptase. FASEB Journal, 1995. 9(15): 1660-1663.
11. Barre-Sinoussi FC, J. C., Rey F, Nugeyre MT, et al., Isolation of a T-lymphotropic retrovirus from a patient at risk for acquired immune deficiency syndrome (AIDS). Science, 1983. 220(4599): 868-871.
12. Beck W, Donahue DJ, Jull AJT, et al., Ambiguities in direct dating of rock surfaces using radiocarbon measurements Science, 1998. 280(5372): 2132-2139.
13. Beebe DC, The use of cell lines to "model ocular tissues: cautionary tales. Investigative Ophthalmology and Visual Science, 2013. 54: 5720.
14. Benveniste J, Platelet-activating factor, a new mediator of anaphylaxis and immune complex deposition from rapid and human basophils. Nature, 1974. 249(5457): 581-582.
15. Benveniste J, Le Couedic JP, Polonsky J, et al., Structural analysis of purified platelet-activating factor by lipases. Nature, 1977. 269(5624): 170-171.
16. Benveniste J, Dr. Jacques Benveniste replies Nature, 1988. 334(6180): 291.
17. Boek E, Chemical and Engineering News. 55.
18. Bostanci A and Vogel G, German inquiry finds flaws, not fraud. Science, 2002. 298(5598): 1531-1532.
19. Broad WJ, Report absolves Harvard in case of fakery. Science, 1982. 215(4534): 874-876.
20. Brodeur P, Annals of radiation. The hazards of electromagnetic fields I-power lines. The New Yorker, 1989: 51.

21. Brodeur P, Annals of radiation. The hazards of electromagnetic fields III-video-display terminals. The New Yorker, 1989: 39.
22. Brodeur P, Annals of radiation. The hazards of electromagnetic fields II-something is happening. The New Yorker, 1989: 47.
23. Brodeur P, Annals of radiation. Calamity on Meadow street. The New Yorker, 1990.
24. Brodeur P, Department of amplification of annals of radiation. The New Yorker, 1990: 134.
25. Brodeur P. Annals of radiation. The cancer at Slater school. The New Yorker. 1992 7 December.
26. Campbell RJ, Wilson LG, Herschman HR, et al., Paradoxical decrease in norepinephrine content of adult mouse spleen and heart after neonatal nerve growth factor treatment. Biochemical Pharmacology, 1975. 24(23): 2213-2216.
27. Chang G and Roth CB, Structure of MsbA from E. coli: a homolog of the multidrug resistance ATP binding cassette (ABC) transporters.[Retraction in Chang G, Roth CB, Reyes CL, Pornillos O, Chen YJ, Chen AP. Science. 2006 December 22;314(5807):1875; PMID: 17185584]. Science, 2001. 293(5536): 1793-1800.
28. Chang G, Roth CB, Reyes CL, et al., Retraction.[Retraction of Pornillos O, Chen YJ, Chen AP, Chang G. Science. 2005 December 23;310(5756):1950-1953; PMID: 16373573].[Retraction of Reyes CL, Chang G. Science. 2005 May 13;308(5724):1028-1031; PMID: 15890884].[Retraction of Chang G, Roth CB. Science. 2001 September 7;293(5536):1793-1800; PMID: 11546864]. Science, 2006. 314(5807): 1875.
29. Chang G, Retraction of "Structure of MsbA from Vibrio cholera: a multidrug resistance ABC transporter homolog in a closed conformation" [Journal of Molecular Biology (2003) 330 419-430].[Retraction of Chang G. Journal of Molecular Biology. 2003 July 4;330(2):419-430; PMID: 12823979]. Journal of Molecular Biology, 2007. 369(2): 596.
30. Chignard M, Le Couedic JP, Tence M, et al., The role of platelet-activating factor in latelet aggregation Nature, 1979. 279(5716): 799-809.
31. Cho A, Neutrinos travel faster than light according to one experiment. Science, 2011. ???
32. Cho A, Once again, physicists debunk faster-than light neutrinos. Science, 2012. 336(6086).
33. Clark A, Tamm ER, Al-Ubaidi MR, et al., Editorial. On the use of immortalized ocular cell lines in vision research: The unfortunate story of RGC-5. Experimental Eye Research 116:433, 2013. Experimental Eye Research, 2013. 116: 433.
34. Cohen J, John Crewdson: science journalist as investigator. Science, 1991. 254(5034): 946-949.

35. Cohen J, Intellectual property. Dispute over lab notebooks lands researcher in jail. Science, 2011. 334(6060): 1189-1190.
36. Cohen J and Enserink M, False positive. Science, 2011. 333(6050): 1694-1701.
37. Crewdson J. The great AIDS quest. Chicago Tribune. 1989 November 19.
38. Culliton B, Errors in cell paper acknowledged. Science, 1988. 242(4883): 1240.
39. Culliton BJ, A bitter battle over error (II). Science. 241(4861): 18-21.
40. Culliton BJ, A bitter battle over error. Science. 240(4860): 1720-1723.
41. Culliton BJ, Panel completes interviews in "Baltimore case". Science, 1988. 241(4863): 286.
42. Culliton BJ, Whose notes are they? Science, 1989. 244(4906): 765.
43. Culliton BJ, NIH goes "extra mile" on Gallo. Science, 1990. 247(4945): 908.
44. Culliton BJ, Inside the Gallo probe. Science, 1990. 248(4962): 1494-1398.
45. Culliton BJ, Gallo inquiry takes puzzling new turn. Science, 1990. 250(4978): 202-203.
46. Culliton BJ, Gallo reports mystery break-in. Science, 1990. 250(4980): 502.
47. Culliton BJ, Gallo inquiry takes puzzling new turn. Science, 1990. 250(4978): 202-203.
48. Dalton R, International recruitment highlights need to track scientific behavior. Nature, 1996. 383(6596): 107-108.
49. Dalton R, Geographer sues crutics of is rock-dating methods. Nature, 1999. 401(6752): 419.
50. Davenas E, Beauvais F, Amara J, et al., Human basophil degranulation triggered by very dilute antiserum against IgE. Nature, 1988. 333(6176): 816-818.
51. Dawson RJP and Locher KP, Structure of a bacterial multidrug ABC transporter. Nature, 2006. 443(7108): 180-185.
52. DiCara LV, A rare case of spontaneous hemothorax. American Review of Tuberculosis, 1955. 71(5): 755-761.
53. DiCara LV and Miller NE, Instrumental learning of systolic blood pressure responses by curarized rats: dissociation of cardiac and vascular changes. Psychosomatic Medicine, 1968. 30(5): 489-494.
54. DiCara LV and Miller NE, Changes in heart rate instrumentally learned by curarized rats as avoidance responses. Journal of Comparative and Physiological Psychology, 1968. 65(1): 8-12.
55. DiCara LV and Miller NE, Instrumental learning of vasomotor responses by rats: learning to respond differentially in the two ears. Science, 1968. 159(3822): 1485-1486.
56. DiCara LV and Wolf G, Bar pressing for food reinforcement after lesions of efferent pathways from lateral hypothalamus. Experimental Neurology, 1968. 21(2): 231-235.

57. DiCara LV and Miller NE, Transfer of instrumentally learned heart-rate changes from curarized to noncurarized state: implications for a mediational hypothesis. Journal of Comparative and Physiological Psychology, 1969. 68(2): 159-162.
58. DiCara LV and Weiss JM, Effect of heart-rate learning under curare on subsequent noncurarized avoidance learning. Journal of Comparative and Physiological Psychology, 1969. 69(2): 368-374.
59. DiCara LV, Learning in the autonomic nervous system. Scientific American, 1970. 222(1): 30-39.
60. DiCara LV, Role of postoperative feeding experience in recovery from lateral hypothalamic damage. Journal of Comparative and Physiological Psychology, 1970. 72(1): 60-65.
61. DiCara LV, Braun JJ, and Pappas BA, Classical conditioning and instrumental learning of cardiac and gastrointestinal responses following removal of neocortex in the rat. Journal of Comparative and Physiological Psychology, 1970. 73(2): 208-216.
62. DiCara LV and Stone EA, Effect of instrumental heart-rate training on rat cardiac and brain catecholamines. Psychosomatic Medicine, 1970. 32(4): 359-368.
63. DiCara LV, Learning of cardiovascular responses: a review and a description of physiological and biochemical consequences. Transactions of the New York Academy of Sciences, 1971. 33(4): 411-422.
64. DiCara LV, Weaver L, and Wolf G, Comparison of DC and RF for lesioning white and grey matter. Physiology and Behavior, 1974. 12(6): 1087-1090.
65. Dickson D, Benveniste criticism is diluted. Science, 1989. 245(4915): 248.
66. Disney MJ, Book Review. Hubble: toil and trouble. American Scientist, 2009. 97(January-February): 75-76.
67. Doepp F, Paul F, Valdueza JM, et al., No cerebrocervical venous congestion in patients with multiple sclerosis. Annals of Neurology, 2010. 68(2): 173-183.
68. Doty P, Responsibility and Weaver et al. Nature, 1991. 352(6332): 183-184.
69. Dworkin BR and Miller NE, Failure to replicate visceral learning in the acute curarized rat preparation. Behavioral Neuroscience, 1986. 100(3): 299-314.
70. Ehrlich PR, Harte J, Harwell MA, et al., Long-term biological consequences of nuclear war. Science, 1983. 222(4630): 1293-1300.
71. Eisen HN, O'Toole's charges. Science. 245(4923): 1166-1167.
72. Elmer-DeWitt P. Thin skins and fraud at M.I.T Time. 1991 April 1
73. Franzen S, Cerruti M, Leonard DN, et al., The role of selection pressure in RNA-mediated evolutionar materials synthesis. Journal of the American Chemical Society, 2007. 129(49): 15340-15346.
74. Friedly J, How congressional pressure shaped the 'Baltimore case'. Science, 1996. 273(5277): 873-875.

75. Galllo, Nature, 1991.
76. Gallo R, Crash development of AIDS test nears goal. Science, 1984. 225: 1128-1131.
77. Gallo RC, Sarin PS, Gelmann EP, et al., Isolation of human T-cell leukemia virus in acquired immune deficiency syndrome (AIDS). Science, 1983. 220(4599): 865-867.
78. Gallo RC, Salahuddin SZ, Popovic M, et al., Frequent detection and isolation of cytopathic retroviruses (HTLV-III) from patients with AIDS and at risk for AIDS. Science, 1984. 224(4648): 500-503.
79. Gallo RC, Virus Hunting AIDS, Cancer, and The Human Retrovirus: A Story of Scientific Discovery. 1991: Perseus Book Group.
80. Gelmann EP, Popovic M, and al. e, Proviral DNA of a retrovirus, human T-cell leukemia virus, in two patients with AIDS. Science, 1983. 220(4599): 862-865.
81. Goldberg JS, The "Baltimore" affair. 1986, Lincoln, NE: University of Nebraska Press
82. Grosschedl R, Weaver D, Baltimore D, et al., Introduction of a mu immunoglobulin gene into the mouse germ line: specific expression in lymphoid cells and synthesis of functional antibody. Cell, 1984. 38(3): 647-658.
83. Gugliotti LA, Feldhelm DL, and Eaton BE, RNA-mediated metal-metal bond formation in the synthesis of hexagonal palladium nanoparticles. Science, 2004. 304(5672): 850-852.
84. Gynaecology RCoOa, Report of the independent committee of inquiry into the circumstances surrounding the publication of two articles in the British Journal of Obstetrics and Gynaecology in August 1994. 1995, London: Royal College of Obstetrics and Gynaecology. pp. 26.
85. Hamilton DP, Hints emerge from the Gallo probe. Science, 1991. 253(5021): 728-729.
86. Hendry LB, Wichmann. J.K., Hindenlang DM, et al., Evidence for the origin of insect sex pheromes presence in food plants. Science, 1975. 188(4183): 59-63.
87. Hetherington NSE, Cosmology: Historical, Literary, Philosphical, Religious, and Scientific Perspectives. Garland Reference Library of the Humanities Vol. 1634. 1993, New York and London: Garland Publishing, Inc. pp. 631.
88. Hilts PJ, Nobelist caught up in fraud case resigns as head of Rockefeller U. New York Times, 1991: A1.
89. Hindenlang DM and Wichmann JK, Reexamination of teradecenyl acetates in oak leaf roller sex pheromone and in plants. Science, 1977. 195(4273): 86-89.
90. Huang CA, Henry C, Iacomini J, et al., Adult bone marrow contains precursors for CD5+ B cells. European Journal of Immunology, 1996. 26(10): 2537-2540.

91. Imanishi-Kari T, Reis MH, Weaver D, et al., Altered repertoire of endogenous immunoglobulin gene expression in transgenic mice containing a rearranged mu heavy chain gene.[Erratum appears in Cell 1988 November 18;55(4):541]. Cell, 1988. 55(4): 541.
92. Imanishi-Kari T, Weaver D, and Baltimore D, On the specificity of BET-1 antibody. Cell, 1989. 57(4): 515-516.
93. Imanishi-Kari T, Bandyopadhay S, Busto P, et al., Endogenous Ig production in mu transgenic mice. I. Allelic exclusion at the level of expression. Journal of Immunology, 1993. 150(8 Pt 1): 3311-3326.
94. Imanishi-Kari T, Huang CA, Iacomini J, et al., Endogenous Ig production in mu transgenic mice. II. Anti-Ig reactivity and apparent double allotype expression. Journal of Immunology, 1993. 150(8 Pt 1): 3327-3346.
95. Jaroff L, Nash M, and Thompson D. Crisis in the labs. Time Magazine. 1991 August 26:45-51.
96. Kaiser J and Marshall E, Imanishi-Kari ruling slams ORI. Science, 1996. 272(5270): 1864-1865.
97. Kerr RA, Requiem for life on Mars? Support for microbes fades. Science, 1998. 282(5393): 1398-1400.
98. Kevles DJ, ed. The Baltimore Case. A Trial of Politics, Science, and Character. 1998, W.W.Norton: New York. pp. 509.
99. Kish L, Statistical medicine. Science, 1994. 265(5172): 591.
100. Krishnamoorthy RR, Clark AE, Daudt D, et al., A forensic path to RGC-5 cell line identification: lessons learned. Investigative Ophthalmology and Visual Science, 2013. 54: 5712-5719.
101. Kugler A, Stuhler G, Walden P, et al., Regression of human metastatic renal cell carcinoma after vaccination with tumor cell-dendritic cell hybrids. [Retraction in Kugler A, Stuhler G, Walden P, Zoller G, Zobywalski A, Brossart P, Trefzer U, Ullrich S, Muller CA, Becker V, Gross AJ, Hemmerlein B, Kanz L, Muller GA, Ringert RH. Nature Medicine 2003 September 9(9):1221; PMID: 12949529]. Nature Medicine, 2000. 6(3): 332-336.
102. Kugler A, Stuhler G, Walden P, et al., Retraction: Regression of human metastatic renal cell carcinoma after vaccination with tumor cell-dendritic cell hybrids.[Retraction of Kugler A, Stuhler G, Walden P, Zoller G, Zobywalski A, Brossart P, Trefzer U, Ullrich S, Muller CA, Becker V, Gross AJ, Hemmerlein B, Kanz L, Muller GA, Ringert RH. Nature Medicine. 2000 March; 6(3):332-6; PMID: 10700237]. Nature Medicine, 2003. 9(9): 1221.
103. Lassmann H, Multiple sclerosis pathology: evolution of pathogenetic concepts. Brain Pathology, 2005. 15(3): 217-222.
104. Leary WE. Research labs in the U.S. feel chill of secret service 'witch hunt'. Durham Morning Herald 1989 May 14.
105. Li J, Fernandez L, O'Connor KC, et al., The rearranged V(H) domain of a physiologically selected anti-single-stranded DNA antibody as a precursor

for formation of IgM and IgG antibodies to diverse antigens. Journal of Immunology, 2001. 167(7): 3746-3755.
106. Liu MB and Davis K, Lessons from a horse named Jim: A Clinical Trials Manual from the Duke Clinical Research Institute 2nd Edition. 2010: Wiley, John and sons.
107. Lombardi VC, Ruscetti FW, Das Gupta J, et al., Detection of an infectious retrovirus, XMRV, in blood cells of patients with chronic fatigue syndrome. Science, 2009. 326(5952): 585-579.
108. Lombardi VC, Ruscetti FW, Das Gupta J, et al., Detection of an infectious retrovirus, XMRV, in blood cells of patients with chronic fatigue syndrome. Science, 2009. 326(5952): 585-589.
109. Lombardi VC, Ruscetti FW, Das Gupta J, et al., Detection of an infectious retrovirus, XMRV, in blood cells of patients with chronic fatigue syndrome. [Retraction in Alberts B. Science. 2011 December 23;334(6063):1636; PMID: 22194552]. Science, 2009. 326(5952): 585-589.
110. Ma C and Chang G, Structure of the multidrug resistance efflux transporter EmrE from Escherichia coli. Proceedings of the National Academy of Sciences of the United States of America, 2004. 101(9): 2852.
111. Maddox J, Randi J, and Stewart WW, High-dilution" experiments a delusion. Nature, 1988. 334(6180): 287.
112. Marshall E, When does intellectual passion become conflict of interest? Science, 1992. 257(5070): 620-623.
113. Matakoff D, Cleared of misconduct, geoscientist sues crtics. Science, 1999. 286(5441): 883-885.
114. Maugh THI. Physicist is found guilty of misconduct. Los Angeles Times. 2008 19 July.
115. McKay DS, Gibson EK, Jr., Thomas-Keprta KL, et al., Search for past life on Mars: possible relic biogenic activity in martian meteorite ALH84001. Science, 1996. 273(5277): 924-930.
116. Miller G, Scientific publishing. A scientist's nightmare: software problem leads to five retractions. Science, 2006. 314(5807): 1856-1857.
117. Nardone RM, Eradication of cross-contaminated cell lines: a call for action. Cell Biology and Toxicology, 2007. 23(6): 367-372.
118. Neff J. College research found to be false. Project begun at NCSU appears to be target. The News and Observer. 2014 January 30.
119. Newman J. 20 of the greatest blunders in science in the last 20 years. What are they thinking? Discover Magazine. 2000 1 October.
120. Nowak R, Problems in clinical trials go far beyond misconduct. Science, 1994. 264(5165): 1538-1531.
121. O'Toole M, The Dingell investigation. Science, 1989. 244(4910): 1243.
122. Ovelgönne JH, Bol AW, Hop WC, et al., Mechanical agitation of very dilute antiserum against IgE has no effect on basophil staining properties. Experientia, 1992. 48(5): 504-508.

123. Packham NTC, Wolf KL, Wass JC, et al., Production of tritium from D2O elecrolysis at a palladium cathode. Journal of Electroanalytical Chemistry, 1989. 270: 451-458.
124. Palca J, Article on Gallo prompts inquiry. Science, 1990. 247(4938): 19.
125. Pennisi E, Microbiology. Concerns about arsenic-laden bacterium aired. Science, 2011. 332(6034): 1136-1137.
126. Pool R, Cold fusion: only the grim remains. Science, 1990. 250(4982): 754-755.
127. Pool R, Cold fusion at Texas A&M: problems, but not fraud. Science, 1990. 250(4987): 1507-1508.
128. Popovic M, Sarngadharan MG, Read E, et al., Detection, isolation, and continuous production of cytopathic retroviruses (HTLV-III) from patients with AIDS and pre-AIDS. Science, 1984. 224(4648): 497-500.
129. Pornillos O, Chen Y-J, Chen AP, et al., X-ray structure of the EmrE multidrug transporter in complex with a substrate. Science, 2005. 310(5756): 1950-1953.
130. Rindfleisch E, Histologisches Detail zur grauen Degeneration von Gehim und Rückenmark. Arch Pathol Anat Physiol Klin Med (Virchow), 1863. 26: 474-483.
131. Sarasohn J, The David Baltimore Affair: Science on Trial. The Whistle-blower, the Accused, and the Nobel Laureate. 1993, New York St.Martin's Press.
132. Sarngadharan MG, Popovic M, Bruch L, et al., Antibodies reactive with human T-lymphotropic retroviruses (HTLV-III) in the serum of patients with AIDS. Science, 1984. 224(4648): 506-508.
133. Schupbach J, Popovic M, Gilden RV, et al., Serological analysis of a subgroup of human T-lymphotropic retroviruses (HTLV-III) associated with AIDS. Science, 1984. 224(4648): 503-505.
134. Seife C, 'Bubble fusion' paper generates a tempest in a beaker. Science, 2002. 295(5561): 1808-1809.
135. Service RF, Researchers raise new doubts about 'bubble fusion'reports. Science, 2006. 311(5767): 1532-1533.
136. Service RF, New Purdue panel faults bubble fusion pioneer. Science, 2008. 321(5886): 473.
137. Silverman RH, Das Gupta J, Lombardi VC, et al., Partial retraction. Detection of an infectious retrovirus, XMRV, in blood cells of patients with chronic fatigue syndrome. Science, 2011. 334(6053): 176.
138. Simmons G, Glynn SA, Komaroff AL, et al., Failure to confirm XMRV/MLVs in the blood of patients with chronic fatigue syndrome: a multi-laboratory study. Science, 2011. 334(6057): 814-817.
139. Simpson CW and Dicara LV, Estradiol inhibition of catecholamine elicited eating in the famale rat. Pharmacology, Biochemistry and Behavior, 1973. 1(4): 413-419.

140. Simpson CW, Dicara LV, and Wolf G, Glucocorticoid anorexia in rats. Pharmacology, Biochemistry and Behavior, 1974. 2(1): 19-25.
141. Sundstrom P, Wahlin A, Ambarki K, et al., Venous and cerebrospinal fluid flow in multiple sclerosis: a case-control study. Annals of Neurology, 2010. 68(2): 255-259.
142. Taleyarkhan RP, West CD, Cho JS, et al., Evidence for nuclear emission during acoustic cavitation. Science, 2002. 295(5561): 1868-1873.
143. Taleyarkhan RP, West CD, Lahey RTJ, et al., Nuclear emissions during during self-nucleated acoustic cavitation. Physical Review Letters, 2006. 96(3): 034301-034301 to 034301-034304.
144. Taubes G, Cold fusion conundrum at Texas A&M. Science, 1990. 248(4961): 1299-1304.
145. Taubes G, Bad Science: The Short Life and Weird Times of Cold Fusion. 1993, New York: Random House, Inc.
146. Turco RP, Toon OB, Ackerman TP, et al., Nuclear winter: global consequences of multiple nuclear explosions. Science, 1983. 222(4630): 1283-1292.
147. Turco RP, Toon OB, Ackerman TP, et al., Climate and smoke: an appraisal of nuclear winter. Science, 1990. 247: 166-176.
148. Unknown, Science, 1991.
149. Unknown, Science, Unknown: 507.
150. van Bergen NJ, Wood JPM, Chidow G, et al., Recharacterization of the RGC-5 retinal ganglion cell line. Investigative Ophthalmology and Visual Science, 2009. 50: 4267-4272.
151. Weaver D, Reis MH, Albanese C, et al., Altered repertoire of endogenous immunoglobulin gene expression in transgenic mice containing a rearranged mu heavy chain gene. [Retraction in Weaver D, Albanese C, Costantini F, Baltimore D. Cell. 1991 May 17;65(4):536; PMID: 2032282]. Cell, 1986. 45(2): 247-259.
152. Weaver D, Albanese C, Costantini F, et al., Retraction: altered repertoire of endogenous immunoglobulin gene expression in transgenic mice containing a rearranged mu heavy chain gene. [Retraction of Weaver D, Reis MH, Albanese C, Costantini F, Baltimore D, Imanishi-Kari T. Cell. 1986 April 25;45(2):247-59; PMID: 3084104]. Cell, 1991. 65(4): 536.
153. White-Scharf ME and Imanishi-Kari T, Genetic basis for altered idiotype expression in the hyperimmune response to (4-hydroxy-3-nitrophenyl) acetyl hapten. Journal of Immunology, 1986. 137(3): 887-896.
154. Wiegant FA, Memory of water revisited. Nature, 1994. 370: 322.
155. Wilson JR and Dicara LV, Effects of previous curare-immobilization on Pavlovian conditioned heart decelerations in the curarized rat. Physiology and Behavior, 1975. 14(3): 259-264.
156. Wilson LM, Wilson JR, and Dicara LV, Facilitation of Pavlovian conditioned cardiodecelerations following preshock in immobilized rats. Physiology and Behavior, 1975. 15(6): 653-658.

157. Wolf G and DiCara LV, A third ascending hypothalamopetal pathway. Experimental Neurology, 1971. 33(1): 69-77.
158. Wolf G, Dicara LV, and Simpson W, The contact method: a simple technique for electrical self-stimulation without external leads. Physiology and Behavior, 1973. 11(5): 721-723.
159. Wolf G and Dicara LV, Impairments in sodium appetite after lesions of gustatory thalamus: replication and extension. Behavioral Biology, 1974. 10(1): 105-112.
160. Wolf G, McGovern JF, and Dicara LV, Sodium appetite: some conceptual and methodologic aspects of a model drive system. Behavioral Biology, 1974. 10(1): 27-42.
161. Wolfe-Simon F, Switzer Blum J, Kulp TR, et al., A bacterium that can grow by using arsenic instead of phosphorus. Science, 2011. 332(6034): 1163-1166.
162. Xu Y and Butt A, Confirmatory experiments for nuclear emissions during acoustic cavitation. Nuclear Engineering and Design, 2005. 235(11-12): 1317-1324.
163. Zamboni P, The big idea: iron-dependent inflammation in venous disease and proposed parallels in multiple sclerosis. Journal of the Royal Society of Medicine, 2006. 99(11): 589-593.
164. Zamboni P, Menegatti E, Bartolomei I, et al., Intracranial venous haemodynamics in multiple sclerosis. Current Neurovascular Research, 2007. 4(4): 252-258.
165. Zamboni P, Galeotti R, Menegatti E, et al., Chronic cerebrospinal venous insufficiency in patients with multiple sclerosis. Journal of Neurology, Neurosurgery and Psychiatry, 2009. 80(4): 392-399.
166. Zamboni P, Menegatti E, Galeotti R, et al., The value of cerebral Doppler venous haemodynamics in the assessment of multiple sclerosis. Journal of the Neurological Sciences, 2009. 282(1-2): 21-27.
167. Zamboni P, Menegatti E, Weinstock-Guttman B, et al., The severity of chronic cerebrospinal venous insufficiency in patients with multiple sclerosis is related to altered cerebrospinal fluid dynamics. Functional Neurology, 2009. 24(3): 133-138.
168. Zamboni P, Regarding "no cerebrocervical venous congestion in patients with multiple sclerosis. Intraluminal jugular septation". Annals of Neurology, 2010. 68(6): 969; author reply 970.
169. Zamboni P, Menegatti E, Weinstock-Guttman B, et al., CSF dynamics and brain volume in multiple sclerosis are associated with extracranial venous flow anomalies: a pilot study. International Angiology, 2010. 29(2): 140-148.
170. Zamboni P and Carinci F, Face, brain, and veins: a new perspective for multiple sclerosis onset. Journal of Craniofacial Surgery, 2011. 22(1): 376.

171. Zamboni P, Menegatti E, Weinstock-Guttman B, et al., Hypoperfusion of brain parenchyma is associated with the severity of chronic cerebrospinal venous insufficiency in patients with multiple sclerosis: a cross-sectional preliminary report. BioMed Central Medicine, 2011. 9: 22.
172. Zimmerman R, The Universe in a Mirror: The Saga of the Hubble Space Telescope and the Visionaries Who Built it. 2008: Princeton University Press. 320.

CHAPTER 14

Scientific Misconduct Prior to the USA Outbreak of Fraudulent Research in the 1970s and 1980s

Since the beginning of science, fraudulent research has taken place, but it was not exposed or publicized as it is today. Charles Babbage (1791-1871)—the distinguished English mathematician, philosopher, inventor, and mechanical engineer who is acknowledged as the father of the computer—was appalled by the decline in the honesty of scientists in 1830 when he wrote his famous book on the subject [4]. He recognized three main varieties of fraudulent science. The first was labeled "forging," which indicated total fabrication, namely, the documentation of findings that were never observed. Forging was the most defenseless form of fraudulent research to expose because it depended entirely on made-up data, and it was most likely to appear phony to persons knowledgeable about the subject matter. It is also exposed by failures to replicate studies. The second type referred to as trimming was characterized by data manipulation to make them appear superior than they were in reality. Trimming sometimes remained concealed for years particularly if a talented dishonest researcher made noteworthy guesses to reach acceptable conclusions. Babbage's third type of fraudulent research was called cooking, and it consisted of selecting data that fits a researcher's hypothesis while rejecting observations that do not support it. Cooking also goes undetected for long periods of time because only parts of the findings are true.

Accusations of Scientific Misconduct by Famous Scientists

Many scientists in past generations have been accused of trimming, cooking, and other types of research misconduct, which was not suspected during their natural lives. The previously held high esteem of these scientists became damaged after flaws were uncovered in their research after amazingly long periods during which it was skillfully concealed.

Claudius Ptolemy. Claudius Ptolemy (c.AD 90/100-c. AD168/170), the famous Greek astronomer in the Roman era of Egypt, is commonly regarded as the greatest astronomer of antiquity. During the second century of the Christian era, he authored a mathematical and astronomical treatise in Greek known as the Almagest. This opus recorded the noticeable movements of the stars and planetary paths and is widely acclaimed to be one of the greatest scientific documents ever written. It stressed the geocentric theory that located Earth in the center of the universe with the sun, planets, and heavenly bodies revolving around it. This geocentric model was acknowledged as the final word for some 1400 years from its conception in about AD 150 in Alexandria, the Hellenistic center of Egypt, until Nicolas Copernicus(1473-1543) in the early Renaissance. Robert Hall Newton, a faculty member of the applied physics laboratory at Johns Hopkins University (JHU) (http://www.jhu.edu/), accused Ptolemy of inventing data to prop up his own astronomical theories. He regarded Ptolemy not as the supreme astronomer of ancient times but as the perpetrator of a colossal triumphant fraud in the annals of science by methodically fabricating or doctoring the data of previous astronomers to sustain his own hypotheses. Newton verified the calculations in the Almagest and presented the evidence for his allegation in a book titled The Crime of Ptolemy [78]. Mistakes by Ptolemy have been detected by others, but differences of opinion exist regarding their explanation. Some are considered unintentional errors, such as discrepancies in his data that can be explained if Ptolemy's watch was constantly30 minutes slow so that his recorded observations at noon were actually made at 12:30 p.m. Much data within the Almagest can be precisely predicted by figuring out the expected solution based on the theory of Ptolemy. Using observations at the equinox, Ptolemy confirmed precisely the findings of Hipparchos (Hipparchus) (c.190 BC-c.120 BC), an earlier astronomer who determined the duration of a year and the autumn equinox. Newton argues that instead of assembling observations on his own, Ptolemy must have toiled in reverse from the conclusion that he was attempting to confirm. Newton has also drawn attention to numerous findings by Ptolemy, which were virtually the same as those of the Alexandrian sage desired to prove and differed markedly from what he should have noted on back computations from modern-day data. Newton is persuaded that the only rationalization is intentional fraud, and he discards the possibility that Ptolemy had a deceitful, disloyal subordinate. Newton suggests that Ptolemy was driven to fraud because of an ambition to be recognized as a famous astronomer. If the work of Ptolemy was so conspicuously fraudulent as pointed out by Newton, it is surprising that his dishonesty was not exposed until centuries after his death. Although Ptolemy clearly made mistakes in his written documents, not everyone attributes his errors to fraud [39]. Owen Jay Gingerich (1930-), a professor emeritus of astronomy and history science at Harvard University (Harvard) (http://www.harvard.edu/), concurs that mistakes exist in Ptolemy's records, but he does not regard them as intentional errors [40]. Several curiously suspicious numbers are present in the Almagest, but Gingerich rejects fraud as the interpretation. He proposes that Ptolemy only documented selected observations in the Almagest that seemed to best fit his theory, and this falls within the current definition of scientific misconduct.

Sebastian Cabot. Sebastian Cabot (c. 1474-c. 1557), a talented explorer of the sixteenth century with a deep knowledge of the North Atlantic Ocean, supposedly reached Hudson Bay during a 1508-1509 expedition before Henry Hudson (1565-1611), the English explorer who explored the bay in 1610. By keeping this knowledge to himself, he was able to play England, Spain, and Venice, the major naval forces of the time, against one another. Historians have doubted whether his vast expeditions left England or even existed. The handful of surviving records indicate that Cabot was a skilled conspirator, whose exploits in the North Atlantic fooled historians for 3 centuries [94].

Sir Isaac Newton. Sir Isaac Newton (1642-1727) was a remarkable English genius who practiced as a physicist, mathematician, astronomer, natural philosopher, alchemist, and theologian. Many consider him the greatest and most influential scientist to have ever lived. In 1687, he published *Philosophiæ Naturalis Principia Mathematica*, one of the most significant scientific tomes to which pen has been put to paper [77]. It formed the basis for classical mechanics, and it documented universal gravitation and the three laws of motion. For three centuries, this book ruled the scientific assessment of the physical universe, and it demonstrated that objects on Earth and in extraterrestrial locations moved by the identical natural laws of motion. Newton showed the constancy between his theory of gravitation and the laws of planetary motion, which appeared in three books.recognized by Johannes Kepler (1571-1630), the German mathematician, astronomer, and astrologer. Among Newton's achievements were calculations of the speed of sound, the precision of the equinoxes, and the orbit of the moon. Unlike scientists of today Newton produced a colossal bulk of handwritten laboratory notes and correspondence. Despite his scientific achievements, his research illustrates indisputable falsification by a variant of what Babbage called trimming. Perhaps because of Newton's unmatched intelligence and the high esteem in which he was held, a quarter of a millennium was needed to uncover his lack of objectivity [108] in developing theories to explain his observations. He manipulated observations to fit the theories that he was convinced were true. Newton was convinced that he knew the answers to questions, before he made observations. As a genius Newton did not follow the standard procedures of the scientific enterprise but made the data conform to what he thought he knew to be so. He frowned upon hypotheses and during his time stated "hypotheses non fingo" and "hypotheses nonsequor" (I do not pursue hypotheses).

Charles Robert Darwin. The English naturalist Charles Robert Darwin (1809-1882) was undoubtedly one of the most influential persons in the history of mankind, and during his lifetime he produced an enormous volume of written notes and correspondence. In 1858, while preparing the text of his theory that would shake the world, the British naturalist Alfred Russel Wallace (1823-1933) forwarded an essay to Darwin documenting the identical concept, and this triggered the urgent combined publication of both of their theories [8]. To explain the diversity of living creatures, Darwin generated a hypothesis that has withstood the test of time and which deserves the right to be regarded as a fact. According to Darwin, all species descended eventually from common ancestors by a procedure dubbed natural selection by him. His theory of evolution was published in 1859 in On the Origin of Species by Natural Selection with persuasive substantiation [23]. Much data to support the theory was gathered on his five-year voyage on HMS Beagle (1831-1836), which also set him up as a distinguished geologist. By the 1870s the scientificenterprise and a vast number of the general population believed that evolution did indeed take place. In 1871 Darwin published his second book on evolutionary theory [25]; and a year later came The Expression of the Emotions in Man and Animals [26], which contained photographs of facial expressions in Europeans showing a wide variety of human emotions characteristic of grief, fear, horror, shame, anger, joy, contempt, disgust, and surprise. Such expressions were already well recognized in adults and even some children. Darwin pointed out that these emotions were common to all humans worldwide as well as animals, such as elephants that also express emotions. Some photographs were manipulated and by contemporary standards would be labeled as scientific misconduct, and some have made this accusation. But the art of publishing photographs in books was still in its infancy when Darwin penned this book, which was one of the first scientific texts illustrated with photographs. In 1889 seven years after Darwin's death, Sir Francis "Frank" Darwin (1848-1925)—a botanist and the third son of Charles Darwin and his wife, Emma—published a second edition of The Expression of the Emotions in Man and Animals, and it included new material and annotations that Charles Darwin had made in the margins of his copy of the first edition. The Expression of the Emotions in Man and Animals was reprinted as a third edition in 1998 with a commentary by Paul Ekman, a social psychologist at the University of California at San Francisco (UC San Francisco) *(http://www.ucsf.edu/)*, who was an expert in Charles Darwin. The English art historian Phillip Prodger also contributed to the third edition [84]. This version of the book included additional information from the records of this genius including previously unpublished photographs. Ekman noted that some of Darwin's photographs had been altered apparently because he had been unable to obtain acceptable pictures because photographic plates at the time could not record short-lived facial appearances, such as the change in a baby from fear to tears. From an analysis of material in the Darwin achieves and his correspondence, Ekman discovered that Darwin had altered images to a large extent. Some of the images had been created by stimulating facial muscles with electrodes and then removing the electrodes from the engravings. The most famous photograph of a weeping baby was in reality a drawing by Oscar Gustave Rejlander (1813-1875), a pioneering Victorian art photographer who created a drawing to look as if it was a photograph. Despite the criticisms made against Darwin and his photographer, they were basically making illustrations to stress a point at a time when photography had not yet reached its heyday. The rules about photographic impartiality had not yet been established, and there was little guidance with regard to using photographs as scientific evidence. Thus, the difference between illustration and objective evidence was hazy. It was only later when scientists started to document evidence photographically that demands for accuracy became essential. As a mark of respect for Darwin's supremacy as a scientist, he was honored with a formal funeral in Westminster Abbey, and his body was left to rest near the tombs of Sir Isaac Newton (1642-1727) and Sir John Frederick William Herschel (1792-1871), two outstanding British scientists.

Gregory Mendel. Gregory Mendel (1822-1884), the Augustinian monk, established the fundamental basis of genetics in breeding experiments carried out on peas in a tiny garden in front of the monastery at Brünn (currently Brno) in Austria. In essence he found that some characteristics of the pea plants were inherited as dominant or recessive traits. He described his findings at two meetings of the Brünn Natural History society on February 8 and March 8 in 1865 and published a paper on the subject a year later [71]. Despite being widely distributed as all students of genetics know, the paper was largely ignored until 1901 after the death of Mendel. At the turn of the century, naturalists rediscovered the rules of heredity that Mendel had established and posthumously brought him fame as the father of the gene theory of heredity. In 1911, when Sir Ronald Aulmer Fisher (1890-1962), one of the founders of population genetics, was an undergraduate student at Cambridge University (Cambridge) *(http://www.cam. ac.uk/)*, he calculated that the Mendel's original results all fell within the limits of probable error. He also found that if the experiments were repeated, the odds of achieving comparable results were approximately 16:1. When Mendel performed his experiments, the science of statistics was rapidly advancing. Fisher tabulated the thousands of experiments performed by Mendel and found that Mendel's actual data were incredulously close to the theoretical anticipation. The later experiments seemed to be biased toward what was expected by Mendel with his theory. He was clearly not familiar with the precautions that are currently needed for a completely objective analysis of the type of experiments that he performed with peas in his garden. The conclusions on heredity from the experiments of the Austrian scientist were obviously correct, but he probably manipulated the results to make them conform precisely to his theory of heredity. In 1936 Fisher accused Mendel of fudging parts of his data while simultaneously praising him for his thoughtful work and his discovery [27]. The data from the experiments of Mendel illustrate a variant of trimming or falsification as they were statistically unbelievable. A statistical analysis of his published data signifies that the odds of obtaining them in a real experiment are extremely low (10, 000:1). A number of admirers of Mendel have come to his defense and pointed out potential explanations for the bias in his research.

Louis Pasteur. The exceptionally renowned French scientist Louis Pasteur (1822-1895) was born five months after Mendel (1822-1884), and he specialized in chemistry and founded microbiology with Ferdinand Cohn (1828-1898) and Robert Koch (1843-1910). Pasteur left a legacy of numerous scientific contributions that included extraordinary advances related to the cause and prevention of infectious diseases. After becoming aware that microorganisms cause disease in both animals and humans, Pasteur's research focused on typhoid fever, rabies, anthrax, chicken cholera, and other infectious diseases. His research verified the germ theory of disease and diminished the mortality of puerperal fever. He also invented a way to prevent the transmission of disease by milk, and the process became known as pasteurization in his honor. Pasteur made profound contributions to infectious diseases, but his research extended far beyond microbiology, and he furthered knowledge in chemistry and discovered the molecular basis for the asymmetry of some crystals. Following Pasteur's doctoral thesis on crystallography, he was appointed professor of chemistry at the University of Strasbourg in 1849. During the same year he got married, and five children were born from this mating. Three of them died in childhood. Two succumbed from typhoid, and this heartbreaking loss motivated Pasteur to find cures for typhoid and other infectious diseases. Another of his discoveries was that microorganisms cause fermentation and that the proliferation of bacteria in nutrient broths results from biogenesis and not from spontaneous generation. Pasteur revealed that the spoiling of beverages such as beer, milk, and wine were caused by the growth of microorganisms. This led to his proposal that infectious diseases could be prevented by blocking the entry of microorganisms into the body. Being aware of infections, following surgery, Lord Joseph Lester (1827-1912) took this concept further and developed antiseptic techniques in surgery. Pasteur even did research on the silkworm. In 1865 the silk industry was devastated by the deaths of enormous numbers of silkworms in Alais (now Ales). Pasteur investigated the epidemic and established that a microbe attacked the silkworm eggs and that by eliminating this pathogen within silkworm nurseries this lethal disease could be exterminated. Pasteur even established that certain microbes can remain alive in the absence of air or oxygen. Before Pasteur's birth, it was known that vaccination could diminish the severity of certain infections and that it was an important therapeutic procedure. The most notable example was the extremely fatal smallpox. It had become established that inoculation with smallpox reduced its mortality and caused considerably less scarring in contrast to the usual infection. In 1796 Edward Jenner (1749-1823) established the efficacy of inoculating cowpox lesions into the skin of human subjects for protecting them against smallpox. Pasteur developed the first vaccine for rabies and also made a vaccine for anthrax.

During his life Pasteur received many awards including the Leeuwenhoek Medal that was established in 1877 by the Royal Netherlands Academy of Science in honor of Antonie Philips van Leeuwenhoek (1632-1723), the famous Dutch scientist who became widely known as the father of microbiology. Another major award was the Grand Croix of the French Legion of Honor established by Napoleon Bonaparte. Following his death Pasteur was initially buried in the Cathedral of Notre Dame, but his remains were later reinterred in a magnificent crypt beneath the Pasteur Institute in Paris, which was named after him in addition to the Université Louis Pasteur. In his honor many communities throughout the world have streets named after him.

In 1878 Pasteur instructed his family to never let anybody see his laboratory notebooks. His family adhered to these wishes until 1964 when his last remaining grandson gave the documents to the Bibliothèque Nationale in Paris. Ten years later a handful of historians were allowed to examine them in a limited way. After going through hundreds of pages of notes written in teeny words by Pasteur, evidence of potential research misconduct and ethically questionable human studies emerged. Gerald L. Geison (1943-2001), a historian at Princeton University (Princeton) (http://www.princeton.edu/), drew attention to these surprising revelations at a meeting of the American Association for the Advancement of Science (AAAS) [2]. Geison provided examples of Pasteur breaking standards of behavior that were not even appropriate in the midnineteenth century. On May 16, 1995, a hundred years after the death of Pasteur, the New York Times published an article titled "Revisionist History Sees Pasteur as Liar who Stole Rival's Ideas, " reporting the findings of Geison [1]. After meticulously going through the laboratory notes of Pasteur, Geison asserted that this legendary figure of science with impeccable credentials provided a deceiving portrayal as to how the anthrax vaccine was prepared in an experiment at Pouilly-le-Fort. According to official publications, Pasteur inoculated half of a flock of sheep with an anthrax vaccine that he had created and then exposed all animals to anthrax. The vaccinated sheep survived, and Pasteur became widely acclaimed for this therapeutic success, but Geison discovered that the vaccine used was not made by inactivating pathogenic anthrax bacilli with oxygen as documented in publications by Pasteur but with a potassium dichromate according to the method of his competitor Jean Joseph Henri Toussaint (1847-1890) [37]. Pasteur's oxygen technique was ultimately used to create an anthrax vaccine but not until Pasteur had obtained a patent on its production. Aside from Pasteur's questionable behavior with anthrax vaccine, his experiments with rabies vaccine were also problematic from an ethical standpoint. In experiments related to the development of a rabies vaccine, Pasteur injected the vaccine into two individuals suspected of having symptomatic rabies despite having never tested it in symptomatic cases. Pasteur also inoculated two young boys with an experimental rabies vaccine without first carrying out trials in animals.

Sigmund Freud. The Austrian neurologist Sigmund Freud (1856-1939) performed research on cerebral palsy, aphasia, and neuroanatomy before extending his effort into studies of the unconscious mind and generating theories about how it functions. He then launched the discipline of psychoanalysis from which he attained fame. From 1887 to 1904 Freud wrote numerous letters to Wilhelm Fleiss (1858-1928), a German physician and intimate friend living in Berlin, who exchanged ideas with Freud. This correspondence remained unpublished until 1950 when a German edition of Sigmund Freud, Aus den Anfägen der Psychanalyse edited by three admirers of Freud was published in London [34]. The editors included Marie Bonoparte Anna Freud (1868-1938), the third daughter of Sigmund; Martha Freud, who followed her father's footsteps into a career as a psychoanalyst; and Ernst Kris. An English translation of the book appeared in 1954 [35]. The letters in this edition were markedly censored as became apparent when they were published in full by another editor in 1985 [63]. One theory proposed by Fleiss was that illnesses of the nose cause psychosexual disorders because of a connection between the genitals and the nose ("nasal reflex neurosis"). This disorder needed to be treated by nasal cauterization and the application of cocaine to bones of the nasal passages. Freud became convinced about the validity of this weird entity and referred Emma Erkstein, an attractive young woman, to Fleiss for therapy. Fleiss treated her according to his recommended therapy but left half a meter of gauze in the surgical wound. After Eckstein came back to Vienna, a serious infection and almost-lethal hemorrhage developed. For Freud the diagnosis was obvious. She was "bleeding for love" for him. While her life was saved after the unfortunate complications, Erkstein was eternally disfigured. Freud hypothesized that the principal force motivating human life was sexual drives. He took advantage of remedial methods like the use of free association, uncovered the phenomenon of transference in the therapeutic association, and determined its cardinal function in the method of analysis, and deciphered dreams as a way of gaining insight into the unconscious yearnings. In addition he was a productive author drawing on psychoanalysis to provide him with a place in history. Throughout his career, Freud was constantly adamant that his theories and techniques were based on his personal experience and research with authentic patients. He observed that female hysterics were victims of sexual seduction by adults, and in a presentation to neurologists and psychiatrists in 1896, he claimed that this observation was authenticated by "some eighteen cases of hysteria, " which he successfully treated. However, in correspondence to Wilhelm Fleiss (mentioned above), Freud admitted that none had been treated successfully. A year later Freud dropped the necessity for seduction in childhood by an adult. He then faulted women for having desires of being seduced by their fathers during infancy or childhood. He then explained the neurotic behavior during adulthood on these repressed unrecognized sexual fantasies. The desire of an infant to seduce a parent of the opposite gender together with repression became the initial tenant of psychoanalytical theory. As pointed out by Adolf Grünbaum, Malcolm Macmillan, Frank Sulloway, and other critics of Freud [22, 46, 60, 102] later modifications to his theory were subjective and lacked the objectivity of science. Early critics of Freud questioned the validity of his methods as really being science. Also, the not unimportant question was raised as to whether his treatment produced cures. Undoubtedly Freud believed intensely in his theories. During his life he was not deterred by his critics, and they attacked him even after his death. In 1979 Sulloway convincingly defended Freud as a scientist in a book titled Freud, Biologist of the Mind: Beyond the Psychoanalytic Legend [102]. For this persuasive analysis the History of Science Society awarded Sulloway a prize. After the publication of this opus, Sulloway continued to doggedly analyze the works of Freud; and just over a decade later, he renounced his earlier opinion in a second edition of one of his books [103] and in journal publications. He considered all case histories of Freud useless because of fabrications and falsifications. Between 1993 and 1994 Frederick Crews published a series of papers in The New York Review of Books between 1993 and 1994 that drew attention to the fact that Freud had fabricated and fudged the evidence for his theories. A handbook containing these papers and embittered arguments for and against Freud were published [22]. Crews summarized the situation, stating, "Freud's theories of personality and neurosis—derived as they were from misleading precedents, vacuous pseudophysical metaphors, and a long concatenation of mistaken interferences that couldn't be subjected to empirical review—amounted to castles in the air." Another critic of Freud in relatively recent times was Grünbaum, a professor of philosophy and research in psychiatry at the University of Pittsburgh (Pitt), who concluded that Freud's clinical theory cannot to be true [46]. Another scrutinizer of the methods and practices of Freud and his disciples was Macmillian, an Australian professor of psychology at Monach University [60]. Like many before him, Macmillian concluded that psychoanalysis was not a science. The fraudulent behavior of Freud was covered up for almost a century in an extremely skillful manner. Authentic case histories on which Freud based his theories and his audacious claims at obtaining cures were an astonishingly small number. Sulloway could only identify eight in 1991[103].

Ernst Heinrich Phillip August Haeckel. Before the publication of The Origin of Species by Natural Selection, or the Preservation of Favoured Races in the Struggle for Life by Darwin on 24 November 1859 [24], biologists were aware that that different species are frequently more akin to each other as embryos than in adulthood. Karl Ernst von Baer (1792-1876), a member of the Russian Academy of Sciences and a founding father of embryology, made numerous notable observations on embryos, and these became known as the von Baer laws. One was that the embryo of a higher animal never resembles the adult of a higher animal but only its embryo. Also all early embryos of reptiles, birds, and mammals develop comparable characteristics that become lost as these embryos proceed to term. The German biologist Ernest Heinrich Phillip August Haeckel (1834-1919) extended these observations *(http://en.wikipedia.org/wiki/Erst_Haeckel)* and hypothesized that from fertilized egg to entire body development, the organism goes through a series of brief condensed repetitions of phases through which the succeeding predecessor evolved. In 1866 Haeckel, an enthusiastic supporter of Darwin's theory of evolution, defined his famous biogenetic law for which he coined the phrase "ontogeny recapitulates phylogeny." Haeckel realized that the recapitulation was in exact and that evolutionary phases could disappear from the developmental stages. The developmental stages were not precise copies of evolutionary periods as proposed by Haeckel. His fabrications misled embryologists and evolutionary biologists up the garden path for decades. Haeckel was another recently exposed dishonest scientist who achieved fame for his work. In 1874 Haeckel published a collection of virtually identical drawings of vertebrate embryos that were claimed to been created by directly examining the samples. He provided illustrations of comparable stages in embryonic life of creatures as diverse as calf, chick, fish, hog, rabbit, salamander, tortoise, and man. These illustrations were provided as evidence to support Haeckel's appeal that ontogeny recapitulates phylogeny. According to this hypothesis, embryos of all higher vertebrates pass through morphological phases of development that they share with other more primitive animals. Higher vertebrates only diverge by the addition of new stages later in development. Haeckel implied that during development, an animal succeeds during embryonic forms of all of its ancestors consecutively from the beginning of the earliest cell to the present. From knowledge about embryology, Haeckel assumed that the phylogenetic history of a species could be predicted. From the time of Haeckel's annunciation that ontogeny recapitulates phylogeny was regarded as fact and hence as a biological law, and it was reiterated in textbooks of biology from that time onward as one of the most renowned tenets in biology. Yet before the twentieth century, it was obvious that Haeckel's naïve scheme was flawed and that organisms diverge in some respects in their development compared to Haeckel's documentation. Von Baer, a contemporary of Haeckel, noted flaws on Haeckel's reasoning and provided examples of embryos that failed to adhere to the biogenetic law of Haeckel. Chick embryos, for example, lack the scales, swim bladders, and fin rays of adult fishes. Adult attributes of ancestors hardly ever emerge as transitional stages in the ontogeny of a progeny. Haeckel was only to some extent correct in that an early embryo does not represent any previous ancestral adult forms. It resembles embryos of evolutionary predecessors. Not only have biologists knownfor a long time that Haeckel's idea was wrong, but the drawings of embryos used by him to support his line of reasoning were inaccurate. In August 1997, when he was still at St.George's Hospital Medical School in London, the British embryologist Michael K. Richardson published important papers with his colleagues from Australia, Canada, France, and the USA [90, 91, 92] indicating that Haeckel fabricated his illustrations to assemble the early embryonic stages of the different vertebrates to seem more identical than they were in reality. In pointing out the inconsistencies, they published photographs of the vertebrate embryos showing that they were distinctly different. Notable errors on Haeckel's part were the fact that he omitted limb buds in human embryos and inserted a tail on a chick embryo. He also drew all specimens to the same scale despite the existence of tenfold differences in size. Despite the fact that Haeckel's colleagues not only spotted the fraudulent observations when he was alive and managed to get an admission from him, modifications of Haeckel's illustrations persisted for more than a century in virtually all textbooks of biology and comparative embryology. As pointed out by Richardson, Haeckel may have created the most famous fake in biology [82].

Robert E. Peary. In 1909, when Admiral Robert E. Peary (1856-1920) (http://en.wikipedia.org/wiki/Robert_Peary), a USA navy surveyor, started his eighth Arctic expedition, he was 53 years old, and all his toes except two had been lost to frost bite. By this time Peary had mastered the skill of dog sledding and was convinced that he could achieve his longtime goal of reaching the North Pole. Peary realized that this trip would be his last opportunity to accomplish this objective. To reach the North Pole many teams would be involved, and they would be arranged in camps that created a pyramid of supplies northward over the ice and sequentially move back to Ellesmere Island. From the final campsite, one party would press on to the North Pole. While proceeding north, Captain Robert "Bob" Abram Bartlett (1875-1946), the Arctic explorer, broke the trail while Peary followed some distance behind. At 87°47' north Peary pulled alongside Bartlett and informed him that he must retreat to safeguard food and fuel. A terribly saddened Bartlett obeyed his commander. Peary continued northward with Matthew Henson and four Inuits, and on April 6, 1909, he notified them that they had reached the North Pole at 89°57' north. Since only Peary could read the precise coordinates of the pole with the relevant equipment, no one in the group could confirm that the North Pole had been reached. To reach that location, the Peary team would have had to travel more than 25 miles per day, which was much farther than the usual average daily mileage of 9.3 traveled in a day. More unbelievable was the speed needed for Peary's return to the camp stationed at 87°47 '[62]. He would have had to travel 133 miles in two and a quarter days. This was considerably faster than the all-time record (30 miles per day) on packed ice claimed by Peary in 1906. Nevertheless, after Peary's return home, the National Geographic Society without any delay endorsed Peary's claims, and an upsurge of public praise for a hero brushed skeptics under the rug. Meticulous scholars had no doubt that Peary's alleged feat could not have been accomplished, and the National Geographic Society was criticized for being too impetuous in accepting Peary's claim of reaching the North Pole. By sending Bartlett back, Peary eliminated the only person who could have readily exposed his records that he had presumably chosen to fabricate.

After an 80-year-old heated discussion about the credibility of Peary and the brushing aside of the claim by American surgeon-explorer Frederick Albert Cook (1865-1940) *(http://en.wikipedia.org/wiki/Frederick_Cook)*, the surgeon on Peary's 1891-1892 Arctic expedition claimed that he researched the North Pole first before Peary on April 21, 1908, the issue seems to have become resolved.

Cook's reputation as a cheater and liar had been documented previously. During an Antarctic expedition in 1898, he met the Anglican missionary Thomas Bridges (ca.1842-1896) shortly before Bridges's death and brought back a Yamana dictionary prepared by Bridges and tried unscrupulously to publish it as his own *(http://en.wikipeia.or/wiki/Frederick_Cook)*. During a second expedition to Alaska in 1906, Cook claimed that he reached the summit of Mount McKinley; but although he did take photographs from the alleged summit, it only appeared as a tiny peak. The object of this fraudulent photograph became known as Fake Peak. Even though others on the expedition doubted his accomplishment, his assertion was not contested in public until his confrontation with Peary over the first to reach the North Pole [10, 48]. A team led by retired admiral Thomas D. Davies and financed by the National Geographic Society, which commissioned Peary's controversial Arctic expedition of 1909, reexamined the data from Peary's alleged trip to the North Pole. Dennis Rawlins (1937-), an independent astronomer, indicted Peary largely on the records of the position of the sun and certain key stars in publications. Rawlins initially interpreted one record as evidence of fraud but later admitted that he had misread it [89]. It is generally accepted that nobody set foot on the North Pole until the USA Air Force pilots Joseph Otis Fletcher (1920-2008) and William Pershing Benedict (1928-1974) achieved this on May 3, 1952, with the geophysicist Albert Paddock Crary (1911-1987) [93].

Robert Andrews Millikan. In 1895 the American experimental physicist Robert Andrews Millikan (1868-1953) was the first to earn a PhD from the department of physics at Columbia University (Columbia) *(http://www.columbia.edu/)*. In 1907 he started performing experiments at the University of Chicago (UChicago) *(http://www.uchicago.edu/)* with oil drops in which he measured the charge of a single electron (e). Following his first publication on the subject in 1909 a bitter controversy broke out with the Austrian physicistFelix Ehrenhaft (1879-1952) who believed that e was only an average value of charges that varied continuously over a wide range. Millikan is widely believed to have been blocked from getting the 1920 prize for physics because of assertions by Ehrenhaft that he had measured charges smaller than Millikan's elementary charge. After constructing a special better piece of equipment for his experiments a strong electric field could be turned on or off between a horizontal pair of brass plates, and each solitary electrically charged minute drop of oil could be observed at a constant temperature and air pressure through a short focus telescope with a scale in the eyepiece. Millikan published his epoch-making research in 1913 in which he found that all electrons have the identical value and he measured this with considerable accuracy [73]. Even the infrared light, which could affect the temperature, was filtered out from the light that illuminated the drops of oil. Each drop would fall slowly, but this would be much slower when the electric field was turned on. The charge on the drops was changed with a beam of X-rays. Hundreds of observations were recorded by Millikan in notebooks, and he calculated the charge on each drop. The charges were always multiples of a precise minimum quantity. From 1920 Millikan published a series of papers on these experiments. In 1923 Millikan was awarded the Nobel Prize in Physics *(http://www.nobelprize. org/nobel_prizes/physics/laureates/1923/millikan-photo.html)* for determining the charge on a single electron and for his research on the photoelectric effect.

In 1917 Millikan became convinced by George Ellery Hale (1868-1938), one of the world's most famous solar astronomers, to start coming to Throop College of Technology in Pasadena, California, for a few months each year. Shortly thereafter this small academic institution became known as the California Institute of Technology (Caltech) *(http://www.caltech.edu/)*, and Millikan departed from UChicago to become Caltech's first chairman of the executive council, de facto itspresident. He retained that prominent title from 1921 to 1945. While at Caltech, Millikan's research concentrated mainly on cosmic rays, a label that he invented, but his concept of them being photons of new atoms turned out to be incorrect. Cosmic rays were shown to be charged particles deflected by the earth's magnetic field. During World War I Millikan served as vice chairman of the National Research Council and assisted in the creation of antisubmarine and meteorological equipment.

In his seminal paper of 1913 [73] Millikan stated, "It is to be remarked, too, that this is not a selected group of drops but represents all the drops experimented on during 60 consecutive days, during which time the apparatus was taken down several times and set up anew." In his autobiography [75], he went on to stress, "These drops represent all those studies for 60 consecutive days, no single one being omitted." A quarter of a century after Millikan's death, Gerald Holton (1922-), a historian of science who was also a physicist at Harvard University (Harvard) *(http://www.harvard.edu/)*, analyzed the notebooks of Millikan, which were available in the archives at Caltech, and found inconsistencies with what he documented in the literature. Millikan discarded many of his oil-drop observations when calculating e. In his notebooks he included comments as to whether certain experiments should or should not be included in publications, such as "Perfect publish, " "Publish this beautiful one, " "Agreement poor. Will not work out, " "Error high will not use." According to Holton the fifty-eight drops reported in Millikan's 1913 paper were selected from approximately one hundred and forty. Millikan's intention was to accomplish enormous accuracy. According to the terminology of Babbage, described at the beginning of the chapter, Millikan achieved his goal by cooking his data. He selected which observations to discard and which to report. The analysis of Millikan's research by Holton stirred up a controversy with many defending Millikan. One of his defenders was Allan D. Franklin, a physicist and philosopher-historian of science at Cornell University (Cornell) *(http://www.cornell.edu/)*. Franklin also reviewed the records documented in Millikan's notebooks [31]. He assumed that by the time Millikan had mastered his equipment and was able to get reliable results, he had tested 68 oil drops. The initial experiments were, in fact, a trial period in which the reliability of the experiment was being tested. In his mind Milligan was not guilty of considerable cooking but of very little trimming to "gain a reputation for extreme accuracy in making observations." Another Millikan defender who had also examined his notebooks was David Goodstein (1939-), a physicist at Caltech [42, 43]. Ullica Segerstråle [100] pointed out that the Millikan case has become an example of either scientific misconduct or of good scientific judgment depending on the view point of the author. Nevertheless, Millikan's research was clearly biased by his preconceptions, and he pointlessly discarded data that he did not like. He also deliberately lied again and again.

Even though Milligan's research on the electron was performed with Harvey Fletcher (1884-1981) including portions that lead to his Nobel Prize, Millikan unethically took all the credit according to in a private arrangement, which Fletcher did not reveal it until many years after Millikan's death [28]. In exchange for credit in the electron measurements, Fletcher was provided full authorship on an associated study for his PhD thesis. Despite his imperfections, Millikan received many honors and awards. Throughout his career, Millikan was fervent about education, and he coauthored more than one well-liked and high-ranking entry-level textbooks [74]. It was hence not surprising that Millikan had two schools in California named after him (Millikan Middle School in suburban Los Angeles and Robert A. Millikan High School in Long Beach). The tallest building on the Caltech campus was also named in his honor (Millikan Library). In 1953 Millikan died from a heart attack at 85 years of age. He was laid to rest in the Court of Honor at Forest Lawn Memorial Park Cemetery in Glendale, California.

Paul Kammerer. In contrast to the genetic theory of Mendel, another theory proposed that at least some acquired characteristics could be inherited to the parents' offspring. Nearly all toads possess black scaly bumps on their hind limbs known as nuptial pads, which enable them to grab hold of the wet slippery female so that they can hang on to each other and mate in water and later lay their fertilized eggs in an aquatic milieu. However, unlike nearly all other frogs and toads, the midwife toad *(Alytes obstetricans)*, which mates on land, lacks these pigmented nuptial pads. After the eggs are released from the female midwife toad, the male wraps the eggs around his legs and transports them until the embryos emerge as tadpoles. In the early 1900s the Austrian biologist Paul Kammerer (1880-1926) was a firm believer in Lamarckism named after the French biologist Jean-Bapitste Lamarck (1744-1829). According to Lamarckism traits attained during the life of an organism can be transmitted to upcoming generations. Kammerer thought that if he could get the midwife toad to mate in water, it would ultimately develop pigmented nuptials like other water-mating toads and that their offspring would inherit them. Kammerer wanted to determine if he could induce these special structures the midwife toad lacks. To do this he confined midwife toads to a dry overheated environment, forcing them to move into water to copulate and lay eggs. Most of the eggs died, but 3-5% of the offspring continued to exist, but they lost the territorial habits of the parents and opted to mate and deposit their eggs in water. This change in phenotype persisted for the duration of Kammerer's experiments over 6 generations. By the third generation the hind limbs were stated to have thickened areas, and two generations later distinct nuptial pads were noted. Eventually midwife toads were born with black scaly marks on the hind limbs allegedly proving to the scientific populace that Lamarckian inheritance was possible. Kammerer believed that the newly formed genes were transmitted to the offspring. But there was something most extraordinary about these amphibians that brought fame to Kammerer. Other features desirable for living in an aqueous environment, such as a thicker jelly coating of the eggs and smaller amounts of yolk as well as larger tadpole gills, did not develop. Furthermore, in experiments mating water toads with terrestrial toads, the traits of the water toads occurred in the proportions expected for Mendelian inheritance. The research of Kammerer was severely criticized and not accepted by the scientific community. William Bateson (1861-1926), a distinguished senior geneticist who coined the word "genetics" in 1906, was the archetypical nemesis. He knew Kammerer's evidence for Lamarckian inheritance of nuptial pads in the midwife toad must be amiss and asserted pressure in public until Hans Leo Przibram (1874-1944) and Gladwyn Kingsley Noble (1894-1940), curator of reptiles and a herpetologist at the American Museum of Natural History in New York, examined them in 1926 and delivered a devastating blow after examining the last residual specimen of Kammerer's water toads and reported that he had not found pads on the hind limbs but India ink within the relevant parts of their limbs. Instead of black scaly marks on their hind limbs, Noble found black spots where India ink had been injected subcutaneously [79]. Kammerer was markedly embarrassed, but he denied any wrongdoing and blamed his laboratory assistants. Nevertheless, Kammerer sparked a long-lasting controversy and received all the fallout from this fraudulent episode; and with his reputation ruined, he committed suicide less than two months later in the Austrian mountains. Was he a brilliant scientist who went astray, perhaps because he became overcommitted to the wrong hypothesis like many other scientists? Koestler and others attribute the India ink incident to an overzealous laboratory assistant who set Kammerer up as a victim of forgery. Some even maintain that he was framed by political or scientific opponents. Nevertheless, Kammerer was depicted a victim in a TV documentary movie and in a book by Arthur Koestler [56]. Alexander Vargas, an evolutionary developmental biologist at the University of Chile in Santiago *(http://www.uchile.cl/)*, recently made a case that the experiments of Kammerer may have demonstrated epigenetic effects and many of his supporters believe that this caused his downfall. Was Kammerer a victim of a scientific society that found his work incomprehensible because it had not yet learned about epigenetics? A new analysis of his data in August 2009 even suggested to one author, who was obviously not familiar with all the evidence, that his renowned research may have revealed the phenomenon of epigenetics of recent times [83, 106]. Chemical modifications of DNA can be passed from generation to generation and can even be augmented in succeeding generations. While Kammerer will probably go down in history as the person responsible for one of the most renowned scientific frauds in the early twentieth century, his research was not a fatal blow to Lamarckian inheritance, which obtained a resurge in popularity due to the discovery of epigenetics. Because of this, some acquired traits seem to be inherited. Vargas also drew attention to the fact that Kammerer noticed a parent of origin effect in which a trait appeared to only be transmitted if coming from a particular parent. If the father was a "water" toad, 100% of the first generation and three quarters of the next generation were "water toads." When the father maintained the terrestrial tradition, the reverse was true. This strange finding only complicated the situation leading to a greater suspicion of fraud. Yet similar outcomes have been detected in mice with regard to the inheritance of coat colors and are attributed to epigenetic changes in a gene variant called Aguiti variable yellow [107]. According to Vargas changes in the jelly coat around the eggs of Kammerer's toads may have led to anomalous methylation of some genes. The eggs of all vertebrates have a jelly coating, and if it is removed immediately after fertilization, DNA methylation can be decreased in primitive embryos. Not all are convinced about Vargas's arguments that Kammerer's finding could be due to epigenetics, and many find it hard to believe that he was anything but a failure as a scientist. Time will tell, and eventually someone is bound to try to repeat the experiments of Kammerer in view of the door that Vargas has opened to potentially restore the credibility of Kammerer, who was preoccupied in trying to prove the Lamarckian theory of evolution. If his claims are true, the experiments still need to be replicated [83, 104, 105].

Cyril Lodowic Burt. Sir Cyril Lodowic Burt (1883-1971) was a British educational psychologist with considerable political standing where his research supported the contentions of both the snobby elitists and racists. In 1907 Burt assisted in a national survey of physical and mental qualities of the British population as suggested by the famous Sir Francis Galton (1822-1911), and he standardized psychological tests. During this assignment, Burt came in contact with Charles Edward Spearman (1863-1945), the English psychologist known for his seminal work in statistics and models of human intelligence. He also became acquainted with Karl Pearson (1857-1936), the English mathematician who founded the statistics department at University College London (UCL) *(http://www.ucl.ac.uk/)*. This was the first deparment of this nature created at any university. In 1908 while a lecturer at Liverpool University (http://www.liv.ac.uk/), Burt was influenced by Sir Charles Scott Sherrington (1857-1952), the 1932 Nobel laureate in physiology or medicine *(http://www.nobelprize.org/nobel_prizes/medicine/laureates/1932/sherrington-bio.html)*. In 1931 Burt was appointed professor and chair of psychology at UCL. His other duties included being a consultant on committees that developed the Eleven Plus examination in the United Kingdom (UK) that is sometimes administered to students in the final year of their primary education, governing admission to various types of secondary school. Throughout a career that covered more than five decades, Burt was extensively honored for his contributions to psychological testing and for facilitating educational opportunities. He was also elected president of the British Psychological Society, and in 1946 Burt was the first British psychologist to be knighted. Burt proposed that an international organization composed of members with extremely high intelligence quotients (IQs) be formed; and in 1946 Mensa was founded in the UK by Roland Berrill, an Australian barrister, and Dr. Lancelot Ware (1915-2000), a British scientist and lawyer. For admission the only required qualification was an IQ at or above the 98th percentile on an approved IQ test. In 1960 Burt was elected honorary president of Mensa. For the initial half of the twentieth century, Burt was the dominant figure in research on intelligence, and he tested it in exceptionally bright as well as substandard children. He was the acknowledged authority in statistical techniques for studying intelligence, and he competently portrayed the conclusions of his research to audiences. Burt published hundreds of documents in journals and books for the general public on the subject. Being employed in London, Burt had contact with numerous children and their school documents. In a 1909 paper he was adamant that the offspring of aristocratic families in privileged preparatory schools fared better that students in the usual elementary schools and that the diversity was inborn [11]. His research focused on the heritability of intelligence, and his studies made use of comparisons between monozygotic (identical) and dizygotic (fraternal) twins. Twins that are derived from the same egg are genetically indistinguishable, whereas those coming from different eggs contain different genes. When identical twins are separated soon after birth and raised apart, differences in them can be attributed to environmental factors. In about 1913 Burt started to collect data on twins but did not publish anything on twin studies until 1943 [12] when he reported an analysis of 62 pairs of monozygotic twins and 156 pairs of dizygotic twins. Fifteen of the identical twins had been brought up separately. The correlation between IQs, a measure of intelligence originally created by the German psychologist William Stern (1871-1938), of non-separated twins was 0.86, and with separated identical twins the correlation dropped to 0.771. With dizygotic twins the correlation was even less (0.54). These observations indicate that intelligence, as reflected in IQ scores, has a considerable genetic influence. Surprisingly Burt did not divulge how he obtained the twins, and his raw data was not made public. In 1955 Burt reexamined the topic of twin studies and IQs and published another paper on the subject titled "Evidence for the Concept of Intelligence" [13]. In this more up-to-date publication the number of twins had increased because of the assistance of Jane Conway. Amazingly the correlations were still exactly the same as those published by Burt in 1943. In the following year Burt published an additional paper on the topic [14], but no raw data was provided, and details were insubstantial, but correlations for separated twins were provided to four decimal places. In 1957 Burt created the British Journal of Statistical Psychology under the sponsorship of Sir Godfrey Hilton Thomson (1881-1955), another English educational psychologist and pioneer in intelligence research. The journal did not attract much attention, and less than one hundred individuals contributed to it.

Particularly after the death of Thomson, Burt became the only editor, and he published a major part of his research in this journal without undergoing peer review. Jane Conway published papers in this journal [19] [20], which was eventually renamed the British Journal of Mathematical and Statistical Psychology in 1965 testifying that the number of identical twins reared apart had reached 42, twice the number reported by Burt in 1966 [15]. By now the number of identical pairs reared apart had reached 53. Despite increased numbers in the different twin categories, Burt's correlation coefficients for IQs for twins raised apart (0.771) and those brought up together (0.999) did not change when additional twins were studied. The correlations to three decimal places remained as previously documented, but it was extremely unlikely for this to occur statistically. Burt's work went unchallenged while he was alive, but his research particularly in his later years contained blatant errors, but they were ignored. The preposterous statistics in Burt's papers were there for everyone to notice, but nobody apparently paid attention to the divulging evidence. His publications should have provoked disbelief in the integrity of his research. Other educational psychologists noted questionable aspects of Burt's research but little other concerns. Gradually the conclusions from his data became more and more contentious politically. Conservative reporters and politicians leaped on Burt's findings that indicated that the IQ differences between whites and persons of color were predominantly innate and genetically determined. If Burt's conclusions were correct, education could do little to elevate persons of color to the level of whites. This issue was debated in the literature [51, 58]. When Burt died, his critics had barely started to attack his flawed research [47][54]. The development of hypotheses is often the outcome of the prevailing beliefs of society at the time. A scientist wedded to the wrong hypothesis or to a politically incorrect hypothesis can lead a scientist down the wrong garden path as occurred with Burt. In the early 1900s racism was still prevalent, and Caucasians were believed to be more intelligent than other races because of their genetic makeup. Even in Caucasians those in high society were believed to be more intelligent than the peasants. This widely held perception protected Burt from being exposed as a fraud during his natural life [80]. He hypothesized that intelligence is influenced partly on heredity, and he supported his theory with observations carried out in studies on identical twins. He was able to promote his theory while serving as the advisor to the British government in the 1930s and 1940s. Because of his high esteem, he was able get a school system established in which children were allocated to one of three educational levels based of a test taken at 11 years of age. Burt's work confirmed what many people wished to accept as true, or feared to be true, and there was no evidence to dispute it.

Ultimately, after a period of unchallenged influence for almost 4 decades, Burt's data was discovered to be peppered with improbabilities and fundamental technological errors. Some investigators even decided that Burt may have manipulated studies on identical twins or even concocted his compilation of IQ data [72]. Arthur Robert Jensen (1923-2012), a former professor of educational psychology at the University of California-Berkeley (UC Berkeley) *(http://www.berkeley.edu/)*, suggests that Burt was an eccentric and sloppy scientist rather than a fraudulent investigator. Jensen pointed out that Burt was a sophisticated statistician, and it is extremely unlikely that someone with his knowledge would attempt to fake the data. Why he would report precisely the same correlations three times in succession was problematic [9], but it might have been the result of Burt becoming wedded to his hypotheses, which he was convinced were true regardless of evidence. The twin studies of Burt that demonstrated unchanging correlations between IQs of identical twins reared apart even with more than tripling of the sample is generally no longer accepted as valid.

Soon after his death, it became recognized that all of Burt's documents had been burned, but flaws in Burt's data were suspected by Leon J. Kamin (1927-), a psychologist at Princeton University (Princeton) *(http://www.princeton.edu/)*, who later became the chairman of the department of psychology at Northeastern University *(http://www.northeastern.edu/)* [53]. The likelihood of data fabrication first surfaced when Kamin drew attention to the extremely improbable correlation coefficients of monozygotic and dizygotic twins IQs reported by Burt. They were the same to three decimal points across articles even when new data was twice added to the sample of twins. Another blow to Burt's credibility occurred in 1976 when the front page of the London Sunday Times on October 24 contained a report by its medial correspondent Oliver Gillie pointing out that Burt had concocted his research collaborators, Ms. Jay Conway and Ms. Margaret Howard [38]. Conway and Howard supposedly assisted Burt in gathering data on the twins, but Jensen doubted any fabrication on their part during the era when racism was particularly rife and when it was commonly believed that genes influenced the IQ more than the environment. Fifteen years ago, few geneticists would have questioned that IQ is influenced more by the genotype than the environment. Studies carried out between 1955 and 1966 by Burt on identical twins appeared to solidify the belief that genes influenced the IQ more than the environment. The IQ scores of monozygotic twins were invariably closely matched regardless of whether the twins grew up apart or together. Particularly remarkable about Burt's studies was his alleged success. No other researcher has been so successful in tracing twins who had been brought up in different surroundings after becoming separated at birth. In 1979 Leslie Hearnshaw [47] verified these allegations after scrutinizing his personal diaries. Burt did not personally perform a lot of his documented studies. Recent researchers have confirmed an association between genetic and IQ variation but not to the extent previously claimed by Burt. Evidence for fraudulent research by Burt are presented in Cyril Burt: Fraud or Framed [59] and in *The Mismeasure of Man* [44] as well as in other books and journal articles.

His earlier work is frequently regarded as legitimate, but there is evidence that his research data later in life was falsified. Hearnshaw considered the majority of Burt's research after World War II untrustworthy or falsified. In

1997 Tucker maintained that "a comparison of his data on twins with that from other acknowledged investigations left little uncertainty that Burt's work was fraudulent" [105]. Aside from books critical of Burt, others have praised and defended him [29, 30, 52].

Richard Meinertzhagen. Col. Richard Meinertzhagen (1878-1967) was a man whose background and achievements were the envy of many. His father was an extremely wealthy socially connected British banker second only in importance to the Rothschilds, and his mother cofounded the London School of Economics. Many of his relatives were rich and influential with British titles. He received an excellent education as one would expect for a person born with a golden spoon in his mouth. He attended the world-renowned English independent Harrow School for Boys, commonly known just as Harrow, in which his education overlapped with Sir Winston Churchill (1874-1965). Another friend of the Meinertzhagen family was Charles Robert Darwin (1809-1882), who is discussed earlier in this chapter. Meinertzhagen was an exceptionally talented artist and drew maps as well as drawings of landscapes and wildlife. He had no interest in a banking career and became a military man during the glory days of the British Empire serving in India and East Africa. During these assignments he participated in big-game hunting and earned a reputation as a fierce and ruthless soldier. While in Kenya in 1905, he crushed a major revolt by murdering the leader of the uprising (Nandi Orkoiyt Koitatel Arap Samoei) while shaking his hand after he came to negotiate and then orchestrated a cover-up of the incident. During his travels, this British warrior recorded considerable detail in his diaries and wrote four books based on their content, but doubt has been raised about the validity of some of the material [66, 67, 69]. Meinertzhagen had a passion for bird watching and shooting and regarded himself as a scientist-explorer rather than a soldier. He was an avid bird collector with a enormous compilation of bird and birdlike specimens and was recognized in his time as the most distinguished ornithologist in the UK. Meinertzhagen became the chairman of the British Ornithologist Club and was a recipient of the most prestigious Godman-Salvin Gold Medal of the British Ornithologists' Union named after Frederick DuCane Godman (1834-1919) and Osbert Salvin (1835-1898). The British Museum of Natural History even named one of its rooms after him.

In 1914 Meinertzhagen claimed that he had collected and stuffed a forest owlet (Blewitt owl). This large-eyed brown 8-inch high bird has distinctive bands on its wings and underside. The first recorded sighting of this bird was in the 1870s, and during the next few years, several of them were seen. Despite being idolized by many fans, Meinetzhagen's claim turned out to be his Achilles' heel leading to an unequivocal exposure of his fraudulent life. Pamela Rasmussen (1959-), an internationally renowned ornithologist formerly of the Smithsonian National Museum of Natural History in Washington DC, studied the specific bird that Meinertzhagen had allegedly preserved; but after examining the specimen, his claim was found to not to be true *(http://en.wikipedia.org/wiki/*

Pamela_C._Rasmussen) [36][85, 86]. The stuffed owlet had, in fact, been stolen from the British museum where it had originally been conserved in the 1880s by an ornithologist known for a unique method of preservation.

In 1967, Meinertzhagen inspired three biographies after his death. The novelist Brian Francis Wynne Garfield (1939) [36] regarded Meinertzhagen as a liar, charlatan, and gigantic fraud. Meinertzhagen became internationally famous after sighting the giant African forest hog (Hylochoerus meinertzhageni) which he killed, stuffed, and transported back to London. He edited Nicolle's Birds of Egypt in two volumes [65] following the death of the author in 1925. In 1948 and 1949 Meinertzhagen escorted Dr. Philip Alexander Cancey (1917-2001), a foremost expert on South African birds, on an ornithological expedition to many countries. Meinertzhagen's magnum opus Birds of Arabia [70] is thought to be based on the unpublished script of the American naturalist George Latimer Bates (1863-1940), who was not adequately acknowledged. In the 1990s a scrutiny of the Meinertzhagen bird collection at the Walter Rothschild Zoological Museum (currently known as Natural History Museum) in Tring in the UK disclosed considerable theft and falsification. The British ornithologist Alan G. Knox discovered the fraud [55] and stated that Meinerertzhagen stole the finest specimens from the collections of others and fabricated information about them. More current research by Rasmussen and Robert Prys-Jones, the bird collection manager, department of zoology at the National History Museum in the UK, revealed that the fraudulent work was even more wide ranging than initially suspected. Numerous specimens contributed to museums by him were actually discovered to be have gone astray from the Natural History Museum, where they had been acquired from other collectors such as the English ornithologist Hugh Whistler (1889-1943) [16].

James Shearer. On September 30, 1916, the British Journal of Medicine (BMJ) contained a three-page report by James Shearer titled "Dispatches from the Western Front" [101]. In 1916, while British troops were fighting the Germans in Europe during World War I, Shearer, a virtually unknown American medical doctor who was a sergeant in the Royal Army Medical Corps, documented a new technique for portraying the internal organs with an electrical method carried out in the dark in 60 seconds. His procedure allegedly worked when X-rays were unable to detect images below the surface of the body. Flashing lights or crackling spark gaps were needed, and a record of the organ could be conveyed to a wax sheet before being printed [17] [76]. Because the delineator described by him was stated to not only have the approval of the director of medical services in France but could also potentially transform the therapy of injured soldiers, the British military enthusiastically started an expensive program to test Shearer's method in reality. But it did not take long to realize that Shearer's procedure was merely a fabrication; and on learning about the dishonesty, the BMJ swiftly retracted the paper, and it is no longer at the website of the journal. Shearer was court-martialed and condemned to death by a firing squadron. Countless individuals considered the reaction by the military court to Shearer's fabrication just as inappropriate. Fortunately for him this extremely harsh sanction was commuted, but he died a year later in prison from tuberculosis.

Bernard Kettlewell. During the Industrial Revolution when steam power was fueled principally by coal, buildings, trees, and other structures darkened. Prior to this technological change in society, most peppered moths (*Biston betularia*) were light colored in parts of the UK; but over the next century, the pigmented variety became more numerous and outnumbered the typical pale moths. Although natural selection was a cardinal part of Darwin's theory of evolution, on no occasion did he observe an unambiguous example before he departed his biological life. It became apparent by 1886 that the Industrial Revolution affected the ratio of light—to dark-colored peppered moths. The melanic moth was thought to receive protection against its predators while resting on a darkened background compared to the less dark background, which offered camouflage to the remaining light-colored moths. Being an apparent classic example of Darwin's evolution by natural selection [24], the tale of the peppered moth established a place in biology textbooks throughout the world as well in science museums. However, it later became one of science's utmost humiliations in the eyes of the public. In 1953 the Oxford University (Oxford) *(http://www.ox.ac.uk/)* School of Ecological Genetics under the leadership of Edward Brisco "Henry" Ford (1901-1988), the founder of ecological genetics, employed Henry Bernard Davis Kettlewell (1907-1979), the British geneticist, to test the idea experimentally. For consecutive summers, Kettlewell camped out in the neighborhood of industrialized Birmingham or in rural Dorset. He let light and dark moths go free and counted the living ones that he could trap back. He hypothesized that in polluted regions dark moths would survive more readily than in unspoiled areas and that light-colored moths would do better in unspoiled Dorset than in Birmingham. If this hypothesis was correct, the selective interaction between predators and prey could be explained, and indeed this is what Kettlewell found. He observed that in Birmingham twice as many dark months survived than light ones and three times as many light months survived compared to dark ones in Dorset. This was a fantastic new discovery for Darwin supporters, and the observations were acclaimed as "Darwin's missing evidence" despite notable flaws in the experimental design creationists needed to find fault in the research. A provocative book entitled Of Moths and Men was penned by the American journalist Judith Hooper (1949-) [50]. The text unambiguously condemned the Kettlewell studies aimed at authenticating evolution experimentally as fraudulent. Hooper expressed concerns about the scientific techniques and accused Kettlewell of not only being guilty of fraud but also of making lackadaisical mistakes. She criticized the methods of Kettlewell, which included gluing moths to portions of tree bark where they would not be anticipated to rest in nature. By providing abundant moths for birds to eat, nobody would expect feeding to occur at the tree bark. Hence, increasing the number of captured moths falsely biased the project subconsciously in support of an anticipated result. The biology historian David W. Rudge at the Western Michigan University (WMU) *(http:// www.wmich.edu/)* also reviewed the evidence that Hooper used to come to her conclusion, and in his opinion Hooper's historical research was mediocre and her text displayed a deep-seated confusion about the characteristics of science [95]. Fault has also been found in Hooper's book by numerous other critics. According to Jerry Allen Coyne (1949-), an American professor of biology, Hooper had a "flimsy conspiracy theory of ambitious scientists who ignore the truth for the sake of fame and recognition" [21]. In a review of Hooper's book, Bruce S. Grant, a member of the department of biology at the College of William and Mary, pointed out the scientific assessment of the evidence of natural selection in the peppered moth much of which caused doubt by the authors persistent mistrust in the validity of the research [45]. Bryan Campbell Clarke (1932-), a British geneticist and professor emeritus at the University of Nottingham *(http://www.nottingham. ac.uk/)*, who worked with Kettlewell in Oxford, regarded Hooper's book "as a treasure of insinuations worthy of an unscrupulous newspaper" [18]. Michael Eugene Nicolas Majerus (1954-2009), another authority on peppered moth evolution at the University of Cambridge (Cambridge) *(http://www.cam.ac.uk/)*, maintained that The Peppered Moth was "enlisted with errors, misrepresentations, misinterpretations, and falsifications, and published research specifically proving that some of Hooper's assertions were false" [61]. The characteristic moth closely resembled lichen in color, and when on trees, their camouflage concealed them from predators; but in industrialized areas, the lichen fell off, and coloring was no longer beneficial. Aside from Hooper, others have also confronted the experiments of Kettlewell, but most comments in Hooper's book come straight from the mouths or pens of others. Other exposés have been documented in New Scientist (1987), Whole Earth (1999), and other magazines. In 1969 Ted Sargent at the University of Massachusetts (UMass) at Amherst *(https://www.umass.edu/)* drew attention to the fact moths do not settle on a color that resembles their own color [97, 98, 99]. Kettlewell was not masked to what he was quantifying. It was he alone who determined how to classify a moth while simultaneously being familiar with the anticipated results. He also failed to establish that birds ate the most noticeable moths selectively. Peppered moths do not unambiguously relax on the trunks of trees. Probably because the conclusions of his experiments were what the journal editor and the reviewers would have liked to accept as true, the defects in Kettlewell's research slipped through the peer review process into the literature. It is also noteworthy that his research on this subject became accepted as the gospel truth even though nobody attempted to replicate the research. Notwithstanding the shortcomings of Kettlewell's research and the unverified accusations of scientific misconduct in his research damage to the reputation of not only Kettlewell but also science in general was generated. Moreover, it provided missiles for individuals who sincerely had reservations about Darwin's theory of evolution. The story of the peppered moth precipitated controversy between evolutionists and creationists, and the research damaged whatever scientific relevance the experiments had.

John William Heslop Harrison. In the 1940s Professor John William Heslop Harrison (1920-1998), a distinguished British botanist and member of the Royal Society, was at the University of Newcastle *(http://www.ncl.ac.uk/)* before it became a university in 1963. He hypothesized that species of plants off the west coast of Scotland survived the last ice age. To advance his favorite theory that lacked the support of credible scientific evidence, Harrison deceitfully established aliens' edges on the Hebridean Isle of Rum. He firmly believed in his flawed theory despite evidence that ice sheets during the ice age extended well south of the mainland of Scotland. To support his controversial theory, he claimed that he had proof with regard to plants and grasses that grew on the island of Rum. For years he astounded naturalists with a major theoretical implication about plants never before identified in the UK. Some suspected fraud, but the professional botanists of the day preferred not to provoke their colleague with such an allegation. Harrison contended that unlike the mainland portions of Scotland, the Hebrides avoided the ice age. He based this hypothesis on his finding on Rum of a dozen or so species of lichens and other plants that were practically nonexistent on the mainland where such plants were presumably destroyed during the ice age.

In the late 1940s John Raven (1914-1980), a British classics scholar who was a passionate amateur botanist, took on the task of evaluating the Harrison data by boldly questioning whether these grasses were indigenous to the island or planted there after being transported from somewhere else. Overwhelming evidence against Harrison existed. Raven found just under a dozen specimens of the sedge Carex bicolor enlarging tidily in a cleaned upsite but not in another part of the isle. Two "surviving" species were discovered in a solitary bundle and yet again not elsewhere on the island. Furthermore, the earth from one of Harrison's specimens differed from that where it allegedly grew. Over time Raven unraveled a tale about rival scientists and exposed the fraudulent research of Harrison based on circumstantial evidence after obsessively investigating his research. After Raven reported several of his findings in Nature [87] without making an accusation against Harrison, several species of plant were delisted in the botany registers. Harrison was offended by Raven's comments and accused him of disloyalty and conspiracy in a letter, yet Harrison preserved his professorship and much of his status as an elder statesman in biology. Raven documented his findings in a blistering report that remained sealed in the library of the King's College in Cambridge University *(www.cam.ac.uk)* [88]. The latter was presumably done so that the dirty linen that was exposed would not become public knowledge. Raven's extensive report concluded: "That the Professor is deliberately indulging in the most culpable dishonesty in order to secure for himself an immediate reputation and an immortal place in the annals of British botany." Raven's following words mirrored the class facet of the tragedy: "The Professor is clearly a man of very humble origin: he is in fact, I have been told credibly enough, the son of a miner. He is obviously able and almost equally obviously ambitious." The Raven report was effectively buried because Harrison had threatened legal action on several occasions to private accusations. The shame was referred to far and wide as a Rum affair. Raven built a robust argument implicating the participation of Harrison, faking botanical data, including his finding of the large blue and water beetles. After the death of Harrison, Karl Sabbagh exposed the five-decade-old fraud in a book entitled A Rum Affair [96], making a compelling case of scientific fraud based on circumstantial evidence.

Margaret Mead. When an anthropologist travels to a distant land about which little is known, the person can return to civilization with tales that become accepted as fact without confirmation by others. Such a situation arose in 1928 when Margaret Mead (1901-1978) at the young age of 26 years published a book entitled Coming of Age in Samoa [64]. The book described a captivating narrative about a tranquil standard of living on the tropical Samoan islands when they were controlled by New Zealand, and it supported the belief in the plasticity of human characteristics. Mead described a primitive society in which love was freely available unburdened by the pace of modern Western societies. The portrayal of the Samoan's sexual practices by Mead was the most renowned portion of her celebrated book, and it unquestionably promoted the sales. The Samoans expressed an extremely tolerant stance toward sex and putting up with free premarital lovemaking among adolescents. Marriage lacked claims for faithfulness, and envy was extremely rare. This publication immediately made Mead a household name, and this book rocketed her into becoming one of the most renowned global figures in anthropology and an idol in the eyes of the public, which has always found books on sex appealing *(http://en.wikopedia.org/wik/Margaret_Mead)*. As a well-respected academic cultural anthropologist, she commonly communicated with the public as both a speaker and a writer during the 1960s and 1970s, but she was controversial. In a monograph published in 1969, Freeman and others drew attention to some of Mead's mistakes in the 1950s and 1960s. Her descriptions of the mind-set of primitive traditional societies in the South Pacific and Southeast Asian cultures toward sex helped fuel the America sexual revolution in the 1960s. Her work was supported by the paintings in the early 1880s in Tahiti by the post-impressionist artist Eugène Henri Paul Gauguin (1848-1903).

Mead became a well-respected scientist as reflected in her election to the presidency of the American Association for the Advancement of Science (AAAS). Virtually alone Mead convinced Congress that anthropology was a topic worthy of taxpayer financial support, and she assured careers in anthropology for many of her associates in the same discipline. She advocated broadened sexual mores within a frameworkof time-honored Western spiritual life. On behalf of Jimmy Carter (1924-), the thirty-ninth president of the USA, Andrew Jackson Young (1932-), United Nations ambassador for the USA, delivered the Presidential Medal of Freedom to Mary Catherine Bateson, the daughter of Mead, and her third husband as a posthumous honor to Mead for her notable contributions to anthropology. After her death in 1978 serious doubt was cast on the validity of Mead's work when books by the Australian anthropologist John Derek Freeman (1916-2001) hit the bookstands [32, 33]. After reading the Coming of Age in Samoa, Freeman became fascinated by Samoa and regularly visited the islands. He not only studied the customs of the people and their language but even participated in the local politics. Freeman achieved an excellent reputation in the eyes of this peers and was well respected by the Somoans, but he also drew attention to inadequacies in Coming of Age in Samoa [41, 57, 81, 109]. In February 1983 Freeman ignited a controversy with a shocking criticism of the validity of Mead's studies on Samoa. He believed that Mead may have misapprehended Samoan life to some extent because she was not permitted to attend meetings of the all-male community council. Freeman's research not only brought the credibility of a world-renowned matriarch in cultural anthropology into question but also drew attention to the need for major changes in the manner in which anthropological data is collected and analyzed in projects that involve the accumulation of data in field trips. Freeman [32] confronted the truthfulness of Mead's research, and he often debated relative contributions due to nature or nurture in determining physical and behavioral traits. After reading a copy of Freeman's book prior to its publication, Ernst Walter Mayr (1904-2005), a former professor of zoology at Harvard University (Harvard) (http://www.harvard.edu/), was so convinced by Freeman's facts critical of Mead that he encouraged Harvard Press to publish the Freeman book, which allegedly lacked obvious factual errors, but Freeman seems to have an axe to grind for Mead. Was Freeman's book written to attract the attention of the public about a scandal involving a well-known public figure so that it could potentially become a best seller and generate significant income for both the author and the publisher?

Two decades prior to Freeman's controversial book, Lowell D. Holmes (1925-2010), an anthropologist at Wichita State University (WSU) *(http://www.wichita.edu/)* Kansas, spent several years in Samoa restudying the work of Mead. He wrote a doctoral thesis on the subject that remained unpublished until after the storm brought on by Freeman [49]. Holmes found numerous errors in Mead's work that are cited by Freeman. Some of Freeman's peers were convinced by the devastation of Mead's classic best seller. Holmes's comments are especially noteworthy because he spent a number of years in Samoa and was the first scientist to meticulously study the work of Mead after her. Initially Holmes feared being a critic of Mead would jeopardize his ability to get grant support. According to Freeman, Mead perceived Samoa as a functional society but not as it actually existed. He pointed out that Mead lived with an American family in Samoa for nine months and was not eloquent in the Samoan language. In the opinion of Freeman, female virgins are substantially appreciated in Samoa, and he thinks that Mead was hoodwinked by the adolescents that she questioned. Being uncomfortable about answering her questions connected to sex, male adolescents may have told her concocted amusing tales. Freeman estimated that forcible rape in Samoa in 1966 was double that in the USA and 20 times that in the UK. Particularly common in Samoa was an opportune type of rape referred to as moetotolo that occurs at night when young men attempt to destroy the virginity of sleeping girls. Freeman completed the research for his book prior to Mead's death and discussed his concerns about her work with her. As a courtesy to Mead, the publication of Freeman's whistle-blowing report was delayed until after her death so that her reputation would not be ruined while she was alive. However, despite the claims of Freeman, some question the validity of his allegations and wonder whether the Mead evaluation typifies the case of he says she says.

Harold M. Bates. Prior to the scandals of scientific misconduct in the 1970s and 1980s, instances of scientific dishonesty were considered an internal issue and an appropriate institutional committee would often evaluate the allegations and deal severely with a guilty offender. In the late 1950s a young biochemistry PhD graduate student Harold M. Bates working in laboratory of Melvin V. Simpson at Yale University (Yale) *(http://www.yale.edu/)*, rapidly made noteworthy progress in the cell-free synthesis of cytochrome c, an important protein involved in cellular energy releasing responses. This was the first time that a single protein had been synthesized outside of a cell, and the triumphant experiments were documented in two papers authored by Bates, Simpson, and other coworkers [5, 6]. This remarkable study propelled Bates to a postdoctoral fellowship with Fritz Albert Lipmann (1899-1986), the codiscoverer of coenzyme A and the recipient of the 1953 Nobel Prize in Physiology or Medicine *(http://www.nobelprize.org/nobel_prizes/medicine/laureates/1953/lipmann-lecture.html)* with Sir Hans Adolf Krebs (1900-1981) *(http://www.nobelprize.org/nobel_prizes/medicine/laureates/1953/krebs-lecture.html)*.

Meanwhile, Simpson went to the UK on a sabbatical leave for several months. On returning he started additional experiments to extend the successful work with cytochrome c, but to his dismay his efforts were fruitless. Researchers in the Lipmann laboratory also had trouble duplicating the research. Bates was requested to return to Yale and instructed to repeat the cyctochrome c experiment. After diligently working day and night under supervision, he too could not replicate his findings. Unquestionable fraudulent research was suspected, and Bates was instructed to give up research in general. One of his papers was coauthored with Lipmann [7]. After the exposure of his research misconduct, these papers were allegedly retracted, but they remain detectable in the USA National Library of Medicine (MEDLINE) and are still present within the journals without any comment about the retraction. Fortunately none of them have been cited. After the discovery of Bates's misconduct, a review of his past record uncovered the fact that his undergraduate college in Massachusetts had no record of him graduating. Unlike today the facts about the student's research misconduct remained within the affected laboratory. If this had occurred today the media would probably have jumped on the opportunity to publicize the scientific misconduct now that science and society are much more blended together.

References

1. Altman LK. Revisionist history sees Pasteur as liar who stole rival's ideas. The New York Times. 1996 May 16.
2. Anderson C, Pasteur notebooks reveal deception. Science, 1993. 259(5098): 1117.
3. Aronson LR, The case of the midwife toad. Behavior Genetics, 1975. 5(2): 115-125.
4. Babbage C, Reflections on the Decline of Science in England, and on Some of Its Causes. 1830, London: Fellowes.
5. Bates HM and Simpson MV, The net synthesis of cytochrome c in calf-heart mitochondria. Biochimica et Biophysica Acta, 1959. 32: 597-599.
6. Bates HM, Craddock VM, and Simpson MV, The biosynthesis of cytochrome c in cell-free systems. I. The incorporation of labeled amino acids into cytochrome c by rat liver mitochondria. Journal of Biological Chemistry, 1960. 235: 140-148.
7. Bates HM and Lipmann F, Identification of gamma-glutamylcysteinyl ribonucleic acid as intermediary in glutathione synthesis. Journal of Biological Chemistry, 1960. 235: PC22-23.
8. Beddall BG, Wallace, Darwin, and the theory of natural selection. Journal of the History of Biology, 1968. 1(2): 261-323.
9. Broad WJ, Fraud and the structure of science. Is fraud a trivial excrescence on the process of science or do the recent cases have deeper roots Science, 1981. 212(4491): 137-141.
10. Bryce RM, Dr. Cook-Mt. McKinley controversy closed. DIO. The International Journal of Scientific History, 1997. 7(No.s 2-3): 38-99.
11. Burt C, Experimenal tests of general intelligence. British Journal of Psychology, 1909. 2(1-2): 94-177.
12. Burt C, Ability and income. British Journal of Psychology, 1943. 13(2): 83-98.
13. Burt C, Evidence for the concept of intelligence. British Journal of Experimental Psychology, 1955. 25(3): 158-177.
14. Burt C and Howard M, The mutifactorial theory of inheritance and its application to intellegence. British Journal of Statistical Psychology, 1956. 9(2): 95-131.
15. Burt C, The genetic determination of intelligence: a study of monozygotic twins reared together and apart. British Journal of Psychology, 1966. 57(1-2): 137-153.
16. Capstick PH, Warrior: The Legend of Colonel Richard Meinetzhagen. 1998, New York: St. Martin's Press.
17. Charrow RP, Bench notes Judgments. Scientfic misconduct: Sanctions in research of procedures. The Journal of NIH Research, 1990. 2: 91-92.
18. Clarke B, Heredity—The art of innuendo. Heredity, 2003. 90(4): 279-280.

19. Conway J, The inheritance of intelligence and its social implications. British Journal of Statistical Psychology, 1958. 11(2): 171-190.
20. Conway J, Class differences in general intelligence. II A reply to Dr.Halsey British Journal of Statistical Psychology, 1959. 12(1): 5-14.
21. Coyne J, Evolution under pressure. Review of Judith Hooper: "Of Moths and Men: Intrigue, Tragedy and the Peppered Moth". Nature, 2002. 418(6893): 19-20.
22. Crews FC, The Memory Wars: Freud's Legacy in Dispute. 1995, New York: New York Review Books.
23. Darwin C, On the Origin of Species by Natural Selection. 1859, London: John Murray.
24. Darwin C, On the Origin of Species by Means of Natural Selection, or the Preservation of Favoured Races in the Struggle for Life. 1859.
25. Darwin C, The Descent of Man, and Selection in Relation to Sex. 1871, London: John Murray.
26. Darwin C, The Expression of the Emotions in Man and Animals. 1872, London: John Murray.
27. Fisher RA, Has Mendels's work been rediscovered? Annals of Science, 1936. 1: 115-137.
28. Fletcher H, My work with Millikan on the oil-drop experiment. Physics Today, 1982. 35(6): 43-47.
29. Fletcher R, The doubtful case of Cyril Burt. Social Policy and Administration, 1987. 21(1): 40-57.
30. Fletcher R, Science, Ideology, and the Media: The Cyril Burt Scandal. 1991, New Brunswick NJ: Transaction Publishers.
31. Franklin AD, Millikan's published and unpublished data on ol droplets. Historical Studies in the Physical Sciences, 1981. 11(2): 185-201.
32. Freeman D, Margaret Mead and Somoa: The Making and Unmaking of an Anthropological Myth. 1983, Cambridge, MA: Harvard University Press
33. Freeman D, The Fateful Hoaxing of Margaret Mead. A Historical Analysis of her Samoan Research. 1998, Boulder, CO Westview (Perseus). pp. 291.
34. Freud S, ed. Sigmund Freud, Aus den Anfägen der Psych-analyse: Briefe an Wilhelm Fleiss (Marie Bonoparte, Anna Freud, Ernst Kris editors). 1950, S. Fleischer: London
35. Freud S, The Origin of Psycho-Analysis: Letters to Wilhelm Filess, Drafts and Notes 1887-1902. (Marie Bonaparte, Anna Freud, Ernest Kris, editors). 1954.
36. Garfield B, The Meinertzhagen Mystery: The Life and Legend of a Collosal Fraud. 2007, Washington DC: Potomac Books. 353.
37. Geison GL, The Private Science of Louis Pasteur. 1995, Princeton, NJ: Princeton University Press. 378 pp.
38. Gillie O. Crucial data was faked by eminent psychologist. The Sunday Times. 1976 October 24.

39. Gingerich O, Was Ptolemy a fraud?. Quarterly Journal of the Royal Astronomical Society, 1980. 21: 253-266.
40. Gingrich O, Review. On Ptolemy as the greatest astronomer of antiquity. Science, 1976. 193(4252): 476-477.
41. Goodenough W, Margaret Mead and cultural anthropology. Science, 1983. 220(4600): 906-908.
42. Goodstein D, Scientific fraud. The American Scholar, 1991. 60(4): 505-515.
43. Goodstein D, In defense of Robert Andrews Milikan. Engineering and Science, 2000. 4: 30-38.
44. Gould SJ, The Mismeasure of Man. 1981, New York: W. W. Norton and Company. pp.352.
45. Grant BS, Sour grapes of wrath. Science, 2002. 297(5583): 940-941.
46. Grünbaum A, The Foundations of Psychoanalysis: A Philosophical Critique. 1984, Berkeley: University of California Press.
47. Hearnshaw LS, Cyril Burt: Psychologist. 1979, New York: Cornell University Press.
48. Henderson B, True North: Peary, Cook, and the Race to the Pole. 2005, New York: W.W. Norton and Company.
49. Holmes LD, Quest for the Real Samoa. The Mead/ Freeman Contoversy and Beyond. 1986: Bergin and Garvey.
50. Hooper J, Of Moths and Men: An Evolutionary Tale. The Untold Story of Science and the Peppered Moth. 2002, New York: W.W. Norton and company. pp. 397.
51. Jensen AR, How much can we boost I.Q. and scholastic achievement. Harvard Education Review, 1969. 39(1): 1-123.
52. Joynson RB, The Burt Affair. 1989, New York: Routledge.
53. Kamin LJ, The Science and Politics of IQ. 1974, Potomac MD: Lawrence Erlbaum Associates.
54. Kamin LJ, Heredity, intelligence, politics and society presented to the Thirteenth International Congress of Genetics, 1973, reprinted in The IQ Controversy:Critical Readings, Ned Block and Gerald Dworkin (editors). 1997, London: Quarter Boks.
55. Knox AG, Richard Meinertzhagen—a case of fraud examined. Ibis, 1993. 135(3): 320-325.
56. Koestler A, The case of the midwife toad. 1971, London: Hutchinson, Random House.
57. Levy RI, The attack on Mead. Science, 1983. 220(4599): 829-832.
58. Lewontin RC, Race and intelligence. Bulletin of the Atomic Scientists: Science and Public Affairs, 1970. 26(3): 2-8.
59. Mackintosh NJ, Cyril Burt: Fraud or Framed. 1996, Oxford: Oxford University Press.
60. Macmillian M, Freud Evaluated: The Completed Arc. 1996, Cambridge, MA: MIT Press. 786 pp.

61. Majerus MEN, Industrial melanism in the peppered moth, "Biston betularia": An excellent teaching example of Darwinian evolution in action Evolution Education and Outreach, 2009. 2(1): 63-74.
62. Marshall E, Peary's North Pole claim reexamined. Science, 1989. 243(4895): 1131-1132.
63. Masson JM, The Complete Letters of Sigmund Freud to Wilhelm Fleiss 1887-1904. Vol. 492. 1985, Cambridge, MA: The Belknap Press of Harvard University Press.
64. Mead M, Coming of Age in Samoa. A Psychological Study of Primitive Youth for Western Society. 1928: William Morrow and Company. pp. 297.
65. Meinertzhagen R, Nicholl's Birds of Egypt. 1930, London: H. Rees Ltd.
66. Meinertzhagen R, Kenya Diary. 1957, Edinburgh: Oliver and Boyd.
67. Meinertzhagen R, Middle East Diary. 1959, Edinburgh: Oliver and Boyd.
68. Meinertzhagen R, Army Diary. 1960, Edinburgh: Oliver and Boyd.
69. Meinertzhagen R, Diary of a Black Sheep. 1964, Edinburgh: Oliver and Boyd.
70. Meinertzhagen R, Birds of Arabia. 1980, London: Henry Sortheran.
71. Mendel JG, Versuche über Pflanzenhybriden. Verhandlungen des naturforschenden Vereines in Brünn, 1866. 4: 3-47.
72. Miller DJ, Hersen M, and Editors, eds. Research Fraud in the Behavioral and Biomedical Sciences. 1992, John Wiley and Sons: New York.
73. Millikan RA, On the elementay electrical charge and the Avogadro constant. Physical Review, 1913. 2(2): 109-143.
74. Millikan RA, Gale HG, and Davis IC, Exercises in Laboratory Physics. 1925, Oxford, United Kingdom: Ginn and company.
75. Millikan RA, The Autobiography of Robert A. Millikan. 1950, New York: Prentice Hall. 342.
76. Mould RF, A Century of X-Rays and Radioactivity in Medicine with emaphass on photographic records of the early years. 1993, Boca Raton FL: CRC Press.
77. Newton I, Philosophiæ Naturalis Principia Mathematica. 1687, London: Apud Guil and Joh. Innys.
78. Newton RR, The Crime of Claudius Ptolemy. 1977: Johns Hopkins Press.
79. Noble GK, Kammerer's alytes. Nature, 1926. 118(2962): 209.
80. Olsen James L J, A neglected literature. Science, 1976. 194(4268): 894.
81. Orans M, Mead misrepresented. Science, 1999. 283(5408): 1649-1650.
82. Pennisi E, Haeckel's embryos: fraud rediscovered. Science, 1997. 277(5331): 1435.
83. Pennisi E, History of science. The case of the midwife toad: fraud or epigenetics? Science, 2009. 325(5945): 1194-1195.
84. Prodger P, Photography and The Expression of the Emotions, in Charles Darwin The Expression of the Emotions in Man and Animals, 3rd edition. New York: Oxford University Press.

85. Rasmussen PC and King BF, The rediscovery of the forest owlet Athene (Heteroglaux) blewitti. Forktail, 1998. 14: 53-55.
86. Rasmussen. P.C and Ishtiaq F, Vocalizations and behavior of forest spotted owlet Athene blewitti. Forktail, 1999. 15: 61-66.
87. Raven JE, Alien plant introductions on the Isle of Rhum. Nature, 1949. 163(4133): 104-105.
88. Raven JE, John Raven's report on his visit to the Hebrides, 1948 (edited by C.D. Preston) Watsonia, 2004. 25: 17-44.
89. Rawlins D, Peary at the North Pole: Fact or Fiction?. 1973, Washington: Luce.
90. Richardson MK, Hanken J, Gooneratne ML, et al., There is no highly conserved embryonic stage in the vertebrates: implications for current theories of evolution and development. Anatomy & Embryology, 1997. 196(2): 91-106.
91. Richardson MK, Hanken J, Selwood L, et al., Haeckel, embryos, and evolution. Science, 1998. 280(5366): 983.
92. Richardson MK and Keuck G, A question of intent: when is a 'schematic' illustration a fraud? Nature, 2001. 410(6825): 144.
93. Roberts D. Great discovery hoaxes. Many of history's explorers actually explored nothing more than the public's gullibility. Travel and Leisure Magazine. 1984:106-110.
94. Ruddock AA, The reputation of Sebastian Cabot. Historical Research, 2007. 47(115): 95-99.
95. Rudge DW, Did Kettlewell commit fraud? Re-examining the evidence. Public Understanding of Science, 2005. 14(3): 249-268.
96. Sabbagh K, A Rum Affair A True Story of Botanical Fraud. 2001: Da Capo Press.
97. Sargent TD, Behavioral adaptations of crytic moths. II. Experimtal studies on bark-like species. Journal of the New York Entomological Society, 1969. 77(2): 75-79.
98. Sargent TD, Bchavioral adaptions of cryptic moths. III: Resting attitudes of two bark-like species, Melanophia Canafaria and Catocala ultronia. Animal Behavior, 1969. 17(4): 670-672.
99. Sargent TD and Keiper RR, Behavioral adaptations of cryptic moths. I. Preliminary studies on bark-like species. Journal of the Lepidopterists' Society, 1969. 23(1): 1-9.
100. Segerstråle U, Science and Engineering Ethics, 1995. 1.
101. Shearer J, Despatches from the Western Front. September 30, 1916 British Medical Journal, 1916. 2(2909).
102. Sulloway FJ, Freud, Biologist of the Mind: Beyond the Psychoanalytic Legend. 1979, London: Burnett Books. 612 pp.
103. Sulloway FJ, Reassessing Freud's Case Histories: The Social Construction of Psychoanalysis. 1991: Burnet Books.

104. Svardal H, Can epigenetics solve the case of the midwife toad?—a comment on Vargas. Journal of Experimental Zoology, 2010. Part B. Molecular and Developmental Evolution. 314(8): 625-628.
105. Tucker WH, Re-reconsidering Burt: Beyond a reasonable doubt. Journal of the History of the Behavioral Sciences, 1997. 33(2): 145-162.
106. Vargas AO, Did Paul Kammerer discover epigenetic inheritance? A modern look at the controversial midwife toad experimints. Journal of Experimental Zoology Part B. Molecular and Developmental Evolution, 2009. 312B(7): 667-678.
107. Waterland RA, Travisano M, and Tahiliani KG, Diet-induced hypermethylation at agouti viable yellow is not inherited transgenerationally through the female. FASEB Journal. 21(12): 3380-3385.
108. Westfall RS, Never at Rest: A Biography of Isaac Newton. 1983, Cambridge: Cambridge University Press.
109. Young RE and Juan S, Freeman's Margaret Mead myth: The Ideological virginity of anthroplogists. Australian and New Zealand Journal of Sociology, 1985. 21: 64-82.

CHAPTER 15

Modern American Examples of Scientific Misconduct

Fraudulent Research Prior to the Creation of the Office of Research Integrity

In the 1970s and 1980s several major examples of fraudulent research were identified in biomedical research. The perpetuators included William L. Summerlin, John Rowland Darsee, Elias A.K. Alsabti, Mark J. Straus, John C. Long, Vijay Soman, Stephen C. Breuning, Robert Slutsky, Eric Poehlman and Mark Spector. These cases stirred up hornet nests at prominent institutions in the USA when stories about them came to the attention of the public. Since then research misconduct has become well-recognized and numerous cases are recognized annually.

William L. Summerlin. Over many years numerous researchers have contributed to the science of transplantation immunology [21] and this has led to the successful transplantation of the liver, kidney, heart, lungs, pancreas, intestine and bone marrow. These achievements saved countless lives and for research contributions on the subject Sir Frank Macfarlane Burnet (1899-1985) (*http://www.nobelprize.org/nobel_prizes/medicine/laureates/1960/burnet-lecture.html#*) and Sir Peter Brian Medawar (1915-1987) (*http://www.nobelprize.org/nobel_prizes/medicine/laureates/1960/medawar-lecture.html*) shared the Nobel Prize Nobel in physiology or medicine in 1960.

In 1974 in an extension of this work Dr. William L. Summerlin (1938—) a young physician in training performed research at the highly regarded Sloan-Kettering institute for cancer research in New York under the mentorship of Robert A. Good (1922-2003). Good was widely regarded as the founder of modern immunology and he had carried out the first successful human bone marrow transplant for a disease other than cancer. Summerlin coauthored publications with his mentor [197, 199]. Summerlin firmly believed that he had made an astonishing new advance in allogeneic transplantation by transplanting tissue from unrelated mice successfully without immunosuppression by retaining the tissue initially in culture for four to six weeks [196, 198]. As a way of determining that skin grafts can be carried out without immunosuppression pieces of skin were transferred from black mice to white mice. Summerlin claimed that such patchwork grafts could be successful in mice if the donor tissue was first incubated in an appropriate solution, but his research came under attack because the work could not be replicated by others. Unsuccessful experimental disappointments did not deter Summerlin's belief that he was on to something important. One day in 1974 when visitors were expected in the laboratory and triumphant results needed to be demonstrated Summerlin wanted to demonstrate the success of his research and this was done by painting a black square on the skin of a white mice with a felt-tipped pen to resemble a successful skin graft [31]. The deception of the painted mice was readily detected when a laboratory assistant removed the black ink with a ball of cotton soaked in a tad of ethanol. The scandal created by scientific misconduct received much publicity. This incident made front page news for Summerlin, not as the discoverer of something new, but as a person responsible for scientific fraud. This fraudulent research shocked the world when the public's attention was drawn to it by Jane Brody in the *New York Times* [27]. Brody felt that the patchwork mouse reflected "dangerous trends in current efforts to gain scientific acclaim and funds for research" and Good was accused of "manipulating national attention and attracting an enormous amount of money for the institute" [27]. An entire book was published on *The Patchwork Mouse* by Hixson [60]. On the day that Summerlin admitted fakery he was relieved of his post and an investigation into all of his work began. Summerlin was undoubtedly influenced by the image of his productive mentor who had 341 publications (an average of 68 per year) during the 5 years prior to Summerlin's resignation. Shortly after the fraud was exposed Good's role as President of the Memorial Sloan-Kettering cancer center ended and so did most transplantation without immunosuppression. The basic concept in Summerlin's research led to the possibility of performing corneal grafts with tissue that was first stored as organ cultures. Such grafts between species could survive longer than usual perhaps because of reduced immunogenicity [45]. In contrast to the slow pace at which Harvard University (Harvard) (*http://www.harvard.edu/*) responded to the fraudulent research by Darsee the institutional responses to Summerlin was more prompt [25]. As a result of an unbearable clinical and experimental work load and his research misconduct Summerlin suffered from mental and physical exhaustion.

John Rowland Darsee. John Rowland Darsee had impeccable credentials and was predicted to have a bright future. He attended Notre Dame University (ND) (*http://www.nd.edu/*) as an undergraduate and then went to the medical school of

Indiana University (IU) (*http://www.indiana.edu/*) until 1974. He then moved to Emory University (Emory) (*http://www.emory.edu/*) in Atlanta to attend a variety of residencies including a stint as the chief medical resident at Grady Memorial hospital. Among the people Darsee worked for was Donald O. Nutter, executive associate dean of the school of medicine. Nutter was most impressed with him and considered him the most outstanding person with whom he had worked with during his tenure at Emory. In July 1979, Darsee joined the cardiac research laboratory of Eugene Braunwald (1929—) at Harvard University *(Harvard)(http://www.harvard.edu/)*. Braunwald was an extremely productive eminent cardiologist and a member of the *National Academy of Sciences* (*NAS*). Darsee was dedicated to a career as a clinician-scientist. He chose performing research in a prestigious internationally respected laboratory over money. He was paid considerably less than what he could have received in private practice and he worked steadfastly to move his career forward spending most of his time in the laboratory. In 1980 he received a small annual stipend of $22, 000 at Harvard from a fellowship from the National Institutes of Health (NIH). During relatively short time over 2 years in the superb environment at Harvard Darsee thrived and published almost 100 papers and abstracts, and six were coauthored with Braunwald when he was his mentor [35-37] but most of them contained fabricated data. Braunwald regarded Darsee highly and considered setting him up with a separate laboratory at Beth Israel hospital.

With research funding from the NIH four research groups in heart disease agreed to participate in a prospective collaborative multicenter clinical trial designed to evaluate in a masked fashion whether ibuprofen and verapamil could protect the heart in dogs from damage caused by ischemia following coronary artery occlusion. The participating researchers were from Johns Hopkins University (JHU) *(http://www.jhu.edu/)*, the Braunwald laboratory, the Robert Jenning laboratory at Duke University (Duke)*(http://duke.edu/)*, and the Joseph Greenfield laboratory at the Durham veterans administration (VA) center. One of the involved researchers in the Braunwald laboratory was Darsee. All data in the multicenter study were to be sent to all participating laboratories after the experimental code was broken. After his fraudulent research was unearthed it was found that Darsee had not even examined the hearts that should have been studied. The preserved hearts from earlier experiments were still in the freezer and had not been injected with radioactive isotopes as required by the experimental protocol. Only Darsee got positive results with the test drugs. Darsee shook the academic community by surprise in 1981 when it became apparent that he had faked data while performing research in the Braunwald laboratory. It was difficult to imagine how these shocking events could take place in well-respected laboratories of a leading biomedical scientist at a premier research institution in the USA. A comprehensive investigation followed the discovery of this fraudulent research. A record of repetitive fraudulent behavior by Darsee was esblished. He fabricated data not only while at Harvard, but also when he attended Emory, and even when he was an undergraduate student at ND. In common with many other researchers who fabricate data, Darsee accumulated numerous publications (47) over relatively few years. In keeping with publications of other junior investigators, the papers were coauthored typically by established scientists who enhanced the credibility of his research.

After the discovery of Darsee's fabricated research he was found to have an outstanding track record as an ambitious phony. On May 18 1981, when Darsee was 32 years old, two young research fellows and a technician, who worked daily with him in the Braunwald laboratory and were distrustful of his enormous productivity, surprisingly watched the brazen Darsee gather hemodynamic data within the laboratory complex from a dog in the multicenter study and mark the chart day 1, day 2, day 3 and so on to give the impression that the recordings had been made over a period of a several days. The research fellows reported this fakery to Robert A. Kloner, MD, PhD chief of the Braunwald laboratory at the Peter Bent Brigham and women's hospital in Boston. Kloner looked into the matter and found irregularities in the records and requested the raw data. After being challenged soon thereafter Darsee confessed to this one case of scientific misconduct, but stated that he had only done it because he had discarded the authentic data. Four days later Kloner notified Braumwald about the shocking discovery. As one would expect this fraud was a shock to the 51 year old Braunwald, Darsee's mentor and main coauthor.

The Darsee case raised the issue of how mentors, coauthors and institutions should react to allegations of scientific misconduct. Since scientific falsification and fabrication was not an issue that Harvard had previously encountered institutional policies did not exist and Braunwald and Harvard officials made several serious mistakes after detecting Darsee's fraud. They considered it a single stupid act and wanted to be fair to Darsee and to ensure due process. Harvard did not notify all relevant parties. But this was not a trivial matter it involved an extremely large expensive multicenter research project of considerable importance to public health with many co-investigators. Harvard terminated Darsee's fellowship on June 30 1982 without providing the NIH with the reason, but from that date Braunwald continued to employ Darsee while paying him from an unrestricted fund code even though he was no longer a member of the Harvard staff. Moreover Darsee was retained on the multiinstitutional project because he had already played an important role in this soon to be completed study, but his fraudulent research had already invalidated the research and Harvard was denied a request to do the study over. Also, Darsee continued to publish papers as if nothing had happened. An appraisal disclosed that the original data from many of Darsee's experiments were missing without explanation. A faculty position offered to Darsee for the following year was promptly abandoned.

In May and June of 1981 after recovering from the shock of Darsee's admitted fabrication Braunwald and Kloner were under the impression that they were dealing with a solitary weird act by an immature researcher who had until now functioned superbly.

Towards the end of September 1981 data from the study was sent to the NIH for analysis by each participating institution after breaking the code that presented the investigators from knowing whether particular dogs received a test drug or the placebo. Regrettably, Leonard Holman Darsee's coworker in the division of nuclear medicine, was not informed about his faked research or the termination of Darsee's appointment until November. The Harvard data did not require a rocket scientist to detect the fraud. As Keith A. Reimer MD, PhD (1945-2002) a cardiac pathologist at Duke pointed out "the data were too clean, and when you started plotting them to see what types of correlations would show up, based on the work of other laboratories, the correlations did not show up. There were a number of inconsistencies. The Harvard people had no explanation, aside from the fact that Darsee had been involved in the collection of the data." In November 1981 after an unbelievable delay of 6 months from the time of Darsee's confessed fraud, Harvard publically admitted the scientific misconduct, but only after pressure from the NIH compelled it to explain the scientific inconsistencies in this $724, 154 research project [26]. In a brief press report Harvard provided a concise statement about Darsee's acknowledged falsifications and announced the withdrawal of Darsee's findings from the multicenter study and anticipated presentations at a professional meeting. The NIH appointed a panel of outside scientists, headed by Howard E. Morgan of Pennsylvania State University (Penn State) *(http://www.psu.edu/)* college of medicine, to investigate Darsee in December 1981. The panel also looked into how the laboratory activities of Braunwald were supervised. There was an obvious concern about how these unacceptable acts could be permitted on NIH supported research in a major laboratory. For his fraudulent behavior Darsee was debarred from getting NIH funds or for sitting on review or advisory committees for 10 years [33]. The panel detected significant deficiencies in the supervision of the Braunwald laboratory with his numerous responsibilities and it requested an on-site reassessment of the supervisory and research methods in about a year to make sure that they had become acceptable. Braunwald was extremely unhappy with the panel's evaluation of the situation. He did not believe that it was feasible for him to conduct a tight control over the supervision of persons working under him. Furthermore he did not believe that Darsee's fraudulent behavior was caused by working in a high-pressured laboratory, but rather to Darsee's personality defects. Nobody at Harvard saw a need to stop Darsee's pending publications, or to retract those that were in print, or to stop him from publishing more papers. Ten abstracts authored with Darsee were accepted by *Circulation* for the American heart association meeting in Dallas (November 16-19, 1981). Although the abstracts were printed Darsee was prevented from presenting the material coming from Harvard, but did present the information stemming from Emory. New papers coauthored by Darsee appeared in journals such as the *American Journal of Cardiology* [38, 40-43] and *Proceedings of the National Academy of Science* [73] after his scientific misconduct was exposed.

The Darsee affair was aggravated by the failure of Harvard to act promptly and reasonably after it became obvious that Darsee was guilty of scientific misconduct. Harvard started an official investigation into the embarrassing incident in December 1981 when Daniel Tosteson the dean of Harvard medical school appointed an eight member blue-ribbon committee chaired by Richard S. Ross, dean of JHU school of medicine, to evaluate the matter further.

Perhaps not surprising a month later the Ross committee did not fault Harvard for allowing Darsee to continue performing research for 6 months after he admitted fabricating data. The Ross committee believed that Darsee was treated reasonably. For several months after the accusations surfaced Darsee was treated appropriately for his crime on the assumption that it was merely a single unique act of data fabrication. The Ross committee made two suggestions for enhancing the institutional reaction in dealing with delicate situations of this nature. The committee felt that advice should have be sought immediately after misconduct was discovered from a small committee of senior professors from within Harvard, or from elsewhere, but not within the same department as Darsee.

Following a confession of dishonest research Darsee was relieved of his academic title by Braunwald, but still worked for him and prepared additional papers for publication. Even though Darsee admitted only a single act of fakery, this deed alone warranted a prompt reaction. Harvard probably accepted Darsee's explanation but after the serious allegations of scientific misconduct were raised he should never have been allowed to continue working on research funded by an NIH grant. Darsee illustrates a problem that faces senior faculty with large research programs. As pointed out by Braunwald it not possible to hold a fellow's hand at every turn. The role of postdoctoral research fellows in research projects varies with their experience. It is essential that senior scientists trust their postdoctoral fellows as well as their technicians, and students [33, 74].

In what was clearly an oversight the Ross report failed to recommend that Harvard should have informed the funding agencies about Darsee's fabrication. The only criticism rendered by the NIH was that Harvard failed to alert them about the issue. William Raub, head of extramural affairs at NIH stressed this important omission on a Columbia broadcasting system (CBS) television programs that Harvard had a responsibility to notify the NIH straight way about the confessed fabrication. Braunwald allowed Darsee to continue working on the NIH supported randomized project under close guidance for a few more months because the project was profoundly reliant on persistent procedures and methods and to alter an investigator in one of the participating multicenter laboratories might put the entire study at risk eventhough the Darsee input had already severely jeopardized the randomized masked project that was being performed concurrently in four autonomous laboratories. A comparison of results from different laboratories made it readily apparent that the Harvard data had been mishandled. The retention of Darsee's participation in the project after discovering his research misconduct destroyed both the morale and research output of the participants. Braunwald should have notified all of Darsee's collaborators about the discovered fabrication of his research, but Braunwald was unaware of the Darsee-Holman collaboration [26].

Almost 100 papers and abstracts written by Darsee were considered acceptable by the Ross committee, which clearly did not evaluate them in as much detail as Walter Stewart and Ned Feder who found mistakes in many of them [195]. The entire Ross committee was also satisfied that a review by Braunwald and Kloner disclosed no widespread fraudulent behavior and none of Darsee's untrustworthy work had been published and pending papers and abstracts containing concocted research data were retracted. Braunwald was clearly sympathetic to Darsee's singular admitted indiscretion and believed that an exposé in the public domain based on a solitary incident would have devastated Darsee for the rest of his life so to avoid this they adopted a guarded position. Administrators at Emory were eventually informed. On June 8, 1983 the *New England Journal of Medicine* retracted two fraudulent papers on heart studies authored by Darsee [44]. The research fabrication of Darsee also compromised the integrity of eight papers published by him when he was located at Emory and nine more at Harvard, which were also eventually retracted.

Stewart a biomedical basic scientist at the NIH shared a laboratory with Feder, a long time friend, mentor, and research collaborator. At one time they studied the nervous system of snails using a research technique developed by Stewart in 1978 [194]. During the spring of 1983, Stewart and Feder abandoned their customary laboratory research at the NIH and began an extraordinary scrutiny of the authenticity of the scientific literature. They desired to determine whether coauthors, editors and reviewers of scientific papers were meeting the accepted standard of scientific writings. After critically reviewing the Darsee publications Stewart and Feder launched their new careers as self-appointed fraud busters. They found Darsee's papers noticeably defective with obvious errors that should have been detected by his coauthors, peer reviewers of the papers as well as the journal editors. Stewart and Feder analyzed 18 papers, 88 abstracts, and 3 book chapters authored by Darsee and 47 coauthors at Emory and Harvard [32]. In 1983 they prepared a manuscript on their study and submitted it unsuccessfully to *Nature*, but it was later withdrawn because the authors considered the required alterations needed to prevent possible litigation weakened their message too much. Reviewers of the Stewart-Feder paper felt that they renounced their coworkers and their profession. Potential libel suits blocked the publication of the paper for almost 3 years. They concluded that more than 50% (35 of 47) of the coauthors were involved in haste, inattentiveness and other questionable behavior. Stewart and Feder found that papers authored by Darsee were loaded with mistakes and irregularities that the coauthors should have detected with greater care during the preparation of the papers for publication. Stewart and Feder also claimed that journal editors and manuscript reviewers failed to detect noticeable mistakes.

Approximately a quarter of the coauthors had taken part at least once in various degrees of research misconduct by permitting information that they knew or should have known to be inaccurate. Stewart and Feder considered some errors an insult commonsense, such a report on a curious familial heart disease affecting a 17 year old man with 4 children, including one who was 9 years old [39].

Subsequently Stewart and Feder tried to get their analysis of Darsee publications accepted by several other journals, including *Science* and the *New England Journal of Medicine* which rejected it for editorial reasons. In letters made public through the *Freedom of Information Act* in response to requests for legal and editorial documents on the Stewart-Feder manuscript, Daniel Koshland Jr. the editor of *Science* informed them in July 1985 "I do not concur with you that details in this case are extraordinary important". Arnold Relman the editor of the *New England Journal of Medicine* also rejected a version of the Stewart-Feder paper stating "What was true was not new, and what was new was not true, or in any event very misleading." Other editors considered the article significant to the public domain, but potentially too costly to deal with because of its libelous nature. Eventually *Nature* published an extensively edited rendering of the Stewart-Feders report on January 13, 1987 [195]. The publication was associated with an editorial titled "fraud, libel and the literature" [6] and a refutation of criticisms by Braunwald Darsee's mentor infuting the accusation that he and his coworkers permitted the documentation of shoddy research. Braunwald [20] recalled "When we discovered the ultimately exposed all of Darsee's misconduct, we were staggered as its extent" and "perhaps from positive outcome of the unfortunate Darsee affair is that it has helped focus attention on fraud in science, in part promoting both research institutions and granting agencies to develop more formal rules and procedures for dealing with suspected fraud."

Stewart and Feder sought extensive publicity on their paper and reported the core of their study to a congressional subcommittee on 26 February 1986, when they drew attention to the fact that fear of libel was thwarting publication. They also submitted copies to every member of the *NAS*. Stewart and Feder pointed out in their infamous *Nature* paper that their study raised the question about what fraction of papers in the biomedical sciences are not supported by primary data at the time of publication. They conceded that a large number of mistakes and discrepancies were minor. The editorial emphasized that Stewart and Feder had by now acquired notoriety comparable with the Darsee affair itself." Maddox went on to state "Stewart and Feder complained that they could not publish what they chose because of threats of libel."

John C. Long. In 1970 John C. Long started a residency at the world-renowned Massachusetts general hospital (Mass General), which attracts physicians who hope to pursue careers in the medical profession. While there he performed research from 1972 to 1974 in the laboratory of Paul Zamecnik, a prominent member of the prestigious *NAS*, who sought to culture the typical cells of Hodgkin disease and to determine if this disease was caused by a virus.

Most cultures from tissue with Hodgkin disease only survived for a short period of time, but Long is alleged to have been able to establish permanent cell lines with the characteristic cell type of this disease and he documented this with Zamecnik in papers published in the *Proceedings of the National Academy of Sciences of the United States of America* between 1973 and 1977 [78, 79, 222]. The cell lines derived from patients with Hodgkin disease were regarded as potentially valuable in research on this disorder as well as in supporting the hypothesis that Hodgkin disease stems from macrophages. In his early academic career Long obtained grants from the NIH in 1976 ($209, 000) and 1979 ($550, 000) and supervised two research fellows. Long established a reputation in Hodgkin disease and his research was cited in a prominent textbook on this disease [69]. Long also coauthored a paper with David Baltimore the 1975 Nobel laureate in physiology or medicine *(http://www.nobelprize.org/nobel_prizes/medicine/laureates/1975/baltimore-bio. html)* [75]. Because of his research productivity, grant support and reputation he was promoted from assistant professor to associate professor of pathology in 1979. Long falsified data related to a contaminated cell line in research on Hodgkin disease [210].

While Long was working at the Mass General Steven Quay became a resident at the same hospital. Quay trained under Har Gobind Khorana (1922-2011), a scientist who shared the Nobel Prize in physiology or medicine in 1968 *(http://www.nobelprize.org/nobel_prizes/medicine/laureates/1968/khorana.html)* with Robert William Holley (1922-1993) *(http://www.nobelprize.org/nobel_prizes/medicine/laureates/1968/holley.html)* and Marshall Warren Nirenberg (1927-2010) *(http://www.nobelprize.org/nobel_prizes/medicine/laureates/1968/nirenberg.html)* for research on the interpretation of the genetic code and its function in protein synthesis. In 1979 Quay coauthored a paper with Long on the immune complexes formed by Hodgkin cell lines derived from Hodgkin disease [80]. Long collaborated with Quay in an experiment in which Quay determined by ultracentifugation that the density of the immune complexes was much smaller than expected. The journal to which this paper was submitted rejected it pointing out that a reviewer had recommended that the anomalous measurements be better substantiated. While Quay was on vacation Long allegedly repeated the experiment, obtained the expected results and submitted the paper with the additional data to the *Journal of the National Cancer Institute* and in April 1979 the paper appeared in print [80]. Having trained under a Nobel laureate Quay was a meticulous experimentalist, and he was surprised that Long was capable of carrying out the complicated measurements while he spent two weeks on vacation. Long's results concerned him, but it was not until the fall of 1979 that Quay decided to explore the size of immune complexes in Hodgkin disease once more using another method. He requested to see Long's raw data, but each time Long told him that they had been lost. Later that year Quay began to become suspicious that Long had not carried out the reported experiment. Moreover Long became defensive and furiously asked Quay whether he was aware of the serious implications of his request. Quay blew the whistle on his senior colleague and drew the matter to the attention of Robert T. McCluskey (1923-2006), chairman of the pathology department about the concerns. McCluskey instructed him to prove his allegation. For two months Quay pursued the issue in agony and frequently considered withdrawing his accusation. During this phase of mental turmoil Long delivered a notebook full of the reported data to Quay, who looked it over and regretted doubting the integrity of Long. Being panic-stricken about his seemingly mistaken allegations he apologized to Long, who courteously accepted his apology. A couple of weeks later, while studying the notebook Quay observed a page to which photographs of Ouchterlony plates had been taped. The lighting of the room disclosed a ridge beneath the photographs. Quay removed the picture and discovered an additional piece of tape beneath it suggesting that another photograph had been inserted with adhesive tape. This finding motivated Quay to scrutinize the rest of the data in detail and he found evidence that the notebook had been forged. Quay then started to evaluate the past research contributions of Long on Hodgkin disease and other topics.

In January 1980 Quay provided McCluskey with evidence of Long's dishonesty and McCluskey confronted Long with the allegations, but Long denied the charges. He provided the ultra-centrifugation logbooks to document that he had performed the relevant experiments, while Quay was convinced that the logbook had been written over McCluskey considered the evidence insufficient to doubt Long's integrity. From information recorded in the logbooks Quay calculated the expected rotor count had Long performed the experiments, but found the recorded count to be much less than it should have been and could not be explained by the machine shutting down from overheating.

After Long's contribution to the dishonest paper was recognized it remained to be determined whether other aspects of Long's research were valid or not. This unpleasant task fell upon Nancy Harris Long's research associate, a pathologist who joined the Mass General in 1978 with virtually no research experience. She soon became aware of something weird about the cell lines used in Long's research. The three cell lines expressed the type A form of glucose-6-phosphate dehydrogenase (G-6-PD) which is typical of cells arising from persons of color, such as the immortalized HeLa cell line derived from the cervical cancer of the African American woman Henrietta Lacks (1920-1951) by George Otto Gey (1899-1970) in 1951 [110]. The patients from who Long established his three Hodgkin disease cell lines were Caucasians and this marker is extremely rare in individuals of this ethnicity. Malignant HeLa cells are infamous for polluting other cell lines and overgrowing cultures. Long ignored the possibility that Hela cells might have contaminated his cell lines because other recognized HeLa cell markers were lacking. Long's four cultures derived from patients with Hodgkin disease were designated FQ, RB, SpR, and RY. He thought that the patients might have had both the "black" and "white" variants of the enzyme G-6-PD and indeed one patient (FQ) was heterozygous for this enzyme. Lines FQ, RB, SpR had the HeLa cell marker. RY was different, but it had been founded by Zamecnik. During the fall of 1978 when Long was attempting to set up new cultures of Hodgkin disease he requested Harris to take over the assignment. She noticed a growth of groups of malignant cells in a flask which in due course occupied the entire culture. From records in the laboratory notebooks Harris discovered that on that day Long had subcultured the cells and had worked with the RY cell line. Harris established the karyotype of the new cell line and found it to resemble closely that of RY. Long ignored the possibility of contamination, but Stephen J. O'Brien, PhD the head of the genomic diversity laboratory at the NCI and a cell culture expert confirmed that the suspected new cell line was merely a case of contamination with RY cells. By January 1979 Harris completed identifying all of Long's cell lines and found that the FQ, RB and SpR lines were strikingly similar hinting at a common origin, but none of their chromosomes resembled any human chromosome. Harris convinced Long to submit cell lines and blood to O'Brien for analysis, but when the results came back Long kept the findings to himself. After Long admitted data fabrication in January1980 Harris contacted O'Brien and learnt that FQ, RB and SpR were all derived from the same individual and the FQ line could not have come from the patient, whose blood sample was submitted for analysis. Further, the patient was not heterozygous for G-6-PD as previously asserted by Long. O'Brien analyzed the lines for six enzymes and concluded that they were of human origin. Considering HeLa contamination, Harris forwarded them to Walter Nelson-Reese, an expert in HeLa cells at the naval biosciences laboratory in Oakland, who pointed out that he could not identify the karyotype, but it was not human and O'Brien using a larger panel of enzymes confirmed this conclusion. Bharati Hukku, an associate of Ward Petterson, of the child research center in Detroit observed similarities between the chromosomes of FQ and those of the northern Colombian brown-footed owl monkey. Subsequently the New England retinal primate research center established that FQ and the two other lines stemmed from this monkey [51]. By scrutinizing the laboratory notebooks of Zamechnik and Long, Harris established that Long had been toiling with an owl monkey cell line (OMK-210) while establishing the FQ cell line. In all likelihood the existence of the type A G-6-PD HeLa cell marker reflected the fact that the owl monkey has this form of G-6-PD. Long's RY cell line was human but not a Hodgkin disease cell line as it was derived from a patient without tumors. It remains an open question as to whether the contamination of the other three lines with OMK-210 was deliberate or unintentional. As pointed out elsewhere cell cultures frequently become contaminated and this provided an acceptable explanation for what is known about Long's research involving cell cultures and this severely damaged Long's career.

The only aspect of Long's research that was established to be fraudulent and to which Long confessed related to his measurements of immune complexes. His serious lapse of judgement involved only several paragraphs in a nine page paper prepared under pressure for a grant application. However, other aspects of his research were not beyond criticism. While unintentional and not falling in the serious category of scientific misconduct they represented sloppy science and ignored red flags to which attention should have been paid. He inappropriately suppressed the report from O'Brien from his collaborators and incorrectly informed them that FQ was heterozygous for type A G-6-PD. Other questions also emerged from an analysis of Long's research. In 1977 Long published a paper on Hodgkin disease on the nature of the tumor generated when the FQ type cells where inoculated into mice. Long maintained that the tumors were of a cell type and perhaps related to Hodgkin disease. However, other pathologists who reviewed the original material disagreed and maintained that the tumors were not of the Hodgkin type. Long clearly should not have misinterpreted the findings. As pointed out by McCluskey Long was a sufficiently good pathologist to be able to recognize the difference. But as elaborated scientists often subconsciously misjudge their findings to support what they wish to believe because they somehow become so wedded to their hypothesis that they lose objectivity. Another problem with Long's research stems from immunofluorescence data which were provided to Baltimore of the MIT cancer center. The data in his laboratory notebook did not match what he published. A third glitch relates to all publications by Long and his colleagues which used the contaminated cell lines allegedly derived from the splenic tumors from patients with Hodgkin disease. According to the hospital records tumors were not observed in the spleens of patients FQ and RY.

Unfortunately the only act of scientific misconduct required Long to resign from the Mass General on 31 January 1980 ending his promising research career provided that he terminate his research efforts and express remorse for his misconduct. When asked whether the pressure of competitive research contributed to his data fabrication, he replied that he acted in haste and made a mistake for which he paid a high price. He did not think that there was any justification for his behavior. He expressed remorse for being unfair to his coauthors of the paper. He attributed the HeLa type marker to heterozygosity of the patients based on an incorrect assumption. Long lost his objectivity in not following-up on the HeLa type maker that was a clue pointing to something being wrong with the cell lines. Long confirmed that the patients from whom the FQ and RY lines were derived did not come from splenic tumors, but were assumed to have come from such sites and earlier papers clearly indicated that tumors had not been observed. With a firm belief in his hypothesis from the beginning Long had immense faith in his cell lines and questions have been raised as to why it took several years for their true nature to become detected, especially since his research related to an important highly visible topic. The contamination of cell cultures is an extremely common grave widely publicized problem. It is noteworthy that Long's grant application to the NIH drew attention to the HeLa cell marker in the cell lines, but this information slipped through the peer review progress without any members of the Study Section pointing out the need to follow-up this unusual finding. His flawed research probably escaped detection for years because Long was a respected researcher working at the internationally famous Mass General rather than at an obscure institution. By March 1981 Long was working as a pathologist in a Midwestern hospital.

Mark Spector. As a graduate student in the biochemistry laboratory of Ephraim Racker (1913-1991) at Cornell University (Cornell) *(http://www.cornell. edu/)* Mark Spector was found to commit scientific misconduct in 1981 to the surprise of Racker, who founded the department of biochemistry at Cornell. Racker described Spector as an outstanding teacher who delivered marvelous lectures. He was found to work hard for many hours in the laboratory and on more than one occasion Racker found it necessary to get him to slow down. After becoming aware of his dishonesty Racker considered Spector a professional cheater and a master in deception who was difficult to catch. Spector hypothesized that a cascade of more than several protein tyrosine kinases accounted for sodium potassium adenosine triphosphatase (ATPase) in Ehrlich tumor cells. The downfall of his research was utilizing radioactive iodine rather than phosphorus in his experiments. One of Spector's research preparations was inadvertently shielded by a glass plate, which still allowed the transmission of radioactivity through the glass. Glass would have prevented the passage of radioactive phosphate, but not radioactive iodine.

When Spector was questioned about suspicious experiments, he disavowed misconduct and argued that radioactive iodine must have been inserted into his samples by another laboratory worker. Racker requested that Cornell retract Specboc's doctoral degree and as well as previously published paper in Science [95]. Spector was requested to voluntarily resign, but at first he refused and threatened litigation. However, following a discussion between Racker and Spector's mother, Spector was eventually convinced to withdraw from the PhD degree program.

Robert Slutsky. At one time when Robert Slutsky was a young hyperproductive cardiac-radiologist at the University of California, San Diego (UCSD) *(http://www. ucsd.edu/)* school of medicine he churned out a new research article every 10 days!!! Such productivity by an author is a red flag suggestive of scientific misconduct. An investigation into this extraordinary researcher disclosed that he had altered data and lied about the methods used. He even concocted data related to studies that were not carried out. To provide his papers with an appearance of authenticity Slutsky frequently swayed scientists more prominent than he to coauthor his articles. Even by today's criteria he was consistently prolific. Over a period of 6 years he mass produced 161 research papers. After his remarkable productively was discovered to stem from fraudulent research he resigned from the school of medicine in 1986. Investigators said 13 of his publications were unquestionably fraudulent and the legitimacy of 55 were dubious. By the conclusion of a UCSD investigation into Slutsky'research towards the end of 1986 Slutsky was found to have fabricated data in 3 papers. At least 10 of Slutsky's 135 papers contained fraudulent material. Slutsky did not admit the fraud, but retracted 15 of his papers [82, 111-184] by informing the journal editors that there were critical questions about their validity. Slutsky did not go sufficiently far in his response according to the university, which wanted the papers not only retracted, but accompanied by a written statement pointing out that UCSD had detected evidence of fraudulent research during its evaluation of his research. In 1986 UCSD forwarded its request to all journals that published his papers and even to those that apparently published legitimate papers. Paul J. Friedman, MD, the associate dean of academic affairs at UCSD at the time, attempted to condemn Slutsky's fake and dubious publications and to get them retracted or flagged in the professional literature. Friedman's campaign met with a mixed response. Of the 13 journals that published only valid papers by him, 5 issued a statement listing papers by Slusky retracted in other journals. Friedman's effort to get journal retractions was highly variable. Fourteen journals listed the retractions under their table of contents. One journal published an editorial about the retraction [48]. Numerous retractions were published in the letter sections of journals. Some were provided at the end of the journal. The editor of one journal notified Friedman that a retraction had been printed, but Friedman could not locate it.

Vijay Soman. Vijay Soman was trained in India and received a fellowship in endocrinology at Yale University (Yale)(*http://www.yale.edu/*) school of medicine in 1975. While in the USA he established a notable research record as a laboratory researcher and eventually received a faculty appointment at Yale. At 39 years of age he was an associate professor studying the binding of insulin to the human blood cells. He published papers with Yale collaborators in reputable journals, such as *Nature* [190], the *New England Journal of Medicine (NEJM)* [191, 192] and the *Journal of Clinical Investigation* [187]. In 1973 Soman was responsible for one of the earliest documented examples of scientific misconduct when he violated the confidentiality of a competitor's manuscript that was undergoing peer review for the *NEJM*. His failure to respect the traditional peer review process used in evaluating manuscripts submitted to journals led not only to his academic downfall, but also to the disgrace of Philip Felig MD, his mentor who had complete faith in him. Felig, his coauthor remained in a managerial and ethical mess [23]. The indiscretions of Soman and Felig disclosed many lessons.

No attempt was made to evaluate the extent of Soman's research misconduct until a year after he was found guilty of plagiarism and data fabrication. In 1980 following the audit of his records Soman confessed that he had falsified work reported in one paper. All of Soman's research data was immediately seized for an audit by a researcher from another institution [23, 187-189, 191]. The inspection of his records disclosed a paucity of unprocessed data and eleven papers by Soman and Felig were retracted [187-189], mainly because of uncertainties relating to deception rather than proof. As in the Darsee case in which Harvard officials neglected to examine all raw experimental data without delay Yale officials delayed the initiation of the Soman investigation. Felig was severely penalized because of his connection in the unpleasant disclosure of Soman's misconduct [24, 25].

Following a meeting with Roth, Felig came back to Yale and met head-on with Soman who confessed that he copied the Wachslicht-Rodbard paper and lifted several expressions since he felt ill at ease with the English language and was under stress to publish his paper in the near future. He also acknowledged using equations from the Wachslicht-Rodbard paper for his calculations. Despite the exposure of Soman's research misconduct Felig continued to consider him honest without a blemished reputation. Wachslicht-Rodbard was outraged by what she considered fraudulent research and felt that that Soman should be sternly chastised for this wrongdoing.

In 1980 *Science* reported that Wachslicht-Rodbard, an endocrinology researcher at the NIH had protested to her supervisor and others that Felig and Soman stole words and ideas from an unpublished paper that she had submitted to the *New England Journal of Medicine*. In addition she wrote to Felig, Soman's superior, as well as Robert W. Berlinger (1915-2002), the Yale school of medicine dean at that time, about this incident expressing her disbelief about the legitimacy of the Soman-Felig research in the disputed paper and requested an investigation. She did not question the assumptions documented in the Soman-Felig manuscript that blood cells attach more insulin per cell than normal in anorexia nervosa, and that the eating of patients improves with therapy. Wachslicht-Rodbard wanted verification that the Felig-Soman studies had been performed and she insinuated that the reported studies had not been carried out.

Soman provided a list of patients in the project with names and dates, but not medical record data sheets, and figures assembled from analyses on the blood on six patients. Also, six female patients in the study were anorexic or lacked menses, but got them back after gaining weight, a clinically unlikely outcome.

Felig reassured Berlinger that the studies had been carried out as documented and in a letter Berlinger notified Wachslicht-Rodbard that this was unquestionable and he desired that she would now regard the dispute as over, but she was not convinced.

Soman and Felig planned to present their study at a forthcoming meeting of the *American Federation of Clinical Research*. Wachslicht-Rodbard called Roth in an outrage and he agreed to act without delay. As documented later in *Science*, Wachslicht-Rodbard daringly intimated to Roth that she was considering humiliating them in public at that meeting if their work was no investigated. Probably because of this threat Roth and Felig agreed to request an impartial evaluation of the allegations. Felig recommended a potential auditor, but this individual declined the assignment apparently because of insufficient available time. But subsequently Roth selected Jeffrey Flier an assistant professor at Harvard to audit the Felig-Soman paper. Flier was acceptable to both sides of the controversy as an unbiased evaluator of the data and he was chief of the diabetes metabolism unit at Beth Israel hospital in Boston. He was a young 31 year old physician, but thoroughly familiar with a special method of studying insulin metabolism. Flier had been trained in diabetes research at the NIH. He spent hours scrutinizing the laboratory records of Soman related to an accusation of fraudulent research on insulin metabolism. Flier examined the hospital records on particular patients.

Instead of six patients information was only available on five and Soman failed to explain the missing case. Felig indicated that the diagnoses of all were anorexia nervosa. In 1980 the resignation of Soman was requested.

A report on this case was published in *The New York Times* [62].

After months of no activity the NIH and Yale both seemed content to let the matter dissipate. Felig had ample grounds to want the incident to disappear, because he had been contacted by the search committee of the prestigious college of physicians and surgeons of Columbia University (Columbia) *(http://www.columbia.edu/)* in the Spring of 1979 to be a candidate for the chairmanship of their department of medicine and in December. After being offered the position he accepted the chair and the promised Samuel Barr distinguished professorship of in medicine. This was an opportunity to advance his career substantially as that department of medicine, was renowned for its most important role in medical education in the USA. At the time accusations by Wachslicht-Rodbard remained secret and word about them had not reached Columbia. Indeed Felig was so convinced that the rage had settled that he brought Soman with him to Columbia in January 1980. He even introduced Soman to the officials and faculty and recommended that he be appointed assistant professor. During the same month the Felig-Soman paper came out in the *American Journal of Medicine (Am J Med)* [47] with Felig listed in the esteemed position as first author.

At the time assessments of research laboratories for research misconduct were extremely rare and established procedures for performing them did not exist. Near the beginning of February, 1980 Flier traveled to Yale for the special assignment. Flier considered Soman to be an earnest researcher, but his work was not particularly ground-breaking. The paper under assault was coauthored with Felig, the 43 year old chief of endocrinology research at Yale. Felig had a reputation as a highly productive researcher and his curriculum vitae listed about 200 papers between 1975 and 1980 or an average of 31.8 papers per year and most had coauthors. Nevertheless the number of his publications was considerably more than what most researchers author during their entire careers. In retrospect this high output put an excessive pressure on Soman who hoped to emulate the role model of his supervisor. Flier believed that it was highly improbable that a respected medical researcher of the caliber of Felig would participate even marginally in scientific misconduct.

On going through the data with Soman Flier found all of them defective in some way and the composite curve in the paper could not have been obtained from the data. Also, the published data did not correspond to the data Soman showed Flier. During the audit Soman appeared more and more disturbed and blamed a technician for errors, but Flier asked why Soman would publish the flawed data because of a technician. Soman pointed out that he had been under considerable pressures to publish immediately to enhance a funding priority. The Felig laboratory was geared towards output and achievement and these attributes are now known to predispose towards scientific misconduct. When Felig questioned how the audit went Soman informed Felig that Flier was prejudiced against the two of them. On hearing this Felig promptly telephoned Flier who informed him that Soman had unquestionably falsified data in the recently published paper that Felig coauthored with him. Appreciating that his own reputation and to some extent that of Yale were at risk. Felig met with Flier and Berlinger and a decision was made to fire Soman and the Felig-Soman paper would need to be retracted. Soman was summoned to Felig's office and informed that his best option was to resign and stop doing research. Soman agreed. Felig and Berlinger took possession of all of Soman's records and recruited Jerrold M. Olefsky MD, an endocrinology researcher at the University of Colorada (UC) *(http://www.colorado.edu/)*, to come to visit Yale in late March. Felig wrote remorseful letters to both Roth and Wachslicht-Rodbard. Felig knew he had to inform Columbia about the scandal and in late February while giving a seminar at Columbia Felig met with Donald Fraser Tapley (1927-1999), the dean of the medical school at Columbia at the time. Felig informed him about the scientific misconduct of Soman and the review of his laboratory research and pointed out that Soman was no longer suitable for a faculty position at Columbia.

Back at Yale a second audit on the research of Felig and Soman was taking place and it revealed that raw data for 9 of Soman's papers were missing. Olefsky found extra defects with the available data. In one paper 25-50% of the data could not be found and the interpretations were unjustifiable. Wachslicht-Rodbard called Olefsky in April after initially learning about the second audit and the absent data. Berliner received the findings of the audit on May the 12th after returning from a month of travel. Wachslicht-Rodbard called Olefsky in April and initially learnt about his audit and the absent data. Wachslicht-Rodbard called Olefsky in April after learning about the second audit and the absent data. She wrote to Berliner and insisted that the paper be retracted, but he did not receive her letter until the same date that he saw the findings of the audit, but other individuals at Yale heard about Wachslicht-Rodbard's demand.

Late in March Olefsky arrived at Yale expecting to audit fourteen Soman papers, including one that had not yet been published. Surprisingly most needed material was missing and Soman claimed that he had disposed of numerous research records during the prior year. Olefsky spent two days examining the relevant material and in performing calculations from the available data for the five papers supported by the records. Olefsky was unable to locate insulin binding data for all patients in any publication. The conclusions in the papers usually appeared reasonable, but a significant amount of data was absent in all case and could not be matched to the content of the publications. There was a general inclination to even out the data. Because of missing material the scientific misconduct might have been considerably more widespread than Olefsky could determine. Olefsky could find no fault in only two papers. The others were discarded because of questionable or falsified raw data. The majority of papers by Soman were coauthored with Felig and other colleagues.

By March the scandal had spread like wildfire and it was too widely known to persist in secret. Columbia faced a dilemma about about what to do with their newly appointed chairman-elect of the department of medicine and Samuel Barr professor-to-be. A senior advisory faculty committee of Columbia requested a meeting with Felig in May to clarify the situation. He met with them, but seemingly confined his remarks to the falsifications of Soman. Those members of the faculty present later confirmed that they had heard nothing from Felig about the plagiarism or the other wrongdoings of Soman. A member of the committee proposed that Felig retract the papers that were either not backed up with raw data, or for which raw data was missing. Shortly letters were submitted to several journals withdrawing twelve papers stemming from Yale and eight were coauthored by Felig [187-189, 191]. Felig moved with his family to New York to accept his new job in June, but by way of the grapevine rumors arrived before him. When the vice-president of Columbia for health sciences Paul Marks returned from his holiday in July the rumors about Felig and his associate were already rampant and he rapidly appreciated that Columbia was confronting a predicament. Marks contacted Tapley, who found it necessary to shorten his vacation and come back home before he originally intended.

What started in 1978 as a confrontation to publish a research observation before competitors escalated over the next year into a troublesome out of control nightmare. Soman acknowledged lifting about 60 words from manuscript written by an opponent. This act precipitated a demand for an investigation into whether the Soman-Felig study at Yale did or did not take place.

During the second half of 1979, an auditor had not been selected. In February 1980 numerous weaknesses in the research of Soman were unearthed. The scandal was particularly vulnerable because of the sensitive issue of Felig's recruitment into a prominent faculty position at Columbia. Felig was prepared to abandon a chairmanship at Yale to assume the position as chairman of medicine at Columbia with a distinguished professorship. Just over 5 months later Felig was forced to resign from Columbia, to a certain extent because he failed to report the entire tale from the outset. Felig did not consider the plagiarism issue sufficiently important to notify officials at Columbia, but Columbia considered the topic grave despite the fact that nobody at Columbia observed the five dozen or so plagiarized words. Because of several poor judgmental decisions the scandal shattered the career of Felig, who mistakenly considered the errors of Soman as trivial. Felig asked the *Am J Med to* publish the Felig-Soman paper [47]. All but two passages were copied verbatim from the Wachslicht-Rodbard manuscript. Despite numerous imperfections some wandered what really went wrong in the Soman scandal. Felig coauthored several papers with Soman and failed to oversee the raw research data for which he was responsible.

On 5 February, an auditor from Harvard pointed out that the concern with Soman extended further than a short segment of plagiarism. Soman had fudged much of the data for the *Am J Med* paper. On February 14, Felig and officials at

Yale requested the resignation of Soman and decided that an even wider ranging audit was indicated.

When Marks visited Yale he paid his respects to Dean Berliner and they discussed Felig among other topics, but Berliner did not bring up concerns about Soman. Because neither Felig nor Berlinger informed him about the Soman issue by Marks called an urgent meeting of the search committee to request the resignation of Felig. On February 27 after giving a seminar at Columbia Felig told Tapley about the audit and the doctored data. Felig felt that it would have been inappropriate to discuss the situation with the vice president without first telling the dean. While precisely correct with regard to academic protocol, common sense suggested that an exception be made. Tapley assumed that the audits were internal matters at Yale, but he did not realize that Soman's books had already been scrutinized by Flier from Harvard or that Olefsky from UC had been recruited to do the same. The scientific community at large learnt more about the Soman affair before officials at Columbia and this was why the judgment of the Columbia committee was so rapid and definitive. After Felig left Columbia on his return to Yale, Tapley contacted Berliner, who admitted Soman's indiscretions, but he tenaciously endorsed the integrity of Felig. When Tapley started to hear gossip about the Soman-Felig affair he attempted to contact Berliner to find out what was happening. At this time Felig was consuming a couple of days each week at Columbia reorganizing his future department of medicine. Felig met with five senior professors in the department of medicine at Columbia and informed them about the findings of the Flier and Olefsky audits. In June, Felig's employment as chairman of medicine at Columbia officially started, but the beginning of the end came during the next month. After being out of the country for about a month Marks came back to Columbia on Sunday, 20 July. On the following day he became aware of rumors about Felig and acted promptly. He talked to Isidore S. Edelman (1921-2005), chairman of the department of biochemistry who contacted Berliner at Yale on 22 July. Berliner provided Edelman with his account of the Soman-Felig saga and volunteered to forward a letter that he had sent to Wachslicht-Rodbard. Edelman recommended that Berliner transmit it straight to Marks and this was done on July 23 with an enclosure of 26 other letters and documents recounting the entire issue, but he failed to include the two thick papers depicting plagiarism. The package reached Marks on Friday, 25 July and a committee composed of 6 members of the faculty (including 4 of the original search committee that picked Felig), while not being previously aware of the Soman affair during the selection procedure. At his time officials at Columbia became aware for the first time of the plagiarism, struggle for a research priority, missing research data and the interruption in the scrutiny that lasted a year.

Felig should not have permitted Soman to evaluate the Wachslicht-Rodbard manuscript [23]. He should have personally investigated Soman's data in person as soon as Wachslicht-Rodbard's allegations were raised in early 1979. Felig should probably not have taken Soman to Columbia in January 1980 as a nominee for an assistant professorship. More significantly many members of the search committee believed that Felig failed to express concerns about Soman to appropriate persons at Columbia. After becoming aware of the search committee's adverse response, Felig offered a 9 page letter of denial drawing attention to information that the search committee failed totake into account. He had been honest and frank about bringing Columbia up to date on all pertinent viewpoints of the Soman scandal. Felig was apologetic for not verifying Soman's data when the accusations initially came to light and confessedthat he should have done a better job in supervising Soman. Felig appreciated that his new faculty position was in jeopardy and he tried unsuccessfully to justify his situation. It was troublesome that Felig did not notify Paul Marks about all known facts about the Soman affair. After Felig retuned to Yale Tapley summoned six senior faculty members to a meeting in his office to talk about bothering aspects of Felig and as well as give advice about a course of action Dr. Henry Holt Bendixen (1923-2004), chairman of the committee, recommended that Felig resign. By Friday the committee provided a seven page summary of transactions over the preceding 19 months. Amongst the considerations over that time period the committee pointed out that Felig inefficiently communicated with officials at Columbia incriminating himself as being responsible of inadequate reasoning in his inattention and the way he dealt with the confidential assessment of the Wachslicht-Rodbard paper and his fiasco to suitably supervise Soman's research after she made her charges and his inadequate reasoning in proposing employment at Columbia for Soman, being aware of his unacceptable actions. The committee presumed from the detailed information and Felig's mind-set when queried about these issues were sign of ethical insensitivity and unfavorable standards in science. The committee decided that Felig should neither maintain the the pit seemed that position as chair of the department of medicine at Columbia or the endowed professorship. On the same day, Tapley notified Felig of the effect and presented him with a replication of the committee report. For Felig it seemed as if his career had suddenly disintegrated and over the weekend he composed a nine page letter to Michael Sledern, president of Columbia in which he offered a detailed refutation of the committee report. Berlinger and other supporters of Felig acknowledged that some of these acts were ill-advised. At the worst, he may have been culpable of dishonorable attempt to scoop Wachslicht-Rodbard with the publication of an article, highly competed with fake data. People interrogated at Columbia by *The New York Times* ignored possible fraud on the part of Felig. The basic issue as viewed by Tapley basically was that the chairman of the department of medicine and a Samuel Barr professor directly has an effect on each student admitted to the medical school at Columbia and that Felig's poor judgment in failing to make known the incidents of the Soman scandal that made him not good enough to serve as a role model for the trainees. Judgment is an essential characteristic for somebody to chair an academic department at Columbia or anywhere else. Felig evidently did not realize the consequences of numerous interacting matters in this complex case that demanded a thorough considered opinion.

A sad aspect of this example of scientific misconduct is the detrimental effect that this case had on the three main players Wachslicht-Rodbard, Soman and Felig. In April 1980 Soman departed from Yale and later in that year he returned to Pruna, India with his family. Wachslicht-Rodbard completed a residency in internal medicine and then went into private practice stating that should probably never return to research. Wachslicht-Rodbard kept quiet but soon thereafter being disheartened with the life of researchers she resigned from the NIH, and started a residency in Washington. She then went into private practice stating that she would probably never return to research. Bendixen expressed the opinion that search committees for departmental chairs that in the past were courteous and commonly asked mild questions, but now the questions are much more difficult. After a distressing analysis of the Felig-Soman affair that lingered for three months Yale reappointed Felig as a tenured professor but his endowed chair was not restored. Most of Felig's academic effort was devoted to research and he spent little time teaching. He took part in several collaborative research assignments. Felig was an esteemed clinician-scientist and he was the recipient for many academic honors and he had considerable grant support from the NIH, the March of Dimes Foundation, and the American Diabetes Association. Despite being directly answerable for the misconduct of Soman Felig had his main NIH grant renewed and continued to. The Soman affair caused a tremendous fall from grace for Felig's and his life underwent major changes. Felig had other offers to head many other significant departments of medicine in the USA, but none were ever as renowned as the one that occurred at the time of his academic downfall.

Stephen E. Breuning. Stephen E. Breuning PhD (1953—) worked at the Oakland regional center for developmental disabilities in Michigan for a year after receiving a PhD in psychology from the Illinois Institute of Technology in 1973. In 1978 he relocated to the Coldwater Michigan regional center. He was considered a brilliant psychologist by many colleagues and while working there Robert Sprague selected him in 1979 as an on-site investigator for a clinical trial accessing the effects of neuroreactive drugs on the retarded that was being financially supported by the National Institute of Mental Health (NIMH). Breuning became director of the John Merck program at the Western psychiatric institute of the University of Pittsburgh (Pitt) (*http://pitt.edu/*) in January 1981 until 1984. He continued to compose manuscripts on his research studies at Coldwater and acquired a grant from the NIMH as principal investigator (PI) to study the effects of stimulant medication on retarded individuals. After resigning from Pitt he became assistant director of clinical services at the Polk center in Pennsylvania's department of public welfare. At that me his alleged scientific misconduct surfaced and an investigation into it was in progress at Pitt and it progressed at a glacial pace by the NIMH in December 1986 [61]. Breuning authored papers on some important studies using psychoactive medications on institutionalized patients with mental retardation and despite few publications he strongly influenced research on the mentally impaired. Sprague began to become skeptical of the legitimacy of Breuning's research when unbelievably high reliability ratings with the use of the tardive dyskinesia rating scale. Sprague analyzed Breuning's publications and the progress reports on his NIMH research grant. He observed numerous features that were implausible or not in agreement.

In 1983 Sprague received an abstract from Breuning summarizing a study that Breuning hoped to coauthor with Gualtieri at the annual scientific meeting of the *American College of Neuro-Psycho-Pharmacology* (ACNP) in 1983. It provided an additional 2 year follow-up of 45 of the 57 Coldwater patients with tardive dyskinesia after the withdrawal of narcoleptics. By contacting Neal Davidson, the director of psychology at Coldwater and learnt that Davidson had no knowledge of research being performed by Breuning on any patients at Coldwater. Breuning claimed that he was could not locate much of the old Coldwater data, but an analysis of these disclosed that they were unfeasible. When confronted with these facts Breuning agreed that all raw data was unavailable, and that it would not be appropriate to present the findings at the ACNP. The legitimacy of his research was initiated in December 1983 by two of his former collaborators. These colleagues were Sprague of the University of Illinois (U of I) *(http://www.uillinois. edu/)* and Thomas Gualtieri of the University of North Carolina (UNC) *(http:// unc.edu/)*, who coauthored papers on persons affected with tardive dyskinesia with him in affected persons from North Carolina and at the Coldwater Michigan regional center Coldwater with him. Sprague and Gualtieri informed the NIMH about discrepancies in Breuning's research and alleged that some of the research documented in his scientific reports had not been performed.

In 1985 the NIMH appointed a panel of experts to audit the research of Breuning because of the allegations against his research [61]. Despite participating in serious scientific misconduct Breuning denied taking part in fraudulent research and pointed the finger at the panel for its use of threat and intimidation. After greater than three years the NIMH issued a preliminary report of its findings and in March 1987, Breuning reviewed this document and stalwartly disavowed taking part in any unlawful activity despite some negligible mistakes. A final NIMH report was released in May 1987. After a long-drawn-out analysis of the validity of research performed by Breuning, the NIMH found this former grantee, guilty of knowingly, willingly, and repeatedly engaging in misleading and deceptive practices in reporting the results of his research. Many studies reported in articles or progress reports for grants by Breuning were not performed.

It considered Breuning as a perpetrator of long-term premeditated fraudulent research. The required complicated painstaking techniques were not utilized. In his long research career Breuning reported the research findings from the mid 1970's in Chicago until his resignation from Pitt despite not performing the studies. Breuning gained substantial distinction in his studies that signified the overuse of anti-psychotic drugs in the mentally impaired and that stimulant drugs were more effective in the therapy of such individuals that were hyperactive. After discussions between Breuning and Gualtieri Sprague documented his criticisms in writing and turned them over to the head of his department and to the NIMH, which dispatched them to Pitt school of medicine. Sprague discovered contradictory or improbable data concerning the number of participants in the studies and with regard to time spans, travel and the research techniques employed. The records at Coldwater also had noticeable gaps. In an attempt to explain questionable findings Breuning claimed that some raw data that was more than a few years old and had been destroyed to safeguard patient privacy.

Donald MacDonald, chief of the alcohol, drug abuse and mental health administration at NIMH passed the case against Breuning on to the USA justice department with a recommendation of potential prosecution even though no prior perpetrator of scientific misconduct had faced a criminal trial [93]. Moreover MacDonald advocated that Breuning not be allowed to receive any grants or contracts from the Department of Health and Human Services (DHHS) for a decade [93]. The career of Breuning was allegedly characterized by widespread fraudulent research. As a researcher he experimented with psycho-active drugs in the treatment with mentally retarded.

Breuning claimed that some data from his experiments were collected in Oakdale. A book titled *Drugs and Mental Retardation* co-edited by Breuning and Alan D. Poling was published in 1982 [22]. The computer resource that Breuning claimed to use had no record of him actually using it.

Breuning was only one of a handful of researchers world wide making significant input to knowledge about psychoactive drugs with institutionalized mentally impaired individuals with high rates of emotional, behavioral and brain disorders. NIMH officials requested that Pitt perform a couple of investigations into the integrity of research performed by Breuning. He contended that stimulant drugs are often more effective, and his studies seemed to substantiate this contention. In the summer of 1984 Pitt notified the NIMH that it had found no evidence to implicate Beuning in research misconduct while he worked at Pitt. Larry B. Silver, the acting director of the NIMH, informed Sprague that the NIMH would perform an extensive investigation into the allegations of research by Breuning. Lorraine Torres, the NIMH official responsible for looking into the allegations against Breuning, asked James Shriver, a former NIH official, to investigate the situation on behalf of the NIMH. In January 1985 administrators at Pitt started a second investigation, into Breuning, apparently after being nudged by Schriver to examine Breuning's research at the Pitt Western Psychiatric institute and clinic. Breuning's studies at Pitt during the period 1982 to 1984 altered the therapy of markedly retarded children. His research apparently drew attention to weaknesses in the frequently used tranquilizers and disclosed more encouraging effects in treating these individuals with stimulants. Following all investigations into allegations of data fabrication by Breuning a federal grand jury indicted him, making Breuning the initial researcher to be unlawfully arraigned.

In 1986, Breuning admitted submitting fake research information in applications for federal research grants. After pleading guilty Breuning was sentenced on November 10, 1988 to 60 days in a work-release program. He was required to serve 250 hours of community service and stay out of the discipline of psychology for a five year period of probation. He was also required to pay back $11, 321 in salary to Pitt [28, 29]. By blowing the whistle on a previous colleague this case profoundly effected Sprague [193].

Research Misconduct Investigations Performed by the Office of Scientific Integrity

The Office of Scientific Integrity (OSI) was created in 1989 and it investigated cases until 1992 when the OSI was replaced by the Office of Research Integrity (ORI).

Margit Hamosh. Margit Hamosh (1933-2011) was born in Germany and survived the holocaust. She then spent years in Romania before immigrating to Israel in 1950 where she obtained a PhD at Hadassah University in Jerusalem. In 1960 she moved to the USA [9] and became a professor of pediatrics and head of the division of developmental biology and nutrition at Georgetown University (Georgetown) (*http://www.georgetown.edu/*) in Washington, DC. Hamosh was charged by Dr. Lois Freed, her former laboratory research assistant, whom she dismissed, with exaggerating the status of her research in a 1985 grant application. Freed, who was a senior staff scientist at the National Institute of Aging (NIA) in 1991, charged Hamosh with eight counts of possible misconduct for making false claims in two grant proposals to the NIH [204] (*http://www.nytimes.com/1991/03/23/us/report-says-scientists-used-false-claims-for-grant.html*) Georgetown fired Hamosh and dismissed the case in 1989 maintaining that a full investigation was unnecessary. A civil suit filed by Hamosh contested her dismissal from Georgetown and she retained her faculty position. Not being satisfied Freed reported the case to the OSI which decided to consider only four of the allegations. Two were thought to lack proof of misconduct and the residual two accusations related to preliminary data and an intended animal model in two different grant applications of which only one had been submitted to the NIH; the other application went to the USA department of agriculture an agency beyond the jurisdiction of the OSI. The NIH's punishment of Hamosh was extraordinarily rapid and harsh. In May 1990, her appointment on a study section reviewing grant applications on maternal and child health research was abruptly terminated by the Eunice Kennedy Shiver National Institute of Child Health and Human Development (NICHHD) and the funding of a recently approved grant application was deferred effectively bringing her laboratory to a standstill. Why the OSI should have hurriedly responded in this way is obscure because the NIH usually puts a researcher's funds on hold while an investigation into scientific misconduct is still active.

Hamosh was investigated by both the OSI and the ORI. The OSI found Hamosh guilty of scientific misconduct and faltered Georgetown for not investigating the accusation fully [205]. The OSI also accused Hamosh of lying and submitting fake data in grant applications to the department of agriculture and the NIH. In one grant application Hamosh was charged with asserting that she was exploiting an experimental model for total parental nutrition based on research using newborn rabbits. Hamosh allegedly confessed that she had never used such a model. Hamosh had also provided data from preliminary studies on the stomach emptying times of various dogs in a grant application to the USA department of agriculture. This was evidently fabricated or falsified. A report on Hamosh first appeared in the *Washington Post* and criticized the original Georgetown inquiry into allegations made against Hamosh. Georgetown did not bother to question obvious witnesses but again and again took debatable testimonials at face value. On October 6 1993, the ORI stopped its prosecucion of Hamosh. Like the Imanishi-Kari case the OSI's handling of the Hamosh case had a comparable beginning and detrimental consequences.

Research Misconduct Beyond the Jusisdiction of the Office of Scientific Integrity and the Office of Research Integrity

Jan Hendrik Schön. Jan Hendrik Schön of Bell laboratories (formerly known as AT&T Bell Laboratories), the world's most well-known commercial research unit, had many failures with his experiments, but the successful ones made this young researcher one of the brightest stars in nanotechnology perhaps on the path to a future Nobel Prize. In one paper after the other he documented how he could transform poor electrical conductors into semiconductors, metals of even superconductors. By the beginning of 2002 Schön had penned 15 papers in the high impact journals *Science* [98, 106] and *Nature* [2, 3, 99-104] and in many other professional journals. His productivity was remarkable. His curriculum vitae listed more than 90 papers between 1998 and May 2002 and many of them were in the top 1% of all physics citations, yet key details were commonly neglected. Other physicists became jealous of his achievements and most of his intellectually challenging findings could not been replicated by independent investigators in spite of months of struggle at a cost of millions of dollars. Then the biggest fraud in physics became exposed in May 2002. Bell Laboratories appointed an outside committee to investigate his work after researchers drew attention to apparent duplicated data in multiple publications by Schön [107]. According to some scientists the suspicious data made all of his research questionable. Schön's 2000 paper in *Science* [105, 106]was a sensation in the physics community and was extensively cited and reached the top 0.01% of all physics papers published in the same year. Several of his papers with questionable data were retracted in 2003. Articles from many journals including both *Science* [98, 105, 106] and *Nature* [99-104]. The Schön affair was reported in a book by Eugenie Samuel Reich [97], but this text did not receive raving reviews [46, 108].

Zhiwen Zhiwen Zhang. Standard proteins are composed almost entirely of twenty amino acids that are programmed by the DNA code, yet hundreds of other amino acids occur in nature. Peter Schultz, a well-known chemist, and his colleagues at the Scripps research institute in San Diego California carried out ground-breaking research on the assimilation of non-native amino acids into proteins. This work provided cell biologists and pharmaceutical companies with a unique method to assimilate nonstandard amino acids into proteins where they could be used as markers in studies of proteins of interest [215]. The research was extended in studies in which the unnatural amino acid was preloaded with a specific sugar moiety [223]. Because sugars are components of glycoproteins a new approach to evaluate how proteins become modified emerged and another paper documented the integration into a protein of a sugar loaded amino acid [221]. Assisting with this research was Zhiwen Zhang, a post-doctoral fellow, who subsequently left the Schultz laboratory to become an assistant professor of pharmacology at the University of Texas (UT) *(http://www.utexas.edu/)* in 2004. Eric Tipman, who became a member of Schultz's laboratory as a postdoctoral assistant, before Zhang's departure to UT, became interested in Zhang's work because others in the laboratory were having difficulty replicating his, work including what was reported in *Science*, moreover Zhang's laboratory notebooks documenting the details of his experiments could not be found. After considerable effort most of the research could be reproduced but not what had been reported in the two 2004 papers. These papers were retracted in 2009 during a time of exceptional intrigue involving attempted extortion and a threat of suicide. One had been published in *Science* [224] the other in the *Journal of* the *American Chemical Society* [50]. The paper in *Science* was retracted on November 27 2009 [225]. In the middle of the investigation of Zhang received an e-mail from someone who he did not know named Michael Pemulis claiming that he had discovered fraud in multiple papers and that if Zhang failed to send $4, 000 by overnight mail to a post office box in San Diego the faculty at both the Scripps Institute and UT Austin would be informed about the fraud. The e-mail also stated that all of Zhang's postdoctoral work would be retracted and that he would be fired before he gets tenure. The thought of this terrified Zhang. Without delay he notified Schultz who contacted Richard Lerner, President of Scripps, and the San Diego police department. The police referred the case to their electronic crimes unit. Within a month, the law enforcement had a suspect, but Zhang declined to press charges hoping that the state of affairs would end abruptly, but it did not. An unsigned letter, claiming to be from the Schultz laboratory, was received by officials at a number of institutions including Scripps, UT Austin, University of California at Berkeley (UC Berkeley) *(http://www.berkeley.edu/index.html)*, as well as to the editorial department of *Science*. The letter stated that the 2004 paper in *Science* was a fake, but also mentioned "I feel like leaving science or committing suicide". During the next two years, Zhang received several anonymous phone calls in which the caller said nothing and just hanged up. Zhang tried to contact the caller using the number that appeared on his telephone during these episodes, but only received a recording from Mississippi. The incident distressed Zhang as well as his family and they feared for their safety. The constant worry caused Zhang's wife and children to move from Texas and go into virtual hiding. In October 2009, Zhang learnt he had been denied tenure at UT Austin [109, 214].

Stephen Arnold. A potentially important paper on the effects of hormone like chemicals on health was published in *Science* in 1996 by endocrinologist John A. McLachlan PhD, a former scientific director of the National Institutes of Environmental Health Sciences (NIEHS) in North Carolina, a component of the NIH, and colleagues maintaining that combinations of pesticides could have powerful hormonal consequences [10, 86]. In this study yeast cells with a human estrogen receptor were used to evaluate the possible estrogenic effects of distinct compounds. Pairs of some pesticides were one thousand times more powerful at setting off estrogenic activity than single chemicals. The possibility that pesticides could imitate estrogen alarmed environmentalists and toxicologists and at the time it facilitated efforts to persuade the USA Congress to include prerequisites in laws forcing makers of chemicals to monitor thousands of those on the market for estrogenic activity. Considerable attention was paid to the paper because of the outstanding reputation of its senior author McLachlan. Within months other laboratories were unable to replicate the findings and an investigation of scientific misconduct was initiated by the Tulane University (Tulane) *(http://tulane.edu/)* medical center where McLachlan was located at the time. Because of an uproar over the paper, McLachlan retracted it during the following year [87]. Tulane medical center, however, found that the data of Stephen Arnold, a researcher in the study, did not verify the most important conclusions of the manuscript. Arnold resigned from Tulane in 1997 and later worked at Rosewell Park Cancer Institute before attending business school. Tulane cleared McLachlan of wrongdoing [68] and in defending the reputation of McLachlan the chancellor of Tulane, John C. LaRosa penned a letter adamantly stated that McLachlan did not commit or have knowledge of any scientific misconduct related to the paper [76].

Research Misconduct Investigated by the Office of Research Integrity

Fraudulent research of individuals found guilty of scientific misconduct by the ORI has taken place in the most highly regarded medical centers, institutions and universities in the country. Persons of all academic ranks have been included and the guilty parties have been from a wide spectrum of disciplines and departments.

From 1994 to 2006 the confirmed cases of scientific misconduct detected by the ORI were summarized but the names and institutions of the respondents were kept confidential. Because a discussion of all cases of scientific misconduct is beyond the scope of this book only a sample of the research misconduct are provided.

Eric Poehlman. Between 1987 and 2001 Eric Poehlman (c.1956—) held faculty positions at the University of Vermont (UVM) *(http://www.uvm.edu/)* college of medicine and the University of Maryland (UM)*(http://www.umd.edu/)* and rose though the academic ranks to a full professor. He established a place for himself in history as the first academician to spend time in prison for falsifying data on a grant application. During his career Poehlman's human research was funded predominantly by the NIH and the USA Departments of Agriculture and Defense. His fraudulent research focused on human obesity and his publications asserted that injections of hormone replacement were beneficial for menopause despite having no substantiated evidence. He afterward worked at UM for three years and was then recruited back to the UVM as a full professor. Poehlman developed a reputation as an authority on age-related metabolism, particularly during menopause. Over the course of twenty years Poehlman published more than 200 journal articles mostly on the effect of exercise and genetics on obesity. In his research he documented physiological changes. Eventually Walter PeNino, Poehlman's former laboratory technician at the UVM accused him of scientific misconduct. On March 17, 2005, the USA attorney's office for the district of Vermont, the DHHS, the office of the inspector general and the ORI announced that Poehlman, a former tenured research professor at UVT agreed to a far-reaching criminal, civil and administrative settlement related to falsifying and fabricating research data in seventeen grant applications to the NIH and ten professional publications between 1982 and 2000. On June 28, 2006, Poehlman was sentenced to a year and a day in a federal jail for data falsification in federal grant applications. According to government prosecutors Poehlman defrauded the NIH of $2.9 million. Submitting a federal grant application and having the plan funded despite fraudulent data diverts large quantities of taxpayer money from the funds available to support other valid scientific applications. Poehlman fraudulently drained off millions of dollars from the federal government in the name of research and this deprived honest scientists of funding for valid projects. The prosecution of Poehlman set an example to those who might consider comparable deplorable behavior. Prior to sentencing Judge William Sesions pointed out that Poehlman had violated the public trust. Besides spending time in prison Poehlman was eternally barred from getting future federal research grants and was ordered to write letters of retraction and correction in relevant scientific journals. Poehlman voluntarily pleaded guilty in submitting false statements in federal research grant application to the NIH in April, 1999 for which he was funded $542, 000. In the settlement Poehlman consented to pay $180, 000 to resolve a civil complaint related to many false grant applications he filed with

UVM. Moreover, he paid $16, 000 in attorney fees to the counsel of PeNino, who blew the whistle on Poehlman for scientific misconduct. In addition Poehlman agreed to be barred for life from asking for or getting financial support from any federal agency in years to come. Poehlman also agreed to be permanently excluded from participation in all federal health care programs. Poehlman admitted that he alone was responsible for the research misconduct and in filing fake grant applications. Between 1992 and 2000, Poehlman submitted 17 grant applications to federal agencies or departments containing falsified or fabricated data related predominantly to preliminary observations and overall he applied for about $11.6 million in research funding, but many of the grant applications were not funded. Poehlman also published papers on the metabolism in Alzheimer disease [94].

Luk van Parijs. The Belgian born Luk Van Parijs (c1970—) obtained his undergraduate education at Cambridge University *(http://www.cam.ac.uk/)* in England. Subsequently he became a graduate student in the department of pathology at Harvard working in the laboratory of professor Abul Abbas at the Brigham women's hospital before earning a PhD in immunology from Harvard in 1997. He subsequently received a postdoctoral fellowship with David Baltimore, the 1975 Nobel laureate in physiology or medicine *(http://www.nobelprize. org/nobel_prizes/medicine/laureates/1975/baltimore-autobio.html#)* at MIT and the California institute of technology (Caltech) (1998-2000). Following this experience he joined the biology department at the MIT in 2000. In the following year van Parijs received a three-year appointment as the Ivan R. Cottrell career development assistant professor of immunology at MIT. Following this he was promoted to associate professor of biology without tenure in July 2004 at the MIT center for cancer research. The focus of van Parijs' research was the application of short-interference RNA (RNAi) in disease processes, particularly with regard to autoimmune diseases. He studied normal and defective immune cell function in these cells in the pathogenesis of diseases. He had a robust publication record and was regarded as a rising star. A month after being promoted to an associate professor MIT started an investigation into the van Parijs' laboratory after coworkers accused him of fraudulent research. Van Parijs fabricated or falsified data in NIH grant applications, published papers, and in multiple submitted manuscripts and after being caught he confessed to these indiscretions. Within a month MIT put him on paid administrative leave and banned him from his laboratory. After an investigation into scientific misconduct lasting more than a year van Parijs was fired by MIT on 27 October 2005 [30]. While at MIT van Parijs admitted falsifying and fabricating data in a publication, as well as in numerous presentations and in a grant application to the NIH. On 6 October 2005 an inquiry into van Parijs was started Caltech on the recommendation of Baltimore after *New Scientist* drew attention to flaws in papers that van Parijs wrote while at Caltech [15, 96]. On 28 October 2005: coverage by the news media began. Harvard/BWH and Caltech work questioned by *New Scientist;* Abbas determining "course of action" in light of *New Scientist* enquiries. 26

January 2006: Data in 2 Caltech patent applications having inventors van Parijs and others questioned in press. In March 2007 one investigation by Caltech determined that van Parijs was guilty of research misconduct and that four of his publications needed correction. On 23 January 2009 the ORI's findings of scientific misconduct at Harvard/BWH, Caltech, and MIT as well as van Parijs' voluntary exclusion agreement published in the *Federal Register* [1]: Two of van Parijs's papers [207, 208] were scrutinized for *The New Scientist* and found to have extremely similar figures despite captions indicating that they were derived from different mice, but van Parijs disputed the findings of experts hired by *The New Scientist* (Reich ES (2005, October 28). Massachussetts Institute of Technology (MIT) professor sacked for fabricating data. (NewScientist.com. *http://www. newscientist.com/article.ns?id=dn8230.Accessed January 2006)*.

In March 2011, van Parijs pleaded guilty in a district court in Boston to a single count of falsifying a statement in a federal grant application. Since the fraud was serious and implicated a $2-million grant the government requested Judge Denise Casper to sentence van Parijs to six months in prison, but following an appeal for leniency by several prominent researchers van Parijs was sentenced to six months of home confinement with electronic survalence and 400 hours of community service. Moreover MIT needed to return the spent funds to the NIH. Van Parijs falsified data in seven publications and in three additional papers submitted to journals. He falsified the expression of IFNY and KJ-126 INFY in flow cytometry and the expression of different proteins in flow cytometry while at Caltech. He used parts of the identical dot blot images to represent T cell populations in the expression of injected compounds. He falsified images in grant applications to the NIH and admitted falsifying or fabricating data in different manuscripts. He fallaciously declared that the bifunctional lentiviral vectors had been created, when they had not been made. A falsified contention was that the introduction of a lentivirus into non-obese diabetic mice diminished the number of CD8+ T-cells and reduced the incidence of diabetes mellitus. Van Parijs also claimed that he had generated transgenic mice with the mono-functional lentiviral vectors with c-Myc, Ras or Akt, when this was not true. He also made false claims in presentations and manuscripts that mice had been inoculated with specific plasmids when this had not taken place. He also lied about claming that particular compounds were useful.

Evan B. Dreyer. Evan B. Dreyer MD, PhD was born in the Netherlands and grew up in New York City. After graduating summa cum laude and phi beta kappa from Columbia college Dreyer received the dual doctoral degrees from the joint Harvard-MIT program in 1984. This was followed by an internship in internal medicine at NYU, a residency in ophthalmology at the Massachusetts Eye and Ear Infirmary (MEEI), and a glaucoma fellowship at the same institution. Later Dreyer became director of the glaucoma services at MEEI in 1991. The misfortunes of Dreyer started in 1995 when he agreed to investigate whether glutamate might be responsible for the untreatable form of deafness known as

Meniere disease with Dr. Cliff Megerian a junior researcher. The collaborative research pair compared an analysis of fluid from the inner ears of guinea pigs suffering from a Menierelike disease to that from normal guinea pigs, which contained much less glutamate in the inner ears. This suggested that glutamate might play an important role in the disease pathogenesis. In September 1995 Megerian attempted to get NIH grant support to study Meniere disease based on the research of Dreyer, but the grant application was rejected mainly because of insufficient supporting data to indicate that the project was realistic or that the glutamate hypothesis was compelling. Nonetheless, the concept was patented by Dreyer and Megerian during the spring of 1996. The research team then approached private philanthropy for research funds and obtained start up funds from a family having a daughter with Meniere disease. In June 1996 Megerian extracted a small volume of glutamate containing inner ear liquid from 12 guinea pigs and thus obtained potential preliminary data that might support an NIH grant application. It would also support the patent application aspired towards a potentially profitable market in the therapy of Meniere disease. Megerian transmitted ear fluid samples to Dreyer who measured amino acid levels by high performance liquid chromatography (HPLC). Subsequently Dreyer told Megerian that the glutamine levels were higher in the treated animals compared to the controls and Dreyer asked Megerian to incorporate the new information into a revised NIH grant application. In addition Dr. Joseph Madil, a senior MEEI scientist considered the data so heartening that he recommended that Dreyer and Megerian prepare an abstract on the novel work for presentation at an approaching professional meeting. In December 1996, Michael McKenna MD, a specialist in otology and neurology at the MEEI became skeptical of the HPLC findings. They seemed too constant from animal to animal and some quantities were conspicuously comparable to data documented by different earlier researchers. McKenna and Sewell reported their unease to Nadall, chairman of the department. In a tense meeting Dreyer could not provide the raw data or notebooks to substantiate his biochemical findings. Nadall volunteered to report the event as foreseeable research misconduct, but Dreyer requested more time to search for the raw data. The following day Dreyer turned over a magneto-optical computer disk supposedly holding the missing data. Subsequently additional findings were reviewed by Sewell and others and were found to be falsified. The PHS intended to deprive Dreyer of federal research funding for five years because of his untrustworthy research [11]. The ORI found Dreyer guilty of faking or fabricating experimental results to support the hypothesis that elevated levels of glutamate contributed to Meniere disease. The documents supporting this charge were NIH grant applications in which the experimental results for 19 amino acids were reported. In addition the fabricated matter was presented in an abstract presented to the Triological Society Eastern division meeting in early February 1997. A disk contained 21 chromatograms of amino acid patterns, which Dreyer confessed were fabricated. Subsequently in 1997 Dreyer moved to the Sheie

Eye Institute of the University of Pennsylvania (UPenn)*(http://www.upenn.edu/)* in Philadelphia as an associate professor of ophthalmology. In 2000 the *Boston Globe* pointed out that some experimental data from this rising star in glaucoma research was too good to be true. As indicated by information made public by a federal inquiry Dreyer fabricated experimental findings without performing the research. He was forced to fight for his reputation and to evade a five year cut in research funding from the PHS. A hearing was set to start on October 30, 2000. The PHS accused Dreyer of fabricating data and becoming involved in deceitful behavior in keeping with the desire to defraud. A combined announcement from Harvard and the MEEI asserted that Dreyer had abused the institutional policies on research misconduct. The punitive actions proposed by the PHS concerned UPenn as Dreyer had recently joined its faculty a few months after allegedly faking research data. In October 2000 a UPenn representative stated that UPenn was unaware of the accusations against Dreyer when they employed him. Both MEEI and Harvard presumed that Dreyer had performed scientific misconduct. Dreyer maintained conflicting descriptions of his experimental research and tried to dodge accountability by placing fault on others. Dreyer was highly respected as a gifted extremely bright well-educated glaucoma researcher with very high ethical standards who had made research contributions to glaucoma. He allegedly demonstrated that glutamate was released from a small population of retinal cells and that the release of this amino acid caused the death of other cells. Dreyer also discoverd that a glutamate blocking drug stopped eye damage in animal experimental models and this effect had been evaluated in an international human clinical trial. Initially Dreyer appealed against the accusations to the federal appeals board (FAB) but on November 13, 2000, he entered into a voluntary exclusion arrangement with the PHS in which he did not divulge that he falsified or fabricated the findings at issue, but Dreyer acknowledged that if the case advanced to closure, there was enough proof to reach a ruling of research misconduct. Dreyer consented to eliminate himself from being part of a contract with any USA federal organization for ten years and also from being a mentor of any person who applies or receives federal funding. He also became ineligible to serve the PHS in any capacity including, but not limited, to service on a PHS advisory committee, board, and/or peer review committee, or as a consultant. After his turbulent academic beginning Dreyer joined Glaucoma-Cataract Consultants, Inc in 2001 where he directed their glaucoma services taking care of patients in Western Pennsylvania. On January 17, 2010, Dreyer was in Pittsburgh, PA. Surprisingly none of the publications by Dreyer were retracted. By 2010 he had authored more than 150 publications on glaucoma. He was best known for introducing the hypothesis of neuroprotection in glaucoma, which focused glaucoma therapy not on elevated intraocular pressure (IOP) but directedly on the optic nerve. Dreyer was innovative and held seven patents and was on the editorial boards of more than 25 professional journals. He also advised a number of pharmaceutical companies.

Amitav Hajra. It did not take a rocket scientist to identify Amitav Hajra as the dishonest student, who embarrassed Francis Sellers Collins (1950—) the director of the National Institutes of Health (NIH) for coauthoring fraudulent publications. This MD/PhD graduate student worked in the Collins laboratory when his scientific advisor was on the faculty of the University of Michigan (UM)(*http://www.umich.edu/*) before accepting the directorship of the National Human Genome Research Institute (NHGRI). Hajra fabricated data on the effect of leukemia genes in a paper coauthored by Collins that was submitted to *Oncogene* in August 1996. A meticulous peer reviewer detected research misconduct. Immediately after learning about the misconduct Collins swiftly confronted the wrongdoer and informed the public about problems with the data. An investigation into the data related to this case led to the discovery that the student had had difficulty generating a control cell line for tests to determine whether a gene on chromosome 16 would trigger cancer in murine tissue cells. Supposedly the student made up the data for the control while frustrated. Another study involving the same thing allegedly took place when a different gene was transfected into mouse cell lines. In an additional experiment the nucleotide sequence in stretches of human DNA were concocted using mouse DNA as a guide. Following the discovery of these fraudulent papers several extra papers authored by Hajra were retracted or some material was withdrawn or corrected with an erratum [77, 219, 222]. When Collins learnt about the episode in August 1996, almost two months before the public, he took care of the matter swiftly and with great precision, unlike previous scientists who had discovered that they had been party to scientific misconduct in publications without suspecting a cheater among their trusted research collaborators. Collins contacted the suspected offender, under his supervision at the UM before Collins left Ann Arbor in 1993, about the problematic data in this paper.

Despite his role as a coauthor of the paper and as a mentor of the culprit Collins fared well in this episode. As soon as Collins was convinced the fraud had been perpetrated, he alerted the relevant authorities at the UM, where the student was registered, and the ORI at the DHHS, where cases of scientific misconduct involving NIH grantees are investigated and ruled upon. He also notified the director of the NIH at the time Harold E. Varmus, who shared the 1989 Nobel Prize in physiology or medicine *(http://www.nobelprize.org/Harold E. Varmus)* with J. Michael Bishop *(http://www.nobelprize.org/nobel_prizes/medicine/laureates/1989/bishop-bio.html)*. He also obtained a written confession from Hajra. After discussing the matter with associates Collins wrote a quasi-public letter delineating what had transpired and mailed it to 100 researchers on a need-to-know basis about the tainted areas of research on October 1, 1996. He did not make a public announcement because the government lawyers forbade him doing so.

He also absolved many coauthors by name. In retrospect, Collins was not sure how this dishonesty could have been discovered earlier. Some scientists were concerned that Collins' busy schedule may have precluded him from monitoring his graduate student as carefully as he should have. According to Varmus it would be incorrect to suppose from the Collins case that scientists have to mistrust everything or require duplication of every significant result in the laboratory. According to Collins it is naïve to try to create a fail-safe mechanism to prevent fraud; "deceit and betrayal are part of the price; we must pay for a free system. If research is going to be open, if it is going to be creative; if you are going to allow people with talent to explore the unknown, there are people who are going to take advantage of that freedom and abuse it. Collins believes that the only remedy may be to do science with eyes more open [83]

(*http://ori.dhhs.gov/documents/newsletters/vol5_no4.pdf.Accesssed January 2006*)

Marc D. Hauser. The American evolutionary biologist Marc D. Hauser (1959—) received a PhD from UCLA. Because of this research productivity and established reputation he rose through the academic ranks to professor in the departments of psychology, organismic and evolutionary biology, and biological anthropology at Harvard University (Harvard)*(http://www.harvard.edu/)*. He became the co-director of a program on the mind, brain, and behavior at Harvard as well as the director of the cognitive evolution laboratory. He was also appointed adjunct professor in the graduate school of education and in the neurosciences program. The focus of his research was at the boundary between cognitive neuroscience and evolutionary biology. After reaching the pinnacle of his academic career his repute came crashing down after research misconduct was detected by his students. The misconduct involved three papers related to federally funded grants. On August 19, 2010, *The Chronicle of Higher Education* [12] documented that a former research assistant of Hauser had alleged that Hauser falsely coded videotapes of monkey behavior, and ignored requests by his research assistants and students to have them recoded by an independent observer. Moreover he pressured his students to accept his analysis of the data. Without obtaining permission from Hauser the students went ahead and recoded the data on their own. Allegedly they found little relation between Hauser's coding system and what was evident on the tapes. After the students accused Hauser of data falsification Harvard investigated the Hauser laboratory [206] (*http://www.newscientist.com/article/dn19293-misconduct-found-in-harvard-animal-morality-profs-lab.html. Retrieved 2010-08-12*). According to the *The Chronicle of Higher Education* other members of the Hauser laboratory had comparable conflicts with him [12]. In August 2010, following a three year investigation Michael D. Smith dean of the faculty of arts and sciences at Harvard publicly confirmed that Hauser was solely accountable for eight counts of unspecified scientific misconduct [14, 90, 218] (*http://chronicle.com/article/Marc-Hauser-Resigns-From/128296/*). Hauser was a highly productive internationally reknowned scientist with more than 240 publications.and the allegations against him shocked the scientific community particularly researchers opposed to his research. On July 20, 2011, the *Boston Globe* announced that Hauser was resigning from Harvard, effective August 2, 2011. Nicholas Wade summarized the situation in *The New York Times* as follows "There is a wide spectrum of scientific sins, ranging from wrist-slap offenses like bad data storage at one end, to data fabrication at the other. He has admitted only to unspecified "mistakes, " not to misconduct [212]. One paper published in 2002 in *Cognition* was retracted [57]. Two other papers published in 2007 had been "corrected"; Hauser and Kralik responded to criticisms of these papers and clarified their coding criteria [53]. Harvard cooperated with independent further investigations by the ORI, the NSF, the office of the inspector general and the USA attorney's office for the district of Massachusetts [13] *(http://chronicle.com/blogPost/Harvard-Confirms-Hausergate/26198/. Retrieved 2012-08-20)*. In 2010 an article in *Cognition* was retracted *(http://retractionwatch.wordpress.com/2010/08/10/monkey-business-2002-cognition-paper-retracted-as-prominent-psychologist-marc-hauser-takes-leave-from-harvard/. Retrieved 2010-08-26)* [65]. In this paper, Hauser and his co-workers concluded that the small New World monkey cotton-top tamarin money (*Saguinus oedipus*), also known as the Pinché tamarin, can learn simple rule-like patterns. In an earlier controversial paper published in 1995 Hauser asserted that members of this species of monkey can identify themselves in a mirror [59] *(http://www.pubmedcentral.nih.gov/articlerender. fcgi?tool=pmcentrez&artid=40702)*. This paper may have been precipitated the research misconduct investigation of Hauser [64]. In two more publications some field notes or video recordings were "incomplete", although Hauser and Wood replicated the experiments [52]. The *Proceedings of the Royal Society* published the replication of the missing data in an addendum to one paper [52] *(http://rspb. royalsocietypublishing.org/content/early/2010/07/22/rspb.2010.1441. Retrieved 2010-08-26)*. In April 2011 Hauser and Wood (coauthor of the original paper) replicated the results of the 2007 *Science* study [91] *(http://news.sciencemag.org/ scienceinsider/2011/04/science-publishes-replication-of.html?ref=hp. Retrieved 3 May 2011)* [220]. *Michael Tomasello* a well-respected animal cognition researcher, claimed that some of Hauser's past students had informed him that there "was a pattern and they had specific evidence" [213] *(http://www.nytimes.com/2010/08/14/ education/14harvard.html?pagewanted=2)*. This statement was later confirmed by the dean of Harvard [209]. Several different sources documented that Hauser has been found guilty of eight counts of scientific misconduct by Harvard *(http:// content.usatoday.com/communities/sciencefair/post/2010/08/marc-hauser-harvard-science-misconduct-/1. Retrieved 2010-08-26)*. As Frans de Waal informed Dan Vergano of *USA Today*: "But it leaves open whether we in the field of animal behavior should just worry about those three articles or about many more. From my reading of the dean's letter, it seems that all data produced by this lab over the years are potentially in question." [209] *(http://content.usatoday.com/communities/ sciencefair/post/2010/08/marc-hauser-harvard-science-misconduct-/1. Retrieved 2010-08-26)*. Vaux and Watumull worked with Hauser after the start of the Harvard investigation and their opinions were not without *self*-interest [63] *(http:// www.boston.com/news/local/breaking_news/2010/08/journal_editor.html. Retrieved 2010-08-27) (http://homepage.mac.com/gerry_altmann/me/mind/blog/blog.html.*

Retrieved 2010-08-27). [88] *(http://news.sciencemag.org/scienceinsider/2010/08/ journal-editor-says-he-believes.html. Retrieved 2010-08-27)* [211] *(http://www. nytimes.com/2010/08/28/science/28harvard.html?hp. Retrieved 2010-08-27) (http:// homepage.mac.com/gerry_altmann/me/mind/blog/blog.html. Retrieved 2010-08-27).* In the experiments it would have been more appropriate for Hauser to interchange the control and experimental conditions and to videotape the findings, but Hauser only videotaped the experimental conditions *(http://www.nytimes.com/2010/08/28/ science/28harvard.html?hp. Retrieved 2010-08-27).* It seemed that the control experiments were not performed, but "Altmann told *Science* that there may yet be an alternative explanation for this discrepancy. I know that the investigation was rigorous to the extreme [88] *(http://news.sciencemag.org/scienceinsider/2010/08/ journal-editor-says-he-believes.html. Retrieved 2010-08-27).* Altmann considered Hauser guilty of data fabrication and that the data reported in *Cognition* had no valid basis. Another option to explain the dispute was a series of mistakes. An article at the *Science* website indicated that the investigation of Hauser by the USA attorney's office was "somewhat unusual" and "historically they've done so in cases in which the misconduct was particularly egregious and significant amounts of money were involved." [89] *(http://news.sciencemag.org/scienceinsider/2010/08/ hausergate-scientific-misconduct.html. Retrieved 2010-08-26).* The article also pointed out that it was still not precisely known what Hauser did and whether it was severe enough to justify a jail sentence. *(http://www.nytimes. com/2010/08/28/science/28harvard.html) Retrieved 2010-08-27. The New York Times* pointed out that data fabrication typically results in the expulsion of the scientist from the research community. On April 21, 2011, the *Boston Globe* reported that Hauser had been prohibited from teaching during the 2011-2012 academic year by the psychology faculty. Michael D. Smith, dean of the faculty of arts and sciences at Harvard agreed and extended it to other arts and sciences departments [66]*(http://articles.boston.com/2011-04-21/news/29459892_1_marc-hauser-scientific-misconduct-embattled-harvard).* Gordon G. Gallup Jr PhD (1941—) a psychologist who became famous developing the mirror self-recognition test which gauges self-awareness in certain animals had reservations about Hauser's findings after examining some video recordings of his research and observed no evidence to support the notion that tamarins had discovered how to accurately interpret mirrored information about themselves. Gallup wanted to examine additional videotapes, but he was told that other tapes had been stolen [7] *(http://www.economist.com/ node/16886218?story_id=16886218&CFID=141739115 &CFTOKEN=94865933. Retrieved 2010-08-28).*

Gallup and Anderson published critiques of Hauser's article [3] in which they stated that the described coding criteria for the monkey's behavior were documented in insufficient detail and that in their assessment cotton-top tamarins did not demonstrate what they regarded as evidence for mirror recognition in great apes. In their response Hauser and Kralik provided their coding criteria

[53]. However, Hauser subsequently reported that later attempts to replicate the experiments were unsuccessful [54, 55]. An indication of Hauser's research is provided in some of his intellectual publications [56, 58]. It is noteworthy that his research misconduct Hauser makes a point that humans have evolved a universal moral instinct propelling humans to deliver judgments of right and wrong [58] *(http://www.thecrimson.com/article/2011/7/19/marc-hauser-resigns-psychology-harvard/)*. Numerous individuals confirmed that Hauser had been found guilty of scientific misconduct at Harvard, including Michael D. Smith dean of the faculty of arts and sciences [8] the *(http://harvardmagazine.com/breaking-news/harvard-dean-details-hauser-scientific-misconduct. Retrieved 2010-08-26)*

(http://retractionwatch.wordpress.com/2010/08/10/monkey-business-2002-cognition-paper-retracted-as-prominent-psychologist-marc-hauser-takes-leave-from-harvard/. Retrieved 2010-08-26.) [52]. In 2007 Wood and colleagues reported that cotton-to-tamarins, chimpanzees and rhesus macaques can make inferences about the goals of human experimenters [220]. In 2011 the journal *Science* published a replication of some the reported observations in the controversial 2007 publication from the Hauser laboratory by issuing an erratum [92] ((*http://news.sciencemag.org/scienceinsider/2011/04/science-publishes-replication-of.html?ref=hp. Retrieved 3 May 2011)*[220].

Eric J. Smart. The focal point of the research of Eric J. Smart, a former professor in the departments of pediatrics and physiology at the University of Kentucky (UKY) *(http://www.uky.edu/)* was to understand atherosclerosis, hypertension, diabetes mellitus and other diseases affecting the cardiovascular system. After allegations of research misconduct by him surfaced an extensive investigation of Smart was undertaken by UKY and it involved interviews with 27 witnesses [49]. Smart was found to have committed research misconduct for at least a decade (1998-2008). He falsified or fabricated images on numerous occasions involving knockout mice that did not exist. Ten were in published papers [18, 50, 72, 84, 186]. One was in a paper submitted for publication. Three were in progress reports for grants and seven involved grant applications. These papers and others by Smart were eventually retracted [19, 50, 71, 85, 185, 200, 201, 203, 216, 227]. He concealed his research misconduct by playing an active role in the design, analysis and the reporting of his research findings. For his experiments he depended on a group of junior technicians and each of them were kept in the dark about the work of others and by and large about his research program. A handful of graduate students and postdoctoral fellows contributed to the preparation and review of manuscripts, but all images were finalized by Smart who was in complete control of the preparation of figures for publications and grant applications. Many figures were created by copying single bands and inserting them into multiple lanes. Without a meticulous forensic program the manipulated images were difficult to detect, but common pixilation patterns within individual bands indicated that they were duplicated, but by cropping and modifying the outer edges they gave the false impression that they were different.

As Garfinkel has pointed out [49] the misconduct of Smart provided several red flags to alert individuals to dishonest behavior. Firstly one individual should not control all raw data and it is advisable that all data and research findings get reviewed by more than a single person, particularly with regard to the content of potential manuscripts and grant applications. Image manipulation is a common form of research misconduct and members of a research team should not permit a single individual from finalizalizing images. The whistle-blower in the smart case was William Everson who discovered while preparing a presentation for a professional meeting that a 2005 grant application by Smart had stated that his laboratory had specific genetically engineered mice, but this was a lie as the laboratory did not receive such mice until 2007 or 2008 [17].

Other. An electronic bulletin board of the PHS (*http://silk.nih/public/ CBZiBJE.@WWW.ORDTLS.html*) lists individuals who have had administrative actions imposed against them by the ORI until March 9, 2007. These researchers supported by grants from the NIH are listed. Although the names of graduate students and postdoctoral fellows involved in scientific misconduct are often not revealed by the institution where the dishonest research was performed they can often be identified from knowledge of the names of the authors of retracted papers. A bulletin board alerts the public to individuals found guilty of scientific misconduct and to the sanctions imposed on them in public records (*http:// www.ori.dhhs.gov/misconduct/AdminBulletinBoard.shtml*). Each guilty person is listed by name, date of birth, type of research misconduct, and the name of the institution that conducted the investigation is also provided. In addition the enforced administrative actions and their expiration date are summarized.

Research Misconduct by Scientists Not Overseen by the Office of Research Integrity

Scientific misconduct has been performed by numerous researchers whose research was not funded by federal agencies that required an investigation by the ORI. Those listed below are examples.

Victor Ninov. Victor Ninov, a Bulgarian nuclear chemist, trained at the Gesellschaft für Schwerionenforschung (GSI) in Germany before being recruited to the Lawrence Berkeley National Laboratory (LBNL) (*http://en.wikipedia. org/wiki/Victor_Ninov*). He made a name for himself by being involved in the discovery of elements 110 (darmstadium), 111 (roentgenium), 112 (copernicum), 116 (ununhexiun) and 118 (ununoctium). He was widely regarded as an expert at using the complex software needed to identify the decay chain of the unstable elements of transuranium. The findngs of some of his reported experiments could not be replicated at the LBNL and elsewhere. An investigation into the reason for the unrepeatable studies disclosed fraudulent research by Ninov [34]. This led to a major scandal. Ninov was the only member of a large research group involved in converting raw computer data into results that could be deciphered by humans and he is alleged to have used his computer knowledge to insert falsified data. Ninov vehemently denied any wrong-doing and claimed that a defect in his complicated equipment caused the faulty data. The LBNL was not convinced by Ninov's explanation and he was fired in 2001. The following year his report on the discovery of elements 116 and 118 was retracted [67, 93].

Arthur H. Hale. On October 18, 1981, the *Durham Morning Herald* [4] announced that Bowman Gray school of medicine had begun an investigation in late March 1981 into Dr. Arthur H. Hale, an assistant professor of microbiology and immunology, on its faculty for research dishonesty. The cancer researcher faced possible dismissal and was suspended with pay until the investigation was completed. The research by Hale came under scrutiny when he could not prove that he had performed experiments that he claimed that he had. Officials at the Wake Forest University (WFU) *(http://www.wfu.edu/)* asserted that inadequate laboratory records made it unfeasible to replicate the experiments he had reported in publications. They also contended that Hale lied to investigators evaluating earlier allegations of falsifying data. Moreover, Hale was accused of using radioactive iodine when it was not available to him. Although he could not provide proof, Hale claimed that he did have the isotope. Initially he was also charged with using a strain of laboratory mouse which had never set foot on the campus of WFU, but this charge was dropped by the investigators. Hale acknowledge that his record keeping was not ideal but while standing by his work he believed that he would probably be out of science regardless of the findings of the investigation into his research. This indeed seems to be case as he had no more publications in peer reviewed journals after 1981. On November 7, 1981 the *Durham Morning Herald* [5] reported that dean Richard Janeway had recommended to the President of WFU that Hale "be terminated from the faculty for cause", but Hale chose to resign his academic position effective October 30, 1981 rather than appeal. In his letter of resignation Hale denied any wrongdoing or incompetence and did not admit any liability or wrongdoing. Janeway lamented the need for this action by an otherwise brilliant and capable young scientist but there can never be a question about the integrity of science conducted at WFU". The investigation by Bowman Gray School of Medicine concluded that experiments with at least one research project were not performed as documented in a manuscript prepared for publication, but never published. Investigators found that Hale did not have adequate records to substantiate data or other experiments. Hale's research, which focused on the body's defense system, particularly on the lymphocyte that kills specific viruses, was relevant to a breakdown in the body's defenses that are suspected of causing some forms of cancer.

Michael Bellesiles. Social scientists have entered political disputes on either side of debates, such as the battle over the right to bear arms even when concealed. Kennedy drew attention to a humorous saga related to gun control in academics and politics [70] Michael Bellesiles, a former professor of colonial and legal history at Emory University (Emory)(*http://www.emory.edu*) supported the gun control case in his book *Arming America* [16]. He contended that guns were rare in early USA history, but when confronted on that assertion, he was unable to produce the data, claiming that his records were demolished when his office flooded. A scholarly committee investigated Bellesiles on behalf of Emory and its report cast doubt on the scholarly integrity of Bellesiles, who then resigned from the university and the Bancroft Prize awarded for his book was revoked. This provided ammunition for the pro-gun bloc. John Lott of the American enterprise institute and formerly of the University of Chicago (UChicago) *(http://www.uchicago.edu/)* law school claimed in the book *More Guns, Less Crime* [81] that when the vast majority of citizens (98%) use guns defensively, the simple manufacture of a weapon makes criminals desist. According to Lott his conclusion was based on about 2000 interviews that he conducted. But when pressed for the survey data, Lott gave the increasingly accustomed explanation that a computer crash had whipped out his data on the hard drive. The dispute related to gun laws overflowed from scholarly journals to the internet and Lott became vehemently defended by Mary Rosh against his academic critics, including John Donohue at Stanford University(Stanford) (*http://www.stanford.edu/*) law school, but Rosh turned out to be none other than Lott himself. *More Guns, Less Crime* unfortunately retains an important place in the gun control debate. For social science issues as grave at this society needs to receive sound data that is not fabricated political propaganda. Death from shooting remains a significant national problematic public health issue that demands a logical scientific solution.

References

1. Federal Register, 2009. 74(14): 4201-4202.
2. Anderson C, NIH fraudbusters get busted. Science, 1993. 260(5106): 288.
3. Anderson JR and Gallup Jr GG, Self-recognition in Saguinus? A critical essay Animal Behavior, 1997. 54(6): 1563-1567.
4. Anonymous. NEED TITLE. Durham Morning Herald. 1981 October 18.
5. Anonymous. WFU Professor resigns; denies falsifying project. Durham Morning Herald. 1981 November 7.
6. Anonymous, Fraud, libel and the literature. Nature, 1987. 325(6101): 181-182.
7. Anonymous. Scientific misconduct: Monkey business? Allegations of scientific misconduct at Harvard have academics up in arms. The Economist. 2010 August 26.
8. Anonymous, FAS Dean Smith confirms scientific misconduct by Marc Hauser. Harvard Magazine, 2010.
9. Anonymous. Margit Hamosh obituary. Washington Post. 2011 November 5.
10. Arnold SF, Klotz DM, Collins BM, et al., Synergistic activation of estrogen receptor with combinations of environmental chemicals.[Retraction in McLachlan JA. Science. 1997 July 25;277(5325):462-463; PMID: 9254413]. Science, 1996. 272(5267): 1489-1492.
11. Associated Press. Doctor appeals accusation of fraud. Boston Globe. 2000 October 3.
12. Bartlett T. Document sheds light on investigation at Harvard. The Chronicle of Higher Education. 2010 August 19.
13. Bartlett T. Harvard confirms 'Hausergate' The Chronicle of Higher Education. 2010 August 20.
14. Bartlett T, Marc Hauser resigns from Harvard. The Chronicle of Higher Education, 2011.
15. Beckett LE, MIT professor fired for faking data. MIT biologist and HMS grad may also have falsified data in work at Harvard. The Harvard Crimson, 2005.
16. Bellesiles MA, Arming America: The Origins of a National Gun Culture. 2000, New York: Alfred A. Knopf.
17. Blackford LB. Whistle-blower in UK research-fraud case: 'The system is badly broken'. Faculty researcher blew the whistle on his boss and then lost his job. Kentucky.com. 2012 December 9.
18. Bradshaw EL, Li X-A, Guerin T, et al., Nucleoside reverse transcriptase inhibitors prevent HIV protease inhibitor-induced atherosclerosis by ubiquitination and degradation of protein kinase C.[Retraction in American Journal of Physiology—Cell Physiology. 2013 April 15;304(8):C811; PMID: 23588579]. American Journal of Physiology—Cell Physiology. 291(6): C1271-1278.

19. Bradshaw EL, Li X-A, Guerin T, et al., Nucleoside reverse transcriptase inhibitors prevent HIV protease inhibitor-induced atherosclerosis by ubiquitination and degradation of protein kinase C.[Retraction in American Journal of Physiology—Cell Physiology 2013 April 15;304(8):C811; PMID: 23588579]. American Journal of Physiology—Cell Physiology. 291(6): C1271-1278.
20. Braunwald E, On analyzing scientific fraud. Nature, 1987. 325(6101): 215-216.
21. Brent L, A History of Transplantation Immunology, Academic Press. 1997: 482.
22. Breuning SE and Poling AD, Drugs and Mental Retardation. 1982: Charles C. Thomas.
23. Broad WJ, Imbroglio at Yale (I): emergence of a fraud. Science, 1980. 210(4465): 38-41.
24. Broad WJ, Fraud and the structure of science. Is fraud a trivial excrescence on the process of science or do the recent cases have deeper roots Science, 1981. 212(4491): 137-141.
25. Broad WJ, Coping with fraud. Science, 1982. 215(4532): 479.
26. Broad WJ, Harvard delays in reporting fraud. Science, 1982. 215(4532): 478-482.
27. Brody J. Charge of false research data stirs cancer scientists at Sloan-Kettering. The New York Times. 1974 April 18.
28. Byrne G, Breuning sentenced. Science, 1988. 242(4881): 1004.
29. Byrne G, Breuning pleads guilty. Science, 1988. 242(4875): 27-28.
30. Couzin J, Scientific misconduct. MIT terminates researcher over data fabrication. Science, 2005. 310(5749): 758.
31. Culliton BJ, The Sloan Kettering affair (II): an uneasy resolution. Science, 1974. 184(4142): 1154-1157.
32. Culliton BJ, Coping with fraud: the Darsee case. Science, 1983. 220(4592): 31-35.
33. Culliton BJ, Fraud inquiry spreads blame. Science, 1983. 219(4587): 937.
34. Dalton R, Misconduct: the stars who fell to earth. Nature, 2002. 420(6917): 728-729.
35. Darsee JR, Kloner RA, and Braunwald E, Early recovery of regional performance in salvaged ischemic myocardium following coronary artery occlusion in the dog.[Retraction in Braunwald E, Kloner RA. Journal of Clinical Investigation 1982 October;70(4):following 915; PMID: 6765636]. Journal of Clinical Investigation. 68(1): 225-239.
36. Darsee JR, Kloner RA, and Braunwald E, Time course of regional function after coronary occlusions of 1—to 120-min duration.[Retraction in Braunwald E, Kloner RA. American Journal of Physiology 1983 March;244(3):H470; PMID: 6101122]. American Journal of Physiology. 240(3): H399-407.

37. Darsee JR, Kloner RA, and Braunwald E, Demonstration of lateral and epicardial border zone salvage by flurbiprofen using an in vivo method for assessing myocardium at risk.[Retraction in Braunwald E, Kloner RA. Circulation. 1982 November;66(5):1136; PMID: 6751590]. Circulation. 63(1): 29-35.
38. Darsee JR and Kloner RA, The no reflow phenomenon: a time-limiting factor for reperfusion after coronary occlusion?.[Retraction in Kloner RA. American Journal of Cardiology. 1982 October;50(4):929; PMID: 7124650]. American Journal of Cardiology, 1980. 46(5): 800-806.
39. Darsee JR, Miklozek CL, Heymsfield SB, et al., Mitral valve prolapse and ophthalmoplegia: a progressive, cardioneurologic syndrome.[Retraction in Annals of Internal Medicine. 1983 August;99(2):275-276; PMID: 6349461]. Annals of Internal Medicine, 1980. 92(6): 735-741.
40. Darsee JR, Mikolich JR, Walter PF, et al., Transcutaneous method of measuring Doppler cardiac output—I. Comparison of transcutaneous and juxta-aortic Doppler velocity signals with catheter and cuff electromagnetic flowmeter measurements in closed and open chest dogs.[Retraction in Walter PF, Schlant RC, Glenn JF. American Journal of Cardiology. 1983 July;52(1):220; PMID: 6400477]. American Journal of Cardiology, 1980. 46(4): 607-612.
41. Darsee JR, Walter PF, and Nutter DO, Transcutaneous Doppler method of measuring cardiac output—II. Noninvasive measurement by transcutaneous Doppler aortic blood velocity integration and M mode echocardiography. [Retraction in Glenn JF, Nutter DO, Walter PF. American Journal of Cardiology1983 July;52(1):220; PMID: 6400542]. American Journal of Cardiology, 1980. 46(4): 613-618.
42. Darsee JR and Kloner RA, Dependency of location of salvageable myocardium on type of intervention.[Retraction in Kloner RA. American Journal of Cardiology. 1982 October;50(4):929; PMID: 7124650]. American Journal of Cardiology, 1981. 48(4): 702-710.
43. Darsee JR, Kloner RA, and Braunwald E, Time course of regional function after coronary occlusions of 1—to 120-min duration.[Retraction in Braunwald E, Kloner RA. American Journal of Physiology. 1983 March;244(3):H470; PMID: 6101122]. American Journal of Physiology, 1981. 240(3): H399-407.
44. Darsee JR, A retraction of two papers on cardiomyopathy. New England Journal of Medicine, 1983. 308(23): 1419.
45. Doughman DJ, Miller GE, Mindrup EA, et al., The fate of experimental organ-cultured corneal xenografts. Transplantation. 22(2): 132-137.
46. Eigler D, Neglecting the crucial "why"?. Science, 2009. 325(5939): 395.
47. Felig P and Soman V, Insulin receptors in diabetes and other conditions. American Journal of Medicine, 1979. 67(6): 913-915.

48. Frye RL, Vlietstra RE, Anderson RH, et al., The problem of deception within medical science. International Journal of Cardiology, 1988. 18(2): 123-124.
49. Garfinkel S. A research misconduct case: 10 years of bad behavior. Office of Research Integrity Newsletter. 2012 December:1-8.
50. Graf GA, Roswell KL, and Smart EJ, 17beta-estradiol promotes the up-regulation of SR-BII in HepG2 cells and in rat livers.[Retraction in Journal of Lipid Research. 2013 April;54(4):1151; PMID: 23515499]. Journal of Lipid Research, 2001. 42(9): 1444-1449.
51. Harris NL, Gang DL, Quay SC, et al., Contamination of Hodgkin's disease cell cultures. Nature, 1981. 289(5795): 228-230.
52. Hauser M and Wood J, Replication of 'Rhesus monkeys correctly read the goal-relevant gestures of a human agent'. Proceedings of the Royal Society, 2010. B 278(1702): 158-159.
53. Hauser MD and Kralik J, Life beyond the mirror: a reply to Anderson and Gallup. Animal Behaviour, 1997. 54(6): 1568-1571.
54. Hauser MD, Wild Minds: What Animals Really Think (illustrated by Ted Dewan). 2001, New York: Henry Holt.
55. Hauser MD, Miller CT, Liu K, et al., Cotton-top tamarins (Saguinus oedipus) fail to show mirror-guided self-exploration. American Journal of Primatology, 2001. 53(3): 131-137.
56. Hauser MD, Chomky N, and Fitch WT, The faculty of language: What is it, who has it, and how did it evolve?. Science, 2002. 298 (5588): 1569-1579.
57. Hauser MD, Weiss D, and Marcus G, RETRACTION: Ruleleraning by cotton-top tamarins. Cognition, 2002. 86(1): 816-822.
58. Hauser MD, Moral Minds: How Nature Designed a Universal Sense of Right and Wrong. 2006, Ecco NY 2: Harper Collins.
59. Hauser MJ, Kralik C, Borro-Mahan M, et al., Self-recognition in primates: phylogeny and the salience of species-typical features. Proceedings of the National Academy of Sciences USA, 1995. 92(23): 10811-10814.
60. Hixson JR, The Patchwork Mouse. 1976, Garden City, New York: Anchor Press. x, pp. 228.
61. Holden C, NIMH review of fraud charge moves slowly. Science, 1986. 234(4783): 1488-1489.
62. Hunt M. A fraud that shook the world of science The New York Times. 1981 November 1.
63. Johnson C, Journal editor questions Harvard researcher's data. The Boston Globe, 2010.
64. Johnson CY. Author on leave after Harvard inquiry. Investigation of scientist's work finds evidence of misconduct, prompts retraction by journal Boston Globe. 2010 August 10.

65. Johnson CY, Author on leave after Harvard inquiry. Investigation of scientist's work finds evidence of misconduct prompts retraction by journal. The Boston Globe, 2010.
66. Johnson CY. Embattled Harvard professor barred from teaching. Role at university still undefined. The Boston Globe. 2011 April 21.
67. Johnson G. At Lawrence Berkeley, physicists say a colleague took them for a ride. The New York Times. 2002 October 15.
68. Kaiser J, Tulane inquiry clears lead researcher. Science, 1999. 284(5422): 1905.
69. Kaplan HS, Hodgkin's Disease (2nd edition). 1980: Harvard University Press.
70. Kennedy D, Research fraud and public policy. Science, 2003. 300(5618): 393.
71. Kincer JF, Uittenbogaard A, Dressman J, et al., Hypercholesterolemia promotes a CD36-dependent and endothelial nitric-oxide synthase-mediated vascular dysfunction. [Retraction in Journal of Biological Chemistry 2013 March 1;288(9):6584; PMID: 23457389]. Journal of Biological Chemistry, 2002. 277(26): 23525-23533.
72. Kincer JF, Uittenbogaard A, Dressman J, et al., Hypercholesterolemia promotes a CD36-dependent and endothelial nitric-oxide synthase-mediated vascular dysfunction. Retraction in Journal of Biological Chemistry. 2013 March 1;288(9):6584; PMID: 23457389]. Journal of Biological Chemistry, 2002. 277(26): 23525-23533.
73. Kloner RA, DeBoer LW, Darsee JR, et al., Recovery from prolonged abnormalities of canine myocardium salvaged from ischemic necrosis by coronary reperfusion.[Retraction in Brawnwald E. Proceedings of the National Academy of Sciences of the United States of America. 1982 October;79(20):6390; PMID: 6765471]. Proceedings of the National Academy of Sciences of the United States of America, 1981. 78(11): 7152-7156.
74. Knox R, The Harvard fraud case: where does the problem lie? Journal of the American Medical Association, 1983. 249(14): 1797-1807.
75. Kung PC, Long JC, McCaffrey RP, et al., Terminal deoxynucleotidyl transferase in the diagnosis of leukemia and malignant lymphoma. American Journal of Medicine, 1978. 64(5): 788-794.
76. LaRosa JC, Tulane investigation completed. Science, 1999. 284(5422): 1932.
77. Liu PP, Hajra A, Wijmenga C, et al., Molecular pathogenesis of the chromosome 16 inversion in the M4Eo subtype of acute myeloid leukemia. Blood, 1995. 85(9): 2289-2302.

78. Long JC, Aisenberg AC, Zamecnik MV, et al., A tumor antigen in tissue cultures derived from patients with Hodgkin's disease. Proceedings of the National Academy of Sciences of the United States of America, 1973. 70(5): 1540-1544.
79. Long JC, Aisenberg AC, and Zamecnik PC, An antigen in Hodgkin's disease tissue cultures: radioiodine-labeled antibody studies. Proceedings of the National Academy of Sciences of the United States of America, 1974. 71(7): 2605-2609.
80. Long JC, Dvorak AM, Quay SC, et al., Reaction of immune complexes with Hodgkin's disease tissue cultures: radioimmune assay and immunoferritin electron microscopy. Journal of the National Cancer Institute, 1979. 62(4): 787-797.
81. Lott J, More Guns, Less Crime: Understanding Crime and Gun Control Laws (3rd Edition). 2010, Chicago: University of Chicago Press.
82. Mancini GB, Peck WW, and Slutsky RA, Analysis of phase-angle histograms from equilibrium radionuclide studies: correlation with semiquantitative grading of wall motion. American Journal of Cardiology, 1985. 55(5): 535-540.
83. Marshall E, Fraud strikes top genome lab. Science, 1996. 274(5289): 908-910.
84. Matveev SV and Smart EJ, Heterologous desensitization of EGF receptors and PDGF receptors by sequestration in caveolae.[Retraction in American Journal of Physiology—Cell Physiology 2013 April 15;304(8):C810; PMID: 23588578]. American Journal of Physiology—Cell Physiology. 282(4): C935-946.
85. Matveev SV and Smart EJ, Heterologous desensitization of EGF receptors and PDGF receptors by sequestration in caveolae.[Retraction in American Journal of Physiology—Cell Physiology 2013 Apr 15;304(8):C810; PMID: 23588578]. American Journal of Physiology—Cell Physiology, 2002. 282(4): C935-C946.
86. McLachlan JA, Synergistic effect of environmental estrogens: report withdrawn. [Retraction of Arnold SF, Klotz DM, Collins BM, Vonier PM, Guillette LJ Jr, McLachlan JA. Science. 1996 June 7;272(5267):1489-92; PMID: 8633243]. Science, 1997. 277(5325): 462-463.
87. McLachlan JA, Synergistic effect of environmental estrogens: report withdrawn.[Retraction of Arnold SF, Klotz DM, Collins BM, Vonier PM, Guillette LJ Jr, McLachlan JA. Science. 1996 June 7;272(5267):1489-92; PMID: 8633243]. Science, 1997. 277(5325): 462-463.
88. Miller G, Journal editor says he believes retracted Hauser paper contains fabricated data. ScienceInsider, 2010.
89. Miller G, Hausergate: scientific misconduct and what we know we don't know. ScienceInsider, 2010.

90. Miller G, Harvard dean confirms misconduct in Hauser investigation. ScienceInsider, 2010.
91. Miller G, Science publishes replication of 2007 Hauser study. ScienceInsider., 2011.
92. Miller G. Science publishes replication of 2007 Hauser study. ScienceInsider. 2011 April 25.
93. Norman C, Prosection urged in fraud case. Science, 1987. 236(4805): 1057.
94. Poehlman E, Test of friendship. A short story. RN, 1971. 34(10): 42-43 passim.
95. Racker E and Spector M, Warburg effect revisited: merger of biochemistry and molecular biology.[Retraction in Racker E. Science. 1981 September 18;213(4514):1313; PMID: 6765631]. Science, 1981. 213(4505): 303-307.
96. Reich ES, Scientific misconduct report still under wraps. NewScientist, 2007(2631): 16.
97. Reich ES, Plastic Fantastic: How the Biggest Fraud in Physics Shook the Scientific World. 2009, London: Macmillan.
98. Schön JH, Meng H, and Bao Z, Field effect modulation of the conductance of single molecules. Science, 2001. 294(5549): 2138-2140.
99. Schön JH, Dodabalapur A, Bao Z, et al., Retraction: Gate-induced superconductivity in a solution-processed organic polymer film.[Retraction of Schon JH, Dodabalapur A, Bao Z, Kloc C, Schenker O, Batlogg B. Nature. 2001 March 8;410(6825):189-92; PMID: 11242074]. Nature, 2003. 422(6927): 92.
100. Schön JH, Dorget M, Beuran FC, et al., Retraction: Superconductivity in CaCuO2 as a result of field-effect doping.[Retraction of Schon JH, Dorget M, Beuran FC, Zu XZ, Arushanov E, Deville Cavellin C, Lagues M. Nature. 2001 November 22;414(6862):434-436; PMID: 11719801]. Nature, 2003. 422(6927): 92.
101. Schön JH, Kloc C, and Batlogg B, Retraction: Superconductivity in molecular crystals induced by charge injection.[Retraction of Schon JH, Kloc C, Batlogg B. Nature. 2000 August 17;406(6797):702-704; PMID: 10963589]. Nature, 2003. 422(6927): 93.
102. Schön JH, Kloc C, and Batlogg B, Retraction: Superconductivity at 52 K in hole-doped C60. [Retraction of Schon JH, Kloc C, Batlogg B. Nature. 2000 November 30;408(6812):549-552; PMID: 11117735]. Nature, 2003. 422(6927): 93.
103. Schön JH, Kloc C, Bucher E, et al., Retraction: Efficient organic photovoltaic diodes based on doped pentacene.[Retraction of Schon JH, Kloc C, Bucher E, Batlogg B. Nature. 2000 January 27;403(6768):408-410; PMID: 10667788]. Nature, 2003. 422(6927): 93.

104. Schön JH and Meng HB, Zhenan, Retraction: Self-assembled monolayer organic field-effect transistors.[Retraction of Schon JH, Meng H, Bao Z. Nature. 2001 October 18;413(6857):713-716; PMID: 11607026]. Nature, 2003. 422(6927): 92.
105. Schön JH, Berg, S., Kloc, Ch. Batlogg, B, Ambipolar pentecene field-effect transistors and inventers. Science, 2000. 287(5456): 1022-1023.
106. Schön JH, Kloc, Ch., Batlogg, B., Fractional quantum hall effect in organic molecular semiconductors. Science, 2000. 288(5463): 2338-2340.
107. Service RF, Pioneering physics papers under suspicion for data manipulation. Science, 2002. 296(5572): 1376-1377.
108. Service RF, Winning streak brought awe, and then doubt. Science, 2002. 297(5578): 34-37.
109. Service RF, Scientific integrity. A dark tale behind two retractions. Science, 2009. 326(5960): 1610-1611.
110. Skoot R, The Immortal Life of Henrietta Lacks. 2010, New York: Broadway Paperbacks.
111. Slutsky R and Froelicher VF, The electrocardiographic response to dynamic exercise. Exercise and Sport Sciences Reviews, 1978. 6: 105-123.
112. Slutsky R, Curtis G, Battler A, et al., Effect of sublingual nitroglycerin on left ventricular function at rest and during spontaneous angina pectoris: assessment with a radionuclide approach. American Journal of Cardiology, 1979. 44(7): 1365-1370.
113. Slutsky R, Gordon D, Karliner J, et al., Assessment of early ventricular systole by first pass radionuclide angiography: useful method for detection of left ventricular dysfunction at rest in patients with coronary artery disease. American Journal of Cardiology, 1979. 44(3): 459-465.
114. Slutsky R, Karliner J, Battler A, et al., Reproducibility of ejection fraction and ventricular volume by gated radionuclide angiography after myocardial infarction. Radiology, 1979. 132(1): 155-159.
115. Slutsky R, Karliner J, Ricci D, et al., Left ventricular volumes by gated equilibrium radionuclide angiography: a new method. Circulation, 1979. 60(3): 556-564.
116. Slutsky R, Karliner J, Ricci D, et al., Response of left ventricular volume to exercise in man assessed by radionuclide equilibrium angiography. Circulation, 1979. 60(3): 565-571.
117. Slutsky R, Battler A, Gerber K, et al., Effect of nitrates on left ventricular size and function during exercise: comparison of sublingual nitroglycerin and nitroglycerin paste. American Journal of Cardiology, 1980. 45(4): 831-840.
118. Slutsky R, Battler A, Gerber K, et al., A simplified method for the calculation of left ventricular volume by equilibrium radionuclide angiography. Catheterization and Cardiovascular Diagnosis, 1980. 6(1): 49-60.

119. Slutsky R, Battler A, Karliner JS, et al., First-third ejection fraction at rest compared with exercise radionuclide angiography in assessing patients with coronary artery disease. Radiology, 1980. 136(1): 197-201.
120. Slutsky R, Hooper W, Gerber K, et al., Assessment of right ventricular function at rest and during exercise in patients with coronary heart disease: a new approach using equilibrium radionuclide angiography. American Journal of Cardiology, 1980. 45(1): 63-71.
121. Slutsky R, Karliner J, Gerber K, et al., Peak systolic blood pressure/end-systolic volume ratio: assessment at rest and during exercise in normal subjects and patients with coronary heart disease. American Journal of Cardiology, 1980. 46(5): 813-820.
122. Slutsky R, Karliner JS, Battler A, et al., Comparison of early systolic and holosystolic ejection phase indexes by contrast ventriculography in patients with coronary artery disease. Circulation, 1980. 61(6): 1083-1090.
123. Slutsky R, Pfisterer M, Verba J, et al., Influence of different background and left-ventricular assignments on the ejection fraction in equilibrium radionuclide angiography. Radiology, 1980. 135(3): 725-730.
124. Slutsky R, Hemodynamic effects of inhaled terbutaline in congestive heart failure patients without lung disease: beneficial cardiotonic and vasodilator beta-agonist properties evaluated by ventricular catheterization and radionuclide angiography. American Heart Journal, 1981. 101(5): 556-560.
125. Slutsky R, Response of the left ventricle to stress: effects of exercise, atrial pacing, afterload stress and drugs. American Journal of Cardiology, 1981. 47(2): 357-364.
126. Slutsky R, Ashburn W, and Karliner J, A method for the estimation of right ventricular volume by equilibrium radionuclide angiography. Chest, 1981. 80(4): 471-477.
127. Slutsky R, Hooper W, Ackerman W, et al., Evaluation of left ventricular function in chronic pulmonary disease by exercise gated equilibrium radionuclide angiography. American Heart Journal, 1981. 101(4): 414-420.
128. Slutsky R, Hooper W, Gerber K, et al., Left ventricular size and function after subcutaneous administration of terbutaline. Chest, 1981. 79(5): 501-505.
129. Slutsky R, Watkins J, Peterson K, et al., The response of left ventricular function and size to atrial pacing, volume loading and afterload stress in patients with coronary artery disease. Circulation, 1981. 63(4): 864-870.
130. Slutsky R, Pulmonary blood volume: the response to afterload in normal subjects and patients with previous myocardial infarction. Investigative Radiology, 1982. 17(3): 241-248.
131. Slutsky R, Bhargava V, Dittrich H, et al., Comparison of single-plane and biplane contrast analyses of right ventricular function and size. American Heart Journal, 1982. 104(1): 100-104.

132. Slutsky R, Higgins C, Bhargava V, et al., Pulmonary blood volume: correlation of equilibrium radionuclide and dye-dilution estimates. Investigative Radiology, 1982. 17(3): 233-240.
133. Slutsky R, Hooper W, Ackerman W, et al., The response of right ventricular size, function, and pressure to supine exercise: a comparison of patients with chronic obstructive lung disease and normal subjects. European Journal of Nuclear Medicine, 1982. 7(12): 553-558.
134. Slutsky R, Berger F, and Garver P, The effect of abstinence on left ventricular performance in asymptomatic chronic alcoholics. Cardiovascular and Interventional Radiology, 1983. 6(3): 154-159.
135. Slutsky R, Higgins C, Costello D, et al., Mechanism of increase in left ventricular end-diastolic pressure after contrast ventriculography in patients with coronary artery disease. American Heart Journal, 1983. 106(1 Pt 1): 107-113.
136. Slutsky R, Mancini GB, Costello D, et al., Radionuclide analysis of pulmonary blood volume: the response to spontaneous angina pectoris and sublingual nitroglycerin in patients with coronary artery disease. American Heart Journal, 1983. 105(2): 243-248.
137. Slutsky R, Watkins J, and Costello D, Radionuclide evaluation of the systolic blood pressure/end-systolic volume relationship: response to pharmacologic agents in patients with coronary artery disease.[Retraction in Brown MD. American Heart Journal. 1986 March;111(3):623; PMID: 3513502]. American Heart Journal, 1983. 105(1): 53-59.
138. Slutsky RA, Ackerman W, Karliner JS, et al., Right and left ventricular dysfunction in patients with chronic obstructive lung disease. Assessment by first-pass radionuclide angiography. American Journal of Medicine, 1980. 68(2): 197-205.
139. Slutsky RA, Carey PH, Bhargava V, et al., A comparison of peak-to-peak pulmonary transit time determined by digital intravenous angiography with standard dye-dilution techniques in anesthetized dogs. Investigative Radiology, 1982. 17(4): 362-366.
140. Slutsky RA, Carey PH, and Higgins CB, Effects of acute incremental volume overload on cardiac chamber size, function, and the pulmonary circulation: analysis by digital intravenous angiography. American Heart Journal, 1982. 104(2 Pt 1): 254-262.
141. Slutsky RA, Gerber KH, and Higgins CB, Digital intravenous ventriculography: analysis and validation of wall thickness changes during ischemia and inotropic stimulation in dogs. American Journal of Cardiology, 1982. 50(4): 874-880.
142. Slutsky RA, Relationship of radionuclide indexes of cardiac function during interventions: volume loading, afterload stress, exercise, and pacing. Cardiovascular and Interventional Radiology, 1983. 6(1): 7-13.

143. Slutsky RA, Radionuclide evaluation of cardiocirculatory performance at rest and with exercise. Cardiology Clinics, 1983. 1(3): 441-455.
144. Slutsky RA, Reduction in pulmonary blood volume during positive end-expiratory pressure. Journal of Surgical Research, 1983. 35(3): 181-187.
145. Slutsky RA, Regional blood volume changes during acute pericardial tamponade.[Retraction in Peters RM, Friedman PJ. Critical Care Medicine 1987 March;15(3):288]. Critical Care Medicine, 1983. 11(9): 721-725.
146. Slutsky RA, Phenylephrine stress in the evaluation of patients with congestive cardiomyopathies.[Retraction in Slutsky RA.Investigative Radiology. 1986 February;21(2):97; PMID: 3514538]. Investigative Radiology, 1983. 18(2): 122-129.
147. Slutsky RA, On the analysis of left ventricular volume from gated radionuclide ventriculograms. Chest, 1983. 84(1): 2-3.
148. Slutsky RA, Bhargava V, and Higgins CB, The accuracy of radionuclide ventriculography in the analysis of left ventricular ejection fraction. Computers and Biomedical Research, 1983. 16(3): 234-246.
149. Slutsky RA, Bhargava V, and Higgins CB, Pulmonary circulation time: comparison of mean, median, peak, and onset (appearance) values using indocyanine green and first-transit radionuclide techniques. American Heart Journal, 1983. 106(1 Pt 1): 41-45.
150. Slutsky RA, Dittrich H, and Peck WW, Radionuclide analysis of sequential changes in central circulatory volumes: inspiration, expiration, and the Valsalva maneuver.[Retraction in Peters RM, Friedman PJ. Critical Care Medicine. 1987 March;15(3):288; PMID: 3545676]. Critical Care Medicine, 1983. 11(12): 913-917.
151. Slutsky RA, Gerber KH, and Higgins CB, The effect of ischemia and isoproterenol on the pulmonary circulation in dogs. Analysis by digital intravenous angiography. Investigative Radiology, 1983. 18(1): 27-32.
152. Slutsky RA, Hackney DB, Peck WW, et al., Extravascular lung water: effects of ionic and nonionic contrast media. Radiology, 1983. 149(2): 375-378.
153. Slutsky RA and Higgins CB, Intravascular and extravascular pulmonary fluid volumes II. Response to rapid increases in left atrial pressure and the theoretical implications for pulmonary radiographic and radionuclide imaging. Investigative Radiology, 1983. 18(1): 33-39.
154. Slutsky RA and Higgins CB, Intravascular and extravascular pulmonary fluid volumes. III. Response to sustained left atrial hypertension. Cardiovascular Research, 1983. 17(2): 113-121.
155. Slutsky RA and Higgins CB, Analysis of the pulmonary circulation using digital intravenous angiography. Radiology, 1983. 146(1): 219-221.
156. Slutsky RA, Mancini GB, Gerber KH, et al., Radionuclide analysis of ejection time, peak ejection rate, and time to peak ejection rate: response to supine bicycle exercise in normal subjects and in patients with coronary heart disease. American Heart Journal, 1983. 105(5): 802-810.

157. Slutsky RA, Mancini GB, Gerber KH, et al., Analysis of ventricular emptying and filling indexes during acute increases in arterial pressure. American Journal of Cardiology, 1983. 51(3): 468-475.
158. Slutsky RA, Mancini GB, Norris S, et al., Digital intravenous ventriculography. Comparison of volumes from mask-mode and nonsubtracted images with thermodilution and sonocardiometric measurements. Investigative Radiology, 1983. 18(4): 327-334.
159. Slutsky RA, Mattrey RF, Long SA, et al., In vivo estimation of myocardial infarct size and left ventricular function by prospectively gated computerized transmission tomography. Circulation, 1983. 67(4): 759-765.
160. Slutsky RA, Peck WW, and Higgins CB, Pulmonary edema formation with myocardial infarction and left atrial hypertension: intravascular and extravascular pulmonary fluid volumes. Circulation, 1983. 68(1): 164-169.
161. Slutsky RA, Shabetai R, LeWinter M, et al., Pulmonary blood volume: analysis during exercise in patients with left ventricular dysfunction. European Journal of Nuclear Medicine, 1983. 8(12): 523-527.
162. Slutsky RA, Thallium pulmonary scintigraphy. Relationship to pulmonary fluid volumes during left atrial hypertension and the acute release of pressure.[Retraction in Slutsky RA. Investigative Radiology. 1986 February; 21(2):97; PMID: 3514538]. Investigative Radiology, 1984. 19(6): 510-516.
163. Slutsky RA, Andre MP, Mattrey RF, et al., In vitro magnetic relaxation times of the ischemic and reperfused rabbit kidney: concise communication. Journal of Nuclear Medicine, 1984. 25(1): 38-41.
164. Slutsky RA, Brown JJ, Peck WW, et al., Effects of transient coronary ischemia and reperfusion on myocardial edema formation and in vitro magnetic relaxation times.[Retraction in Journal of the American College of Cardiology 1987 April;9(4):973; PMID: 3549839]. Journal of the American College of Cardiology, 1984. 3(6): 1454-1460.
165. Slutsky RA, Brown JJ, and Strich G, Extravascular lung water: effects of using ionic contrast media at varying levels of left atrial pressure and during myocardial ischemia.[Retraction in Radiology. 1986 January;158(1):279]. Radiology, 1984. 152(3): 575-578.
166. Slutsky RA and Higgins CB, Thallium scintigraphy in experimental toxic pulmonary edema: relationship to extravascular pulmonary fluid. Journal of Nuclear Medicine, 1984. 25(5): 581-591.
167. Slutsky RA, Long S, Peck WW, et al., Pulmonary density distribution in experimental noncardiac canine pulmonary edema evaluated by computed transmission tomography. Investigative Radiology, 1984. 19(3): 168-173.
168. Slutsky RA and Mattrey RF, Pulmonary edema formation after myocardial infarction and coronary reperfusion: intravascular and extravascular pulmonary fluid volumes. IV. Circulatory Shock, 1984. 13(2): 183-191.

169. Slutsky RA and Olson LK, Intravascular and extravascular pulmonary fluid volumes during chronic experimental left ventricular dysfunction.[Retraction in Brown MD. American Heart Journal 1986 March;111(3):623; PMID: 3513502]. American Heart Journal, 1984. 108(3 Pt 1): 543-547.
170. Slutsky RA, Olson LK, Costello D, et al., Extravascular lung water in patients with mitral stenosis: relationship to pulmonary capillary wedge pressure and Kerley B lines.[Retraction in Radiology. 1986 January;158(1):279; PMID: 3330619]. Radiology, 1984. 153(2): 317-320.
171. Slutsky RA, Peck WW, Higgins CB, et al., Pulmonary density distribution in experimental and clinical cardiogenic pulmonary edema evaluated by computed transmission tomography.[Retraction in Brown MD. American Heart Journal 1986 March111(3):623; PMID: 3513502]. American Heart Journal, 1984. 108(2): 401-407.
172. Slutsky RA, Peck WW, and Mancini GB, The effect of pulmonic and aortic constriction on regional left ventricular thickening dynamics, geometry and the radius of septal curvature. Analysis by gated computed transmission tomography. Investigative Radiology, 1984. 19(5): 374-379.
173. Slutsky RA, Peck WW, Mancini GB, et al., Myocardial infarct size determined by computed transmission tomography in canine infarcts of various ages and in the presence of coronary reperfusion.[Retraction in Journal of the American College of Cardiology 1987 April;9(4):973; PMID: 3549839]. Journal of the American College of Cardiology, 1984. 3(1): 138-142.
174. Slutsky RA, Peck WW, Mancini GB, et al., Intravascular and extravascular pulmonary fluid volumes during acute experimental pericardial tamponade. American Heart Journal, 1984. 108(1): 90-96.
175. Slutsky RA, Watkins J, and Engler R, The systolic arterial pressure/end-systolic volume relationship in patients with severe left ventricular dysfunction.[Retraction in Cardiovascular and Interventional Radiology. 1987;10(1):52; PMID: 3105885]. Cardiovascular and Interventional Radiology, 1984. 7(2): 59-64.
176. Slutsky RA, Computed tomography of the heart and great vessels. Chest, 1985. 87(3): 392-393.
177. Slutsky RA and Brown JJ, Chest radiographs in congestive heart failure: response to therapy in acute and chronic heart disease.[Retraction in Radiology. 1986 January;158(1):279]. Radiology, 1985. 154(3): 577-580.
178. Slutsky RA and Higgins CB, In vivo validation of the thermal-green dye technique for measuring extravascular lung water.[Retraction in Peters RM, Friedman PJ. Critical Care Medicine. 1987 March;15(3):288; PMID: 3545676]. Critical Care Medicine, 1985. 13(5): 432-435.

179. Slutsky RA and Murray M, Computed tomographic analysis of the effects of hyperosmolar mannitol and methylprednisolone on myocardial infarct size. [Retraction in Petersdorf RG, Leopold GR. Journal of the American College of Cardiology. 1985 December;6(6):1440; PMID: 3905919]. Journal of the American College of Cardiology, 1985. 5(2 Pt 1): 273-279.
180. Slutsky RA, Olson LK, and Dittrich HC, Furosemide during sustained left atrial hypertension in functionally anephric dogs: intravascular and extravascular pulmonary fluid volumes. V. Journal of Surgical Research, 1985. 38(3): 216-223.
181. Slutsky RA and Peck WW, Effects of beta-adrenergic blockade on the natural progression of myocardial infarct size and compensatory hypertrophy. [Retraction in Petersdorf RG, Leopold GR. Journal of the American College of Cardiology 1985 December ;6(6):1440; PMID: 3905919]. Journal of the American College of Cardiology, 1985. 5(5): 1132-1137.
182. Slutsky RA, Peterson T, Strich G, et al., Hemodynamic effects of rapid and slow infusions of manganese chloride and gadolinium-DTPA in dogs. Radiology, 1985. 154(3): 733-735.
183. Slutsky RA and Strich G, Extravascular lung water: effects of intravenous ionic and non-ionic (Iopamidol) contrast media during ischemia.[Retraction in Radiology. 1986 January;158(1):279; PMID: 3330619]. Radiology, 1985. 155(1): 11-14.
184. Slutsky RA, Notice of retraction.[Retraction of Slutsky RA. Invest Radiol. 1983 March-April;18(2):122-129.[Retraction of Slutsky RA. Investigative Radiology. 1984 November-December;19(6):510-516]. Investigative Radiology, 1986. 21(2): 97.
185. Smart EJ, De Rose RA, and Farber SA, Annexin 2-caveolin 1 complex is a target of ezetimibe and regulates intestinal cholesterol transport.[Retraction in De Rose RA, Farber SA. Proc Natl Acad Sci U S A. 2013 Jan 8;110(2):797; PMID: 23267065]. Proceedings of the National Academy of Sciences of the United States of America. 101(10): 3450-3455.
186. Smart EJ, De Rose RA, and Farber SA, Annexin 2-caveolin 1 complex is a target of ezetimibe and regulates intestinal cholesterol transport.[Retraction in De Rose RA, Farber SA. Proceedings of the National Academy of Sciences of the United States of America 2013 January 8;110(2):797; PMID: 23267065]. Proceedings of the National Academy of Sciences of the United States of America, 2000. 101(10): 3450-3455.
187. Soman V and Felig P, Glucagon and insulin binding to liver membranes in a partially nephrectomized uremic rat model.[Retraction in Felig P. Diabetes. 1980 August;29(8):672; PMID: 7002682]. Journal of Clinical Investigation, 1977. 60(1): 224-232.

188. Soman V and Felig P, Glucagon binding and adenylate cyclase activity in liver membranes from untreated and insulin-treated diabetic rats.[Retraction in Felig P. Diabetes. 1980 August;29(8):672; PMID: 7002682]. Journal of Clinical Investigation, 1978. 61(3): 552-560.
189. Soman V and Felig P, Regulation of the glucagon receptor by physiological hyperglucagonaemia.[Retraction in Felig P. Diabetes. 1980 August;29(8):672; PMID: 7002682]. Nature, 1978. 272(5656): 829-832.
190. Soman V and Felig P, Regulation of the glucagon receptor by physiological hyperglucagonaemia.[Retraction in Felig P. Diabetes. 1980 August ;29(8):672; PMID: 7002682]. Nature, 1978. 272(5656): 829-832.
191. Soman V, Tamborlane W, DeFronzo R, et al., Insulin binding and insulin sensitivity in isolated growth hormone deficiency.[Retraction in Felig P. New England Journal of Medicine 1980 November 6;303(19):1120; PMID: 6999347]. New England Journal of Medicine, 1978. 299(19): 1025-1030.
192. Soman VR, Koivisto VA, Deibert D, et al., Increased insulin sensitivity and insulin binding to monocytes after physical training. New England Journal of Medicine, 1979. 301(22): 1200-1204.
193. Sprague RL, Whistleblowing: a very unpleasant avocation. Ethics and Behavior, 1993. 3(1): 103-133.
194. Stewart WW, Functional connections between cells as revealed by dye-coupling with a highly fluorescent naphthalimide tracer. Cell, 1978. 14(3): 741-759.
195. Stewart WW and Feder N, The integrity of the scientific literature. Nature, 1987. 325(6101): 207-214.
196. Summerlin WT, Allogeneic transplantation of organ cultures of adult human skin. Clinical Immunology and Immunopathology, 1973. 1(3): 372-384.
197. Summerlin WT, Broutbar C, Foanes RB, et al., Acceptance of phenotypically differing cultured skin in man and mice. Transplantation Proceedings, 1973. 5(1): 707-710.
198. Summerlin WT, Miller GE, and Good RA, Successful tissue and organ transplantation without immunosuppresion. Journal of Clinical Investigation, 1973. 52: 34a.
199. Summerlin WT, Miller GE, Harris JE, et al., The organ-cultured cornea: an in vitro study. Investigative Ophthalmology, 1973. 12(3): 176-180.
200. Uittenbogaard A, Everson WV, Matveev SV, et al., Cholesteryl ester is transported from caveolae to internal membranes as part of a caveolin-annexin II lipid-protein complex.[Retraction in Journal of Biological Chemistry 2013 March 1;288(9):6586; PMID: 23457391]. Journal of Biological Chemistry. 277(7): 4925-4931.

201. Uittenbogaard A, Ying Y, and Smart EJ, Characterization of a cytosolic heat-shock protein-caveolin chaperone complex. Involvement in cholesterol trafficking.[Retraction in Journal of Biological Chemistry 2013 March 1;288(9):6587; PMID: 23457392]. Journal of Biological Chemistry. 273(11): 6525-6532.
202. Uittenbogaard A and Smart EJ, Palmitoylation of caveolin-1 is required for cholesterol binding, chaperone complex formation, and rapid transport of cholesterol to caveolae.[Retraction in Journal of Biological Chemistry 2013 March 1;288(9):6585; PMID: 23457390]. Journal of Biological Chemistry, 2000. 275(33): 25595-25599.
203. Uittenbogaard A and Smart EJ, Palmitoylation of caveolin-1 is required for cholesterol binding, chaperone complex formation, and rapid transport of cholesterol to caveolae.[Retraction in Journal of Biological Chemistry 2013 March 1;288(9):6585; PMID: 23457390]. Journal of Biological Chemistry, 3004. 275(33): 25595-25599.
204. Unknown. NEED TITLE. The New York Times. 1991 March 23.
205. Unknown. NEED TITLE. Washington Post. 1991 March 22.
206. Unknown, Misconduct found in Harvard animal morality prof's lab—life. NewScientist, 2010.
207. Van Parijs L, Peterson DA, and Abbas AK, The Fas/Fas ligand pathway and Bcl-2 regulate T cell responses to model self and foreign antigens. [Retraction in Van Parijs L, Peterson DA, Abbas AK. Immunity. 2009 April 17;30(4):611; PMID: 19378493]. Immunity, 1998. 8(2): 265-274.
208. Van Parijs L, Refaeli Y, Lord JD, et al., Uncoupling IL-2 signals that regulate T cell proliferation, survival, and Fas-mediated activation-induced cell death.[Retraction in Van Parijs L, Refaeli Y, Lord JD, Nelson BH, Abbas AK, Baltimore D. Immunity. 2009 April 17;30(4):611; PMID: 19382300]. Immunity, 1999. 11(3): 281-288.
209. Vergano D. Updated: Harvard says Marc Hauser guilty of science misconduct. USA Today. 2010 July 20.
210. Wade N, A diversion of the quest for truth. Science, 1981. 211(4486): 1022-1025.
211. Wade N. Harvard researcher may have fabricated data. The New York Times. 2010 August 27.
212. Wade N. Harvard finds scientist guilty of misconduct. The New York Times. 2010 August 20.
213. Wade N. Inquiry at Marc Hauser's Harvard lab, a raid and then a 3-year wait. The New York Times. 2010 August 13.
214. Wang L, Brock A, Herberich B, et al., Expanding the genetic code of Escherichia coli. Science, 2001. 292(5516): 498-500.
215. Wang L, Brock A, Herberich B, et al., Expanding the genetic code of Escherichia coli. Science, 2001. 5516(498-500): 498.

216. White J, Guerin T, Swanson H, et al., Diabetic HDL-associated myristic acid inhibits acetylcholine-induced nitric oxide generation by preventing the association of endothelial nitric oxide synthase with calmodulin. [Retraction in American Journal of Physiology—Cell Physiology 2013 April 15;304(8):C812; PMID: 23588580]. American Journal of Physiology—Cell Physiology. 294(1): C295-305.
217. White J, Guerin T, Swanson H, et al., Diabetic HDL-associated myristic acid inhibits acetylcholine-induced nitric oxide generation by preventing the association of endothelial nitric oxide synthase with calmodulin.[Retraction in Journal of Biological Chemistry 2013 April 15;304(8):C812; PMID: 23588580]. American Journal of Physiology—Cell Physiology. 294(1): C295-305.
218. White WN. Harvard faculty dean confirms misconduct in Hauser's laboratory. Michael S. Smith confirmed eight instances of scientific misconduct. The Harvard Crimson. 2010 August 20.
219. Wijmenga C, Gregory PE, Hajra A, et al., Core binding factor beta-smooth muscle myosin heavy chain chimeric protein involved in acute myeloid leukemia forms unusual nuclear rod-like structures in transformed NIH 3T3 cells.[Erratum appears in Proceedings of the National Academy of Sciences of the United States of America 1996 December 24;93(26):15522]. Proceedings of the National Academy of Sciences of the United States of America, 1996. 93(4): 1630-1635.
220. Wood JN, Glynn DD, Phillips BC, et al., The perception of rational, goal-directed action in nonhuman primates.[Erratum appears in Science. 2011 April 29;332(6029):537]. Science, 2007. 317(5843): 1402-1405.
221. Xu R, Hanson SR, Zhang Z, et al., Site-specific incorporation of the mucin-type N-acetylgalactosamine-alpha-O-threonine into protein in Escherichia coli.[Retraction of Xu R, Hanson SR, Zhang Z, Yang YY, Schultz PG, Wong CH. J Journal of the American Chemical Society 2004 December 8;126(48):15654-15655; PMID: 15571382]. Journal of the American Chemical Society, 2009a. 131(38): 13883.
222. Zamecnik PC and Long JC, Growth of cultured cells from patients with Hodgkin's disease and transplantation into nude mice. Proceedings of the National Academy of Sciences of the United States of America, 1977. 74(2): 754-758.
223. Zhang Z, Gildersleeve J, Yang Y-Y, et al., A new strategy for the synthesis of glycoproteins.[Retraction in Zhang Z, Gildersleeve J, Yang YY, Xu R, Loo JA, Uryu S, Wong CH, Schultz PG. Science. 2009 November 27; 326 (5957):1187; PMID: 19965450]. Science, 2004. 303(5656): 371-373.
224. Zhang Z, Gildersleeve J, Yang Y-Y, et al., A new stategy for the synthesis of glycoproteins. Science, 2004. 303(5656): 371-373.

225. Zhang Z, Gildersleeve J, Yang Y-Y, et al., Retraction. A new strategy for the synthsis of glycoproteins [Retraction of Zhang Z, Gildersleeve J, Yang YY, Xu R, Loo JA, Uryu S, Wong CH, Schultz PG. Science 2004 January 16; 303 (5656): 371-373; PMID: 14726590) Science, 2009.
226. Zhu W and Smart EJ, Myristic acid stimulates endothelial nitric-oxide synthase in a CD36—and an AMP kinase-dependent manner. Journal of Biological Chemistry. 280(33): 29543-29550.
227. Zhu W and Smart EJ, Myristic acid stimulates endothelial nitric-oxide synthase in a CD36—and an AMP kinase-dependent manner.[Retraction in Journal of Biological Chemistry 2013 March 1;288(9):6583; PMID: 23457388]. Journal of Biological Chemistry. 280(33): 29543-29550.
228. Zhu W and Smart EJ, Myristic acid stimulates endothelial nitric-oxide synthase in a CD36—and an AMP kinase-dependent manner.[Retraction in Journal of Biological Chemistry. 2013 March 1;288(9):6583; PMID: 23457388]. Journal of Biological Chemistry, X. 280(33): 29543-29550.

CHAPTER 16

Modern Non-American Research Misconduct

Research Misconduct in Australia

William Griffith McBride. In a letter to the editor of *The Lancet* William Griffith McBride (1927—), one of Australia's most prominent gynecologists and obstetricians, was the first to alert the world of the dangers of thalidomide (Distaval) [144]. His astute anecdotal observations indicated that this sedative prescribed for anxiety and morning sickness was causing infants to be born with severe limb deformities. Later more detailed investigations proved him right and this drug was withdrawn from the market preventing additional births of children with severe shortening or complete absence of the limbs [143]. Two decades later, McBride published a report on experiments with hyoscine, a component of the morning-sickness drug called Debendox, claiming that it clearly caused birth defects in rabbits. Because this obstetrician became famous for his original observations related to the dangers of thalidomide, prompt attention was paid to his remarks and the pharmaceutical company Merrell Dow removed Debendox from the market in 1983 amid a flood of lawsuits [145]. Unfortunately for McBride one of his assistant's resigned after discovering that his boss had changed some original figures and altered data in research carried out by assistants. Four years later the row went public when a radio program by the Australian broadcasting corporation accused McBride of scientific fraud [129, 210]. An initial informal inquiry into McBride's research in 1988 found him guilty of scientific misconduct, but the investigating body was powerless to inflict a punishment. Nevertheless his professional reputation was severely damaged and his medical practice deteriorated. The New South Wales health department held a hearing in 1989, but after perhaps the longest inquiry in medical history McBride fell dramatically from grace in 1993 when he was found guilty by the medical tribunal of New South Wales in Australia of scientific fraud in four of seven allegations against him [148]. (http://www.independent.co.uk/news/world/thalidomide-doctor-guilt-of-medical-fraud-wil . . . 1.30/2010).

He published statements that he knew were false and misleading and after the inquiry into his fraudulent behavior started he was forced to sell properties to pay for his defense. This beloved member of Sydney society was humiliated and struck off the Australian medical register at the age of 65 years, but he was reinstated six years later in 1998.

Ronald Wild. Professor Ronald Wild the dean of social sciences at La Trobe University (http://www.latrobe.edu.au/) in Australia published a textbook in 1985 [219]. Soon thereafter major portions of the book were found to be plagiarized from 10 separate sources without giving credit to the original authors. Under mounting pressure because of the obvious plagiarism the publisher withdrew the book and La Trobe University investigated the allegations. Before the investigations were complete Wild resigned from the university in 1986. His dishonest behavior was not detrimental to his career as he rapidly obtained a lucrative academic position in Hedland college of technical and further education in northwest of Australia because of his "academic experience and his leadership" [88].

Ashoka Prasad. Ashoka Prasad a psychiatrist at the Victoria mental health institute in Australia fabricated data on one thousand patients with schizophrenia allegedly determining that the incidence was higher in the winter. A committee of inquiry reported its findings to the Victoria state parliament.

Other. Another Australian researcher Michael Briggs performed fraudulent research in Australia involving human subjects.

Research Misconduct in China

With the exception of the USA China has more publications in international journals than any other country, but a very high percentage of them do not withstand a critical evaluation and many have been plagiarized. In China scientific and academic misconduct, and particularly plagiarism, runs rampart and a detailed review of it is beyond the scope of this book. Ethical interruptions s are also extremely widespread according to Professor Rao Yi, dean of the life sciences school at Peking University (http://english.pku.edu.cn/) in the capital.

Since the early 1990s the Chinese government has provided its academic community considerable financial assistance, but a considerable number of its researchers have been accused of cheating as a way to acquire coveted positions. The dishonesty has ranged from fudging resumes to fabricating data and Chinese leaders tolerate misconduct [231].

Like academicians elsewhere those in China are under pressure to publish for promotion and the salaries of Chinese professors are often linked to research grants. In sharp contrast to other countries plagiarism, ghostwriting and the fabrication of research findings are out of control in China and have prospered [58, 227].

The works of others are translated into Chinese as their own without attribution. An example in that country was exposed in the *Journal of Dialectics of Nature* (before early 1996) and later in *Science* [230]. Cong Cao of the department of sociology at Columbia University (Columbia) (*http://www.columbia.edu/*) became aware of this open secret while performing research in Beijing. According to Cao pressure had been asserted to not publicly discuss the incident and indeed three journals declined to publish an article bringing the case to public attention. The fraudulent research in China is so widespread that a British journal needed to retract 70 papers by the same two senior authors in December 2009 because of fabricated research at a Chinese university [226]. When Chinese professors seek academic promotion they frequently contact Chinese professional writers, such as the ex-school teacher Lu Keqian, who provides ghost written papers for anyone prepared to pay his fee usually 300 yuan ($45) [226]. According to Gillian Wong, a writer for Associated Press [226] professor Shen Yang at the Wuhan University (http://en.whu.edu.cn/) discovered a fivefold increase in ghostwritten papers in 2009 compared to 2007 numbers rather than quality in academic promotions as well as the weak penalties. Keqian does not consider it wrong to write papers for others who need assistance. Nearly 1 billion yuan (more than $145 million), was consumed on academic publications in 2009, up fivefold from 2007, according to an analysis by Shen Yang, a professor at Wuhan University (http://en.whu.edu.cn/).

A company providing ghostwriting in the southern industrial Chinese city of Liuzhou includes the Lu Ke Academic Center, which includes a network of graduate students and professors willing to write on a wide variety of topics ranging from computer technology to military affairs.

Clients contact the Lu Ke Academic Center mainly by way of Internet chat programs and Lu maintained that on one occasion 10 students from the same college class collectively applied for him to compose their papers. Perhaps not surprising the products of the lucrative ghost writing business are frequently plagiarized texts. The Wuhan study found that more than 70% of the papers sold in 2007 were plagiarized.

Early in 2009, internet users discovered that the deputy principal of Anhui Agricultural University (http://www.ahau.edu.cn/ plagiarized more than 20 papers [228]. The university relieved him of this academic position, but still permitted him to continue teaching.

The Chinese have started to retract flawed publications. An example is a 2011 paper on the chloroplast in wheat in *Acta Biochimica et Biophysica Sinica* [61], which was retracted [62].

After learning about the finding of major data misrepresentation of data in an investigation by the Shangai Institute of biological sciences, Chinese academy of sciences the *Journal of Neuroscience* retracted an unreliable paper by Chao Guo and coauthors [80].

The Israeli professor Dan Ben-Canaan taught at Heilongjiang University (*http://www.at0086.com/hlju/*) in the Chinese city of Harbin recalls stories that suggest cultural difference between the Chinese and Western civilizations. In 2008 a colleague contacted him about a manuscript that he had written about a Jewish musician who was kidnapped and murdered in 1933 in Harbin during the Japanese occupation. This person had the audacity to deliver this paper as his own without embarrassment at a conference organized by Ben-Canaan. On another occasion Ben-Canaan provided a document to a scientist at the esteemed Chinese academy of social sciences and was flabbergasted to later be given a book written by this individual that was predominantly a translation of the information Ben-Canaan had given him, but in the absence of acknowledgements.

Research Misconduct in the Congo

Serge Valentin Pangou. Serge Valentin Pangou an ecology researcher in the Congo found at least seven of his papers that have either been retracted or whose editors have agreed to retract because of plagiarism [158, 161] *"Meaning of tree size on the reproductive phenology in Pterocarpus soyauxii Taubert (Fabaceae) in Mayombe rain forest, Congo,"* published in 2011 in the *International Research Journal of Agricultural Science and Soil Science* (removed from the journal's site) [159]. *"Tree selection for fruit/seed production in Carapa procera D.C. from wild stands in the rain forest of Congo,"* published in 2011 in the *International Research Journal of Plant Science* (removed from the journal's site) [157, 160] The downfall of Pangou occurred after Patrick Jansen an ecologist at Wageningen University (http://www.wageningenur.nl/en/wageningen-university.htm) in *the Netherlands* received a manuscript to review for the *International Journal of Biodiversity and Conservation*. The paper was strikingly similar to one that Jansen had coauthored with Pierre-Michel Forget in *Conservation Biology* [68]. To confirm plagiarism Jansen compared their paper to the one by Pangou with the software package Turnitin (http://turnitin.com/) and found the papers 90% identical. As pointed out by Michael Balter [20] Pangou unfortunately plagiarized a paper by Forget the son of a "well-known private detective in France". Together with Jansen Forget scrutinized other publications by Pangou and discovered other examples or plagiarism.

Research Misconduct in Germany

Friedhelm Herrmann and Marion Brach. The most infamous scandal encompassing scientific fraud in Germany, and indeed in Europe, involved Friedhelm Herrmann and Marion Brach. By March 1999 these two prominent German molecular biologists published numerous papers containing seemingly fabricated data [1]. Institutional committees pronounced strong doubts about the validity of 47 papers by these two prominent cancer researchers because of falsified data. A figure in one of Herrmann and Brach's articles led to the detection of their scientific misconduct [4, 5, 14, 21-23, 26-56, 64-66, 70-79, 81-86, 116-119, 121, 138, 139, 156, 168, 169, 174, 181, 182]. Changes related to those here [82-86]. The figure contained four columns that separated the three elements at four different time points on an autoradiograph. The images clearly showed duplicate data. Lane 1 was identical to lane 4 and lane 2 contained the same images as lane 3, but part of the figure was turned around. The whistle blower in this scandal was Earhardt Hildt, a young postdoctoral fellow who worked with these two scientists at the Max Delbrück center for molecular medicine in Berlin. The PhD mentor of Hilt in Munich played a crucial role in toppling the reputation of Herrmann and Brach. German research organizations assigned much energy to restoring the harm done to the science community because of the fraud of Hermann and Brach. Both the Deutshe Forschungsgemeinschaft (DFG) and the Mildred Scheel foundation funded the research of Herrmann and Brach and an investigation by these two agencies tried to unravel the complex mixture of truths and falsehoods. Software was used to scrutinize portions of their enormous bibliography of 550 papers and 82 book chapters. While the panel was investigating the Hermann-Brach scandal the DFG made recommendations as to how alleged scientific misconduct should be conducted and what should be done to foster ethical research. The DFG suggested that research funds should be denied to institutions that failed to adopt effective procedures [122]. During an investigation drawn out over 18 months under the leadership of Ulf Rapp a professor of molecular and cell biology at the University of Würzburg *(http://www.uni-wuerzburg.de/en/ueber/university/)* many journal papers and book chapters written by this cancer research team sometimes with collaborators were found to contain deceptively fabricated data. Data manipulation was evident in 94 papers and in 121 papers may have contained defective information. Many more scientists than Hermann and Bach may have been involved in the fraudulent research [7]. Only 132 of 347 papers published by Hermann between 1985 and 1996 were cleared of any suspicion of fraud.

[3]. Hermann denied any knowledge of the falsification, but he was barred from participating in DFG advisory boards and he was provisionally removed from the university. Brach, who admitted falsifying data in two or three publications, lost his professorship at Lübeck.

Inge Czaja. Shortly after the emergence of the Hermann-Bach scandal Germany faced another scientific humiliation when Inge Czaja, a research technician at the Max Planck institute for plant breeding in Cologne confessed to fraud in a scientific paper. Because of concern that her fraudulent behavior might have involved experiments performed over 6 years scientists at the Max Planck Institutes for Plant Breeding repeated all investigations documented in manuscripts using assays performed by Czaja [2]. Within several months the chore was concluded and a list of non-replicated papers was published in March 1999 in the *Plant Journal* [24].

Research Misconduct in India

Pattium Chiranjeevi. An investigation of Pattium Chiranjeevi, an Indian chemistry professor at Sri Venkateswara University (SVU) (*http://www.svuniversity.in/*) in Tirupati in India was initiated after a reviewer of a manuscript submitted by him to *Analytica Chimica Acta* (*ACA*) noted that it was almost identical to one already published by different authors in another journal [177]. After receiving this information the editor of *ACA* notified the chair of the SVU chemistry department about this plagiarism. A three-member university panel investigated the matter and found Chiranjeevi guilty of publishing dozens of unethical and fraudulent research papers. Chiranjeevi also cited the use of scientific equipment that did not exist at his university and included unjustified persons as coauthors. Gary Christian, one of the editors-in-chief of the *Talant*a investigated papers by Chiranjeevi in that journal and this led to the retraction of five papers published between 2003 and 2007. Many of his publications were retracted [18, 19, 120, 123-128, 163, 165, 166, 179, 204-209, 216]. Elsevier journals retracted another 8 of his publications. *Environmental Monitoring and Assessmen* suspended the publication of 6 papers by Chiranjeevi that were in press and investigated other publications by him in that journal. SVU did not fire Chiranjeevi, but punished him for unethical behavior involving dozens of papers by not allowing him to hold an administrative position or to mentor students. SVU also denied him pay raises. Chiranjeevi stopped receiving research funding. Chiranjeevi denied any wrong doing and it is noteworthy that Chiranjeevi was senior author of the articles. A red flag pointing suspicion to Chiranjeevi's deceitful research was his extremely large list of publications (66) between 2004 and 2007.

Viswa Jit Gupta. Viswa Jit Gupta, a professor at Punjab University *(http://www.pu.edu.pk/)*, not only plagiarized the work of others, but also fabricated geological findings in the Himalayas [203]. Gupta was exposed by John A. Talent, an Australian scientist, who found that many of the fossilized animals alleged to be found in the Himalayas are actually found in other parts of the world and are most unlikely to be endogenous to this mountain range [203].

Reserach Misconduct in Iran

Top Iranian government scientists, including Kamran Daneshjou, a former minister of science were accused of including plagiarized material in four scientific papers [11] describing experiments related to tungsten alloy rods ricocheting off steel plates. One Iranian paper coauthored by Kamran Daneshjou, a professor in the school of engineering at the Iran University of Science and Technology (UST) in Tehran and Majid Shahravi, his former PhD engineering student reported by *Nature* [57] and that journal discovered plagiarism in two more articles by the same authors. Major parts of an article by Daneshjou and Shahravi in Engineering and Computers [63] were identical to an article reported seven years early by South Korean scientists. Plagiarism has also detected in another 2006 paper coauthored by the Iranian roads and transport minister and the former PhD advisor of president Mahmud Ahmadinejad. Four of these papers were retracted [162] and the incident created an outrage by émigré scientists from Iran and this enhanced Daneshjou's role in the widely rejected allegedly fraudulent Iranian re-election of president Ahmadinejad in 2009. The scapegoat in this episode became Shahravi and he is stated to have admitted his guilt and will face punishment. However, Shahravi denied plagiarism in an interview with Iranian media and pointed out that papers cite the work of others in footnotes.

Research Misconduct in Japan

Remarkably few examples of scientific dishonesty have been recognized in Japan and two explanations have been proposed. Some ascribe the paucity of scientific dishonesty to an inherent truthfulness of Japanese scientists. Others suspect that it just mirrors the Japanese culture in which individuals look the other way to situations that may provoke social frictions. The hierarchal configuration of faculties, especially within traditional universities and the dictatorial powers of the professor act as disincentives for researchers to become whistle-blower after witnessing questionable scientific activities. Indeed in Japan whistle blowing is rare and those who have ventured into this activity have not escaped unharmed and some have even lost their jobs. Young investigators working in a Japanese research laboratory have little if any hope in influencing questionable scientific practices. An authoritative professor dictates the research procedures and junior members of the team have virtually no say in how it is performed.

While the incidence of scientific misconduct appears to be extremely low in Japan scientists in that country are perceived by the public as being less ethical than desired according to a public survey by the Kanazawa Institute of Technology. In the late 1990s the public trust in science plunged in Japan after several nuclear accidents and succeeding cover-ups as well as a scandal involving the fabrication of research data at nuclear fuel recycling plants [173].

Naoki Mori. Professor Naoki Mori, a Japanese virologist received numerous awards for his research on *human T-lymphotropic virus type I* (HTLV-I), a *retrovirus* which causes *adult T-cell leukemia*. Mori published a paper with image manipulation. As punishment for this deed the *American Society of Microbiology* which owns the journal banned further publications in their official journal for 10 years. Other papers by Mori published in *Leukemia Research, Biochemical and Biophysical Research, Infection and Immunity, International Journal of Cancery and Blood* {} have been retracted. As of October 2011, thirty papers coauthored by Mori have been retracted, Marcus, Adam (12 October 2011). For his research dishonesty the University of the Ryukyus (http://www.u-ryukyu.ac.jp/en/) in the town of *Nishihara* on *Okinawa Island* fired Mori only to rehire him again at a later date. His publications between 2000 and 2009 documented research on *H. pylori*. (*http://news.sciencemag.org/scienceinsider/2011/01/japanese-virologist-loses-job.html*) (*http://retractionwatch.wordpress.com/2010/12/24/japanese-virologist-hit-with-publishing-ban-after-widespread-data-manipulation/*). He was determined that he had manipulated data and images in a number of articles *"Japanese Virologist Loses Job, Gets Publishing Ban for Image Manipulation". Science. http://news.sciencemag.org/scienceinsider/2011/01/japanese-virologist-loses-job.html*. "That's a Mori! Seven more retractions brings latest count to 30". (*http://retractionwatch.wordpress.com/2011/10/12/thats-a-mori-seven-more-retractions-brings-latest-count-to-30/#more-4770*). Mori took "full responsibility" for "multiple inaccurate and inappropriately duplicated" images *"'Manipulation' nets virologist a 10-year journal ban". Times Higher Education. (http://www.timeshighereducation. co.uk/story.asp?sectioncode=26&storycode=414802&c=1)*., and was dismissed from his university post in August 2010. *Science. (http://news.sciencemag.org/scienceinsider/2011/01/japanese-virologist-loses-job.html)*. This disgraced Japanese researcher *lost more than 23 paper*s to retraction for image manipulation and duplication.**Shigeaki Kato.** In January 2012 an outside whistleblower raised questions about the validity of 24 papers by the Japanese endocrinology researcher Shigeaki Kato at the Institute of Molecular and Cellular Biosciences at the University of Tokyo *(http://www.u-tokyo.ac.jp/en/)*. This precipitated an investigation into scientific misconduct by him and one of his papers was retracted in March 2012.

Yoshitaka Fujii. At various times seven different Japanese institutions employed the anesthesiologist *Yoshitaka Fujii* despite almost two decades of questionable data. A panel investigating the validity of his research found 172 of his fabricated papers over the 19 years. If all of these flawed papers get retracted he would hold the world record in the number of retractions by a single author [13]. The committee on publications ethics (COPE) informed officials at the institutions where he performed his research that unless they validate his 193 papers they will be retracted. Fuji was fired by Toho University *(http://www.toho-u. ac.jp/english/)* in February 2012 allegedly for not obtaining ethical approval of several research projects, but when his position was terminated journal editors were concerned that the university would not investigate the integrity of Fujii's data. In 2012 the British anesthesiologist John Carlisle published a *statistical meta-analysis* of randomized clinical trials on the prevention of postoperative nausea and vomiting and compared studies by Fujii and his colleagues with other authors. The analysis disclosed unusual distributions and cast grave distrust on the authenticity of his research [59].

Research lead by the Japanese chemist Kazunari Taira could not be replicated. Concerns about the inability of numerous researchers to replicate the studies were drawn to the attention of the RNA Society of Japan during the Spring of 2005, which reported the problematic research to Tokyo University initiating an investigation into scientific misconduct. The lack of replicability was apparently attributed to the improper recording of the experimental data. Questions were raised about the research content in 11 papers published between 1998 and 2004 in Nature, Nature Biotechnology and the Proceedings of the National Academy of Sciences and other scientific journals [153]. First author on 10 of the 11 papers was Hiroaki Kawasaki, a research associate working in the Taira laboratory. The corresponding author on nine articles was Taira and the other two articles listed Taira and Kawasaki as coauthors [108-113]. Yoichiro Matsumoto, a mechanical engineer in the Graduate School of Engineering at Tokyo University was selected to head the investigative panel looking into the Taira affair and the selected four papers for detailed scrutiny, but found that raw data could not be produced to substantiate the research. Taira claimed that he could replicate the experiments so the panel requested him to do so [152]. Kawasaki claimed that he had replicated the experiments in one paper but the materials used were not as documented in the manuscript. Taira attempted to repeat the experiments for the panel, but was unable to do so within a reasonable time frame. Following this first ever investigation into research misconduct at Tokyo University both Taira and Kawasaki faced disciplinary action and the university considered establishing a permanent committee or an office to investigate research misconduct.

Research Misconduct in the Netherlands

Died*erik Stapel*. Diederik Stapel, a highly regarded Dutch social psychologist started working at Tilburg University *(http://www.tilburguniversity.edu/)*, which specializes in the social sciences, in the southern Netherlands in 2006 and became head of the Tilburg institute for behavioral economics research. He gradually became a prolific investigator and flourishing fund raiser. His research generated innovative ideas as to how the brain functions and his findings attracted the attention not only of the media, but also of policy makers. He found that individuals are more prone to stereotype or discriminate in chaotic environments [202]. In contrast to the usual custom of researchers not speaking to representatives of the media until research findings are published, or at least in press, Stapel was a publicity seeker and on one occasion made headlines with a press release alleging that "meat brings out the worst in people". This claim was made even though the research project on which this conclusion was based had not passed peer review. Indeed a draft manuscript had not even been completed. Junior investigators working in the Stapel laboratory reported him to Philip Eijlander the rector of the university accusing Stapel of using fabricated data in his research and Staple acknowledged his use of fabricated data in some papers [12](*http://scim. ag/_Stapel*). Immediately after Tilburg University found him guilty of fabricating data in some studies the Triburg University fired him on September 7, 2011 and removed his webpage. The rector pointed out that having admitted to using faked data Stapel would not be permitted to return to the university and his "tainted papers" will be retracted. Having already acted on the allegations the university requested Willem Levelt, a former president of the royal Netherlands academy of arts and sciences, to chair an inquiry into the scope of his fraudulent research. The allegations against Stapel were particularly shocking as he was widely regarded as the most eminent social psychologist in Europe and served on the editorial boards of six different professional journals. He was also the recipient of the 2007 "early career award" from the international society for self and identity (*http:// persistentastonishment.blogspot.com/2011/09/dieterik-stapel-and-frequency-of.html*). The allegations of research misconduct by Stapel were extensively investigated by three separate committees—one from each of the universities where Stapel worked. The committees were from the Free University of Amsterdam *(http:// www.vu.nl/en/privacy/index.asp?Referer=/en/index.asp)* Tilburg University *(http:// www.tilburguniversity.edu/)* and University of Groningen *(http://www.rug.nl/kvi/)*. The three committees announced their findings on 28 November 2012 [67]. They not only found Stapel guilty of research misconduct, but faulted his coauthors as well as the reviewers of his papers and the editors of the journals in which his work was published. They found that "The critical function of science has failed on all levels". Aside from flagrant research misconduct other errors reflected unintentional mistakes, such as sloppy research, errors in statistical evaluation, the removal of participants from data for questionable reasons, and the failure to report parts of experiments. When Stapel's career collapsed his family was distraught and he contemplated suicide. In the final analysis 55 of Stapel's 137 publications contained fabricated data and the same situation was found in 10 PhD theses of students supervised by him. In ten other papers research misconduct could not be excluded without reservations, because of missing original data and a failure of Stapel to acknowledge fraud. On the same day that the three committees announced the conclusions of their investigations the publicity seeking Stapel issued a video statement that had been prepared by a Dutch television station at his home. He deeply regretted his wrongdoing and announced that he had written an autobiography titled *Ontsporing (*Derailment) explaining his downfall [185]. The book appeared the next day causing a media outburst. Two days later the book was on the internet at a file sharing web site as a 324 page portable document format (PDF) for a short period of time, before it reappeared at a new site. To capitalize on personal research misconduct by writing a book did not go down well. In his autobiography Stapel does not deny his fraudulent research, but points out that he did not know how to stop after this became a way of life. "I had become addicted to the rhythm of digging, discovering, testing, publishing, and applause". His book documents how he planned experiments for PhD students and colleagues. Careful questionnaires ended in the trash heap. He insisted on collecting the data himself usually at a high school or university where he had good contacts [15-17, 25, 69, 105-107, 130-133, 141, 142, 150, 167, 170-172, 175, 176, 183, 184, 186-202, 212-215].

Don Poldermans. Before November 16, 2011, Don Poldermans was an esteemed extremely productive scientist at the Erasmus Medical Center in *Rotterdam* when allegations about his research integrity were raised. His research focused on complications of cardiovascular surgery and he authored about 600 papers (Marcus A. November 17, 2011 Retractionwatch.com *http:// retractionwatch.wordpress.com/2011/11/17/breaking-news-prolific-dutch-heart-researcher-fired-over-misconduct-concerns/*). Sixteen of his publications had been cited at least 100 times and one study had more than 700 citations [89] (*http:// www.forbes.com/sites/larryhusten/2011/11/20/more-details-emerge-about-the-dutch-research-scandal/*), (*http://www.dutchnews.nl/news/archives/2011/11/erasmus_ medical_centre_suspend.php*) (*http://www.nrc.nl/nieuws/2011/11/17/nieuw-geval-van-wetenschapsfraude-hoogleraar-erasmus-mc-ontslagen/*)(*http://www.erasmusmc. nl/perskamer/archief/2011/3488672/*). After a committee at the Erasmus medical center found Poldermans guilty of several episodes of research misconduct (van der Maas, P.J, Löwenberg B, Peters R.J.G, Rabelink A.J., Oosting J.M, Juttmann R.E., Struhkamp R.M. 16 Nov 2011. Report Summary [Internet]. Erasmus MC, University Medical Center Rotterdam: Committee for the Investigation of Scientific Integrity; *http://www.erasmusmc.nl/5663/135857/3397899/report_ summary_investigation_integrity?lang=en*) the medical center fired him. The first of his projects questioned related to the dutch echocardiographic cardiac risk evaluation applying stress echocardiography (DECREASE) designated DECREASE II, III, IV and VI (Reducing cardiac risk in non-cardiac surgery: evidence from the DECREASE studies *http://eurheartjsupp.oxfordjournals.org/ content/11/suppl_A/A9.full*). Poldermans failed to obtain acceptable written consent from the human subjects and his actions violated scientific behavior and patient trust. None of the participants in the studies were injured, but he could have potentially harmed their health with his illegal and indefensible activities. Moreover Poldermans failed to gather data as defined in the research protocols. In two projects DECREASE II and DECREASE VI, serious errors existed because of a failure to follow procedures. The studies were also subject to biases created by Poldermans and his associates. The committee found that Poldermans fabricated data in a number of cases. No other researchers were involved in this scandal. In view of his esteemed status junior coworkers did not probe his know-how and took for granted that he was behaving correctly. As an additional violation of correct scientific behavior the committee also concluded that Poldermans intentionally submitted manuscripts for publication while being aware that the data integrity was not trustworthy. His ongoing project DECREASE VI was aborted because it was considered unethical to have patients participate in a clinical trial in which they had not consented. Poldermans accepted the committee's assessment, but he maintained that his research misconduct was unintentional.

Dirk Smeesters. The unique Uri Simonsohn method for detecting suspicious statistical patterns disclosed questionable publications of Dirk Smeesters of Erasmus University in Rotterdam and at least one of his papers was retracted [104].

Research Misconduct in Norway

The Norwegian oncologist Jon Sudbø participated in scientific misconduct involving patients with oral cancer

Research Misconduct in Poland

Andrzej Jendryczko. An example of plagiarism was discovered accidentally by Jan Fallingborg when he searched the medical literature for papers on selenium concentrations in ulcerative colitis. He had been the senior author of a paper on the subject in a Danish journal [149] and to his surprise he found that along with his own 1989 abstract on the subject, the computer coughed up an almost precise replica of publications by Andrzej Jendryczko, a chemical engineer and former professor of biochemistry at the Silesian Medical University (SMU) *(http://www.slam.katowice.pl/)* of Katowice in Poland, and three other Polish authors in 1992 [99]. Dr. Marek Wronski, a researcher in neuro-oncology at the Staten Island University in New York *(http://hospitals.nyhealth.gov/browse_view.php?id-233)*, had never heard of Jendryczko prior to becoming aware of his plagiarized paper in 1997 checked his publications and unearthed a vast body of plagiarized papers by Jendryczko. Several years later other authors summarized the scandal in the *British Medical Journal* while it was being investigated by two Polish institutions [232]. Jendryczko, mass produced plagiarized papers over many years. Following an investigation by the university more than forty papers were declared to be plagiarized by SMU [229].

Wronski pointed out that Jendryczko was one of the most bizarre plagiarists. The vast majority of his plagiarized publications spanned a wide diverse spectrum of subspecialties in medicine. They covered such diverse topics as lead poisoning [102], mitochondrial aging [98], atherosclerosis and myocardial infarction [103], neonatal growth{}, systemic lupus erythematosus [94], the composition of milk

[95], hepatocytes and the liver [92], zinc and copper in cancer tissue [91, 96], cholesterol and hypertension{}, enzymes in the placenta [93], cancer of the cervix (), menopause [101], the effects of selenium [100], and ionizing radiation [97]. Publications on these subjects would raise a red flag for any physician who would dare to list such diverse original papers in curriculum vitae.

What benefit could a chemical engineer hope for with these medical publications? If it were not for Wronski and Medical Literatures Analysis and Research System (MEDLINE) on the internet he might not have been exposed. In one paper Jendryczko reported data on 300 patients in a single paper {} and another paper documented studies on one thousand patients without providing credit to the medical doctors who must have taken care of them. In his plagiarized publications Jendryczko listed about a dozen coauthors, including some who were full professors or heads of university departments. It is not apparent whether they were factious or given honorary authorship without their permission. At least 30 biomedical research papers repeated verbatim passages from other authors without appropriately acknowledging the primary sources. Wronski found it extremely disturbing at what he considered a mild initial response by the Polish authorities to the plagiarism of the Danish text. After a 13 months inquiry behind closed doors the Polish university officials decided not to punish Jendryczko because his indiscretion had passed the Polish law on higher education had passed the three year time limit of the statute of limitations. Following Jendryczko's resignation from SMU in early 1997, he was appointed professor at the polytechnic institute of Czestochowa in Poland. Wronski claimed that several individuals were familiar with the plagiarized papers and nobody took action, because the exposure of scientific misconduct was prohibited by an "old boys" network. He was told to keep quiet, because news about research misconduct would set back Polish science. Very few of the fraudulent papers by Jendryczko were retracted. Wronski contacted colleagues in Poland for photocopies of Jendryczko's papers and within a week, he received 90 of them in a package.

Research Misconduct Singapore

In Singapore accusations of research misconduct are treated exceedingly gravely and are investigated at a senior level.

Alirio Melendez. At least three papers by *Alirio Melendez*, a former researcher with expertise in immunization and infection at the National University of Singapore (NSU) were retracted after an investigation indicated that some of his published data in scientific journals were manipulated or copied from other work. One of this papers in the *Journal of Cellular Physiology* was retracted by the journal editor in chief, Dr. Gary S. Stein, and by Wiley Periodicals, Inc. the publisher (*http://www.liverpoolecho.co.uk/liverpool-news/local-news/2011/10/12/*

liverpool-university-academic-professor-alirio-melendez-suspended-over-fake-research-investigation-100252-29580675/#ixzz1rwiaEAX).

The possibility of scientific misconduct by Melendez was precipitated by a persuasive anonymous email that raised serious doubts about 70 papers that he had authored with a group of collaborators. He was forced to retract a paper in *Nature Immunology* [164] and another paper by him was listed as an Expression of Concern in *Science*. An investigation into the validity of the research by Melendez involved eight researchers from NUS and the Defense Science Organization (DSO) national Laboratories that included scientists, academics, research fellows and students. A panel of professors experienced in the investigation of fraudulent research from different countries was appointed to look into the matter. After being discharged from the SNU Melendez became employed in the department of molecular and clinical pharmacology at the University of Liverpool *(http://www.liv.ac.uk/)* in August 2011. This university suspended him without prejudice, awaiting the conclusion of the inquiry. According to Retraction Watch on October 12, 2011 Melendez informed Richard van Noorden an assistant news editor of *Nature* that he was currently unwell and on sick leave, and that he was conducting his own investigation into other papers which he agreed did contain "questionable data", but which, he asserted, were not his fault.

Research Misconduct in South Africa

An extremely controversial study involving patients with breast cancer was carried in South Africa by Werner Bezwoda.

Research Misconduct in South Korea

Woo Suk Hwang. A team of scientists in South Korea under the leadership of Woo Suk Hwang, a veterinary cloning expert, and the gynecologist Shin Yong Moon pulled off two remarkable scientific feats that many anticipated as achievable. They apparently overcame impediments that hampered other researchers and managed to produce human embryonic stem cells and clone human cells. On February 12, 2004, in an article in *Science* Hwang and colleagues claimed in an article that they had established a human embryonic stem cell line from a cloned blastocyst after inserting the nucleus of a human cumulus cell (a cell surrounding the developing eggs in the ovary) into a human egg *(www.sciencemag.org/cgi/content/abstract/1094515)* [217]. They used patient-specific embryonic stem cells derived from human SCNT blastocysts This was a major advance since the birth of the cloned sheep Dolly in 1996 and it made Hwang a national idol in South Korea and researchers gathered at Seoul National University (SNU) *(http://www.useoul.edu/)* to be taught how to do it. A few months later an allegation of unethical behavior focused on the investigators related to the oocyte donation [8]. In a reaction to the allegations Hwang confessed to unethically obtaining oocytes from his research group and other paid donors. The scientific goals of Hwang were admirable and if his experiments had been performed honestly and ethically he could have brought therapeutic cloning into reality by allowing patients to generate tissues matched directly to their genetic makeup. The 2004 Hwang communication was followed by a report on May 19 2005 of the creation of 11 different stem cell lines derived from human patients by transferring somatic nuclei [90]. On August 4, 2005, the Hwang research months team reported the first cloning of an Afghan hound called Snuppy [9, 134] and this cloning was confirmed [135]. A few later Korean televisions reported evidence of unethical oocyte donation. By early December 2005 the scientific community was skeptical about the validly of papers published by Hwang. An anonymous Korean researcher reported that the data to some images in the 2005 *Science* paper were not valid, but duplicated images and that debatable DNA fingerprint data were evident in his articles. Later that month Hwang and Gerald Schatten a coauthor retracted the May 2005 *Science* paper in *Science* because of problematic data. Hwang persisted in endorsing the soundness of the observations. Shortly thereafter SNU investigated the seminal paper in *Science* in which Hwang reported the creation human stem cell lines [90]. After the possibility of fraud emerged a committee at SNU investigated the allegations and found fabricated data [151]. The investigating SNU committee concluded after a far-reaching review that the 2005 paper did not provide evidence of patient-specific stem cells, but contained extensive fabricated data, relating to DNA fingerprinting, photographs of a teratoma, embryoid bodies, MHC-HLA isotype matches and karyotyping [Seoul National University Investigation Committee, (2006, January 10). Summary of the Final Report on Hwang's Research Allegation. SNU News. (*http://www.snu.ac.kr/engsnu/*). SNU also concluded that the other papers in *Science* were fraudulent, but that Snuppy had indeed been cloned. On January 12th 2006 Donald Kennedy the editor of *Science* officially retracted both papers in *Science* by Hwang and colleagues [114]. Soon after the SNU committee report Hwang resigned the directorship of his research unit. The loss of his reputation as a scientist did not end Hwang's woes. A few months later after being fired by SNU Hwang obtained a position in veterinary science at Sooam biotech research foundation where he continued to clone dogs and publish papers. More than three years after a criminal trial lasting 40 months a South Korean court convicted Hwang of embezzlement and the bioethics accusations and sentenced him to a 2 year suspended prison term on 26 October 2009 [154]. A few days later, *Science* documented the story of Hwang's fall from grave. Other questions about the validity of his work steadily accumulated. Because of Hwang's ethical lapses some scientists were afraid that his research would be used by anti-stem cell proponents against those pursuing this line of research.

Six months after the papers by Hwang were retracted Shinya Yamanka of Kyoto University introduced four genes into mouse skin cells transforming them into so-called induced pluripotent stem (iPS) cells that closely mimicked embryonic stem cells. Since then, some investigators have even reported the establishment iPS cells from patients with different diseases.

Despite his unethical and fraudulent work Hwang still has numerous supporters and many stood by him throughout his ordeal in court. Hwang was even awarded the Jang Yeong-il Memorial Foundation medal for scientific excellence early in 2009, before his trial in the court that convicted him had reached a decision.

Professor Gerald P. Schatten of the University of Pittsburgh medical center (UPMC) was a coauthor of papers on stem cells with the disgraced Hwang, but he subsequently admitted that he did not contribute anything to the high profile publications of Hwang. This did not restore his reputation because in April 2004 he attempted to patent technology to create stem cells with two UPMC researchers (Calvin Simerly and Christopher Navara), in an application that failed to include Hwang with whom he had previously worked (*Pittsburgh Tribune Review*)

(Kolata G. 2005, December 16) Clone scandal: 'a tragic turn', *The New York Times*, pA6).

Tae Kook Kim. Tae Kook Kim obtained a PhD in the USA from Rockefeller University (Rockefeller) *(http://www.rockefeller.edu/)* and then underwent post-doctoral training at both Harvard University (Harvard) *(http://www.harvard.edu/)* and the SNU's institute for molecular biology and genetics. Having gained this superb experience he accepted a faculty position at the Korea advanced institute of science and technology (KAIST), which was established by the South Korean government in 1974. While at KAIST he developed a novel method of identifying drug targets by following protein movements in living cells. The technique was called magnetism-based interaction capture (MAGIC) and it involved getting magnetized nanoparticles coated with extremely small particles of interest to interact with proteins in living cells. The labeled smidgens become engulfed by the cell compelling the nanoparticle to move and an observer can determine whether a particular fluorescent labeled protein moves with it. The work was documented in *Science* [221] and in *Nature Chemical Biology* [225]. To commercialize this method Kim founded a company (CGK Co) in Daejeon, South Korea in 2004. Jin Hwan Kim became the chief executive officer (CEO) of CGK Co. and to fund the company, 2.5 million dollars was raised from three Korean venture capitalists in 2006. Scientists at CGK were unable to replicate the MAGIC technology. CGK encountered several difficulties with MAGIC. The published images could not be generated with magnetized nanoparticles used in the study. Kim's mentors were surprised and amazed to learn about his scientific misconduct. According to a molecular biologist, who supervised his post-doctoral training in the 1990s, he was an extremely hard worker, very efficient, and extremely focused. Yong-Weon Yi, a former PhD student of Kim and a coauthor on both papers, contacted the two journals in December 2007 requesting that his name be withdrawn from the papers. KAIST was notified about the inability to repeat the studies and appointed an internal institutional committee to investigate the matter. The investigators found that the published papers lacked scientific truth and steps were taken to retract the 2008 papers. Although the technique of MAGIC may be valid the papers contained serious fabricated data that damaged the integrity of the entire papers. The validity of the research was also seriously weakend by an inability of the authors to provide notebooks or the raw data from the reported experiments. Jaejoon Won, the first author on both papers, submitted a written declaration admitting serious scientific misconduct in the two papers. After learning about the fraudulent findings the president of KAIST promptly suspended Kim and informed the editors of the journals, who reviewed the peer reviewed critiques that the articles garnered during the peer review process. The reviews were enthusiastically positive. Within days *Science* posted an editorial expression of concern [115] and the aftermath of this fraudulent scandal for Korean scientists was devastating. Kim was fired from KAIST and was sued for criminal fraud by the company that he founded. In addition won Yi brought criminal defamation charges against five members of the investigating committee of KAIST. *Science* generally wants all authors to agree with a retraction, but the journal encountered difficulties in locating Neoncheol Jung, one of the authors, and when he was found he stated that he had nothing to do with the fabrication and claimed that his name was added to the paper without his consent. In an unprecedented situation *Science* retracted the paper in *Science* without the consent of all authors [6]. The paper in *Nature Chemical Biology* was retracted in July 2006 [223]. In July 2008 the journal editor retracted the paper [10]. In the same copy of the journal a *Nature Chemical Biology* signed by all 9 authors except Kim retracted the paper. Despite the evidence of scientific misconduct in the two retracted papers Kim contemplated legal action against several parties for defamation and libel.

During 2007 and 2008 Kim became entangled in a legal disagreement between KAIST and CGK related to patent disputes over the rights to MAGIC. CGK asserted that it had an agreement with KAIST to allocating intellectual property and commercial rights of the MAGIC technology to CGK. On December 9, 2009 a Korean court restored the MAGIC patent rights to KAIST, but the Korean intellectual property office proposed on January 19 2010 that the patent be rejected. On March 13, 2009 KAIST issued a press release stating that the preliminary investigation alleged that Won Yi and Kim were all involved or aware of the misconduct, but the final report issued later did not designate responsibility for the assumed misconduct. Following the exposure of the two fraudulent papers a dispute emerged focusing on the patent rights of MAGIC because the methodology relevant to intellectual proper was still believed to be valid [60, 222].

After publishing an article on the technique Won and colleagues published a corrigendum drawing attention to the fact that they could be perceived as having a conflict of interest that may have influenced their interpretation of the results [224] [155].

Research Misconduct in Spain

José Luis Ortiz Moreno. The astronomer José Luis Ortiz Moreno, the former vice-director of technology at the Instituto de Astrofísica de Andalucia in Spain *(http://www.iaa.es/)*, led a group of astronomers studying minor solar system objects at the Sierra Nevada Observatory in Granada, Spain. On July 29, 2005 Morenco announced the discovery of a dwarf planet that was given the temporary designation of 2003 EL_{61}. However, a team of astronomers at California Institute of Technology (Caltech) under the leadership of Michael E. Brown (1965—), a professor of planetary astronomy at Caltech, had been tracking this planet for a longer period, but were delaying publication of their discovery until they secured its orbital parameters. Because of the delayed publication Brown and colleagues were scooped and they acknowledged that the Spanish team deserved credit for the discovery. His backing was withdrawn when Brown learnt that the Ortiz team used the Caltech public observation logs and research prior to their announcement of the discovery *(http:/www.nytimes.com/2005/09/13/science/13plan.htm)*. Ortiz claimed that he only accessed this information to make sure that they were tracking the same object *(http://www.newscientist, com/article.ns?id=dn8033)*. Eventually the international astronomical union (IAU) resolved the dispute by giving the place of discovery as the Sierra Nevada Observatory in Granada, Spain and the name of the dwarf planet as Haumea (the Caltech recommendation) rather than Ataecina (the Spanish preferred designation) *(http://en.wikipedia.org/wiki/Jos%C3%A9José_Luis_Moreno.*

Research Misconduct in the United Kingdom

Until relatively recently many medical institutions in the United Kingdom (UK) traditionally dealt with scientific misconduct by brushing it under the carpet and hounding the whistleblower out of the person's job [136]. In the UK the general medical council (GMC) has legal powers over the medical profession and several medical doctors have had their names removed from the UK medical register and have lost their ability to practice medicine after being found guilty of fraud in clinical drug trials and it other inappropriate research [218]. By 1998 research misconduct was virtually unknown in the UK and its existence in that country was questioned [136]. Since then several notable medical doctors in the UK have become involved in fraudulent research affecting patient care: John

Anderson, Mark H.Williams [220], Malcolm Pearce [137] (Locke S. Lessons from the Pearce affair: handling scientific fraud. BMJ 310:1547-1548, 1995). Andrew J. Wakefield, and Peter Nixon. The Royal college of obstetrics and gynecology (RCOG) appointed a committee to investigate the situation related to the two papers published in the *British Journal of Obstetrics and Gynaecology* in August 1994. London: RCOG, 1995) to one new case per million population every year (P Riis, personal communication, 1995

Robert Davies. Robert Davies, a professor of respiratory medicine in London attempted to cover up for slip-ups in a clinical trial by a drug company [180].

Sir Samuel Roy Meadow. The distinguished British pediatrician professor Sir Samuel Roy Meadow (1933—) became famous after writing a paper in 1977 on the Munchausen syndrome by proxy [146], which later became controversial. He became well-known because of a campaign against parents who deliberately injury or kill their offspring. He introduced the dictim that *"one sudden infant death is a tragedy, two is suspicious and three is murder, until proved otherwise"*, which became widely accepted as the Law of Meadow and it was once embraced by social workers and child protection agencies in the UK and his textbook became the bible on child abuse [147]. Meadow co-founded London's royal college of pediatrics and child health. For his important contributions Meadow was knighted by Queen Elizabeth II in 1997 the year that his book was first published. Distorting statistics inappropriately to make a point is a common offence, but the offender often gets away with the crime. Meadow was an expert witness in a criminal trial suspected of involving murder and child abuse in 1999. He got into unsuspected trouble trying to help find the mother guilty of killing her two sons by overstating the low probability of two infants from the same family dying rapidly and unexpectedly. An inquiry found the most egregious error in Meadow's testimony to be his declaration that the risk of two infants in the same family dying of the sudden infant death syndrome (SIDS) was 1 in 73 billion. He reached this high probability by squaring the stated frequency of single cases of unexplained deaths due to natural causes in infants (1 in 8543). The mother, Sally Clark an attorney, was seemingly incorrectly convicted in 1999 of murdering her two infant sons, mainly because of evidence provided by Meadow and ended up in prison for three years until her husband, also an attorney, and other members of the defense team were able to cast doubt on the expert's evidence. Frank Lockyer, the father of Sally, who was also a retired senior police officer, filed a grievance to the GMC accusing Meadow of grave professional misconduct. *("BBC: Sir Roy Meadow struck off by GMC". BBC News.* 15 July 2005 (*http://news.bbc.co.uk/2/hi/health/4685511.stm*).

According to the appeals court medical details were withheld and the jury was influenced by Meadow's evidence. A professional panel chaired by Mary Clark-Glass, a former law lecturer and lay member of the GMC in the UK evaluated the case and considered his statistical blunder serious and on July 15, 2005, when Meadow was 72 years old and retired, the panel ruled that Meadow abused his position as a doctor by providing misleading evidence, albeit unintentionally, and that he extended his professional competence beyond its limits by wandering into the area of statistics. The fame and influence of Meadow carried such considerable weight that it accentuated the errors in his court testimony. The panel found Meadow's probability of risk incorrect because in reality the risk of a second child dying of SIDS is elevated when one family member has died from SIDS. Aside from condemning Meadow for providing incorrect numbers the panel also found fault in his metaphor implying that the probability of a family having two children dying in the identical manner would be like selecting the winner in the Grand National horse race in four consecutive years. The panel found Meadow, "guilty of serious medical misconduct" and imposed its severest penalty, namely a recommendation that his name be removed from the register of physicians in the UK for offering "erroneous" and "misleading" evidence in the Sally Clark case [211] (*http://society.guardian.co.uk/health/news/0, 8363, 1110264, 00.html*) Despite the shortcomings of Meadow's testimony many members of the medical profession were horrified that the GMC would become involved in this case and some believed that the GMC was exceptionally harsh in making Meadow the scapegoat for flaws in the legal system related to child abuse cases. A comment on Meadow's defense was published in the *Lancet* by Richard Horton [87]. Nevertheless the case stresses that expert witnesses should adhere to impartiality and integrity and not provide unsound statistical opinions. The decision by the GMC panel raised serious concerns about whether individuals will be reluctant to offer expert testimony as witnesses in the future. Not being satisfied with the GMC verdict Meadows appealed to the court of appeal and it ruled in his favor in February 2006. In October 2006 by a majority decision the court of appeal upheld the decision of the high court ruling that Meadow's misconduct was not sufficiently serious to merit the punishment which he had received [140]. Clark acquired a grave psychiatric disorder with alcohol dependency and died in 2007 [178] (http://en.wikipedia.org/wiki/Sally_Clark).(*http://web.tiscali.it/human*rights/articles/meadow77.html). Roy Meadow (*http://en.wikipedia.org/wiki/Roy_Meadow*)

References

1. Abbott A, Fraud claims shake German complancency. Nature, 1997. 387(6635): 750.
2. Abbott A, German technician's confession spurs check on suspect data. Nature, 1998. 393(293): 293.
3. Abbott A, Science comes to terms with the lessons of fraud. Nature, 1999. 398(6722): 13-17.
4. Ahlers A, Belka C, Gaestel M, et al., Interleukin-1-induced intracellular signaling pathways converge in the activation of mitogen-activated protein kinase and mitogen-activated protein kinase-activated protein kinase 2 and the subsequent phosphorylation of the 27-kilodalton heat shock protein in monocytic cells. Molecular Pharmacology, 1994. 46(6): 1077-1083.
5. Ahlers A, Engel K, Sott C, et al., Interleukin-3 and granulocyte-macrophage colony-stimulating factor induce activation of the MAPKAP kinase 2 resulting in in vitro serine phosphorylation of the small heat shock protein (Hsp 27).[Retraction in Blood. 1999 May 15;93(10):3573; PMID: 11012249]. Blood, 1994. 83(7): 1791-1798.
6. Alberts B, Retraction. Science, 2009. 324(5926): 463.
7. Anonymous, Fraud investigation concludes that self regulation has failed. British Medical Journal, 2000. 321(7253): 72.
8. Anonymous, Ethics of therapeutic cloning. Nature, 2004. 429(6987): 1.
9. Anonymous, A dog's life. Nature, 2005. 436(7051): 604.
10. Anonymous, Correcting the scientific record. The retraction of a Nature Chemical Biology paper is a step toward a full accounting of a case of scientific misconduct. Nature Chemical Biology, 2008. 4(7): 381.
11. Anonymous, Iran science officials in plagiarism flap. Science, 2009. 326(5950): 211.
12. Anonymous, Psychologist sacked over faked data. Science, 2011. 333(6049): 1556.
13. Anonymous, By the numbers. Science, 2012. 337(6091): 140.
14. Asano Y, Brach MA, Ahlers A, et al., Phorbol ester 12-O-tetradecanoylphorbol-13-acetate down-regulates expression of the c-kit proto-oncogene product. Journal of Immunology, 1993. 151(5): 2345-2354.
15. Avramova YR and Stapel DA, Moods as spotlights: the influence of mood on accessibility effects. Journal of Personality and Social Psychology, 2008. 95(3): 542-544.
16. Avramova YR, Stapel DA, and Lerouge D, The influence of mood on attribution. Personality and Social Psychology Bulletin, 2010. 36(10): 1360-1371.

17. Avramova YR, Stapel DA, and Lerouge D, Mood and context-dependence: Positive mood increases and negative mood decreases the effects of context on perception. Journal of Personality and Social Psychology, 2010. 99(2): 203-214.
18. Babu SH, Suvardhan K, Kumar KS, et al., The facile flow-injection spectrophotometric detection of gold(III) in water and pharmaceutical samples using 3, 5-dimethoxy-4-hydroxy-2-aminoacetophenone isonicotinoyl hydrazone (3, 5-DMHAAINH).[Retraction in Journal of Hazardous Materials 2008 April 1;152(2):892; PMID: 18459183]. Journal of Hazardous Materials, 2005. 120(1-3): 213-218.
19. Babu SH, Kumar KS, Suvardhan K, et al., Preconcentration technique for the determination of trace elements in natural water samples by ICP-AES. Environmental Monitoring and Assessment, 2007. 128(1-3): 241-249.
20. Balter M, Scientific misconduct. Reviewer's deja vu, French science sleuthing uncover plagiarized papers. Science, 2012. 335(6073): 1157-1158.
21. Belka C, Ahlers A, Sott C, et al., Interleukin (IL)-6 signaling leads to phosphorylation of the small heat shock protein (Hsp)27 through activation of the MAP kinase and MAPKAP kinase 2 pathway in monocytes and monocytic leukemia cells. Leukemia, 1995. 9(2): 288-294.
22. Belka C, Brach MA, and Herrmann F, The role of tyrosine kinases and their substrates in signal transmission of hematopoietic growth factors: a short review. Leukemia, 1995. 9(5): 754-761.
23. Belka C, Wiegmann K, Adam D, et al., Tumor necrosis factor (TNF)-alpha activates c-raf-1 kinase via the p55 TNF receptor engaging neutral sphingomyelinase. EMBO Journal, 1995. 14(6): 1156-1165.
24. Bibikova TN, Blancaflor EB, and Gilroy S, Microtubules regulate tip growth and orientation in root hairs of Arabidopsis thaliana. Plant Journal, 1999. 17(6): 657-665.
25. Blanton H and Stapel DA, Unconscious and spontaneous and ... complex: the three selves model of social comparison assimilation and contrast. Journal of Personality and Social Psychology, 2008. 94(6): 1018-1032.
26. Brach M, Klein H, Platzer E, et al., Effect of interleukin 3 on cytosine arabinoside-mediated cytotoxicity of leukemic myeloblasts. Experimental Hematology, 1990. 18(7): 748-753.
27. Brach MA, Cicco NA, Riedel D, et al., Mechanisms of differential regulation of interleukin-6 mRNA accumulation by tumor necrosis factor alpha and lymphotoxin during monocytic differentiation.[Retraction in Mertelsmann R. FEBS Lett. 1998 Jun 16;429(3):426; PMID: 9687234]. FEBS Letters, 1990. 263(2): 349-354.
28. Brach MA, Lowenberg B, Mantovani L, et al., Interleukin-6 (IL-6) is an intermediate in IL-1-induced proliferation of leukemic human megakaryoblasts. Blood, 1990. 76(10): 1972-1979.

29. Brach MA, Riedel D, and Herrmann F, Induction of monocytic differentiation and modulation of the expression of c-fos, c-fms and c-myc protooncogenes in human monoblasts by cytokines and phorbolester. [Erratum appears in Virchows Archiv. B. Cell Pathology 1999 July;435(1):76]. Virchows Archiv. B. Cell Pathology, 1990. 59(1): 54-58.
30. Brach MA, Riedel D, Mertelsmann RH, et al., Synergistic effect of recombinant human leukemia inhibitory factor (LIF) and 1-beta-D-arabinofuranosylcytosine (Ara-C) on proto-oncogene expression and induction of differentiation in human U 937 cells. Leukemia, 1990. 4(9): 646-649.
31. Brach MA, Gauer E, Ludwig WD, et al., Expression of the c-fes proto-oncogene in granulocyte-macrophage colony-stimulating factor-dependent acute myelogenous leukemia cells grown autonomously. International Journal of Cell Cloning, 1991. 9(1): 89-94.
32. Brach MA, Henschler R, Mertelsmann R, et al., To overcome pharmacologic and cytokinetic resistance to cytarabine in the treatment of acute myelogenous leukemia by using recombinant interleukin-3? Seminars in Hematology, 1991. 28(3 Suppl 4): 39-43.
33. Brach MA, Henschler R, Mertelsmann RH, et al., Regulation of M-CSF expression by M-CSF: role of protein kinase C and transcription factor NF kappa B. Pathobiology, 1991. 59(4): 284-288.
34. Brach MA and Herrmann F, Hematopoietic growth factors: interactions and regulation of production. Acta Haematologica, 1991. 86(3): 128-137.
35. Brach MA, Buhring HJ, Gruss HJ, et al., Functional expression of c-kit by acute myelogenous leukemia blasts is enhanced by tumor necrosis factor-alpha through posttranscriptional mRNA stabilization by a labile protein. [Retraction in Blood. 1999 May 15;93(10):3573; PMID: 11012249]. Blood, 1992. 80(5): 1224-1230.
36. Brach MA, de Vos S, Arnold C, et al., Leukotriene B4 transcriptionally activates interleukin-6 expression involving NK-chi B and NF-IL6. European Journal of Immunology, 1992. 22(10): 2705-2711.
37. Brach MA, deVos S, Gruss HJ, et al., Prolongation of survival of human polymorphonuclear neutrophils by granulocyte-macrophage colony-stimulating factor is caused by inhibition of programmed cell death. Blood, 1992. 80(11): 2920-2924.
38. Brach MA, Gruss HJ, Asano Y, et al., Synergy of interleukin 3 and tumor necrosis factor alpha in stimulating clonal growth of acute myelogenous leukemia blasts is the result of induction of secondary hematopoietic cytokines by tumor necrosis factor alpha. Cancer Research, 1992. 52(8): 2197-2201.
39. Brach MA, Gruss HJ, Riedel D, et al., Effect of antiinflammatory agents on synthesis of MCP-1/JE transcripts by human blood monocytes. Molecular Pharmacology, 1992. 42(1): 63-68.

40. Brach MA, Gruss HJ, Riedel D, et al., Activation of NF-kappa B by interleukin 2 in human blood monocytes. Cell Growth and Differentiation, 1992. 3(7): 421-427.
41. Brach MA and Herrmann F, Interleukin 6: presence and future. International Journal of Clinical and Laboratory Research, 1992. 22(3): 143-151.
42. Brach MA, Herrmann F, and Kufe DW, Activation of the AP-1 transcription factor by arabinofuranosylcytosine in myeloid leukemia cells.[Retraction in Blood. 1999 May 15;93(10):3573; PMID: 11012249]. Blood, 1992. 79(3): 728-734.
43. Brach MA, Herrmann F, Yamada H, et al., Identification of NF-jun, a novel inducible transcription factor that regulates c-jun gene transcription. EMBO Journal, 1992. 11(4): 1479-1486.
44. Brach MA, Kharbanda SM, Herrmann F, et al., Activation of the transcription factor kappa B in human KG-1 myeloid leukemia cells treated with 1-beta-D-arabinofuranosylcytosine. Molecular Pharmacology, 1992. 41(1): 60-63.
45. Brach MA, Mertelsmann RH, and Herrmann F, Hematopoietins in combination with 1-beta-D-arabinofuranosylcytosine: a possible strategy for improved treatment of myeloid disorders. Seminars in Oncology, 1992. 19(2 Suppl 4): 25-30.
46. Brach MA, Arnold C, Kiehntopf M, et al., Transcriptional activation of the macrophage colony-stimulating factor gene by IL-2 is associated with secretion of bioactive macrophage colony-stimulating factor protein by monocytes and involves activation of the transcription factor NF-kappa B. Journal of Immunology, 1993. 150(12): 5535-5543.
47. Brach MA, Gruss HJ, Kaisho T, et al., Ionizing radiation induces expression of interleukin 6 by human fibroblasts involving activation of nuclear factor-kappa B. Journal of Biological Chemistry, 1993. 268(12): 8466-8472.
48. Brach MA, Gruss HJ, Sott C, et al., The mitogenic response to tumor necrosis factor alpha requires c-Jun/AP-1. Molecular and Cellular Biology, 1993. 13(7): 4284-4290.
49. Brach MA and Herrmann F, The mitogenic response of AML blasts to tumor necrosis factor-alpha requires functional c-jun/AP-1. Leukemia, 1993. 7 Suppl 2: S22-S26.
50. Brach MA, Mertelsmann RH, and Herrmann F, Modulation of cytotoxicity and differentiation-inducing potential of arabinofuranosylcytosine in myeloid leukemia cells by hematopoietic cytokines. Cancer Investigation, 1993. 11(2): 198-211.
51. Brach MA, Sott C, Kiehntopf M, et al., Expression of the transforming growth factor-alpha gene by human eosinophils is regulated by interleukin-3, interleukin-5, and granulocyte-macrophage colony-stimulating factor. European Journal of Immunology, 1994. 24(3): 646-650.

52. Brach MA, Kiehntopf M, and Herrmann F, [Gene therapy in oncology. A model of change in pharmacology?]. Deutsche Medizinische Wochenschrift, 1995. 120(4): 113-120.
53. Brach MA, Sott C, Belka C, et al., [Molecular basis of tumor formation. Significance for the diagnosis of malignant diseases]. Deutsche Medizinische Wochenschrift, 1995. 120(3): 73-79.
54. Brach MA and Herrmann F, [Principles and strategies of gene therapy]. Internist, 1996. 37(4): 343-349.
55. Brach MA, Kauer M, and Herrmann F, Contribution of transcription factors to oncogenesis. Cytokine and Molecular Therapy, 1996. 2(2): 81-87.
56. Brennscheidt U, Riedel D, Kolch W, et al., Raf-1 is a necessary component of the mitogenic response of the human megakaryoblastic leukemia cell line MO7 to human stem cell factor, granulocyte-macrophage colony-stimulating factor, interleukin 3, and interleukin 9. Cell Growth and Differentiation, 1994. 5(4): 367-372.
57. Butler D, Iranian ministers in plagiarism row. Nature investigation reveals duplications in papers by science and transport chiefs Nature, 2009. 461(7262): 568-579.
58. Cao C, Plagiarism in China. Science, 1996. 274(5294): 1820-1825.
59. Carlisle JB, A meta-analysis of prevention of postoperative nausea and voitinig: randomized controlled trials by Fujii et al. compared wth other authors. Anesthesia and Analgesia, 2012. 67(10): 1076-1090.
60. Couzin J and Normile D, Scientific misconduct. Two papers from Korean lab found to lack 'scientific truth'. Science, 2008. 319(5869): 1468-1469.
61. Cui C, Song F, Tan Y, et al., Stable chloroplast transformation of immature scutella and inflorescences in wheat (Triticum aestivum L.). Acta Biochimica et Biophysica Sinica, 2011. 43(4): 284-291.
62. Cui C, Song F, Tan Y, et al., Stable chloroplast transformation of immature scutella and inflorescences in wheat (Triticum aestivum L.).[Retraction in He G. Acta Biochimica et Biophysica Sinica (Shanghai). 2012 April;44(4):373; PMID: 22467140]. Acta Biochimica et Biophysica Sinica, 2012. 43(4): 284-291.
63. Daneshjou K and Shahravi M, Retracted article: Analysis of critical ricochet angle using two space discretization methods. Eng, Comp., 2009. 25(2): 191-206.
64. de Vos S, Brach MA, Asano Y, et al., Transforming growth factor-beta 1 interferes with the proliferation-inducing activity of stem cell factor in myelogenous leukemia blasts through functional down-regulation of the c-kit proto-oncogene product. Cancer Research, 1993. 53(15): 3638-3642.
65. de Vos S, Brach M, Budnik A, et al., Post-transcriptional regulation of interleukin-6 gene expression in human keratinocytes by ultraviolet B radiation. Journal of Investigative Dermatology, 1994. 103(1): 92-96.

66. Engel K, Ahlers A, Brach MA, et al., MAPKAP kinase 2 is activated by heat shock and TNF-alpha: in vivo phosphorylation of small heat shock protein results from stimulation of the MAP kinase cascade. Journal of Cellular Biochemistry, 1995. 57(2): 321-330.
67. Enserink M, Final report on Stapel also blames field as whole. Science, 2012. 338(6112): 1270-1271.
68. Forget P-M and Jansen PA, Hunting increases dispersal limitation in the tree Carapa procera, a nontimber forest product. Conservation Biology, 2007. 21(1): 106-113.
69. Gordijn EH and Stapel DA, Behavioural effects of automatic interpersonal versus intergroup social comparison. British Journal of Social Psychology, 2006. 45(Pt 4): 717-729.
70. Gruss HJ, Brach MA, Mertelsmann RH, et al., Interferon-gamma interrupts autocrine growth mediated by endogenous interleukin-6 in renal-cell carcinoma. International Journal of Cancer, 1991. 49(5): 770-773.
71. Gruss HJ, Brach MA, Drexler HG, et al., Expression of cytokine genes, cytokine receptor genes, and transcription factors in cultured Hodgkin and Reed-Sternberg cells. Cancer Research, 1992. 52(12): 3353-3360.
72. Gruss HJ, Brach MA, Drexler HG, et al., Interleukin 9 is expressed by primary and cultured Hodgkin and Reed-Sternberg cells. Cancer Research, 1992. 52(4): 1026-1031.
73. Gruss HJ, Brach MA, and Herrmann F, Involvement of nuclear factor-kappa B in induction of the interleukin-6 gene by leukemia inhibitory factor.[Retraction in Blood. 1999 May 15;93(10):3573; PMID: 11012249]. Blood, 1992. 80(10): 2563-2570.
74. Gruss HJ, Dolken G, Brach MA, et al., High concentrations of the interleukin-1 receptor antagonist in serum of patients with Hodgkin's disease. The Lancet, 1992. 340(8825): 968.
75. Gruss HJ, Dolken G, Brach MA, et al., The significance of serum levels of soluble 60kDa receptors for tumor necrosis factor in patients with Hodgkin's disease. Leukemia, 1993. 7(9): 1339-1343.
76. Gruss HJ, Dolken G, Brach MA, et al., Serum levels of circulating ICAM-1 are increased in Hodgkin's disease. Leukemia, 1993. 7(8): 1245-1249.
77. Gruss HJ, Brach MA, Schumann RR, et al., Regulation of MCP-1/JE gene expression during monocytic differentiation. Journal of Immunology, 1994. 153(11): 4907-4914.
78. Gruss HJ, Scott C, Rollins BJ, et al., Human fibroblasts express functional IL-2 receptors formed by the IL-2R alpha—and beta-chain subunits: association of IL-2 binding with secretion of the monocyte chemoattractant protein-1. Journal of Immunology, 1996. 157(2): 851-857.

79. Gruss HJ, Ulrich D, Dower SK, et al., Activation of Hodgkin cells via the CD30 receptor induces autocrine secretion of interleukin-6 engaging the NF-kappabeta transcription factor.[Retraction in Blood. 1999 May 15;93(10):3573; PMID: 11012249]. Blood, 1996. 87(6): 2443-2449.
80. Guo C, Qiu H-Y, Shi M, et al., Lmx1b-controlled isthmic organizer is essential for development of midbrain dopaminergic neurons.[Retraction in Journal of Neuroscience 2011 September 28;31(39):14047; PMID: 21957266]. Journal of Neuroscience, 2008. 28(52): 14097-14106.
81. Henschler R, Lindemann A, Brach MA, et al., Expression of functional receptors for interleukin-6 by human polymorphonuclear leukocytes. Downregulation by granulocyte-macrophage colony-stimulating factor. [Retraction in Mertelsmann R, Lindemann A, Henschler R, Mackensen A. FEBS Lett. 1998 Jun 16;429(3):426; PMID: 9687233]. FEBS Letters, 1991. 283(1): 47-51.
82. Herrmann F, Gebauer G, Lindemann A, et al., Interleukin-2 and interferon-gamma recruit different subsets of human peripheral blood monocytes to secrete interleukin-1 beta and tumour necrosis factor-alpha. Clinical and Experimental Immunology, 1989. 77(1): 97-100.
83. Herrmann F, Lindemann A, Cannistra SA, et al., Monocyte interleukin-1 secretion is regulated by the sequential action of gamma-interferon and interleukin-2 involving monocyte surface expression of interleukin-2 receptors. Haematology and Blood Transfusion, 1989. 32: 299-315.
84. Herrmann F, Andreeff M, Gruss HJ, et al., Interleukin-4 inhibits growth of multiple myelomas by suppressing interleukin-6 expression.[Retraction in Blood. 1999 May 15;93(10):3573; PMID: 11012249]. Blood, 1991. 78(8): 2070-2074.
85. Herrmann F, De Vos S, Brach M, et al., Secretion of granulocyte-macrophage colony-stimulating factor by human blood monocytes is stimulated by engagement of Fc gamma receptors type I by solid-phase immunoglobulins requiring high-affinity Fc-Fc gamma receptor type I interactions. European Journal of Immunology, 1992. 22(7): 1681-1685.
86. Herrmann F, Kiehntopf M, Brach MA, et al., Marker gene transfer into leukapheresis preparations containing hematopoietic progenitor cells: application in high-dose therapy rescued by reinfusion of peripheral blood hematopoietic progenitors in patients with multiple myeloma. Journal of Molecular Medicine, 1995. 73(4): 197-203.
87. Horton R, In defence of Roy Meadow. The Lancet, 2005. 366(9479): 3-5.
88. Howard J. Dr Ronald Wild takes college job in far North-west. The Australian. 1986 July 16.
89. Husten L. More details emerge about the Dutch research scandal. CardioBrief. 2011 November 20.

90. Hwang WS, Roh SI, Lee BC, et al., Report. Patient-specific embryonic stem cells derived from human SCNT blastocysts.[Erratum appears in Science. 2005 December 16;310(5755):1769].[Retraction in Kennedy D. Science. 2006 January 20;311(5759):335; PMID: 16410485]. Science, 2005. 308(5729): 1777-1783.
91. Jendryczko A, [Effect of zinc on immunoproteins in women with uterine myomas]. Wiadomosci Lekarskie, 1983. 36(24): 2039-2041.
92. Jendryczko A and Drozdz M, Metabolism of hydralazine by rat hepatocytes. Acta Physiologica Polonica, 1984. 35(2): 185-187.
93. Jendryczko A and Drozdz M, [Activity of enzymes of the methionine cycle in human placentas in various periods of pregnancy]. Ginekologia Polska, 1984. 55(9): 687-689.
94. Jendryczko A and Drozdz M, [Proline levels and collagen metabolism in patients with systemic lupus erythematosus]. Przeglad Lekarski, 1984. 41(9): 563-565.
95. Jendryczko A and Drozdz M, [Picolinic acid level in human milk]. Pediatria Polska, 1985. 60(7): 515-518.
96. Jendryczko A, Drozdz M, Tomala J, et al., Copper and zinc concentrations, and superoxide dismutase activities in malignant and nonmalignant tissues of female reproductive organs. Neoplasma, 1986. 33(2): 239-244.
97. Jendryczko A and Drozdz M, [Late sequelae of small doses of ionizing radiation in humans]. Wiadomosci Lekarskie, 1989. 42(12): 815-819.
98. Jendryczko A and Drozdz M, [Free radical-related processes of mitochondrial aging]. Wiadomosci Lekarskie, 1989. 42(2): 110-113.
99. Jendryczko A, Drozdz M, Kozlowski A, et al., [Concentration of selenium in serum of patients with ulcerative colitis](Polish). Przeglad Lekarski, 1992. 49(9): 292-293.
100. Jendryczko A, Drozdz M, Kozlowski A, et al., [Concentration of selenium in serum of patients with ulcerative colitis]. Przeglad Lekarski, 1992. 49(9): 292-293.
101. Jendryczko A and Tomala J, An imbalance between the excretion of thromboxane and prostacyclin metabolites in women after the menopause. Zentralblatt fur Gynakologie, 1993. 115(4): 163-166.
102. Jendryczko A, [Involvement of free radicals in lead poisoning]. Medycyna Pracy, 1994. 45(2): 171-175.
103. Jendryczko A, [Iron as yet another risk factor in heart diseases?]. Wiadomosci Lekarskie, 1994. 47(3-4): 118-121.
104. Johnson CS, Smeesters D, and Wheeler SC, Visual perspective influences the use of metacognitive information in temporal comparisons.[Retraction in J Pers Soc Psychol. 2012 Oct;103(4):605; PMID: 23002957]. Journal of Personality and Social Psychology. 102(1): 32-50.

105. Johnson CS and Stapel DA, No pain, no gain: the conditions under which upward comparisons lead to better performance. Journal of Personality and Social Psychology, 2007. 92(6): 1051-1067.
106. Johnson CS and Stapel DA, It depends on how you look at it: being versus becoming mindsets determine responses to social comparisons. British Journal of Social Psychology, 2010. 49(Pt 4): 703-723.
107. Joly JF, Stapel DA, and Lindenberg SM, Silence and table manners: when environments activate norms. Personality and Social Psychology Bulletin, 2008. 34(8): 1047-1056.
108. Kawasaki H, Onuki R, Suyama E, et al., Identification of genes that function in the TNF-alpha-mediated apoptotic pathway using randomized hybrid ribozyme libraries.[Retraction in Onuki R, Suyama E, Taira K. Nat Biotechnol. 2006 Sep;24(9):1170; PMID: 16964231]. Nature Biotechnology. 20(4): 376-380.
109. Kawasaki H, Suyama E, Iyo M, et al., siRNAs generated by recombinant human Dicer induce specific and significant but target site-independent gene silencing in human cells.[Retraction in Suyama E, Iyo M, Taira K. Nucleic Acids Res. 2006;34(8):2153; PMID: 16670428]. Nucleic Acids Research. 31(3): 981-987.
110. Kawasaki H and Taira K, Induction of DNA methylation and gene silencing by short interfering RNAs in human cells.[Retraction in Taira K. Nature. 2006 Jun 29;441(7097):1176; PMID: 16810259]. Nature. 431(7005): 211-217.
111. Kawasaki H and Taira K, Retraction: Hes1 is a target of microRNA-23 during retinoic-acid-induced neuronal differentiation of NT2 cells. [Retraction of Kawasaki H, Taira K. Nature. 2003 Jun 19;423(6942):838-42; PMID: 12808467]. Nature. 426(6962): 100.
112. Kawasaki H and Taira K, Identification of genes by hybrid ribozymes that couple cleavage activity with the unwinding activity of an endogenous RNA helicase.[Retraction in Taira K. EMBO Rep. 2006 Jul;7(7):746; PMID: 16819465]. EMBO Reports. 3(5): 443-450.
113. Kawasaki H and Taira K, Hes1 is a target of microRNA-23 during retinoic-acid-induced neuronal differentiation of NT2 cells.[Retraction in Kawasaki H, Taira K. Nature. 2003 Nov 6;426(6962):100; PMID: 14603326]. Nature. 423(6942): 838-842.
114. Kennedy D, Editorial retraction. Science, 2006. 311(5759): 335.
115. Kennedy D and Alberts B, Editorial expression of concern. Science, 2008. 319(5868): 1335.
116. Kiehntopf M, Brach MA, Licht T, et al., Ribozyme-mediated cleavage of the MDR-1 transcript restores chemosensitivity in previously resistant cancer cells.[Retraction in EMBO J. 1997 Jul 1;16(13):4153; PMID: 9280722]. EMBO Journal, 1994. 13(19): 4645-4652.

117. Kiehntopf M, Esquivel EL, Brach MA, et al., Ribozymes: biology, biochemistry, and implications for clinical medicine. Journal of Molecular Medicine, 1995. 73(2): 65-71.
118. Kiehntopf M, Esquivel EL, Brach MA, et al., Clinical applications of ribozymes. The Lancet, 1995. 345(8956): 1027-1031.
119. Kiehntopf M, Herrmann F, and Brach MA, Functional NF-IL6/CCAAT enhancer-binding protein is required for tumor necrosis factor alpha-inducible expression of the granulocyte colony-stimulating factor (CSF), but not the granulocyte/macrophage CSF or interleukin 6 gene in human fibroblasts.[Retraction in Adler G. J Exp Med. 1997 Jul 7;186(1):171; PMID: 9244806]. Journal of Experimental Medicine, 1995. 181(2): 793-798.
120. Kiran K, Suresh Kumar K, Suvardhan K, et al., Preconcentration and solid phase extraction method for the determination of Co, Cu, Ni, Zn and Cd in environmental and biological samples using activated carbon by FAAS.[Retraction in J Hazard Mater. 2008 Aug 15;156(1-3):624; PMID: 18630349]. Journal of Hazardous Materials, 2007. 147(1-2): 15-20.
121. Klein H, Becher R, Lubbert M, et al., Synthesis of granulocyte colony-stimulating factor and its requirement for terminal divisions in chronic myelogenous leukemia. Journal of Experimental Medicine, 1990. 171(5): 1785-1790.
122. Koenig R, Panel proposes ways to combat fraud. Science, 1997. 278(5346): 2049-2050.
123. Kumar KS, Rao SP, Krishnaiah L, et al., Detection of lead in vegetables with new chromogenic reagent by spectrophotometry. Environmental Monitoring and Assessment, 2004. 98(1-3): 191-199.
124. Kumar KS, Krishnaiah L, Babu SH, et al., Spectrophotometric determination with a chromogenic reagent of lead in vegetables. Chemistry and Biodiversity, 2005. 2(3): 386-391.
125. Kumar KS, Swaroop BL, Suvardhan K, et al., Facile and sensitive spectrophotometric determination of synthetic pyrithroids in their formulations, water and grain samples. Environmental Monitoring and Assessment, 2006. 122(1-3): 1-8.
126. Kumar KS, Suvardhan K, Rekha D, et al., Development of simple and sensitive spectrophotometric method for the determination of bendiocarb in its formulations and environmental samples. Environmental Monitoring and Assessment, 2007. 127(1-3): 67-72.
127. Kumar KS, Suvardhan K, Rekha D, et al., Facile and sensitive spectrophotometric determination of carbosulfan in formulations and environmental samples.[Erratum appears in Environmental Monitoring and Assessment 2008 April;139(1-3):369-370]. Environmental Monitoring and Assessment, 2007. 129(1-3): 271-276.

128. Kumar SK, Rao PS, Krishnaiah L, et al., Determination of nickel in alloys and plant leaves with salicylaldehyde 3-oxobutanoylhydrazone by spectrophotometry. Analytical Sciences, 2004. 20(6): 951-953.
129. Lagan B, Brown M, and Jamrozik W. Dr McBride's rise and falter: From fame to controversy. Sydney Morning Herald. 1987 December 19.
130. Lammers J and Stapel DA, How power influences moral thinking. Journal of Personality and Social Psychology, 2009. 97(2): 279-289.
131. Lammers J, Stoker JI, and Stapel DA, Differentiating social and personal power: opposite effects on stereotyping, but parallel effects on behavioral approach tendencies. Psychological Science, 2009. 20(12): 1543-1549.
132. Lammers J, Stoker JI, and Stapel DA, Differentiating social and personal power: opposite effects on stereotyping, but parallel effects on behavioral approach tendencies. Psychological Science, 2009. 20(12): 1543-1549.
133. Lammers J, Stapel DA, and Galinsky AD, Power increases hypocrisy: moralizing in reasoning, immorality in behavior. Psychological Science, 2010. 21(5): 737-744.
134. Lee BC, Kim MK, Jang G, et al., Dogs cloned from adult somatic cells. [Erratum appears in Nature. 2005 August 25;436(7054):1102], [Erratum appears in Nature. 2006 Mar 9;440(7081):164], [Erratum appears in Nature. 2006 October 12;443(7112):649 Note: Shamim, M Hossein [corrected to Hossein, M Shamim]]. Nature. 436(7051): 641.
135. Lee JB, Park C, and Seoul National University Investigation C, Molecular genetics: verification that Snuppy is a clone. Nature, 2006. 440(7081): E2-3.
136. Lock S, Misconduct in medical research: does it exist in Britain? British Medical Journal, 1988. 297(6662): 1531-1535.
137. Lock S, Lessons from the Pearce affair: handling scientific fraud. British Medical Journal, 1995. 310(6994): 1547-1548.
138. Mantovani L, Henschler R, Brach MA, et al., Differential regulation of interleukin-6 expression in human fibroblasts by tumor necrosis factor-alpha and lymphotoxin.[Retraction in Mantovani L, Mertelsmann R, Henschler R. FEBS Letters 1998 June 16;429(3):426; PMID: 9687235]. FEBS Letters, 1990. 270(1-2): 152-156.
139. Mantovani L, Henschler R, Brach MA, et al., Regulation of gene expression of macrophage-colony stimulating factor in human fibroblasts by the acute phase response mediators interleukin (IL)-1 beta, tumor necrosis factor-alpha and IL-6.[Retraction in Mantovani L, Mertelsmann R, Henschler R. FEBS Lett. 1998 June 16;429(3):426; PMID: 9687235]. FEBS Letters, 1991. 280(1): 97-102.
140. Marshall E, Flawed statistics in murder trial may cost expert his medical license. Science, 2005. 309(5734): 543.

141. Marx DM, Stapel DA, and Muller D, We can do it: the interplay of construal orientation and social comparisons under threat. Journal of Personality and Social Psychology, 2005. 88(3): 432-446.
142. Marx DM and Stapel DA, Distinguishing stereotype threat from priming effects: on the role of the social self and threat-based concerns. Journal of Personality and Social Psychology, 2006. 91(2): 243-254.
143. McBride W, Health of thalidomide victims and their progeny. The Lancet, 2004. 363(9403): 169.
144. McBride WG, Thalidomide and congenital abnormalities. The Lancet, 1961. 278(7216): 1358.
145. McBride WG, Debendox withdrawn. Medical Journal of Australia, 1984. 140(7): 445.
146. Meadow R, Munchausen syndrome by proxy. The hinterland of child abuse. The Lancet, 1977. 2(8033): 343-345.
147. Meadow R, Mok, Jacqueline Rosenberg, Donna, ABC of Child Abuse. 3rd ed. 2007, London: BMJ Books. pp. 85.
148. Milliken R. Thalidomide doctor' guilty of medical fraud: William McBride, who exposed the danger of one anti-nausea drug, has been disgraced by experiments with another. The Independent 1993 Saturday February 20.
149. Mortensen PB, Abildgaard K, and Fallingborg J, Serum selenium concentration in patients with ulcerative colitis. Danish Medical Bulletin. 36(6): 568-570.
150. Noordewier MK and Stapel DA, Affects of the unexpected: when inconsistency feels good (or bad). Personality and Social Psychology Bulletin, 2010. 36(5): 642-654.
151. Normile D and Vogel G, Korean Univesity will investigate cloning paper. Science, 2005. 310(5755): 1748-1749.
152. Normile D, Tokyo professor asked to redo experiments. Science, 2006. 309(5743): 1973.
153. Normile D, Scientific conduct. Panel discredits findings in Tokyo University team. Science, 2006. 311(5761): 595.
154. Normile D, Scientific misconduct. Hwang convicted but dodges jail; stem cell research has moved on. Science, 2009. 326(5953): 650-651.
155. Normile D, Scientific misconduct. Science retracts discredited paper; bitter patent dispute continues. Science, 2009. 324(5926): 450-451.
156. Oster W, Brach MA, Gruss HJ, et al., Interleukin-1 beta (IL-1 beta) expression in human blood mononuclear phagocytes is differentially regulated by granulocyte-macrophage colony-stimulating factor (GM-CSF), M-CSF, and IL-3.[Retraction in Blood. 1999 May 15;93(10):3573; PMID: 11012349]. Blood, 1992. 79(5): 1260-1265.
157. Pangou SV, Lechon G, Bouki T, et al., Retracted: Characteristic of natural regeneration of Aucoumea klaineana (Pierre) in Mayombe rain forest, southern Congo. African Journal of Ecology, 2007. 45(2): 156-164.

158. Pangou SV, Kampe J-P, De Zoysa M, et al., Effects of soil properties on growth of young tree seedling in logged-over tropical rain forest in Mayombe, Congo. American Journal of Plant Sciences, 2011. 2(3).
159. Pangou SV, Neela DZ, and Gema L, Comparison between field performance of cuttings and seedlings of Carapa procera D.C. (Meliaceae). International Research Journal of Plant Science, 2011. 2(9): 281-287.
160. Pangou VS, Zassi-Boulou AG, Bouki AG, et al., Evaluation of seed rain from remnant trees in fields for food crops in the tropical wet forest of Mayombe (Central Congo). Candollea, 2009. 64(2): 219-235.
161. Pangou VS, Jean-Pierre K, Lenoir C, et al., Evaluation of high-value indigenous trees for the rehabilitaion of deforested areas in Mayombe rain forest, Southern Con. International Research Journal of Plant Science, 2011. 2(9): 288-293.
162. Philip J H, 2 Universities to Pay U.S. $1.6 Million in Research Fraud Case. New York Times, 1994: 9.
163. Priya BK, Subrahmanayam P, Suvardhan K, et al., Cloud point extraction of palladium in water samples and alloy mixtures using new synthesized reagent with flame atomic absorption spectrometry (FAAS).[Retraction in J Hazard Mater. 2008 May 30;153(3):1320; PMID: 18543408]. Journal of Hazardous Materials, 2007. 144(1-2): 152-158.
164. Puneet P, McGrath MA, Tay HK, et al., The helminth product ES-62 protects against septic shock via Toll-like receptor 4-dependent autophagosomal degradation of the adaptor MyD88.[Retraction in Nat Immunol. 2011 August;12(8):804; PMID: 21772286]. Nature Immunology, 2011. 12(4): 344-451.
165. Rekha D, Suvardhan K, Kumar JD, et al., Solid phase extraction method for the determination of lead, nickel, copper and manganese by flame atomic absorption spectrometry using sodium bispiperdine-1, 1'-carbotetrathioate (Na-BPCTT) in water samples.[Retraction in J Hazard Mater. 2008 June 30;155(1-2):393; PMID: 18584810]. Journal of Hazardous Materials, 2007. 146(1-2): 131-136.
166. Rekha D, Suvardhan K, Kumar KS, et al., Spectrophotometric determination of V(V) in environmental, biological, pharmaceutical and alloy samples by novel oxidative coupling reactions.[Retraction in J Hazard Mater. 2008 April 15;152(3):1340; PMID: 18461699]. Journal of Hazardous Materials, 2007. 140(1-2): 180-186.
167. Renkema LJ, Stapel DA, Maringer M, et al., Terror management and stereotyping: why do people stereotype when mortality is salient? Personality and Social Psychology Bulletin, 2008. 34(4): 553-564.
168. Riedel D, Lindemann A, Brach M, et al., Granulocyte-macrophage colony-stimulating factor and interleukin-3 induce surface expression of interleukin-2 receptor p55-chain and CD4 by human eosinophils. Immunology, 1990. 70(2): 258-261.

169. Riedel D, Brennscheidt U, Kiehntopf M, et al., The mitogenic response of T cells to interleukin-2 requires Raf-1. European Journal of Immunology, 1993. 23(12): 3146-3150.
170. Ruys KI and Stapel DA, Emotion elicitor or emotion messenger? Subliminal priming reveals two faces of facial expressions. Psychological Science, 2008. 19(6): 593-600.
171. Ruys KI and Stapel DA, How to heat up from the cold: examining the preconditions for (unconscious) mood effects. Journal of Personality and Social Psychology, 2008. 94(5): 777-791.
172. Ruys KI and Stapel DA, The secret life of emotions. Psychological Science, 2008. 19(4): 385-391.
173. Saegusa A, Japanese scandals raise public distrust. Nature, 1999. 398(6722): 14.
174. Schumann RR, Nakarai T, Gruss HJ, et al., Transcript synthesis and surface expression of the interleukin-2 receptor (alpha-, beta-, and gamma-chain) by normal and malignant myeloid cells. Blood, 1996. 87(6): 2419-2427.
175. Schwinghammer SA, Stapel DA, and Blanton H, Different selves have different effects: self-activation and defensive social comparison. Personality and Social Psychology Bulletin, 2006. 32(1): 27-39.
176. Sei Jin K, Judd CM, and Stapel DA, Stereotyping based on voice in the presence of individuating information: vocal femininity affects perceived competence but not warmth. Personality and Social Psychology Bulletin, 2009. 35(2): 198-211.
177. Service RF, Chemist found responsible for ethical breaches. Science, 2008. 319(5867): 1170-1171.
178. Shaikh T. Sally Clark, mother wrongly convicted of killing her sons, found dead at home. The Guardian. 2007 March 17.
179. Sivani C, Ramakrishna Naidu G, Narasimhulu J, et al., Determination of Co(II) in water and soil samples using spectrophotometry coupled with preconcentration on 4-amino methyl pyridine anchored silica gel column. [Retraction in J Hazard Mater. 2008 Jul 15;155(3):610; PMID: 18642438]. Journal of Hazardous Materials, 2007. 146(1-2): 137-141.
180. Smith R, Research misconduct: the poisoning of the well. Journal of the Royal Society of Medicine, 2006. 99(5): 232-237.
181. Sott C, Dorner B, Karawajew L, et al., Transforming growth factor-beta relieves stem cell factor-induced proliferation of myelogenous leukemia cells through inhibition of binding of the transcription factor NF-jun. [Retraction in Blood. 1999 May 15;93(10):3573; PMID: 11012249]. Blood, 1994. 84(6): 1950-1959.
182. Sott C, Herrmann F, and Brach MA, The NF-jun transcription factor in the hematopoietic response to mitogenic signals. Immunobiology, 1995. 193(2-4): 149-154.

183. Spears R, Gordijn E, Dijksterhuis A, et al., Reaction in action: intergroup contrast in automatic behavior. Personality and Social Psychology Bulletin, 2004. 30(5): 605-616.
184. Stapel D and Suls J, Method matters: effects of explicit versus implicit social comparisons on activation, behavior, and self-views. Journal of Personality and Social Psychology, 2004. 87(6): 860-875.
185. Stapel D, Ontsporing ("Derailment"). 2012, Amsterdam: Prometheus Books. pp. ill, .315.
186. Stapel DA and Koomen W, Social categorization and perceptual judgment of size: when perception is social. Journal of Personality and Social Psychology, 1997. 73(6): 1177-1190.
187. Stapel DA, Koomen W, and Zeelenberg M, The impact of accuracy motivation on interpretation, comparison, and correction processes: accuracy x knowledge accessibility effects. Journal of Personality and Social Psychology, 1998. 74(4): 878-893.
188. Stapel DA and Koomen W, How far do we go beyond the information given? The impact of knowledge activation on interpretation and inference. Journal of Personality and Social Psychology, 2000. 78(1): 19-37.
189. Stapel DA and Koomen W, Distinctness of others, mutability of selves: their impact on self-evaluations. Journal of Personality and Social Psychology, 2000. 79(6): 1068-1087.
190. Stapel DA and Koomen W, I, we, and the effects of others on me: how self-construal level moderates social comparison effects. Journal of Personality and Social Psychology, 2001. 80(5): 766-781.
191. Stapel DA and Tesser A, Self-activation increases social comparison. Journal of Personality and Social Psychology, 2001. 81(4): 742-750.
192. Stapel DA, Koomen W, and Ruys KI, The effects of diffuse and distinct affect. Journal of Personality and Social Psychology, 2002. 83(1): 60-74.
193. Stapel DA and Blanton H, From seeing to being: subliminal social comparisons affect implicit and explicit self-evaluations. Journal of Personality and Social Psychology, 2004. 87(4): 468-481.
194. Stapel DA and Koomen W, Competition, cooperation, and the effects of others on me. Journal of Personality and Social Psychology, 2005. 88(6): 1029-1038.
195. Stapel DA and Koomen W, When less is more: the consequences of affective primacy for subliminal priming effects. Personality and Social Psychology Bulletin, 2005. 31(9): 1286-1295.
196. Stapel DA and Koomen W, When less is more: the consequences of affective primacy for subliminal priming effects. Personality & Social Psychology Bulletin, 2005. 31(9): 1286-1295.

197. Stapel DA and Van der Zee KI, The self salience model of other-to-self effects: integrating principles of self-enhancement, complementarity, and imitation. Journal of Personality and Social Psychology, 2006. 90(2): 258-271.
198. Stapel DA and Marx DM, Distinctiveness is key: How different types of self-other similarity moderate social comparison effects. Personality and Social Psychology Bulletin, 2007. 33(3): 439-448.
199. Stapel DA and Marx DM, Distinctiveness is Key: How Different Types of Self-Other Similarity Moderate Social Comparison Effects. Personality & Social Psychology Bulletin, 2007. 33(3): 439-448.
200. Stapel DA and Semin G, The magic spell of language: linguistic categories and their perceptual consequences. Journal of Personality and Social Psychology, 2007. 93(1): 23-33.
201. Stapel DA, Joly JF, and Lindenberg SM, Being there with others: how people make environments norm-relevant. British Journal of Social Psychology, 2010. 49(Pt 1): 175-187.
202. Stapel DA and Lindenberg S, Coping with chaos: how disordered contexts promote stereotyping and discrimination. Science, 2011. 332(6026): 251-253.
203. Stevens WK. Scientist accused of faking findings. The New York Times. 1989 April 23.
204. Subrahmanyam P, Krishnapriya B, Suvardhan K, et al., Simple, selective and sensitive spectrophotometric determination of fenitrothion using novel chromogenic reagent.[Retraction in J Hazard Mater. 2008 Jun 15;154(1-3):1210; PMID: 18508008]. Journal of Hazardous Materials, 2007. 146(1-2): 51-57.
205. Suvardhan K, Kumar KS, Krishnaiah L, et al., The determination of nickel(II) after on-line sorbent preconcentration by inductively coupled plasma atomic emission spectrometry using Borassus Flabellifer inflorescence loaded with coniine dithiocarbamate.[Retraction in J Hazard Mater. 2008 Mar 21;152(1):450; PMID: 18572427]. Journal of Hazardous Materials, 2004. 112(3): 233-238.
206. Suvardhan K, Babu SH, Kumar KS, et al., Inductively coupled plasma-optical emission spectrometry for on-line determination of multi trace elements in environmental samples after preconcentration by using Borassus flabellifer inflorescence loaded with 2-propylpiperidine-1-carbodithioate. Chemistry and Biodiversity, 2005. 2(4): 477-486.
207. Suvardhan K, Kumar KS, and Chiranjeevi P, Extractive spectrofluorometric determination of quinalphos using fluorescein in environmental samples. Environmental Monitoring and Assessment, 2005. 108(1-3): 217-127.

208. Suvardhan K, Krishnaiah L, Suresh Kumar K, et al., Studies of selenium in environmental samples and synthetic mixtures by spectrophotometry. [Retraction in Chemosphere. 2008 May;71(11):2199; PMID: 18584777]. Chemosphere, 2006. 62(6): 899-904.
209. Suvardhan K, Rekha D, Kumar KS, et al., Novel analytical reagent for the application of cloud-point preconcentration and flame atomic absorption spectrometric determination of nickel in natural water samples.[Retraction in J Hazard Mater. 2008 May 1;153(1-2):899; PMID: 18456906]. Journal of Hazardous Materials, 2007. 144(1-2): 126-131.
210. Swan N. The Man who stopped thalidomide accused of fraud. Sydney Morning Herald. 1987 December 14.
211. Taylor M. Cot death expert to face investigation. The Guardian. 2003 December 18.
212. Teunissen PW, Stapel DA, Scheele F, et al., The influence of context on residents' evaluations: effects of priming on clinical judgment and affect. Advances in Health Sciences Education, 2009. 14(1): 23-41.
213. Teunissen PW, Stapel DA, van der Vleuten C, et al., Who wants feedback? An investigation of the variables influencing residents' feedback-seeking behavior in relation to night shifts. Academic Medicine, 2009. 84(7): 910-917.
214. Trampe D, Stapel DA, and Siero FW, On models and vases: body dissatisfaction and proneness to social comparison effects. Journal of Personality and Social Psychology, 2007. 92(1): 106-118.
215. van den Bos A and Stapel DA, Why people stereotype affects how they stereotype: the differential influence of comprehension goals and self-enhancement goals on stereotyping. Personality and Social Psychology Bulletin, 2009. 35(1): 101-113.
216. Venkata Mohan S, Mohanakrishna G, Chiranjeevi P, et al., Ecologically engineered system (EES) designed to integrate floating, emergent and submerged macrophytes for the treatment of domestic sewage and acid rich fermented-distillery wastewater: Evaluation of long term performance. Bioresource Technology, 2010. 101(10): 3363-3370.
217. Vogel G, Human cloning. Scientists take step toward therapeutic cloning. Science, 2004. 303(5660): 937-939.
218. Wells F, The British pharmaceutical industry's response. In Fraud and Misconduct in Research, Lock S and Wells F (Eds). 1993, London: British Medical Journal.
219. Wild R, An Introduction to Sociological Perspectives. 1985, Crows Nest, New South Wales, Australia: Allen and Unwin.
220. Williams MH and Bowie C, Evidence of unmet need in the care of severely physically disabled adults. British Medical Journal, 1993. 306(6870): 95-98.

221. Won J, Kim M, Yi Y-W, et al., A magnetic nanoprobe technology for detecting molecular interactions in live cells. Science, 2005. 309(5731): 121-125.
222. Won J, Kim M, Yi Y-W, et al., A magnetic nanoprobe technology for detecting molecular interactions in live cells.[Retraction in Alberts B. Science. 2009 Apr 24;324(5926):463; PMID: 19390025]. Science, 2005. 309(5731): 121-125.
223. Won J, Kim M, Kim N, et al., Small molecule-based reversible reprogramming of cellular lifespan. Nature Chemical Biology, 2006. 2(7).
224. Won J, Kim M, Kim N, et al., Corrigendrum: Small molecule-based reversible reprogramming of cellular lifespan. Nature Chemical Biology, 2007. 3: 126.
225. Won J, Kim M, Kim N, et al., Retraction: Small molecule-based reversible reprogramming of cellular lifespan. Nature Chemical Biology, 2008. 4: 431.
226. Wong G. Rampant cheating hurts China's research ambitions. Associated Press. 2010 April 10.
227. Wong G. Academic fraud runs rampart in China. News and Observer 2010 April 11.
228. Wong G. China research hurt by plagiarism, faked results. Associated Press. 2010 Sunday April 11.
229. Wronski M, [Plagiarism in publications by Dr. Andrej Jendryczko](in Polish). Przegl Lek., 1998. 55(11): 629-633.
230. Xiguang L and Lei X, Chinese researchers debate rash of plagiarism cases. Science, 1996. 274(5286): 337.
231. Xin H, Scientific misconduct. Scandals shake Chinese science. Science, 2006. 312(5779): 1464-1466.
232. Zawadzki Z and Abbasi K, Polish plagiarism scandal. British Medical Journal, 1998. 316: 645.

CHAPTER 17

Effects of Scientific Misconduct

Research misconduct is not a trivial inconsequential act performed by dishonest researchers that can be ignored because of a lack of significant importance to humanity. Aside from affecting the perpetrator it has serious repercussions for society and other scientists.

Embarrassment of Mentors and Collaborators

Deliberate fraudulent behavior and blunders by a relatively small number of researchers have had a detrimental effect on most disciplines to the embarrassment of honest hardworking scientists. Research misconduct not only has a harmful effect on the perpetrator, but also on individuals responsible for supervising or collaborating with that person. For example, Philip Felig (due to Vijay Soman see chapter 14) and Joseph Nevins (due to Antil Potti see chapter 14). Some hardworking prominent scientists have been total surprised, to learn about misconduct by a graduate student or postdoctoral fellow, who they were mentoring but nobody has apparently fallen into this trap more than once. Coauthors of publications should have learnt the important lesson of not allowing their names to be associated with papers authored by others if they are not fully conversant with the raw experimental data and the validity of the research. Scientists who have embarrassed to discover this dishonest behavior by a trainee were initially struck in misbelief, have included Robert Good (1922-2003) (see William L. Summerlin, Eugene Braumwald (1929—)(see John Darsee), Ephraim Racker (1913-1991) (see Mark Spector chapter 14) Robert Good (1922-2003), Francis Sellers Collins (1950—), the current director of the National Institutes of Health (NIH). (see Amitav Hajra chapter 14), Dr. Cameron Bowie, emeritus director of public health in Somerset, England, among many others. The gloom of a scientist who

495

discovers scientific misconduct by a trusted colleague in a coauthored publication is difficult to imagine.

Collins was a coauthor of fraudulent papers by one of his students. This leading genetic researcher was the previous director of the National Center for Human Genome Research (NCHGR) and he was an ardent defender of research funding. In October 1996 the *Chicago Tribune* broke the news to the public about alleged scientific fraud in papers coauthored by this prominent scientist. The exposure of this renowned scientist gave the NIH and the scientific enterprise another black eye based on papers that needed to be retracted. Many prominent scientists who have coauthored papers with mentees have found themselves coconspirators in research misconduct by allowing their names to be included in the authorship. Many have previously held the cheater in extremely high regard and had complete confidence in them. This happened with anonymous mentees of the Czech biochemist and pharmacologist Carl Ferdinand Cori (1896-1984) the 1947 Nobel laureate in physiology or medicine (*http://www.nobelprize.org/nobel_prizes/medicine/laureates/1947/cori-cf.html*) with his wife Gerty Theresa Radnitz Cori (1896-1957) (*http://www.nobelprize.org/nobel_prizes/medicine/laureates/1947/cori-gt-bio.html*) and the Argentine physiologist Bernardo Alberto Houssay (1887-1971) in 1947 (*http://www.nobelprize.org/nobel_prizes/medicine/laureates/1947/*) when he was on the faculty at Washington University in Saint Louis. It also affected Dr. David Green in Medicine, and also occurred with Francis Collins (1950—), Ephraim Racker (1913-1991), Melvin Simpson (Mark Spector) at Yale University (Yale) (*http://www.yale.edu/*) and later with Fritz Albert Lipmann (1899-1986) the 1953 Nobel laureate in physiology or medicine (*http://www.nobelprize.org/nobel_prizes/medicine/laureates/1953/lipmann.html*) at Rockefeller University. As soon as reputable scientists (Cori, Simpson, Lipmann, and Collins) disclosed research misconduct they retracted the flawed publications and Green even went so far as to withdraw the work at a major meeting on the Federation of American Societies for Experimental Biology (FASEB).

Scores of prominent scientists who coauthored papers with mentees have found themselves participants in scientific misconduct by allowing their names to be included in the authorship.

Loss of Confidence in the Scientific Enterprise

Scientists are still highly regarded perhaps second to fire fighters in status, but the unacceptable deeds of research misconduct have diminished the prestige of scientists in the eyes of the public. Mere allegations of scientific misconduct blemish the image of the scientific enterprise and weaken its trustworthiness.

Dishonest or untrustworthy researchers develop a reputation and become known to other researchers in their subspecialty by word of mouth and other mechanisms. Such individuals get kept under control by becoming banished socially and by being denied letters of reference for future employment, and the withholding of research resources. Steps like these can discourage and castigate dishonorable actions within research institutions.

Many careers in science have been destroyed because the lure to falsify data was too great. Even accusations of misconduct can unjustly affect the reputations of innocent scientists. Eminent scientists became tragic casualties of scientific misconduct for debatable reasons but such cases shared common features. They usually managed huge research enterprises that were too large for them to pay close attention to the detailed research of those under their care and with whom they coauthored papers. Some scientists are concerned about having extremely busy researchers monitor trainees when they are managing major organizations. However, most researchers do not believe that administrators of research should become isolated from basic laboratory investigations, as the prevention of scientific misconduct cannot be completely avoided.

An amplified alertness to research misconduct has upset the scientific enterprise and caused much money and effort to be devoted to clearing up the chaos caused by publications based on fraudulent data.

Detrimental Impact of Research Misconduct on the Credibility of Science and Scientists

The impact on the perpetrator of the crime or on the individual who should have overseen the research is trivial compared to the effect that scientific misconduct has on the mother ship *Science*. Scientific misconduct elicits considerable publicity in television, radio, the printed press and on the internet. By chipping away at the public's trust fraudulent studies have an extremely bad effect on science. When those empowered to perform research funded by government grants or other forms of the taxpayer money do not perform honestly the public loses trust. They can also set back knowledge in some fields enormously especially if the culprit is a prominent scientist whose work has been accepted for years. When the scientific misconduct has involved important topics unnecessary effort has had to be used in attempts to replicate the results. When research deals with topics that relate to politically sensitive issues the scientist's research commonly loses its objectivity and the tested hypothesis is retained as in the case of Cyril Lodowic Burt and his studies and the role of genetics in the intelligence quota (IQ). Deliberate fraudulent behavior and blunders by a small number of researchers have had a detrimental effect on most scientific disciplines. Even an extremely low incidence of fraudulent research can be gravely injurious to science. False science can be costly to the scientific enterprise. Dishonest inappropriate behavior by scientists weakens the connection between science and society.

Anyone thinking of performing scientific misconduct had better give the matter serious thought as the penalties to perpetuators are most severe. Those who commit misconduct frequently lose their academic faculty position and their reputation is ruined and tarnished for life. The chastisement of those who commit scientific misconduct is stern and commonly results in a total derailment of the perpetrator's career in research. The penalties for research misconduct by any offender are severe and they typically derail the person's career completely. By participating in scientific misconduct researchers ruin their reputation as being truthful. Those found guilty of research misconduct face public humiliation as information about their misconduct is published by the Department of Health and Human Services (DHHS) and the Office of Research Integrity (ORI) on the internet. The emotional effect on persons accused of scientific misconduct is tremendous even if found innocent of the charges. Investigations into alleged scientific misconduct are time consuming and decisions are often not rendered until many years after the accusation first surfaces. The case of Theresa Ishimari-Kari lingered on for years. Robert B. Fogel was an example of someone who falsified and fabricated data in an abstract in 2001 and in a paper in 2003. Allegations against him lingered on for years before the ORI found him guilty of research misconduct on April 9, 2009. After the reputation of a scientist becomes damaged, even if it stems from being the principal investigator (PI) of a grant in which unidentified members of a research team commit scientific misconduct, the research remains questionable from then onwards.

Individuals determined accountable of scientific misconduct lose their credibility and this seriously affects their career in science. Almost invariably the person responsible for misconduct gains no financial or other tangible reward, but the penalties are severe. Some liberal minded individuals feel that penalties to brilliant young researchers who commit scientific misconduct is too harsh and that they should be receive a second chance, especially since the incentives to misconduct are not entirely self-determined. The responsible young researcher working in the Cori laboratory was provided a second opportunity to survive in science and his career advanced to a professor of public health at Johns Hopkins University (JHU) *(http://www.jhu.edu/)* until an early death apparently after some marvelous but unverified publications. Regardless of its cause scientific misconduct cannot be tolerated and must be harshly punished. Numerous universities and other institutions have a policy of terminating positions of persons found guilty of research misconduct. It is a pity that persons, who commit scientific misconduct commonly, waste the education and experience that they have accumulated prior to their fatal mistake. Much talent goes down the drain, but hopefully some of it can be salvaged. On January 8, 2013 James Dubois started a pilot study four health scientists to determine if they could be rehabilitated [2]. The spectrum of research misconduct is broad and so is experience of those found guilty of it. The sanctions inflicted vary from case to case and from institution to institution. Also, cases where the research is funded in the USA by the National Institutes of Health (NIH) the ORI imposes variable sanctions on those cases where guilt has been found. When the substantiation stands up in court, conviction and goal sentences are in order for certain types of scientific misconduct. Extenuating situations like unjustified pressure and a declaration of guilt may deserve milder reactions. Loss of both a job and reputation is by itself a brutal penalty. The fraudulent research by gifted young researchers has not only wrecked their careers. It has also brutally injured prominent scientists in whose laboratories they have worked because their supervisor/mentor was responsible for supervising the research of their trainees, who carried out these despicable acts. To coauthor a paper that allegedly contains defective research can affect the career of a scientist as it did with David Baltimore. These events have undoubtedly had a major psychological impact on all involved persons and only those who have been directly information affected can fully appreciate the effect.

Effect on the Careers of Innocent Participants

Probably the most severe consequence of fraudulent research is its effect on young students, post doctoral fellows, and junior faculty who innocently coauthor a paper based on fraudulent data derived from a dishonest colleague. After the misconduct is discovered the flawed paper causes an unwarranted guilt by association and the paper is virtually always retracted. Junior researchers lose publications from a developing curriculum vitae and the persons effort in its preparation goes down the drain. False science can be costly as it goes down the drain and careers may become deteriorates by unwarranted guilt by association. The retraction of flawed papers affects the publication record of all authors. This is particularly important for non-tenured faculty whose tenure decision may be effected by the retractions. To be found guilty of fraudulent research is a ground for dismissal at most universities, but proving it may be difficult, expensive, and fraught with potential litigation.

Retraction of Papers Due to Research Misconduct

Most papers containing defective data because of scientific misconduct are retracted when the dishonesty is recognized have had their papers retracted, but sometimes the papers are not retracted. Students and junior faculty without tenure are particularly severely hit when they are tarnished by the fraudulent content of a study in which they innocently participated. Publications in high impact journals are difficult to acquire and their loss weakens the person's curriculum vitae significantly.

Effects of Plagiarism in Scientific Misconduct

Plagiarism sometimes precipitates litigation because words and phrases fall under the umbrella of IP and laws that govern it. For allegations of plagiarism to end in court potential proof that a significant piece of text has been copied inappropriately without due recognition of the original source is needed. A copyright infringement involving multiple people can lead to a class action suit, such as the one that brought against Google claiming violations of "copyrights of authors, publishers, and other copyright holders by scanning in copyright Books and Inserts and displaying excepts without permission". In the case brought against Google, the parties agreed to a settlement if approved by court. The settlement will "authorize Google to scan in—copyright books and inserts to the USA and maintain an electronic database of Books. For out-of-print books, and if authorized by rights holders of in-print books, Google will sell right of entry to single books. An institutional subscription to the database, and the placement of advertisements on any page dedicated to the book and makes other profitable uses of the Books. Through a Book Rights Registry, Google will compensate Rights holders 63% of revenues from these uses (*http://www.* Googlebooksettlement. com). Despite denying the claims of copyright infringement, Google agreed to pay $34.5 million to establish the fund to the initial operations of the registry.

Jay Gunasekera had been chairman of the department of mechanical engineering at Ohio University from 1987, but in 2004 he became entangled in a plagiarism scandal. A graduate student discovered plagiarized text within the theses of several candidates for master or doctoral degrees. A university committee investigated the matter and Gunasekera and two other members of the faculty were blamed for allowing this cheating. They were reprimanded for "either failure to monitor the writings in advisee's theses or simply ignored academic honesty, integrity and basically supported academic fraudulence" [5]. Following this incident Gunasekera stepped down as the department chairman and he was dismissed from his role as a graduate student advisor. He then sued the University for damaging his reputation, particularly since the university had no evidence that he was aware of the plagiarism. In particular he sued dean Dennis provided Irwin and the former executive vice president Kathy Krendl because he was not an opportunity to exonerate his name. The lawsuit was settled on March 2011 in favor of Gunasekera who received $32, 501 in the settlement in addition to $118, 238 that was awarded to cover the fees of his attorney [3].

As long as plagiarism steals honest research findings it usually has no significant effect on the body of scientific knowledge. This is in sharp contrast to other forms of scientific misconduct. However, plagiarized clinical studies duplicates data and significantly over represents it in meta-analyses. If this occurs there is not only an additional cost to the research community but also potentially an adverse affect on the therapy of patients.

Most damage from plagiarism is to the ego of the plagiarized person, but some authors consider the plagiarism of their work a sincere form of flattery. Others just ignore it, while some writers become extremely upset and report the plagiarized text to the relevant authorities and even take legal action against the perpetrator. The time and costs in bringing a case to court is usually not worth the effort.

Many researchers mentor young graduate students and postdoctoral fellows. This responsible task cannot be taken lightly and it is essential that the mentor not fall into the trap of becoming too busy to supervise the research of the trainee or to become overconfident in the research conducted by the apprentice.

Prominent scientists who have supervised developing researchers have found themselves participants in flawed publications by allowing their names to be included in the authorship. This has happened with an anonymous mentee of Carl Ferdinand Cori (1896—1984). It also occurred with Mark Spector with Melvin Simpson at Yale University.

Impact on Human Health

Between 1975 and 1984 clinical investigators received administrative sanctions because of evidence collected by inspectors under the bioresearch monitoring program, which protects human research participants by continuous monitoring those involved in clinical drug research [6]. These individuals were prohibited from evaluating investigational drugs or were restricted in using such drugs. Inadequately carried out or fraudulent clinical trials not only place the subjects at needless risk, but may result in a far greater danger should the drug be endorsed based on insufficient safety and efficacy data.

Fortunately most instances of dishonest research have not impacted the treatment of patients, but fraudulent biomedical research and unintentional scientific errors is particularly grave when it affects the lives and health of human subjects. Errors in clinical research is commonly weeded out because of the customary practice of replicating experiments in independent laboratories before acknowledging the validity of the findings. Clinical trials by the pharmaceutical industry have an incentives to intentional bias and hence a danger of human injury should misconduct takes place. Manufacturers of drugs have a financial interest in having their drugs approved and then prescribed and sold. Fabrications and falsifications in clinical trials can lead to invalid conclusions and endanger patient safety. Moreover participants in fraudulent research may be put unnecessarily at risk especially when potentially dangerous treatments are involved. Healthcare resources are wasted when clinical trials need to be repeated when they are not performed correctly. Also, drugs may become approved by the FDA when they should not.

Dissemination and Reiteration of Misinformation

While scientific misconduct eventually becomes exposed in the long run the numerous detrimental effects of it persist for years. Misinformation is spread not only by papers containing the products of research misconduct, but also unknowingly by publications with unintentional errors that cite the flawed research.

Papers based on falsified and fabricated errors are usually retracted over time, but this often occurs many years after they appear in print and such articles are often cited numerous times in lectures, journals and textbooks. It takes time for the fraudulent nature of a particular paper to become widely known and this often does not occur until the paper is retracted, but this does not usually take place soon after the flawed paper is released. Indeed some papers with errors are retracted years later. Indeed the longest delay between the publication of a flawed paper and its retraction seems to be twenty five years. This paper was authored by I.E. Swift and V.E. Milborrow [8]. The dissemination of this misinformation leads to further costly unnecessary studies by others.

Unfortunately, many perpetrators of scientific misconduct have published dishonest manuscripts for years before they get exposed. The social psychologist Dierick Stapel performed fraudulent research for 15 to 20 years before his research was recognized as being dishonest [7]. When such a prolific researcher deceitfully mislead fellow scientists for such a long time the bottom line of their papers gets perpetuated exponentially as in compound interest.

Lost Time and Effort by Other Researchers

Publications with fake information cause other honest researchers, who trust in peer reviewed publications to waste time, effort and money tying to replicate or extend the phony work of those who have committed research misconduct. When this is done much effort becomes diverted from meritorious research tasks so that measures can be undertaken to investigate the fraudulent undertakings and maintain the integrity of science.

Criminal Prosecution for Research Misconduct

Scientific misconduct is illegal when the research is funded by federal grants and criminal prosecution can follow this dishonest behavior. Ross S. Laderman of the office of compliance, center for drugs and biologics of the Federal Drug Administration (FDA) has pointed out that the FDA took action through the justice department against several clinical research investigators who violated sections of the *Food, Drug and Cosmetic Act* or Title 18 of the USA code which deals with the submission of false data to the government, conspiracy, and obstruction of justice among other issues [6]. After 1975 the FDA submitted twenty cases of fraud and other criminal violations related to the clinical evaluation of new investigational drugs to the justice department. The investigators were convicted in thirteen cases and some received prison terms.

Effect on Society

Some research misconduct is relatively inconsequential on society particularly when the lives or available resources of society are not endangered. More importantly cheaters in research threaten the credibility of science in the eyes of the general public. Scientific misconduct is of particular concern because it damages the bond between science and society and misleads readers into not believing the scientific literature. Publicity about scientific misconduct and undisclosed COI in the media has weakened the trustworthiness of the whole scientific community. The adverse exposure about unacceptable deeds of scientists weakened knowledge about the work of scientists impairing the formation of well-informed policies. Stories of scientific misconduct and mere allegations of this behavior in the public arena have blemished the image and weakened the trustworthiness of the whole scientific enterprise. The debate between evolution and creationism is a notable example of the lack of confidence is science by the general public. Despite overwhelming evidence that the Darwinian theory of evolution is now an established fact rather than just a theory some extremely religious individuals find the concept totally unacceptable because of a profound belief in the divine creation according to the literal word of the Bible. They adhere to their conviction notwithstanding the fact that history has not been documented in writing since the origin of human life. Fraudulent science has made some political groups highly skeptical about the importance and power of science. This has led to rampant campaigns to lower the authority of science and scientific discoveries in the educational system of schools. In this regard it became official in 1995 that Colorado students would not be tested on evolution. Kansas followed this practice a few years later in 1999. Even more devastating to the teaching of science Mississippi and Tennessee do not teach the subject at all, and school curricula in Florida and South Carolina touch on the subject only lightly. The exposure of the public to dishonest research undoubtedly gives the impression that tax payer money is being misused and claims of science cannot be believed. Reports of scientific misconduct have also raised questions about whether scientists are indeed doing as much as necessary to keep their house in order. A prime example is the theory of evolution and fraudulent instances in this regard have been fodder for the creationists. The shortcomings of Kettlewell's research and the allegations of misconduct in his research on the peppered moth not only damaged his reputation, but also science in general. Moreover, it strengthened reservations about the Darwin theory of evolution.

Costs of Research Misconduct

Research misconduct is a significant financial burden on society. The amount is impossible to estimate, but just to maintain the ORI the American taxpayers dole out $9 million per year. Institutional inquiries and investigations into accusations of scientific misconduct are expensive especially in terms of the hours of manpower needed to carry out the required tasks efficiently. Scientific misconduct carried out by recipients of funded grants deprive other worthy grant applicants with research funds. Equally important are the expenses caused by the research misconduct *per se*. Eric Poehlman, who pleaded guilty to criminal and civil charges in resulted in several noteworthy costs, which the ORI needed to help evaluate for the injured parties. The University of Vermont cooperated with the ORI and of justice department in deriving this estimate. Poehlman's research involved hundreds of participants, who underwent extensive procedures. When it became available knowledge about the falsified research of Poehlman hindered the university's capability to recruit additional participants. His research misconduct had an effect on numerous collaborators and coauthors in more than two hundred publications. The coauthors included young clinician-scientists and their association with Poehlman made others mistrust their research publications and this seriously impaired their ability to obtain research positions.

It is impossible to determine the costs related to attempts by numerous research laboratories to replicate concocted studies before it becomes publically appreciated that misconduct has taken place. Sometimes fraudulent data enters the literature and becomes widely cited by readers who accept the work on face value. Graduate students and postdoctoral fellows bear the brunt of the costs related to trying to replicate and extend fraudulent research, because they cannot afford to spend valuable time pursuing unsuspected fraudulent research findings.

A vast body of at the University of Alabama at Birmingham (UAB) *(http://www.uab.edu/)* led to 16 retractions of papers by Drs. Judith Thomas and Juan Contreras and four of them from the Thomas Laboratory were withdrawn—for fake assertions. The misconduct resulted in debarment to Thomas and Contreras of 10 and 3 years respectively.

When the money of tax payers is wasted supporting dishonest research the used funds go down the drain. More importantly in an attempt to verify or extend fraudulent findings the talent and time of other scientists become diverted with wastage of additional financial resources. For example, the fraudulent work of the graduate student of Simpson caused him to spend a year untwisting the confusion on cytochrome *c* and this unpleasant task caused him to lose ground in another area of biochemistry.

Positive Effects of Research Misconduct

The emergence of fraudulent science during the misconduct revolution of the 1970s and 1980s not only had a negative effect on science, but also a positive one. Indeed it led to considerable improvements in the manner in which research is taught, performed, and published. It resulted in a vastly improved way in which trainees in science are mentored and learn research integrity and ethics. The philosopher Thomas C. Kuhn partitions the history of science into times of normal and revolutionary goings-on [9]. During normal times variances found by scientists need to be ignored or repressed. Paul Feyerabend, another philosopher of science argues that small scale cheating is vital to the progression of science [1]. He claims no theory, no matter how good, accounts for all observations in a field. Of necessity scientists must ignore certain facts or neutralize them within a hypothesis. Scientific misconduct has also had a positive effect on professional journals by making them evaluate submitted papers much more closely. Some cases of fabrication or falsification could not have been detected before publication because of fraudulent images, but software was developed under a grant from the Federal Bureau of Investigation (FBI) to detect such fraudulent images [4]. Some journals, such as *Science* and the *Journal of Cell Biology* now routinely screen images in papers that they accept for publication with a sophisticated digital process for uncovering image-fraud. Journals have also strengthened their policies for acceptance by requiring statements of COI, approval of research projects by Institutional Review Boards (IRBs) and animal care committees and input from statisticians and epidemiologists when indicated and relevant. Many journals now require each author to identify his/her specific contribution to the manuscript documenting the research findings and demand that all coauthors sign off on the final manuscript. Researchers are now more devoted to appropriate record keeping of laboratory activities and to meticulous managing of research funds. Because the culture of science has traditionally been based on trust and not suspicion, fraudulent research has made scientists more cautious in accepting the results of students, fellows and collaborators.

References

1. Broad WJ, Paul Feyerabend: science and the anarchist. Science, 1979. 206(4418): 534-537.
2. Cressey D, 'Rehab' helps errant researchers return to the lab. Nature, 2013. 493(4731): 147.
3. Hoke F, Veteran Whistleblowers advise other would-be 'ethical resisters' to carefully weigh personal consequences before taking action. The Scientist, 1995. 9(8): 1.
4. Kennedy D, Good news-and bad Science, 2006(3111): 145.
5. Labowsky M, Mann M, Meng C-K, et al. Yale's patent lawsuit was terrible mistake. Yale Daily News. 2005 February 25.
6. Laderman RS, Letters. Scientific fraud and prosecution. Science, 1987. 236(4809): 1613.
7. Stapel DA and Lindenberg S, Retracation. Science, 2011. 334(6060): 2011.
8. Swift IE and Milborrow BV, Retention of the 4-pro-R hydrogen atom of mevalonate at C-2, 2' of bacterioruberin in Halobacterium halobium. [Retraction in Swift IE, Milborrow BV. Biochem J. 2005 August 1;389(Pt 3):919; PMID: 16047404]. Biochemical Journal, 1980. 187(1): 261-264.
9. Wade N, Thomas S. Kuhn: revolutionary theorist of science. Science, 1977. 197(4299): 143-145.

CHAPTER 18

Correction of Errors in Science

Like all other human endeavors scientists inevitably make mistakes and they may result of from honest" unintentional errors or less desirably from research misconduct. Scientists also sometimes change their minds about the understanding of research data. When this occurs they are not embarrassed to bring new interpretations to the attention of others. Since the early days of science mistakes have been corrected by withdrawing former statements or retracting or correcting errors. The latter is an important aspect of science and as Confucious is stated to have said "A man who has committed a mistake, and doesn't correct it, is committing another mistake".

Spectrum of Error Correction

Once published the literature gets purged of flawed papers in more than one way. Mistakes discovered after publication are typically corrected by drawing them to the attention of the journal editor, who has several options. The journal may publish a brief comment under the heading "corrections and clarifications" or "errata". Of these choices errata are most frequently issued soon after the original publication. When mistakes are considered important they can be corrected by submitting them to the editor, but sometimes all authors cannot be traced or some authors refuse to give consent. Under such circumstances some journal editors may refuse to retract the paper, but the editor of other journals may retract papers based on his/her evaluation of the evidence. As recently as 1990 even when obvious reasons for retracting papers with scientific misconduct exist it is not always possible to enforce a retraction as the dean of academic affairs at the University of California San Diego (UCSD) *(http://ucsdnews.ucsd.edu/)* School of Medicine discovered in trying to get fraudulent papers by Robert Slusky MD retracted. After getting editors to retract papers due to editorial policies, Friedman recommended that journals establish a standard written policy about how to cope with retractions and errors in content [37]. Subsequently the journal editors in the United Kingdom (UK) established a Committee on Publication Ethics (COPE), which revealed, marked variations in the policies and procedures of different journals before establishing guidelines for the retraction of papers in 2009.

Expression of Concern

Some journals publish editorial expressions of concern about suspicious papers [9,85]. Several weeks after the publication of a paper by Beloqui and colleagues entitled "Reactosome array: forging a link between metabolome and genome [18] an editorial expression of concern was published to alert readers about serious questions raised about the methods and data presented in paper [2]. After this initial red flag the paper was retracted in November 2010 [19]. When reported research or the contents of the manuscript is questionable sometimes print editorial expressions on concern when referring to a disputable paper [2-4,10,28,55-57,84]. This practice usually takes place when the authors fail to satisfactorily rectify discrepancies, the editor or professional society governing the journal in which the paper was published by publishing an expression of concern. For example the *American Heart Association* published expressions of concern about five publications coauthored by the prominent Japanese cardiologist Hiroaki Matsubara of Kyoto Prefectural University following allegations of image manipulation in the papers [5,38,51,69,73,75,86]. The *Journal of Clinical Oncology* issued an expression of concern over similarities in images published by a Spanish research group in an article published in April 15, 2003 by Roman-Gomez et al. [83].

Most blunders are discovered and dealt with before papers go to press. Some mistakes pass the numerous gatekeepers of the scientific literature and get published. Once a paper is accepted for publication after surviving a rigorous peer review safety measures continue to protect against mistakes discovered by the authors and readers of the text. Minor mistakes in a publication can be fixed in journals in a section designated "corrections". More serious errors are dealt with by retracting the paper. Retractions of manuscripts in journals are becoming more frequent [35].

When significant errors or disputed interpretations or questionable research techniques are detected readers can write to the editor and the confronted embarrassed author is usually given an opportunity to comment on the real or alleged errors. When mistakes are found the authors and editor can be made aware of them. Journals publish letters to editors and this allows readers to point out mistakes and fraudulent material that slips past the peer review process. Other investigators can also openly draw attention to differences of opinion in personally written manuscripts. Such journals usually allow authors to respond to the criticisms as which are sometimes hypercritical especially if stemming from competitors. Nevertheless this creates a forum in which the merits of the paper and its interpretation can be debated and overtime the truth emerges as errors get corrected [11].

Alternatively colleagues familiar with the subject can correct errors in the discussion portion of subsequent new publications on a relevant topic. Even then most flawed information in papers gets purged from the literature. Relatively minor errors missed during the proofreading of a manuscript, or that get advertently inserted during editing are commonly trivial and inconsequential and not worth further action. Mistakes of this type often involve the innocent mislabeling of illustrations and errors in syntax. Occasionally the names of some coauthors are accidentally omitted or incorrectly spelt.

Retractions

A common way of dealing with seriously flawed scientific papers is have them retracted. This is usually performed by the editor of the journal with the consent of all authors. Before 1970 scientific papers were virtually never retracted, and as recently as 1990 journals rarely had written procedures on how manuscripts should be removed because of research misconduct. The retraction system is an expeditious method of exposing uncovered false knowledge [40].

The earliest known documented retraction seems to be a notice in the *Philosophical Transactions of the Royal Society* made by Benjamin Wilson (1721—1783) on June 24, 1756 (Ref), a distinguished scientist known for controversial disputes with the renowned Benjamin Franklin (1706-1790). Wilson was made a fellow of the Royal Society in 1751 based on his research including his *Treatise on Electricity* [107] and in 1756 that distinguished society awarded him their most prestigious Copley Medal created by the wealthy English landowner Sir Godrey Copley FRS (1653-1709). Other recipients of this medal were Michael Faraday FRS (1791-1867), Charles Darwin FRS (1809-1882), Dorothy Crowford Hodgkin FRS (1910-1994), the recipient of the 1964 Nobel Prize in chemistry *(http://www.nobelprize.org/nobel_prizes/chemistry/laureates/1964/hodgkin-bio.html)* and Francis Crick FRS (1916-2004), who shared the 1962 Nobel Prize in physiology or medicine *(http://www.nobelprize.org/nobel_prizes/medicine/laureates/1962/crick-bio.html)* with James Watson (1928—) *(http://www.nobelprize.org/nobel_prizes/medicine/laureates/1962/watson.html)*.

It was discovered by Jeffrey L. Furman, Kyle Jensen, and Fiona Murray while preparing a paper on retractions [40] Wilson stated: "Gentlemen, I think it necessary to retract an opinion concerning the explication of the Leyden experiment, which I troubled this Society with in the year 1746, and afterwards published more at large in a *Treatise upon Electicity*, in the year 1750; as I have lately made some farther discoveries relative to that experiment, and the minus electricity of Mr. Franklin, which shew I was then mistaken in my notions about it. What I mean by the minus electricity of Mr. Franklin, regards the minus electricity of the Leyden experiment only, which that gentleman discovered. I shall be very glad to have this acknowledgment made public, and to answer that end the effectually, I wish that it may have a place in the Transactions of the Royal Society". *The Philosophical Transactions of the Royal Society, Abridged,* 1755 to 1763, XI (London, 1809, p. 15).

By tradition journals had different policies related to the correction of published papers and many were reluctant to correct or retract them unless requested to do so by all coauthors. For retractions to be approved some journals, such as the *Journal of Clinical Oncology,* required all authors to sign not only a statement approving the retraction, but also the wording of the explanation. Now without significant hesitation journals retract hundreds of flawed publications. Papers can be retracted by the authors or by the journal in which they have been published. From the standpoint of the authors it is preferable that the process be initiated by them as they have the right to explain the reason for the retraction and how the flaws were detected. When the journal retracts the paper the situation is more serious and usually because errors have been detected by others, as in the finding of research misconduct by an institution or an agency like the Office of Research Integrity (ORI). Also, available evidence suggests that retractions made by authors are looked upon more favorably than those initiated by journals. Lu and coauthors [65] analyzed the effects of 667 retractions occurring mainly after 2000 on the citations of the earlier work of the authors. When journals retract the papers the number of annual citations of earlier papers drop by 6.9% on average, but this fall was not detected with author initiated retractions suggesting that the scientific community had more confidence in the research of researchers who cared about getting their papers correct.

Reasons for Retractions

Papers reporting flawed research unfortunately sometimes fall through the cracks in the peer process and manage to get published despite excellent reviewers and journal editors [102]. Papers are retracted for fraudulent research, unintentional mistakes and political reasons. Intentional errors due to research misconduct account for most retractions (67% 73.5%). The rest are accounted for by unintentional "honest" mistakes or when the validity of the study is questionable or not replicable. Fabrication or falsification account for the rest (26.6%).

Multi-authored papers can be retracted if a single author does not consent to its submission for publication. For example Antonia D. Asencio and two coauthors published an online paper in *Journal of Phycology* on August 22, 2011, on the Wiley Online Library (*http://www.wileyonlinelibrary.com*)[17], but one coauthor (Ferran Garcia-Pichel) did not agree to the submission so the journal editor (Robert Sheath), together with the publisher (Wiley Periodicals Inc) retracted the paper.

The reasons include: inappropriately manipulated digital images, unreliable reported findings, to confirm original results, a cDNA library found to be derived from another organism, impure critical reagent, and plagiarism. A survey by the Committee on Publication Ethics (COPE) of Medline retractions from 1988-2004 found that 40 percent of retractions were for honest mistakes or non-replicable findings. Other causes were research misconduct (28%), redundant publications (17%), other or unstated (15%) [103]. Retractions are still relatively rare and affect approximately once in every 10, 000 publications.

The main cause of retractions in the life sciences is research misconduct [26]. Most retractions are due to intentional fabrication, falsification, or plagiarism. Retractions have also taken place because of an acknowledgment of considerable manipulated or fabricated data [21,90,95,96] of "data irregularities"[58]. Other retractions are due "honest" unintentional errors and duplicate publications. For example the journal *Blood Cells, Molecules, and Diseases* retracted a 2011 paper [43,44] after discovering that the data had already been published (elsewhere). Also, *Neurological Sciences* retracted a 2009 paper by a Korean scientist after realizing that the article contained parts of a different papers have been retracted by authors from many different countries and some have been collaborative studies involving multidisciplinary investigators from diverse parts of the world.

In recent years the number of retractions has increased immensely and in part the explanation seems to be because journals are acting more swiftly [89].

When publications are involved in research misconduct the PHS or ORI requests that the sanctioned person to retract the papers with an explanation in letters to the editors of journals in question identifying missing data as well as material that was falsified or fabricated in the paper. Also relevant manuscripts in press need to be retracted. Individuals found guilty of research misconduct by the ORI are required to contact the editors of journals in which fabricated or falsified material is published. Unless the publications can be corrected with errata retractions are required with detailed explanations in letters approved by the ORI.

Examples of this involved a paper by Atsushi Handa and Kenin E. Brown in the *Journal of General Virology* [45]. When this paper was retracted by Brown, the senior author, he pointed out that he had learnt that the first author had fabricated one of the figures and that the supporting data was lost so that the accuracy of the research could not be verified.

In a review of 2, 047 biomedical and life-science research articles indexed by PubMed as retracted Fang and coauthors found on May 3, 2012 that the majority (67.4 %) of the retracted scientific publications were the direct result of research misconduct (fraud, suspected fraud, and plagiarism) [36]. In October 2011 Fang and co-workers [35] reported that over four decades the journal *Infection and Immunity* published 28, 000 articles, but only fifteen papers were retracted over the same period, yet six were withdrawn in 2011 from the same laboratory [7,8,29,35,52,70,72,74-76,80,82,87,92,93].

Statistics from the National Library of Medicine indicate that retractions have increased immensely in recent years. Retractions have sky rocketed tremendously in recent years by increasing tenfold [59]. For 2012 retractions increased 37-38 % over 2011 *(http://www.nhl.nih.gov/)*. The increase in retractions aised the question of whether scientists have become more dishonest [81,88]. The number of publications relative to retractions also provides an indication of the flawed scientific publications. Numerous highly respected scientists, including Nobel laureates, have retracted publications from their laboratories after discovering manufactured findings by a collaborator. Because of the common misconception that retraction equates to research misconduct, some researchers are reluctant to withdraw publications with "honest" unintentional errors because of possible shame, but distinguished Nobel laureates, such as David Baltimore the 1975 Nobel laureate in physiology or medicine *(http://www.nobelprize.org/nobel_prizes/medicine/laureates/1975/baltimore-lecture.html)* Von Parris have not refrained from retracting papers when necessary. Baltimore was required to retract a paper from the Baltimore affair [100,101,106]. He also has had an academic connection with a third individual John Long[22].

Linda Brown Buck (1947—) the 2004 Nobel laureate in physiology or medicine *(http://www.nobelprize.org/nobel_prizes/medicine/laureates/2004/buck-lecture.html)* retracted three papers unrelated to her Nobel Prize winning research on the sense of smell that were coauthored by her postdoctoral fellow Zhihua Zou. These retracted papers dealt with the olfactory system in mice were retracted because she and others could not repeat the experiments [20][60][110] *(http://www.nytimes.com/2010/24/science/24retraction.html?r=1)*. Despite denying any wrongdoing Zou signed the retraction of one paper, but refused to provide consent to retract the other two manuscripts [109]. Retraction [60][108,110-112].

Errata

Some errors in publications can be corrected by preparing errata and sometimes errata progress to retractions. The rate of errata since 2002 has remained comparatively constant at about 1%, in contrast to vast increase in retractions.

By custom science has been regarded as self-correcting because blemished information about important topics eventually becomes forgotten and stops being cited. Many papers with significant errors unfortunately become embedded in the scientific literature; especially in textbooks and become cited on vast occasions before the desired outcome of a retraction takes effect [35].

Their work was featured in the *New York Times* in April 2012. Publications in high impact journals do not necessary translate into high quality arenas. A *Wall Street Journal* study of 11, 600 peer reviewed journals disclosed 22 retractions in 2001 Ref. By 2006 retractions had increased to 139 and in 2010 there were 339 [24, 25, 32-34, 71].

Competition for research funds forces scientists to try to produce publishable results at all costs. Also an incentive exists to produce positive results as negative findings are less likely to be publishable. Competitive academic environments, especially in the USA, increases the productivity of researchers as well as their bias [30,27,88].

In a review of 742 retracted papers Steen found that the retractions had increased 10 fold during the past decade [88] and he suggests that many of these papers were written with a purposeful intent by the author to deceive the readers [89].

A recent editorial in *Infection and Immunity* [35] showed that high impact journals retract more publications than low impact journals [13]. Based on retractions in 17 journals Fang and Casadevall created a *retraction index*. For each journal it is the number of retractions between 2001 and 2010 multiplied by 1, 000 and divided by the number of published articles with abstracts. The *retraction index* correlated robustly with the journal *impact factors*. They have suggested that high impact journals encourage authors to take more risk with their research. The more prestigious the journal the more probable a researcher will cut corners or fudge data. Also journals like *Science* and *Nature* require crisp, clean scientific publications despite the fact that "everyday science is often a messy affair" and they suggest that such requirements may inspire investigators to manipulate their data to meet this expectation. Other factors that may contribute to retractions include, the greater scrutiny that papers in these high-impact journals may make them more prone to the detection of unintentional errors and phony data than articles in less influential journals. Perhaps others detect flaws in the papers or cannot replicate the findings. On August 10 2011 two days after the Fang and Casadevall paper a report by Gayam Naik in the *Wall Street Journal* drew attention to the increased number of retracted papers in various journals [77] (*http://www.burrillreport.com/articlefraud_and_errors_fuel_research_journal_retractions.html* . . . 8/31/2011). It pointed out that during the past decade the number of published papers has risen 44%, but that retractions had risen 15 times.

Time Delay in the Retraction of Papers

Unfortunately flawed papers are often retracted years after their publication and the defective information may already have been cited and widely disseminated. The retraction of a paper does not signify its fatality as defective papers every so often keep on being quoted especially if the retraction is delayed.

Retracted papers mostly take place within less than two years, but some retractions take an extremely long time to take place. Journals very rarely become engrossed in a contested article many years after its publication. For example the notorious 1998 flawed article by Dr. Andrew Wakefield and colleagues [104] that linked the measles, mumps and rubella (MMR) vaccine to autism was only retracted 12 years after it appeared in *The Lancet* on February 3 in 2010.

Ivan Oransky and Adam Marcus launched *RetractionWatch* (*http://retractionwatch.worldpress.com*) on August 3 2010. According to *RetractionWatch* a paper by Swift and Milborrow [91] holds a record twenty five years for the longest time between the publication of a paper and its retraction. After such a long time the retraction is probably caused by staunch criteria that journals have about manuscript retractions and a fear of litigation by authors who firmly adhere to the validly of their paper. This paper was published in *Science* on August 19, 2011 [14].

Once fraudulent material becomes entrenched in the literature, especially within textbooks it is difficult to flush it from public knowledge. This makes it is particularly difficult for young scientists who are trying to absorb a vast body of information about different topics. The important lesson is not to believe everything that is read and to challenge all existing information as a healthy doubter.

When readers discover questionable material in published papers they are free to draw the dubious content to the attention of the corresponding author of the paper, or the journal editor. The designations for rectifying mistakes in publications range in trivial corrections such as errata to retractions the most severe form of dealing with a flawed manuscript. Sometimes errata are published before the more serious retraction is assigned as illustrated by a paper authored by Wang and colleagues [105]. An example of this type was made by Hor. For data fabrication by Chan Struhl the senior author retracted a paper titled

"Evidence that armadillo transduces wingless by mediating nuclear export or cytosolic activation of pangolin" [21,90].

Errors Created by Journal Editors

Journal editors have considerable freedom on how to deal with questionable manuscripts. When Sir *John Maddox* (1925-2009) was editor of *Nature* Benveniste submitted a controversial paper to that high-impact journal. The paper would normally have been rejected because of unbelievable findings, but Benveniste was an excellent researcher with a superb reputation. Hence Maddox wondered if this renowned scientist had made a remarkable new discovery and decided to accept the paper on two conditions that Benveniste accepted. Maddox wanted the experiments authenticated in other impartial laboratories and after publication the Benveniste team needed to carry out the study in front of witnesses chosen by *Nature*. The required replication was easily taken care of by the addition of more coauthors to the paper. In publishing the manuscript Maddox highlighted it with a perspective and editorial proviso [6,15]. After the paper was published Maddox arranged for an independent confirmation of the experiments in front of James Randi, a former magician, the American fraud buster Walter Stewart and him, but Benveniste was unable to convince the skeptic trio. A retraction was then requested and after Benveniste's refused to retract this paper a critique of the study was written pointing out the flaws in the experiments [66]. The incident was an embarrassment to the research team and Philippe Lazar, director of the Institut National de la Santé la Recherche Médicale (INSERM), felt that *Nature* had dealt badly with the Benveniste controversy and had publically humiliated the coauthors excessively.

Examples of Retractions

Jean-Calvin Nguele a former graduate student of at the University of Houston (UH) *(http://www.uh.edu/)* published a severely fraudulent paper with fabricated data based on considerable plagiarized published and unpublished research data from the Masaya Fujita laboratory without appropriate authorization. Sections were pilfered from Fujita's grant. Nguele forged the signatures of three alleged coauthors (Prahathees Eswaramoorthy, Mou Bhattacharya, and Masaya Fujita). A fifth author named Edouard Ngou c Milama could not be identified and was not known to the Fujita laboratory. The editorial staff of the journal took the drastic action of retracting the paper without the consent of the authors [12] after receiving evidence from Fujita and letters from the available named coauthors who neither participated in the study nor agreed to an authorship. According to *Retraction Watch* the fraudulent paper was surprisingly discovered by Eswaramoorthy during a literature search. Shimomura the senior author of a paper. [58] retracted it because a questionable transgenic mouse never existed and all gel pictures were forged by one of his colleagues.

Who retracts papers.

A paper by Deb and co-workers [30] was retracted because Deb intentionally falsified and fabricated digital images. Numerous manuscripts are retracted by coauthors, after the discovery of fraudulent research by other authors. When the retraction of papers is demanded some authors still maintain that the thrust of their paper is valid [49]. After publishing a paper Horsewill and colleagues retracted it after discovering in further studies [48] that the sample used by them contained residues of dichloromethane which may have affected their results.

Geremia and colleagues published a paper in 2004 [41] and which then subsequently found to be flawed and four years later it was retracted.

Three papers by R.B. Tracy were retracted and Tracy. The editor of the *Journal of Gastrointestinal Surgery* retracted a 2009 article [67,68] for plagiarism [62].

Retractions are a blow to the career of distinguished scientists [23][53][73]. A questionable paper allegedly disclosed that antibiotics might sometimes be safe in treating appendicitis rather than the traditional method of removing the appendix surgically. Several surgeons from Bologna, Italy, confronted the Indian authors in correspondence to the academic journal [16]. The manuscript was retracted by the editor of the *Journal of Gastrointestinal Surgery and* the board of directors of the *Society for Surgery of the Alimentary Tract (SSAT)* representing the organization managing this journal because significant parts of the paper had previously been published by other authors [31,50]. "A paper by Lipardi and Paterson in the Proceedings of the National Academy of Sciences in the USA [63,64] was retracted by the authors who provided a detailed explanation for the retraction and rescinded their original interpretation [64] due to an unintentional error [46,47,54,79,94,98,99]. This surge in the retraction of papers highlighted weaknesses in the system for handling them.

However, a vast number of them (about 200) become targeted for the rectification of mistakes or become recognized as being notably flawed so that they become withdrawal from the literature. A scientifically sound paper may be withdrawn from the scientific literature because data has already been published [97].

More than 50% of retractions are thought to be because of falsified or fabricated research or plagiarism. Some falsified research leading to retraction has involved doctored images. Retractions have also occurred because of the submission of the same paper to more than one journal. Other scientific publications are retracted for unintentional mistakes, such as artifacts of analysis, an agreement between the journal editor and one or more coauthors, inappropriate duplication of images and/or previously published text, political reasons, scientific misconduct, and because studies cannot be replicated.

RetractionWatch

When Steven Shafer was editor-in-chief of *Anesthesia & Analgesia* he took part in two important misconduct investigations. One dealing with 11 journals culminated in the retraction of 90 papers. Soon thereafter on August 2010, Adam Marcus, managing editor at Anesthesiology News, co-founded the blog *Retraction Watch* with Ivan Oransky the executive editor at Reuters Health. This site on the internet site (*http://retractionwatch.wordpress, com/*) tracks new retractions soon after they are announced and indicates how often these papers have been cited according to the Thomson Scientific Web of Knowledge. Regardless of who retracts a paper Retraction Watch recommends that the reason be stated mainly to safeguard blameless authors who need to be encouraged to retract papers when honest errors are discovered. In the USA when allegations of scientific misconduct are confirmed by the ORI in NIH funded research publications the dishonest research is invariably retracted.

Multiple Retractions

Because multiple retractions due to unintentional mistakes by a specific researcher are extremely uncommon the presence of more than one retracted paper by a particular author is strongly suggestive of scientific misconduct or sloppy science and the caliber of that person's research becomes extremely questionable. The current leader in multiple retractions is Yoshitaka Fujii who currently has 172 retracted papers. Second place goes to Joachim Boldt, who is far behind and just short of one hundred (90). Diederick Stapel has twenty eight retractions and Antil Potti has more than 18.

Other notable examples with multiple retractions are John Darsee, Schön, and Scott S. Reuben MD.

Frequency of Retractions

Considering the vast number of papers published each year the number of retractions is low.

Announcements of retractions have increased with amazing frequency in recent years (von Noorden) and the reason remains uncertain particularly in high impact journals. During the early 2000s only about 30 papers were retracted each year, but during the past decade the number of retracted papers has grown exponentially and risen from 0.001% of all publications to about 0.02%.

Now the Web of Science (thompsonreuters.com/products_services/science/science_ products/a-z/web_of_science/) keeps tracks more than 400 retractions (Rise of the retractions).

Without doubt the progressively expanding number of readers of scientific papers has led to the detection of more flaws in the scientific literature. A rise in "honest" inadvertent errors may be a factor due to the increased awareness of the importance of keeping the literature accurate. Perhaps part of the growth in retractions reflects an increased recognition of scientific misconduct due to regulatory bodies such as the Office of Research Integrity (ORI) in the USA.

Effects

The retraction of papers clearly is detrimental to junior investigators who are developing their credentials and curriculum vitae.

Explanations

When papers are retracted from the scientific literature they are frequently accompanied by the reason. It is commonly prepared by the authors and relates to the discovery of mistakes by the authors, but purposefully conceals the authentic reason without indicating whether they reflect research misconduct or unintentional "honest" mistakes. Many journals now strongly encourage an explanation for retractions and discourage authors from volunteering to have any of their own papers retracted. Authors are reluctant to retract papers particularly when explanations are required, unless they are forced to because of scientific misconduct. When research misconduct is established by the ORI respondents are generally required to write to the journal editors giving the precise reasons why the flawed papers are being withdrawn. The wording of the letters also needs to be approved b the ORI. Journals distinguish retractions based on "honest errors" from those precipitated by scientific misconduct, but some journals do not insist on a clear-cut explanation for the retraction. When provided the wording of the explanation is important. Some editors prefer not to retract publications and even when they do all journals do not provide a clear explanation for this drastic action. When the explanations are drafted by editors and the ambiguity of the reasons every now and then suggests an attempt to avoid a lawsuit. Ideal retractions are those that provide a clear explanation for why the paper is being withdrawn. Unless the explanation honestly and explicitly states why the article is being retracted the reader is left uncertain about the true reason. To just state that a paper is being retracted because the authors wanted to do so is unethical and gives the misconception that an honest unintentional error was made. Researchers guilty of wrongdoing may seek a retraction as soon as possible before questionable papers face a forced withdrawal so that the retraction does not mention the misconduct. To avoid this possibility COPE recommends that papers do not become retracted during investigations of research misconduct, but only after the investigations are concluded. During this interim period an "expression of concern" may be appropriate.

It is difficult to flush the literature of publications containing errors, but when severely defective information is discovered in scientific articles the entire paper can be retracted. This frequently used method together with other procedures is part of the ongoing sanitization of the scientific literature. Many contemporary errors in science occur because of increasingly complicated data. Small errors in computer programs can generate substantial mistakes. Research in the social sciences and humanities is more difficult and less precise than in other sciences and replications of such research are uncommon yet retractions seldom occur. Regardless of the reason why a paper is retracted the author's ego and reputation is affected.

An analysis of 724 papers in medicine and biology that were retracted between 2000 and 2010 related to scientific misconduct then increased 7 times between 2004 and 2009 (published in the *Journal of Medical Ethics*).

For an article to be retracted from the universal body of knowledge journals have traditionally required explicit permission of all authors. This policy has challenged journal editors when all authors are reluctant or unavailable to consent to a retraction of a blemished paper. Some editors have retracted papers on the request of the senior author even if other authors refuse to agree. Some papers are retracted in response to requests by coauthors as occurred with Herrmann and Brach, but because of a lack of a uniform policy regarding paper retractions at that time the editors were uncertain about how to react to flawed publications in their journals.

After appropriate consultation some journals, such as *Nature* publish anything they can about publications verified as suspect or false. *Nature* attempts to sway authors to retract tainted papers formally and as soon as possible, but sometimes coauthors disagree among themselves. Not all journal editors are persuaded to extend their activities into the domain of law enforcement and consider it inappropriate for them to function as a judge or jury with regard to disputed publications. It was questionable whether editors should report suspect manuscripts submissions to the appropriate authorities. Journals have different editorial policies on how to respond when reviewers detect content suspicious of scientific misconduct. A response to this is more difficult than dealing with paper retraction. Journals can draw the attention of their readership to publications with questionable content and publish remarks by authors while simultaneously making it obvious where the parties disagree.

Errors picked up by knowledgeable reviewers can be corrected before going to press.

At one time it was debatable as to whether editors should retract articles containing fraudulent data even when at least one author did not agree with an investigation the committee's decision.

Certain journals, such as the *Journal of Immunology* are prepared to notify institutions about probable scientific misconduct when it is detected during peer review if the authors do not satisfactorily clarify questionable data. On the other had Magne Nylenna, a former editor of the *Journal of the Norwegian Medical Association* regards the submission of manuscripts to journals extremely confidential and that the editorial staff should not act as part of the secret police [1].

When the products of scientific misconduct are published or the contents of papers are proven inaccurate steps need to be taken to withdraw the information as soon as possible from the public domain. In general the scientific community, institutions and editors of professional journals agree that such papers need to be retracted and this is most frequently achieved by having the authors withdraw the paper.

To remove a flawed paper from the universal body of knowledge journals have traditionally required explicit permission of all authors. It matters not whether the retraction is based on scientific misconduct, sloppy research, or honest errors in the performance or interpretation of the research. With most retracted articles the reason is not made public and only the editor is provided with an explanation.

The Committee for Research Integrity (CRI) at the Osaka University Graduate School of Medicine investigated biochemical experiments reported by Fukuhara et al. (Science) that focused on the interaction of visfatin (also known as pre-B cell-colony enhancing factor), a protein secreted by visceral fat, with the insulin receptor. Because of the findings of this committee the authors agreed to retract the paper, but stood by their conclusions and pointed out that many publications have shown that human plasma visfatin levels correlate with metabolic states.

Regardless of whether published misinformation is caused by scientific misconduct, unintentional "honest mistakes" caused by sloppy research or other technical errors in the research several ways of correcting flawed information exist. If only minor imperfections are present the authors can draw these errors to the attention of the editors and if deemed significant a brief comment may be published sometimes as a letter to the editor. Some journals publish letters to editors and this allows readers to draw attention to errors and fraudulent material that slipped past the peer review process

References

1. Abbott A, Science comes to terms with the lessons of fraud. Nature, 1999. 398(6722): 13-17.
2. Alberts B, Editorial expression of concern. Science, 2010. 330(6006): 912.
3. Alberts B, Editorial expression of concern. Science, 2010. 327(5962): 144.
4. Alberts B, Editorial expression of concern. Science, 2011. 333(6038): 35.
5. Amano K, Matsubara H, Iba O, et al., Enhancement of ischemia-induced angiogenesis by eNOS overexpression.[Erratum appears in Hypertension. 2004 January 19;43(2):e9]. Hypertension, 2004. 41(1): 156-162.
6. Anonymous, When to believe the unbelievable. Nature, 1988. 333(6176): 787.
7. Anonymous, Retraction. Activation of intercellular adhesion molecule 1 expression by Helicobacter pylori is regulated by NF-kB in gastric epithelial cancer cells.[Retraction of Mori N, Wada A, Hirayama T, Parks TP, Stratowa C, Yamamoto N. Infection and Immunity 2000 April;68(4):1806-14; PMID: 10722567]. Infection and Immunity, 2000. 79(1): 542.
8. Anonymous, Retraction. Directional gene movement from human-pathogenic to commensal-like streptococci.[Retraction of Kalia A, Enright MC, Spratt BG, Bessen DE. Infection and Immunity 2001 August;69(8):4858-69; PMID: 11447161]. Infection and Immunity, 2001. 77(10): 4688.
9. Anonymous, Editorial expression of concern. Journal of Pharmacology and Experimental Therapeutics, 2009. 329(2): 848.
10. Anonymous, Editorial expression of concern. Journal of Pharmacology and Experimental Therapeutics, 2009. 329(2): 848.
11. Anonymous, Retraction--Ileal-lymphoid-nodular hyperplasia, non-specific colitis, and pervasive developmental disorder in children.[Retraction of Wakefield AJ, Murch SH, Anthony A, Linnell J, Casson DM, Malik M, Berelowitz M, Dhillon AP, Thomson MA, Harvey P, Valentine A, Davies SE, Walker-Smith JA. The Lancet. 1998 February 28;351(9103):637-641; PMID: 9500320]. The Lancet, 2010. 375(9713): 445.
12. Anonymous, Retraction of "Genetic and biochemical analyses of sensor kinase A in Bacillus subtilis sporulation" by J.C. Nguele, P. Eswaramoorthy, M. Bhattacharya, E. Ngou-Milama and M. Fujita. Genetics and Molecular Research. 9 (1): 573-590 (2010).[Retraction of Nguele JC, Eswaramoorthy P, Bhattacharya M, Ngou-Milama E, Fujita M. Genetics and Molecular Research. 2010;9(1):573-90; PMID: 20391342]. Genetics and Molecular Research, 2011. 10(3): 2257.
13. Anonymous, Higher Impact, More Retractions. Science, 2011. 333(6045): 924.
14. Anonymous, Corrections and clarifications. Science, 2011 333(6045): 937.

15. Anonymous., Editorial reservation: readers of this article may share the incredulity of the many referees Nature, 1988. 333(6176): 818.
16. Ansaloni L, Catena F, Coccolini F, et al., Letter to the editor. Re: Conservative management of acute appendicitis. Journal of Gastrointestinal Surgery, 2010. 14(5): 931-932; author reply 933.
17. Asencio AD, Garcia-Pichel F, and Hoffmann L, RETRACTED. Carotenoids, mycosporine-like amino acid compounds, phycobiliproteins, and scytonemin in the genus Scytonema (cyanobacteria): a chemosystematic study. Journal of Phycology, 2011. 47(5): 1228.
18. Beloqui A, Guazzaroni M-E, Pazos F, et al., Reactome array: forging a link between metabolome and genome.[Retraction in Beloqui A, Guazzaroni ME, Pazos F, Vieites JM, Godoy M, Golyshina OV, Chernikova TN, Waliczek A, Silva-Rocha R, Al-Ramahi Y, La Cono V, Mendez C, Salas JA, Solano R, Yakimov MM, Timmis KN, Golyshin PN, Ferrer M. Science. 2010 November 12;330(6006):912; PMID: 21071648]. Science, 2009. 326(5950): 252-257.
19. Beloqui A, Guazzaroni ME, Pazos F, et al., Retraction.[Retraction of Beloqui A, Guazzaroni ME, Pazos F, Vieites JM, Godoy M, Golyshina OV, Chernikova TN, Waliczek A, Silva-Rocha R, Al-Ramahi Y, La Cono V, Mendez C, Salas JA, Solano R, Yakimov MM, Timmis KN, Golyshin PN, Ferrer M. Science. 2009 October 9;326(5950):252-7; PMID: 19815770]. Science, 2010. 330(6006): 912.
20. Buck LB, Retraction.[Retraction of Zou Z, Buck LB. Science. 2006 March 10;311(5766):1477-1481; PMID: 16527983]. Science, 2010. 329(5999): 1598.
21. Chan S-K and Struhl G, Evidence that Armadillo transduces wingless by mediating nuclear export or cytosolic activation of Pangolin.[Erratum appears in Cell. 2003 July 25;114(2):267].[Retraction in Struhl G. Cell. 2004 February 6;116(3):481; PMID: 15025097]. Cell, 2002. 111(2): 265-280.
22. Chang K. Nobel laureate retracts two papers unrelated to her prize. The New York Times. 2010 September 23.
23. Chen L and Woo SLC, Complete and persistent phenotypic correction of phenylketonuria in mice by site-specific genome integration of murine phenylalanine hydroxylase cDNA.[Retraction in Woo SL. Proceedings of the National Academy of Sciences of the United States of America. 2010 August 10;107(32):14514; PMID: 20622151]. Proceedings of the National Academy of Sciences of the United States of America, 2005. 102(43): 15581-15586.
24. Christiansen P and Mazak JH, A primitive late Pliocene cheetah, and evolution of the cheetah lineage.[Retraction in Mazak JH. Proceedings of the National Academy of Sciences of the United States of America. 2012 September 11;109(37):15072; PMID: 22908293]. Proceedings of

the National Academy of Sciences of the United States of America. 106(2): 512-515.
25. Christiansen P and Mazak JH, A primitive late Pliocene cheetah, and evolution of the cheetah lineage. Proceedings of the National Academy of Sciences of the United States of America. 106(2): 512-515.
26. Corbyn Z, Misconduct is the main cause of of life-sciences retractions. Opaque announcements in journals can hide fraud, study finds. Nature, 2012. 490: 21.
27. Couzin-Frankel J, Shaking up science. Two journal editors take a hard look at honesty in science and question the ethos of their profession. Science, 2013. 339(6119): 386-389.
28. Cozzarelli NR, Editorial expression of concern. Proceedings of the National Academy of Sciences of the United States of America, 2003. 100(20): 11816.
29. Cue DR and Cleary PP, High-frequency invasion of epithelial cells by Streptococcus pyogenes can be activated by fibrinogen and peptides containing the sequence RGD.[Retraction of Cue DR, Cleary PP.Infection and Immunity 1997 July;65(7):2759-64; PMID: 9199447]. Infection and Immunity. 66(9): 4577.
30. Deb K, Sivaguru M, Yong HY, et al., Cdx2 gene expression and trophectoderm lineage specification in mouse embryos.[Retraction in Roberts RM, Sivaguru M, Yong HY. Science. 2007 July 27;317(5837):450; PMID: 17656701]. Science, 2006. 311(5763): 992-996.
31. Eriksson S and Granstrom L, Randomized controlled trial of appendicectomy versus antibiotic therapy for acute appendicitis. British Journal of Surgery, 1995. 82(2): 166-169.
32. Fanelli D, How many scientists fabricate and falsify research? A systematic review and meta-analysis of survey data. PLoS ONE, 2009. 4(5): e5738.
33. Fanelli D, Do pressures to publish increase scientists' bias? An empirical support from US States Data. PLoS ONE [Electronic Resource], 2010. 5(4): e10271.
34. Fanelli D, "Positive" results increase down the hierarchy of the scieces. PLoS ONE, 2010. 5(4): e10068.
35. Fang FC, Casadevall A, and Morrison RP, Retracted science and the retraction index. Infection and Immunity, 2011. 79(10): 3855-3859.
36. Fang FC, Steen D, Grant R, et al., Misconduct accounts for the majority of retracted scientific publications. Proceedings of the National Academy of Sciencs of the United States of America, 2012. 109(42): 17028-17033.
37. Friedman PJ, Correcting the literature following fraudulent publication. Journal of American Medical Assocation, 1990. 263(10): 1416-1419.
38. Fujiyama S, Matsubara H, Nozawa Y, et al., Angiotensin AT(1) and AT(2) receptors differentially regulate angiopoietin-2 and vascular endothelial growth factor expression and angiogenesis by modulating

heparin binding-epidermal growth factor (EGF)-mediated EGF receptor transactivation. Circulation Research. 88(1): 22-29.

39. Fukuhara A, Matsuda M, Nishizawa M, et al., Retraction.[Retraction of Fukuhara A, Matsuda M, Nishizawa M, Segawa K, Tanaka M, Kishimoto K, Matsuki Y, Murakami M, Ichisaka T, Murakami H, Watanabe E, Takagi T, Akiyoshi M, Ohtsubo T, Kihara S, Yamashita S, Makishima M, Funahashi T, Yamanaka S, Hiramatsu R, Matsuzawa Y, Shimomura I. Science. 2005 Jan 21;307(5708):426-30; PMID: 15604363]. Science, 2007. 318(5850): 565.

40. Furman JL, Jensen K, and Murray F, Governing knowledge in the scientific community: exploring the role of retractions in biomedicine. Research Policy, 2012. 41(2): 275-290.

41. Geremia JM, Stockton JK, and Mabuchi H, Real-time quantum feedback control of atomic spin-squeezing. Science, 2004. 304(5668): 270-273.

42. Geremia JM, Stockton JK, and Mabuchi H, Retraction.[Retraction of Geremia JM, Stockton JK, Mabuchi H. Science. 2004 April 9;304(5668):270-273; PMID: 15073372]. Science, 2008. 321(5888): 489.

43. Gilligan DM, Finney GL, Rynes E, et al., Comparative proteomics reveals deficiency of NHE-1 (Slc9a1) in RBCs from the beta-adducin knockout mouse model of hemolytic anemia.[Retraction in Blood Cells Molecules and Diseases 2012 February 15;48(2):145; PMID: 22393569]. Blood Cells Molecules and Diseases, 2011. 47(2): 85-94.

44. Gilligan DM, Finney GL, Rynes E, et al., Comparative proteomics reveals deficiency of NHE-1 (Slc9a1) in RBCs from the beta-adducin knockout mouse model of hemolytic anemia. Blood Cells Molecules and Diseases, 2011. 47(2): 85-94.

45. Handa A and Brown KE, GB virus C/hepatitis virus replicates in human haematopoietic cells and vascular endothelial cells [Retracted by K.E. Brown]. . Journal of General Virology, 2000. 81(10): 2461-2469.

46. Harland CW, Bradley MJ, and Parthasarathy R, Phospholipid bilayers are viscoelastic.[Retraction in Harland CW, Bradley MJ, Parthasarathy R.Proceedings of the National Academy of Sciences of the United States of America. 2011 August 30;108(35):14705; PMID: 21817064]. Proceedings of the National Academy of Sciences of the United States of America, 2010. 107(45): 19146-19150.

47. Harland CW, Bradley MJ, and Parthasarathy R, Retraction for Harland et al., "Phospholipid bilayers are viscoelastic".[Retraction of Harland CW, Bradley MJ, Parthasarathy R.Proceedings of the National Academy of Sciences of the United States of America 2010 November 9;107(45):19146-50; PMID: 20974934]. Proceedings of the National Academy of Sciences of the United States of America, 2011. 108(35): 14705.

48. Horsewill AJ, Jones NH, and Caciuffo R, Evidence for coherent proton tunneling in a hydrogen bond network. Science, 2001. 291(5501): 100-103.

49. Horsewill AJ, Jones NH, and Caciuffo R, Retraction.[Retraction of Horsewill AJ, Jones NH, Caciuffo R. Science. 2001 January 5;291(5501):100-103; PMID: 11141555]. Science, 2002. 298(5596): 1171.
50. Horton MD, Counter SF, Florence MG, et al., A prospective trial of computed tomography and ultrasonography for diagnosing appendicitis in the atypical patient. American Journal of Surgery, 2000. 179(5): 379-381.
51. Iba O, Matsubara H, Nozawa Y, et al., Angiogenesis by implantation of peripheral blood mononuclear cells and platelets into ischemic limbs. Circulation. 106(15): 2019-2025.
52. Ismail SO, Skeiky YA, Bhatia A, et al., Molecular cloning, characterization, and expression in Escherichia coli of iron superoxide dismutase cDNA from Leishmania donovani chagasi.[Retraction of Ismail SO, Skeiky YA, Bhatia A, Omara-Opyene LA, Gedamu L. Infection and Immunity1994 February;62(2):657-64; PMID: 8300222]. Infection and Immunity. 63(9): 3749.
53. Kaiser J, Smart vector restores liver enzyme. Science, 2005. 306(5729): 1756.
54. Kawakami H, Tomita M, Okudaira T, et al., Inhibition of heat shock protein-90 modulates multiple functions required for survival of human T-cell leukemia virus type I-infected T-cell lines and adult T-cell leukemia cells.[Retraction in International Journal of Cancer 2011 December 1;129(11):2762-3; PMID: 21960263]. International Journal of Cancer, 2007. 120(8): 1811-1820.
55. Kennedy D, Editorial expression of concern. Science, 2006. 314(5799): 592.
56. Kennedy D, Editorial expression of concern. Science, 2006. 311(5757): 36.
57. Kennedy D and Alberts B, Editorial expression of concern. Science, 2008. 319(5868): 1335.
58. Komazawa N, Matsuda M, Kondoh G, et al., Enhanced insulin sensitivity, energy expenditure and thermogenesis in adipose-specific Pten suppression in mice.[Retraction in Komazawa N, Matsuda M, Kondoh G, Mizunoya W, Iwaki M, Takagi T, Sumikawa Y, Inoue K, Suzuki A, Mak TW, Nakano T, Fushiki T, Takeda J, Shimomura I. Nature Medicine. 2005 June;11(6):690; PMID: 15937475]. Nature Medicine, 2004. 10(11): 1208-1215.
59. Krueger J. What do retractions tell us? Office of Research Integrity Newsletter. 2012 2012:1-6.
60. Li F and Buck LB, Retraction for Zou et al., Odor maps in the olfactory cortex.[Retraction of Zou Z, Li F, Buck LB. Proceedings of the National Academy of Sciences of the United States of America. 2005 May 24;102(21):7724-7729; PMID: 15911779]. Proceedings of the National Academy of Sciences of the United States of America, 2010. 107(40): 17451.

61. Lieber M, Transcription-dependent R-loop formation at mammalian class switch sequences.[Retraction of Tracy RB, Lieber MR. EMBO Journal. 2000 March 1;19(5):1055-1067; PMID: 10698946]. EMBO Journal, 2001. 19(17): 4855.
62. Linn JG, Neff JA, Theriot R, et al., Reaching impaired populations with HIV prevention programs: a clinical trial for homeless mentally ill African-American men. Cellular and Molecular Biology, 2003. 49(7): 1167-1175.
63. Lipardi C and Paterson BM, Identification of an RNA-dependent RNA polymerase in Drosophila involved in RNAi and transposon suppression. [Retraction in Lipardi C, Paterson BM. Proceedings of the National Academy of Sciences of the United States of America 2011 September 6;108(36):15010; PMID: 21821790]. Proceedings of the National Academy of Sciences of the United States of America, 2009. 106(37): 15645-15650.
64. Lipardi C and Paterson BM, Retraction for Lipardi and Paterson, "Identification of an RNA-dependent RNA polymerase in Drosophila involved in RNAi and transposon suppression".[Retraction of Lipardi C, Paterson BM. Proceedings of the National Academy of Sciences of the United States of America. 2009 September 15;106(37):15645-15650; PMID: 19805217]. Proceedings of the National Academy of Sciences of the United States of America, 2011. 108(36): 15010.
65. Lu SF, Jin GZ, Uzzi B, et al., The retraction penalty: evidence from the web of science. Scientific Reports, 2013. 3(3146).
66. Maddox J, Randi J, and Stewart WW, High-dilution" experiments a delusion. Nature, 1988. 334(6180): 287.
67. Malik AA and Bari S-u, Conservative management of acute appendicitis.[Retraction in Journal of Gastrointestinal Surgery2011 December;15(12):2302; PMID: 21975685]. Journal of Gastrointestinal Surgery, 2009. 13(5): 966-970.
68. Malik AA and Bari S-u, Conservative management of acute appendicitis. Journal of Gastrointestinal Surgery, 2009. 13(5): 966-970.
69. Mano A, Tatsumi T, Shiraishi J, et al., Aldosterone directly induces myocyte apoptosis through calcineurin-dependent pathways. Circulation, 2004. 110(3): 317-823.
70. Marcato P, Mulvey G, and Armstrong GD, Cloned Shiga toxin 2 B subunit induces apoptosis in Ramos Burkitt's lymphoma B cells.[Retraction of Marcato P, Mulvey G, Armstrong GD. Infection and Immunity. 2002 March;70(3):1279-86; PMID: 11854211]. Infection and Immunity. 71(8): 4828.
71. Mazak JH, Retraction for Christiansen and Mazak. A primitive Late Pliocene cheetah, and evolution of the cheetah lineage.[Retraction of Christiansen P, Mazak JH. Proc Natl Acad Sci U S A. 2009 Jan 13;106(2):512-5; PMID: 19114651]. Proceedings of the National Academy of Sciences of the United States of America. 109(37): 15072.

72. McNally A, Roe AJ, Simpson S, et al., Differences in levels of secreted locus of enterocyte effacement proteins between human disease-associated and bovine Escherichia coli O157.[Retraction of McNally A, Roe AJ, Simpson S, Thomson-Carter FM, Hoey DE, Currie C, Chakraborty T, Smith DG, Gally DL. Infection and Immunity 2001 August;69(8):5107-14; PMID: 11447192]. Infection and Immunity. 73(4): 2571.
73. Miller G, Scientific misconduct. Misconduct by postdocs leads to retraction of papers. Science, 2010. 329(5999): 1583.
74. Mori N, Retraction. Essential role of transcription factor nuclear factor-kB in regulation of interleukin-8 gene expression by nitrite reductase from Pseudomonas aeruginosa in respiratory epithelial cells.[Retraction of Mori N, Oishi K, Sar B, Mukaida N, Nagatake T, Matsushima K, Yamamoto N. Infection and Immunity. 1999 August;67(8):3872-8; PMID: 10417151]. Infection and Immunity, 1999. 79(8): 3473.
75. Mori N, Ueda A, Geleziunas R, et al., Retraction. Induction of monocyte chemoattractant protein 1 by Helicobacter pylori involves NF-kB.[Retraction of Mori N, Ueda A, Geleziunas R, Wada A, Hirayama T, Yoshimura T, Yamamoto N. Infection and Immunity 2001 March69(3):1280-6; PMID: 11179289]. Infection and Immunity, 2001. 79(1): 543.
76. Mori N, Krensky AM, Geleziunas R, et al., Retraction. Helicobacter pylori induces RANTES through activation of NF-kB.[Retraction of Mori N, Krensky AM, Geleziunas R, Wada A, Hirayama T, Sasakawa C, Yamamoto N. Infection and Immunity 2003 July;71(7):3748-56; PMID: 12819056]. Infection and Immunity, 2003. 79(1): 544.
77. Naik G. Mistakes in scientific studies surge. Wall Street Journal. 2011 August 10.
78. Nguele JC, Eswaramoorthy P, Bhattacharya M, et al., Genetic and biochemical analyses of sensor kinase A in Bacillus subtilis sporulation. [Retraction in Genetics and Molecular Research. 2011;10(3):2257; PMID: 21968764]. Genetics and Molecular Research, 2010. 9(1): 573-590.
79. Okudaira T, Hirashima M, Ishikawa C, et al., A modified version of galectin-9 suppresses cell growth and induces apoptosis of human T-cell leukemia virus type I-infected T-cell lines.[Retraction inInternational Journal of Cancer 2011 December 129(11):2762-2763; PMID: 21960263]. International Journal of Cancer, 2007. 120(10): 2251-2261.
80. Orme IM, Furney SK, Skinner PS, et al., Inhibition of growth Mycobacterium avium in murine and human mononuclear phagocytes by migration inhibitory factor.[Retraction of Orme IM, Furney SK, Skinner PS, Roberts AD, Brennan PJ, Russell DG, Shiratsuchi H, Ellner JJ, Weiser WY.Infection and Immunity. 1993 January;61(1):338-342; PMID: 8418058]. Infection and Immunity. 62(5): 2141.
81. Resnick DB and Dinse GE, Scientific retractions and corrections related to misconduct findngs. Journal of Medical Ethics, 2013. 39(1): 46-50.

82. Reynaud A, Federighi M, Licois D, et al., R plasmid in Escherichia coli O103 coding for colonization of the rabbit intestinal tract.[Retraction of Reynaud A, Federighi M, Licois D, Guillot JF, Joly B.Infection and Immunity 1991 June;59(6):1888-1892; PMID: 2037350]. Infection and Immunity. 61(10): 4533.
83. Roman-Gomez J, Castillejo JA, Jimenez A, et al., Cadherin-13, a mediator of calcium-dependent cell-cell adhesion, is silenced by methylation in chronic myeloid leukemia and correlates with pretreatment risk profile and cytogenetic response to interferon alfa. Journal of Clinical Oncology, 2003. 21(8): 1472-1479.
84. Schekman R, Editorial expression of concern for Choi et al., Use of combinatorial genetic libraries to humanize N-linked glycosylation in the yeast Pichia pastoris. Proceedings of the National Academy of Sciences of the United States of America, 2010. 107(15): 7113.
85. Schekman R, Editorial expression of concern for Choi et al., Use of combinatorial genetic libraries to humanize N-linked glycosylation in the yeast Pichia pastoris. Proceedings of the National Academy of Sciences of the United States of America, 2010. 105(15): 7113.
86. Shibasaki Y, Matsubara H, Nozawa Y, et al., Angiotensin II type 2 receptor inhibits epidermal growth factor receptor transactivation by increasing association of SHP-1 tyrosine phosphatase. Hypertension. 38(3): 367-372.
87. Shin JJ, Bryksin AV, Godfrey HP, et al., Retraction. Localization of BmpA on the exposed outer membrane of Borrelia burgdorferi by monospecific anti-recombinant BmpA rabbit antibodies.[Retraction of Shin JJ, Bryksin AV, Godfrey HP, Cabello FC. Infection and Immunity. 2004 April;72(4):2280-2287; PMID: 15039353]. Infection and Immunity. 76(10): 4792.
88. Steen RG, Retractions in the scientific literature: is the incidence of research fraud increasing? . Journal of Medical Ethics, 2011. 37(4): 249-253.
89. Steen RG, Casadevail A, and Fang FC, Why has the number of scientific retractions increased? . PLOS One 8: e68397, July 8 2013), 2013. 8: e68397.
90. Struhl G, RETRACTION.Evidence that Armadillo transduces wingless by mediating nuclear export or cytosolic activation of Pangolin.[Retraction of Chan SK, Struhl G. Cell. 2002 October 18;111(2):265-280; PMID: 12408870]. Cell, 2004. 116(3): 481.
91. Swift IE and Milborrow BV, Retention of the 4-pro-R hydrogen atom of mevalonate at C-2,2' of bacterioruberin in Halobacterium halobium. [Retraction in Swift IE, Milborrow BV. Biochemical Journal 2005 August 1;389(Pt 3):919; PMID: 16047404]. Biochemical Journal, 1980. 187(1): 261-264.
92. Takeshima E, Tomimori K, Teruya H, et al., Retraction. Helicobacter pylori-induced interleukin-12 p40 expression.[Retraction of Takeshima

E, Tomimori K, Teruya H, Ishikawa C, Senba M, D'Ambrosio D, Kinjo F, Mimuro H, Sasakawa C, Hirayama T, Fujita J, Mori N. Infection and Immunity 2009 April;77(4):1337-48; PMID: 19179414]. Infection and Immunity, 2009. 79(1): 546.
93. Tomimori K, Uema E, Teruya H., et al., Retraction. Helicobacter pylori induces CCL20 expression.[Retraction of Tomimori K, Uema E, Teruya H, Ishikawa C, Okudaira T, Senba M, Yamamoto K, Matsuyama T, Kinjo F, Fujita J, Mori N. Infection and Immunity. 2007 November;75(11):5223-32; PMID: 17724069]. Infection and Immunity, 2007. 79(1): 545.
94. Tomita M, Kawakami H, Uchihara J-n, et al., Curcumin (diferuloylmethane) inhibits constitutive active NF-kappaB, leading to suppression of cell growth of human T-cell leukemia virus type I-infected T-cell lines and primary adult T-cell leukemia cells.[Retraction in International Journal of Cancer. 2011 December 1;129(11):2762-2763; PMID: 21960263]. International Journal of Cancer, 2006. 118(3): 765-772.
95. Tracy RB, Hsieh CL, and Lieber MR, Stable RNA/DNA hybrids in the mammalian genome: inducible intermediates in immunoglobulin class switch recombination.[Retraction in Tracy RB, Hsieh CL, Lieber MR. Science. 2000 August 18;289(5482):1141; PMID: 10970226]. Science, 2000. 288(5468): 1058-1061.
96. Tracy RB and Lieber MR, Transcription-dependent R-loop formation at mammalian class switch sequences.[Retraction in Lieber M. EMBO Journal. 2001 September 1;19(17):4855; PMID: 12134869]. EMBO Journal, 2000. 19(5): 1055-1067.
97. Trives R, Palestis BG, and Zaatari D, The Anatomy of a Fraud: Symmetry and Dance 2009: TPZ Publishers.
98. Uchihara J-N, Krensky AM, Matsuda T, et al., Transactivation of the CCL5/RANTES gene by Epstein-Barr virus latent membrane protein 1.[Retraction in International Journal of Cancer. 2011 Dececember 1;129(11):2762-2763; PMID: 21960263]. International Journal of Cancer, 2005. 114(5): 747-755.
99. Uchihara J-N, Matsuda T, Okudaira T, et al., Transactivation of the ICAM-1 gene by CD30 in Hodgkin's lymphoma.[Retraction in Int J Cancer. 2011 Dec 1;129(11):2762-3; PMID: 21960263]. International Journal of Cancer, 2006. 118(5): 1098-1107.
100. Van Parijs L, Refaeli Y, Abbas AK, et al., Retraction. Autoimmunity as a consequence of retrovirus-mediated expression of C-FLIP in lymphocytes. [Retraction of Van Parijs L, Refaeli Y, Abbas AK, Baltimore D. Immunity. 1999 December;11(6):763-770; PMID: 10626898]. Immunity, 2009. 30(4): 612.
101. Van Parijs L, Refaeli Y, Lord JD, et al., Retraction. Uncoupling IL-2 signals that regulate T cell proliferation, survival, and Fas-mediated activation-induced cell death.[Retraction of Van Parijs L, Refaeli Y, Lord JD, Nelson

BH, Abbas AK, Baltimore D. Immunity. 1999 Sep;11(3):281-8; PMID: 10514006]. Immunity, 2009. 30(4): 611.
102. Wade N. It may look authentic. Here's how to tell that it isn't. The New York Times. 2006 December 16.
103. Wager E and Williams P, Why and how do journals retract articles? An analysis of Medline retractions 1988-2008. Journal of Medical Ethics, 2011. 37(9): 567-570.
104. Wakefield AJ, Murch SH, Anthony A, et al., Ileal-lymphoid-nodular hyperplasia, non-specific colitis, and pervasive developmental disorder in children.[Retraction in The Lancet. 2010 February 6;375(9713):445; PMID: 20137807]. The Lancet, 1998. 351(9103): 637-641.
105. Wang F, Okamoto Y, Inoki I, et al., Sphingosine-1-phosphate receptor-2 deficiency leads to inhibition of macrophage proinflammatory activities and atherosclerosis in apoE-deficient mice.[Erratum appears in Journal of Clinical Investigation 2012 January;122(1):419]. Journal of Clinical Investigation, 2010. 120(11): 3979-3995.
106. Weaver D, Albanese C, Costantini F, et al., Retraction: altered repertoire of endogenous immunoglobulin gene expression in transgenic mice containing a rearranged mu heavy chain gene. [Retraction of Weaver D, Reis MH, Albanese C, Costantini F, Baltimore D, Imanishi-Kari T. Cell. 1986 April 25;45(2):247-59; PMID: 3084104]. Cell, 1991. 65(4): 536.
107. Wilson B, A Treatise on Electricity. Medical Science and Technology. 2010, Farmingham Hills, Michigan: Gale ECCO, Print edition.
108. Zou Z, Horowitz LF, Montmayeur J-P, et al., Genetic tracing reveals a stereotyped sensory map in the olfactory cortex.[Retraction of Zou Z, Horowitz LF, Montmayeur JP, Snapper S, Buck LB. Nature. 2001 November 8;414(6860):173-9; PMID: 11700549]. Nature, 2001. 452(7183): 120.
109. Zou Z, Li F, and Buck LB, Odor maps in the olfactory cortex.[Retraction in Li F, Buck LB. Proc Natl Acad Sci U S A. 2010 Oct 5;107(40):17451; PMID: 20864630]. Proceedings of the National Academy of Sciences of the United States of America, 2005. 102(21): 7724-7729.
110. Zou Z, Horowitz LF, Montmayeur JP, et al., Genetic tracing reveals a stereotyped sensory map in the olfactory cortex.[Retraction in Zou Z, Horowitz LF, Montmayeur JP, Snapper S, Buck LB. Nature. 2008 March 6;452(7183):120; PMID: 18322536]. Nature, 2008. 414(6860): 173-179.
111. Zou Z and Buck LB, Combinatorial effects of odorant mixes in olfactory cortex.[Retraction in Buck LB. Science. 2010 September 24;329(5999):1598; PMID: 20929829]. Science, 2010. 311(5766): 1477-1481.
112. Zou Z, Li F, and Buck LB, Odor maps in the olfactory cortex.[Retraction in Li F, Buck LB. Proceedings of the National Academy of Sciences of the United States of America. 2010 October 5;107(40):17451; PMID: 20864630]. Proceedings of the National Academy of Sciences of the United States of America, 2010. 102(21): 7724-7729.

CHAPTER 19

Reasons for Scientific Misconduct

The reason why some researchers become involved in misconduct has been a topic of considerable speculation. Is dishonesty a fundamental trait of human nature in which any individual is prepared to stretch the truth and become dishonest and unethical under particular circumstances? Or is it an inherent or acquired peculiarity of only certain people who stray from the expected attributes of civilized society in Science? An analysis of cases of scientific misconduct suggests that the personality of the researcher and the environment in which the person performs the research contribute to this research misconduct. Important factors promoting this unacceptable behavior include dwindling available research funds, and an excessive emphasis on productivity by universities and research institutions in promotion, salary and tenure decisions. An amplified competition in research performance promotes fraudulent research include the pressure to publish and obtain grant support during times, too little communication between young researchers with their peers and mentors and an intensifying desire by young ambitious researchers to create names for themselves in their areas of specialization through publication. All of these factors generate additional nervous tension and this makes it more likely for some researchers to perform scientific misconduct. Several motives to fabricate and falsify research data have been identified and they differ considerably amongst scientists.

Research misconduct is a product of a wide spectrum of researchers with dissimilar backgrounds and at separate stages of their careers. Some are young and are still developing their careers, other scientists have reached the pinnacle of distinguished careers, or are just average researchers. The higher degrees range from MD, PhD, or MD/PhD. Regardless of status the researchers perform scientific misconduct when they presuppose that their dishonest work will not be detected. Some students, fellows and technicians have performed fraudulent research under the noses of eminent scientists, who did not detect or suspect their mentee of misconduct. How and why young promising scientists and prominent established investigators start fabricating and/or falsifying data or begin plagiarizing the work of others is a question of profound importance, but it astounds many who find it difficult to comprehend. What makes them go astray and ruin the reputation that they have built over many years or have the potential to build?

An understanding of the reasons that show the way to research misconduct could hopefully lead to the promotion of research integrity and a diminution, if not a abolition of scientific misconduct. To understand what coerces a scientist to cheat requires an appreciation of why some individuals in all human endeavors behave unethically and unscrupulously, as discussed in elsewhere. Such an understanding is important if this type of behavior is to be controlled [7].

Fang estimated that almost all scientists are driven by fear because of the pressure to get a research grant to pay their salaries and those of their research staff or other publications to enhance the probability of obtaining a research grant [12].

Pressure to Publish Research Findings to Support Grants

Some scientists have succumbed to scientific misconduct when cornered under the pressure to publish research findings to support a grant application. Notable examples are John Long and Vijay Soman.

An Overemphasis on Publications

Universities and research institutions provide an environment in which researchers face a pressure to publish. Professional societies, such as the Institute of Medicine (IOM) have noticed that researchers are under pressure to write numerous publications and that this productivity, which influences scholarly promotions and wage advancements is dependent on the number of papers published in scientific journals. It is a reality of life that publications are essential to develop reputations on specific topics. Papers in reputable journals are needed to acquire research grants and to become promoted to move up the academic ladder and obtain the desired job security provided by tenure. Promotions hinge on the number and magnitude of research grants in many institutions. These realities transform the goal of research laboratories and departments from scientific endeavors to a yardstick counting publications and the measuring of the total grants awarded as well as the dollar amounts of research support. The more desirable goal of furthering knowledge through discovery gets displaced and becomes a secondary issue. When rewards from research almost come within the grasp of some researchers they become tempted to cut corners, manipulate data, or even to fabricate some of it. Exceptionally susceptible to these persuasions are ambitious young students and researchers in large institutions with inadequate direct supervision. Teaching and the quality of science receive lip service at universities, which are deeply dependent on the revenue derived from indirect costs that come with grants to support their infrastructure. Contemporary biomedical research has become dependent on research grants particularly from agencies, such as the National Institutes of Health (NIH) and the National Science Foundation (NSF), which filter funding applications by vigorous peer review and publications.

Particularly in the past academicians needed to publish numerous original papers in professional journals and this was overly stressed relative to their quality when scholarly progress was measured and this affected funding, promotions and tenure decisions. Until relatively recently productivity in publications was taken to reflect a notable achievement. The volume of journal articles was emphasized over quality in tenure and promotion decisions and also by funding agencies. Prior to the 1980s and 1990s a cliché of the intellectual community was realistically publish or perish. An irrefutable enticement for numerous inappropriate authorship and for research misconduct is the embedded academic mind-set from the "publish or perish" era that towers over academic careers in universities. Over time the axiom changed to publish and perish because it became recognized that the pressure to publish promoted carelessness, shortcuts, and scientific misconduct. This way of thinking promotes high volume publications regardless of quality and provides insignificant motivation to teach the research scientists of the future. As the editors of *Nature* have contended, the most momentous pressure to publish stems from less eminent institutions. Top notch scientists stress quality and publish variable numbers of papers each year. The forces to print or give up the ghost in academia unquestionably promote research misconduct.

Tenure and promotion committees at academic institutions used to devote attention to the weight of the publications even without reading them to judge their content. Total papers were a major consideration of committees evaluating individuals for promotion and for giving high priorities in applications for peer reviewed grant support. Perhaps greater importance was paid to the quantity of the publication record in evaluating noteworthy learned accomplishments and funding decisions.

Personality Defect. Some researchers, who publish falsified data, have a warped sense of realism and stretch the truth because of a personality disorder comparable to compulsive gamblers. Unfortunately some such researchers believe that they can get away with it and are beyond detection.

Psychopathology. Psychopaths, sociopaths, criminals and others with antisocial behavior constantly exhibit unethical dishonest behavior and have no respect for others and the rules and regulations of civilized society. Some scientific misconduct seems to arise from the psychological makeup of the perpetrator in an environment in which the individual performs the research. The psychological factors that permit the misconduct include a self deception of the research and a denial or reluctance to judge that their dishonest feats will be both discovered and penalized. Woolf [30] recognizes a variety of personalities in researchers prone to scientific misconduct: (i) individuals who are negligent and desperate, (ii) unskilled and desperate, (iii) cases that do not seem to have done as they should have been and then data is fabricated in an attempt to cover their traces. These cases are most often detected in federal drug administration (FDA) audits and often the individuals have not known what they were supposed to do.

Some individuals found accountable for fabricating, falsifying or distorting research findings have not realized that what they did was unacceptable, unethical and below the standard expected in science. Some such individuals have not been appropriately mentored and brought up to practice research integrity in an honest ethical way.

Professional Cheater. The mentors of some students and postdoctoral fellows, who committed research misconduct, have regarded their mentees as extremely intelligent, skilled experimenters with knowledge about their research arena. They have commonly appreciated unanswered challenges which required major leaps forward. These bright researchers enlisted in laboratories under the leadership of respected scientists. This category of mentee has been called a professional cheater in contrast to the amateurish committer of fraudulent research that fabricates data on non-existent patients or allegedly performs experiments that are as a rule readily uncovered by fellow students and colleagues. The professional phonies probably realize that if a key breakthrough stems from their research attempts will be swiftly made to replicate their experiments and it will not take long to expose their fraudulent research. The professional cheaters are psychologically disturbed and subconsciously desire that their fraudulent work will be detected. They are rarely psychotic and not mentally ill in the eyes of the law, but to some extent they resemble Richard Milhous Nixon (1913-1994), the 37th President of the USA in their attitude. They do not desire or warrant institutionalization in a psychiatric hospital, but they are emotionally and psychologically unsettled and commonly seek self-destruction.

Some misconduct is attributed to the psychopathology of the cheater. In testifying before the Gore subcommittee in 1981 about the falsified data of Dr. John Long. Ronald W. Lamont-Havers, the chief of research at the Massachusetts General Hospital (Mass Gen) at the time, expressed the opinion that a realistic person appreciates that falsified data will eventually become detected. Some analysts of scientific misconduct contend that psychopathology makes fraudulent research likely, but not unavoidable.

Time and time again individuals found guilty of misconduct have been described by their mentors as brilliant young researchers with the potential for an outstanding career. If they had only been honest and patient their futures would have been bright. This scenario was repeated with William L Summerlin, John Darsee, John C. Long, Robert Slutsky, Elias A.K. Alsabti, Mark Spector, van Parjis, Amitav Hajra and others, who caught their mentors completely by surprise when their fraudulent research was discovered. People are more likely to cheat if they become convinced that free will is an illusion.

Antisocial Personality Disorder. The American psychiatric association has published a handbook known as *Diagnostic and Statistical Manual of Mental Disorders* that characterizes the various distinct psychiatric disorders [2]. One recognized entity is the antisocial personality disorder (ASPD). Sociopaths with this disorder are manipulative and consider their self-serving behavior permissible. They are also characterized by a grandiose sense of self, a lack of remorse, shame or guilt, shallow emotions, callousness and lack of empathy. Some researchers that partake in scientific misconduct undoubtedly have ASPD and some are charlatans who cleverly deceive and readily explain why certain experiments only work for them. Some people found guilty of scientific misconduct, such as John Darsee, have a long history of fraudulent behavior dating back many years even though their fraudulent past record only became apparent in retrospect.

Creative People. Creative people are not necessarily cheaters and unethical and no relationship has been established between them and scientific misconduct. While imaginative scholars do not necessarily pursue a dishonest career path divergent thinkers tend to find "original ways to bypass moral rules." However, studies at the business schools of Harvard University (*http://www.huffingtopost.com/2011/01/31/harvard-study-creative-pe_n_816137.html*) and Duke University (*http://www.zawya.com/story.cfm/sidZAWYA20110727093300/Creative_people_more_pro . . . 8/4/2011*) found that people with a creative mindset are more likely to cheat than non-creative individuals [13].

Risky Behavior. On January 22 2013 Fang, Bennett, and Casadevall published an analysis of the files of 72 faculty members found guilty of scientific misconduct by the Office of Research Integrity (ORI) and noted that the male female ratio was 63:9 (7 males to:1 female) (reported in the American society of microbiology open access publication known as *mBio (http://mbio.asm.org)* [12]. To explain this marked gender difference Fang and collaborators postulated that men are more likely to engage in risky behavior than women. This belief is supported by well documented social science literature.

Stupidity. Why some researchers are dishonest and resort to unethical or untruthful behavior seems strange when truth, honesty, and ethics are cardinal features of science. One would have anticipated that the public humiliation from participating in scientific misconduct, unethical research or other inappropriate research together with the consequential sanctions and scarred reputation would be a sufficient deterrent to dishonest research. Hence perhaps the main reason for research misconduct is stupidity as the consequences of being detected are extremely serious and almost always lead to the termination of a career in research. A well-known quotation of Albert Einstein (1879-1955), one of the most famous scientists of all time, aptly applies to scientific misconduct. Einstein stated "one can be sure of two infinite events; the universe and human stupidity and I'm not sure about the universe".

Flaws in Academic Environment

Ambitious university administrators who seek to improve the ranking of their departments and institutions relative to others desire that their research faculty publish outstanding papers of stellar quality in high impact papers. This unintentionally inflicts additional pressure on their faculty who wish to satisfy the heads of their departments and their deans. Many researchers as well the deans of several medical schools, including Henric H. Bendixen (1923-2004), the former dean and vice president of the college of physicians and surgeons at Columbia University (Columbia) have been of the opinion that the high pressure research environment at major universities distinctly contributes to scientific misconduct. Scientists are often forced to face increased pressure to accumulate new data for their research publications and grant applications. Writing in *The New York Times* about the fabrication of data by John Long of the Massachusetts General Hospital (Mass Gen), Dr. Robert H. Ebert (1915—1996), a former dean of the Harvard University (Harvard) *(http://www.harvard.edu/)*, stressed that a ferocious competition, was financially imposed on medical schools and academic institutions to support their laboratories financially. Ebert believed that cheating was caused in part by increasing pressure from the competition [7] and he contended that it would be a mistake to consider Long merely as an example of human weakness. Pressures to attract grants, develop drugs and publish new findings create an environment conducive to fraudulent research

Lack of Confidence in the Reproducibility of Publications

Because of a lack of confidence in the reproducibility of publications in psychology a group of more than 50 academic psychologists calling themselves the Open Science Collaboration (OSC) has undertaken a bold major initiative to determine if they can replicate published psychological experiments in leading journals [9].

Fierce Competition Among Peer

In 1985 a panel appointed by the *American Associated for the Advancement for Science* (AAAS) concluded that fraudulent research was becoming more common mainly because of the expanding competition for research grants and promotions, both of which depend on the output of publications. A significant number of examples of fraudulent research have been pinned on graduate students, postdoctoral fellows, research technical support, well-established scientists, and academic faculty with titles of assistant professor, associate professor, or full-professor (Table 1 chapter 14). Junior investigators may cheat because of a desire to satisfy supervisors and mentors with productivity. Pressure exerted by their supervisor in charge of them. Misconduct is mainly committed by junior investigators rather than senior scientists. Cases in point are John Darsee, Stephen Breuning, and Amitav Hajra who were turned over by senior faculty.

Academic Research Environment

A major motivation towards cheating in science seems to be the environment in which scientists find themselves. Arturo Casadevall [10] suggests that prizes are detrimental to science because of their focus on winner takes all culture rather than on cooperation. A detrimental research milieu in some departments at universities and research institutions are not conducive to an honest collegial academic research environment. Academic freedom at universities and research institutes has developed a permissive nonjudgmental research milieu that tolerates sloppy or careless slap dash research practices. The extremely lenient research milieu at universities and research institutions, fails to discourage slipshod practices and correct early deviant research behavior. Without an appropriate infrastructure early unacceptable research behavior cannot be recognized and rectified. A toxic academic environment can create dishonesty out of otherwise normal individuals, unless it is closely supervised by knowledgeable scientists. It is extremely important that early deviant behavior be detected and corrected before it gets out of hand. Crucial institutional factors that promote shoddy procedures include the excessive pressure that is inserted indirectly on faculty to further knowledge and obtain research funding and an excessive stress to publish peer reviewed articles, particularly in high impact journals, as the foremost criterion for academic promotion.

Lack of Expertise in all Facets of Research Project

Collaborative interdisciplinary complex research projects result in papers with multiple coauthors. When none of the researchers in such multi-authored research projects are proficient in all facets of the research the potential risk of scientific misconduct exists, especially when busy scientists have their fingers in too many projects. When no investigator in a research team has proficiency in every aspect of a research program a single collaborator may commit scientific misconduct and taint the reputation of other innocent members of the research team, who become co-conspirators in their role as coauthors in the resultant publications. To prevent all coauthors should clearly define their role in the project.

Even though cheating in research has existed throughout the ages the temptation to deceive has never been been never been more profound as researchers struggle for research grants, kudos, awards, and jobs. Numerous studies by Dan Ariely of Duke University (Duke) *(http://duke.edu/)* point to conditions that lead to cheating. Mazar and colleagues have provided evidence that dishonest behavior seems to be based to some extent on external rewards and partially on internal rewards. When people recognize benefit from some level of dishonesty they will engage in it [21]. It is common to point the finger at those who commit scientific misconduct and claim that they are bad apples, but it is perhaps more idealistic to believe that society creates environments that lure people and expect them to behave themselves. For example students and faculty are required at many universities and research institutions to take mandatory courses and pass examinations on subjects that often include information of no obvious significant value to the person required to pass the quiz. These time consuming exercises encourage faculty to save time and effort by using dishonest techniques.

Competition for Insufficient Research Funds

The game of academic research survival depends on the ability to raise money through grants in the face of fierce competition with numerous professional scientists for restricted funds. In 1960 half or more grant applications were funded, but now research funding is scarcer than ever. Successful research depends on adequate financial resources and when money is in short supply researchers may move toward unethical conduct and research misconduct. Drops in funding from the NIH and in its purchasing power from 1979 resulted in an increased scrutiny of research grant applications by reviewers as well as more fierce competition for funds and an apparent increase in fraudulent research. Insufficient research funds seems to be a potential factor in at least some research misconduct, such as the case of Mark J. Straus. Some researchers and administrators of research have blamed the shrinkage of federal grant support in the 1970s and the tight competition for research misconduct and the tendency to exaggerate research findings. While condemning the excessive competition for grant support because of limited available financial resources some scientific leaders were not convinced that more scientists became engaged in greater deception than usual. In 1981 David Baltimore stated "There is no question that the pressure on research workers grows because of the limitation of funds and the increasing formalization of the academic world, with its demands to produce and appear successful, and I am sure that everyone has a cracking point. But whether any of this has to do with John Long or not, I have no idea."

With the vast number of active researchers, the increasing costs of research and the financial restraints created by cuts in research funding competition for research awards, and academic positions has never been keener. Unfortunately many researchers cannot compete on their own abilities with the research competition and resort to other tactics. The vast majority of known cases of research misconduct involve biomedical research, perhaps because that is where most research funds are available. When setbacks emerge in the research of scientists who are markedly dependent on research grants they safeguard their careers by covering up the drawbacks. When research funds become increasingly difficult to acquire and when competition for them becomes fierce the research environment deteriorates and enters a phase prone to scientific misconduct. Competition for research funds from the NIH, NSF and other research funding sponsors has been increasing every year and has now reached an unparalleled level, and shows no sign of abating. As pointed out by Dr. William F. Raub, the NIH associate director for research and training "Never before has so many established investigators faced so much uncertainty about their longevity as active scientists. Never before have so many novices faced so many disincentives to entering or continuing a research career" *(deainfo.nci.nih.gov/advisory/ncab/138_0606/presentations/Zerhouni.pdf)*. After remaining virtually steady for the past decade the number of NIH funded researchers has dramatically dropped from 22, 116 to 21, 511 [1]. Time will tell whether this will become a detrimental loss of productive researchers and a discouragement of recruitment into careers in science, or an appropriate overdue constraint of science within an affordable budget. It could also enhance research misconduct.

There can be little doubt that research has become increasingly more competitive for available research funds, but convincing evidence that competitive pressures have made the research environment distinctly more prone to misconduct than in the past do not exist [7, 16]. Unfortunately many cannot surpass their peers and reach their goals on their own abilities and need to resort to other tactics. At universities naïve students are influenced by captivating faculty who influence career development. Because graduate students and post-doctoral fellows contribute immensely to the productivity of research laboratories they are often recruited to perform research in the mentor's discipline regardless of whether a PhD or postdoctoral training is likely to lead to a reliable independent career in that branch of learning. Many individuals who receive one or more doctoral degree are not capable of surviving as an independent investigator, particularly when obligated to compete with the best and the brightest. Available evidence strongly suggests that modern society is responsible to a major extent for the increased scientific fraud that has been identified during the past four decades.

Because the productivity of researchers is dependent on grant support researchers are under pressure to obtain funding and this force promotes research misconduct and shoddy research practices particularly in persons having difficulty in being competitive in grant support. Many members of the scientific community are convinced that the number of dishonest scientists is growing as a result of the progressively more brutal competition for grant support that is essential for scholarly survival. Grants are awarded more often to researchers who deliver or contend that they can generate the most notable new discoveries in the timeliest fashion.

Research by Overcommitted Busy Scientists

Productive scientists rely on vast numbers of persons to perform their ongoing research projects and they generally have insufficient time to mentor, supervise and oversee their research personnel or who fear not to become too involved. To oversee individual researchers in greater detail would have a detrimental effect on the morale of research staff and on their ability to perform research. Some individuals concerned about the reasons for research misconduct believe that some scientists who mentor graduate students and postdoctoral fellows are too busy to supervise their mentees and especially in overseeing their research. In his comments about Amitav Hajra the MD PhD candidate of Francis Collins who was found guilty of research misconduct Varmus expressed the opinion that it would be incorrect to suppose that scientists have to mistrust everything or require duplication of every significant result in the laboratory. Also, Collins felt that it be naïve to try to create a fail-safe mechanism to prevent fraud; "deceit and betrayal are part of the price; we must pay for a free system. If research is going to be open, if it is going to be creative; if you are going to allow people with talent to explore the unknown, there are people who are going to take advantage of that freedom and abuse it". Collins believed that the only remedy may be to do science with eyes more open [20]. While the remarks of these two leaders of the NIH may be true, it is clear that an inadequate supervision of research trainees is going to provide an opportunity for them to exploit misconduct when their research faces difficulties that they cannot overcome. Busy scientists who participate in many projects are particularly prone to induce unsuspected scientific misconduct if they had only had intermittent contact and insufficient time to supervise the research in detail. This is particularly true for persons with heavy administrative duties, such as Charles J. Glueck MD because of an overcommitment or reliance on others to perform the actual research.

Rewards of Research

The products of research affect the careers of scientists. If the outcomes of a researcher meant nothing scientific misconduct would probably not exist as it would be meaningless. But this is not the case except in countries that are not under communist rule. Aside from promotion and job offers research can lead to fame in an area of specialization.

Research Funding

An Undue Emphasis for Researchers to Publish In the past the number of publications was a gauge of scholarly progress and it was an important criterion used by funding agencies to enhance the judgment of the caliber of new grant applications as well as the quality of renewal grant applications. The "publish or perish" attitude encouraged suboptimal research and the pressure for academic advancement drove some individuals into scientific misconduct.

Other

Untruthfulness and deceitfulness in research were once considered rare and a peculiarity of only a handful of scientists, but this is clearly not true. Examples of people doing bad things that nobody would ever have imagined them doing are heard about almost every day on television as well as in the newspapers.

As discussed elsewhere the attributes of dishonesty and unethical behavior affect virtually all human endeavors. An understanding of why these individuals behave unethically and unscrupulously is crucial to the appreciation of why some researchers resort to scientific misconduct. Such information is important if this type of behavior is to be eliminated.

Undue Effort to Emulate an Exceptionally Prominent Scientist

Some young researchers found guilty of scientific misconduct have had prominent scientists as role models and mentors. Individuals in this category include William Summerlin, John Rowland Darsee, Vijay Soman, and Amitav Hajra. Such mentees become attempted to emulate the success of the leaders who guide them. Some people who have committed scientific misconduct share a strikingly common work ethic with their mentors, who have been extremely busy, hard working and highly productive and as role models they have placed pressure perhaps unintentionally on the careers of the young developing scientists. Summerlin who established notoriety for his *patchwork mouse* was watched over by Robert Alan Good (1922—2003), a giant in biomedical research who founded modern immunology. Likewise Eugene Braunwald (1929-) the mentor of Darsee was an extremely productive American cardiologist with more than one thousand publications in peer reviewed journals. At one time he seemingly authored about 28.5 publications per year.

Even people who consider themselves honest are dishonest under certain circumstances.

Broad and Wade [8] suggest that the conceit of scientists somehow renders them particularly vulnerable to fraud, but they do not make a sound case for their contention.

The motivation for falsification and fabrication often stems from a conflict of interest (COI) caused by an intellectual, financial and social bias in which the investigator is determined to find experimental evidence to prove a theory, especially one with commercial ramifications.

The Vague Boundary Between Unacceptable and Acceptable Research

The boundary between research dishonesty and research integrity is not always as clear cut as black and white. They have firmly believed that their activities were not wide of the accepted standard of research. To justify their behavior a frequent excuse is that everyone does it. Some modern research activities lie at the interface of tolerable research endeavors and numerous scientists do not consider such practices unacceptable and negligent. Examples of such research practices include misconduct with publications, such as honorary authorships in multi-authored papers, repetitive publications of the same data, and salami publications.

Internal and External Pressures

Competition for Insufficient Research Funds

Ongoing research projects are attempting to elucidate factors that sway professional conduct in the contemporary research milieu. Scientists face a pressure to not only obtain research grants, but also to publish. This together with the paucity of research funds amplifies the tolerance of sloppy research and lackadaisical thought in interpreting the results of experiments that are often unrepeatable. These external pressures can coerce researchers into fraudulent behavior.

Weak Mentors

Misconduct by junior researchers often reflects inadequate guidance and mentoring by personnel working in the trenches of a research laboratory. Neglected supervision by mentors of research and of the publications stemming from it by graduate students and post doctoral fellows is another cause of research misconduct that could be nipped in the bud prior to its submission to journals.

Traditions of Science and Research Incentives

The nervous tension that a scientist faces dissipates soon after each new discovery is documented. The scientific community is too large, too entrepreneurial, and its long term custom of emphasizing the first to discover something is overly stressed. The first researcher or laboratory to report a new significant discovery receives the credit in the standard tradition of winner takes all. This custom of science makes researchers want to generate new information as soon as possible and to publish it before rival laboratories. By giving undue credit to the first person to make new discoveries has placed scientists working on the same topic under unnecessary stress. Unfortunately the race between laboratories has diminished the guidance of junior researchers and lessened mutually respectful discussion, frank assessments of the research and a trend towards secrecy and a failure to share data.

Money and Personal Profit

Some people clearly sell their souls for money. The truism that money is the source of all evil is supported by many well documented examples of scientific misconduct involving clinician-scientists that receive vast amounts of money from pharmaceutical companies. The potential for faking data exists when an incentive for making personal profit exists. When scientists, or other professionals, become expert witnesses in legal disputes there is a danger of them not interpreting the evidence objectively. A disconcerting aspect of some allegations of research misconduct is the rapid defense that sometimes follows even before an objective inquiry takes place [27].

Reward Incentives

Policies with incentives to direct research endeavors probably promote research misconduct as it did in the high school teaching scandal in the USA state of Georgia.

Survival in a Competitive Society

What individuals learn as children from their parents and peers clearly influences their future behavior. Society is undoubtedly also responsible to some extent in developing personal behavior and the religious upbringing of children play an important role in this regard. Ambition is good, but not when it leads to greed, dishonesty, unethical behavior and shortcuts. With regard to culprits like John Darsee and Robert A. Slutsky Mario Biagioli, a professor of the history of science at Harvard University (Harvard) *(http://www.harvard.edu/)* has pointed out that the psychological profile of these people is usually a B-plus, A-minus scientist who gets into a hyperproduction mode".

In highly competitive societies, such as the USA, bright individuals learn early in life that they can reach the pinnacle of their dreams if they can excel and stand head and shoulders above their peers. Society has preconditioned individuals on how to survive in a competitive environment. In the competition to forge ahead students need to prepare papers on various topics and pass examinations that rank participants. These assignments are often too much for some young individuals, particularly when parents establish standards that are beyond their reach. They hence resort to the methods of cheaters, unless they have been brought up not to lie and cheat and strongly believe that honesty is the best policy. An advisory group that examined the scandal in stem cell research caused by Hwang Woo-sek warned that the increasing competiveness and incentives to over claim or even cheat exists. Powerful competitive forces undoubtedly contribute to fraudulent science. These forces start in the formative years of future researchers while they are at high school and in college and they gain momentum from then onwards in highly competitive lines of work, such as architecture, medicine, theatre, music, and professional athletics. From the standpoint of research medical schools and academic research institutions have unintentionally promoted an extremely, brutal competition. The struggle to gain admission into medical school is fierce and so is the obtaining of scholarships that support the financial expenses of the education. This practice has generated an atmosphere that compensates success even if publications result from unethical behavior and research misconduct and often the wrongdoing remains undetected. panel of academic officials and scientific journal editors at the annual meeting of the AAAS on May 29, 1985 found the growth of professional incentives for frequent scientific publishing an incentive for unethical conduct.

Growth of Departments

In the past papers containing fraudulent material were weeded out before publication by colleagues and by departmental heads. At one time the heads of departments had the desire and sufficient time to time to read and endorse what their staff put in writing, but departments have grown immensely with the multiplication of scholarly specialties. Because of this it has become unrealistic for chairs of departments to oversee manuscripts coming from their specialties prior to their submission to journals.

Large Laboratories and Demanding Task Masters

Large research laboratories headed by a prominent senior scientist prosper by attracting considerable amounts of grant money which enables numerous technicians, graduate students, and post-doctoral fellows to be hired. With such a laboratory growth the leaders become progressively more distant from the actual research and the participating personnel. The research projects are performed largely by junior colleagues who also contribute intellectually to the documentation of the findings, but the senior investigator may not necessarily appreciate the specific technical details of the research. This has increased the opportunity for fraudulent research by junior colleagues because statements in the publications cannot always be verified by the principal Investigator (PI) responsible for the research.

Another apparent factor in the modern fraud factory was the creation of what Broad and Wade called the "research mills". Moreover extremely busy scientists who lead large laboratories are almost invariably unable to adequately oversee and supervise all of the ongoing research for which they are responsible. Also, in papers in which their names are listed as coauthors some researchers often lack the time to go over the manuscripts generated by the research with a fine tooth comb. Examples of research misconduct by trainees stemming from large laboratories headed by prominent scientists included those under the directorship of Robert Good, Eugene Braunwald, and Francis Collins (chapter 14). The heads of some large successful research laboratories are overly ambitious demanding task masters, who consciously or subconsciously, place excessive pressures on junior members of the research squads. The leader of these groups are probably ignorant of the detailed research of some members of their research groups, who sometimes manipulate data and perform experiments that they do not discuss at regular laboratory meetings. The pressure was presumably overwhelming when the laboratory leader expected results and laboratory teams provided what was wanted even if the results were dishonest [16]. In other cases the academic researcher is clearly acting in response to social pressures, such as getting promotions or peer reviewed research funding.

Pressures for Research Progress Reports

Researchers are under extreme pressure to write frequent publications with new findings that are preferably positive and to obtain research grants to achieve an academic promotion and tenure in universities. Reviewers of papers submitted to journals play a part in this misbehavior by preferring comparatively concise manuscripts and the prompt publications of research projects by renowned laboratories, to expand knowledge, to obtain and retain research grants. Some researchers who depend on research grants for survival become lured into cutting corners to produce manuscripts that can not only pass the eyes of a journal editor, but also the peer reviewers. When this becomes difficult some engage in untruthful scientific behavior by committing research misconduct. As pointed out by Broad and Wade [8] some researchers found guilty of scientific fraud have attempted to be successful in a system demanding a continuous high rate of publication of necessity by cheating.

To keep up with professional demands when progress is often extremely slow some scientists are forced to cut corners, stretch the truth, adjust data and if not corrected may partake in fabrication, falsification, and plagiarism. Scientists jostle for space and grants to perform their research with diminishing available funds to an increasing number of investigators. The latter is partly caused by the scientists themselves as they go out of their way to train graduate students and postdoctoral fellows to assist in their research enterprises. By these actions researchers have created a double edge sword that is an incentive to scientific misconduct unless strongly counteracted by mentoring about the extreme importance of scientific integrity.

The Population Effect

The contemporary situation with regard to research funding has become comparable to what the British scholar Reverend Thomas Robert Malthus (1766-1834) predicted in a theory about the control of populations that he developed in *An Essay on the Principle of Population*, first published in 1798 [19]. Malthus thought that the populace eventually gets checked by famine, disease, and widespread mortality. In an analogous way the population of practicing scientists becomes controlled by veracious competition and those who cannot make the grade either change their career goals or try to survive in a research career through dishonesty. Some researchers alter data and now and then develop it into obvious fabrication and falsification. When their findings fit the current conventional system of belief, they may go unchallenged, particularly when the research was performed at a well-regarded institution and the research paper was coauthored by at least one renowned researcher. The proposal of Malthus is equivalent to lectures on population growth delivered at Oxford University by the Reverend William Forester Lloyd [18] and a paper by Garrett Hardin [14]. Lloyd suggested that if pasteurlands were overgrazed with herds of cattle the time would come when "the unmanaged commons would be ruined by overgrazing". Hardin stressed that when a resource is open to everyone it is available to nobody.[14]. When dealing with cattle and wild animals limited resources can be overcome by culling different overpopulated inhabitants. Unless more financial resources are provided for research the comparable action would be restrict the number of research trainees and the encouragement of existing scientists to pursue non-scientist initiated research careers.

Over time the life of a researcher has become increasingly competitive, particularly in complex experiments that are not easy to replicate. Currently with a vast number of active scientists competing for prestige and research funds that are insufficient to support all meritorious projects, various types of scientific misconduct have become more apparent. Analysts of this despicable phenomenon have noted that there is not only a motivation from the academic research environment to falsify and fabricate data, but also to cut corners in coming to conclusions that the cheat is convinced is true.

The distortion of ethics seems to result from competition as indicated by the sociological studies of Barber and colleagues [3]. A determination by senior investigators to retain acceptable funding levels to keep a research laboratory operational together with a demand for high quality researchers by university officials has amplified the incidence of fraudulent research at some institutions.

For many years the scientific literature has expanded annually at a remarkable rate and the deluge of papers submitted to journals has become so enormous that the peer review system is stumbling under the load. A person who fabricates research data believes that it is not unlikely to be discovered.

Pressure to Recruit for Clinical Trials

Researchers involved in clinical trials agree to recruit a definite number of human subjects with a particular disease who meet specific criteria, but who must be excluded if other criteria are present. When the enrollment has been met prestige and money are at stake and the patient records may be altered to ensure that the committed number of patients gets recruited.

A Rigid Belief in the Wrong Hypothesis and an Inability to Evaluate Research Data Objectively

As mentioned in chapter 2 scientists commonly develop a hypothesis before performing experiments to test it but they need to remain objective in evaluating the observations. Unfortunately some researchers believe so deeply in their hypothesis that they lose objectivity and become trapped strongly into believing an unproven theory despite support against its authenticity. They ignore contradictory evidence and decide perhaps subconsciously that it is justifiable to take risks to convince others in the validity of the theory. Such researchers get forced into misconduct because of a need to not lose face by aborting their long held hypothesis. In a classic transcript of a talk in 1953 Irving Langmuir pointed out how one can fool oneself [17]. If a scientist pursues an idea and strongly believes that he or she is onto something important the researcher may fail to notice the accumulated evidence and end up bending the truth due to a misconception in which real observations are ignored. This is particularly true when scientists embark on a path with potential financial rewards.

Personal Internal Pressure and Ambition

At least in biology, physics and academic medicine a major motive for scientific misconduct seems to be career advancement, particularly by weak researchers who have difficulty keeping up with the crowd. Some individuals anxiously desire by personal ambition and internal pressure to rapidly climb the academic ladder.

Deception by the Government. Political reasons also contribute to research dishonesty. In May 1953 when the USA atom bomb testing program in Nevada was vulnerable fallout from two atomic blasts killed thousands of sheep within a short time period as radioactive material fell on herds of sheep grazing near where the bombs were blasted. Shortly after the blast about an eighth of the ewes and a quarter of the lambs died often with signs of radiation. Soon thereafter news about the deaths rapidly disseminated amongst the small local communities in southern Nevada and western Utah and people complained to the Atomic Energy Commission (AEC), which initiated a wide-ranging investigation involving veterinarians as well as nuclear scientists. They soon discovered significant information implicating the atomic bombs. Surviving sheep manifested lesions that closely resembled those caused by intense beta radiation. Ewes that ingested high levels of radioiodine produced stillborn or defective lambs as reported by the ranchers. At the conclusion of the investigation the AEC wanted to prevent widespread panic by the general public and decided not to reveal the critical findings for fear of causing a widespread panic. Because at least one veterinarian believed that radiation contributed to the loss of these animals the AEC inserted considerable pressure on the individual through letters and personal visits in an attempt to make him revise his opinion. With this strategy the AEC prevented a public uproar and avoided having to pay damages to the sheep owners.

In 1956 the animal owners demanded compensation in a lawsuit against the AEC. The trial was heard by federal Judge Albert Sherman Christensen (1905-1996), who rejected the owners demand for damages referring to data in the public summary of the AEC investigation and pointed out that one veterinarian had altered his opinion. The Judge also stated that "some of the best informed experts in the country were convinced that radiation could not have caused or contributed to the ovine disaster". This dishonesty remained concealed for three decades until Harold A. Knapp (1924-1989), a high-level defense department analyst blew the whistle after carefully reviewing the public records. The extent of the AEC's dishonesty did not surface until 1979 when Scott Milne Mathesum (1929-1990), the governor of Utah, achieved the release of formerly classified federal documents on the sheep deaths. More information became accessible through a sequence of congressional hearings in the same year.

In 1980 Knapp documented further evidence of the concealed deception in the conclusions of the experiment and sheep report [28]. When the fraud became exposed the sheep's owners renewed their claim for damages before Judge Christiansen, who by this time was 77 years old and still presiding over the federal court in Salt Lake City. On August 24, 1982 in his ruling the Judge pointed out the AEC had undeniably misrepresented the facts and he identified several attorneys and scientists who were responsible for committing the fraud upon his court [29]. The Judge ordered a new trial, which the Justice Department appealed.

The Spectrum of Morality

The spectrum of morality is wide and ranges from the saint-like behavior of certain religious people to truly evil people. The wide range of unethical behavior varies from the telling of minor untruths to the activities of the evil. Likewise all dishonest research is not the same.

Evil People. Evil and wicked people have been recognized for eons so perhaps those who commit research misconduct represent part of the spectrum of these misfits in society. Their existence has attracted the attention of such famous writers as William Shakespeare (1564-1616) and Fyodor Dostoevsky (1821-1881). Shakespeare recognized evil as he made Anthony state in Act 3 scene 2 of his play Julius Caesar "The evil men do lives after them; The good is oft interned with their bones". The Russian author Dostoevsky wrote about it in his book *The Brothers Karamazov* [11]. "In every man a demon lies hidden-the demon of rage, the demon of lustful heat at the screams of the tortured victim, the demon of lawlessness let off the chain." Some wicked people genuinely exist, but the evils of many cannot be accounted for solely as the acts of a few psychopaths.

Lucifer Effect or "Evil People"

The psychology of evil has been extensively studied by social psychologists, such as Philip Zimbardo (1933—). Social forces can bring forth social evil. Some scientists who are unable make original new discoveries and compete satisfactorily for research funds may become converted from good into evil by the so called "Lucifer effect", named after the biblical transformation of God's favorite angel Lucifer into Satan. The Stanford prisoner experiment of

Zimbardo is a critical example of how situations may influence behavior more than the personality traits of the involved individuals. Zimbardo showed that situations of madness can create insane behavior even in normal people. Perhaps under certain circumstances there is a breaking point in which a previously honest individual may resort to dishonesty. The conversion from good to evil was demonstrated by Zimbardo, when he was a 38 year-old professor in the psychology department at Stanford University (Stanford) *(http://www.stanford.edu/)*. His best-known experiments became an important corner stone of social psychology when it was carried out innocently in 1971 during an experiment. By today's standards the experiment is considered unethical and it would almost certainly have not been approved by any institutional review board (IRB) for human experimentation. Contemporary IRBs forbid professors from performing research on their students even with a signed informed consent. Zimbardo established a simulated prison at the university and sequestered twenty four educated, college-age men as participants for the study. These men were normal according to clinical interviews and personality tests. Each individual received $15 for participating in the research. By flipping a coin each subject was assigned to be a guard or a prisoner, and Zimbardo appointed himself warden. When the experiment started measurable personality differences were not detected between the two groups and Zimbardo was concerned that subjects would not genuinely participate in the project, but surprisingly the two groups soon began to act like real persons in their assignments. The prisoners became downhearted and lost control. One needed to be discharged in less than 38 hours, one had to be released because of marked depression, disorganized thinking, out of control crying and episodes of anger. Over the next three days, another three prisoners needed to be released because of symptoms of anxiety. The entire body of a fifth prisoner becomes covered in a psychosomatic rash and the person needed to be dismissed after his appeal to the mock parole board was rejected.

The guards all exerted their power and required the prisoners to obey frivolous, often incoherent rules and compelled them to carry out monotonous, meaningless tasks, like moving cartons from one cupboard to another or incessantly extracting thorns from blankets. The convicts were required to sing songs or laugh or stop smiling when so directed. They were also expected to curse and slander one another in public and to cleanse toilets with their bare hands. They were required to call out their numbers repetitively and to perform continuous push-ups.

Behaving as if they no longer realized that they were participants in an experiment the prisoners often became overwhelmed in the circumstances and during the mock parole board hearings, most would rather give up funds owed to them in exchange for their liberation. Even Zimbardo became embroiled in his position as warden becoming concerned about malingering convicts and the avoidance of prison breaks rather than the madness precipitated by his experiment. Zimbardo did not realize how the experiment had affected his personality until a girlfriend pointed this out to him. The experiment was planned to continue for two weeks, but it had to be terminated after six days because the prisoner abuse got out of hand. The results of these experiments were not published after they were performed, but became well-known by word of mouth and through videotapes.

Thirty five years after his notorious Stanford prisoner experiment Zimbardo documented this research in more detail in a book entitled *The Lucifer Effect: Understanding How Good People Turn to Evil* [31]. He also explored contemporary small and large scale evils, such as the deceitful behavior of executives at Enron and WorldCom, the sexual abuse of parishioners by Catholic priests *(http://en.wikipedia.org/wiki/Catholic_sex_abuse_cases)*, the *My Lai massacre* in Vietnam *(http://en.wikipedia.org/wiki/My_Lai_Massacre)*, systematic programs of police and military torture in several countries, the mass suicides at Jonestown *(http://history1900s.about.com/od/1970s/p/jonestown.htm)*, and the genocides in Rwanda *(http://en.wikipedia.org/wiki/Rwandan_Genocide)* and elsewhere. Zimbardo persuasively made it clear how each of these evils typifies the lessons of the Stanford prison experiment and points out that if the lessons had been effectively learnt perhaps some of these events could have avoided at least in part. The *Lucifer effect* is activated by a far-reaching list of noxious forces and psychological reactions to them. His book recounts tragedy after tragedy in which apparently good people give in to the psychological forces created by the circumstances with the nastiest of results. In each instance, those in power consistently come to the incorrect assumption that the abnormal behavior is caused by "bad apples" when indeed the larger problem was the makeup of the barrel in which they were situated. This type of mistaken belief is extremely common and known to social psychologists as the "fundamental attribution error". It occurs when the behavior of others is explained, particularly when it shows the way to no good, and overvalues the significance of personality traits while undervaluing the strength of the forces of the situation.

Zimbardo considers the infamous sadistic acts performed by the USA military personnel in Abu Ghraib prison *(http://en.wikipedia.org/wiki/Abu_Ghraib_torture_and_prisoner_abuse)* as an example of how a military reservist with an impeccable military record with more than a dozen medals and records can change so radically under the stress or war. Zimbardo in fact testified in this regard as an expert witness at the court-martial of Staff Sergeant Chip Frederick the scapegoat in the Abu Ghraib prison scandal who became converted from idealistic soldier to abuser. Nevertheless the Army sentenced Frederick to prison for eight years, gave him a dishonorable discharge and stripped him of both his medals and his pension [25].

The personality of some scientists predisposes them to scientific misconduct, especially when they have excessive traits of ambition, publicity seeking, and a desire for fame.

Some psychologists regard the form of research misconduct known as plagiarism an intentional theft and a desperate effort to get caught and punished. It is an attempt to rescue self-esteem as it stems from a lack of confidence and an impression that researchers do not warrant the achievement for which they are striving. According to Dr. Arnold Cooper, a past pesident of the American Psychoanalytic Association, unconscious cribbing is more common. There is an inner need for public punishment and in some way the person invites shame.

Most well known evil individuals have only achieved their inappropriate goals with the active involvement of vast numbers of ordinary individuals. Hitler for example could not have accomplished his massive exterminations of the Jews without the involvement of vast numbers of everyday German citizens. Also, Hutus slaughtered with machetes an estimated 900, 000 of their neighboring Tutsis in Rwanda in central Africa over a three-month period in 1994.

The tendency of individuals to deviate from morally unacceptable behavior depends on their upbringing and the customs of the society in which the person was raised. Research in social psychology has convincingly established that even slight aspects of a situation often bring to surface the most awful attributes of humanity. Humans predictably behave badly under particular conditions. The 1961 trial of the Nazi Adolf Eichmann (1906-1975) influenced at least two thoughtful individuals who wanted to understand how such an evil person could carry out his deeds. Hannah Arendt (1906-1975), a German political theorist, who covered the Eichmann trial for *The New Yorker*, labeled the crimes of Eichmann as "the banality of evil" in the title of her renowned 1963 book about the Eichmann trial titled *Eichmann in Jerusalem: A Report on the Banality of Evil* [4]. Arendt believed that the huge wicked acts of the past and especially the Nazi Holocaust were not performed by sociopathic personalities or extremists, but by everyday individuals who agreed with the assertions of their country and consequently involved themselves believing that their acts were ordinary. In accounting for this behavior Edward S. Herman [15] stressed the significance of "normalizing the unthinkable." In his opinion "doing terrible things in an organized and systematic way rests on 'normalization.' The American social psychologist Stanley Milgram '(1933-1984) made unspeakable acts routine and as 'the way things are done. He made a name for himself by his infamous controversial experiments on obedience to authority [6]. His "Milgram experiment" was performed in the 1960s while he was a professor at Yale University (Yale) *(http://www.yale.edu/)* [6]. The happenings of the Nazi Holocaust during World War II stimulated Milgram to investigate the relationship between obedience and authority. Milgram reported the findings of his extremely controversial experiments in 1963 [22] and a decade later he published a book on *Obedience to Authority* [23] which received the annual social psychology award by the *AAAS*. Aroused partly by the 1961 trial of Eichmann Milgram's models could explain the 1968 *My Lai Massacre*. In this experiment, 37 out of 40 participants received a full range of shocks up to 450 volts. This bears a resemblance to real-life happenings in which people consider themselves as just components in an organization carrying out their part and being permitted to shun accountability for the effects of their deeds. The shocks themselves were a sham. The person taking part as the 'learner' in the study was in reality a compensated actor or pretending to respond to the effects of the shock at different voltages. Milgram became infamous for this ploy. It did not take long for these stressful experiments by Milgram to be labeled unethical. The Milgram experiments became one of the most discussed psychological experiments in current times and Milgram found himself in the core of public interest. The psychology community was divided on the usefulness of the Milgram studies to interpret the actions of the partakers in the Nazi Holocaust. Milgam's experiment on *Obedience to Authority* have been depicted in songs, novels and short films [5, 6, 24, 26].

References

1. Anonymous, Has the cull begun in biomedicine? Science, 2014. 243(6176): 1183.
2. Association AP, Diagnostic and Statistical Manual of Mental Disorders, Fourth Edition—Text Revision (DSMIV-TR). 4th ed. 2011, Arlington, VA: American Psychiatric Association.
3. Barber B, Lally JJ, Makarushka JL, et al., Research on Human Subjects: Problems of Social Control in Medical Experimentation. 1973, New York: Russell Sage Foundation.
4. Bird D. Hannah Arendt, political scientist, dead. The New York Times. 1975 6 December.
5. Blass T, Obedience to Authority: Current Perspectives on the Milgram Paradigm. 2000, Mahwah, New Jersey: Kawrence Eribaum Associates.
6. Blass T, The Man Who Shocked the World: The Life and Legacy of Stanley Milgram. 2004, New York: Basic Books.
7. Broad WJ, Fraud and the structure of science. Is fraud a trivial excrescence on the process of science or do the recent cases have deeper roots Science, 1981. 212(4491): 137-141.
8. Broad WJ and Wade N, Betrayers of Truth. 1983, New York: Simon and Schuster. 256.
9. Carpenter S, Psychology research. Psychologist's bold initiative. Science, 2012. 335(6076): 1558-1561.
10. Couzin-Frankel J, Shaking up Science. Two journal editors take a hard look at honesty in science and question the ethos of their profession. Science, 2013. 339(6118): 386-389.
11. Dostoevsky F, The Brothers Karamazov (translated by Constance Garnett with a new afterword by John Bayley), ed. Komroff M. 1880, New York: Signet Classic.
12. Fang FC, Bennett JW, and Casadevall A, Males are overrepresented among life science researchers committing scientific misconduct. mBio, 2013. 4: e00640-00612.
13. Gino F and Ariely D, The dark side of creativity: original thinkers can be more dishonest. Journal of Personality and Social Psychology, 2012. 102(3): 445-459.
14. Hardin G, The tragedy of the commons. Science, 1968. 162(3859): 1243-1248.
15. Herman ES, The Banality of Evil, Chapter 13 in Triumph of the Market: Essays on Economics, Politics, and the Media. 1995, Boston, MA: South End Press. 97-102.
16. Kennedy D, Mixed grill. Science, 2007 317(5842): 1145.
17. Langmuir I and Hall RN, Pathological science. Physics Today, 1989. 42(10): 36-48.

18. Lloyd RWF, Two Lectures on the Checks to Population Delivered Before The University of Oxford in Michaelmas Term 1832. 1833, Oxford.
19. Malthus TR, An Essay on the Principle of Population at it Affects the Future Improvement of Society with Remarks on the Speculations of Mr. Godwin, M. Condorcet, and Other Writers (orginally Anonymous). 1798, London: J. Johnson.
20. Marshall E, Fraud strikes top genome lab. Science, 1996. 274(5289): 908-910.
21. Mazar N, Amir O, and Ariely D, The dishonesty of honest people: A theory of self-concept maintenance. Journal of Marketing Research, 2008. 45(6): 633-644.
22. Milgram S, Behavioral study of obedience. Journal of Abnormal and Social Psychology, 1963. 67: 371-378.
23. Milgram S, Obedience to Authority: An Experimental View. 1974, New York: Harper and Row.
24. Milgram S, The Indvidual In A Social World: Essays and Experiments, 3rd expanded edition. 2010: Pinter and Martin Ltd.
25. Miller G, Social psychology. Using the psychology of evil to do good. Science, 2011. 332(6029): 530-532.
26. Millgram S, Obedience to Authority; An Experimental View. 1974: HarperCollins.
27. Mishkin B, Responding to scientific misconduct. Due process and prevention. Journal of American Medical Association, 1988. 260(13): 1932-1936.
28. Smith RJ, Atom bomb tests leave infamous legacy. Science, 1982. 218(4569): 266-269.
29. Smith RJ, Scientists implicated in atom test deception. Science, 1982. 218(4572): 545-547.
30. Woolf PK, Deception in scientific research. Jurimetrics, 1988. 29(1): 67-95.
31. Zimbardo P, The Lucifer Effect: Understanding How Good People Turn to Evil. 2007, New York: Random House Rider Books.

CHAPTER 20

Non-Scientific Frauds and Scandals

Dishonesty is not a trait expressed only by some scientists. Countless scoundrels exist in virtually all walks of life and in probably all professions. An understanding of why they resort to this behavior is clearly needed if research misconduct is to be potentially prevented.

Cheating in Institutions of Education

Clarence Mumford Sr. American teacher certification examinations are written and administrated by the educational testing service. A series of these tests known as the praxis tests are generally required before, during and after teaching training courses in the USA. Clarence Mumford Sr., a longtime educator, discovered an opportunity to generate extra money helping teachers in the southern states of Arkansas, Mississippi, and Tennessee, pass the praxis tests that prove that they are qualified to teach. Until his scheme was discovered after 15 years he sent other people to take the tests for his clients. Each time he received a fee of between $1, 500 and $3, 000. One of Mumford's test ringers would take the praxis test with fake identifications, and his clients would receive a passing grade enhancing their teaching careers. The victims in this scam were hundreds, if not thousands, of innocent public school children who received an education from unqualified teachers. After the fraud was exposed Mumford was indicted on about 60 charges including mail, wire and social security fraud, in addition to identity theft [40].

Online Essay Mills. In the competition to forge ahead students need to prepare papers on various topics. These assignments are often too much for some young individuals who choice cheating over honesty. To combat this concern, high schools, colleges, and universities have honor systems that hopefully keep cheaters at bay. Nevertheless when papers need to be written a market is available to help those who desire others to prepare the assignments. Online essay mills, such as echeat *(www.echeat.com)* will generate papers on any requested subject for those prepared to pay a fee. Dan Ariely, a James B. Duke professor of behavioral economics at the Fuqua School of Business at Duke University (Duke) *(http://duke.edu/)*, carried out a study with Aline Grüneisen examining four essays that they purchased from writers of these essays allegedly prepared to assist students in the writing of their papers. Ariely and Grüneisen found significant mistakes and garbled passages in the essays [31].

Public Schools. Immediately after George Walker Bush (1946—) took office as the 43rd President of the USA on January 20, 2001 his administration introduced a bill on *No Child Left Behind* which received tremendous bipartisan support in Congress. There was a widespread consensus that the education of children in public schools was unacceptable and the *No Child Left Behind Act of 2001* was based on the belief that setting high standards and launching quantifiable goals can improve individual outcomes in education. The purpose of the legislation was to improve the academic achievements of disadvantaged children and steps were taken to measure the acquired knowledge of the students and the quality of the teachers based on objective measurements of the students that they taught. *The Act* required that if states were to receive federal funding for schools individual states need to develop assessments in basic skills to be given to all students in certain grades. *The Act* provided monetary incentives to states and to participants in the educational system of public schools. Unfortunately it also promoted dishonesty as illustrated by the scandal in the state of Georgia. Surveys of high school students indicate that most students cheated and that honor codes hopefully may be an effective strategy for decreasing the amount of cheating. Proactive/preventative measures seem to be more effective than punitive actions against dishonest behavior [34]. Extensive cheating took place in certain public schools. Being concerned about this undesirable issue the State of Georgia investigated the public schools in Atlanta, Georgia and released its report on July 5, 2011. The study unearthed a "culture of fear, intimidation and retaliation" and extensive cheating was discovered in more than 78% of schools investigated. The full report in three volumes is available on the internet *(http://www.myfoxatlanta.com/dpp/news/seen_on_tv/Atlanta-Public-Schools-Investgation—* . . . *8/23/2011)* and *(http://www.myfoxatlanta.com/dpp/news/local_news-New-Details-Emerge-in Atlanta-Publ* . . . *8/23/2011)*. Teachers and administrators in the Atlanta public school system were involved in widespread standardized test cheating. Test answers were corrected using erasures. At one of the schools a newly appointed principal Christopher Walter improved the test scores of students in the criterion references competency test (CRCT) from 50-81%. For this remarkable achievement he was widely acclaimed as a role model for other teachers and he received awards and even a financial incentive to remain at his school. However he achieved improved teaching statistics by encouraging cheating in the standardized tests.

The shocking State of Georgia findings were turned over to the prosecutors and the state professional standards commission for action. The cheating in the Atlanta public schools seemed to be a direct effect of the *No Child Left Behind Act.*

Business Schools. Some ambitious business students also have a temptation to side sweep the truth and embark on dishonest and unethical behavior. A survey of 5, 300 students (including 623 business students) from 32 USA and Canadian institutions during 2002 and 2004 was headed by Donald McCabe, a professor at Rutgers University (Rutgers) (*http://www.rutgers.edu/*). It found that 56 % of Master of Business Administration (MBA) students admitted cheating [34]. This was more than the 47% acknowledged by students in other disciplines. In May 2007 the Fuqua school of business at Duke faced a major scandal when it discovered 34 of its first-year MBA students cheating [21]. The majority of the culprits behaved inappropriately when given a take-home examination intended to be taken individually. The students were punished and some of the cheaters faced expulsion. Concerned individuals pondered over the reason behind the unethical student cheating. The high standard of MBA programs together with the rivalry between students for both grades and jobs was considered an important driving force. Unfortunately students are reluctant to blow the whistle on their colleagues and faculty is often hesitant to report detected cheating [7].

United States Military Academy. The USA military academy at West Point faced a notorious humiliation involving most of the army cadets from the Black Knights football team in 1951 when accusations that football players were distributing unauthorized academic information surfaced *(en.wikipedia.org/wiki/1951)*. In 1966 another scandal at the military academy resulted in the expulsion of 40 cadets, but that one was barely mentioned in the press and at meetings with West Point officials. The cadets were instructed not to write to parents or friends about the scandal because of a concern about it becoming widely known. In 1976 a major scandal tarnished the military academy at West Point about violations in its honor system. Information about the scandal was widely dispersed in newspapers, such as *The New York Times*, and on television *(www.west-point.org/publications/borman.html)*. According to Lucian K. Truscott, a former West Point cadet the honor code was secret and the key was obedience *(http://aliciapatterson.org/APF001976/Truscutt/Truscutt02/Truscott02.html)*. The cadets were required to not lie, cheat, or steal and for abiding they were assured of graduation. They were required to turn in their best friends or roommates should they commit an honor violation. Many cadets rebelled against the honor system especially when Major General Samuel W. Koster (1919-2006), who commanded the *My Lai* massacre in Vietnam and its cover-up, was appointed superintendent of the military academy. When this was revealed he resigned his position of superintendent because of his major effect in disintegrating the honor code. Based on interviews with cadets Truscott concluded that cadets do not cheat for profit, but because it feels good.

USS Memphis. A lapse of integrity with cheating also took place aboard the nuclear submarine USS Memphis. After being detected it was rapidly dealt with by firing the commanding officer and 10% of the crew. Some former officers maintained that this cheating was not unique to this submarine and that it was not uncommon for sailors to be given answers or other hints before taking training examinations. Apparently the examinations had not only become extremely difficult, but also detached from really needed skills. *Associated Press* obtained a report on the investigation through the *Freedom of Information Act.* Answers to the questions were emailed to the sailors prior to qualification examinations which were undertaken in the absence of proctors, and officers were allowed to answer questions asked by the sailors (*http://www.forbes.com/feds/ap/2011/08/15/general-us-submarine-cheating-scandal_8624* . . . 8/23/2011).

The Sale of Corpses and Body Parts

When entrepreneurs develop new ideas to generate money nothing remains off limit to some of them even if the idea is undoubtedly unethical. For example Michael Mastromarino a New York businessman and former dentist sold parts of human cadavers without consent in a program that he masterminded with the assistance of a group of "cutters" who pilfered pieces of corpses from deceased humans under the care of three funeral directors (Louis Garzone, Gerald Garzone, and James McCafferty). The funeral directors were charged by a grand jury of selling two hundred and forty four bodies to Mastromarino who managed *Biomedical Tissue Services,* a company that resold the often-diseased body parts at around one thousand dollars per specimen for the therapy of medical indications, such as burns and fractured bones. Kidneys, eyes, bones, skin and other body parts were cut from dead bodies and sold to persons having a need. Sales of body parts on the black-market took place from at least February 2004 through September 2005. Some organs were transplanted into patients for therapeutic reasons worldwide without the recipient or the surgeon being aware of the origin of the human material. Soon after the discovery of this dishonest unethical behavior a grand jury in Philadelphia heard the case and charged Mastromarino and Lee Cruceta, and a former nurse for running the cutting crew. *Biomedical Tissue Services* was accused of plundering 1, 077 corpses in New York and Philadelphia. A duty of the participating undertakers was to obtain consent and they allegedly falsified death certificates and indicated that the donors were for the most part in good physical shape, but had succumbed to heart attacks or blunt-force trauma. The funeral directors were charged with a vast number of wrong doings that included theft of body parts, forgery, and managing a fraudulent business. Evidently Mastromarino only gathered the organs and tissues and forwarded them to the next processing unit. Most of the deceased were poverty-stricken and their families were unaware that the corpses of their recently departed relatives were being plundered. They were under the misconception that the bodies were being cremated, but instead the corpses frequently remained unrefrigerated for days, every so often in lanes adjacent to the funeral home, pending the arrival of the designated cutter. In giving testimony about this scandal one cutter described the place where the tissues were obtained as dirty and comparable to the back of a butcher shop. One donor actually suffered from cancer and the body was infected with both hepatitis C and the human immunodeficiency virus (HIV). According to the indictment the names of donors were usually concocted and family consent forms were forged. To make the human material ideal for transplantation the ages of donors were lowered and the dates of death were altered to give the impression that the tissues were fresher than they were in reality. Because of falsified documents it was virtually impossible to link unauthorized donors with recipients. The true identity of only 48 of the 244 corpses could be established. The probability of complications occurring after organ and tissue transplantation carried out under the appalling outlined conditions was extremely high, but the magnitude of them remains unknown. One female patient from Philadelphia became infected with hepatitis following a transplant from a putative contaminated piece of the body instigated a civil suit. The Garzone brothers eventually relinquished their state funeral licenses. Aside from the undertakers involved in the Mastromarino scam seven funeral directors in New York pleaded guilty, including one whose funeral home allegedly stole pieces of the body of Alistair Cooke (1908-2004) the host of the Public Broadcasting Service (PBS) "Masterpiece Theater" from 1971-1992.

(*http://www.msnbc.msn.com/id/21140879/ns/health-health_care/t/charged-selling-body-parts/*).

Art and Artists

The paintings of masters have been, and continue to be, copied by many individuals, but this is not regarded as plagiarism or theft of IP. Sometimes the copying is performed as an exercise to acquire skills in the art; on other occasions it is done by highly talented artists to mimic the original. When skillfully copied some paintings are difficult to distinguish from the original, especially to the eye of persons who are not skilled art scholars. Such paintings are what con artists sell to gullible customers under the pretense that they are the real thing. Such fake paintings can generate considerable income for the rogues who sell them.

Early in the 1880s Eugène Henri Paul Gauguin (1848-1903) a leading post-impressionist artist travelled from Paris to Polynesia and remained there for the rest of his life except for a few trips elsewhere. While in Tahiti he made ceramics, wood carvings and paintings with exotic bare-breasted native women. With his art Gauguin depicted the primitive island inhabitants in a way that he had hoped they would be, but by the time of Gauguin's arrival the Tahitians had been thoroughly

Christianized and colonized by the French and the women did not walk around half-naked, but tended to wear Christian missionary gowns

(http://www.npr.org/2011/03/15/134537646/gauguins-nude-tahitians-give-the-wrong-impre . . . 3/16/2011).

While artists have a right to depict images in an imaginative way Gauguin insisted that his illustrations of life in the Polynesian colonies were truly factual, but in reality this was an unquestionable lie. A co-conspirator was the French tourist industry which wanted to promote tourism in Tahiti by Europeans. Despite this dishonesty history recorded Gauguin merely as the creator of an intentionally perpetuated myth, rather than as a dishonest artist.

Politicians

As everyone knows many politicians are not known for their integrity and power commonly leads many to treachery. Honest democratic elections are desirable in modern countries, but in certain countries those in control as well as their supporters take steps to rig elections to retain power. Notable examples of this are Zimbabwe, Afghanistan, Syria, and Iran. As pointed out by Galbraith, a former top United Nation's official, fraud was witnessed during the 2009 presidential election in Afghanistan [23]. Some provinces counted more votes than were actually cast. Even in relatively minor elections some politicians have a reputation for unethical fraudulent behavior and for bending the truth. This is to be expected as their jobs are dependent on getting sufficient votes to become elected and to remain in office. To do this they often make pledges to support legislature and when it comes to the floor for a vote they turn the other way. Sometimes this is a change of opinion based on evidence that had not been previously considered and at times it results from political pressure from their constituencies as reflected in surveys of public opinion. Their supporters go out of the way to raise enormous amounts of money for their campaigns because without sufficient funds nobody can hope to be elected. Space will not permit a detailed review of bad behavior by politicians. Fortunately the USA Congress and Senate have ethics committees that define the rules and they throw the book at individuals who overstep the line in which Congress and the Senate consider inappropriate. In 2010 a panel of the USA Congress voted 9-1 in favor of a resolution to censure representative Charles Bernard Rangel (1930-)(Democrat, New York) before the full house for not paying taxes on unreported income from his rental property in the Dominican Republic. When this occurred Rangel was 80 years old and had served the district of Harlem in Congress for 40 years. He was censured before the entire house for desecrating more than a dozen congressional rules and other offenses including the misuse of congressional staff and supplies [5]. This censure was just short of the most harsh penalty available to Congress namely expulsion from the house.

Athletics

Violations of the National Collegiate Athletic Association. At least in the USA College athletes can pursue an extremely profitable career if they make it to the professional level. This is especially true for football, basketball, baseball and golf. Top notch student athletes can receive scholarships to support their college or university education. Many colleges set a high academic standard and expect their athletes to reach an acceptable scholarly level to be admitted to their programs. Some superbly talented athletes are exceptionally good at their sport, but cannot make the admission grade at highly prestigious universities. Moreover some that get admitted to institutions of higher learning are unable to keep up with the academic requirements, because of the vast amount of time that these athletes must devote to the sport for which they were recruited where the object of the game is winning and enhancing the reputation of the university and keeping the alumni happy. The National Collegiate Athletic Association (NCAA) has regulations to prevent abuses, but this is not sufficient to deter violators. Many college football programs have been investigated and penalized by the NCAA for violations of expectations. For example in 2010 several football players at the University of North Carolina (UNC) at Chapel Hill *(http://unc.edu/)* ran into trouble with the authorities for not complying with the rules. The scandal started with a tweet from Marvin Austin one of the UNC's football players with the potential to become a professional [29]. It arrived in the early morning of May 18, 2010 and stated "I live in club LIV so get the tenant rate. Bottles comin (sic) like it's a giveaway." When UNC became aware of wrongdoings in its football program it imposed a $50, 000 fine on itself and forfeited all won games during the 2008 and 2009 football seasons. After an investigation by the NCAA's committee on infractions a final verdict was announced and it led to the humiliation of the UNC football program and the university's academic reputation. [11]. This disgrace precipitated the retirement of Dick Barbour, the former Tar Heel director of athletics, and the firing of Butch Davis an ex-football coach. Eventually Holden Thorp the chancellor of UNC also fell from grace because of the scandal. The penalties were severe. UNC was placed on three years of probation and was banned from postseason games in 2012, including the Atlantic Coast Conference (ACC) championship and a bowl game. Fifteen scholarships were lost over three years. John Blake, an ex-assistant coach was banned from recruiting and probably from coaching for three years. Some players accepted more than $10, 000 in impermissible benefits. Several major infractions in NCAA rules were detected in the UNC football program. They included impermissible benefits from agents, failure to monitor the football ball program and even academic fraud. For the violations the guilty parties received a variety of penalties such as being required to miss a certain number of games. Some were suspended for a season or even dismissed from the program. A few were even declared permanently ineligible to play college football by the NCAA.

Even coaches are not all honest and some receive "gifts" and "loans" from sports agents having an interest in eventually recruiting talented individuals into professional teams. Players may not receive gifts and benefits from agents, and they often need tutors to keep up with their class work and tutors have stepped out of line by preparing papers for the studies. Such academic misconduct is clearly unacceptable. When schools commit major violations of NCAA rules the institutions sometimes impose sanctions on themselves, as mentioned above with UNC, to avoid more serious penalties. The fact that the NCAA needs to establish rules and monitor their use indicates that unmonitored misconduct is a significant problem. Even states have laws, such as the *Uniform Athlete Agent Act* of North Carolina, that require agents to work with a license and not provide benefits to students

Providing Gifts to College Football Players. Nevin Shapiro (1969—), who is currently serving 20 years in a federal prison for money laundering and securities fraud related to a massive Ponzi scheme, participated in widespread violations of NCAA rules. Shapiro singled out numerous current and past Miami University (Miami) *(http://www.miami.edu/)* football players to whom he provided money, cars, jewelry, televisions and other gifts. He picked up the bill for sex parties, meals in restaurants, amusement at nightclubs, and strip joints. He also gave players access to his multimillion dollar home and yacht and even claimed that he paid for 39 different football players to have sex with prostitutes. He allegedly even paid for the abortion of a woman who became impregnated by a football player. At least one former Miami football player accepted $1, 000 from Shapiro when he started college. Shapiro also maintained that he was a part-owner in an agency, *Axcess Sports and Entertainment LLC* in Ponte Verde Florida that channeled money to athletes to sign with its agents, an allegation denied by others in the agency. Shapiro informed YAHOO!Sports that he provided financial compensation to potential recruits on behalf of the sports agency. At least one linesman received $50, 000 to sign with the agency as a first-round pick. The allegations against Miami set off investigations by the NCAA related to college football.

Shapiro even offered a five thousand dollar prize to any player who could knock Chris Rix (1961) the former quarterback for the Florida State University (FSU) *(http://www.fsu.edu/)* Seminoles out of the game. Shapiro asserted that he recruited numerous football players to Miami and even paid some recruits $10, 000. Shapiro was an ex-Miami booster connected to the Miami football program for about a decade. For a 10-year pledge to Miami the university named its student-athlete lounge in Shapiro's honor, but only until 2008 when he defaulted on a pledged payment. Perhaps his fall from grace in his dishonest financial matters played a role.

Sexual Abuse. Not all football scandals at universities focus on violations of NCAA regulations. One of the most disgraceful events related to college football programs centered on sexual abuse in boys and young men and the cover-up after it was discovered. Gerald Arthur "Jerry" Sandusky (1944—) a 67 year old admired assistant football coach of the University of Pennsylvania (Penn State)*(http:// www.upenn.edu/)* founded an organization called *The Second Mile* in 1977 as a foster home for troubled youths. Over the years at least some of these boys went on trips to professional and college football games. Gifts of computers, clothes, and cash were also doled out to them. While this behavior might have given the impression of generosity Sandusky had an ulterior motive in helping unfortunate youngsters. In addition to his role as coach and mentor Sandusky became a sexual predator of young boys and frequently molested them individually in the bedroom of his home or in a school workout room when nobody was around to bear witness. In 1998 the mother of one victim called the Penn State police after her son arrived home with wet hair after showering naked with Sandusky, who hugged the boy in the shower. In 2002 Michel McQueary, a graduate assistant, witnessed a young boy being assaulted in the showers of Penn State's football facility by Sandusky. Joe Paterno (1926-2012) Penn State's legendary head football coach was informed of this incident and it was reported to his superiors, but the police were allegedly not notified. The fact that a prominent university would keep allegations about unacceptable sexual behavior concealed under a cloak of secrecy for years without permitting law enforcement agents to investigate the matter is inexcusable. Clearly Penn State did not want the adverse publicity about its prominent football program. The first alleged victim to report this despicable behavior to the police prompted an investigation into the coach and this led to him being charged with 40 counts of sexual abuse [28]. When the story broke in 2011 it caused the biggest dishonor in American college football history and it came to a head after a two year long investigation by a grand jury in Pennsylvania found Sandusky guilty of numerous sexually abused youngsters. After Sandusky was arrested and charged with 52 counts of sexual abuse involving young boys over fifteen years old. Four charges were dropped, but he was found guilty and he was convicted of 45 of the charges by a jury in a federal court. On October 9 2012 Sandusky received a prison was sentence of thirty to sixty years. Because he was 68 years old at the time he is expected to spend the rest of his life in prison. A former athletics director of Penn State (Timothy M. "Tim" Curley (1954—) and the university's senior vice president for business and finance (Gary Schultz) faced perjury charges for denying to the grand juror that they knew anything about the 2002 episode.

On November 2, 2011 the nine member Penn State's board of trustees fired Panero and forced the resignation of President Graham Basil Spanier (1948—). Curley was placed on administrative leave pending his trial and the university did not renew his contact when it expired in June 2013. Penn State funded an inquiry into the matter chaired by Louis Freeh, a former director of the Federal Bureau of Investigation (FBI) and after seven months it found that Paterno, Spanier and other former administrators tried to protect the institution from bad publicity by not reporting the Sandusky episode. After this report was released a statue of Paterno was removed from Penn State football stadium and the NCAA withdrew

111 wins from the Penn State football record. The net effect was that Paterno no longer held the record as the major football coach with the most wins. The family of Paterno commissioned its own inquiry by a group that included Dick Thornburg (1932—) the former attorney general of the USA. It considered the Freeh report factually incorrect and "fundamentally flawed" [33].

Professional Athletes. Today many people in western societies enhance their perception by means of drugs, such as Marijuana, Ritalin (methyl phenidate), caffeine, and ethanol. It is hence perhaps not surprising that professional athletes resort to methods to enhance their athletic skills. The skill of an athlete depends on an inherent talent that presumably relates to a genetic predisposition, an active devotion to the sport, and from an early age excellent mentoring and coaching, a financial sponsor, and a considerable amount of practice. Top notch athletes are extremely competitive and thrive to be the best in their chosen sport. Obviously they can eventually only reach the limit of their capability by a given age. Each sport creates its own rules and regulations, but some talented athletes try to push their capability using means that have been made illegal and unethical by the administrators of their sport. The governing organization of some sports, such as those that take place in the Olympic Games, take special steps to expel dishonest athletes, such as those who use enhancement drugs. Examples of individuals going beyond the legal limits are known in different sports. An example will be provided in baseball and cycling.

Lance Edward Armstrong (born Lance Edward Gunderson) (1971—) was regarded as the greatest cyclist of all time after winning the *Tour de France*, the most prestigious race in cycling seven times in a row (1999-2005). This number of wins was more than anyone else had achieved. His accomplishment was considered unbelievable by other cyclists. His triumph was even more astonishing because he had survived a testicular seminoma that was diagnosed in 1996 and Armstrong became an inspiration to sufferers of different types of cancer. Unfortunately for Armstrong his reputation collapsed like a pack of cards because his cycling success was exposed in what the USA Anti-Doping Agency (USADA) described as "the most sophisticated, professionalized and successful doping program that sport had ever seen". For this the International Cycling Union (ICU) stripped him of all his *Tour de France* titles and banned him from cycling events for life. On October 10 the USADA published its report on the scandal that Armstrong elected not to contest [49] *(www.cnn.com/2012/10/10/sport/armstrong-report-highlights.index. html)*. Armstrong's former teammates testified against him and for doing so they received less severe sanctions. Initially Armstrong vehemently denied doping and while he was racing he never failed a drug test. For his indiscretion Armstrong not only destroyed his reputation, but also financial support from sponsors worth millions of dollars and the chairmanship of *Livestrong*, a cancer charity that he founded in 1997. Despite the overwhelming evidence against Armstrong his lawyer labeled the USADA report as a "one-sided hatchet job" and a "government-funded witch hunt". Eventually in a vague confession Armstrong admitted cycling while under the influence of doping in a live interview on television, but he did not apologize for vicious attacks that he made on whistle-blower that drew attention to his dishonest behavior [4].

William Roger Clemens (1962—), one of the greatest baseball pitchers, became the subject of a scandal on December 13, 2007 when allegations were made that his former strength and conditioning coach injected him with Human Growth Hormone (HGH) and anabolic steroids over a three year period (1998-2000) [16]. The allegations were made in a report informally referred to as the Mitchell Report [35] (*www.mlb.com; accessed on 12/13.2007*). Clemens steadfastly denied the accusations under oath facing Congress, and because of suspected perjury his case was referred to the USA department of justice. A federal grand jury indicted Clemens on six felony counts related to false statements, perjury and obstruction of Congress. The case went to trial, but prosecutorial misconduct resulted in a mistrial. In a second trial Clemens was found not guilty of lying to Congress. While doping could not be unconvincingly pinned on Clemens other members of the baseball community participated in drug enhancement.

Olympic Games. The modern Olympic Games are divided into separate winter and summer games. They are major international events in which athletics from all countries compete in a wide variety of sports to determine the individuals who are the best in the world in particular sports. The games are organized by the International Olympic Committee (IOC) which sets the standard for all athletic activities. To enhance their chances of winning some athletes use enhancing drugs even though they are banned by the IOC. To prevent cheating with the use of drugs that are known to enhance physical activity the IOC requires all participants to be tested to make sure that they have not used the banned drugs. Nevertheless at almost all Olympic Games some individuals are disqualified for taking enhancing drugs.

Businesses

Unethical behavior and cheating by members of the business community have tainted their profession and numerous examples have received publicity in the media. Particularly notable examples involved the major companies of Enron, Tyco, Global Crossing Limited, and WorldCom. The tradition of tying pay to accomplishment is a recipe for bright individuals to become cheaters and scandals have involved former Chief Executive Officers (CEOs) and Chief Financial Officers (CFOs) of major corporations. In the USA fraudulent "*creative accounting*" has been the downfall of some companies typically after the initiation of investigations by *government oversight* agencies, such as the *Securities and Exchange Commission* (SEC). An exceptional number of corporate scandals emerged in the early part of this millennium. Together these companies shared neglected commercial authority, bookkeeping exploitations, and blatant greed.

Enron. Enron, an energy trading corporation with headquarters in Houston Texas, was once one of the biggest companies in the world and its collapse destroyed the lives and the retirement investments of thousands of individuals. Months before the company went bankrupt Kevin Kindall a mid-level analyst exposed the financial rot that was destroying the company. Before its downfall Enron was claimed to be the corporate triumphant of the decade, but it ignored the basic principles of business by pursuing an ambitious quest of profit due to greed. Kindall relayed his concerns to his boss, but the warning went no further. His superior was the recipient of more than $1 million from his own wrongdoing and he had no intention of bringing the Enron money machine to a stop. In the business model of Enron employees were judged by the number of "deals" that they made for the company. Compensation and bonuses to employees were based on how much growth the person generated. Many of the business deals caused flashy financial losses that could be covered up with tricky creative accounting. Enron collapsed because of one of the largest audit malfunctions and ambiguities using the General Accepted Accounting Principles (GAAP). A scoundrel in the Enron scandal was the businessman Andrew Stuart Fastow (1961—) who served as its CFO until the SEC investigated him and Enron for financial irregularities in 2001. For his participation in a complex maze of creative accounting that concealed massive financial losses he went to prison for 6 years. Jeffrey Skilling the former President of Enron was convicted on 19 felony counts, including conspiracy, securities fraud and insider-trading charges in October 2006. For his greed Stilling was sent to the Colorado federal prison for 24 years. The downfall of Enron also had a detrimental effect on Arthur Andersen one of the big five auditing firms in 2002 which was responsible for overseeing its records. This accounting giant was indicted and found guilty criminally, but the verdict was reversed in 2005 by the USA Supreme Court. The book "Conspiracy of Fools" describes the Enron fiasco in detail [17].

Tyco International. A security systems company Tyco International Ltd. was founded by Arthur J. Gandua in 1960. It focused on security solutions, fire protection and flow control. The business was incorporated in Switzerland, with operational headquarters in Princeton, New Jersey in the USA. Initially it dealt mainly with government research and military experiments, but subsequently it grew with stockholder equity and an elaborate acquisition of companies. Dennis Koslowski (1946—) joined Tyco in 1975 and eventually became the CEO in 1992. With an aggressive acquisition strategy more than 1000 companies were acquired and Tyco underwent a name change from Tyco to Tyco International Ltd in 2007. In 1996 Tyco was added to the Standard and Poor's (S&P) composite index. With its acquisitions Tyco suffered extensive losses. During 2002 Tyco became entangled in a scandal created by the greed of Kozlowski and his senior management. Kozlowski the villain of Tyco supposedly received $81 million in unauthorized bonuses and purchased art worth $14, 725 million. In addition an investment banking fee of $20 million was paid by Tyco to Frank Walsh Jr

(1941—), a former Tyco director, for helping to broker an acquisition. This payment to Walsh led to the resignation of Kozlowski and more than two dozen class-action lawsuits against Tyco and former officers of the company.

In 2005 Kozlowski was convicted of his crimes and sentenced to 8.33 to 25 years in prison. He now resides in the mid-state correctional facility in Marcy, New York and cannot be released until he becomes eligible for parole on January 17, 2014. Walsh avoided prison by agreeing to return the $20 million together with a fine of $2.5 million [44].

Global Crossing Limited. Gary Winnick (1942—), the financier who founded Global Crossing Limited in 1997, served as its initial chairman until 2002 when the company filed for bankruptcy. This company laid the earliest privately funded undersea fiber optic cable network across the Atlantic Ocean and it went public in 1998. Before the collapse of the company Winnick made about $734 million from the sale of his shares. Because many of the employees lost money in their 401k retirement accounts Winnick donated $25 million to them. Shareholders considered the financial activities of Global Crossing Limited fraudulent, and brought a class-action lawsuit against company executives. The case was settled without anyone admitting unlawful activity, but as part of the deal Winnick consented to pay $55 million.

WorldCom. Bernard J. Ebbers (1941—) was the former CEO of WorldCom the onetime telecommunications giant. He made a fortune and by June 1999 he was billionaire. By 2002 he resigned as CEO and after WorldCom collapsed Ebbers was left penniless with nothing to show for the company he established. Dennis J. Moberg the new corporate CEO ended up in prison after being convicted of conspiracy, security fraud and making false filings with the SEC. He was answerable for an $11 billion fraud in WorldCom. A prompt loss in the company's revenue together with a bookkeeping disgrace generated an illusion of billions of dollars in earnings and more than $107 billion were listed in its assets when WorldCom unnerved the telecommunications business when it filed the largest claim for corporate bankruptcy in the history of the USA after the accounting irregularities were discovered. WorldCom divulged in late June 2002 that it had inappropriately reported more than $3.8 billion [41]. WorldCom had considerably more assets than Enron, which filed for insolvency before it (*http://www.scu.edu/ethics/dialogue/candc/cases/worldcom.html*) [12].

Other factors leading to the collapse of WorldCom were a strategic growth through acquisition and the awarding of loans to its senior executives based on friendships between those in the corporate power rather than on objective criteria. The idol in exposing the collapse of WorldCom was Ms. Cynthia Cooper, who blew the whistle on the inappropriate dealings of this corporate giant. She exposed deceptive accounting fraudulent wrongdoings [13]. Cooper and her fellow internal auditors distrusted unusual accounting irregularities that were approved by Arthur Andersen but which ultimately contributed to the WorldCom downfall [18]. Together with her fellow auditors she exposed fake accounting entries and other skillful methods planned to conceal nearly $4 billion in misallocated expenses [38]. Cooper first suspected wrongdoings in March 2002 when a manager informed her supervisor Scott Sullivan, the CFO, that a $400 million reserve account that had been established as a shield to protect against unexpected losses of revenue had been taken over. Cooper unsuccessfully brought her concerns to the Andersen accounting firm so she took them to the board's audit committee, but this created tension between her and Sullivan who warned her not to become involved in the disputed concerns. Cooper continued to investigate questionable auditing procedures secretly and discovered that the WorldCom financial analyst had been sacked during the previous year for not accepting the accounting chicanery [37]. Over time she and her fellow auditors exposed a $2 billion financial entry for unauthorized capital expenditures. To collect additional confirmation about the wrongdoings, the Cooper team illicitly explored WorldCom's computerized accounting information system and discovered additional fraudulent data. When Sullivan learnt about the audit, he asked Cooper to postpone announcing her findings until the third quarter, but she refused and reported them to the board's audit committee. As a consequence Sullivan and a couple of other members of WorldCom were fired in June 2002. For her remarkable detective work *Time* magazine highlighted Cooper as one of three "Persons of the Year" on December 21, 2002. Using a generous interpretation of bookkeeping rules for fiscal records the company gave the false impression that the earnings were improving. WorldCom was apparently trying to misrepresent operating costs as capital expenditures to give the illusion that the corporation was more profitable that it really was. The attainment of Microwave Communications Inc (MCI)(now Verizon) gave WorldCom an additional opportunity for creative accounting. After the accounting irregularities were discovered the corporation filed for bankruptcy in July 2002. The inappropriate accounting of operating expenses as capital expenses infringed upon the GAAP. Not surprisingly the Andersen accounting firm of WorldCom, which also served Enron, stood by scores of the bookkeeping indiscretions that contributed to the WorldCom downfall [18]. When the *Glass-Steagall Act* was revoked by William Jefferson "Bill" Clinton (1946—), the 42nd President of the USA in 1999, financial organizations could offer a greater variety of financial services to their clients and this was considered an important cause of the financial crisis of the late 2000s.

HealthSouth Corporation. Richard M. Scrushy, the founder of HealthSouth Corporation, served at one time as its CEO, but a federal judge found him guilty of contributing $500, 000 to a foundation formed by the democratic politician Don Eugene Siegelman (1946—), the 51st Governor of Alabama from 1999 to 2003. As consequence he spent six-and-a-half years in prison for a controversial bribery conviction, which was upheld on appeal in a trade for a place on the Alabama hospital regulatory board. He was sent to prison for 7 years for bribery in June 2007.

Samsung. Lee Kun-hee, chairman of Samsung the largest company in South Korea, was suspected of corruption for many years. In the 1990s he was found guilty of paying off politicians and was also charged with: (i) having power over a colossal slush account from which he purportedly funded bribes, (ii) evading billions of dollars in taxes, (iii) concealing a vast amount of money in about 1, 200 accounts under different names, (iv) putting together a sale of shares in Samsung companies to his son at falsely low values to shift control of Samsung to his successor. Evidence in support of bribery was inadequate and he was helped in this regard by the expiration of the statute of limitations. By South Korean standards, these were drastic moves, but the whole family will still be Samsung's largest shareholder. Lee theoretically faced a life sentence in prison, but he circumvented punishment, because as recent as 2008 the judicious system in South Korea devoted little effort to prosecuting white-collar crime. *The Economist* pointed out that the higher up the corporate level the more likely that a person would get a suspended prison sentence for the benefit of the nation's economy, especially if the culprit donated part of his/her wealth to a charitable organization [2]. On April 21, 2008 on live television Lee astounded South Koreans by resigning from the business his father created seven decades earlier. He stated that he would take legal and moral responsibility for his actions. On April 17, 2009 Lee was charged by prosecutors for tax evasion and breach of trust. On December 29, 2009 he was pardoned by the South Korean government to aid their country's bid to host the winter Olympic Games in 2012 at Pyeongchang, but the IOC suspended his right to participate in meetings of its commission for 5 years *(http://news.bbc.co.uk/2/ hi/asia-pacific/735999.2.stm).*

It was presumed that Lee's heir would become the chairman of Samsung after the public anger subsided, but Lee returned to his position on March 24, 2010.

While he was in charge the sales of Samsung, encompassing 59 companies covering the well-known electronic division, ship construction, sports teams, and shopping malls grew to multibillion dollar enterprises, estimated to be worth about $160 billion.

Lies About Type of Writing

A big difference exists between the writings of fiction and non-fiction. Novels are typically fiction and the author's imagination reigns supreme, but in non-fiction the author has an obligation to document the truth. As pointed out in Rotvik's book *In His Image. The Cloning of a Man* was alleged to be a non-fiction tale [42], but it turned out to be a factious story. In her popular television show on October 26, 2005 Oprah Winfrey declared that she had chosen the talented American writer James Christopher Frey's non-fictional memoir about the author's battle with alcoholism, crime, and drug addiction and his treatment at a rehabilitation center in Minnesota titled *A Million Little Pieces* [22] for her influential book club. Winfrey very enthusiastically appraised the supposedly factual account of his life in this book and so did book critics. The book sold more than 3.5 million copies, but an examination of the content by *The Smoking Gun* based on discussions with relevant people and an examination of a variety of sources including law enforcement and court documents revealed major flaws in the factual information in *A Million Little Pieces*. Frey's book was a lie and almost pure fiction based on fantasy. Nevertheless despite evidence against him Frey defended his book as non-fictional and pointed out that minor details in the contents of the account were modified for literary effect [10].

Financial Advisors and Stock Brokers

Many financial advisors and stock brokers have been guilty of fraudulent financial dealings with their clients. Examples are provided here.

Bernard L. Madoff. The most famous dishonest financial advisor is Bernard L. Madoff (1932—), who founded the Bernard L. Madoff Investment Securities LLC firm in 1960 after graduating from law school. This Wall Street firm traded securities, money markets and advised wealthy clients. At the New York Stock Exchange Madoff had influence at the NASDAQ (National Association of Securities Dealers Automated Quotations) stock market and he was chairman of the board of directors of NASDAQ in 1990, 1991 and 1993. On December 10, 2008 Madoff allegedly confessed to his sons, Mark and Andrew that the asset management arm of his investment advisory business was "a giant Ponzi scheme". The sons turned Madoff in to the USA authorities and two days later the FBI arrived at his apartment in Manhattan in New York and arrested him. Madoff remained chairman of his firm until he was arrested and charged with one count of securities fraud by the FBI on December 11, 2008. This occurred after Harry Markopolos a financial analyst blew the whistle on Madoff and informed the SEC that he did not think that Madoff could achieve his reported financial gains legally and mathematically. After appearing in court he was freed on a $10 million bond. After a trial in which Madoff pleaded guilty to securities fraud, wire fraud, mail fraud, money laundering, making false statements, perjury, theft from an employee benefit plan, as well as making false filings with the SEC and other federal offenses he was found guilty and sentenced to 150 years in a federal prison (*http://www.bloomberg.com/apps/news?pid=20670001&refer=home&sid=a tUk.QnXAvZY* 12/24/2008).

He ended up being incarcerated in the federal correctional institution at Butner in North Carolina registered as #61727-054. Madoff's life there was undoubtedly a new experience for him. His first incident in a yard fight with another elderly inmate was documented on October 13, 2009 [26] *(bostonherald. com/news/national/northeast/view/20091013madoff_wins_prison_yard-scuffle-over-stick-market)* and a couple of months later he was hospitalized because of serious facial injuries, broken ribs and a collapsed lung after he allegedly landed on his face after falling out of his prison bed *(www.dailymail.co.uk/news/article-1238450/ Fraudster-Madoff-hospital-facia-injuries-fallig-prison-bed.html)*.

Who would have suspected Madoff of being a crook? He was extremely wealthy and a prominent philanthropist who contributed significant amounts of money to charities. Through a $19 million private Madoff family foundation, Madoff made philanthropic gifts. Some of his major charitable contributions were with an ulterior motive. For example he donated $6 million to lymphoma research, only after his son Andrew was found to have that type of cancer. Madoff served on boards of nonprofit institutions, but he benefitted from his role as many of them entrusted him with their endowments. Greed seems to be a major factor driving Madoff to his unethical fraudulent behavior as it motivates many others on Wall Street.

The fraud of Madoff allegedly swindled particularly his fellow Jews and destroyed the fortunes on many Jewish charities and institutions. The losses to the Jewish philanthropic world were devastating and greater than what anyone could remember. The psychological consequences were also calamitous. His customers as well as firms lost vast amounts of money estimated at about $50 billion. Perhaps the biggest loser was Walter Noel's Fairfield Greenwich Group, which invested $7 billion with Madoff's firm (en. wikipedia.org/wiki/Fairfield_Greenwich_Group). The Madoff scandal also affected research directly [14]. The endowments of some nonprofit foundations, such as Picower foundation, the Robert I. Lappin charitable foundation, and the JEHT foundation based in New York City and named with the acronym that stands for the core values of the foundation's mission: "Justice, Equality, Human dignity and Tolerance" that provided grants to fund research were managed by Madoff's firm. Following the exposure of Madoff's disgrace these foundations were forced to close. Picower Foundation, which had assets of almost $1 billion before its financial collapse, ceased issuing grants and stopped funding research on consortia studying Parkinson disease, diabetes mellitus, and obesity as well as graduate fellowships *(http://en.wikipedia. org/wiki/Bernard_Madoff)*.

When the fraudulent finances of Madoff were exposed it was clear that such vast amounts of money could not have been lucrative without the knowledge of others. After an investigation of the Maldoff scandal nine people pleaded guilty to criminal charges in the government probe lasting more than five years. Those that pleaded culpable included Irwin Lipkin and his son Eric Lipkin and Peter B. Maldoff the brother of Bernard Maldoff [1].

Marc M. Harris. Marc M. Harris became an offshore financial authority after graduating with honors from the University of North Carolina at Chapel Hill (UNC-CH)(http://unc.edu/). He was driven by greed and a need to impress and fled Florida in 1989 after earning a reputation for money laundering, tax fraud, and trying to bring into the USA banned Freon (the trade name of DuPont), that includes the organic compound chlorofluorocarbon (CFC). This organic compound is composed of *carbon*, *chlorine*, and *fluorine* and is produced as a *volatile* derivative of *methane* and *ethane*. Many chlorofluorocarbons were extensively used as *refrigerants*, propellants (in aerosol applications), and solvents until it was realized that they were responsible for the depletion of ozone. After this adverse effect became established the manufacture of these compounds was phased out by an international treaty that countries started signing on September 16, 1987. Harris created the Marc Harris organization and became ensnared in a diplomatic struggle between USA and the Panamanian authorities, who were suspected of protecting him.

In the 1990s Harris was a partner in a suspicious offshore financial advisory group in Panama City known as *Trust Services, S.A* Arturo Paz Guzmán, who was a public housing official under Pedro Juan Rosselló González (1944—) the physician who served as the Governor of the Commonwealth of Puerto Rico from 1993-2001 received a one year prison sentence for money laundering making use of *Trust Services, S.A*. With promises of offshore investments Harris deceived retirees and others of millions of dollars. In November 2000 Harris was expelled from Panama and fled to Nicaragua after legal and political pressures on him reached a boiling point. While on his way to immigration hearings in Managua, the capital city in Nicaragua, with his Panamanian-born wife Harris was stopped and transferred to USA Marshals, who were waiting for him in Nicaragua. He was handcuffed and transported to Miami Florida, where his attention was drawn to a 13-count indictment charging him as well as Aurelio and Joseph Vigna, his co-conspirators, with failing to pay $6.2 million in excise taxes related to the importation of the banned Freon between 1993 and 1994. Harris was supposedly sheltered between 1994 and1999 under Ernesto Pérez Balladares González-Revilla (1946—) by the Panamanian Governor. After his conviction Harris chose to work with the USA government in exchange for a milder prison sentence. He could almost certainly provide polished if not humiliating information about several prominent international public figures. At present Harris is serving a prison sentence of 17 years at a federal medical center outside of Fort Worth, Texas for tax fraud and money laundering and a plot to sneak banned Freon into the USA. His release from prison is scheduled for 2018.

Samuel Israel III. Samuel Israel III (1959—) was another, but less infamous swindler. He was from a prominent New Orleans family and founded the company Bayou Management. He owned a large Tudor mansion located 30 miles north of New York that had at one time been the property of Henry John Heinz (1844-1919), the wealthy food manufacturer. Israel started deceiving investors from the outset after he created a hedge fund in 1996 [20]. As a charming hedge-fund manager he stole $300 million from stockholders in the USA over time. He also pleaded guilty to 3 counts of mail fraud, investment advisor fraud, and conspiracy. Daniel E. Marino, Israel's deputy and Bayou Management's CFO pleaded guilty to the same three crimes as well as wire fraud. Israel admitted that he had deceived investors by informing them that Bayou Management was doing better than it really was. The poor performance of the hedge fund was concealed in regular client mailings. Factitious reports kept the investors happy and new funds attracted hundreds of millions of dollars. Fraudulent audited financial statements sent to investors included documents with colossal gains and high total assets. This was despite millions of dollars in losses. In 2004 Israel invested $120 million in a fraudulent program promising very high yields that jumped between banks in New York, London, Hamburg, and Hong Kong. Finally, unluckily for Israel this vast amount of money set down in a bank in Arizona only to be seized by state regulators participating in prime bank instrument fraud. In April 2008 for his crimes he was sentenced to 20 years in prison. Because he did not show up at the prison on an assigned day and later faked a suicide while in prison his sentence was increased to 22 years. One investor appeared in Stamford Connecticut at the headquarters of the hedge fund to meet head-on with Bayou's managers. He found a vacant office and an apparent suicide node by Marino, allegedly describing Bayou as a fraud, but Marino did not kill himself [19].

Scott W. Rothstein. Scott W. Rothstein (1962—) is a former managing shareholder, chairman and CEO of a large law firm known as Rothstein Rosenfelt Alder. The firm with 150 employees and 70 lawyers had offices in various parts of Florida (Boca Raton, West Palm Beach, Fort Lauderdale, Miami and Tallahassee), but also in New York and Caracas, Venezuela. In October 2009 Rothstein's conjured financial kingdom collapsed as soon as numerous investors started to insistent on withdrawing millions of dollars invested in fake employment-discrimination and sexual-harassment settlements. The firm became non-existent after he filed for bankruptcy. From his now non-operational law firm in Fort Lauderdale, Florida he sold concocted legal settlements to rich financiers. He was also disbarred as a lawyer for his wrongdoings following his massive $1.2 billion Ponzi scheme that resulted in him being sentenced on June 9, 2010 to 50 years in prison. The shamed disbarred lawyer is currently serving a prison sentence for both conspiracy and racketeering under the federal witness protection program. In an attempt to lesson his sentence Rothstein helped the FBI in a sting operation that led to the arrest of the Miami Beach wine merchant Roberto Settineri, who was suspected by USA and Italian authorities of being an intermediary between the Gambino crime family and the Santa Maria de Gesu clan from Cosa Mostra or the Sicilian Mafia. The scandal about his scam investments stunned numerous communities in South Florida. Rothstein lost about $360 million belonging to some 320 clients and investors. (*http://www.miamiherald.com/2011/12/12/2543724/imprisoned-ponzi-schemer-scott.html#ixzz1gRMsuWCE*).

Nevin Shapiro. Another mastermind of a $930 million Ponzi scheme was Nevin Shapiro (1969—), who was a former booster for the Miami University (Miami) *(http://www.miami.edu/)* athletic department. After his fraudulent activity was discovered Shapiro ended up in prison with a twenty year sentence. In addition he was instructed to reimburse his prey more than $82 million. Between 2002 and 2010 Shapiro provided considerable benefits to Miami football players in violation of NCAA rules (*http://www.miamiherald.com/2011/12/12/2543724/ imprisoned-ponzi-schemer-scott.html#ixzz1gRMLlJkN*)

Raj Rajaratnam. Another scoundrel in the finance business is the former hedge fund titan Raj Rajaratnam, who was born in Sri Lanka. He founded the Galeon hedge fund in 1997 and acquired a fortune from persuading corporate tipsters to provide him illegally with confidential information about companies so that he could have an upper hand in trading stocks. Like many others on Wall Street his undoing was greed and corruption. He was convicted of insider trading and sentenced on July 29, 2011 to more than 11 years in prison [27, 39]. This was the longest jail term for insider trading. To convict him the prosecutors recorded his wheeling and dealing with corrupt executives and consultants on wiretaps and other methods used against drug lords, mobsters, and other big fish in the crime world. In sentencing him the USA district Judge Jed S. Rakoff (1943—) stated that the crime of Rajaratnam demanded a fine that would deprive him of "a material part of his fortune". He was ordered to pay the highest recorded financial penalty in civil case brought by the SEC ($92.8 million). In addition he was ordered to pay a $10 million fine and forfeit $53.8 million in profits. The monetary sanctions imposed on Rajaratnam were enormous from the standpoint of the average citizen, but for a billionaire it was a slap on the wrist. The cost of SEC civil case together with and criminal cases cost Rajaratnam more than $156.6 million [9]. Aside from his fall from grace Rajaratnam suffers from diabetes type 2 and had a severe stroke in February 2007. His physicians hope to prolong his life with a kidney transplant, but this desire raises the important ethical question of whether special consent should be obtained from the donor or the donor's next of kin. Would an average citizen be prepared to give a kidney to this greedy bastard? Also after the drainage of his financial resources in his lawsuit would Rajaratnam have sufficient money left over to buy a kidney at a black market price?

Insurance Industry

The sale of annuities brings a considerable amount of money into the insurance industry. In 2012 annuities totaled $24.8 billion nationwide. With annuities contracts between an insurance company and the purchaser provide money to the buyer at predetermined time intervals. The cost is a major worry for families of elderly investors, but also for state and federal regulators. Regrettably certain unscrupulous insurance agents take advantage of the elderly and investors may be prevented from gaining access to their personal assets for decades. On July 28, 2008 Thomas Goldsmith a staff writer of the *News and Observer* reported an outlandish insurance scam touching the elderly [25]. He drew attention to the swindle of Rosalie Whittington, a 80 year old federal retiree living in Raleigh North Carolina. An insurance agent sold this resident with dementia more than $200, 000 worth of annuities so that she would receive a monthly income.

But she would only start receiving installments when she was 105 years old. During late 2006 and early 2007 the same agent sold her a still bigger policy that would begin distributing monthly payments when she reached 90 years of age. How could anyone sell an annuity to a person of her age? This complicated product with very high fees is as a rule only the proper product for very few investors. Whittington accumulated more than $1 million from 1960 and put the assets into complex policies referred to as equity indexed annuities. This type of investment is known to regulators as one of the most all-encompassing products implicated in senior citizen investment fraud. Grievances by relatives of Whittington brought these most unusual annuity sales to the attention of the North Carolina State department of insurance. The department took no action against Edward Reinheimer, the insurance agent who sold the policies, but the two insurance companies returned Whittington's investments within weeks knowing that she was not competent to make sensible financial choices, when the policies were purchased. Early withdrawals could have taken place, but would have resulted in extra fees and tax consequences. The department of insurance does not take the bull by the horns against insurance agents except when they disobey a law or statute. Between 2005 and 2008, more than 400 North Carolinians have protested to the state with reference to the sales and practices of selling annuities. In early 2007 $125 million was obtained in refunds for thousands of senior citizens in Minnesota after law suits were settled against the insurance companies Allianz Life and American Equity for using deceptive sales practices in selling equity—index annuities.

Fraudulent Dentistry

Most dentists are honest and ethical and give their patients the appropriate compassionate treatment expected of their profession, but some dentists participate in swindles and deceive their patients. The non-profit coalition against insurance fraud (CAIF) (*www.insurancefraud.org*) is an organization dedicated to fighting insurance fraud and it was founded in 1993. A small but alarming number of dentists are dishonest according to complaints made by some dental patients (*http://www.avvo.com/legal-guides/ugc/dental-scams-and-dentist-fraud-rip-offs-and-dental-malpractice-in-california*). Some appear to participate in fraudulent, neglectful and unscrupulous therapy. According to *R. Sebastian Gibson* the types of dishonest behavior that have been reported by dentists include providing worthless or unnecessary treatment, over-charging for routine services such as teeth cleaning, inflating estimates for dental work, pushing dental insurance plans that may or may not be valid, operating dental clinics without the proper precautions for safety and hygiene, operating without sufficient or proper training, questionable billing practices charging for services not performed. Some dentists simply provide negligent treatment, but others sell phony dental plans, or advocate disfiguring surgical procedures that are not shielded by insurance policies. Some inform patients that they suffer from diseased gums and need costly mouth rinses and antibiotics, and dental cavities that need fillings or fillings need to be removed and substituted with pricey dental posts, teeth extractions, deep cleaning and sometimes gum treatment. Because of a desire to generate more income some dentists resort to disgraceful unnecessary expensive procedures, especially in the care of children's teeth, such as multiple root canals.

Legal Profession

The law profession also has a reputation for having more than its fair share of dishonest unethical lawyers. Many have been sent to prison for breaking the law despite their professional knowledge of the subject. Joseph P. Collins, for example, was sentenced to seven years in prison after he was convicted in July 2009 for aiding owners and executives of the extinct Refco Inc in stealing more than $2 billion by concealing its financial woes. After the jury found Collins guilty of conspiracy, security and wire fraud he was sentenced to 7 years in prison, but his lawyer, William J. Schwarz, appealed the conviction [8] and on January 9, 2012 the conviction was overruled, because the trial judge had improper discussions with a juror [32].

Wine Traders

Wealthy wine connoisseurs have always cherished precious pricey wines, but oenophiles have doubts about whether the wine label reflects the contents of the bottle. Research has shown that the label on a bottle of wine strongly influences how the taste is judged [47]. For example in an experiment, the same wine was given to two groups of individuals. One group received wine from a bottle identified from a factious North Dakota winery and the other received the wine with a bogus Californian label. The wine with the North Dakota label rated much lower that the California wine. Counterfeiters enable the taste of cheap wines to appear much better than they are by putting desirable forged labels on the bottles and selling them at a price appropriate for the real wine. This way the placebo effect of wine labels enables more profit to be made by selling inexpensive wines. The incentive for fraudulent wines that are supposedly blue-chip has perhaps increased as the sales of fine wines escalates when the wealthy can afford to cast bids at Sotheby for wines that mere mortals can only dream about.

In 1985 Hardy Rodenstock, a German music promoter and wine merchant came into the possession of several bottles of 1787 Château Lafite wine declared to have formerly belonged to Thomas Jefferson (1743-1826) the third President of the USA. They had been found in Paris in a walled-up underground room where Jefferson lived when he was the USA ambassador to France (1785-1789). Rodenstock was not prepared to divulge the name of the person who sold him the bottles which were all engraved with the letters Th.J. Other details were distrustfully blurred. Rodenstock had no difficulty getting buyers for the wines. At a Christie auction in London in 1985, Malcolm Stevenson Forbes (1919—1990) paid $156, 450 and set a record price for a single bottle of wine. In 1988 the shyster tricked billionaire William Ingraham Koch (1943—) into purchasing four bottles for half a million dollars. Koch tried to validate their origin in 2005 and sadly discovered that Jefferson, a meticulous record keeper, had made no documentation about these bottles. To pursue the matter further Koch employed investigators to appraise the wines and they concluded that the Jefferson initials on the bottles were forged [30, 45]. This provoked Koch to take legal action against Rodenstock for fraud. The wine world was stunned because Rodenstock was one of its most important participants in the wine industry. Allegations against Rodenstock were tried in a German court which found him guilty in 1992 of knowingly selling adulterated wine. An appeal was settled out of court.

Another law suit reached a federal court in New York on January 31, 2007, while Rodenstock was seemingly in Europe and the court was asked to dismiss the allegation by Koch [24]. On August 14, 2007 a default judgment was recommended against Rodenstock by the judge, but Rodenstock turned out to be an alias as his birth name was Meinhard Goerke. The story about how this swindler hoodwinked wine experts was published as a book entitled *The Billionaire's Vinegar* [46] which precipitated a lawsuit from Mr. Broadbent a former head of Christie's wine department against the publisher, but not the author. Broadbent claimed that the book falsely depicted him as complicit in the crime. The lawsuit was settled out of court because of the UK's plaintiff-friendly libel suits [3]. A movie based on the book by Wallace has an expected release date in 2014 [36].

Journalism

Ideally journalism recounts the news accurately and impartially. Those who participate in plagiarism, fabrication and the omission of facts violate both the law and the rules of ethics. Image manipulations and other dishonesty in reporting stories in newspapers and the television medium chip away at the public's confidence. Several steps in the editing of articles in journalism offer an opportunity to catch flaws in stories before their release, but senior editors sometimes fail to detect bias, potentially libelous statements or fabrication within stories created by reporters. Appropriate checks and balances are occasionally neglected because of the need to rapidly release a crucial new story to the radio, television or press. The general public relies on the credibility of reputable newspapers and magazines, but sometimes unscrupulous members of their staff go overboard and do not report the truth or over interpret their conclusion.

As in the review of scientific journal articles it is virtually unfeasible for an editor to unearth all first-rate lies. The extreme desire to become a Pulitzer Prize winner has sometimes provoked journalists to report untrue stories. This coveted Prize was established in 1917 in the will of the Hungarian born American publisher Joseph Pulitzer (1847-1911). Occasionally Pulitzer Prizes have been erroneously been awarded for inaccurate stories released in newspapers and have commonly been withdrawn with an expression of regret. The journalists have usually been withdrawn from their positions, and often the misrepresented businesses or individuals sue for considerable amounts of money.

Walter Duranty. The British journalist Walter Duranty (1884-1957), who served as the Moscow bureau chief of *The New York Times* (1922-1936), received the coveted Pulitzer Prize in 1932 for untrue stories written in 1931 denying widespread famine in the Soviet Union and especially the Ukraine mass starvation *(http://en.wikipedia.org/wiki/Walter_Duranty)*

Brian Walski. On April 1 2003 Brian Walski lost his job at the *Los Angeles Times* for dishonestly joining two digital images obtained at some stage in the USA led military coalition in Iraq known as *Operation Iraqi Freedom (en.wikipedia.org/wiki/Brian_Walski)*. Walski maintained that he was only attempting to establish a more persuasive photograph. The flaw in the image was discovered when some individuals were found twice in the photograph.

Chris Cecil. In June 2005 Chris Cecil, a 28-year-old associate managing editor of the *Daily Tribune News* in Cartersville, Georgia was laid off after this daily newspaper with circulation of eight thousand discovered that he had plagiarized at least 8 columns written by the Pulitzer Prize winning columnist Leonard Pitts, Jr (1957—). The dishonesty was exposed to Pitts by a reader of the Carterville newspaper [48].

Evan Hajj. The freelance photographer Evan Hajj had 920 photographs pulled by Reuters on the 2006 Israeli Lebanon war. A number of high profile photographs had been markedly manipulated using the software graphics editing computer program published by Adobe Photoshop to embellish the destruction by Israeli violence a.

Allan Detrich. Allan Detrich, a 1998 Pulitzer Prize finalist, was the photographer for the newspaper *The Blade*. He resigned from *The Blade* shortly after the newspaper ran a front page article with a doctored photograph by him depicting the Bluffton University (Bluffton) *(http://www.bluffton.edu/)* baseball team praying for five teammates who had passed away in a bus accident in Atlanta, Georgia. Detrich removed the legs of photographed subject digitally using the software package Photos. Detrich was adamant that he had submitted the image erroneously from his private collection of photographs. The fact that the image had been altered was initially realized when photographers from competitive newspapers discovered that photographs from almost the same position contained a pair of legs behind a banner hanging on the outfield fence that the Detrich photograph did not show. Dietch had polished the image aesthetically without distorting the information relevant to the news story [43].

Jayson Blair. While attending the University of Maryland (UMD)*(http:// www.umd.edu/)* Jason Blair became editor-in-chief of The Diamondback the university's student newspaper. While holding this position thirty members of the Diamondback's staff recognized that he had significant flaws in his character and signed a letter drawing attention to serious mistakes published in that newspaper. Subsequently in 1999 he was hired by The New York Times despite considerable faults in The Diamondback under his watch. On April 18 2003 Macarena Hernandez, who had previously been an intern at *The New York Times* wrote an article for the *San Antonio Express-News*. A few days later the senior editor of that newspaper noticed a remarkably close resemblance between the article authored by Hernandez for the *San Antonio Express-News* and a column prepared in *The New York Times* by Blair. On April 28, 2003, *Jim Roberts*, the national editor of *The New York Times* confronted Blair about the striking similarities between what he had recently written and the story penned in the *San Antonio Express-News* by Hernandez. When Blair's plagiarism to *The New York Times* was recognized his journalistic misdeeds were reported in an extraordinary 7, 239-word front-page story on May 11, 2003 [6]. For *The New York Times* this embarrassing scandal was "a low point in the 152-year history of the newspaper." *The New York Times* remained unfaltering in its determination to produce a newspaper that adheres to the highest standards of integrity in journalism. The Blair scandal not only had an immediate effect on the reputation of the newspaper and Blair's career, but it caused the resignation of the executive editor (Howell Raines) and managing editor (Gerald Boyd) of *The New York Times* and the publisher (John Schulzberger) accepted them in the best interest of *The New York Times* (*http:// msnbc.com/news/922400/asp*).

On May 1 2003 Blair tendered his resignation from *The New York Times* after that newspaper discovered recurrent exploits of *plagiarism* and fabrication in reports authored by him. The final straw that broke the camel's back were charges of plagiarism related to a story about a family of an American soldier in Iraq (*www. pbs.org/newshour/media/media_ethics/casestudy_blair.php*).

Consumer Reports

When consumers desire information on different products they often seek the opinion of companies that are believed to provide objective opinions, but in February 2007 *Consumer Reports* documented that just two of the tested infant car seats were safe in high speed automobile accidents for infants. The publication maintained that other versions of the car seats occasionally broke away from their bases or undersized mannequins went flying out of the seats, but after the National

Highway Traffic Safety Administration (NHTSA) appraised the other infant car seats they functioned as expected in collisions at 38.5 miles per hour (mph). Later it was established that a business recruited by *Consumer Reports* to evaluate the car seats performed the tests at 70 mph and not at the standard speed of 38 to 40 mph. On January 18 2008 *Consumer Reports* partially withdrew the flawed tests by pointing out that the seats were going to be retested.

Medicare Fraud

Criminal healthcare fraud and abuses are common and involve fraudulent charges made to health insurance companies as well as *Medicare* in an attempt to gather *Medicare* health care compensation by deception [15]. Numerous varieties of this *Medicare* fraud exist and they siphon money unlawfully from the *Medicare* program, which was established in 1965. The amounts of money derived from *Medicare* fraud are vast and over the years annual losses gradually totaled many billions of dollars. By the end of 1998 the USA attorney's office obtained 326 convictions and 107 civil cases resulted in the recovery of $480 million from damages or settlement fees. During 2010 improper payments were estimated to be about $47.3 billion according to the Office of Management and Budget (OMB). Initially the USA government took wide-ranging action against healthcare fraud that came to its attention, but the authorities had difficulty dealing with Medicare fraud. Eventually rewards for whistleblowers and the sanctions in fines and jail sentences of those prosecuted controlled the crime to some extent. Whistleblowers had a significant financial incentive to turn in culprits guilty of healthcare fraud and abuse. Other cases were reported by the *Medicare* contractor for different districts. Bills submitted by medical doctors that fall outside the usual expected range signified a red flag worthy of investigation. In the June 1999 copy of the *EyeNet Magazine* Susan E. Davis drew attention to some examples of individuals found guilty of fraudulent charges in the healthcare system [15]. The cases which she summarized involved ophthalmologists, psychiatrists, a dentist, a podiatrist and a variety of other medical specialists from different states. After billing *Medicare* for $43.6 million from 1988 to 1992, an ophthalmologist in San Diego California was convicted on more than two hundred counts of falsifying records, up coding and performing uncalled for surgical procedures. An appeal eliminated some charges, but he ended up with a $5 million fine and 5 years in prison. In the biggest healthcare fraud payment against a solitary doctor in the USA another ophthalmologist in Massachusetts settled a $750, 000 civil claim for submitting bogus claims according to wire reports. In a guilty plea to three criminal charges an ophthalmologist in Connecticut settled a $700, 000 lawsuit. After this eye doctor shone lights into the eyes of patients he informed them that he performed laser surgery and *Medicare* was billed $640, 000 for trabeculoplasties, iridotomies, iridectomies, and photocoagulations. Fraudulent bills from this ophthalmologist went to Medicare for $640, 000. A psychiatrist in Florida billed *Medicare* and *Medicaid* (a special health program for low income families funded jointly by the states and federal government) for $4.5 million under the pretense of psychotherapy services by showing movies to residents in a nursing home and teaching them to play cards and how to make clothespin dolls and playing cards. He ended up with 6 years in a federal prison and had to reimburse more than $1.6 million for the fraudulent federal healthcare charges. Another psychiatrist, this time from California, spent 18 months in prison for billing *Medicare* more than $100, 000 for treatment sessions to nursing home patients that did not take place. A Californian podiatrist billed *Medicare* for patients that he treated in nursing homes for surgical procedures that he carried out. In reality he cut toenails and his lucrative conspiracy might have gone undetected, but his greed enticed him to re-bill more than 1, 750 surgical procedures over a 6 month period, which provoked the curiosity of the USA government. The foot doctor received 18 months in jail and was required to pay more than $350, 000 to defray the government's civil case. A pediatric dentist near Sacramento, California pleaded guilty to submitting claims to the government for procedures that were neither indicated nor appropriate. His penalty was one year in prison and a fine of $6, 000 in the civil case. It cost him $500, 000 to resolve a related civil claim. A doctor in Beverly Hills had the audacity to bill *Medicare* $216, 000 for treating persons who were already deceased, incarcerated, or where living too far away for him to treat. This upper class doctor ended up shelling out more than $1.5 million in repayment. In Seattle Washington a doctor and his wife provided acupuncture to patients and generated $375, 000 from *Medicare*, by billing for physical therapy and in a few instances by submitting a bill to the government without providing services. After the couple's deceitful scheme was discovered they were required to reimburse $520, 000 in restitution and damages to the government and an additional $117, 000 to insurance companies. A hospital for the elderly in Maryland was required to pay the government $800, 000 after it charged the room and board fees to a fund code for supplies, laboratory work, and other services after the hospital had its room-and-board code turned down. A con artist in Nebraska with only a high school diploma asserted that he was a surgeon, a biochemist, and a specialist in molecular medicine. He claimed that he could cure cancer and the acquired immunodeficiency syndrome (AIDS) and to establish his diagnoses he analyzed samples of hair and fingernails. From his diagnostic tests, treatments, and selling of merchandise he made about $1 million. During an investigation of him he was provided with a sample of material from a guinea pig and declared after his analysis that it was from a patient with allergies and organ troubles.

References

1. Albergotti R. Another guilty plea in Maldoff probe. The Wall Street Journal. 2012 Friday November 9.
2. Anonymous. Economist. 2008 April 26:82.
3. Asmov E. 'Billionaire's Vinegar' lawsuit is settled. The New York Times. 2009 October 14.
4. Austen I. Those wronged by Armstrong see little right in interview. The New York Times. 2013 January 18.
5. Barrett D. Rangel pleads for mercy. The Wall Street Journal. 2000 November 19.
6. Barry D, Barstown D, Glater JD, et al. CORRECTING THE RECORD; Times Reporter who resigned leaves long trail of deception. The New York Times. 2003 May 11.
7. Beck R. Cheating scandal sends message: law is needed. News and Observer 2007 May 6.
8. Bray C. Lawyer gets 7-year term in Refco case. The Wall Street Journal. 2010 January 15.
9. Bray C. Rajaratnam is ordered to pay record SEC fine The Wall Street Journal. 2011 November 9.
10. Carr D. How Oprahness trumped truthiness. The New York Times. 2006 January 30.
11. Carter A. The sanctions againt the heels include a one-year postseason ban. News and Observer. 2012 March 13.
12. Colvin G. Bernie Ebbers' foolish faith. Fortune. 2002 November 25:146.
13. Colvin G. Wonder women of whistleblowing is it significant that the prominent heroes to emerge from the two great business scandals of recent years were women? Fortune. 2002 August 12.
14. Couzin J, For many scientists, the Madoff scandal suddenly hits home. Science, 2009. 323(5910): 25.
15. Davis SE. Meet the prosecutor: the government's take on fraud and abuse. EyeNet Magazine. 1999 June:55-57.
16. Edes G. Clemens implicated in steroid scandal by trainer. The Boston Globe. 2007 December 14.
17. Eichenwald K, Conspiracy of Fools. A True Story. 2005, New York: Random House Digital, Inc. pp.742.
18. Elstrom P. How to hide $3.8 billion in expenses. Business Week. 2002 8 July.
19. Farrell G. 2 executives admit guilt in fraud at hedge fund. USA Today. 2005 29 September.
20. Farrell G, Empty promises in hedge fund fraud. USA Today, 2005: 3B.
21. Finder A. 34 Duke business students face discipline for cheating. The New York Times. 2007 May 7.

22. Frey JC, A Million Little Pieces. 2003, New York: Random House Inc.
23. Galbraith PW. How to rig an election: a former top U.N. official recounts the fraud he witnessed during Afghanistan's presidential vote and why he lost his job for speaking up. Time. 2009 October 19:40-43.
24. Goldberg HG. Court asked to dismiss "Jefferson' wine case. Decanter 2007 January 31.
25. Goldsmith T. Seniors' investments leave little time to cash in. News and Observer. 2008 July 28.
26. Grillo T. Madoff wins prison yard scuffle over stock market. Boston Herald 2009 October 13.
27. Hays TN, L. Exhedge fund titan guilty of insider trading. News and Observer. 2011 May 12.
28. Johnson K. Victim 1 USA Today. 2011 11-13 November.
29. Kane D, Blythe A, and Tysiac K. Problems surfaced with tweet, but more had been brewing. News and Observer. 2011 October 23.
30. Keefe PR. The Jefferson bottles. The New Yorker. 2007 3 September.
31. Koelsch A. Purchasing essays online not worth the effort, prof finds The Chronicle. 2010 Tuesday 5 October.
32. Lattman P. Conviction of former Refco lawyer is overturned. The New York Times. 2012 January 9.
33. Maher K. Paterno family fires back at Freeh report. The Wall Street Journal. 2013 Monday 11 February.
34. McCabe D and Katz D, Curbing cheating. Tech Directions, 2009. 69(3): 32-34.
35. Mitchell GJ, Report to the commissioner of baseball of an independent investigation into the illegal use of steroids and other performance enhancing substances by players in major league baseball. 2007: Collins.
36. Neyfakh L. Oenophile row: Brooklyn stoopmates race to produce wine-fraud flick. The New York Observer. 2008 13 February
37. Orey M. Career Journal: WorldCom-inspired 'whistle-blower' law has weaknesses. The Wall Street Journal. 2002 October 1.
38. Pelliam S. Questioning the books: WorldCom memos suggest plan to bury financial misstatements. The Wall Street Journal 2002 July 9.
39. Pulliam S and Chad B. Trader draws record sentence. Rajaratnam slapped with 11-Year prison term for orchestrating insider scheme. The Wall Street Journal. 2011 Friday October 14.
40. Rich M. Educator aided others at cheating, U.S. charges. The New York Times. November 26, 2012.
41. Romar EJ and Calkins M, WorldCom case study update 2006, 2006: Santa Clara University, The Jesuit University in Silicon Valley.
42. Rotvik DM, In his Image: The cloning of a Man. 1978, Philadephia and New York: J.B. Lipinncott.

43. Royhab R. A basic rule: Newspaper photos must tell the truth. The Blade. 2013 Friday August 9.
44. Sorkin AR. Former director of Tyco fined/board member pleads guilty to fraud, settles. The New York Times. 2002 December 18.
45. Steinberger M. Excuse me, waiter, there's fake wine in my glass. Slate Magazine. 2007 September 12.
46. Wallace B, The Billionaire's Vinegar: The Mystery of the World's Most Expensive Bottle of Wine. 2008, New York City: Random House.
47. Wansink B, Mindless Eating: Why We Eat More Than We Think. 2005, New York City: Bantam Dell.
48. Weber HR. Columnist fired for plagiarizing Pulitzer winner's work. Florida Times Union. 2005 Friday 3 June 3.
49. WireStaff C. Highlights of the Armstrong report. 2012 Wednesday October 10.

CHAPTER 21

Detection of Scientific Misconduct

Research misconduct can be detected in many different ways, but many have pointed out that fabrications by sophisticated cheaters are almost impossible to detect [41].

Whistle-blowers

Potential whistle-blowers who become aware of the law suit by Dorn against the whistleblowers who condemned Ronald Dorn for scientific misconduct should be discouraged from making allegations of misconduct in public. Litigation of this nature may dampen debate on controversial research.

The collegiality of the faculty discourages individuals from revealing negative information about a colleague or student. Steps should be taken to prevent retaliation against the whistle blower if the accusations are made in good faith regardless of the outcome of the investigation. Snitching is not a part of many cultures, including our own, as every child learns at school. Nobody likes people who tattletale and many of those who have turned in cases of scientific misconduct as whistle-blower. Researchers who draw attention to scientific misconduct, unethical research and other inappropriate behavior by their colleagues are an important way in which these violations of scientific standards get exposed. Individuals working close to the perpetrator commonly become distrustful and collect the evidence to expose the person. Individuals who turn people in for dishonest research either work in the same laboratory or they are from other institutions. Whistle-blowers with inside information are the commonest way in which scientific misconduct is exposed, but their accusations need to be taken with an open mind and dealt with confidentially.

Some whistle-blowers are naïve and do not appreciate the difference between honest unintentional honest errors and scientific misconduct, moreover some have an ax to grind. LaFollette [21] distinguishes between the whistleblower, which brings attention to the misconduct from the inside and the nemesis, which does it from the outside.

Kyriakie Sarafoglou. Dr. Kyriakie Sarafoglou blew the whistle on an inappropriate use of a five-year $23 million NIH grant by Cornell University (Cornell) *(http://www.cornell.edu/)*. The grant was to perform research on children's diseases and Sarafoglou discovered that several of the projects just existed on paper and many of the participants did not have the relevant diseases and much of the money was used to treat adults. Although it is it illegal for universities to retaliate against whistleblowers that make allegations of misconduct in good faith it is not unusual for them to take steps to deal with the accuser and to insinuate that whistle-blower are psychiatrically unstable. For her role as a whistle blower Sarafoglou maintained that Cornell retaliated against her by ostracizing her, recommending that she obtain a psychiatric examination and not renewing her tenure track faculty appointment when it expired.

Fear of litigation prevents some individuals from blowing the whistle, such as John Raven. Others have discovered, despite the law that is supposed to protect them, that they face many repercussions. At one time scientists were apprehensive about drawing attention to cheating in science, but an ascent in social consciousness has hopefully changed this attitude. Indeed the *National Academy of Sciences (NAS)* in the USA considers it an obligation of researchers to blow the whistle when such behavior is recognized [34]. The Glazer husband-wife team interviewed 64 whistleblowers over 6 years and considers this goal idealist but unrealistic [13].

The turning in of a colleague is extremely stressful and frequently has adverse psychological effects on the whistle-blower and not uncommonly leads marriages into divorce. To assist whistleblowers deal with the stress of blowing the whistle on a colleague for scientific misconduct and in dealing with the hurdles over which they need to jump in seeing the matter through Robert Sprague (the whistle-blower in the case of Robert Breuning and Carolyn Phinney have spear headed a group called *Whistleblowers for Integrity in Science and Education* which counsels a network of informed lawyers and provides access to misconduct cases *(http:www.sciencemag.org/content/277/5332/16112, short)*.

Individuals who have studied whistleblowers consider it inadvisable for potential whistleblowers to accuse someone of scientific misconduct until the relevant evidence has been fully documented and unintentional errors have been excluded. Whistleblowers should be extremely careful not to rush to judgment and attribute to malice what is in reality incompetence or unintentional errors. Moreover the opinion of other experienced individuals should be sought and preferably also the view of past whistleblowers [18] (*http://www.the—scientist. com/images.yr1995/may.prof_950515.html*).

It is particularly risky for a junior inexperienced researcher to blow the whistle, especially since a lot of valuable time will become diverted from meaningful research endeavors [25]. Whistle-blowers now frequently expose fraudulent research by making allegations against individuals whom they suspect as perpetrators. Allegations of scientific misconduct are usually made by a whistleblower who is generally someone very familiar with the research, such as graduate students, postdoctoral fellows, research technicians, and other members of the research staff. Whistleblowers who have made names for themselves include Margot O'Toole who accused Thereza Imanishi-Kari of scientific errors, even though she was not guilty of scientific misconduct Robert Sprague, Dr.Earhard Hildt and Roger Bioscopy. Despite being the whistle blower in the Herrmann-Brach scandal Earhard remained positive about science and in 1999 he was in charge of research team at the University of Munich. Cheating has been exposed from the inside by laboratory assistants (Straus) or scientific competitors as in the research misconduct of William Summerlin and John Long.

The careers of whistleblowers and arch-rivals rarely process smoothly. They are often depicted as troublemakers or disgruntled students, technicians and other employees even if they are not. They become ostracized by their colleagues and some have faced reprisals from the institution in which they work. They have had to pay the consequences in being overlooked for promotions, appointments and other things because of discriminatory reasons that are not revealed. Who would want to hire someone who had turned a prior employee into the federal authorities? For example Daniel Freedman had to make a career change after bringing suit against his previous employer Dr. van Gorp at Cornell University (Cornell).

In reality most whistleblowers seem to believe in honesty, justice morality and abiding by the rules [13]. They are idealistic and believe in the system [49].

The exposure of scientific misconduct by a nemesis leads to criticism of the person and puts his or her career at risk as if the bringing to light of the bad deeds is a bigger problem than the misconduct itself. LaFollette wonders why scientists often censure the nemesis when scientists desire to know the truth. According to her academic institutions seem to value stability and equilibrium over honesty, despite giving lip service to the contrary. In human behavior people generally act in their own perceived best interests and perhaps envision their best interests with the accused rather than with the accuser.

When whistleblowers have reported allegations of misconduct there has frequently not been an open collegial atmosphere in the laboratory or even in the department and at least one of the groups often has an ax to grind or a chip on the shoulder.

Detection of Scientific Misconduct During Peer Review

Professional journals are the gatekeepers to the scientific literature and the detection of fraudulent research misconduct prior to publication is desirable is sometimes detected during the peer-review process, but it is unreasonable to expect all intentional and unintentional mistakes in papers to be detected during their review by journals. Journal reviewers and editors can sometimes detect swindles, but they cannot be expected to detect fraudulent publications. Critics have demanded that journal editors and reviewers take more responsibility for not letting peer reviewed papers slip through the net, but it is impractical to imagine that a journal will invariably detect a fraudulent manuscript. Under exceptional circumstances it does as it did in the exposure of Amitov Haija, a MD/PhD student of Francis Collins.

When fraudulent papers are submitted for publication they commonly get published and remain buried in the scientific literature unless someone later, such as a whistle blower or research colleagues exposes the misconduct after the research is reported.

Peer review can rarely detect research misconduct because reviewers of manuscripts assume that the authors are telling the truth. While the peer review process rarely detects research misconduct it sometimes does as occurred with Amitav Hajra an MD/PhD student of Francis Collins. A meticulous reviewer of one of his manuscripts observed something strange about proteins created by rearranged genes on chromosome 16 following cell transformation resulting from a fusion protein coded by genes on an abnormal chromosome found in acute myeloid leukemia. Two lanes in a Western blot labeled as containing distinct proteins shared the same background artifacts. After finding fraudulent material in this paper Collins uncovered additional fabricated data in "five published research papers, two published review articles, in one submitted but unpublished paper, in the student's doctoral dissertation, and in his submission to the GenBank database" [3].

A figure illustrated a peculiar repetition of "little telltale glitches" and lanes for two different proteins had the same visual flaws. When Collins was informed about the peculiarity discovered by the reviewer for *Oncogene* he was taken by surprise as he had not detected it as a coauthor before the submission of the paper. A painstaking reviewer drew the attention of John Jenkins, editor of *Oncogene*, to something strange in the submitted paper documenting unusual proteins generated by rearranged genes on chromosome 16 in a sequence of articles over a two year period. The paper submitted to *Oncogene* contained an assortment of proteins in columns from Western blot assays. The reviewer surprisingly noted the peculiar recurrence of tiny artifacts that are generally ignored in two separate proteins. It was obvious that a cut-and-paste job had been performed to generate the figure. After Collins received a telephone call from Jenkins, he and a researcher in his laboratory, P. Paul Liu, devoted two weeks going through the freezer, sequencing clones attempting to make sure that all aspects of the research that had been performed by their junior colleague was valid. But this is not what they found. Instead they discovered the graduate student was unable to produce a control cell line for tests to determine whether the gene on chromosome 16 would initiate cancer in mouse cells. Being discouraged, Hajra supposedly created fake data for the control. Similar allegedly fabricated data was created in a different experiment in which another gene was transfected into mouse cells. Sections of human DNA sequence were extrapolated from mouse DNA. Collins pointed out that it was astounding that someone sufficiently knowledgeable to concoct such a fraud did not comprehend it would be discovered. With all evidence in hand Collins took a flight to Ann Arbor and handed over the assembled material documenting scientific misconduct to the young graduate student. Three-and— half hours later according to Collins the student confessed, but was not apparently regretful, although seemly glad that he had been caught. Collins credits the student for sticking to his confusion and putting a description of the fraudulent behavior to pen in a three page document. Collins considered it important to advise investigators in the relevant scientific field as rapidly as possible about inaccurate information in the literature. In his letter Collins retracted two papers entirely and portions of third one [1, 14-17]. If research is a product of fraudulent research subsequent studies will be unable to replicate it and thus expose its lack of validity. The likelihood of fraudulent research being detected correlates directly with the importance of the discovery being reported and the number of researchers investigating the topic.

Editors and manuscript reviewers have sometimes had doubts about the authenticity of the research in papers received for their evaluation. For example, fabrication or falsification can be suspected in studies documenting unreasonable randomized trials [46, 53]. RK Chandra a Canadian researcher published numerous randomized trials alone in major journals over a decade, and such studies [7] were seriously questioned because it is extremely difficult if not impossible for a single researcher to conduct such studies at a single institution. Memorial University in Newfoundland, which employed Chandra, was asked to investigate his credibility, but the university found no evidence of research misconduct in an investigation with a disputed thoroughness. Eventually a paper by him in *Nutrition* was retracted [24, 45]. Rarn B. Singh an Indian researcher published dozens of fraudulent clinical studies in multiple journals and the papers were well cited. The studies were suspicious, but research misconduct was difficult to substantiate [54].

A long-time disputed genetics paper on the detection of gene interactions was published in February 2012 in the high impact journal *Nature* by investigators at the reputable Johns Hopkins hospital. The paper was fully retracted by the journal on November 27, 2013 after the retraction was signed by all twelve authors except the lead author Yu-yi Lin, who apparently committed suicide. The retraction took place because the results could be replicated. The accuracy of the paper had been questioned in August 2012 by Daniel Yuan an employee working in the same laboratory as the researchers. Shortly after Yuan raised doubts about the authenticity of the paper Lin was found dead in a laboratory in Taiwan with pucture marks in his arm skin and empty vials of sedatives and muscle relaxants at the site. The whistleblower had been fired by JHU in December 2011 after working there for a decade.

Detection of Misconduct in Papers Prior to Journal Submission

As pointed out by Bloch and Relman [4] fraudulent research is best uncovered by coworkers and supervisors in the laboratory before papers are submitted to journals. The expected standard of scientific publications has gradually increased over time as demanded by editors and reviewers will hopefully result in an improved detection of mistakes prior to publication. The flaws in the infamous paper in *Cell* that precipitated the Baltimore scandal could have been detected by the reviewers and editors of *Cell* in the unlikely event that these individuals had been familiar with the scientific details and had been able to evaluate objectively the manuscript knowing that one of the authors was a Nobel laureate. When the research of the culprit focuses on a relatively unpopular topic and has unpretentious aspirations, such as a goal to only obtain university tenure, the likelihood of discovery is probably less than with misconduct in an extremely trendy area of research. When the results of the cheater fall within the presently acknowledged dogma the work may go unchallenged.

Simultaneous Discoveries

Two scientists rarely reach similar or identical conclusions simultaneously and independently, but by pursuing different research tracks based on existing knowledge similar breakthroughs may be made. As the axiomatic expression states "all roads lead to Rome (Omnes viae Romam ducunt)" so under extraordinary circumstances it is not surprising that this sometimes happens as it did with Alfred Russel Wallace (1823-1913) the British naturalist and Charles Robert Darwin (1809-1882), who separately proposed a theory of evolution by natural selection. Likewise the 1975 Nobel laureates in physiology or medicine David Baltimore (*http://www.nobelprize.org/nobel_prizes/medicine/laureates/1975/baltimore-autobio.html#*) and Howard Temin (*http://www.nobelprize.org/nobel_prizes/medicine/laureates/1975/temin-lecture.html*) discovered reverse transcription of DNA and RNA template independently at the same time. Because scientists occasionally discover things at the same time as pointed out in chapter 2 it is extremely important that collaborators and competitors are not immediately accused of deliberate plagiarism when they report somewhat similar finding.

Without convincing evidence of plagiarism they should be given the benefit of the uncertainty.

Plagiarism Detection

Plagiarism can be detected in many ways.

Plagiarism Detection by Original Author

In the early days most cases probably went undetected or were discovered accidentally as when the author of the original text uncovers it when recognizing personally written text in something documented by others. In 1974, while attending an international ecology meeting in the Netherlands the attention of Dr. Harold J. Morowitz (1927—) was drawn to a book titled *L'Anti-hasard*, by Ernest Schoffeniels [38]. To his surprise many sentences within the book were French translations from parts of his book *Energy Flow in Biology* that was published in 1968 [26, 27]. On other occasion plagiarism follows a laborious task when specific reasons existed for suspecting it. In the past plagiarized text was difficult to detect unless it reached the eyes of a person familiar with the original text and this was most often the author of the original work as the case of Chaturvedi Sunil and colleagues. This modern informational age has rapidly changed the field.

Plagiarism detection by similarities in stylistic writings

To detect a plagiarist suspected of stealing part of a specific literary contribution of a known author stylistic similarities, such as the frequency in which prepositions, adverbs, and conjunctions are used can be helpful. Such detective work was used to determine the authors essays written by three founding fathers of the USA: Alexander Hamilton (1755/57-1804)(first USA secretary of the treasury), James Madison (1751-1836) (later secretary of state and the fourth President of the USA) and John Jay (1745-1829) (the first chief justice of the USA). These distinguished statesmen took turns over a nine month period (1787-1788) writing 85 essays under the pseudonym of Publicus for a semiweekly newspaper in New York called the *Independent Journal* and in *The New York Packet*. The essays together with eight other letters were jointly incorporated into a book called *The Federalist; or the New Constitution*. These literary contributions advocated the ratification of the USA Constitution before it was approved and the author of each essay was kept secret. Later a dispute arose about who penned which letters (C. Rossiter, Ed, 1962; The Federalist Paper xi) [36]. In the mid-1950s the historian Douglass Greyhill Adair (1912-1968) solved the long-time mystery of who wrote what by analyzing stylist differences between the writings of these authors and found that Hamilton had taken credit for 12 letters that Madison actually wrote.

Exposure of Plagiarism by eTBLAST or Déjà vu

Computer software developed for uncovering plagiarism and image manipulation has disclosed misconduct that would otherwise have gone undetected. Plagiarism has been tagged by eTBLAST or Déjà vu [8]. eTBLAST detects suspicious articles based on comparable titles abstracts and the examined papers can be categorized into those with distinct, sanctioned, update, or duplicated. Suspicious papers detected by *Déjà vu* sometimes elicit acknowledgments of wrongdoings and even retractions.

Today plagiarism can be detected by extremely sophisticated computerized programs. With the assistance of colleagues Harold "Skip" Garner, a computational biologist at the University of Texas (UT) *(http://www.utsystem.edu/)* developed a software program called eTBLAST which is capable of detecting similarities in text as in plagiarism (*http://wikipedia.org/wiki/ETBLAST*) *(invention.swmed.edu)*. Garner originally hoped to use the program to access the literature on specific topics more efficiently, because it was immensely time consuming to do it by hand. eTBLAST can readily detect plagiarism by going through many millions of manuscripts in a database like a medical literature analysis and retrieval system (MEDLINE). Together with his colleagues Garner hunted worldwide for evidence of duplicated material in publications. In 2007 he obtained grant support from the National Institutes of Health (NIH) and the Office of Research Integrity (ORI) to develop eTBLAST and it spawned the program *Déjà vu (spore.swmed.edu/dejavu)* which has impacted the scientific community by identifying thousands of papers with evidence of plagiarism. Both programs are available free, but upgraded versions and customizable computer aided translation systems can be purchased *(http:www.atrile.com/)*. In an analysis of 17 million scientific papers with a sophisticated computer program researchers at UT Southwestern Medical Center identified about 73 papers with an indication of plagiarism [12]. By May 22 2009 *Déjà vu* listed 74, 790 pairs of papers drawn from MEDLINE with remarkably comparable language and content. The notoriety of being listed by *Déjà vu* in the public domain among the tens of thousands of identified papers makes *Déjà vu* particularly controversial because sometimes it points the finger unjustifiably at innocent individuals whose reputations can be unjustly damaged. Being singled out by *Déjà vu* suggests plagiarism, but innocent explanations may exist. Some listings in *Déjà vu* are caused by individuals repeating, or appearing to repeat, their own previously published work. The program does not clarify the reason for the text connection so the onus is placed on the named individual to explain the situation to restore credibility. Showing up on *Déjà vu* is somewhat comparable to being blacklisted in the 1950s in the McCarthy era. Many journals in India, China and Egypt, frequently get listed in duplicate publications. The similarity between some papers was so comparable that journals retracted 48 untrustworthy papers. *Déjà vu* has identified plagiarism in papers authored by investigators in humble institutions throughout the world as well as by Nobel laureates. Knowledge about the existence of Déjà *vu* and eTBLAST and its ability to detect publication duplication should be a deterrent against plagiarism and duplicate publications [8]. *Déjà vu* is also valuable in academic institutions checking papers before making decisions about hiring and promotion. However, it is not without its flaws as papers above suspicion, such as translations, get detected together with the suspect ones. The program has been criticized mainly by individuals who feel that they have been unjustly tagged, such as Lawrence Solin, a radiation oncologist at Albert Einstein Healthcare Network in Philadelphia. Two papers authored by him were tagged by *Déjà vu*. One reported the long-term outcome of ductal carcinoma detected on mammography and treated with conservative surgery followed by breast radiation [47] and a vastly similar paper appeared in 2005 [48], but it was essentially an updated version of the same study with more participants. The other paper documented the proceedings of an international consensus conference on breast cancer risk, genetics and risk management [39]. Solin was not an author of this paper, which was followed by a report of the same data by many more authors presumably to publicize the research finding [40]. The first court case testing a controversial plagiarism detector emerged in 1991 when heirs to the estate of John Marquise Converse (1909-1981), a plastic surgeon, alleged that the publisher W.B. Saunders and Converse's former assistant editor Joseph G. McCarthy copied entire sections of Converse's seven volume plastic surgery textbook *Reconstuctive Plastic Surgery* "almost verbatim" for another edition under McCarthy's name. An attorney for Converse's estate requested a comparison of the original and revised texts and hired Ned Feder and Walter Stewart to use their plagiarism detection program in this case and their analysis picked up a 57% overlap between the texts. An expert witness in the case was C. Kristina Gunsalus a misconduct policy officer at the University of Illinois at Urbana-Champaign (UIUC), who regarded the text scanning plagiarism program of Feder and Stewart valuable in detecting or refuting allegations of plagiarism [52].

Extreme Productivity of the Mentor or Mentee

Over the years it has become apparent that individuals who publish excessive numbers of papers, particularly early in their careers, should be suspected of scientific misconduct. Outstanding research universities in the USA where research superiority is stressed the faculty publish significantly more papers than the academic custom. This is true not only for senior faculty, but also for young investigators who still need to establish their careers seem to court fraudulent research more than older established scholars. Many individuals responsible for scientific misconduct have been brought to the attention of the scientific community have more publications to their credit than usual, particularly during the time period contiguous with the exposure of their research misconduct.

Some highly publicized examples of fraudulent research have occurred in laboratories with a more prolific publication track record than the usual research group. Philip Felig the mentor of Vijay Soman published 201 papers or an average of 31.8 each year between 1975-1980. This observation suggests that trainees working in such high output laboratories might be directly or indirectly under pressure to publish and hence cut corners to achieve this goal. Alternatively, trainees might try to mimic the productivity of the role model mentor. Many prominent scientists, who mentored junior investigators found guilty of research misconduct have been extremely productive such as Robert Good, who published 341 papers at an average annual rate of 68 papers a year during the time that Summerlin was one of his trainees. While there is no evidence of fraudulent research by productive scientists, such role models often inspire young trainees to emulate their idol. To reach a vast publication record per year, particularly during the publish or perish era of yester year authors with an extremely large number of publications have been suspected of scientific misconduct

The research productivity of John Darsee, Robert Slutsky typified some researchers who commit scientific misconduct. While at the University of California San Diego (UCSD)*(http://www.ucsd.edu/)* school of medicine Slutsky churned out a new research article every 10 days!!! This extreme productivity should have been a red flag suggesting that his research may not have been honest. Over two decades Poehlman published over 200 journal articles, namely 10 papers per year. Arnold Relman, editor-in-chief of the *New England Journal of Medicine* and Marcia Angell that journal's executive editor pointed out the merits of having manuscripts reviewed by peers but this system is not infallible [35]. The fact that much fraudulent research gets published in peer-reviewed journals indicates that the peer-review process does not unearth most cases of scientific misconduct. Reviewers of manuscripts are not expected to find perfection, but rather to improve the quality of papers and to get rid of papers that are obviously wrong.

Detection of Scientific Misconduct in Laboratory Meetings

When a member of a research team observes experimental errors and possible research misconduct the witness should bring the suspicious findings before the entire laboratory group and provide the suspected culprit and the entire research team with an opportunity to provide alterative explanations. This is particularly true if the group realizes that everyone makes innocent mistakes and the research task force has frequent laboratory meetings and openly discusses all positive and negative results without criticizing those who make errors.

The Exposure of Research Misconduct by Serendipity. Researchers who perform dishonest research have no conscience and do not see fault in what they do. They do not confess and turn themselves in, but their dishonest deeds are often discovered by serendipity sometimes years after the ends of the lives. The Darwinist Paul Ekman reported photographic manipulations by Charles Darwin. The historian of science Gerald Holton, who was also a physicist, discovered on reviewing both the published papers and saved raw data that Robert Millikan the 1923 Nobel laureate in Physics (*http://www.nobelprize.org/nobel_prizes/physics/laureates/1923/millikan-bio.html*) purposely discarded many of his oil-drop observations when calculating the weight of an electron (e).

Marek Wronski PhD, a cancer researcher, unearthed a scandal of immense proportions involving plagiarism in Poland while gathering data about a wrongdoing related to a translation from a Danish journal into Polish without permission. The Danish committee on scientific dishonesty confirmed the finding of plagiarism when Danish authors of the 1987 paper found a duplicate of the abstract in the medical literature analysis and retrieval system (MEDLINE) under other names. After discovering the plagiarized Danish paper, Wronski screened MEDLINE for publications by Andrzej Jendryczko, an engineer, and surprisingly unearthed about 125 plagiarized medical papers published by him over a 13 year period. The inquiry opened the doors into an extensive example of plagiarism and into a scandal about the inability of the scientific enterprise in Poland to police itself. Wronski discovered 29 plagiarized papers published in Polish that involved text lifted from at least 12 journals including *The Lancet, the Journal of the American Medical Association (JAMA)* [55]. One paper published in 1989 in the *British Medical Journal* was combined with part of a 1992 paper from a *New England Journal of Medicine* to create a composite article published in 1993.

Failure to Replicate Findings in a Study

As pointed out in chapter 2 the gold standard in scientific studies is confirmation of the results by impartial investigators. A failure of others to replicate new studies is the hallmark of suspected research misconduct. Reproducibility is an essential part of scientific investigations and real discoveries get replicated by the original researchers or by others. Complex questions being addressed with new methods, as well as interdisciplinary collaborative research that is often international pose challenges for replication, particularly when the studies are long term and performed and are difficult to replicate. An inability of other scientists to replicate studies does not automatically indicate research misconduct. Indeed many researchers have reported experiments that others were unable to replicate. The complexity of contemporary research makes it tremendously tricky for others to independently authenticate, especially if delicate technical aspects are not documented. To replicate the findings and meet expected customs it is essential that researchers document the specific experimental conditions. Unsuccessful attempts at replication are regarded as spurious and are caused by unintentional errors, an unrecognized quirk in the experimental conditions or fraudulent research. This practice of replication promotes truthfulness in research, because it exposes cheaters and questionable conclusions. Regardless of the reason non-replicated research tarnishes the reputation of scientists. The replication of experimental findings under certain conditions is hindered by a difficulty in defining the precise unique conditions under which the new findings are collected [37].

Failure to Replicate Data

A self-correcting mechanism to validate scientific discoveries is the ability of other researchers to replicate the work. Until experiments are repeated and their findings replicated with the same methods the data are always suspicious and questionable. Indeed if they cannot be replicated with the same techniques intentional or unintentional errors are suspected. The detection of scientific misconduct is particularly high when fame or wealth are potential rewards of the research because other investigators will undoubtedly attempt to replicate the experimental findings. Fraudulent papers are often detected after publication when investigators in other laboratories working on the same general topic cannot replicate the work and occasionally the studies cannot even be confirmed within the laboratory of the principal investigator (PI). Moreover it is important to realize that a failure to replicate does not invariably signify research misconduct as other explanations may exist. A failure of replication is how errors were exposed in the works of Mendel, Newton and Ptolemy and others.

To challenge the results of a colleague, teacher, or mentor on the grounds of scientific misconduct because experiments cannot be replicated takes considerable audacity. As discussed elsewhere under cold fusion alleged replications of research do not always reflect honesty. Without supervision in the performance of trustworthy research students and other junior researchers lack the tools and knowledge to query research protocols or appreciate potentially precarious circumstances where allegations may not be justified.

More than fifty academic psychologists have taken bold steps to share the task of trying to scrutinize the reproducibility of research findings in their discipline [6]. By repeating a sample of published studies they hope to get an indication of how prevalent research misconduct is in their field. This ongoing project will presumably detect some studies that cannot be replicated.

Research Misconduct Detection from Scrutiny of Raw Data and Correspondence

Historians and other scholars have played an important role in exposing inappropriate research by scientists years after their death. By delving in meticulous detail into the archival records of raw data and written communications between scientists and their colleagues unsuspected evidence detrimental to the image of famous figures in science have been unearthed. Thus Gerald Grieson exposed unfavorable information about Louis Pasteur and the lies and deceit of Sigmund Freud was uncovered by X. While studying the archival records of Dr. John Charles Cutler (1915-2003) after his death Wellesley College's historian *Susan Reverby* discovered unethical experiments performed on syphilis in Guatemala.

A Campaign to Disclose Research Misconduct

When a member of the faculty embarrasses an institution because of what that individual states under their freedom of speech that is protected by academic freedom as well as the constitution of the USA other members the faculty who are offended by the remarks may start an investigation unearthing flaws in the credentials of their colleague who was tolerated until that time. This occurred with Ward Churchill and despite his tenured position he was fired.

Detection of Manipulated Images

The manipulation of images with Adobe photoshop is common and such manipulations can be detected. Art experts can usually recognize flaws in counterfeit paintings, but sophisticated methods are now available to recognize their authenticity. One uses a computer for pattern recognition from neuroscience known as sparse coding [19]. Journals can detect manipulated images and some journals now screen all images for this distortion in submitted manuscripts.

The Media and Investigative Reporters

Investigative reporters from newspapers and television stations also expose scientific misconduct and other inappropriate types of research. A major asset that they bring to the table in exposing culprits is their ability to follow-up leaked confidential information from whistle-blower who do not want the expected humiliation and psychological stress from reporting allegations of scientific misconduct directly to the relevant authorities despite their theoretical protection guaranteed by law. They play an important role in bringing fraudulent research to the attention of the public. Two of the most successful investigative reporters who have exposed inappropriate behavior by scientists are John Crewdson, who put the heat on Robert Gallo for his work on AIDS [9] and Brian Deer who not only exposed the unbelievable notorious scandal of Andrew Jeremy Wakefield(1957—), but also the fraudulent work of Michael H. Briggs on the safety of oral contraceptives. Others who have exposed dishonest work are the investigative journalist Duncan Campbell and the British television Channel 4 which exposed the Harley street charlatan Peter G. Nixon despite the threat of litigation.

John Crewdson (1945—) began his career in his early twenties with *New York Times* with the coveted duty of dealing with the Watergate scandal, which he and eight other reporters contested effectively with notes from then President Richard Nixon's re-election committee. Crewdson became a well-known investigative reporter and in 1981 he received a coveted Pulitzer Prize for a string of stories about immigration injustices in the USA. He established a reputation for his determination and assertiveness in pursuing an issue in remarkable detail. Because of his awareness of fraudulent research in the 1970s and 1980s he appreciated that some science is corrupt and assumed a witch-hunt to annihilate Robert Gallo. Aside from the outstanding jobs performed by Crewdson, Deer and Campbell reporters from other newspapers, such as the *Boston Globe*. have served the important role of exposing fraudulent research. The National Cancer Institute (NCI) would probably not have investigated the Straus affair if it had not been for the series of articles published by the *Boston Globe*.

Poor Record Keeping

Good record keeping will not expose fraudulent research, but bad record keeping will draw attention to sloppy research and may bring to light scientific misconduct.

Detection of Scientific Misconduct by Professional Cheaters

Scientific fraud by sophisticated highly intelligent imaginative researchers is extremely difficult to detect. A noteworthy aspect of their misconduct is the fact that even outstanding scientists at renowned institutions, such as Harvard, the Massachusetts Institute of Technology (MIT) and the California institute of technology (Caltech) *(http://www.caltech.edu/)*, can again and again have the wool pulled over their eyes by individuals whom nobody would suspect of falsifying anything.

Astounding Results

There is a truism that if something appears too good to be true it is probably not true. Some research misconduct is discovered when authors document seemingly implausible findings and skeptics look into the data. An example of this is the breast cancer clinical trial of Bezwoda. The results of research studies that conflict with the beliefs of others is suspect especially if it relates to topic of great economic importance.

Failure to Openly Debate Research Results

Intellectual curiosity and open debate over research findings and hypotheses is a widely practiced component of science. It is also an important factor in the detection of research misconduct when researchers fail to freely discuss their research findings and do not volunteer to present them at seminars and conferences where the data can be questioned and openly debated.

Statistical Method of Fraud Detection

Uri Simonsohn, a social psychologist at the Wharton school of the University of Pennsylvania (Penn) *(http://www.upenn.edu/)* developed, a novel statistical method (currently unpublished) that makes it possible to detect misconduct in the literature [11]. The method assumes that if a study measures the identical variable in different groups the same probability of this happening can be determined with a computer simulation. The detection of too little variance is a sign of data manipulation. The Simonsohn method has also flagged another psychology paper, which precipitated another investigation leading to a resignation. In a paper coauthored with two colleagues Simonsohn reported that it is unacceptably easy to prove virtually anything with common methods to massage data proposing that a high percentage of papers may be false positives [42]. He has drawn attention to the not uncommon dubious research practice of doing multiple analyses of data and selecting the most convincing ones, while rejecting certain subjects. A failure to find raw data has often been attributed to a computer crash or the loss of data during an office move. The Dutch psychologist Dirk Smeesters resigned from Erasmus University *(http://www.eur.nl/)* Rotterdam after errors were detected in this publications using the Simonsohn method and he requested the retraction of two of his papers [20, 43, 44].

Ned Feder and Walter Stewart the Fraudbusters

Ned Feder (1929—) joined the NIH in 1967 when he was 38 years old, and later shared his laboratory with Walter W. Stewart (1945/1947—). The Darsee case changed the career pathway of of both Stewart and Feder getting them to enter the fraud game and they embarked on becoming the self appointed fraudbusters and the overseers of scientific honesty. In their role they were frequently referred to as Batman and Robin. While sharing Feder's laboratory, Stewart obtained a chance to confront the scientific establishment. In 1971 he successfully isolated the structure of a tobacco plant disease and reported his results in a controversial paper in *Nature*. He professed to detect a chemical named scotophobin that could transfer learned behavior from one rat to another. *Nature* published the paper along with a critique by Stewart who decisively argued against the existence of scotophobin [50]. The basic research of Steward and Feder focused on dyes that make it possible to visualize branching nerve cells. Stewart established his research reputation with a paper on a Lucifer yellow dye [51], which is still widely cited by researchers in neuroscience. After these researchers probed the nervous system of snails their research productivity declined and for more than a decade neither Stewart nor Feder published a basic research paper. During a broad rearrangement of space and restoration at the NIH in 1985 their laboratory in building 4 was extraordinarily large and of low research productivity. The duo were moved to the basement of building 8 and during that move some of their equipment was inadvertently not marked for relocation and ended up in the dumpster for salvage. This precipitated an email war of words that jumped back and forth between the NIH administration and the two of them. Stewart and Feder interpreted the incident as harassment; the NIH considered their failure to safeguard their equipment a reflection of their insincerity to perform research.

This contentious study was extensively probed, but some scientists apparently replicated Feder's findings. During this turbulent time the extremely notorious Darsee case provoked them to enter the scientific fraud game. These two intramural NIH researchers benefited from considerable autonomy to undertake any projects that they considered significant. The matter was taken to the Office of the NIH director where, in the spring of 1984, Joseph E. Rall, deputy director for intramural research, and others decided to allow Stewart and Feder to finish their analysis of the Darsee publications. In July 1988 the NIH loaned Stewart and Feder to the Dingel congressional subcommittee as full-time employees until June 1990. When Stewart and Feder joined the subcommittee staff, while remaining on the NIH payroll, they were in an ideal position to affect the congressional inquiry. These new duties protected them from potential acts of vengeance by the NIH. By June 29, 1990, after more than a year on Capitol Hill helping Dingell's congressional subcommittee, Stewart and Feder were back in their laboratory in the basement at the NIH in Bethesda. Shortly before their return to NIH, Dingell requested that their time with the House subcommittee on oversight and detail be extended, but his wish was not granted by the Department of Health and Human Services (DHHS). These fraud-busters became despised by the scientific community and their future status was in doubt after they returned to their laboratory in the basement at the NIH in Bethesda. In the remaining time Stewart and Feder would toil with NIH staff in composing a scientific protocol in harmony with the mission of the intramural program of the National Institute of Diabetes and Digestive and Kidney Disease (NIDDK). At that time the NIH had an Office of Scientific Integrity (OSI) and a fraud oversight office.

William F. Raub PhD, the deputy director of the NIH (1986-1991), who was also the acting director of the NIH from 1989 to 1991, participated in the decision as head of the NIH office of research fraud.

A couple of Darsee's coauthors retained legal counsel to block publication of the initial versions of the contentious Stewart and Feder paper, but eventually Stewart and Feder published their analysis of the Darsee papers using a less destructive much more diplomatic prose. Stewart and Feder were not completely pleased with the editorial alterations and some of them were inserted without their approval. The NIH formally rewrote their job description allowing them to devote 20% of their time on fraudulent cases of research. Eventually the workload related to scientific misconduct encompassed all of their time making it impossible for them to perform basic research. Throughout the 1980s, Stewart and Feder's association with the NIH was nerve-racking all around, but their initial assignment to the Dingell subcommittee seemed an ideal solution to the clash over their roles in the intramural program.

By 1989 the subject of scientific misconduct remained emotionally charged and many scientists were ignoring the issue. Stewart strongly maintained that this was a serious problem being ignored by the scientific community. By that year Stewart and Feder were learning about 100 allegations of research misconduct per year, and they believed that this was just the tip of a colossal iceberg [5]. Many scientists were resentful of Stewart because of his role in the Baltimore affair.

From then onwards they were only available for less than 2-3 days a month [10]. Later they embarked on being overseers of scientific honesty and desired to continue research on the integrity of the scientific literature to answer the questions that an analysis of a Darsee paper could not answer.

Why their full-time role on the congressional subcommittee ended abruptly after 2 years is fuzzy and buried in speculation. In that position they were expected to enhance the status of Dingell, but the congressman failed dismally to attain widespread admiration for his investigations of scientific misconduct.

In 1993 the NIH locked Stewart and Feder out of their laboratory and took possession of their files, because these self-appointed fraud busters had steered off target from their assigned activities by scrutinizing the works of the historian Stephen B. Oates (1936—), who wrote biographies of several prominent Americans, including Abraham Lincoln (1809—1865) [29, 30] *Abraham Lincoln: The Man Behind The Myths*), the 16th president of the USA, Martin Luther King,

Jr. (1929—1968)[32], *Clara Barton* (1821-1912) [33] William Faulkner (1897—1968), the 1949 Nobel laureate for literature [31] and *John Brown* (1800—1859) the revolutionary abolitionist who preached and practiced armed revolt to abolish slavery [28].

Oates was accused of plagiarism in several pieces of his written text particularly in his biography of Lincoln, but was cleared by the University of Massachusetts (UMass) (*http://www.massachusetts.edu/*) and the *American Historical Association* (*http://www.hnn.us/articles/648.html*).

Stewart and Feder developed a machine for detecting plagiarism and they started to use it in 1993 to find parts of Oates' writings that had been lifted word-for word from the literature without acknowledgement, but this project fizzled out because it was inappropriate for individuals on the NIH payroll to investigate this matter. By scanning the text of Oates' life history of Abraham Lincoln Feder and Stewart found numerous examples of copied text that had not been attributed to the source. A lack of training in the writings of historical documents impaired Feder and Stewart's ability to assess allegations of plagiarism. While they provided instances of lifted text in Oates' books. Oates used the text of *Plagiarism and Originality* [22] to defend himself. Dingell maintained that "It is one thing to test the so-called plagiarism machine. It is another to undertake an exhaustive inquiry at the government's expense and send the results to some three dozen scholars as well as members of the press". Because Oates was not employed by the NIH and the Feder-Stewart fraud team over stretched their mission in testing plagiarism in the historical works of Oates and they were instructed not to investigate Oates for allegations of plagiarism [2] and those in power stepped in and confiscated data related to Oates. In protesting this incident Stewart started a 33-day hunger strike until June 12, 1993 and Dingle condemned him for introducing a degree of blackmail by adopting such starvation and urged Stewart and Feder to plea by way of accessible channels at NIH to end the hunger strike. At the end Stewart claimed victory, but he failed to achieve his desired aim of remaining a fraud buster [23]. Nevertheless an inquiry into the authenticity of Oates was kicked off by both UMass, where Oates was a former professor, and the *American Historical Association*. After years of squabbling *American Historical Association* came to an indecisive ruling on Oates and decided not to investigate any further allegations of plagiarism because of insufficient resources (*http://www, hnn.us.articles/658.html*). The modern language association (MLA) stresses the importance of using quotation marks to signify quotations that are given verbatim.

References

1. Adelstein RS, Collins FS, Hajra A, et al., The leukemic core binding factor beta-smooth muscle myosin heavy chain (CBF beta-SMMHC) chimeric protein requires both CBF beta and myosin heavy chain domains for transformation of NIH 3T3 cells.[Retraction of Hajra A, Liu PP, Wang Q, Kelley CA, Stacy T, Adelstein RS, Speck NA, Collins FS. Proceedings of the National Academy of Sciences of the United States of America 1995 March 14;92(6):1926-1230; PMID: 7892201]. Proceedings of the National Academy of Sciences of the United States of America, 1996. 93(26): 15523.
2. Anderson C, NIH fraudbusters get busted. Science, 1993. 260(5106): 288.
3. Anonymous, Case Summaries: Amitav Hajra, University of Michigan. (*http://ori.dhhs.gov/documents/newsletters/vol5_no4.pdf.Accesssed*). Office of Research Integrity Newsletter., 2006.
4. Bloch K and Relman AS, Preventing fraud. Science, 1989. 245(4925): 1436.
5. Booth W, A clash of cultures at meeting on misconduct. Science, 1989. 243(4891): 598.
6. Carpenter S, Psychology's bold inititiative. Science, 2012. 335(6076): 1558-1561.
7. Chandra RK, Effect of vitamin and trace-elements supplementation on cognitive function in elderly subjects. Nutrtion, 2001. 17(9): 700-712.
8. Couzin-Frankel J and Grom J, Scientific publishing. Plagiarism sleuths. Science, 2009. 324(5930): 1004-1007.
9. Crewdson JM, Science Fictions: A Scientific Mystery, a Massive Cover-Up, and the Dark Legacy of Robert Gallo. 2002, New York: Little Brown & Co.
10. Culliton BJ, Fraudbusters back at NIH. Science, 1990. 248(4963): 1599.
11. Enserink M, Fraud-detection tool could shake up psychology. Science, 2012. 337(6090): 21-22.
12. Errami M and Garner H, A tale of two citations. Nature, 2008. 451(7177): 397-399.
13. Glazer MP, Glazer, P. M., The Whistleblowers: Exposing Corruption In Government and Industry. 2002, New York: Basic Books.
14. Hajra A, Liu PP, Wang Q, et al., The leukemic core binding factor beta-smooth muscle myosin heavy chain (CBF beta-SMMHC) chimeric protein requires both CBF beta and myosin heavy chain domains for transformation of NIH 3T3 cells.[Retraction in Adelstein RS, Collins FS, Hajra A, Kelley CA, Liu PP, Speck NA, Stacy T, Wang Q. Proceedings of the National Academy of Sciences of the United States of America. 1996 December 24;93(26):15523; PMID: 9340650]. Proceedings of the National Academy of Sciences of the United States of America, 1995. 92(6): 1926-1930.

15. Hajra A and Collins FS, Structure of the leukemia-associated human CBFB gene.[Retraction of Hajra A, Collins FS. Genomics. 1995 April 10;26(3):571-579; PMID: 7607682]. Genomics, 1996. 38(1): 107.
16. Hajra A, Liu PP, Speck NA, et al., Overexpression of core-binding factor alpha (CBF alpha) reverses cellular transformation by the CBF beta-smooth muscle myosin heavy chain chimeric oncoprotein.[Retraction of Hajra A, Liu PP, Speck NA, Collins FS. Molecular and Cellular Biology1995 September;15(9):4980-4989; PMID: 7651416]. Molecular and Cellular Biology, 1996. 16(12): 7185.
17. Hajra A, Liu PP, Speck NA, et al., Overexpression of core-binding factor alpha (CBF alpha) reverses cellular transformation by the CBF beta-smooth muscle myosin heavy chain chimeric oncoprotein.[Retraction of Hajra A, Liu PP, Speck NA, Collins FS. Molecular and Cell Biology. 1995 September;15(9):4980-4989; PMID: 7651416]. Molecular and Cellular Biology, 1996. 16(12): 7185.
18. Hoke F, Veteran Whistleblowers advise other would-be 'ethical resisters' to carefully weigh personal consequences before taking action The Scientist, 1995. 9(8): 1.
19. Hughes JM, Graham Daniel J., and Rockmore D, Quantitation of artistic style through sparse coding anlaysis in the drawings of Pieter Bruegel the elder. Proceedings of the National Academy of Sciences of the United States of America, 2010. 107(4): 1278-1283.
20. Johnson CS, Smeesters D, and Wheeler SC, Visual perspective influences the use of metacognitive information in temporal comparisons.[Retraction in Journal of Personality and Social Psychology 2012 October;103(4):605; PMID: 23002957]. Journal of Personality and Social Psychology, 2012. 102(1): 32-50.
21. LaFolette MC, Stealing into Print. Fraud, Plagiarism, and Misconduct in Scientific Publications. 1992, Berkeley, CA: The University of California Press.
22. Lindley AL, Plagiarism and Originality. 1952, New York: Harper and Brothers.
23. Marshall E, National institutes of health. Fraudbuster ends hunger strike. Science, 1993. 260(5115): 1715.
24. Meguid MM, Retraction [Retraction of Chandra RK Nutrition 2001, 17(9):709-712: September PMID 11527656]. Nutrition, 2005. 21(2): 256.
25. Miceli MP and Near JP, Blowing the Whistle. The Organizational and Legal Implications for Companies and Employees. 1992, New York: Lexington Books
26. Morowitz HJ, Energy Flow in Biology. 1968: Academic Press.
27. Morowitz HJ, The smoking gun. Hosp Pract (Off Ed), 1990. 25(12): 13, 16.
28. Oates SB, To Purge This Land With Blood: A Biography of John Brown 1970: University of Massachusetts Press.

29. Oates SB, With Malice Toward None: A Life Of Abraham Lincoln. 1977: HarperCollins e-books.
30. Oates SB, Abraham Lincoln: The Man Behind The Myths. 1984: HarperColins.
31. Oates SB, William Faulkner: The Man and the Artist. 1987, New York: Harper and Row.
32. Oates SB, Let The Trumpet Sound: The Life Of Martin Luther King, Jr. 1994, New York: HarperCollins. pp.592.
33. Oates SB, A Woman of Valor: Clara Barton and the Civil War. 1994, New York and Toronto: MacMillan.
34. Policy. CoSEaP, On Being A Scientist: Responsible Conduct in Research. Third Edition. 2009, Washington DC National Academy Press
35. Relman AS and Angell M, How good is peer review? New England Journal of Medicine, 1989. 321(12): 827-829.
36. Rossiter C, The Federalist papers; Alexander Hamilton, James Madison, John Jay (republished 1999). New York New American Library. 1961, New York.
37. Ryan MJ, Replication in field biology: the case of the frog-eating bat. Science, 2011. 334(6060): 1229-1230.
38. Schffeniels E, L'Anti-Hasard. 1973: Gauthier-Villars.
39. Schwartz GF, Hughes KS, Lynch HT, et al., Proceedings of the international consensus conference on breast cancer risk, genetics, and risk management, April 2007 Cancer, 2008. 113(10): 2627-2637.
40. Schwartz GF, Hughes KS, Lynch HT, et al., Proceedings of the international consensus conference on breast cancer risk, genetics, and risk management, April, 2007. Breast Journal, 2009. 15(1): 4-16.
41. Shafer SL, Tattered threads. Anesthesia and Analgesia, 2009. 108(5): 1361-1364.
42. Simons JP, Nelson L, D, and Simonsohn U, False-positive psychology undisclosed flexibility in data collection and analysis allows presenting anything as significant. Psychological Science, 2011. 22(11): 1359-1366.
43. Smeesters D and Liu JE, The effect of ocular color (red versus blue) on assimilation versus contrast in prime-to-behavior effects. Journal of Experimental Psychology, 2011. 47(3): 653-656.
44. Smeesters D and Liu JE, Retracted: The effect of ocular color (red versus blue) on assimilation versus contrast in prime-to-behavior effects. Journal of Experimental Psychology, 2013. 49(2): 3i5.
45. Smith R, Education and debate. Investigating the previous studies of a fraudulent author. British Medical Journal, 2005. 331: 288.
46. Smith R, Research misconduct: the poisoning of the well. Journal of the Royall Society of Medicine, 2006. 99(5): 232-237.
47. Solin LJ, Fourquet A, Vicini FA, et al., Mammographically detected ductal carcinoma in situ of the breast treated with breast-conserving surgery and

definitive breast irradiation: long-term outcome and prognostic significance of patient age and margin status. International Journal of Radiation Oncology, Biology, Physics, 2001. 50(4): 991-1002.
48. Solin LJ, Fourquet A, Vicini FA, et al., Long-term outcome after breast-conservation treatment with radiation for mammographically detected ductal carcinoma in situ of the breast. Cancer, 2005. 103(6): 1137-1146.
49. Spraque RL. I trusted the research system. The Scientist. 1987 14 December.
50. Stewart WW, Comments on the chemistry of scotophobin. Nature, 1972. 238(5361): 202-210.
51. Stewart WW, Functional connections between cells as revealed by dye-coupling with a highly fluorescent naphthalimide tracer. Cell, 1978. 14(3): 741-759.
52. Stone R, Court test for plagiarism detector?. Science, 1991. 254(5037): 1448.
53. White C, Suspected research fraud: difficulties of getting at the truth. British Medical Journal, 2005. 331(7511): 281-288.
54. White C, Education and debate. Suspected research fraud: difficulties at getting at the truth. British Medical Journal, 205281. 331: 281-289.
55. Wronski M, [Plagiarism in publications by Dr. Andrzej Jendryczko]. Przeglad Lekarski, 1998. 55(11): 629-633.

CHAPTER 22

Prevention of Scientific Misconduct

Lessons learnt from past cases of research misconduct are particularly useful in gathering information for the prevention of new cases of misconduct, especially when persons with the long term fraudulent behavior is analyzed in detail. Dr. Eric J. Smart, a former professor in the department of pediatrics and physiology at the University of Kentucky (UK) is a good example as his misconduct was known to take place over a decade (1998-2008)). Findings on him appeared in the *Federal Register* on November 20, 2012 [4]. Smart consented to a voluntary agreement for seven years of exclusion from eligibility to apply for, or be supported by, federal research funds. During his research misconduct Smart made false claims of images in 10 publications, in one submitted paper, in seven grant applications and three progress reports. These falsifications related experiments allegedly performed with a stain of mouse that does not exist and could not have been known. Knowledge of this could have prevented the misconduct. Smart was the only important person listed on the publications and grant applications and other personnel provided little or no input into his research projects. Smart was directly involved in his laboratory research and without sharing this duty he was able to control his research and prolong his misconduct. Smart also relied on a small group of junior staff to perform his research, but they provided little or no input into the project. Postdoctoral fellows and graduate students participated in writing and review of early manuscripts, but they were finalized by Smart without the input of others. Other members of the Smart research team played an insignificant role in the generation of images. These were taken care of by Smart and he generated them by copying and pasting single bands into multiple lanes and cropping outer edges to make them differ in appearance. The flawed images would have difficult to detect in the absence of meticulous inspection especially with attention to pixilation pattern. However, more oversight by collaborative researchers would have prevented the long term research misconduct by Smart. Misconduct by

Smart could also have been prevented having all students and research staff oversee the raw data and experimental results and particularly when in manuscripts and grant applications. Special attention to image data must be paid by all members of the research teams.

The prevention of research misconduct requires a cultural change in universities and research institutions. The faculty needs to be persuaded to play a more functional role in rendering the imperfections and scientific flaws of their peers unacceptable. This should hopefully be done in a good-natured non-hostile manner.

Stress Research Integrity

While it is impossible to prevent dishonest individuals from performing fraudulent research steps are clearly needed to ensure the integrity of science. From an analysis of known cases of scientific misconduct it has became apparent that several procedures could help control the flood of bad science.

Despite being relatively uncommon fraudulent research does not seem to be completely preventable. Nevertheless steps are needed to prevent as much plagiarism, fabrication, and falsification of research data as well as other forms of scientific impropriety as possible. Highly intelligent dishonest individuals without scruples are capable of performing fraudulent research without the knowledge of their research colleagues. A way to prevent a talented, dedicated committed liar from performing research misconduct does not exist and such people always get away with it and beat you every time according to Ms. C.K. Gunsalus, associate provost at the University of Illinois, Urbana Champaign (UIUC), who encourages dealing forcefully with scientific misconduct. She has handled many misconduct cases and adds "you cannot afford to write rules in a cooperative community—where the foundation must be trust for the bad actors' [5]. Everyone needs to suspicious of misconduct when the red flags of research misconduct exist. Laboratory personnel should also not hesitate to report suspected cases to the relevant authorities at their institutions. Universities and research institutions need to make it clear that scientific misconduct will not be tolerated and that those found guilty of it will be fired and prosecuted according to the relevant regulations and laws.

Fellowship applications commonly contain fabricated misinformation to enhance the probability of the application being accepted. Sekas and Hudson recommended the following steps to prevent fraudulent information in fellowship applications: i) fellowship programs should require that copies of all publications and letters of acceptance for manuscripts in press be submitted with fellowship applications; ii) applications should contain a statement signed by the applicant stating that the information provided is accurate; iii) persons writing letters of recommendation should verify the information being submitted by applicants; iv) medical students and residents should be taught that embellishment of curriculum vitae constitutes misconduct; and v) institutions and professional organizations should develop policies to deal with this problem [7].

Deemphasis of the Importance of Publications

Particularly in the past the total number of publications was stressed and researchers learnt that the number of peer reviewed publications was taken into account when decisions were made with regard to promotions and tenure decisions. A reasonable list of publications was also necessary to obtain grant support. Researchers learnt that the truism "publish or perish" was true. During the extensive assessment of the reasons for research misconduct it became obvious that publication productivity was an incentive for research misconduct. Hence it is advisable for universities, research institutions, and funding agencies to limit the quantity of papers that are essential in judgments related to career progress and the awarding of research grants. The notion that scientists should be evaluated on a limited number of publications was proposed in 1986 by DeWitt Stetten Jr (1909-1990), a former deputy director of the National Institutes of Health (NIH) and other scientific leaders [9]. In this regard only 12 citations are requested in nominations for the Nobel Prize, the highest available award in science and in nominations for membership into *National Academy of Science (NAS)* in the USA. By 1988, Harvard University *(http://www.harvard.edu/)* medical school took the gallant move of changing its criteria for academic advancement so that someone up for promotion to full professor would be judged on the merit of no more than 10 papers and those being considered for associate professor would be evaluated only on 7 papers [3]. Other universities subsequently adopted similar criteria.

Researchers generally agree that safeguards against the existence of scientific misconduct and unintentional errors warrants considerable attention even though it is probably not possible to eliminate them.

Early Detection of Research Misconduct

Better methods of detecting early misconduct would undoubtedly deter this dishonest behavior. Humans are imperfect and so are scientists. When research is supported by the people's money gathered from taxes and distributed by organizations, such as the National Institutes of Health (NIH) and National Science Foundation (NSF) the government must try to influence conduct that leads to the fraudulent misuse of the funds, but the government is seldom successful in controlling human behavior. Science has two things in its favor: time and truth. In the long run truth always wins in science and other human undertakings. Proposals have been made to deal with the erosion into the integrity of research, such as the development standards of research and a code of ethics for researchers.

An academician who appreciates the reality of research and the difficulties faced in discovering new knowledge is more conducive to research integrity than those at institutions which exert undue pressures on the discovery of new knowledge, and the publication of papers and the ability to obtain grant support.

It is essential that the science community prevails over the tendency to secrecy and the failure to share data by unambiguously delineating an acceptable performance and then making use of these yard sticks to identify researchers who do not reach the expected standard. Dr. St. James-Roberts of the University of London pointed out in the British magazine *New Scientist* that science has few real safeguards against cheating [8]. By establishing clear principles of practicing good science an honest desire has been made to prevent fraudulent research by eliminating both the opportunity and motivation. However, this idealistic view is probably as unrealistic as believing that society can prevent crime, sex abuse and other undesirable aspects of human behavior.

Credential Checking

Without a doubt a check on the academic credentials during the recruitment of graduate students and during the hiring of faculty can reveal flags pointing to dishonest behavior. When individuals apply to college and to different positions in academia it is essential that their credentials be checked and that letters of reference be obtained directly from the referees or from sealed letters that the candidates cannot open and read. Confidential follow-up phone calls are often desirable and worthwhile despite the extra needed effort. Today honest opinions are difficult to get in writing especially is the referee wants to express negative remarks about the applicant. If the credentials had been checked on Harold M. Bates, Melvin V. Simpson's graduate student, Mark Williams, and Anil Potti their dishonesty could have been suspected before their recruitment.

Supervision and Mentoring

Mentors play an essential role in preventing research misconduct and experienced senior faculty are extremely important in the development of the careers of technicians, students, postdoctoral research fellows and junior faculty. Senior scientists with students under their care clearly should devote an appropriate amount of time to educate, mentor, and guide junior coworkers, graduate students and postdoctoral fellows as well as technicians. A considerable amount to time needs to be devoted to mentoring them on how to honestly perform research. Not only does the mentor provide useful guidance in career development but also supervision in the conduct of research and in the writing of publications resulting from the research. Successful mentoring is crucial to accomplishments of science in years to come.

It is generally agreed that young scientists need be taught integrity and trustworthy practices in research as well as dependability in publications of data. As part of the educational responsibility of one or more mentor the research teams should ideally be carefully overseen to detect and correct mistakes. In an attempt to prevent scientific misconduct numerous universities now desire that junior faculty select one or more at least one senior faculty member as a mentor. Because of the wisdom and professional connections that mature over decades of experience mentors bestow good judgment. Thelen and Di Lorenzo [6] and the need for excellent mentorship in nurturing a trustworthy scientific career. Like a good parent, a top-notch mentor provides mentees with considerable guidance in developing research careers with integrity. The success and reputation of research depends on its integrity and effective mentoring that draws attention to honest research is essential.

While mentors play a crucial role in guiding the careers of young investigators mentors do not acquire the necessary skills by osmosis from those who trained them. Mentors need to be instructed as to how to become an advisor, teacher, role model and friend of persons whose careers they influence and guide along a path to honest research and research integrity. An excellent relatively short book on being a mentor to students in science and engineering has been published by the National Academy of Sciences (NAS), National Academy of Engineering (NAE) and the Institute of Medicine (IOM) [2]. It provides an outstanding introduction to the duties of a mentor.

Prevention of Research Misconduct by Professional Journals

Professional journals can help prevent research misconduct by demanding that a computerized copy of pertinent data be provided before or after the evaluation of submitted papers as this would facilitate the evaluation of some papers and it may dissuade fabrication or falsification of research findings.

Restriction of Laboratory Management to Scientists with Appropriate Available Time

Several examples of research misconduct have involved senior experienced researchers with considerable time commitments t administrative commitments. In some cases researchers, such as Robert Good, and Francis Collins have lacked sufficient time to mentor and oversee the research of their trainees and have been alarmed to find the mentee guilty of misconduct. On other occasions senior researchers, such as Glueck, has clearly been unable to keep on top of their own research because of all of their commitments. Because of the association of research misconduct with busy overcommitted scientists some believe that one way to diminish research misconduct is to restrict laboratory research to full-time researchers. Eliot Marshall [5] believes that there is no need to isolate research managers from bench science, but busy scientists especially with considerable administrative duties should probably not play the role of a primary mentor.

Promotion of Research Integrity

The government as well as the scientific community needs to promote ethical research with integrity and with as little unintentional errors as humanly possible and ideally without any fraudulent activity. Integrity in research indicates that the documented findings are truthful and correct and are in keeping with accepted research practices. It is widely believed that an emphasis in research integrity will diminish research misconduct. A time-honored aspect of research is the taking for granted that if mistakes are discovered, they will be rectified. The first world congress on research integrity took place in Lisbon, Portugal on September 16-19, 2007. It was sponsored by the European Science Foundation (ESF) and the Office of Research Integrity (ORI) of the Department of Health and Human Services (DHHS). This conference was followed by the second world congress on research integrity in Singapore on July 21-24, 2010. Following the second congress a statement on Research Integrity was issued with the collective participation and insights of 340 individuals from 51 countries. Contributors included researchers, funding agencies, representatives from universities and research institutions and publishers of research findings. A document produced after the second congress was released on 22 September 2010 and it is referred to as the "Singapore Statement" *(http://www.singaporestatement.org)*. The next world congress on research integrity took place in Montreal Canada (May 5-8, 2013). Educators discussed whether ethics could or should be taught as a separate course in training future scientists. Prominent scientists who have followed discussions on scientific misconduct generally agree that the finest solution for preventing scientific misconduct is stressing the importance of research integrity. Contemporary and future scientists need to appreciate that research integrity is utterly important and a vital element in minimizing research misconduct, and procedures promoting it hopefully have educational as well as deterrent components. Most importantly the public needs to be assured that their tax dollars are appropriately used and that dishonest research is not tolerated at all.

The internet contains numerous resources in English of material on good scientific practice as well as guidelines for handling accusations of scientific misconduct. The information is derived from the Danish Committee on Scientific Dishonesty (*www.forkraad.dk/spec-udv/uvvu/*), Deutsche Forschungsgemeinschaft (DFG) the predominant research-funding organization for protecting research integrity in Germany, *(www.dfg.de/aktuell/self-regulation.htm)*, Max Planck Society (*www.mpg.de/fehlengl.htm*), UK Biotechnology and Biological Sciences Research Council *(www.bbsrc.ac.uk/opennet/struct/hrg/sciconco.htm)*, UK Medical Research Council (*www.mrc.ac.uk/mis-con.pdr* and *www.mrc.ac.uk/w-n1.html*) and the USA Office of Research Integrity (*http://ori.dhhs.gov/regguide.htm*).

Since 1989 all applications for training grants from the NIH have been required to outline required courses in the responsible conduct in research and ethics that trainees are required to take. This will hopefully diminish the frequency of dishonest behavior, but they are unlikely to influence unscrupulous researchers. Educational activities related to research integrity probably play an important role in diminishing scientific misconduct and in promoting research integrity. An example of such a conference on research integrity that took place in Washington, DC (October 31—November 4, 2010) attracted the attention of scientists, journal editors, research administrators, and research partners. Aside from the USA organizations in other countries took steps to protect the integrity of science. In Germany the DFG and the Mildred Scheel Foundation sponsored a project in 1998 to keep the effect of publications with fraudulent content in check. The guidelines of the DFG included the appointment of an institutional ombudsman so that young researchers would have an impartial individual to whom they could talk in secret about possible incidents of dishonesty that come to their attention [1]. In 2004, the Slovak Research and Development Agency embraced the thorough advice of the DFG.

Universities and research institutions are major sponsors responsible for research integrity. They also play a significant role in creating an environment in which mistakes in research are both prevented and detected. In 2006 a Research Integrity Committee was formed by Health Canada (HC), to strengthen the research integrity within its country.

Establishment of Boundaries of Accountability in Multi-Authored Collaborative Research

In multidisciplinary collaborative team research, which is now particularly prominent, require clear boundaries of accountability.

Establishment of Procedures for Whistleblowers to Report Mistakes and Potential Research Misconduct

Young individuals who witness possible fraudulent research fear retribution or consider the risk of reporting what they interpret as minor acts of misconduct too risky to blow the whistle. The head of the laboratory, who is usually the Principal Investigator (PI) should be accountable for the quality of the emerging research from the laboratory and should aggressively disapprove of the submission of more than one abstract on the same topic to meetings of professional societies as well as pieces of specific experiments to different publications. With the increasing awareness of fraudulent research it has become increasingly important that researchers maintain records of raw data that can be readily accessed so that the validity of published findings can be established without difficulty at any time. First-rate laboratory notebooks are desirable for recording and storing data. For the benefit of scientific integrity and the research team working on a project it is important that accurate records of raw data be maintained by each researcher in hard covered notebooks with numbered pages that cannot be removed or replaced as in a loose paged binder. Alternatively they could be retained in computerized databases in which information cannot be changed without establishing a documented record of when and who made changes. The original raw data must be kept by the unit in which they were generated.

Appointment and Promotion Criteria

There is a general consensus that the responsibility for research integrity and for keeping scientific misconduct under control belongs with the universities and research institutions where the scientific work is performed. As pointed out elsewhere universities used to evaluate faculty for promotion and tenure by paying much attention to the total number of publications and that this was an important factor promoting research misconduct. It has now become apparent that there should be less emphasis on the quantity of publications and more on their quality. To help prevent research misconduct the policies and procedures of universities and research institutions that govern appointment and promotion decisions were changed. In particular the quality of publications became considerably more important than their number.

Supervision by Head of Department

Years ago the head of the department in which research was performed read and approved scientific manuscripts written by the department's faculty. Alternatively a member of the faculty was made responsible for approving what was written. To some extent this review by an experienced researcher in the same discipline weeded out fraudulent research prior to its submission for publication. This became increasingly less possible as departments enlarged and more and more academic disciplines emerged to cope with specialization. This duty, however, still remains important though not always feasible with publications by junior researchers, who should preferably, be mentored by more experienced faculty, who are intimately familiar with all aspects of the research preferably as a co-investigator.

Code of Ethics

Different recommendations have been proposed to cope with the eroded integrity of research data. They include the use of a code of ethics by researchers and the creation of research standards. Self-governing data inspectors could check the quality and integrity of research data, but this would increase the cost of research and impose an unnecessary burden on scientists with the hope of preventing fraudulent research.

Scientific Community

The scientific community must demonstrate to students how research is, and is not, performed so that they can make the process better when they hopefully become independent investigators. Different organizations have provided useful guidelines for performing ethical research with integrity in an open manner. Codes of good practice in research are undoubtedly worthwhile, but they clearly will not eliminate scientific misconduct. The scientific community needs to establish rules of conduct especially for graduate students and postdoctoral research fellows who are developing careers and face pressures to accumulate publications on their curriculum vitae, while balancing this desire with ethical behavior and honesty. It is crucial that all researchers become familiar with the essential customs of science and the rules that guide trustworthy practices. The existing system for research is fundamentally sound and self-regulated by mechanisms such as peer review in both publications and in the awarding of research grants, the replication of worthwhile discoveries and the oversight of scientists at their institutions where they work usually in an environment where data is preferably discussed freely without secrecy. Specific formalized rules of research conduct are needed to boost confidence that both "honest" unintentional errors as well as blatant dishonesty in research are recognized, but are not published and if they are they are retracted regardless of the cause of the mistakes.

Peer Review of Manuscripts Submitted for Publication

Safeguards against research misconduct include the peer review of manuscripts before their publication in high quality journals and a similar evaluation of grant applications to fund the research. When possible all authors on papers must be responsible for the entire content of papers that include their name and honorary authorship and other forms of publication dishonesty cannot be tolerated. Reviewers of manuscripts should aggressively draw attention to omissions in submitted papers. If it becomes apparent to editors that certain referees perform a slapdash job in critiquing manuscripts such reviewers deserve to be accused of research misconduct and they should be removed from editorial boards, and lists of reviewers and banned from evaluating papers and grant applications.

Investigative Reporters

Freedom of the press by investigative reporters should be encouraged, but without their freedom to mislead and dream up concocted stories about scientists and the research community.

Stressing Serious Consequences of Misconduct

The consequences of scientific misconduct are extremely serious and can bring a research career to an abrupt end. All individuals who pursue a career in research need to appreciate this message. Perpetrators of research misconduct lose their credibility as well as the trust that others previously had in the researcher. Despite temptations to engage in scientific misconduct most intelligent scientists avoid this wrongdoing because of the serious consequences if the cheating is discovered. To just inform researchers with different levels of experience that they must not be deceptive in their research is as worthless as telling the inner-city youth to "Just Say No" in response to requests by peers to reject drugs, not to steal and to rebuff other inappropriate behavior.

Undesirable Publicity For Cheaters

The undesirable publicity and severe penalties that investigators receive for scientific misconduct will optimistically prevent some researchers from committing scientific misconduct. The threat of bringing the names of individuals found guilty of research misconduct into the open with a condemnation of their behavior will hopefully discourage research misconduct. Already the Office of Research Integrity (ORI) in the USA lists individuals funded by the National Institutes of Health (NIH) or National Science Foundation (NSF) who were found guilty of research misconduct in their database. The ORI also manages the DHHS ALERT system of the Office of Public Health and Science (OPHS) that provides information on all PIs under formal investigation. A further exposure by institutions together with serious sanctions will hopefully enhance the positive effect of this negative publicity.

Replication of Research Results

Established researchers are aware that research findings need to be replicated by others before they become accepted into the general scientific knowledge. To make sure that the experiments are valid significant studies should ideally be repeated by other members of the same laboratory group before publication. Developing researchers need to know that fabricated and falsified research cannot be replicated and if someone tries to do so the misconduct will in all likelihood be exposed. Reputable scientists stress that noteworthy research findings should always be double checked by other members of the research laboratory, but this is often not feasible due to the fact that scientists are under considerable pressure to publish before their competitors and the time pressures of obtaining and retaining grant support.

Sharing Data with Collaborators

Graduate students, postdoctoral fellows, research technicians and junior faculty should be persuaded to collaborate when possible with additional colleagues as on projects of mutual interest. Under such circumstances the data should be openly shared and interpreted after ample discussion of the findings amongst the research team prior to the submission of a paper to the most appropriate professional journal.

Research Protocol

Secrecy about research techniques and data need to be discouraged by researchers, because the absence of a free open discussion about the research can potentially lead to scientific misconduct. However patents and the embarking of some university research into money making endeavors has caused secrecy to creep into the research of academicians. To prevent scientific misconduct a strong collegial interactions between all members of the research team should be developed. A major benefit of created bonds developed amongst research groups is the early recognition of experimental errors and an opportunity to rectify them in a timely fashion. Moreover undesirable personality traits can be recognized and dealt with by the laboratory leader.

Knowledge about Methods of Research Misconduct Detection

Knowledge about the different methods of exposing research misconduct should hopefully deter research misconduct even though it may not entirely prevent it. The fact that sensitive computer programs, such as eTBLAST, Déjà vu and refinements of them, are currently available to easily detect plagiarism will hopefully be a major deterrent to this form of misconduct, and knowledge about this powerful method should help to prevent plagiarism.

Universities

Currently several approaches appear to promote responsible conduct in research. They include the review of manuscripts submitted to professional journals and grant applications to funding agencies by anonymous knowledgeable experts in the subject. An essential component of research that helps prevent scientific misconduct is the free discussion at regular laboratory and departmental research seminars. In an open non-hostile environment the details of all projects can be challenged not only by any member of the research team, but also by others in the audience. At these meetings hypotheses and the raw experimental data can be presented and openly discussed, critically evaluated, and deeply analyzed in preparation for scientific meetings and for publication. In an ideal world research should be performed at institutions where everything is carried out in an open environment.

To enhance the integrity of research in departments and in graduate student programs it is popular to have visiting committees from other institutions evaluate the strengths and weaknesses of departments. Also interdepartmental reviews are also useful in assessing the attributes and weaknesses of graduate educational and research courses and programs. With multicenter clinical research programs coordinating committees and appraisals of data collection in audits helps to maintain the integrity of the research and prevents scientific misconduct.

By stressing quality rather than quantity the faculty learn that the number of publications is not important for their promotion, but that the caliber of the scholarly activities carries more weight. University officials and their appointment, promotion and tenure committees now appreciate that faculty positions should be evaluated by the caliber of the research rather than mainly on the mere number of peer-reviewed publications.

Honor System

To help combat dishonesty in high schools, colleges, universities and the military the time tested honor system is used to keep cheaters in bay. The same method can help prevent scientific misconduct, but the honor code can breakdown as it has in prestigious military academies and universities.

Laws Against Cheating

Important deterrents against cheating in business include laws, such as the *Sarbines-Oxley Act* of 2002 (Pub.L.107-204, 116 Stat.7.45) (also known as *Public Company Accounting Reform and Investigator Protection Act* and *Investor Protection Act* and the *Corporate and Auditing Accountability and Responsibility Act*. This federal law set new standards for USA public company boards, but legislature is difficult to enforce against research misconduct.

Ethical Upbringing

An important measure in scientific misconduct prevention is the teaching of honesty and integrity during childhood in the educational period of life. To promote research integrity and prevent research misconduct honesty needs to become part of the normal character of future scientists long before they develop an interest in becoming a scientist. This ethical nurturing ideally comes from parents and religious affiliations which stress moral behavior, and it needs to become imprinted in children at an early age. Unfortunately many youngsters are deprived of this opportunity.

Acceptance of Mistakes in Science

Mistakes in science are inevitable and they are considered different to those in politics and other fields. Most are unintentional and occur for a variety of reasons, such as accidental blunders, ignorance about certain research techniques, and statistical mistakes. When made and acknowledged they sometimes have a positive effect and push research forward. It is essential that members of research teams accept mistakes, discuss them in a friendly non-hostile environment, learn from them and correct them. By openly discussing errors regardless of their cause junior researchers come to realize that they should not feel pressured to fabricate or falsify data when they are unable to obtain the desired results in a timely manner.

Trust

The world of science depends on trust and most scientists have confidence in their students, postdoctoral fellows, technical support staff and research collaborators and are taken by surprise when their publications turn out to be suspected of alleged scientific misconduct. It does not seem feasible, or worthwhile, to design a system in which every move by research colleagues are not only carefully watched over, but in which their work is double checked to confirm every detail in each paper.

References

1. Abbott A, Science comes to terms with the lessons of fraud. Nature, 1999. 398(6722): 13-17.
2. Anonymous, Adviser, Teacher, Role Model, Friend On Being A Menor to students in Science and Engineering. 1997, Washington, DC: National Academy Press. 84 pp.
3. Culliton BJ, Harvard tackles the rush to publication. Science, 1988. 241(4865): 525.
4. Garfinkel S, A research misconduct case: 10 years of bad behavior. Newsletter. Office of Research Integrity, 2012. 21(1): 1, 7.
5. Marshall E, Fraud strikes top genome lab. Science, 1996. 274(5289): 908-910.
6. Miller DJ, Hersen M, and Editors, eds. Research Fraud in the Behavioral and Biomedical Sciences. 1992, John Wiley and Sons: New York.
7. Sekas G and Hutson WR, Misrepresentation of academic accomplishments by applicants for gastroenterology fellowships. Annals of Internal Medicine, 1995. 123(1): 38-41.
8. St James-Roberts I, Are researchers trustworthy?. New Scientist, 1976. 71: 481-488.
9. Stetten DJ, Publication numbers and quality. Science, 1986. 232(4746): 11.

CHAPTER 23

Investigations of Alleged Scientific Misconduct

Almost anybody accused of research misconduct will attempt to convince those charged with investigating allegations of misconduct that he/she made unintentional "honest" mistakes or was using dubious research practices, because the net result of fraudulent research and unintentional errors in the science is identical. In research stupidity is more desirable than dishonesty and anyone would rather be tagged as a jerk in preference to a crook.

Investigations into scientific misconduct are confidential and handled by an institutional misconduct review officer and/or the dean of the school and forwarded to a committee for inquiry. If deemed necessary following the initial probe a formal investigation is launched. At this point, an *ad hoc* committee consisting of individuals chosen for their expertise and open-mindedness look into the matter and prepare a final report which is acted on by a designated university official. The identity of the committee members is generally kept secret to avoid repercussions. Investigators from within the institution of the accused have advantages and disadvantages. For the reputation of an institution it never looks good to find an employee, especially one with an outstanding track record, guilty of scientific misconduct. An allegation of misconduct against a scientist is extremely serious and cannot be dealt with lightly as it can damage the reputations of innocent scientists. Members of investigating panels as a rule take their task extremely seriously. Examples of uncalled for malicious allegations against scientists have occurred throughout the ages. Such charges have been made against honest researchers by disgruntled technicians, graduate students, post-doctoral fellows, or even competitors for many reasons. Political or religious reasons may also lead to allegations of fraudulent research by scientists. When callous charges have been made against an innocent person the victim can sue the perpetrator for libel and such cases have occurred. Even when an allegation has not been formerly made rumors may start [14].

The prolonged time required to prove guilt of research misconduct has enabled some fraudulent scientists to delay the demise of their academic careers. Even when the allegations of scientific misconduct brought by unhappy students, postdoctoral fellows, technicians and collaborators, and even competitors had no foundation their careers were sometimes ruined by eating into their valuable time while rash charges dangled for years before being discredited. Prior to 1980 most institutions lacked policies on how to investigate allegations of research misconduct. Later each university needed an established well-defined policy on scientific misconduct to encounter such allegations. Ideally the committee that investigates misconduct should not be composed entirely of members of the same university as the accused, because of an unconscious or conscious existence not to find the accused guilty because of a loyalty to the institution employing the faculty member. Institutions abhor bad publicity and nobody wants to throws mud at the organization responsible for a person's livelihood. By maintaining the initial inquiry within institutions the accused receives some protection since organizations are prone to bend over backwards in their evaluation of the evidence so that the accused gets the benefit of any doubt. Allegations of research misconduct place a colossal drain on institutional resources, because investigating panels may need to devote many months interviewing witnesses and scrutinizing raw data in laboratory notebooks. When individuals suspect that the work of colleagues is questionable it is highly advisable that they do not make their allegations public especially if they cannot completely substantiate the charges. Accusers and those looking into allegations of research misconduct need to realize that scientists often make unintentional honest errors and those accusations of wrongdoing can damage reputations, especially if the incident is not dealt with rapidly and appropriately without jumping prematurely to incorrect conclusions. Investigations of alleged scientific misconduct should not rush to judgment, particularly because unintentional errors in science are common.

Persons found not guilty of scientific misconduct

Investigations into allegations of research misconduct can take extended periods of time, but it is essential that the procedure not be unduly rushed so that the accused be treated fairly with due process according to the law. Sometimes allegations are made by a young inexperienced whistleblower who does not appreciate the difference between scientific misconduct and "honest" unintentional mistakes in research. Accusations may also be made maliciously against someone by a competitor or a disgruntled employee so this possibility needs to be kept in mind. An accusation of research misconduct does not indicate guilt. Since 1988, the Public Health Service (PHS) in the USA has contended that if an investigation into scientific misconduct fails to disclose evidence of wrongdoing the identity of the accused should remain confidential but steps have been taken to try to change this. The releasing of findings in such cases could be construed as an uncalled-for incursion of privacy because of possible harm to the repute of the scientists acquitted of the accusation. Lyle Bivens, the former director of the Office of Scientific Integrity Review (OSIR) and Office of Scientific Integrity (OSI), believed that allegations of scientific misconduct damaged the reputation of researchers even if the person was found innocent of wrongdoing on a detailed investigation. Because of this policy the OSI only released the names of those shown to have committed misconduct and the later created federal Office of Research Integrity (ORI) retained this policy. A weakness in this policy was that persons suspected of research misconduct sometimes received much notoriety during the accusation phase prior to being cleared of misconduct. A failure to declare an accused researcher innocent of scientific misconduct leaves the public unaware of the outcome of the investigation and hence the accused remains covered by a cloud of suspicion.

Galileo Galilei. Perhaps the most famous scientist to have his views suppressed was Galileo Galilei (1564-1642). This renowned and thriving researcher was accused in 1632 of engaging in unethical activities and publishing erroneous and deceptive proclamations which he knew or should have known were fictitious. An investigation found the allegations sufficient for a formal inquiry and Galileo was found guilty by the authorities according to the laws at that time because of insufficient evidence to support his assertions. Sanctions were imposed on Galileo who was denied further grant support and forced to publically retract his contention that the earth moved around the sun.

James Bruce. In the early days of traveling around in uncharted or incompletely delved into territories voyagers sometimes returned home to disbelief. The most famous explorer of North Africa and Abyssinia (currently named Ethiopia) in the 18th century was James Bruce (1730-1794), also known as Abyssinian Bruce (*http://en.wikipedia.org/wiki/James_Bruce*). After being fired by the British consul-general in Algiers he decided to see the sights of North Africa alone. After sailing off in 1765 he arrived on the Red Sea coast and ventured inland into the virtually unknown kingdom of Abyssinia in search of the source of the Blue Nile and he became most renowned for his discovery of it. For 4 years nothing was known about him until his return to civilization, when sailing into Marseilles in 1773 with astonishing tales. He reported many potential encounters with death while leading an armed horse guard into numerous battles in Abyssinia and had been made governor of that kingdom. Other stories related to being an eyewitness to barbarities, orgies, and natural phenomena that had never been documented before. Bruce was widely applauded in Paris and London following his return, but it did not take long before distinguished English scholars, such as Samuel Johnson (1709-1784), James Boswell (1740-1795), and Horace Walpole (1717-1797) the fourth earl of Oxford considered his tales too good to be true and pronounced him a charlatan. Bruce became the butt of jokes in London for his journey into Abyssinia. Bruce wrote an extremely popular five volume book of his travels [2, 4], but it only provided food for those who wanted to poke fun at the author. Bruce remained proud of his accomplishments, but was resentful at the lack of their acceptance. His reputation only recovered after his death when later European voyagers in Africa confirmed observations that Bruce had made.

Mark and Linda Sobell. In the early 1970s Mark and Linda Sobell, a husband-wife research team, undertook a two year project to treat alcoholics at the Patton State Hospital in San Diego. The study suggested that some individuals physically reliant on alcohol can be coached to control their alcohol consumption. The Sobell duo documented that 20 alcoholics were taught to control their alcohol intake better than a comparable group whose therapy focused on complete abstinence [17, 18]. Their findings were contrary to conventional wisdom which held, and still generally does, that abstinence is the only successful therapy for physically dependent alcoholics. Their controversial results became widely known igniting a storm among alcohol researchers that precipitated allegations of them fabricating data and other improprieties. In early 1982 the controversy made the headlines in newspapers when a separate team of investigators achieved an effective control of alcohol drinking in one of twenty alcoholics over several years but it was questionable whether this alcoholic had indeed ever been physically dependent on alcohol. A subsequent follow-up of the subjects in the Sobell study by an independent group of researchers disclosed that most participants had a high incidence of re-hospitalization, alcohol-related arrests or bouts of heavy drinking, even while being studied by the Sobells [15]. Pendery and colleagues interviewed participants in the Sobell project and also scrutinized hospital and law enforcement documents. The Sobells attempted to prevent the Pendery team from carrying out their study arguing that confidentiality might be compromised. The published version of the Pendery paper lacked allegations of misconduct by the Sobells, but an earlier unpublished version insinuated that the research was not performed in the manner that the Sobells stated and that they reported the results inaccurately. In June 1982, the *New York Times* cited Malzmannas stating that the Sobell study was undoubtedly fraudulent [9]. By this time the Sobells were working for the addiction research foundation in Toronto. These allegations resulted in the appointment of a four person committee chaired by Bernard Dickens, a professor of law at Toronto University (UToronto) *(http://www.utoronto.ca/)* to investigate the charges. The committee interrogated the Sobells, examined their raw data, and listened to tape recorded interviews on experimental subjects as well as controls that had been made by the Sobells. The committee also obtained sworn affidavits from research assistants who participated in the research. They did not interview the research participants again, because the long term memory of alcoholics is unreliable. It is noteworthy that neither Pendery nor Maltzman testified. On November the 5th 1982 after an inquiry lasting 5 months the Sobells were found not guilty of fabrication except for one lapse of carelessness in which a statistic was estimated, but not calculated [13]. The committee was convinced that the Sobells performed the research as documented by them without faking the results. The Sobells were criticized, however, because of their untrue claim to have interviewed research subjects and individuals connected with them every 3-4 weeks throughout the study. They contacted most participating subjects less often and in some instances the pause between interviews was sometimes 3-6 month. The verdict of the Dickens committee failed to convince the critics who made the original allegations of a fraudulent study. They considered it an outrageous whitewash. Never the less the study by the Sorbels was not carried out according to a standard protocol and the sample size was not large enough to obtain statistically significant meaningful results [3]. Even though research misconduct was not documented the errors fell into the category of unintentional mistakes.

Karl Illmensee. Karl Illmensee received a PhD in 1970 after studying chemistry and biology in Munich. Subsequently he held professorships in Geneva, Salzburg and Innsbrook. While at the University of Geneva *(http://www.unige.ch/)* this embryologist and developmental biologist was the recipient of a research grant from the National Cancer Institute (NCI) for collaborative research with Peter C. Hoppe PhD (1942-2006) of the internationally renowned Jackson Laboratory in Maine (USA). In 1979 Illmensee and Hoppe shocked the scientific establishment by claiming that they had cloned three mice by transferring the nuclei from cancer cells into fertilized eggs from which the nuclei has been eradicated by delicate microsurgery. The eggs with the newly implanted nuclei were transplanted into foster mothers for further development and the results of his cutting edge research were presented at a professional meeting before being documented in *Cell* in 1981 [7]. The prevailing dogma at that time was that the cloning of a mammal was impossible so the Illmensee and Hoppe experiments met with considerable suspicion. The lay press gave the disputed experiment much publicity as this would have been the first time that a mammal had been cloned. Illmensee, the microsurgeon in the experiment, was accused of scientific misconduct by his colleagues. Members of the Illmensee research laboratory accused him of altering experimental protocols after finishing the experiments. These allegations were investigated at both the University of Geneva and at the Jackson laboratory. Illmensee justifiably admitted making the changes. The NCI was notified and it withheld the renewal of a $70, 000 research grant until the investigation was completed. A committee of some worldwide renowned scientists also looked into the matter. While evidence of research misconduct could not be established [10] his experiments were inadequately recorded and found to be scientifically of no value. Several researchers attempted unsuccessfully to replicate their study. Their failure to replicate the research did not automatically indicate that the experiments were fraudulent, because they may have lacked the microsurgical dexterity of Illmensee.

A few years later in 1984 a report in *Science* claimed that it was impossible to clone mammals by nuclear transfer but this [11], belief turned out not to be true and on July the 5th 1996, almost fifteen years after the Illmensee-Hoppe paper, the first living cloned mammal was born from an adult somatic cell, namely the lamb called Dolly [5]. After the negated academic beginning of his career Illmensee worked at the Gynecology Clinic of the University of Innsbruck from 1996 until his retirement in 2005 and he was a scientific adviser to the controversial cloning research of Severino Antinori from early 2001 *(http://science.orf.at/science/search?keyword=antinori&tmp=1419)*, In May 2007 Illmensee terminated his relationship with for the notorious company *REPROG* in Lexington, Kentucky that was owned by Panayiotos Zavos, a biologist specializing in reproductive physiology. The Zavos team claimed that they had cloned the first human embryo *(zavos.org)* after using nine microsurgically enucleated human donor oocytes and fusing them via electrical stimulation and activation with human granulosa cells from a patient desiring to have a child via somatic-cell nuclear transfer (SCNT) [20]. In 2007 he documented experimental cloning studies with human cells. Even though research on this topic was banned in many other countries, he was interested in being the first person to clone human cells [8, 19]. The embryo was allegedly created for reproductive purposes [20].

Robert Gallo and Mikulas Popovic. Accusations of research misconduct were leveled against Robert Gallo and his research team who were seeking the cause of the acquired immunodeficiency syndrome (AIDS). Initially the allegations were about stealing the responsible virus from a French scientist now known as the human immunodeficiency virus (HIV). Gallo also became embroiled in a dispute over a patent for a blood test for AIDS from a French scientist because he apparently used the "stolen virus" to generate the relevant reagents. A member of the Gallo research team, Mikulas Popovic was accused of scientific misconduct and both he and Gallo were found guilty of scientific misconduct by the ORI, but in 1993 the three member federal appeals board (FAB) of the DHHS overruled the ORI's decision on Popovic because of imperfections in the ORI investigation [1]. The overruling of the verdict on Popovic caused the ORI to throw in the towel on its long ongoing investigation of his former supervisor Gallo.

Rameshwar Sharma. In 1989 Rameshwar Sharma of the Cleveland clinic foundation was accused of claiming to have data that he did not possess in two unfunded NIH grant applications. Following an institutional investigation the ORI found him guilty of scientific misconduct and imposed administrative actions. Sharma challenged this decision by requesting a hearing before the research integrity adjudications panel. The charges against Sharma were dropped on August 6, 1993 after the ORI could not prove that by a preponderance of the evidence that Sharma committed scientific misconduct *(http://www.hhs.gov/dab/decisions/dab1431.htlm)*.

Simon Shorvon. In December 2000 Simon Shorvon, a 54 year old distinguished British researcher, was recruited to Singapore as director of the

National Neuroscience Institute (NNI), after leading an epilepsy research team at University College in London (UCL). He had been involved in a $5.6 million clinical study aimed at clarifying the genetic basis of Parkinson disease, epilepsy and tardive kinesia funded by the Singaporean government. Patients were requested to not take medications for Parkinson disease at least twelve hours before an extensive evaluation that included an analysis of blood samples. Then they received L-dopa, a well-known anti-Parkinson disease drug. If the patient responded positively to this drug the diagnosis was considered to be Parkinson disease rather than epilepsy or tardive kinesia. A research grant of approximately £3.5 million was awarded in 2002 and a collaborative research project involving investigators at NNI and UCL began. Shortly thereafter Shorvon became entangled in a dispute with Dr. Lee Wei Ling, who was a neurochemist and the deputy director of NNI. A scrutiny into accusations against Shorvon began in January 2003 after patients complained about their therapy. During that month Lee is stated to have found fault ethical violations in the research, but surprisingly she did not discuss her concerns with Shorvon, her co-principal investigator. Shorvon acknowledged making errors, but he maintained that the appraisal panel used exceptional approaches to explore allegations against him, such as locking him out of his office and reviewing years of e-mails. After an internal inquiry by the NII in March, 2003 Singaporean officials accused Shorvon of numerous ethical transgressions linked with breakdowns to achieve neurological information to acquire consent to change their medication levels. Ling had strong political connections, being the daughter of a senior minister in the Singapore government, as well as the sister of Lee Kuan Yew (1923—) prime minister of Singapore from 1959-1990. Shorvon, was suddenly stripped of his directorship of the NNI on 4 April 2003 the day after an investigative panel found him compromising patient safety and well-being. Shorvon considered the penalties for his alleged misdeeds harsh [6]. Because Shorvon and his colleagues struggled to enroll enough patients for the analysis they obtained lists of patients with Parkinson disease from two Singaporean hospitals and contacted them individually. The panel regarded this as an infraction of confidentiality. Considered just as serious was Shorvon's failure to inform the ethical oversight committee and the involved study subjects that they would be required to donate blood and would need to undergo many tests and be taken off their medicine. The informed consent that the patients were required to sign lacked these pieces of relevant information. The assessment of the patients generated an acute discomfort and even though it was not life-threatening some patients risked complications. The panel felt that the patients were treated like laboratory animals rather than as patients with rights. Despite his denial Shorvon was alleged to have tampered with patient files during the investigation and a police report was filed against him. Being from a different culture he did not comprehend the reason behind some accusations against him. Shorvon did not request ethical approval for procedures that were commonly performed by British tradition. Shorvon decided not to challenge the panel report so that he could depart from Singapore without further delays caused by an additional investigation. The day after Shorvon was fired he departed for the UK to resume his position at the UCL. The UCL planned to perform its own investigation of the allegations. When back in the UK in April 2003 Shorvon requested that the General Medical Council (GMC) review the allegations against him so that his name could be cleared. The GMC evaluated the evidence against him and concluded that he did not fall below the expected standard for human research. Then after obtaining the legal advice of an eminent counsel in regulatory affairs in the UK the GMC found no credible evidence to investigate Shorvon and terminated the inquiry. By 2005, the GMC concluded that the accusations against him by the Singaporean medical council (SMC) could not be substantiated beyond a reasonable doubt and brought its inquiry to an end. Later after reviewing the GMC's decision the SMC challenged the authority of the GMC and maintained that it came to the incorrect conclusion after failing to consult with the SMC. Physicians are rarely pursued by medical councils beyond the boundaries of their own countries, but with Shorvon the chase by the SMC was particularly aggressive. On December 21, 2006, the British High Court ruled that the GMC correctly terminated the investigation of alleged wrongdoings by Shorvon. The High Court rejected the SMC assertion and pointed out that the GMC's assessment of the charges against Shorvon were reasonable. The ruling of the High Court disappointed the SMC, which had found Shorvon guilty of professional misconduct. Shorvon was delighted with the ruling because it destroyed an unwarranted stain on his reputation and he was now legally able to apply for research grants, which he could not do until the dispute over the charges against him were legally settled. However, since the Shorvon argument the GMC made the council duty-bound to accept proclamations of regulatory bodies in foreign countries. The Shorvon dispute illustrates how researchers working in foreign countries need to recognize that regulatory systems in different countries are not identical and dissimilar conclusions can be reached when allegations against misconduct are disputed. In an unprecedented move, the local SMC removed Shorvon, from being a top researcher in the NII and tracked him from the shores of Singapore all the way to the British GMC. SMC considered it proper to go after Shorvon even after the GMC disagreed with its ruling. The SMC took the extraordinary step of taking the Shorvon case to the High Court of Justice in London. The SMC apparently found it essential to penalize Shorvon in its own jurisdiction by striking him off the Singaporean medical register, the maximum sanction available to the SMC. This ended Shorvon's ability to practice medicine in Singapore, but it also attempted unsuccessfully to terminate his practice in the UK (*http://www.nature.com/news/2006/061218/full/061218-15.html*). The SMC considered the GMC conclusion indefensible and in violation of an obligation of impartiality to the SMC by not appropriately checking with it. When the High Court made its ruling Shorvon was a professor and subdean at the Institute of Clinical Neurology at UCL. The Shorvon affair will undoubtedly impact negatively on the recruitment of outstanding renowned researchers into Singapore for the development of its international biomedical research complex known as the Biopolis.

References

1. Anderson C, Popovic is cleared on all charges; Gallo case in doubt. Science, 1993. 262(5136): 981-983.
2. Beckingham CF, Travels. Abridged edition. 1964, New York: Horizon Press.
3. Boffey PM. Alcohol study under new attack. The New York Times. 1982 June 28.
4. Bruce J, Travels to Discover the Source of the Nile. In the Years 1768, 1769, 1770, 1771, 1772 and 1773, Five volumes. 1790, London: GGJ and J. Robinson.
5. Campbell KHS, McWhir, J., Ritchie WA, and Wilmut I, Sheep cloned by nuclear transfer from a cultured cell line. Nature, 1996. 380(6569): 64-66.
6. Enserink M, Singapore trial halted British scientist fired. Science, 2003. 300(5617): 233.
7. Illmensee K and Hoppe PC, Nuclear transplantation in Mus musculus: developmental potential of nuclei from preimplantation embryos. Cell, 1981. 23(1): 9-18.
8. Illmensee K, Mammalian cloning and its discussion on applications in medicine. Journal of Reproductive Medicine and Endocrinology, 2007. 4(1): 6-16.
9. Marlatt GA, The controlled-drinking controversy. A commentary. American Psychologist, 1983. 38(10): 1097-1110.
10. Marx JL, Bar Harbor investigation reveals no fraud. Science, 1983. 220(4603): 1254.
11. McGrath J and Solter D, Inability of mouse blastomere nuclei transferred to enucleated zygotes to support development in vitro. Science, 1984. 226(4680): 1317-1319.
12. Mosher EH, Blair C, and Silver GA, Fraud investigation. Science, 1982. 215(4539): 1456, 1458.
13. Norman C, No fraud found in alcoholism study. Science, 1982. 218(4574): 771.
14. Norman C, How to handle misconduct allegations. Science, 1989. 243(4489): 305.
15. Pendery ML, Maltzman IM, and West LJ, Controlled drinking by alcoholics? New findings and a reevaluation of a major affirmative study. Science, 1982. 217(4555): 169-175.
16. Shaw GB, Doctor's Dilemma Act 1. 1911, New York: Brentano
17. Sobell MB and Sobell LC, Alcoholics treated by individualized behavior therapy: one year treatment outcome. Behaviour Research and Therapy, 1973. 11: 599-618.
18. Sobell MB and Sobell LC, Alcoholics treated by individualized behavior therapy: one year treatment outcome. Behaviour Research and Therapy, 1976. 14: 195-215.

19. Unknown. Unknown. Süddeutsche Zeitung 2007 June 19.
20. Zavos PM and Illmensee K, Possible therapy of male infertility by reproductive cloning: one cloned human 4-cell embryo. Archives of Andrology, 2006. 52(4): 243-254.

CHAPTER 24

COI

An investigation of allegatonss of research misconduct by a member of the faculty of the institution where the person is employed is ideally not appropropriat because financial and public relations conflict of interest (COI). This is comparable to having a jury of foxes judge affect the guilt or innocence of Reynard the Fox in Aesop's fables after he broke into the henhouse and made off with fat hens [12]. As Bernard Shaw (1856-1950) the Irish playwright mentions in his famous play Doctor's Dilemma every profession has a conspiracy against the laity [16] and particularly for this reason it is desirable that nonscientists play a role in selected investigations of research misconduct.

As pointed out in an earlier chapter research at institutions and universities requires a considerable infrastructure to support their resources. One of the necessary components of the institutional research administration is a Research Integrity Office (RIO) that oversees COI among other functions. Intellectual, financial, personal, academic, professional and social COI are common and inevitable [28, 44]. For faculty who receive salary support from grants and other sources a conflict of commitment often occurs due a struggle between the allotment of time between different activities. COI is often linked with entrepreneurial pursuits and plays an important role in several aspects of research. For many years universities and research institutions lacked policies on COI, but now they have guidelines that deal especially with financial conflicts generated by interactions with different industries. Despite not being a component of the accepted definition of scientific misconduct a failure to disclose a COI may be as serious as research misconduct.

COI are unavoidable and extremely important in all walks of life and when indicated the person possessing them needs to be fully transparent and able to act without suspicion and loss of public confidence. They occur in the peer review of journal articles and grant applications, research (research COI), institutions (institutional COI), and research administration (administrative COI). Some COI occur because a desire for recognition, rivalry with a competitor, or the over-enthusiastic fondness for a particular theory. Important COI of faculty relate specifically to their research and particularly to their financial working relationships to industry.

Scientists often overlook COI, but vigorous COI have always stemmed from the intellectual investments of researchers in specific hypotheses that they had pursued. Professional decisions or bias in research can potentially be manipulated by financial gain. It is not only unfeasible to eradicate financial and other COI entirely, but this is also probably not wanted, but COI needs to be disclosed. Academic COI based on professional interests are inescapable. Shared intellectual passions generate a great deal of additional strong COI. Authors of manuscripts from companies that support certain products are commonly biased by a COI and the documented findings need to be taken with a pinch of salt until verified by independent investigators.

Accusations of abused clinical trials by academic researchers because of COI commonly appear in popular newspapers. Researchers are now required by law to disclose COI under all conditions where they potentially exist. This is required at professional society presentations, in the submission of manuscripts for publication and by reviewers of grant applications and articles submitted to journals. Disclosures shelter researchers and institutions from potential repercussions. Some researchers tradionally disclose all financial and commercial COI, but others do not divulge most important investments in companies. Some scientists serve as consultants for one or more company but do not hold equity in any company dealing with their research field.

The development of greater interactions between university faculties and the private sector, most notably small venture capitalists became a new opportunity for researchers. It is unacceptable to gain monetary benefit from research in universities or manipulate the stock market based on research findings. Ideally researchers in academic settings should interact in a collegial way in an impartial environment without being worried about how it affects their financial status. Scientists should not have an actual or perceived COI regarding the outcomes of their research and especially if they are involved in clinical trials. Researchers testing products for companies would be prohibited from receiving honoraria from those companies. Some companies, especially in the pharmaceutical industry, rely heavily on university faculties for guidance and for carrying out clinical trials on their promising new drugs. This relationship creates a COI which seriously hampers the desired interaction between the academia and industry. Fortunately waivers from specific restrictions can be granted by universities provided that they are reviewed by the NIH. A deluge of letters to *Science* addressed the issue [2, 8, 35, 38, 44]. The financial rewards for interacting with pharmaceutical and biotechnology companies became so lucrative that the COI cannot be ignored.

Manuscript and Grant Application Evaluations

Until the research activities of scientists became involved with industry their COI issues focused mainly on the review of manuscripts and grant applications during the peer review process and the presentation of their research in publications and/or at professional scientific meetings. In journal articles authors provide readers with new information that may be biased because of COI related to the research of the authors. It is important that journal editors need to understand the COI related to articles that they agree to publish. Because of the frequency of COI some journal editors, such as those of the *New England Journal of Medicine* [3] and *The Journal of the American Medical Association* [27] wrote editorials on the subject. Those who have discussed the matter have commonly stressed the need for disclosure about both financial interests and sponsorships, but those researchers who have ignored this advice have faced dishonor and humiliation. Scandals about research publications have not been particularly extraordinary because for many years journals passed over the importance of making disclosures obligatory.

In early 2009 NIH peer review had undergone reform *(http://enhancing-peer-review.nih.gov/index.html)*, but additional changes were needed because of the enormous administrative burden of the review process [16]. Measures were taken to understand and curtail conflicts. When members of review panels known as Study Sections seek funding they submit their own grant applications which commonly reach the same committee for review, but the regular committee does not judge the proposal with those of competitors. When a scientist is a member of the regular peer review panel that individual's proposal is judged by a special *ad hoc* committee but sometimes it receives an extensive mail review. The peer review process demands that invited reviewers with expert judgment devote time and energy with some altruism as a civic duty for their own profession. To provide an appropriate review of the work of another scientist requires an incentive and this comes from recognition and mutual tradeoffs. It cannot be justified on a cost benefit analysis as reviewers of grant applications receive only a token honorarium for considerable effort and reviewers of manuscripts for journals are not paid.

Conflict of Interest in Peer Review

Members of the *National Academy of Sciences (NAS)* communicate papers to the *Proceedings of the National Academy of Sciences* the journal of the society and also select individuals to serve as peer reviewers for the manuscript. By 1992, *Science* made all authors and peer reviewers of manuscripts notify the editor of any financial or other association, that could be interpreted as a COI. The *New England Journal of Medicine* began to reject review articles by authors with a financial COI in the topic being reviewed. Cases of scientific misconduct or unintentional scientific error have sparked an upsurge in interest in the peer review process. When COI exist the key is full disclosure by honest individuals. Journal editors and funding agencies can decide whether a particular person should be allowed to evaluate a specific grant application or submitted manuscript.

Before researchers in academia became involved in research endeavors with financial rewards the only COI that they faced related to the peer review of manuscripts, and abstracts submitted to professional societies for scientific meetings as poster or platform presentations and those grant applications submitted to funding agencies. It is imperative that steps be taken to avoid having a COI in the reviewers of the research of others.

For the peer review of grant applications and manuscripts submitted to journals judgment is needed from knowledgeable individuals to evaluate their merit. Equivalent opinions are needed for abstracts submitted to professional societies. Persons invited to review the merit of such contributions may have a COI that leads to a positive or negative impression. It is important that reviewers provide non-biased objective opinions that do not favor friends and colleagues and do not discriminate against competitors. The nature of peer review forces researchers to make judgments that influence the livelihood of both comrades and rivals. It places competitors for sparse funds in opposition to each other in transposable roles. At one time reviewers for grants are arbitrators and on other occasions applicants. Reviewers traditionally sign a document of confidentiality and absent themselves from those parts of committee meetings in which they have a COI, or a potential perceived COI. By being absent from the proceedings at these times the person has no knowledge of the specific reactions of the other committee members to the grant applications of particular individuals with whom they have a conflict. COI cannot always be avoided because relatively small groups of researchers focus on highly specialized areas of research and everyone working on the discipline eventually gets to know each other personally or by reputation. The COI should be disclosed to the editors of journals and the officials involved in the evaluation of grant applications and program committees of scientific meetings.

COI Resulting from Entrepreneurial Research

Before the *Bayh-Dole Act* of 1980 it was extremely rare for scientists involved in biomedical research to have significant patents or equity in a company directly related to the person's research. After the passage of this Act academicians developed working relationships with the corporate world and its financial rewards. Until then academicians rarely had financial COI and editors of professional journals were not concerned with this issue. The *Technology Transfer Act* also provided basic researchers with an opportunity to generate income and for some senior investigators this resulted in a vast wealth, but on the negative side this enhanced the likelihood of scientific misconduct as researchers could place their own interest above that of the community. An increasing number of entrepreneurial researchers develop COI by establishing patents on their intellectual property (IP) and in forming start-up companies that are sometimes eventually sold. Academic research now generates an increasing number of patents and more and more researchers develop business relations with industry and the resultant COI. Financial COI by physicians and clinician-scientists because industrial connections have been considered detrimental to medical doctors because it became difficult for them to place convincingly the interests of their patients ahead of their financial rewards and to exclude bias when arriving at therapeutic decisions that must not be hindered by COI.

In 1988 a major uproar was caused by researchers breaking rules on reporting income from pharmaceutical companies. Allegations of COI in academic research became common and researchers were eventually required to disclose COI by law and this promoted confidence and prevented suspicion in the eyes of the public. Academic COI based on professional interests are unavoidable and by 1989 it became clear that several types of COI exist and that rules were needed to deal with those in scientific research.

After 2000 numerous researchers in medical schools had financial COI mainly because of biomedical connections with industry. When the new ball game started researchers, professional societies and journals moved into unchartered territory and needed to respond to financial COI. Over time more and more researchers had equity in pharmaceutical and biotech startup companies and established researchers became paid consultants for industry because of their expertise in disorders in which they were performing their research. After it became feasible for scientists to generate revenue from their personal research in this new arena Pandora's box was opened largely because of COI.

Research Supported by Pharmaceutical Companies

When the research of scientists is financially supported by a pharmaceutical company an undeniable COI exists as it was in the fraudulent research of Scott S. Reuben that was funded from 2002 to 2007 by the drug company Pfizer. All clinical trials evaluating new potential drugs are not expected to always yield positive results, but Reuben never had a negative clinical trial when he evaluated pharmaceutical agents produced by Pfizer. Also the clinical trials by Reuben involving Pfizer drugs yielded bigger therapeutic benefits than what other researchers obtained in comparative clinical trials. Indeed studies by other investigators sometimes found no significant therapeutic benefit even though Reuben reported a positive effect. Reuben was a member of Pfizer's speaker's bureau and one of his studies was the only one supporting the use of a particular expensive drug combination for spinal surgery patients.

COI Caused by Patents

Institutional patent policies were also needed in conjunction with an infrastructure to aid in the processing of patents and licensing arrangements. An outrageous COI was revealed by the *Sunday Times* and the television station Channel 4 in the United Kingdom (UK) related to Andrew J. Wakefield who was responsible for a vast number of children not being vaccinated for the combined measles, mumps and rubella (MMR) vaccine. In June 1997, almost nine months prior to an announcement by Wakefield at a press conference that single vaccines were probably safer than the MMR vaccine and the publication of his notorious paper in *The Lancet* [46] Wakefield filed for a patent on a single measles vaccine, which had little hope of triumph unless support for MMR was annihilated. Confidential records indicated that a group of companies planned to raise venture capital from professed inventions—including a vaccine, testing methods, and strange potential miracle cures for autism.

COI Related to Business Interest

An example academic scientist venturing into the commercial world is illustrated by the story of Sidney Gilman, a former professor of neurology at the University of Michigan (UM) *(http://www.umich.edu/)* [36]. Gilman graduated from the University of California, Los Angeles (UCLA) *(http://www.ucla.edu/)* with honors and then underwent further training at the Medical School of Harvard University (Harvard) *(http://www.harvard.edu/)* before joining UM. While at UM Gilman helped create a national center for research in dementia and the hospital's neurology service was named after him. Like other researchers studying Alzheimer disease Gilman tried to find a cure for this devastating disease. For most of his career his work beyond the UM campus was with professional journals and national advisory committees and for this he received virtually no compensation. In 2000 when 68 years old Gilman continued to desire recognition and he started consulting for two drug companies. Soon thereafter he was contacted by *Gerson Lehrman* an expert network firm formed in 1998 to unite academicians with financial firms. Investors desire contacts with experts in drug development and particularly persons involved in clinical trials and *Gerson Lehrman* served this important function. Gilman's relationship with this company rapidly put him in contact with more than 40 clients, and this gave rise to his participation in 50-100 meetings a year at about $1,000 per encounter. By 2006 *Gerson Lehrman* was asked by Mathew Martoma, a trader with *SAC Capital*, to find an expert familiar with bapineuzumab, a drug under development for Alzheimer disease. At the time Gilman was chairman of the board monitoring the clinical trials of this drug, and was an ideal clinician-scientist to connect with Martoma and within weeks the two of them were talking. Initially a fund of Martoma called *CR Intrinsic* purchased shares in *Wyeth* and *Elan*, the companies developing bapineuzumab. The owner and founder of *SAC Capital* Steven A. Cohen was also encouraged to buy these shares for the company's hedge fund, but after the sale of large investments in the two pharmacy firms Cohen was required to testify in the insider trading scandal of Martoma [30, 37]. Later Gilman informed Martoma that the findings in the clinical trials on bapineuzumab were not faring as desired so all shares in *Wyeth* and *Elan* were sold. By July 2008 Gilman was not only a researcher studying Alzheimer disease, but also a member of panels of the Food and Drug Administration (FDA) and ideally situated to evaluate clinical trials for promising new drugs for this dementia producing disorder. While serving on *Neurobiology of Disease* Gilman delivered a paper on the clinical trial on bapineuzumab showing that this potential drug made by *Elan* and *Wyeth* was not as effective as originally hoped. Those in the audience included David Markowitz, who immediately recommended that his clients sell their shares in *Elan* and *Wyeth*. Unbeknownst to the audience a draft of Gilman's presentation had been emailed by him to a trader at an affiliate of one of the USA's most prominent hedge funds. Aside from his connections with expert networks, Gilman was hired to serve on the advisory boards of several financial firms, including *Pequot Capital,* which went out of business in 2010, after acquiring $25 million of stock in *Wyeth* and $20 million in *Elan*, while Gilman was overseeing the clinical trials being performed by *Wyeth* and *Elan*. In 2007 the stock was sold before the shares of both companies fell.

In 2010 Gilman received a distinguished achievement award from UM and his life changed dramatically on November 20, 2012 when he discovered that Martoma was arrested on a charge of insider trading. Because of his unethical and illegal behavior UM severed its connection with Gilman and on November 28, 2012 Gilman submitted a letter of retirement from the university. Gilman helped hedge fund investors benefit unlawfully in a profitable $276 million insider trading conspiracy in which his expertise and scientific knowledge played an important role. Gilman was extremely fortunate that prosecutors were not charging him with wrongdoing, because he agreed to cooperate in the prosecution of Martoma. He did, however, have a financial cost and agreed to pay the government $234, 000 under a non-prosecution agreement for what he earned from *Gerson Lehrman* and *Wyeth*. An even bigger impact on Gilman was being ostracized by UM after chairing its departments of neurology for seven years (1977-2004). The Gilman saga illustrates the type of quagmire that a distinguished scientist can get caught up in when money becomes an evil force leading researchers astray. It is despicable that Gilman had the audacity to leak confidential information that influenced the price of shares on the stock market.

Unethical Industry Sponsored Clinical Trial

In 2003 psychiatric researchers at the University of Minnesota were involved in an industry sponsored clinical trial involving antipsychotic drugs [15]. One of the patients in the study named Dan Markingson was recruited by researchers despite the fact that he could not provide valid informed consent and his mother Mary Weiss vehemently objected to his participation. Markingson was acutely psychotic and suffered from delusions involving demons. He was considered incompetent of making decisions for himself and was hospitalized under an involuntary commitment order that legally forced him to obey the recommendations of the psychiatrist who enrolled him into the clinical trial. Despite desperate attempts by the mother to get Markingson removed from the clinical trial the psychotic patient eventually committed suicide by cutting his throat. In addition to the unethical aspects of this case a financial COI also existed because the involved psychiatrists were provided with incentives to maintain subjects in the trial for as long as possible. To add insult to injury the study coordinator falsified the initials of psychiatrists in records of the study and failed to inform Markingson of new risks related to the study.

Research Collaborations Between Academicians and Other Federal Funded Investigators

The National Technology Transfer and Advancement Act (NTTAA) produced a COI between the findings of readily available federal funded research and the aspiration to enhance the ability of industry to be competitive. There is a distinct likelihood of withholding the findings from federally supported research and handing them over to a corporation. Research collaborations between academicians and other federal funded investigators face critical ethical issues. A new treatment can make a fortune for a pharmaceutical company.

Need For Guidelines on Conflict of Interest

By 1989 it had become well established that different types of COI exist and that rules were needed to deal with them in scientific research, but COI had not been extensively evaluated and policies remained to be established. Because COI in science had become a major concern congressional hearings were held in the USA under the chairmanship of Theodore "Ted" Weiss (1927-1992) (Democratic, New York). A consensus was reached that federal funded research cannot be contaminated by potential prejudice caused by financial COI. In July, 1989, Bernadine Healy (1944—2011), director of the National Institutes of Health (NIH) at the time, maintained that guidelines were needed to ensure that personal financial interests do not influence research. After it became reasonable for researchers to interact with industry some, but not all, universities prohibited their faculty from obtaining equity in firms that license technology developed in universities. Some institutions earnestly participated in startup ventures, but with guidelines. By concentrating on the corporations and disregarding universities there was no safeguard against favoring faculty who benefited their institution financially.

Together with the Alcohol, Drug Abuse and Mental Health Administration (ADAMHA) the NIH released guidelines on COI on September 15 1989 and some were controversial. Making the instructions available in this unorthodox way was unusual because by law NIH regulations must be published in the *Federal Register* and should also be signed by the Secretary of Health and Human Resources. All researchers involved in ADAMHA and NIH funded research were required to provide a "full disclosure of all financial interest and outside professional activities to their host institution". Without an institutional COI policy research funding from neither the NIH nor the ADAMHA could be obtained. Researchers testing products for companies would be prohibited from receiving honoraria from those companies and there was concern that this restriction would seriously hamper the desired interaction between the academia and industry. Some businesses, especially in the pharmaceutical industry, rely heavily on the faculty of universities for guidance. Fortunately waivers from specific restrictions could be granted by universities provided that they were reviewed by the NIH. Scientists with connections to contemporary biotech companies, especially those focusing on immunology, antisense technology and neuroscience with potential high payoffs did not agree on how to proceed. Also, the guidelines required COI disclosures related to their spouses, dependent children and other dependents [34] and the disclosures needed to be updated at least once annually. All individuals involved in the funded research were prohibited from having personal holdings or options in any company that would be affected by the outcome of the research or the products being evaluated in the research project. Payments to pediatric psychiatrists from drug companies for giving lectures on their drugs seemed rather high and inductive to getting the message of the pharmaceutical companies accepted. An uproar over COI in research reached a crisis level and if the NIH did not establish strict requirements there was a high likelihood that Congress would get involved with strict legislation. The NIH decided to take the bull by the horns and organized a two day meeting on the subject in 1989 [33]. By 1991 it was clear that all academic and research institutions needed policies related to COI and the creation of appropriate institutional policies needed to consider numerous issues [17], including definitions, what faculty should disclose, who should keep the records, who decides on what should be kept confidential, what disclosures would spark a review of COI, who would conduct the review, what criteria would be used to judge the acceptability of an activity, and what sanctions would be applied for non-compliance. In this complex arena it was important to prevent an overreaction with excessive regulations that would hinder research initiatives and particularly collaborations between academia and industry. In 1992 authors could decide which probable COI to disclose and some individuals failed to make known significant equity investments, but some divulged all business-related connections. Some served as advisors for a number of companies but held no equity in any industries related to researcher's expertise. By February 1993 the NIH had not issued CIO guidelines, but in December 2008 the NIH stiffened the federal COI rules for recipients of NIH grants. Some research institutions stated that they would publically disclose income paid to their faculty from drug companies. By January 2010 the NIH was drafting new harsher requirements for reporting financial COI. After considerable discussions a consensus was reached that disclosure was the ideal method of dealing with conflicts of COI and other types of conflict.

Drug Industry and Pediatric Psychiatry

Atypical antipsychotic drugs became exceptionally trendy in psychiatric disorders as they could control almost any radical conduct, often within minutes particularly to the delight of despondent families. The expanding use of antipsychotic drugs in children was intimately associated with the progressively more frequent controversial diagnosis of the mood disorder known as pediatric *bipolar disorder* that is characterized by aggravation, euphoria, *depression* and, sometimes ferocious flare-ups. Therapeutic agents occasionally designated as major tranquilizers deaden neurons within the brain to gushes of dopamine, a neurochemical that has been implicated in euphoria and psychotic delusions. Pediatric psychiatrists developed a cozy relationship with pharmaceutical companies, particularly in Minnesota and this rapport resulted in the psychiatrists receiving large payments from the drug companies [19]. As a consequence of the payments the medical doctors prescribed off-label "atypical antipsychotic" drugs for a variety of psychiatric disorders in childhood. A noteworthy income to pediatric psychiatrists came from giving lectures on promotional medical education. For example in 2003 Dr. George M. Realmuto was hired by *Johnson and Johnson* to deliver three talks for $5, 000 on their psychostimulator Concerta (methylphenidate) which had been approved by the FDA for treating attention deficit hyperactivity disorder (ADHD). At the time Realmuto's university salary was $196, 310 [19].

An example of using an off-label prescription occurred when the teenager named Anya Bailey developed an *eating disorder* and was referred to a psychiatrist at the *University of Minnesota* (UMN) *(http://www1.umn.edu/)* who prescribed the "atypical antipsychotic" drug Risperdal [19]. This powerful drug was approved by the FDA for treating schizophrenia, but not for the off-label treatment of eating disorders. The average number of prescriptions was high particularly when money was received by the doctors from the drug companies. A common side effect of Risperdal is an increased appetite and because doctors may prescribe off-label FDA approved drugs as they see fit even for entities that have not officially been evaluated. While on Risperdal the Bailey child gained weight and also benefitted from this drug because it subdued occasional angry outbursts and the hearing of voices over the previous five years. Her mother was unaware at the time that this drug had not been approved to treat children and clinical trials only included a handful of participating children. Just as astounding the mother learnt that the UMN psychiatrist responsible for the care of her young daughter received more than $7,000 from *Johnson and Johnson* the manufacturer of Risperdal between 2003 and 2004, as a stipend for lectures on one of the company's drugs. The average number of prescriptions written for children by psychiatrists in Minnesota for a relatively new class of drugs known as "atypical antipsychotics" was high between the years 2000 and 2005. Although medical doctors were commonly on the payroll of drug companies at the time they commonly claimed that despite financial COI payments from drug companies that they received did not influence the medications that they prescribed. Increasing payments to pediatric psychiatrists coincided with an increasing prescription of "atypical antipsychotics" to children. These drugs are known as second generation antipsychotics and major tranquilizers and the best sellers among those prescribed for psychiatric childhood disorder are Risperdal, Seroquel, Zyprexa, Abilify and Geodon were best sellers among the drugs being prescribed for psychiatric disorders during the initial years of this millennium. Between 2000 to 2005 pediatric psychiatrists in Minnesota appeared to have written three times as many prescriptions for "atypical antipsychotic" drugs for children as psychiatrists who received little or no financial compensation from drug companies. Despite shortcomings of obtaining the information from pharmacists the data hints at the prescribing patterns in Minnesota. The manufacturers of drugs command a major role in the prescription of pharmaceutical agents at all levels of care. Firstly they pay doctors who prescribe and recommend their drugs. Secondly they provide teaching sessions about the underlying diseases by paying instructors, who have a COI. The audience is also commonly seduced by a free meal at a notable restaurant. Medical doctors participate in the clinical trials organized by drug companies that document the guidelines for all participating doctors. Financial COI created by the receipt of payments from pharmaceutical companies can affect therapeutic decisions without being obviously appreciated. Money paid by drug companies to medical doctors is difficult to distinguish between the honest practice of medicine. For the well-being of patients this issue eventually became extremely controversial in health care, particularly in psychiatry. Between 2000 and 2005 psychiatrists in Minnesota received more income from drug manufacturers than doctors in any other medical specialty. For individual psychiatrists the imbursements ranged from $51 to over $689,000 and the median compensation was $1,750.

Different psychiatrists at UMN preferred dissimilar "atypical antipsychotic" drugs for psychiatric disorders in children. Johnson and Johnson marketed both Risperdal and Concerta. Realmuto assisted in a study of Concerta which was approved by the FDA for treating ADHD. Realmuto appreciated why some people were concerned that his payments for promotional medical education lectures on Concerta might influence him to prescribe this drug but he thought that this would not persuade him to recommend another drug produced from the identical pharmaceutical company. He conceded, his connection with Johnson and Johnson might encourage him to try other therapeutic agents marketed by that company but only if the other if the other drugs were favorable. As a rule Johnson and Johnson and other pharmaceutical companies choose lecturers to talk about their drugs if they have prescribed them or have either performed research on them or are aware of studies on them. Lectures sponsored by drug companies educate physicians about newly established pharmaceutical agents. Indeed, many medical doctors who prescribe these drugs receive no financial reward from the pharmaceutical business. Virtually all psychiatrists who receive payments from drug companies maintain that they remain independent of the pharmaceutical industry. Medical doctors who receive payments, pharmaceutical industry from drug companies are sometimes suspected of being bought by the drug company to prescribe the company's drugs, but most physicians do not believe that this takes place and there is little if any evidence to substantiate this allegation.

An apparent sudden popularity in the diagnosis of pediatric bipolar disorder coincided with a change in prescriptions from antidepressants, such as Prozac, to the much more pricy "atypical antipsychotics". In just Minnesota alone more than $521, 000 was spent mostly on "atypical antipsychotics" for children on Medicaid during 2000. This expense gradually increased over time and by 2005 it topped over $7.1 million, a 14-fold increase. In 2006 Zyprexa (made by Eli Lilly) had $4.36 billion in sales, Risperdal (made by Johnson and Johnson) $4.18 billion and Seroquel (made by AstraZeneca) $3.42 billion. Numerous psychiatrists in Minnesota are alleged to have stated that manufacturers of drugs were paying them almost entirely to discuss the highly contentious childhood bipolar disorder. One of the first and perhaps most influential studies was financed by the pharmaceutical company AstraZeneca. This study by a research team under the leadership of Dr. Melissa DelBello at the *University of Cincinnati* followed the moods of 30 adolescents diagnosed with a bipolar disorder for 6 weeks. Half of the teenagers received Depakote (divalproek), an antiseizure drug used in the therapy of both bipolar disorder and epilepsy in adults. A combination of Seroquel (xeroquel, ketipiner) and Depakote was administrated to the other 15 adolescents. The two groups responded more or less evenly until the final days of the investigation, when subjects in the Seroquel group achieved less well on a standard measure of mania. In an inconclusive study reported in 2002 DelBello and her coauthors [13]. Seroquel in combination with Depakote was documented to be "more effective for the treatment of adolescent bipolar mania" than Depakote alone. In 2005, a respected committee from across the country examined all clinical trials stated to be of bipolar children on psychotic drugs. The panel concluded that ""atypical antipsychotics" should be considered as a first-line of treatment for some children. The guidelines for such treatments in children and adolescents with bipolar disorder were published in *The Journal of the American Academy of Child and Adolescent Psychiatry* [25]. Some doctors on the panel that reached this knowledgeable opinion served as speakers or consultants to manufacturers of "atypical antipsychotics" according to disclosures that accompanied the guidelines. Despite an obvious COI Kowatch a psychiatrist at Cincinnati Children's Hospital and the lead author of the guidelines stated that the drug makers' support had no influence on the conclusions. AstraZeneca hired DelBello, who earns $183, 500 annually from the University of Cincinnati, and Kowatch to give sponsored talks. They later undertook another study comparing Seroquel and Depakote in bipolar children and found no difference. Many medical doctors on the pay role of pharmaceutical companies were originally reluctant to provide information on how much the drug industry was paying them, but now under the *Sunrise Law* the company is required to disclose this amount by law. Medical doctors commonly claim that payments received by them do not influence the treatment of their patients. Many psychiatrists have received marketing or consulting income from as many as eight drug companies, including all five makers of "atypical antipsychotics". Realmuto had heard DelBello speak several occasions and her talks persuaded him to use combinations of Depakote and "atypical antipsychotics" in bipolar children. Some psychiatrists who advocate the use of "atypical antipsychotics" in children acknowledge that the evidence supporting this use is thin. But they say children should not go untreated simply because scientists have failed to confirm what clinicians already know. "We don't have time to wait for them to prove us right, " said Dr. Kent G. Brockmann, a psychiatrist from the Twin Cities who made more than $16, 000 from 2003 to 2005 delivering drug talks and one-on-one sales meetings, and in 2006 he was a leading prescriber of "atypical antipsychotics' to Medicaid children. Therapy with the "atypical antipsychotics" enabled the Bailey child to regain her appetite, but it also heavily sedated her. Significant side effects of "atypical antipsychotics" are rare and difficult to predict, but they include a rapid weight gain. In 2006, the FDA received reports of more than 29 children dying and at least 165 more suffering serious side effects in which an antipsychotic was listed as the "primary suspect." Unlike most universities and hospitals, the Mayo Clinic restricts doctors from giving promotional medical education lectures. A review of 370 clinical trials funded by for-profit organizations biased the interpretation of the results [1]. Turner and coauthors from the department psychiatry at Oregon Health and Science University (OHSU) (*http://www.ohsu.edu/xd/*) reviewed the data from 74 FDA-registered studies on 12 antidepressant drugs involving 12, 564 patients. For reported trials they compared the published outcomes with the FDA outcomes. Only 74 (31%) of the FDA studies were not published and these were almost entirely negative and the failure to publish them biased the overall evaluation of the drugs to positive effects [45].

Some universities and research institutions permit their faculty to consult for companies but sometimes have restrictions, such as not allowing them to be the President or Chief Executive Officer (CEO) of the company. Researchers frequently serve as paid consultants, advisory board members and on speaker bureaus promoting specific products. Some even agree to be listed as authors on ghost-written publications reporting research on which they participate little or no way. Those who work for pharmaceutical companies earn their income from the revenue produced by the industry [8].

Entrepreneurs Face COI and Conflicts of Commitment.

By 1992, many biomedical researchers had equity in biotech companies, but there was still no consensus as to whether their financial COI warranted attention. Eventually journals, professional societies and universities came up with policies on how to address the issue. Following these episodes a consensus emerged maintaining that authors should not only notify editors of financial interests in papers that they submit for publication, but that this COI must also be passed on to their readers. Today this is standard policy with all reputable journals and professional society meetings. The work done in academic laboratories can affect the destiny of companies depending on whether the research is or is not published in highly ranked journals that gather scientific recognition [5]. The American Council on Science and Health (ACSH) recognized that industry had been a powerful driving force in generating new technology and in supporting research, but since about 1998 extensive merged and undermined the public trust in research. Moreover a handful of scandals involving clinical studies with industry emerged and activists attempted to exclude academic researchers with industry connections from participating in governmental scientific advisory boards. Some even advocated that persons with industrial COI should not write review articles for journals. In a discussion to bring reason to the issue of dealing with COI the ACSH issued an extensive review of the topic [4] *(http://www.acsh.org/wp-content/uploads/2012/04/20080401_scrutinizing1.pdf)*.

Financial COI involving tangible or potential fiscal rewards can bias research and influence professional judgments and prejudice research. The source of funding is an important cause of prejudice, particularly if it comes from a pharmaceutical company with a vested interest in the outcome of the research that it funds. Such financial COI should be disclosed by authors in publications, in lectures and presentations given at professional societies to enable readers or the audience to better understand potential bias in the interpretation of the reported data [48]. In research sponsored by a pharmaceutical company the sponsor's drug consistently fares better than drugs of competitors [11]. Financial COI became an important concern for biomedical journal editors and some journals started to ask authors about financial COI by the mid-1980s so that this information could be transmitted to their readers. Accusations of fraudulent marketing from COI became publically known in 2000. NIH organized meetings and drafted regulations and guidelines. University administrators, bureaucrats and scientists questioned the pursuit of self-interest in research rather than public interest. Longrich started examining accusations of fraudulent research and widened its investigation into whether scientists were profiting from government-funded research through interactions with the private sector.

Fred Hollows Foundation

An article published in the professional journal of the Royal Australia and New Zealand College of Ophthalmologists allegedly evaluated intraocular lenses made of polymethylmethacalate and marketed in India [8, 12]. Ravilla considered the article an example of scientific fraud and the product of a COI [22]. Firstly, three of the authors were employed by the Fred Hollows foundation (FHF) which sponsored the study. Moreover Dr. Mark C. Gilles a founder and director of FHF was editor of *Clinical and Experimental Ophthalmology* when the article was published. The administrative costs of FHF allegedly accounted for 80% of the foundation's expenses. A paper supporting the work was later withdrawn [10].

Spectra Pharmaceutical Services

On August 10, 1987 Alfred Edward Maumenee MD Jr (1913-1998) a former chairman of ophthalmology at the Wilmer Eye Institute of Johns Hopkins University (Johns Hopkins) *(http://www.jhu.edu/)* filed a patent with *Spectra Pharmaceutical Services*. A few years later a controversial clinical study on the treatment of the dry eye syndrome with a vitamin A-enriched ointment was performed at the Massachusetts Eye and Ear infirmary (MEEI) by Dr. Kenneth R. Kenyon, an associate professor of ophthalmology, and Dr. Scheffer C. G. Tseng a trainee supervised by him without government funding [32] *(http://www.The crimson.com/article/1992/4/15/state-cleras-prof-of-ethics.charges/)*. The research was particularly contentious because both investigators owned stock in *Spectra Pharmaceutical Services* which expected to market the ointment if it was proven effectual. Because the study was suspected of being fraudulent Kenyon took a leave of absence from administrative duties at the request of his colleagues [29] *(http://www.thecrimson.com/article/1988/11/8/eye-researcher-takes—pa-harvard/)*. The study was scrutinized by at least nine different committees, including a congressional subcommittee looking into scientific misconduct [9]. The study was faulted by the NIH in 1989 because it deviated from federal regulations related to the protection of human subjects. The IRB's procedures for reviewing human experiments was regarded as sloppy and this panel which was responsible for reviewing protocols did not adequately perform its job. The researchers strayed from their approved study design in several respects. They increased the number of patients in the study, changed the doses of the medication and modified the treatment procedures. The NIH demanded that the MEEI contact all participants in the study to determine if any suffered ill effects during the clinical trial. This command was of considerable concern as it would pointlessly alarm the patients, because the study was of low risk to patients and none were known to have had adverse effects. By calling attention to flaws in the study some angry stockholders filed a class-action lawsuit against Tseng and others accusing them of withholding data implying that the vitamin A was ineffective. The drug company later settled a patient class action suit for up to $135 million. This case occurred at a time when Harvard University (Harvard)*(http://www.harvard.edu/)* lacked an appropriate COI policy. The Massachusetts medical board dismissed all charges against Kenyon and Tseng but had concerns about the behavior of the two doctors. In 1990 largely because of this case Harvard modified its COI policy requiring that doctors obtain approval from a medical center committee if they wish to hold stock in a company for which they are performing research *(http://vpr.harvard.edu/content/conflicts-interest)*.

By 1992 it was difficult to find researcher who did not hold equity in a biotech company and there was considerable uncertainty as to what should be disclosed about COI as with Sekoe and Miller [5].

In October 2009 *The Chronicle of Higher Education* listed 14 physicians from a dozen universities who received more than $400, 000 from the manufactures of Vytorm (an anti-cholesterol drug)(Merck and company and Schering-Plough corporation) during a drive by the drug companies to promote the marketing of this medication [6] *(http://chronicle.com/article/Baylor-College-of-Medicine/63648/?sid=at&utm_source=at&utm_medium=en)*. This information was provided to the USA senate by the pharmaceutical companies. Dr. Christie M. Ballantyne, a professor of medicine at Baylor College of Medicine (BCM) received $34, 472 from the drug companies over a five-month period. The companies cited Ballantyne as indicating that Vytorin was "significantly better that Lipitor" (a drug made by Pfizer). The NIH demanded harsher financial disclosures from BCM because of grave worries about the compliance of institutions with regulations related to corporate COI. High payments were also made by drug companies to scientists working in other academic institutions. On May 25, 2011 a USA senate finance committee expressed concerns about the amount received by Dr. Victor Tapson an expert on thrombosis, who was a member of the Duke University (Duke)*(http://duke.edu/)* faculty in the pulmonary, allergy and critical care medicine division of Duke Medicine. According to the senate committee report Tapson had received $260, 604 for consultant fees from the Pharmaceutical Company Sanofi-Aventis.

Universities are primarily responsible for protecting against financial COI. In January 2010 BCM was required to review the recipients of all NIH grants funded since 2004 for financial relations with drug companies. The NIH would also enforce special conditions on future grants to BCM related to applications and payment processes. All participants in the projects needed to be listed with every possible financial COI.

By 2010 USA federal regulations imposed extra requirements on researchers with NIH grants if they received more than $10, 000 from a company *(http:// chronicle.com/article/Baylor-College-of-Medicine/63648/).*

Considerable attention and several unexpected COI became apparent [23]. Shared intellectual passions generate a great deal of additional strong COI. Some professional journals have advertisements that express opinions in support of articles in the journal, but reputable journals do not permit the advertisers to influence the journal content.

COI Related to Journal Publications

The majority of journals failed to deal overtly with COI until after 1992, when most professional publications were adamant in insisting on a financial disclosure. It is not an author's option to decide whether or not the affiliation and connections influence their opinion of information. By 1992, *Nature* and some other journals decided to disclose relevant financial connections of the authors of published papers. Despite the consequences of corporate connections, journal editors usually attempt to recruit the most preeminent available individual to review manuscripts, prepare annotations or write review articles so that readers are familiarized with the financial COI in the scholarly evaluation of the work. Early in the game the *Journal of the American Medical Association (JAMA)*, the *New England Journal of Medicine (NEJM)*, and other journals established rules making authors disclose financial COI that could sway their opinions, but most important biomedical journals lacked formal COI policies in 1992 and relied on the authors' assessment [24]. By that time most universities were only beginning to consider the issue and offered no guidance to their faculty.

In the past it did not cross the minds of many authors such as Selkoe and Miller, that they should draw the attention of both editors and readers to their COI.

Dennis Selkoe. In the early 1990s Dennis Selkoe a Harvard neurologist studying Alzheimer disease was requested to write reviews on this disease in prestigious journals, such as *Science* [42, 43], *Neuron* [39, 47] and *Scientific American* [40, 41], but in these articles he failed to notify readers about his bias based on his financial COI. Selkoe not only founded the biotech company *Athena Neurosciences,* which carried out research aimed at developing diagnostic tests or therapeutic procedures for Alzheimer disease. Selkoe was one of its largest shareholders. When the company went public in November 1991, his 255, 000 shares were valued at over $3 million. *Science* did not realize that *Selkoe* had founded *Athena Neurosciences* when it published articles by him [18, 42].

Paul Miller. The research of Paul S. Miller PhD, a biochemist at Johns Hopkins was at the cutting edge of antisense technology and in the 1990s there was an intense ongoing dispute about the usefulness of two dissimilar kinds of synthetic DNA analogs in preventing gene expression. In 1991 Miller discussed one class of compounds in a review published in *Bio/Technology* [31] that showed promise for future development as therapeutic agents, but readers were not informed that he possessed patents on the compounds, and that he had cofounded the biotech company *Genta* which had the exclusive license for these compounds.

Scientists with connections to biotech companies focused mainly on immunology, antisense technology and neuroscience with potential high payoffs and did not agree on how to proceed. Without doubt it is important that authors reveal their investments when documenting their research findings because this assists the reader's appraisal of the scientist's point of view, particularly when the text is read by individuals not familiar with the topic. The latter concerns not only new young investigators, but also venture capitalists and investors attempting to make an informed decision about new promising biotech companies many of which may not yet have items for consumption. At an early stage in their formation some biotech companies lack distinct benchmarks to pass judgment on them. Barinaga addressed the problem of COI in 1992 [5].

Max L Birnstiel and Meinard Busslinger. A remarkable paper by a virtually unheard of team of Italian investigators documented a major technical advancement in the high impact journal *Cell* [26]. It reported an astonishingly simple way in which foreign DNA could be inserted into mice. The report attracted a congratulatory mini-editorial highlighting this research and it was supposedly written by two apparently impartial prominent scientists promoting the relevance of the science but this was not the case [7]. The editorial was written by Max L. Birnstiel (1933—) and Meinard Busslinger of the research institute of molecular pathology in Vienna. The institute was established in 1988 by two pharmaceutical companies: Genentech of the USA and Boehringer Ingelhein of Germany. Moreover Birnstiel and Busslinger were involved in the preparation of the manuscript and seemingly in the patent rights with potential financial gain. The research institute of molecular pathology filed patent applications based on this novel method of creating transgenic mice. The endorsement of the papers by the mini-editorial clearly gave the readers an undisclosed biased opinion [14].

Leonard S. Schleife. The company *Regeneron Pharmaceuticals* was founded in 1988 by Leonard S. Schleife MD, PhD after he was an assistant professor in the departments of neurology and neurobiology of the medical center at Cornell University (Cornell) *(http://www.cornell.edu/)*. This fully integrated company discovers, develops, manufactures, and commercializes important new medicines for serious diseases. When its stock was first made available to the public in 1991 as REGN on the National Association of Securities Dealers Automated Quotations (NASDAQ) stock exchange the company netted more than $90 million. That sale of stock occurred shortly after two articles in the extremely high impact journal *Nature* documented that brain-derived neurotrophic factor (BDNF) may improve the continued existence of neurons that deteriorate in Parkinson disease [20, 21]. The paper in *Nature* came with hopeful comments that were repeated in *The New York Times*. While many factors influence stock prices, it seemed highly likely that the favorable report in *Nature* affected the high stock price of *Regeneron Pharmaceuticals*. The editor of *Medical Technology Newsletter* disclosed that all being equivalent scientific publications appeared prior to the completion of the contract. One of its commercial opponents may expect such an individual to disclose that they own prejudice because journals have no way of identifying the bias and at the time journals lacked relevant policies to expose this. George D. Yancopoulos MD, PhD, a coauthor of a paper by Hyman and colleagues [21], was the founding scientist, president and a member of the board of directors of *Regeneron Pharmaceuticals*, but this information was not disclosed.

Bias from COI

When researchers have a COI due to an uncompromising adherence to a specific hypothesis in their research, but especially when ties with industry exist the interpretation of their research data is affected because of bias supporting the findings that industry would like to expect. Because of a loss of complete objectively scientific reports are frequently biased by ignoring data that contradicts the hypothesis being tested. In studies of large populations where generalizations are desired adequate steps need to be taken to make sure that cases examined are truly random and that statistically adequate numbers are sampled.

Intellectual COI

Intellectual COI also exist [28].

COI and their Management

COI disclosures are submitted on special forms and some institutions have a video demonstration on how COI needs to be reported on the required forms. Each individual is accountable for keeping an up to date truthful disclosure of their own COI with their institutional RIO. For research COI an annual disclosure form is required from all faculty and non-faculty participating in research. The RIO reviews all COI forms and potential COI in grants that are awarded to the institution. COI disclosure forms are analyzed to determine if the reported relationships are related to the individual's research and/or administrative responsibilities. The RIO makes sure that COI disclosures are in compliance with the policies and regulations of the Public Health Service (PHS) and other federal agencies. All management arrangements are reviewed with a COI committee. Faculty and staff are required to bring their COI disclosure up to date within 30 calendar days of discovering, acquiring or determining any new noteworthy financial interests or changes in existing important financial interests. Training on COI must be fulfilled before taking on research funded by the PHS and having the COI brought up to date by the institution. Those members of the institutional staff that fail to disclose their financial COI completely and truthfully or do not retain an up to date precise disclosure of their COI or fail to follow federal or other sponsor or institutional requirements related to COI may precipitate disciplinary action. Because of potential COI by institutional high level administrators an annual disclosure on COI is needed from persons independently responsible for making decisions for or on behalf of the institution (executive leaders, senior leaders, high-level administrators, and other relevant individuals). These disclosure forms are examined to establish if revealed associations are connected to the disclosing person's administrative duties. An institutional administrative committee on COI reviews all plans related to the management of COI. The RIO reviews possible institutional COI and manages identified cases in collaboration with the chair of the institutional COI committee. When researchers obtain funding from the PHS all reimbursed travel expenses must be disclosed (*https:// radapps.duke.edu/phs_travel*). Universities and research institutions have assumed individual COI policies that apply to their faculty and staff. Universities and Medical Centers have established policies that deal with potential or authentic COI that may crop up in commercial connections that stretch beyond research. Medical schools and research institutions have established policies that deal with the possibility of COI issues concerning NIH-funded Small Business Innovation Research (SBIR) and Small Business Technology Transfer (STTR) grants.

Members of the faculty and staff need to be careful about providing guidance to financial advisors because of the potential of becoming involved in illegal insider trading. Access to a web site related to the administration of COI is protected by a password that is only available to authorized users, such as those in the RIO and departmental liaisons that coordinate COI reporting forms.

An institutional COI portrays circumstances in which the financial interests of an institution or an institutional official, acting on behalf of the institution, may influence or appear to effect the research, education, clinical care, business transactions, or other actions of the institution. Institutional COI are reviewed and dealt with by an institutional COI committee. When an institution has a potential COI the institution is ruled by its institutional COI policy.

Continuing and Promotional Medical Education

Members of the medical profession are required to remain knowledgeable about their specialty. This is achieved by attending courses at professional medical and scientific meetings, by reading articles in relevant journals, by writing papers and by attending certified lectures. Credit for continuing medical education (CME) is provided by accredited sponsors that adhere to the guidelines of the American Medical Association (AMA). Not all educational activities that are available to the medical profession provide CME. For examples many courses provided at resorts and on cruise ships are not designed for CME as they are intended more as a tax deductable business expense.

Promotional medical education activities for physicians and other health professionals are sponsored by pharmaceutical companies and industry. They focus on specific industrial products, such as a device, drug or therapy and do not offer CME credit. When drugs are discussed the promotion of off-label applications of drugs are strictly limited by the regulations of the FDA. The promotional medical education sessions are directly supported financially by honoraria or grants from industry. Because promotional medical education is biased towards the products of the sponsoring industries some institutions have established policies related to the participation of their faculty in such activities. Now the medical institutions require that for the participation of the faculty the content of the presentation must be created by the faculty member and not by the industry sponsoring the event. Moreover the faculty participant must be a recognized as an expert in the topic being endorsed. Faculty who fail to act in accordance with this policy are subject to sanctions and even dismissal from the faculty.

Many members of the faculty at different institutions have interacted with the pharmaceutical industry and biomedical companies delivering promotional medical education lectures or video presentations. By being on the "speaker's bureaus" of these industries faculty members have been able to earn a substantial stipend. Until relatively recently medical doctors could bring up a discussion of off-label uses of drugs in such lectures, but recently the FDA has changed its rules and now speakers hired from promotional medical education lectures or video presentations must speak from approved slides/videos without deviation. From January 1, 2012 payments given for promotional medical education lectures must be posted on a federal government public website.

Sunshine Act

The *Sunshine Act* was passed in 1976 by the USA federal government to create a greater transparency in the activities of the federal government and various legally established federal bodies and agencies. As a result of the *Sunshine Act* part of every meeting by a federal agency or body is open to the public with ten specified exemptions that include information related to national defense, accusations to individuals of a crime and legal proceedings. Because a disclosure of financial interests that physicians have with pharmaceutical companies was initially based on trust and honesty many physicians did not reliably disclose their COI. To rectify this deficiency the *Physician Payments Sunshine Act*, which is part of the *Affordable Care Act* (ACA) was passed and went into effect on August 1, 2013, but by this time some states already had their own state *Sunshine Laws*. The *Sunshine Act* has obligations for the manufacturers of drugs, medical devices, and biologicals that participate in USA federal health care programs. The involved companies are required to follow and report specified payments and valuable items provided to physicians and teaching hospitals. The manufacturers submit their reports to the Centers for Medicare and Medicaid Services (CMS) annually. Group purchasing organizations and manufacturers must also report particular ownership interests possessed by medical doctors and their immediate family members. Most of the information present in the manufacturers' testimony is accessible to the public at a searchable website. Physicians are able to appraise the reports by the manufacturers and challenge them is they are misleading, inaccurate or false. The AMA provides physicians with information about the *Physician Payment Sunshine Act (www.ama-assn.org/go/sunshine)*.

References

1. Als-Nielsen B, Chen W, Gluud C, et al., Association of funding and conclusions in randomized drug trials: a reflection of treatment effect or adverse events? Journal of American Medical Association, 2003. 290(7): 921-928.
2. Amaral MA, COI. Science, 1992. 258(5089): 1717.
3. Angell M, Is academic medicine for sale? New England Journal of Medicine, 2000. 342(20): 1516-1518.
4. Bailey R. Scrutinizing industry-funded science: the crusade against COI American Council on Science and Health. 2008 March.
5. Baringa M, Confusion on the cutting edge. Science, 1992. 257(5070): 616-619.
6. Basken P, Baylor College of Medicine faces NIH sanctions over financial conflicts. The Chronicle of Higher Education, 2010.
7. Birnstiel ML and Busslinger M, Dangerous liaisons: spermatozoa as natural vectors for foreign DNA? Cell, 1989. 57(5): 701-702.
8. Blythman HE, COI. Science, 1992. 258(5089): 1717.
9. Booth W, Hospital faulted for dry eye study. Science, 1989. 243(4894): 1000.
10. Brian G, Evaluation of Indian intraocular lenses flawed: reply. Clinical and Experimental Ophthalmology, 2002. 30(6): 446-447.
11. Carrier M, Howard D, and Kourany J, The Challenges of the Social and the Pressure of Practice: Science and Values Revisited. 2008, Pittsburgh, PA: University of Pittsburgh.
12. Combe R, Watkins R, and Brian G, Evaluation of the quality of generic polymethylmethacrylate intraocular lenses marketed in India. Clinical and Experimental Ophthalmology, 2001. 29(2): 64-67.
13. DeBello MF, Schwiers ML, Rosenberg HL, et al., A double-blind, randomized, placebo-controlled study of Quefiapino as adjunctive treatment for adolescent mania. Journal of the American Academy of Child and Adolescent Psychiatry, 2002. 41(10): 1216-1223.
14. Dickson D, "Dangerous" liaisons in cell biology. Science, 1989. 244(4912): 1539-1540.
15. Elliottt C. Why the University of Minnesota psychiatric research scandal must be investigated. MinnPost. 2013 March 28.
16. Fang FC and Casadevall A, NIH peer review reform—change we need, or lipstick on a pig? Infection and Immunity, 2009. 77(3): 929-932.
17. Friedman PJ, Scientific research conflict of interest: policies and tests. FASEB Journal, 1991. 5(7): 2001.

18. Hardy J and Selkoe DJ, The amyloid hypothesis of Alzheimer's disease: progress and problems on the road to therapeutics. [Erratum appears in Science 2002 September 27;297(5590):2209]. Science, 2002. 297(5590): 2209.
19. Harris G, Carey B, and Roberts J. Psychiatrists, children an drug industry's role. The New York Times. 2007 May 10.
20. Hohn A, Leibrock J, Bailey K, et al., Identification and characterization of a novel member of the nerve growth factor/brain-derived neurotrophic factor family. Nature. 344(6264): 339-341.
21. Hyman C, Hofer M, Barde YA, et al., BDNF is a neurotrophic factor for dopaminergic neurons of the substantia nigra. Nature, 1991. 350(6315): 230-232.
22. Kasthuri RN, Evaluation of Indian Intraocular lenses flawed. Clinical and Experimental Immunology, 2002. 30(6): 445-446.
23. Koshland DE, Jr., Conflict of interest. Science, 1990. 249(4965): 109.
24. Koshland DE, Jr., Conflict of interest policy. Science, 1992. 257(5070): 595.
25. Kowatch RA, Fristad M, Birmaher B, et al., Treatment guidelines for children and adolescents with bipolar disorder. The Journal of the American Academy of Child and Adolescent Psychiatry, 2005. 44(3): 213-236.
26. Lavitrano M, Camaioni A, Fazio VM, et al., Sperm cells as vectors for introducing foreign DNA into eggs: genetic transformation of mice. Cell, 1989. 57(5): 717-723.
27. Lundberg GD and Flanagin A, New requirements for authors: signed statements of authorship responsibility and financial disclosure. Journal of the American Medical Association, 1989. 262(14): 2003-2004.
28. Marshall E, When does intellectual passion become conflict of interest? Science, 1992. 257(5070): 620-623.
29. Masters BA, Eye researcher takes leave. Tseng's supervisor steps down because of investigation. The Harvard Crimson, 1988.
30. McCoy K. Witness: Tight lips on SAC Capital stock sale. USA Today. 2014 January 29.
31. Miller PS, Oligonucleoside methylphosphonates as antisense reagents. Bio/Technology, 1991. 9(4): 358-362.
32. Murray LM. State clears prof of ethics charges. Case of Med School fellow is also dropped. The Harvard Crimson. 1992 Wednesday April 15.
33. Palca J, Ethics in science. NIH grapples with conflict of interest. Science, 1989. 245(4913): 23.
34. Palca J, Conflict over conflict of interest. Science, 1989. 245(4925): 1440.
35. Pesch P, COI. Letter to the editor. Science, 1992. 258(5089): 1717.
36. Popper N and Vlasic B. Quiet doctor, lavish insider: a parallel life. The New York Times. 2012 December 15.

37. Protess B and Stevenson A. After scandal, SAC Capital begins to fade to black. DealB%k. 2014 February 2.
38. Roses AD, COI. Letter to the editor. Science, 1992. 258(5089): 1717.
39. Selkoe D, The molecular pathology of Alzheimer's disease. Neuron, 1991. 6(4): 487-498.
40. Selkoe DJ, Amyloid protein and Alzheimer's disease. Scientific American, 1991. 265: 68-71, 74-76, 78.
41. Selkoe DJ, Aging brain, aging mind. Scientific American, 1992. 267: 134-142.
42. Selkoe DJ, Alzheimer's disease: genotypes, phenotypes, and treatments. Science, 1997. 275(5300): 630-631.
43. Selkoe DJ, Alzheimer's disease is a synaptic failure. Science, 2002. 298(5594): 789-791.
44. Sullivan JL, COI. Science, 1992. 258(5089): 1717.
45. Turner EH, Matthews AM, Linardatos E, et al., Selective publication of antidepressant trials and its influence on apparent efficacy. New England Journal of Medicine, 2008. 358(3): 252-260.
46. Wakefield AJ, Murch SH, Anthony A, et al., Ileal-lymphoid-nodular hyperplasia, non-specific colitis, and pervasive developmental disorder in children.[Retraction in The Lancet. 2010 February 6;375(9713):445; PMID: 20137807]. The Lancet, 1998. 351(9103): 637-641.
47. Walsh DM and Selkoe DJ, Deciphering the molecular basis of memory failure in Alzheimer's disease. Neuron, 2004. 44(1): 181-193.
48. Wray KB, Financial COI worth knowing. Science, 2010. 327(5962): 144.

CHAPTER 25

Legal Aspects of Research

Scientists and research support staff involved in every type of human research face the danger of litigation. Most federal regulations provoking litigation were not in existence before the mid-1970s. Informed consent is crucial to law suits involving human participants. Researchers on these projects are especially at risk since regulations overseeing the research are complex and commonly subject to change. Not being aware of the current regulations can lead to a costly law suit for the researcher. Lawsuits can also be detrimental to the careers of scientists, even if the investigator is triumphant in the litigation. The outcome of malpractice suits often hinges on whether the therapy was typically up to standard or investigational.

An amplified analysis of research projects by congress and the media has made it essential that researchers be familiar with the latest by laws that impact research and especially with regard to human subjects. Prior to 1985 federal regulations on research were relatively vague, but an astonishing escalation in them altered the state of affairs considerably and they pertain to all research regardless of the source of funding. These federal regulations formed the ground work for numerous court cases.

When human participants are involved in research the regulations that govern it come from many different federal agencies within the Department of Health and Human Services (DHHS), such as the National Institutes of Health (NIH), the Office of Human Research Protections (OHRP) and the Federal Drug Administration (FDA), the bureau of prisons, department of energy, and the consumer products safety commission. The department of defense also has rules for human research developed by supporting associations (medical centers, universities, corporations, research institutes). Particular guidelines relate to *in vitro* fertilizations (IVF), children, students, prisoners, mentally disabled, fetuses, and pregnant women. Accusations by students of research misappropriation by faculty supervisors have sometime ended up in court even though this subject is more appropriately dealt with within universities.

Court Cases Involving Universities

Former PhD students and postdoctoral fellows have accused senior faculty of stealing research data and backed up the allegations with legal action against the professors and the universities because they felt that the university grievance systems did not deal with each of their complaints appropriately. The litigation focused on the ownership of ideas particularly in the unbalanced relationship between professor and student.

The regulations of federal agencies have been at the core of numerous cases involving researchers that ended in court. Numerous laws in the USA, such as the *Freedom of Information Act* and the *False Claims Act* have made an impact on scientific misconduct.

Some whistleblowers have been frustrated with the speed or outcome of the investigations into alleged scientific misconduct. Regardless of who initiated legal action those who filed lawsuits\in federal court against universities and researchers challenging the unfavorable decisions.

Carolyn Phinney. Carolyn Phinney, a psychologist with a PhD from the University of California, Berkeley (UC Berkeley) *(http://berkeley.edu)* focusing her research on personality and cognitive disorders became a senior research associate at the University of Michigan (UM) *(http://www.umich.edu/)* Institute of Gerontology (IOG), while completing a postdoctoral fellowship. Her research went well and Marion Perlmutter suggested that Phinney try to get a research grant to extend the research. Initially an application was submitted to the National Institutes of Health (NIH) and because it was not successful the National Science Foundation (NSF) was tried without any luck. Because of Phinney's lack of success Perlmutter decided to submit her own grant application. In preparation for this application Perlmutter gained access to Phinney's research data and grant proposals by deceptively using her data without permission claiming that it was her own. Phinney's research data were also stolen from her office before being submitted by Permutter in a grant application on her own without Phinney. Other promises were also not kept. Perlmutter, an authority on aging, who was the principal investigator (PI) of the grant that supported Phinney's research attempted to establish not only whether wisdom increased with aging, but also whether it can be measured objectively. Phinney maintained that Richard Adelman, the director of the IOG retaliated against her when she refused to withdraw her allegations against Permutter by harassing her and then suspending her for insubordination [6]. The university was asked to investigate the accusation in 1989 and four faculty panels failed to find evidence of the misappropriation of intellectual property by Permutter. Because Phinney was not satisfied with the response and felt that UM botched the investigation, she sued Permutter, Adelman, and the university in 1990. In May 1993 a seven member jury ruled in Phinney's favor. It found that Adelman had violated the state of Michigan's *Whistleblower Protection Act* and concluded that Perlmutter had committed fraud by making false promises to Phinney related to grants, authorship and employment. The jury awarded $989, 200 to be paid by Adelman and another $130, 300 to come from Perlmutter. Later in 1993 a judge upheld the jury award, but increased it by $126, 000 as fees for interest. $1, 246, 000 to Phinney in payment from UM for her allegations that the university blundered its investigation into her accusations. This was the first time that the court levied a substantial fine against a university in a scientific misconduct case. UM appealed the decision to a higher court, and the case was upheld with an award of $1.2 million by Judge Patrick Conlin in the Washtenaw County Circuit Court in Ann Arbor Michigan in September 1993 [2, 7]. The incident precipitated a profound post-traumatic stress disorder on Phinney whose contract with the university was not renewed and from July 1992 to 1995 she remained unemployed [8]. After receiving what is thought to be the largest amount ever awarded to a whistleblower because of professional misconduct ($1.67 million), Phinney pledged part of her award to support a group known as *Whistleblowers for Integrity in Science and Education* which supports whistleblowers who bring allegations of scientific misconduct against their peers [9]. This case illustrates that institutional investigations into allegations of research misconduct cannot be taken lightly and must be done right or the institution can be sued.

Antonia Demas. In 1970s Antonio Demas developed an interest in nutrition and by 1991 she started working towards a PhD in the department of education at Cornell University (Cornell) *(http://www.cornell.edu/)*. Her research focused on trying to substantiate a way to educate children about different foods after receiving her doctoral degree in 1995 and having her thesis copyrighted she was hired by Cornell as a research associate. On March 29, 1999 Demas filed a lawsuit against Professor David Levitsky, a member of her PhD advisory committee, and Cornell because she maintained that Levitsky misused her concepts on instruction and nutrition to obtain research funding on obesity. Demas allowed him to do this, but she became annoyed when he started taking credit for her research in teaching sessions and discussions without appropriate credit and used the material in applications for research funding without including her in the proposal. Demas complained to the ombudsman at Cornell who recommended that Levitsky should coauthor a publication with her but this never happened. In May 1996 Peter C. Stein, dean of the faculty at Cornell, investigated twenty three complaints by Demas of and dismissed all except one because they were not within the jurisdiction of the Department of Human Health and Services

(DHHS) and various other deans failed to inflict sanctions on them. The other grievance was also rejected. Demas raised the important ethical question of whether a professor can acquire the intellectual property (IP) or concepts of a student for his/her own applications for research funding. Stein considered it wrong and unethical to do so if the student had not yet published the study, but with regard to Demas he maintained that use of her ideas was not research misconduct because the published information in her thesis was in the public domain. By February 14, 2002 the Demas case against Levinsky and Cornell reached the State of New York Supreme Court Appellate Division *(www.nacua. org/documents/Demas_v_Levinsky.pdf)*.

Pamela Berge. As part of her graduate school education as a nutritionist at Cornell Pamela Berge went to the University of Alabama Birmingham (UAB) *(http://www.uab.edu/)* to use their vast cytomegalovirus (CMV) database to develop a thesis linking CMV with low birth weight. Later in 1990 Berge was astounded to hear Karen Fowler, a UAB graduate student, deliver a lecture that appeared to reiterate Berge's personal research and she charged Fowler with plagiarism, but two investigations at UAB disclosed no inappropriate behavior. The DHHS did not pursue the allegation further so the former graduate student instigated a *qui tam* lawsuit in 1993 accusing UAB and four researchers with making fake statements in grant applications to the NIH that funded CMV research. By taking the matter directly to the courts Berge potentially opened Pandora's box over suspected misuse of her nutritional studies and universities were worried about comparable lawsuits being brought against them. Berge triumphed in 1995 when the jury of a federal court in Baltimore Maryland ruled in her favor and awarded her $1.6 million. UAB was directed to pay $1.65 million and the researchers were ordered to pay $10,000, 30% of which went to Berge [13]. The verdict was challenged. The legal tactic used by Berge not only concerned UAB, but also other universities because it was out of line with the accepted practice of having scientific misconduct issues settled within universities. On January 22, 1997 the Fourth Circuit Court of Appeals firmly rejected this decision (Science News Staff. Science $1.6 Million Fraud award overturned *(news.sciencemag.org/sciencenow/1997/01/28-04.html)*. Berge appealed against the overturned verdict because the court ignored certain evidence, but the USA Supreme Court refused to hear Berge's appeal [10].

Sheng-ming Ma. Another university that combated a student in a court of law was Columbia University (Columbia) *(http://www.columbia.edu/)*. Duong H. Phong, the former mathematics department chair at Columbia assigned a mathematical problem to Sheng-ming Ma, a mathematics student. The problem was one that Phong had worked on with Elias Menachem Stein (1931—) a professor emeritus at Princeton University (Princeton) *(http://www.princeton.edu/)* since 1991 and had just solved. They published the solution in *Acta Mathematica* [12]. Ma solved the problem after working on it as a thesis project, and then accused Phong of plagiarism. The dean of the graduate school investigated this allegation and found no evidence of research misconduct. Ma was instructed by the mathematics department to apologize to Phong, but because of his refusal Columbia discharged him from his appointment in 1997. Soon thereafter Ma filed a lawsuit against Phong in the New York Supreme Court in March 1998. Columbia tried to get the suit rejected claiming that it had made a thorough complete and unbiased investigation before rejecting the grievance. In addition the university contended that Ma was making an effort to engage the court in scholarly judgments that were outside legal review. The university also insisted that mathematic principals cannot be plagiarized or copyrighted. In October 1999 Judge Emily Jane Goodman of the New York City Supreme Court ruled in favor of the defendants [3]. In coming to this ruling the judge made the astute observation that neither a judge nor jury could fathom whether a mathematical theorem was correct. The decision in this case supported the widespread belief in academic circles that disputes in universities are best solved in institutions rather than in courts of law.

Jon Kolb. In 1987 Eric Glitzenstein, successfully took legal action against the national science foundation (NSF) on behalf of Jon Kolb, an anthropologist who lost a NSF grant because hearsay inappropriately wrongly branded him an agent of the central intelligence agency (CIA). As a consequence the NSF consented to pay, but refused to release the identity of the reviewers.

Wanda and Robert Henke. When researchers receive an extremely harsh review of their work from peers it is not uncommon for them to feel as if they have been ambushed, but they very rarely take the assault to court. Wanda and Robert Henke, two civil engineers from Lutherville, Maryland, who owned a small research company attempted unsuccessfully to obtain funding from the NSF and the National Institute of Standards and Technology (NIST) to build a potential device for testing earthquake risk in soils. On February 3, 1994 they filed suit under Glitzenstein, the lawyer who successfully dealt with Kolb's lawsuit against the NSF in which compensation was obtained for grievances against NSF. They demanded the identity of the critics who berated their research concept. Both the NIST and NSF declined to announce the identity of the reviewers and NIST even refused to provide the critiques. The Henke couple maintained that under the federal *Privacy Act* of 1974 they had the right to gain the right to all documents related on their grant applications so that they could evaluate bias and possible COI (COI) between the referees. They also demanded the identity of the grant reviewers, but this request was of no avail. By that time the confidentiality on the peer view system had not been directly confronted in a court of law. Initially a NSF review panel found their proposal "not clearly written, not feasible in the timeframe proposed, and not adequately supported by preliminary data".

The Henkes implied malicious remarks by a reviewer with a financial interest in the outcome of the research. Not being satisfied with the unfavorable response by the NSF to their grant application the Henkes appealed to several officials at the NSF to investigate the case: Joseph Bordogna, (assistant director for engineering), Frederick Bernthal Deputy (director) and Linda Sundro (inspector general). After evaluating the situation Bordogna indicated that the reviewers were well qualified and that their opinions seemed reasonable. While pleas by the Henke duo were under inspection, the discontented couple filed a *Freedom of Information* application for all documents related to their grant application. In the Henke's case Glitzenstein needed to convince the court that the NSF must release the names of the Henke grant reviewers, but a successful mission would have been detrimental to the evaluations and administration of manuscripts and grant applications. Confidential discussions during the review process are essential.

James H. Abbs. The investigation of James H. Abbs PhD, neurophysiologist at the University of Wisconsin (UW), stemmed from three graphs comparing the tremor of the lip, tongue and jaw of patients with Parkinson disease with healthy individuals in a paper that Abbs had published in *Neurology* [1]. Steven Barlow, a former graduate student of Abbs noticed that these graphs were strikingly similar to graphs that he published with Abbs in the *Journal of Speech and Hearing Research* [4]. Abbs maintained that despite similarities between the graphs in the two papers they depicted data from different patients, but unluckily for Abbs he had lost the raw data. UW and the NIH could not substantiate Barlow's allegations. That would have been the end of the case, but in April 1988 Charles W. McCutchen, a physicist at the NIH with a special interest in scientific misconduct claimed that the probability of the graphs being different was extremely low. Within a fortnight the NIH started to review McCutchen's analysis and months later an external panel of experts in statistics evaluated the graphs. The panel disagreed with UW and the federal Office of Scientific Integrity (OSI) found Abbs guilty of scientific misconduct.

USA Legislation Relevant to Research

Research in the USA is governed by numerous federal laws. Hence numerous regulations govern research and they have gradually grown in complexity and made researchers legally responsible for intentional or unintentional errors in their professional scientific endeavors indicates potentially liable for different aspects of their research. Lots of researchers now appreciate that regardless of their research discipline, knowledge or the duration of their experience as an investigator they are more prone to legal action than previously appreciated. In human research investigators can become liable for their professional activities.

Public Health Service Act

The USA Public Health Service (PHS) was formed in 1798 and in 1944 it was created by the *Public Health Service Act* as the main division of the department of health, education and welfare (DHEW) under Title 42 of the USA Code "The Public Health and Welfare", Chapter 6A "Public Health Service"(*Public Health*

Service Act, FDA.gov's website, accessed 29 July 2007). The PHS is overseen by the assistant secretary for health and agencies within it include the *Centers for Disease Control and Prevention* (CDC), the FDA, and the NIH. *The Public Health Service Act* has been amended numerous times by the *National Cancer Act* (1971). In 1979 the PHS became the DHHS, which has a detailed code of federal regulations (CFR) related to research on human subjects under title 45 part 46 (45 CFR 46), the office for human research protection (OHRP) and several other offices. Other amendments added to the *Public Health Service Act* include Section 493 (1986), the *Health Insurance Portability and Accountability Act (HIPAA)* (1996), and the *Muscular Dystrophy Community Assistance Research and Education Amendments* (2001).

Federal False Claims Act

The *Federal False Claims Act* of 1863 and 2009 (3.USC $\S\S\S3729$-733) also known as the Lincoln law, and q*ui tam* law refers to the first words in the clause "*qui tam pro domino rege quam pro se ipso sequitur*", which translates as "who as well for the lord the king as for himself sues". The Oxford English Dictionary defines the law as "an action brought on a penal statute by an informer, who sues for the penalty both on his own behalf and on that of the crown. This law was designed for whistle-blower and the statute of the *qui tam* suits provides a significant financial incentive for them to expose scientific misconduct. Those who take allegations of scientific misconduct to the courts by instigating *qui tam* suits receive 15-25% of the monetary settlement if the government enters the case and even more if the government declines. Successful *qui tam* suits can be extremely expensive to defendants not only because of the cost of their legal defense, but also because the judge can impose a sizable fine for each false claim ($5, 000—$10, 000) and the defendants may be forced to pay triple the damages incurred by the government. In the 1980s after dozens of congressional hearings on defense procurement fraud the *Federal False Claims Act* was amended in a way that that could influence accusations of scientific misconduct, but *qui tam* suits were not filed in such cases until the USA department of justice made it known on August 10, 1990 that it would join a *qui tam* suit against a researcher alleged to have falsified research supported by the NIH [11].

The Federal Food, Drug, and Cosmetic Act and the Food, Drug and Device Amendments

The *Federal Food, Drug, and Cosmetic Act* (FFDCA) was passed by the USA congress in 1938 and it provided the FDA with the legal power to watch over the safety of food, drugs, and cosmetics. Prior to the creation of the FDA new drugs were marketed in the USA without evidence that they were both effective and safe. After the passage of this law all drugs needed the approval of the FDA after this agency collected evidence indicating that they were safe and effective against one or more specific diseases. In 1985 this law was amended to include the requisites for determining the safety and efficacy of medical devices, an area where costly and essential tests are needed to be conducted. In 2007 the FFDCA was amended again as the *Food and Drug Administration Amendments Act (FDAAA)*. The development of new drugs for possible therapeutic use must be evaluated in extremely complicated clinical trials regulated by the FDA.

Freedom of Information Act

The *Freedom of Information Act* of 1967 permits the public to gain access to the records of federal agencies with the exception of those that are safeguarded from discovery. This *Act* became a reality because of the possibility of the government suppressing information from the tax payers. For example, it would not be unreasonable for a citizen to request information about the use of isotopes or hazardous biological agents. Citizens deserve the right to be knowledgeable about how the taxpayer's money is used. Under the *Freedom of Information Act* certain information about research can be sought by persons desiring such material. Everyone has the legal right to request any information in a grant application except the budgetary or personal data. For preliminary research to be made available to other investigators before it is published or is at least in press is likely to impart a negative incentive to principal investigators (PIs). Likewise it creates a risk of plagiarism from the theft of original ideas. The DHHS developed procedures that researchers and the general public need to follow to obtain data under the *Freedom of Information Act*. For example, when such a request is made the NIH notifies the PI.

The National Research Act of 1974

On July 12, 1974 the National Research Act was signed into law after being passed by the 93rd USA Congress. This law created the National Commission for the Protection of Human Subjects of Biomedical and Behavioral Research. This law was precipitated partly because of the notorious Tuskegee syphilis study. Research regulations were established by the *National Research Act of 1974*. Largely because of the moral outrage caused by the unethical *Tuskegee syphilis study* that took place from 1932 to 1972 the *National Research Act* of 1974 (Pub. L.93-348) was signed into law by President Richard Milhouse Nixon (1913-1994) shortly before his resignation on August 9, 1974. The *National Research Act* generated most of the relevant federal rules and regulations related to human research in the

USA that still exist today. Additional rules were needed by 1993 for the handling of data obtained by immoral methods. This law created a national Commission for the protection of human subjects in biomedical and behavioral research (1974-1978). Among other things the commission was given the responsibility of determining the ethical principles needed to carry out biomedical and behavioral research on human subjects. The Commission was also to create guidelines in performing these duties and was instructed to consider: (i) the boundaries between biomedical and behavioral research and the practice of medicine, (ii) the task of appraising risk-benefit standards in the establishment of their suitability in human research, (iii) apt instructions for selecting human research participants and (iv) the features and meaning of informed consent in different research settings.

Family Education Rights and Privacy Act

The *Family Education Rights and Privacy Act (FERPA)* also known as the Buckley amendment was enacted as a USA federal law on August 21, 1974 (*Pub.L.* 93-579, 88 *Stat.* 1896, enacted December 31, 1974, *5 U.S.C.§552a*). It deals with the subtle balance between the need of the government to keep information on people and the rights of the community to be protected from an unjustifiable incursion of their privacy. Under this *Act*, an agency is required to print an announcement about its record system when the information concerns a person who is recovered using a private identifier. Among other things this law constrains research with student participants. The regulations under this law were established and reviewed by numerous departments and agencies of the federal government, including the departments of energy, defense, education, and the national institute of handicap research, the PHS and the FDA. After the passage of new bills by congress additional new regulations are frequently generated. The regulations deal with such diverse subjects as prisoners in research, research in clinical studies with therapeutic devices, food or color preservatives electronic inventions, medicinal products, secret restricted data, government ethics. FERPA established a code of fair information practice governing identifiable information about individuals maintained by federal agencies. This *Act* requires that the federal government notify the public about their relevant databases in the *Federal Register*. The *Act* prohibits most records from being disclosed without the written permission of particular individuals. Individuals can amend their records.

The USA *Copyright Act of 1976*

The USA Copyright Act of 1976, which was improved during the Berne Convention of 1989, shifted the legal balance of control from publishers to authors. Until that time the work was only protected by copyright law if the author or publisher registered it and this was usually done by the publisher. Now copyright starts much sooner and it's under the control of the author and in most parts of the world it continues for the life of the author plus 50 or 70 years.

Bayh-Dole Act

On December 12, 1980 the USA legislation passed the *Bayh-Dole Act* (*Patent and Trademark Law Amendments Act*). This *Act* was sponsored by Birch Evans Bayh, Jr (1928—) a former democratic senator of Indiana and Robert Joseph "Bob" Doyle (1923-) a past republican senator and presidential candidate of Kansas. The *Act* allows universities, small businesses and non-profit institutions in the USA to retain the intellectual property rights of inventions developed during federal funded research. By allowing this the products of research are able to reach the market in a timely fashion. The *Act* is codified in *35 USC §200*-212 (USPTO. *"Bayh-Dole Act"*. *Manual of Patent Examining Procedure*. Retrieved 2011-08-19) and implemented by 37 *CFR* 401 (Supreme Court of the United States. *"Board of trustees of the Leland Stanford Junior University v. Roche Molecular Systems, Inc., et al.* p.9. *Archived* from the original on 8 July 2011. Retrieved 2011-08-19).

National Technology Transfer and Advancement Act

The National Technology Transfer and Advancement Act (NTTAA; *United States Public Law* 104-113) (*http://itlaw.wikia.com/wiki/Federal_Technology_ Transfer_Act_of_1986*) became law on March 7, 1996. The Act amended some existing laws and mandated new directions for federal agencies. The purpose of the Act was to pass on novel technology and industrial advances to the market more speedily, to promote collaborative research and development between businesses in the private sector and federal laboratories so that would become easier for businesses to acquire exclusive licenses to capitalize on discoveries that are the end results of research collaborations. Until this *Technology Transfer Act* financial incentives did not exist for researchers in the biomedical sciences. The Act brought researchers closer to industry by creating an additional means for government supported scientists in the USA to work in partnership with private businesses. As a result of this law a company can reach a working agreement with a government funded laboratory for a restricted license and an ability to ultimately share in the derived royalties. The foremost challenge was to determine what was lawful and what was dishonorable. The NTTAA brought researchers closer to industry by creating an additional means for government supported basic biomedical scientists in the USA to work in partnership with private businesses. The NTTAA focused on technology transfer and established consortiums of federal laboratories. It provided financial incentives for the researchers, but it also enhanced the likelihood of scientific misconduct as researchers could place the public interest beneath their own interest. The *Act* provided researchers with an opportunity to generate additional income and some senior investigators could accumulate a vast wealth.

Mental Health Act

The *Mental Health Act* that was passed in 1986 allows for the protection of mentally ill persons. It stipulates the regulations that relate to the participation of the mentally ill in research studies.

Certificates of Confidentiality

USA federal law authorizes the DHHS to award certificates of confidentiality to research institutions, but not to PIs, to protect the privacy and confidentiality of data on human research subjects. The NIH recommends that these certificates be issued for genome-wide association studies and data-sharing to avert the forced disclosure of information gathered during research which can potentially have unfavorable outcomes if disclosed for human subjects such as an effect on their financial position, employability, insurability, or reputation. However, the privacy of research subjects may not be fully protected as suggested by a court case [5].

In 2008 the existing law asserted that with a certificate individuals "engaged in biomedical, behavioral, clinical, or other research may not be compelled in any federal, state, or local civil, criminal, administrative, legislative or other proceedings to identify such individuals". In the People vs. Newman, a certificate effectively stopped a disclosure of the persons participating in a drug treatment program facing a subpoena by a grand jury in a murder inquiry [5]. In a criminal case that reached the North Carolina court of appeals research data was brought together under a certificate of confidentiality but the defense subpoenaed it hoping to impeach the reliability of a prosecution witness. In the early 1990s, scientists at Duke University Health System (DUHS) started a study of psychiatric diseases in urban and rural youngsters. A certificate of confidentiality was acquired from the national institute of mental health (NIMH) as the research team intended to collect information concerning psychosocial hardships, substance abuse, unlawful behaviors, and genetic traits.

In 2004 a criminal investigation challenged the study's certificate of confidentiality. A defendant was incriminated with statutory rape taking indecent liberties with a minor. His lawyer believed that a witness for the prosecution participated in the DUHS project and applied for a court order planning to turn over all DUHS study documents about this eyewitness. The court allowed the defendant access to the records for possible exculpatory facts. Without being aware of the certificate of confidentiality the judge was adamant that the records remain confidential except if used at trial or sentencing. The person whose record was sought was not the suspected victim. Months after the accused was found guilty and convicted of all allegations, the defendant's appellate lawyer filed a motion demanding access to the sealed documents. A hearing before the identical judge resulted in an instruction to provide the defense counsel with copies of the records and to share these with the state. DUHS appealed this ruling emphasizing the certificate of confidentiality and the People vs. Newman case. Nevertheless because of the court order DUHS provided the defendant's counsel with the relevant records.

References

1. Abbs JH, Hartman DE, and Vishwanat B, Orofacial motor control impairment in Parkinson's disease.[Retraction in Neurology. 1996 August;47(2):340; PMID: 8927237]. Neurology, 1987. 37(3): 394-398.
2. Anderson C, Scientific misconduct. Michigan gets an expensive lesson. Science, 1993. 262(5130): 23.
3. Anonymous, Case dismissed. Science, 1999. 286(5443): 1265.
4. Barlow SM and Abbs JH, Force transducers for the evaluation of labial, lingual, and mandibular motor impairments. Journal of Speech and Hearing Research, 1983. 26(4): 616-621.
5. Beskow LM, Dame L, and Costello EJ, Research ethics. Certificates of confidentiality and compelled disclosure of data. Science, 2008. 322(5904): 1054-1055.
6. Charatan FB, Psychologist wins damages over theft of research. British Medical Journal, 1997. 315(7107): 501-504.
7. Hilts PJ. Scholar who sues wins $1.2 million. The New York Times. 1993 September 22.
8. Hoke F, Veteran Whistleblowers advise other would-be 'ethical resisters' to carefully weigh personal consequences before taking action. The Scientist, 1995. 9(8): 1.
9. Kaiser J, Random samples. Home for scientific whistleblowers. Science, 1997. 277(5332): 1611.
10. Kaiser J, Scientific misconduct. $1.6 million fraud award overturned. Science, 1997. 275(5300): 610.
11. Mazzaschi AJ, Qui tam suits: scientific misconduct goes to the jury. FASEB Journal, 1990. 4(14): 3173-3174.
12. Phong DH and Stein EM, The Newton polyhedron and oscillatory integral operators. Acta Mathematica, 1997. 179: 105-152.
13. Taubes G, Plagiarism suit wins: experts hope it won't set a trend. Science, 1995. 268(5214): 1125.

CHAPTER 26

Epilogue

The research performed by scientists is extremely important for furthering knowledge and improving the standard of living of all of mankind, but it needs to be carried out in an honest efficient ethical way to produce meaningful results. Most scientists are honest and play by the rules. They appreciate that honesty and truth are cardinal features of science and everyone hopefully appreciates the high standard to which everyone should strive. Although science is composed of numerous apparently independent disciplines, the scientific enterprise is interconnected and advances in one discipline can enhance knowledge in another. Such advances are unpredictable. The standard of research by some full-time professional scientists is exceptionally high and many of them receive numerous awards for their outstanding original contributions. Every year a few of these remarkable giants in science are singled out for Nobel Prizes. Other researchers are often acknowledged for their noteworthy achievements even though their scientific record fails to reach the remarkable standard of the Nobel laureates. The giants of science provide the scientific community with an excellent reputation in the eyes of the public, but unfortunately a small percentage of those who perform research tarnish the status of this learned highly diverse profession by performing fraudulent research that has become known as scientific misconduct. Researchers in this category are guilty of fabrication, falsification, plagiarism and other dishonest activities in their research. Scientists prone to this type of behavior have existed since the beginning of science and more than a handful of scientists who were highly regarded during their lives have had their honesty and scientific achievements seriously questioned. Examples of such scientists are discussed in elsewhere in other chapter. The expected standards of research change with time and the allegations of research misconduct that have been raised against these specific prominent scientists may not be completely justifiable. The errors made by some of them may have been unintentional and a consequence of their scientific approach. While their accusers have provided convincing evidence to implicate them in their misdeeds, it is debatable whether the accusations against all of these prominent scientists are completely justifiable because the underlying principles and methods of science evolve over time and scientists should not be judged by the harsh scientific standards of today, but rather by those of their eras. More importantly they are not able to defend themselves against their accusers and all relevant records are no longer available. Scandals exposing the dark side of the famous always sell books. Some of the accusers wrote books exposing their provocative allegations and some complainants had a conflict of interest (COI) related to royalties from book sales or had an axe to grind.

Scientific fraud was initially considered uncommon [1, 2] and the reasons why some researchers resort to it is a topic of considerable interest. Based on an analysis of many such individuals by different researchers it has become apparent that research misconduct takes place in multiple countries in different disciplines for dissimilar motives and that it will almost certainly never be eliminated. Psychological factors play an important role in promoting dishonest behavior and many researchers that get involved in misconduct suffer from antisocial personality disorders. A major group of persons performing research misconduct are highly intelligent, ambitious, immoral, professional cheaters who have no respect for the standards of science. These psychopathic personalities are extremely unlikely to abide by the expected behavior of scientists. Some researchers that move along the path of misconduct do so because of stupidity for the reason that fraudulent behavior almost always eventually becomes discovered and ruins careers. Abraham Lincoln's statement that "You can fool all of the people all of the time and some of the people some of the time, but not all of the people all of the time" applies to researchers who perform misconduct. Extremely busy senior scientists with extensive administrative duties appear to predispose to personal research misconduct or misconduct by their trainees whose research is apparently not supervised to the extent that they should be prior to publication. Academic environmental factors also contribute to research misconduct by impacting on the perpetrator in a way that promotes dishonesty. Particularly in the past an overemphasis on the number of scientific publications by faculty and postdoctoral fellows created an incentive for scientific misconduct. Fortunately for science the total number of peer reviewed papers is no longer stressed in academic decisions related to promotions and tenure. The same applies to grant applications. Instead the quality of publications is now stressed as being much more important in academic advancement. Academic environmental factors may lead honest individuals to go astray because of the influence of others as shown in the experiments of Milgram and Zimbardi. The Lucifer effect may be responsible for some cases of research misconduct [6].

An important environmental factor leading to research misconduct is the restricted funding that is available for the vast number of competent well trained scientists. Other researchers embarking on research misconduct are unable to survive in a highly competitive environment in which they must produce new meaningful research findings in the face of insufficient research funds. Pumping more funds into research is unlikely to restrict scientific misconduct because society has produced such a vast number of scientists that it may have approached the capacity to which taxpayers can contribute financial support. As pointed out by the late Reverend Thomas Robert Malthus (1766-1834) [5] and others in different population studies a time is eventually reached when specific groups of organisms can no longer be supported with limited resources.

Those who perpetrate fraudulent research rightly suffer the most for their dishonesty. Their reputations are ruined and they face sanctions which are sometimes serious and may even culminate in criminal prosecution as has occurred with some incidents of fraudulent research. Research misconduct also has a detrimental effect on innocent coauthors of falsified publications. The effort and time devoted by them to the preparation of the flawed papers goes down the drain. For junior researchers the loss of publications from their curriculum vitae is particularly devastating especially if the articles were in high impact journals. The entry of falsified and fabricated data into publications leads to the dissemination of misinformation. Hopefully such flawed data eventually becomes retracted, but it becomes cited for unpredictable time periods that may last decades or longer. Scientific misconduct affects not only the researcher responsible for it, but like an infection all associated with the tainted research. Some research misconduct is relatively inconsequential to society, particularly if it does not endanger lives or effect available resources. However, the most detrimental consequence of research misconduct is its effect on the image that it has on the public about the scientific community. Actual and alleged cases of research misconduct genuinely affect the scientific community in general, no matter who is responsible for the misconduct, because the public loses confidence in the scientific enterprise. Moreover those responsible for funding research endeavors develop reservations about allocating grants to researchers. Another negative aspect of research misconduct is the fact that it makes other researchers waste time and money in trying to replicate experiments unnecessarily. The adverse effects of plagiarism are generally much less serious than those of falsification and fabrication because information entering the scientific literature is duplicated rather than inaccurate. However, under certain circumstances plagiarized data can have serious ramifications. When it occurs in clinical trials and is used in meta analyses the conclusions can lead to a false overrepresentation of particular findings and this can potentially have an adverse effect on therapy. Despite all of the negative aspects of research misconduct positive reactions to the dishonesty also emerged. Universities, research institutions, journal editors and the scientific community have improved the oversight of research particularly with regard to the submission of papers to journals. Because universities and research institutions lacked official policies to deal with research misconduct in the 1970s and 1980s the administrators at some of them rushed to punish individuals suspected of misconduct without appreciating that the topic was serious and that the accused needed to be given due process under the law and the benefit of the doubt until proven guilty. Boston University (BU) (*http://www.bumc.bu.edu/*) acted hastily on Marc Straus when whistle-blower accused him of research misconduct. The university's investigation of allegations against Straus lasted only ten days before forcing his resignation. Also, Ohio University inappropriately penalized Dr. Jay Guneschera for plagiarism by students under his supervision.

Digitally manipulated images can serve numerous diverse endeavors as in art, politics journalism, commerce and entertainment. But when used to document factual information the manipulation of images is taboo. Early photographic images were manipulated to achieve what photographs could not do because of serious technical restraints. More realistic representations could be developed and they could be made more pleasing from an aesthetic standpoint. Irrelevant parts of pictures can be copped. The creativity of artistic photographers delved into surrealistic and bizarre imagery. Manipulated photographs preceded modern technology with Adobe Photoshop. Non-digital photographs have been manipulated from the 1840s through the early 1990s [4].

For years cases of scientific misconduct were not given significant consideration until nudged by Congress or the media. Investigations into research misconduct were often not resolved within a reasonable time period and they frequently lingered on for years. Scientific misconduct in NIH funded research has been committed by investigators at all educational levels: graduate students, postdoctoral fellows, clinical coordinators, technicians or other research support staff and even assistant, associate, and full professors in academic institutions. In an analysis of the files of 72 faculty members found guilty of scientific misconduct by the Office of Research Integrity (ORI) Fang and colleagues found that males were much more frequently involved than females (7 males to:1 female)[3]. To explain this marked gender difference Fang and collaborators postulated that men are more likely to engage in risky behavior than women. This belief is supported by well documented social science literature. Another possible explanation is that there are many more male scientists than female researchers.

Several lessons were learnt from examples of scientific misconduct in the 1970s and 1980s. Faculties need to talk about research ethics to improve the appreciation and identification of moral issues. Dishonesty is not a trait expressed only by some scientists. It occurs in all walks of life depending on the ethical upbringing of the individuals and the capability of the person to compete with peers in a competitive world. Some individuals will engage in scientific misconduct as has occurred since the beginnings of science. More than a casual number of scientists cheat and the precise incidence will probably never be known. Scientific misconduct should be thoroughly condemned as it is dishonest and not beneficial to anyone. Based on what is known about scientific misconduct it is clear that researchers should store raw data indefinitely so that qualified scientists can review the records many years later if the legitimacy of the study becomes questionable. Well-kept laboratory records enhance the defense against scientific misconduct.

Publicity about unethical abused research involving human subjects during the early twentieth century shocked the world and this led to the introduction of new laws and regulations affecting human research. The post-World war II doctor's trial exposed many unethical experiments carried out by the Nazis during World War II. Inhumane studies on syphilis that were performed under the direction of high officials in the Public Health Service (PHS) in Tuskegee and in Guatemala illustrated to the USA public that one of their governmental agencies was prepared to perform unethical research to answer questions of relevance to the public health. Indeed such an approach was extremely comparable to what German doctors did for the third Reich. The Tuskegee and Guatemala syphilis studies made it apparent that research by scientists needed to be not only overseen in peer reviewed publications, but also by independent institutional committees that oversee research on human subjects and animal studies. Hence institutional review boards (IRBs)(also known as independent ethics committees (IECs) and clinical review boards (CRBs) were developed to prevent unethical and cruel research on human subjects. For animal research the comparable committee is the Institutional Animal Care and Use Committee (IACUC).

Human research is exceptionally complex and abundant regulations abide over it. When drug companies need to evaluate new potential drugs in clinical trials they need the cooperation of appropriate medical doctors who have access to patients with the target disease as well as relevant controls. To do this the pharmaceutical company selects specific individuals to take care of the patient recruitment and necessary requirements of the study protocol. Unfortunately some recruited medical doctors ignore the established guidelines and do not adhere to the required protocols. By not being in compliance such medical doctors invalidate the clinical trial and this jeopardizes the clinical trial. The child psychiatrist Barry Garfinkel was convicted of several crimes for not playing by the expected rules.

For many years clinical research has been based on the *Belmont Report*, which has three main components: respect for persons, beneficence and justice. Clinical trials are an essential way in which new therapies are evaluated and a vast number of such studies are controlled by specific regulations of the Federal Drug Administration (FDA). When pharmaceutical companies in the USA develop drugs for therapeutic use they need to test them in at least three separate phases (Phase 1, 2, and 3). Only drugs that successfully complete all phases become approved in the USA for a precise application. Drug companies are not permitted to market them for other off label applications, but medical doctors may prescribe FDA approved drugs for any disease that they wish. Despite being forbidden to promote off-label drugs pharmaceutical companies sometimes violate this rule and market them to the medical profession even expecting to pay colossal penalties.

This is not the only way in which drug companies sometimes act dishonestly. On more than one occasion they have misused scientific evidence to promote their products and they occasionally make medications available to the gullible public using distortions of the truth. New devices are sometimes designed and patented by researchers and their institutions. They are now and then widely used in therapy without being appropriately evaluated before being widely used in therapy.

While making money from research endeavors was not a feasible issue until the passage of the *Bayh-Dole Act* (*Patent and Trademark Law Amendments Act*) and the *National Technology Transfer and Advancement Act* (*NTTAA*) many scientists developed wealth from capitalizing on intellectual property (IP) that they have patented usually with the institution at which they worked. The trend of researchers to interact with industry during more than the past three decades has been a popular enterprise with a double edged sword. Secrecy is a vital component of the corporate world so competitors can be kept at bay. Research leading to patents also demands concealment about the details of the IP until the potentially commercial valuable patents have been obtained. Of necessity scientists have been secretive about patenting their IP. Numerous academicians abhor research at universities that is secret and not openly discussed. Moreover, research that is not freely discussed and evaluated in the usual academic forums of science is prone to research misconduct. On other occasions members of the faculty have created one or more company usually in collaboration with appropriate colleagues and then raised money from venture capitalists to get the company off the ground. Eventually vast amounts of money have been raised by allowing the companies to go public. This raising of funds has enabled companies working on therapeutic drugs to raise sufficient funds to perform expensive clinical trials to test the efficiency of their drug.

References

1. Abbott A, Dalton R, and Saegusa A, Briefing. Science comes to terms with the lessons of fraud. Nature, 1999. 398(6722): 13-17.
2. Charrow R, Scientific misconduct: Sanctions in search of procedures—judgments. The Journal of NIH Research, 1990. 2: 91.
3. Fang FC, Bennett JW, and Casadevall A, Males are overrepresented among life science researchers committing scientific misconduct. mBio, 2013. 4: e00640-00612.
4. Fineman M, Faking it. Manipulated Photography Before photoshop, Metroplitan Museum of Art. 2012, New York: Metroplitan Museum of Art.
5. Malthus TR, An Essay on the Principle of Population at it Affects the Future Improvement of Society with Remarks on the Speculations of Mr. Godwin, M. Condorcet, and Other Writers (orginally Anonymous). 1798, London: J. Johnson.
6. Zimbardo P, The Lucifer Effect: Understanding How Good People Turn to Evil. 2007, New York: Random House Rider Books.

INDEX

A

AAAS (American Association for the Advancement of Science), 11, 26, 28, 51, 128, 371, 389, 538, 546, 555
AAMC (Association of American Medical Colleges), 11, 89, 91-92
AAU (Association of American Universities), 11, 89, 92-93
ABC (American Broadcasting Company), 11, 122, 242
ABC (ATP-binding cassette), 317
ACA (Affordable Care Act), 11, 462, 658
ACC (Atlantic Coast Conference), 11, 565
ACE (American Council on Education), 11, 92
ACNP (American College of Neuro-Psycho-Pharmacology), 11, 420
ACSH (American Council on Science and Health), 11, 650, 659
AD (Anno Domini), 11, 24, 37, 127, 129, 366, 661
ADAMHA (Alcohol, Drug Abuse, and Mental Health Administration), 11, 645
ADVANTAGE (Assessment Differences between Vioxx and Naproxen To Ascertain Gastrointestinal Tolerability and Effectiveness), 186-87
study, 186-87

AEC (Atomic Energy Commission), 11, 550-51
AFIP (Armed Forces Institute of Pathology), 11, 284
AGI (Arizona Glaucoma Institute), 11, 185-86
AhR, 11
AHRQ (Agency for Healthcare Research and Quality), 167
AID (Agency for International Development), 11, 239-40, 243
AIDS (acquired immune deficiency syndrome), 11-12, 16, 39, 101, 132, 216, 240-41, 245, 252, 267-69, 282, 323-30, 573, 585, 632
virus, 323-24, 327-29, 331
AIR (American Institutes for Research), 11, 235
ALG (anti-lymphocyte globulin), 11, 293-94
ALJ (administrative law judge), 11, 96
allegations
of misconduct, 66, 68, 72, 81-86, 91, 96, 99-100, 302, 311, 313, 338, 419, 461, 589-90, 627-28
of scientific fraud, 33, 100, 335
Allen, James, 237
Alzheimer's disease, 188, 283, 427, 643, 653
AMA (American Medical Association), 11, 51, 188, 639, 657, 659

685

American Academy of Pediatrics, 203
American Journal of Cardiology, 403
Am J Med (*American Journal of Medicine*), 11, 148, 414, 416
Anesthesia and Analgesia, 195
Anno Domini, 11
Ardipithecus ramidus, 118, 126
Aryl hydrocarbon receptor, 11
ASM (American Society of Microbiology), 11, 92, 464
ASPD (antisocial personality disorder), 11, 537
ASU (Arizona State University), 11, 333
ATP (adenosine triphosphate), 11, 317
ATPase (adenosine triphosphatase), 411
AU (Alfred University), 288
AUC (American University of the Caribbean), 11, 134
Australia, 193, 199, 286, 374, 457-58
autism, 202-6, 270, 514, 642

B

BALB/c (mouse strain), 11, 334, 338
Baltimore, 70-75, 97, 334, 336-46, 410, 427, 512, 530-31, 594
Baltimore Affair, 10, 70, 72-75, 80, 83, 336-37, 340, 342-43, 512, 605
BBC (British Broadcasting Company), 12, 123, 475
BBP (blood-borne pathogens), 12, 57
BCM (Baylor College of Medicine), 12, 148, 150, 312-13, 652-53
BDNF (brain-derived neurotrophic factor), 12, 655, 660
Bextra, 213
Bezwoda, Werner, 183-84, 470, 603
The Blade, 582
Bluffton University, 12, 582
BMJ (*British Medical Journal*), 10, 12, 51, 124, 126-27, 153, 217, 385, 468, 475, 599, 675
Bores, Leo, 185

Boston Globe, 73, 430, 432, 434, 602
Bowie, Cameron, 207, 495
BRCA1 (breast cancer type 1), 12, 265
BRCA2 (breast cancer type 2), 12, 265
Briggs, Michael H., 193, 458, 602
  research of, 194
Brinkley, John Richard, 214
Bristol-Myers Squibb, 187
British Journal of Obstetrics and Gynaecology, 208-9
British Museum of Natural History, 116-17, 128, 384
BRU (AIDS patient), 12, 16, 268-69, 323-24, 327, 329-30
BU (Boston University), 12, 65, 68, 76-77, 680
BUMC (Boston University Medical Campus), 12, 191-92
BYU (Brigham Young University), 12, 349

C

c. (circa), 8, 11-12, 24, 37, 128, 131, 366-67
CAIF (coalition against insurance fraud), 12, 579
Caltech (California Institute of Technology), 12, 271, 285, 340, 377-78, 427-28, 474, 602
Cambridge University, 12, 208, 369, 387-88, 427
Campbell, Duncan, 216, 602
cancer
  breast, 12, 154, 156, 183-84, 189-91, 199-201, 290, 470, 597, 603
  cervical, 331, 408
  lung, 201
  nasopharyngeal, 154
  oral, 198, 468
  testicular, 184
  therapies, 184, 190-91, 199, 201, 215, 267

CBS (Columbia broadcasting system), 12, 202, 404
CDC (Centers for Disease Control and Prevention), 12, 167, 266, 669
cDNA (complementary DNA), 12, 265
CELG (Celgene Corporation), 12, 274
Cell, 70, 73-74, 83, 148, 157, 334-35, 337-42, 344-46, 594, 631, 635, 659-60
Cell Associates, 234
cell lines, 14, 19, 234, 281-82, 326-27, 330, 407-10
Cello scrotum, 124, 127-28
CEO (chief executive officer), 12, 472, 569-72, 577, 650
CERN (Conseil Européen pour la Recherche Nucléare), 12, 316
CFOs (chief financial officer), 569-70, 572
CFR (Code of Federal Regulations), 12, 68, 74, 168, 170, 172, 189, 669, 672
CFS (chronic fatigue syndrome), 26, 216, 332
CGK Co., 12, 472-73
CGS (Council of Graduate Schools), 12, 35, 92
cheating, 8, 31, 57, 110, 458, 500, 538-40, 548, 559-62, 569, 590-91, 614, 620, 623
Chicago Tribune, 72, 76, 190, 268, 324, 496
children, vaccination of, 202
CIA (Central Intelligence Agency), 12, 286, 667
CITI (collaborative institutional training initiative), 57, 176
clinical trials, 167-68, 170-72, 174-75, 182-91, 194-97, 201, 208-9, 293-94, 501, 549, 638, 641-44, 647, 649, 681
  randomized, 316, 465
clinician-scientists, 8-9, 233, 401, 641
  patient care, 234

CME (continuing medical education), 12, 657
CMS (Center for Medicare and Medicaid Services), 12, 168, 658
CMV (Cytomegalovirus), 12, 666
CNRS (Centre National de la Recherche Scientifique), 12, 99
COGR (Council on Government Relations), 12, 92
COI (conflict of interest), 543, 637-41, 644-47, 650-56, 658, 667, 678
  financial, 639-41, 644, 646-47, 650-51, 653, 656
Columbia University, 12, 241, 338, 376, 414, 459, 538, 666
companies
  biotech, 650, 652, 654
  drug, 58, 156, 173, 183, 186-87, 189, 193-94, 196, 211-13, 475, 642, 645-49, 652-53, 681-82
  insurance, 167, 578-79, 585
  pharmaceutical, 167, 171-73, 186, 193-94, 196, 211-13, 294, 297, 424, 545, 641, 644-50, 652, 657-58, 681
COMS (Collaborative Ocular Melanoma Study), 12, 191
Congress, 24, 26, 32-33, 53, 65, 69-71, 73-76, 81, 96-97, 130, 240, 337-38, 340-42, 564, 569
congressional hearings, 25, 68-74, 191, 325, 336-37, 341, 343-45, 551, 644, 669
congressional subcommittee, 24, 71-74, 85, 190, 250, 326, 342-43, 406, 604-5, 651
contraceptives, 193-95
  oral, 193-94, 602
COPE (Committee on Publication Ethics), 12, 94-95, 464
Cornell University, 12, 241, 318, 377, 411, 590-91, 654, 665

COSEPUP (Committee on Science, Engineering, and Public Policy), 12, 57, 89
Council of Scientific and Industrial Research, 13
CRC (clinical research coordinator), 175
CRCT (criterion references competency test), 12, 560
Crewdson, John, 39, 190, 324, 602
CRI (Commission on Research Integrity), 12, 26, 74-75, 86, 520
CRSO (Clinical Research Support Office), 13, 173
CRU (Climate Research Unit), 13, 27, 174-75
CSIR (Council of Scientific and Industrial Research), 13, 138
CSU (Colorado State University), 13, 238
CT (computerized tomography), 13, 121
CTMS (clinical trials management system), 13, 173
CTQA (clinical trials quality assurance), 13, 173
CU (University of Colorado), 288-89
CU-Boulder (University of Colorado-Boulder), 288

D

DARPA (Defense Advanced Research Projects Agency), 13, 353
Darsee, John R., 311, 400-406, 412, 543
Darwin, Charles, 37, 116, 157, 368, 599
Dawson, Charles, 116, 118-20, 127
DC (District of Columbia), 10, 13, 57, 87, 97, 127, 191, 284, 384, 422, 617, 625
DCAA (Defense Contract Audit Agency), 13, 58
DECREASE (Dutch Echocardiographic Cardiac Risk Evaluation Applying Stress Echocardiography), 13, 467-68
Deer, Brian, 204, 602

DEZYMER, 13, 298, 300
DFG (Deutsche Forschungsgemeinschaft), 13, 99-100, 160, 346, 461, 617
DHEW (Department of Health, Education, and Welfare), 13, 68, 170, 235, 238, 668
DHHR (Department of Health and Human Resources), 148, 239
DHHS (Department of Health and Human Services), 8, 58, 66-68, 74-75, 82-87, 96, 150, 168, 170, 172-73, 189, 268-69, 324, 329-30, 345
Difluoromethylornithine, 13
Dingell, John, 70-76, 84, 97, 242, 298, 326, 334, 337-38, 341-43, 604-6
  committee, 70, 72, 298, 334, 338-39
  subcommittee, 70-74, 80, 83, 85, 97, 192, 242, 336, 338, 341-43, 605
DIO (division of investigative oversight), 13, 87
diseases
  cardiovascular, 194
  inflammatory bowel, 203, 206
DMSR (Division of Management Survey and Review), 13, 235-37
DNA (deoxyribonucleic acid), 12-13, 59, 79, 109, 111, 249-50, 265, 279-80, 291-92, 379-80, 431, 471, 593-94, 654, 659-60
DOD (Department of Defense), 168, 353, 663
DRG (Deutsche Forschungsgemeinschaft), 13, 99-100
drugs, 156, 172-73, 183, 190, 195-97, 211-13, 215, 293-94, 501-2, 568-69, 643, 645-50, 657, 669-70, 681-82
  anti-cancer, 189, 215
  manufacturers of, 211-12, 501, 647-48, 658
DSO (Defense Science Organization), 13, 470
DU (Deakin University), 13, 193-94

DU (Duke University), 9, 13, 70, 198-201, 271-72, 294-95, 299-302, 401, 403, 537, 540, 560-61, 652
DUHS (Duke University Health System), 13, 198, 201, 673-74
DUMC, 13

E

ECG (electrocardiographs), 13, 185
ectopic pregnancy, 208-9
ED (Department of Education), 13, 68
EEG (electroencephalograms), 206
Elizalde, Manuel, Jr., 122-24
EMBL (European Molecular Biology Laboratory), 13, 254
embryos, 208, 249-51, 373-75, 378, 632
EMF (electromagnetic field), 13, 109, 111, 319
Emory University, 13, 401, 438
enterprise, scientific, 40, 43, 49, 73, 336, 367, 496-97, 503, 599, 677, 679
Eoanthropus dawsoni, 116
EPA (Environmental Protection Agency), 13, 75, 110, 238-39, 243, 257
Eppley Institute for Research in Cancer and Allied Diseases, 236
ERB (ethical review board), 13, 168
errors
  scientific, 32, 320, 334, 501, 591, 640
  unintentional, 8, 46, 150, 211, 314-16, 319-20, 322, 333-34, 366, 502, 590, 600, 613, 616, 627-28
ESF (European Science Foundation), 14, 616
European Medicines Agency, 13

F

FAB (federal appeals board), 14, 96, 148, 150, 330-31, 335, 342, 345-46, 430, 632

FAES (Foundation for the Advancement of Education in the Sciences), 14, 297
FASEB (Federation of American Societies for Experimental Biology), 26, 35, 75, 92, 312, 496
fault, experimental, 301
FBI (Federal Bureau of Investigation), 14, 282-84, 286, 293, 298, 505, 567, 574, 577
FDA (Federal Drug Administration), 31, 170-73, 182-83, 185-89, 192, 196-97, 211-13, 293-94, 333, 501-3, 536, 646, 648-49, 657, 669-71
federal agencies, 24, 156, 172-73, 247, 302-3, 427, 436, 656-57, 663-64, 670-72
federal research
  grants, 58, 79, 241, 340, 422
  overhead of, 242
Federal Wide Assurance, 14
FEL (free electron laser), 9, 14, 271-72
Felig-Soman paper, 149, 414-16
FERPA (Family Education Rights and Privacy Act), 14, 671
FFDCA (Federal Food, Drug, and Cosmetics Act), 14, 669-70
FHF (Fred Hollows foundation), 14, 651
FHPT (forced hyperventilation provocation test), 14, 217
Fisher, Bernard, 189-91, 369
5-FU (5-fluorouracil), 11
Flexner, Abraham, 214
Fluorescence-activated cell sorting, 14
FOIA (Freedom of Information Act), 14, 324
fossils, 116-21
FQ (human cell line), 14, 408-10
France, 98-99, 122, 268-69, 328-29, 351-52, 374, 385, 460, 568
fraud

scientific, 7, 23, 33, 65, 67, 69-70,
  72-73, 78, 94, 98, 100-101, 312,
  335, 349, 457-58
securities, 566, 570, 574
Fraudbusters, 604
fraudulent behavior, 28, 31, 69, 94, 195,
  207-8, 373, 405, 458, 462, 544,
  593, 678
fraudulent research
  accusations of, 78, 91, 651
  exposing, 216, 602
  performed, 30, 458, 502, 533, 612, 677
FRS, 14
FSU (Florida State University), 14, 566
funds, legal aid, 204-5
FWA (Federal Wide Assurance), 14, 189

G

GAAP (general accepted accounting
  principles), 14, 570, 572
Gallo, Robert, 602, 632
  laboratory of, 72, 76, 268, 297, 323-29,
  331
GAO (Government Accountability
  Office), 14, 235-36
GCP (Good Clinical Practice), 14,
  173-74
GDR (German Democratic Republic),
  14, 257, 259
Georgetown University, 14, 422-23
GlaxoSmithKline, 14, 61, 157
GLP (Good laboratory practices), 14, 110
Glueck, Charles J., 209-10, 542
GMC (General Medical Council), 14, 94,
  101, 206-8, 216, 474-76, 634
G-6-PD (glucose 6-phosphate
  dehydrogenase), 14, 408-10
grant applications, 41, 52, 66, 101,
  129-30, 139-41, 233-34, 241,
  422-23, 426-29, 435-36, 611-12,
  638-40, 664, 666-68
  federal, 199, 426, 428

grant management, 234-35
grants, 30-31, 53, 79, 100, 140-41,
  211, 233, 235-37, 239-41, 338-39,
  419-21, 426, 435-36, 540-41,
  656-57
  funded, 242, 432
Great Moon Hoax, 114-15
GSK (GlaxoSmithKline), 14, 61, 156-57
GWI (Gulf war illness), 14, 290
GWS (Gulf War syndrome), 14, 216, 290
GWU (George Washington University),
  14, 108

H

HAART, 14
Harvard University, 14, 47-48, 60-61,
  65, 137-38, 150, 235-36, 282-83,
  285-87, 311, 367, 400-405, 427-28,
  432-35, 537-38
HC (Health Canada), 14, 617
HDL2 (high-density lipoprotein 2), 14,
  195
HeLa cells, 14, 253-54, 408-10
Hellinga, Homme W., 298-302
  research of, 299, 301
Hellinga affair, 300-301
HEW (Health, Education, and Welfare),
  14, 236
HGH (Human Growth Hormone), 14,
  271, 569
HHMI (Howard Hughes Medical
  Institute), 14, 52, 60-61, 199
HHS, 14, 34
HIPPA (Health Insurance Portability and
  Accountability Act), 14, 57, 176-77
HIV (human immunodeficiency virus),
  15, 39, 57, 240, 252, 267, 322-25,
  328, 563, 632
HL23 (human retrovirus), 15, 326
HMS *Beagle*, 15, 368
H9 (human cell line), 14, 326-27, 330

hoax, 101, 113, 115-19, 121, 123-27, 249
  moon, 114-15, 126
  stunning, 114-15
Hodgkin disease, 14, 19, 406-10
Houston University, 20, 515
HPLC (high-performance liquid chromatography), 15, 429
HR (Human Resources), 15, 173
HRPP (Human Research Protection Program), 15, 172-74, 178
HTLV-I (human T-lymphotropic virus type I), 15, 323, 464
HTLV-II (human T-lymphotropic virus type II), 15, 323
HTLV-III (human T-lymphotropic virus type III), 15, 323-24
HTLV-IIIB (human T-lymphotropic virus type IIIB), 15, 268-69, 324-25, 327, 329-30
human subjects, 9, 62, 74, 88, 167-78, 184, 206, 250, 303, 314, 370, 467, 501, 669-71, 681
  protection of, 68, 74, 170, 172, 176, 652, 670-71
HUT102, 15, 326
HUT78, 15, 326
hypotheses
  clinical, 186-87
  wrong, 317, 347, 379, 382, 549

I

IACUC (Institutional Animal Care and Use Committee), 62, 303, 681
IBAMA (Instituto Brasileiro do Meio Ambiente e dos Recursos Naturais Renováveis), 15, 291
IBT (Industrial Bio-Test Laboratories), 109-10
ICH (International Conference on Harmonization), 15, 173-74
ICMJE (International Committee of Medical Journal Editors), 15, 93, 95, 161, 171
ICU (International Cycling Union), 15, 568
IDE (investigational device), 15, 171, 185
IDH1 (isocitrate dehydrogenase 1), 15, 274
IDH2 (isocitrate dehydrogenase 2), 15, 274
IEC (independent ethics committees), 15, 168, 681
IEEE, 15
IGSP (Institute of Genome Sciences and Policy), 198
Imanishi-Kari, Teresa, 70-72, 80, 83, 85, 96, 334-42, 344-46, 423
immunology, 81, 246, 329, 336-37, 351, 400, 427, 437, 520, 645, 654
IND (investigational new drug), 15, 168, 171, 293
INPA (Instituto Nacional de Pesquisas da Amazônia), 15, 291
INSERM (Institut National de la Santé la Recherche Médicale), 99, 351-52
insulin, discovery of, 275-77
International Research Journal of Plant Science, 460
Investigative Ophthalmology and Visual Science, 158
IOC (International Olympic Committee), 569, 573
IOG (Institute of Gerontology), 15, 664
IOM (Institute of Medicine), 15, 25, 89-90, 200, 204, 325, 534, 615
IOP (intraocular pressure), 15, 185-86, 430
IP (intellectual property), 59-60, 88, 130-31, 138, 266-67, 269, 274, 279-80, 473, 500, 641, 665-66, 682
IPCC (Intergovernment Panel on Climate Change), 15, 28

IQ (intelligence quotient), 15, 381-83, 497
IRB (institutional review board), 15, 94, 168-70, 172-78, 186, 189, 201, 206, 246, 255, 311, 505, 552, 681
approval, 169-70, 177
IRS (Internal Revenue Service), 15, 235, 293
IT (information technology)), 15, 290
IU (Indiana University), 15, 401
IVF (in vitro fertilization), 15, 248-51, 663

J

JAMA (Journal of the American Medical Association), 15, 36, 51, 140, 188, 599, 639, 653, 660
Japanese Journal of Experimental Medicine, 134
Japanese Journal of Medical Science and Biology, 134-35
JCR (Journal Citation Reports), 55
JEHT, 16, 575
JEHT (justice, equality, human dignity and tolerance), 575
JHU (Johns Hopkins University), 16, 159, 242, 366, 401, 498, 594, 651
Journal of Biological Chemistry, 157
Journal of Clinical Investigation, 412
Journal of Molecular Biology, 299-300, 302
Journal of NIH Research, 35, 73, 83, 683

K

KAIST (Korea Advanced Institute of Science and Technology), 16, 472-73
KKG (Kappa Kappa Gamma), 16, 285
Korea Advanced Institute of Science and Technology, 16
KU (University of Kansas), 16, 20, 135, 283

L

Lacks, Henrietta, 14, 253-54, 408
LAN (local area network), 16, 109
Lancet, The, 184, 198, 202-6, 217, 476, 531, 599, 642
LAV (lymphadenopathy virus), 16, 268-69, 323-24, 326-27, 329-31
laws, copyright, 130-31, 158, 671
lawsuits, civil, 241-42, 332
LSD (lysergic acid diethylamine), 16, 295

M

MACS, 16
MAGIC (magnetism-based interaction capture), 16, 472-73
market, 156, 186-87, 270, 425, 457, 560, 651, 672, 681
Massachusetts General Hospital, 16, 65, 536, 538
mastectomy, 189-90
Mayo Clinic, 242, 649
MBA (Master of Business Administration), 16, 561
MBS, 16
McGill University, 16, 159, 313
MCH (Miami Children's Hospital), 16, 270
MCI (Microwave Communications Inc.), 16, 572
MCSC, 16
MD (doctor of medicine), 16, 26, 35, 61, 75-76, 134-35, 148-49, 183, 195-96, 402-3, 412, 428-29, 533, 542, 654-55
MEDLINE (medical literature analysis and retrieval system), 16, 391, 468-69, 596, 599
MEEI (Massachusetts Eye and Ear Infirmary), 16, 428-30, 651-52
Merck and Company, Inc., 186-87, 234, 419, 652

Mesozoic mammals, 116
Miami University, 16, 566, 577
Millikan, Robert Andrews, 376-78
MIT (Massachusetts Institute of
    Technology), 16, 48, 71-72, 75, 83,
    335-37, 339-42, 346, 410, 427-28,
    602
MLA (Modern Language Association),
    16, 131, 606
MMR (measles, munps, rubella), 16,
    202-7, 270, 514, 642
    vaccine, 202-5, 642
money laundering, 566, 574-76
MOST (Ministry of Science and
    Technology), 16, 101-2
mph (miles per hour), 16, 584
MRC (Medical Research Council), 98,
    101, 278, 617
MRI (magnetic resonance images), 185,
    291
MS (multiple sclerosis), 16, 298
MSN, 16
MSSM, 16
MSU (Michigan State University), 16, 86
MTA (Material Transfer Agreement), 16,
    58, 283
MUSC (Medical University of South
    Carolina), 16, 298

N

NACUA (National Association of College
    and University Attorneys), 16, 92
NAE (National Academy of Engineering),
    16, 25, 89, 615
NAS (National Academy of Sciences),
    16, 25-27, 56, 67, 70, 89-91, 162,
    258-59, 292, 324-25, 401, 406-7,
    590, 615, 639
NASA (National Aeronautics and Space
    Administration), 16, 27, 40, 319-20

NASDAQ (National Association of
    Securities Dealers Automated
    Quotations), 17, 574, 655
NASULGC (National Association of
    State Universities and Land Grant
    Colleges), 17, 92
National Geographic Society, 121, 123,
    375-76
National Research Act, 169, 670
National University of Singapore, 17, 469
natural selection, 368, 373, 386-87, 594
Nature, 34-36, 73, 118, 121, 126, 135,
    199, 280, 300-302, 351-53, 405-6,
    423-24, 604, 655, 660
Nature Chemical Biology, 472-73
Nature Medicine, 200
NCAA (National Collegiate Athletic
    Association), 565-67
NCAB (National Cancer Advisory
    Board), 17, 192, 236, 541
NCBI (National Center for Biotechnology
    Information), 17, 62
NCCAM (National Center for
    Complementary Medicine), 17, 353
NCD (national coverage determination),
    17, 167, 173-74
NCHGR (National Center for Human
    Genome Research), 17, 496
NCI (National Cancer Institute), 17, 39,
    79, 135, 184, 189-92, 215, 236,
    268, 297, 323, 332, 407, 602, 631
NCLS (National Conference of Lawyers
    and Scientists), 17, 97
NCPRE (National Center for Professional
    and Research Ethics), 17, 56
NCSU (North Carolina State University),
    322
ND (University of Notre Dame), 17,
    400-401
NEJM (*New England Journal of Medicine*),
    17, 127, 148-49, 157, 187-88, 198,
    256, 405, 412-13, 598-99, 639, 653,
    659, 661

Neurontin, 213
neuroscience, 460, 601, 604, 645, 654
New Scientist, 36, 387, 427-28, 625
NHGRI (National Human Genome Research Institute), 17, 250, 431
NHTSA (National Highway Traffic Safety Administration), 17, 583-84
NIA (National Institute of Aging), 17, 422
NIAID (National Institute of Allergy and Infectious Diseases), 17, 338-39
NICHHD (National Institute of Child Health and Human Development), 17, 235, 422
NIDDK (National Institute of Diabetes and Digestive and Kidney Disease), 17, 605
NIEHS (National Institute of Environmental Health Sciences), 17, 239, 425
NIGMS (National Institute of General Medical Sciences), 17, 236
NIH (National Institutes of Health), 69-76, 78-88, 148-50, 233-36, 248-50, 324-27, 336-42, 344-45, 401-5, 425-29, 540-42, 604-6, 644-46, 651-53, 668-70
  committees of, 340, 344-45
NIMH (National Institute of Mental Health), 17, 137, 139, 211, 235, 419-21, 673
NIST (National Institute of Standards and Technology), 17, 667
Nixon, Peter G., 216, 602
NNI (National Neuroscience Institute), 633
NOAA (National Oceanic and Atmospheric Administration), 17, 27
Nobel Prize, 38-39, 43, 45-46, 59, 81, 97, 162, 248-49, 271-73, 275-81, 285-86, 300, 331, 334, 509
NPR (National Public Radio), 17, 213

NSABP (National Surgical Adjuvant Breast and Bowel Project), 17, 189-90
NSB (National Science Board), 17, 26
NSERC (Natural Sciences and Engineering Research Council), 17, 98
NSF (National Science Foundation), 17, 26, 52, 54, 65, 78, 292, 312, 322, 333, 535, 613, 621, 664, 667-68
NSTC (National Science and Technology Council), 17, 26
NSU (National University of Singapore), 469
NTTAA (National Technology Transfer and Advancement Act), 644, 672, 682
NWU (Northwestern University), 17, 109, 242
NYMC, 17
NYU (New York University), 18, 98, 295, 428

O

OBE (Order of the British Empire), 18, 279
OCR (Office of Civil Rights), 18, 171-73
OCRC (Office of Corporate Research Collaborations), 18, 173
OESO (Occupational and Environmental Safety Office), 18, 57, 176
Office of Civil Rights Compliance, 18
OHRP (Office for Human Research Protections), 18, 168, 170, 189, 247, 663, 669
OHSU (Oregon Health Sciences University), 81, 649
OIG (Office of Inspector General), 30, 173-74, 268-69, 331
OMB (Office of Management and Budget), 18, 234, 237, 584
OMK-210 (monkey cell line), 409

ONR (Office of Naval Research), 18, 350-51
OPHS (Office of Public Health and Science), 18, 621
ophthalmology, 138, 150, 250, 428, 430, 651
ORA (Office of Research Administration), 18, 173
ORI (Office of Research Integrity), 24-25, 30-31, 83-84, 86-89, 148, 190-91, 312-13, 327-28, 330-31, 422-23, 425-26, 498-99, 510-11, 517-18, 632
ORIR, 18
OSHA (Occupational Safety and Health Administration), 18, 57
OSI (Office of Scientific Integrity), 18, 76, 80-86, 96, 325, 327-30, 339, 344-45, 422-23, 605, 629, 668
investigations, 83-85
OSIR (Office of Scientific Integrity Review), 18, 81, 84, 86, 211, 629
OST (Office of Science and Technology), 18, 173
OSU (Ohio State University), 18, 51
Oxford University, 18, 125, 127-28, 199, 386-87, 548, 630, 669

P

PANAMIN (Presidential Arm for National Minorities/Private Association for National Minorities), 18, 122-23
Pasteur Institute, 72, 76, 267-69, 280, 323-24, 326-28, 330
patents, 14, 19, 58-60, 93, 131, 185, 205, 265-73, 278, 282, 323-24, 330, 371, 640-42, 682
PBS (Public Broadcasting Service), 18, 563
PCB (polychlorinated biphenyls), 18, 239
PCR, 18-19
PCT (Patent Cooperation Treaty), 18, 274
PDF (portable document format), 18, 467
Pearce, Malcolm, 208, 475
PECASE (Presidential Early Career Award for Scientists and Engineers), 18, 317
Penn State (Pennsylvania State University), 18, 318, 403, 567-68
Pfizer, 157, 185, 195-96, 213, 641, 652
PhD (doctor of philosophy), 9, 61, 133-35, 199, 277-78, 294, 296-99, 402-3, 419, 427-28, 431-32, 463, 466-67, 654-55, 664-65
PHI (protected health information), 174, 176-77
PHS (Public Health Service), 18, 25-26, 30, 65-69, 75, 80-82, 84, 86-88, 139, 245-47, 429-30, 629, 656, 668-69, 681
physiology, 38-39, 43, 45-46, 51, 81, 97, 162, 248-49, 275-77, 279-81, 286-87, 336-37, 407, 496, 512
PI (principal investigator), 18, 42, 140-41, 169, 174-75, 210, 233, 247, 272, 315, 419, 498, 547, 600, 618
Piltdown Man, 116-21, 127-28
perpetrator of, 119
plagiarism, 24-26, 88, 129-33, 135-41, 149, 416-17, 458, 460, 462-63, 468-69, 500-501, 511-12, 516, 595-97, 606
allegations of, 133, 500, 606
plagiarists, 130-31, 134, 138, 595
PNAS (Proceedings of the National Academy of Sciences), 18, 150, 160, 258-59, 317, 407, 465, 639
PNT (pneumatic trabeculoplasty), 185-86
device, 185-86
Poe, Edgar Allan, 115, 128
polycystic ovary syndrome, 208
Popovic, Mikulas, 80, 83, 148, 268, 327-31, 632, 635

Potti, Anil, 198, 314
  research of, 199, 201
pregnancy, 115, 126, 203, 208-9, 249, 252
  ectopic, 208
Princeton University, 18, 285, 371, 383, 570, 666
proteins, 280, 282, 286, 298-300, 407, 424, 428, 472, 520, 592
psychiatrists, 188, 212, 256, 258, 372, 458, 584-85, 644, 646-49, 660
Public Health Service Act, 67, 80, 668-69
Purdue University, 18, 256, 350-51

R

RAC (Recombinant Advisory Committee), 86
radiation, 190, 257, 286, 348, 550, 597
Randi, James, 351, 515
RB (human cell line), 19, 408-9
RCA, 19
RCOG (Royal College of Obstetrics and Gynecology), 19, 475
RCR (Responsible Conduct in Research), 77-78, 86, 90
replication, 32, 46-47, 90, 336, 418, 433, 435, 519, 599-600, 619
research
  academic, 272, 641
  basic, 99, 271, 604-5
  behavioral, 69, 86, 168, 170, 670-71
  biomedical, 10, 30, 34, 69, 72-73, 90, 99, 156, 399, 541, 543, 640
  cancer, 134, 191, 252-54, 274, 399, 427
  clinical, 168, 170-71, 183, 189, 247, 413, 501, 681
  dishonest, 23, 25, 28, 30, 32, 65, 108, 150, 158, 302, 436, 501, 503-4, 537, 551
  ethical, 29, 76, 90, 99-100, 461, 616
  fraud, 10, 71, 73, 92, 605, 625
  funded, 30, 66, 79, 161, 189, 237, 312, 645, 680
  human, 9, 62, 167-69, 172-75, 178, 189, 195, 198, 426, 634, 663, 668, 670-71, 681
  laboratory, 75, 405, 415, 611, 616
  medical, 35, 90, 94, 162, 252, 271
  scientific, 24, 26, 40, 46, 71, 94, 641, 644
  sloppy, 8, 150-51, 316, 334, 337, 466, 520-21, 544, 602
  social science, 235
  unethical, 95, 245, 259, 537, 589, 681
researcher, cancer, 189, 437, 599
research funds
  divert university, 239
  federal, 67, 192, 233, 236, 611
research institutions, 30, 52-53, 66-67, 69, 76-77, 91, 99-100, 533-34, 539-40, 612-13, 616-18, 645-46, 654, 656, 679
  laboratories of, 237
research integrity, 8, 12, 24, 34-35, 41, 65, 74-76, 79, 84, 86, 88-89, 92, 94, 614-18, 622-23
research laboratories, 39-40, 57, 66, 94, 134, 293, 295, 299, 326, 414, 534, 541, 544, 549, 621
research misconduct, 24-33, 68-69, 76-81, 87-92, 95-98, 336-38, 465-69, 495-99, 502-5, 509-12, 540-42, 611-13, 615-16, 620-23, 678-80
  accusations of, 69, 83, 87-88, 91, 93, 96, 632
  allegations of, 30, 68, 73, 86-87, 93-94, 177, 288, 435, 466, 545, 605, 628, 665, 677
  alleged, 30, 75, 97, 99
  guilty of, 50, 79, 86-87, 96, 151, 200, 331, 350, 428, 498, 511, 598, 621
research participants, 241, 630
  human, 49, 86, 501

research projects, 50, 167-69, 172-75, 207, 237, 247, 254, 403-4, 437, 464, 466, 505, 539, 542, 547
  human, 168-69, 174-75
  unfunded, 233
research regulations, 42, 95, 670
research reputation, 175, 604, 615
research team, 46, 79, 154, 157, 171, 176, 187, 194, 233, 258, 279, 349-52, 539, 598, 621-22
Reuben, Scott S., 8, 195, 641
  research of, 195-96
RIO (Research Integrity Officer), 19, 87-88, 637, 656
RNA (ribonucleic acid), 19, 97, 279, 322, 336, 427, 465, 594
RNAi (RNA interference), 19, 427
RNase, 19
Rockefeller University, 252, 340, 347, 472, 496
Royal College of Obstetricians and Gynecologists, 208-9
Royal Society of Medicine, 216-17
RRTA (rapid response technical assistance), 19, 88
Rutgers University, 19, 108, 277-78, 561
RY (human cell line), 19, 408-10

S

SAGE (Study of Addiction Genetics and Environment), 19, 258
science, fraudulent, 31, 65, 73, 365, 503, 505, 546
scientific misconduct
  accusations of, 46, 80, 88, 92, 99, 311, 333, 342, 366, 504
  allegations of, 23, 30-31, 67, 73, 78-79, 98-102, 312-13, 334, 337, 343, 402, 517, 591, 601, 628-29
  definition of, 25-26, 91, 98, 367
  guilty of, 79, 83, 108, 153, 160, 329, 335, 350, 404, 423, 425, 435-36, 537, 627-28, 632
  potential, 299, 301, 337, 349-50
  prevention of, 86, 497, 611, 623
  scandals of, 40, 147, 391
  suspected, 75, 88, 311, 339, 344
scientists, 24-32, 37-45, 47-58, 68-73, 75-78, 85-87, 97, 348-50, 495-99, 503-7, 533-34, 542-45, 547-51, 638-41, 677-82
  careers of, 52, 542, 663
  fraudulent, 8, 313, 628
  French, 326, 352, 632
  prominent, 139, 162, 276, 351, 495-97, 501, 543, 547, 598, 616, 677-78
  reputations of, 311, 497, 627
  research of, 8, 541, 641
SCNT (somatic-cell nuclear transfer), 19, 470, 632
SEC (Securities and Exchange Commission), 19, 569-71, 574, 578
self-plagiarism, 95, 129, 147
Selkoe, Dennis, 653
settlements, 191-92, 240-42, 271, 290, 426, 500
Shubick, Phillipe, 236
SIDS (sudden infant death syndrome), 19, 475-76
SMARTS (small and moderate aperture telescope system), 19, 139
SMC (Singaporean medical council), 19, 634
SMU (Silesian Medical University), 19, 468-69
SNU (Seoul National University), 19, 470-71
South Africa, 114, 117-18, 126, 183-84, 470
S&P (standard and poor), 19, 570
species, 118, 368, 373-74, 388, 400, 433
SpR (human cell line), 19, 408-9
SSI (solid-state imaging), 19, 320

Stanford University, 19, 59-61, 77, 136-37, 211, 234, 242, 271, 283, 345, 348, 438, 552
Stewart, Walter, 78, 336, 342, 351, 405-6, 597, 604-6
Straus, Marc J., 8, 126, 191-92, 399, 540, 591, 680
streptomycin, discovery of, 278
Summers, William K., 188
Sunday Times, 194, 203-5, 216, 383, 642
SUNY, 19, 299-300
SUNY-Buffalo (State University of New York-Buffalo), 19, 299-300
SVU (Sri Venkateswara University), 19, 462
syphilis, 182, 189, 245-48, 601, 681

T

Tasaday, 122-24, 126-28
TCDD (2,3,7,8-tetrachlorodibenzo-p-dioxin), 19, 237-39
Temple University, 19, 135, 282
THA (tetrahydroaminocrydine), 19, 188
TNT (trinitrotoluene), 19, 38
Tokyo University, 102, 282, 464-65
TRDRP (Tobacco-Related Disease Research Program), 19, 140
Trinitrotoluene, 19, 38
TU (Towson University), 108
Tufts University, 19, 71, 75, 195, 336-37, 346
Tulane University, 19, 425

U

UAB (University of Alabama-Birmingham), 19, 77, 242, 504, 666
UAH (University of Alabama-Huntsville), 266-67
UC (University of California), 281
UC (University of Cincinnati), 20, 209-10, 281, 415, 417, 648-49
UC Berkeley (University of California-Berkeley), 20, 277, 383, 424, 664
UCDavis (University of California-Davis), 20, 280
UChicago (University of Chicago), 20, 89, 210, 294, 345, 376-77, 438
UCI (University of California-Irvine), 20, 251
UCL (University College London), 20, 380, 633-34
UCLA (University of California-Los Angeles), 20, 297, 432, 642
UConn (University of Connecticut), 20, 61
UCSD (University of California-San Diego), 140, 160, 281-83, 411-12, 598
UCSF (University of California-San Francisco), 59, 76, 137, 139-41, 198, 234, 270-71
UD (University of Delaware), 20, 108
UEA (University of East Anglia), 20, 27
UH (University of Hawaii), 20, 124, 240, 272
UIUC (University of Illinois at Urbana-Champaign), 20, 56, 597, 612
UK (United Kingdom), 27, 94, 98, 101, 108, 116, 193-94, 202, 204-8, 313, 319, 380-81, 384-86, 474-76, 634
UKY (University of Kentucky), 435, 611
UM (University of Michigan), 20, 255, 302, 426, 431, 642-43, 664-65
UMass (University of Massachusetts), 20, 387, 606
UMD (University of Maryland), 20, 426, 583
UMMS (University of Minnesota Medical School), 293
UMN (University of Minnesota), 20, 183, 242, 293, 644, 646-48, 659

UNC (University of North Carolina), 20, 290, 322, 420, 565, 575
UNC (University of Northern Colorado), 20, 77, 290, 292, 420, 565-66
Uniform resource locator, 20
University of Bristol, 207-8
University of Iowa, 20
University of Nebraska, 10, 236
University of Pittsburgh, 18, 189-91, 273, 346, 373, 419-22, 582, 659
University of Rochester, 19, 137, 212, 285
University of Utah, 20, 319
UNO (University of New Orleans), 132
U of I (University of Illinois), 20, 35, 56, 420, 597, 612
UPenn (University of Pennsylvania), 18, 182-83, 191, 274, 318, 403, 430, 567-68, 603
UPMC (University of Pittsburgh Medical Center), 346, 472
URL (uniform resource locator), 20, 176
USA (United States of America), 20, 407
USAAMRIID (USA Army Medical Research Institute of Infectious Diseases), 20, 284
USADA (USA Anti-Doping Agency), 20, 568
USC (University of Southern California), 20, 287, 669, 672
USF (University of South Florida), 20, 266
USPTO (USA Patent and Trademark Office), 20, 60, 265, 267-68, 273, 672
UST (University of Science and Technology), 20, 463
UT (University of Texas), 20, 282, 424, 596
UToronto (University of Toronto), 20, 274-77, 630
UTSA (University of Texas-San Antonio), 20, 282

UTSWMC (University of Texas Southwestern Medical Center), 20, 290-91
UVA (University of Virginia), 20, 285
UVM (University of Vermont), 21, 426-27, 504
UW (University of Wisconsin), 21, 238, 313, 668

V

VA (Veterans Administration), 21, 168, 172, 196, 291, 401
vaccines, 202-5, 254, 266, 270, 314, 346, 371, 514, 642
Vanderbilt University School of Medicine, 21
VIGOR (Vioxx Gastrointestinal Outcomes Research), 186

W

Wachslicht-Rodbard study, 148-49, 412-19
Wakefield, Andrew J., 201-7, 270, 475, 602, 642
research of, 202, 206
Wall Street Journal, 61, 243
Weaver, David, 70, 148, 334-35, 341, 344
WFU (Wake Forest University), 21, 437
WHO (World Health Organization), 21, 193, 204
WI-38 (human cell line), 21, 234
Williams, Mark, 207
WMU (Western Michigan University), 21, 387
Woodward, Arthur Smith, 116
WPI (Whittemore Peterson Institute), 21, 332
WPI (Worcester Polytechnic Institute), 21, 51, 332
WSU (Wichita State University), 21, 390

WUSL (Washington University-St. Louis), 21, 155

X

XMRV (xenotropic murine leukemia retrovirus), 21, 332

Y

Yale University, 65, 115, 132, 148, 187, 210, 253, 273, 292, 298-99, 302, 391, 412-19, 496, 506

www.ingramcontent.com/pod-product-compliance
Lightning Source LLC
Chambersburg PA
CBHW020717180526
45163CB00001B/12